SOLAR-TERRESTRIAL PHYSICS

Principles and Theoretical Foundations

ASTROPHYSICS AND SPACE SCIENCE LIBRARY

A SERIES OF BOOKS ON THE RECENT DEVELOPMENTS
OF SPACE SCIENCE AND OF GENERAL GEOPHYSICS AND ASTROPHYSICS
PUBLISHED IN CONNECTION WITH THE JOURNAL
SPACE SCIENCE REVIEWS

Editorial Board

J. E. BLAMONT, *Laboratoire d'Aeronomie, Verrières, France*

R. L. F. BOYD, *University College, London, England*

L. GOLDBERG, *Kitt Peak National Observatory, Tucson, Ariz., U.S.A.*

C. DE JAGER, *University of Utrecht, The Netherlands*

Z. KOPAL, *University of Manchester, England*

G. H. LUDWIG, *NOAA, Environmental Research Laboratories, Boulder, CO, U.S.A.*

R. LÜST, *President Max-Planck-Gesellschaft zur Förderung der Wissenschaften, München, F.R.G.*

B. M. McCORMAC, *Lockheed Palo Alto Research Laboratory, Palo Alto, Calif., U.S.A.*

H. E. NEWELL, *Alexandria, Va., U.S.A.*

L. I. SEDOV, *Academy of Sciences of the U.S.S.R., Moscow, U.S.S.R.*

Z. ŠVESTKA, *University of Utrecht, The Netherlands*

VOLUME 104
PROCEEDINGS

SOLAR-TERRESTRIAL PHYSICS

Principles and Theoretical Foundations

BASED UPON THE PROCEEDINGS OF THE THEORY INSTITUTE HELD AT BOSTON COLLEGE, AUGUST 9–26, 1982

Edited by

R. L. CAROVILLANO

and

J. M. FORBES

Department of Physics, Boston College, Chestnut Hill, Mass., U.S.A.

D. REIDEL PUBLISHING COMPANY

A MEMBER OF THE KLUWER ACADEMIC PUBLISHERS GROUP

DORDRECHT / BOSTON / LANCASTER

Library of Congress Cataloging in Publication Data

Main entry under title:

Solar-terrestrial physics, principles and theoretical foundations.

(Astrophysics and space science library ; v. 104)

Based on papers presented at the Theory Institute in Solar-Terrestrial Physics.

Includes index.

1. Atmosphere, Upper—Congresses. 2. Magnetosphere—Congresses. 3. Magnetohydrodynamics—Congresses. 4. Space plasmas—Congresses. I. Carovillano, Robert L., 1932– . II. Forbes, J. M. (Jeffrey M.). III. Boston College. IV. Theory Institute in Solar-Terrestrial Physics (1982 : Boston College). V. Series.

QC878.5.S64 1983 551.5'14 83–11101

ISBN 90–277–1632–3

CIP

Published by D. Reidel Publishing Company,
P.O. Box 17, 3300 AA Dordrecht, Holland.

Sold and distributed in the U.S.A. and Canada
by Kluwer Academic Publishers,
190 Old Derby Street, Hingham, MA 02043, U.S.A.

In all other countries, sold and distributed
by Kluwer Academic Publishers Group,
P.O. Box 322, 3300 AH Dordrecht, Holland.

Printed in The Netherlands

TABLE OF CONTENTS

MAGNETOSPHERIC PHYSICS

UPPER ATMOSPHERIC PHYSICS

PREFACE

The Theory Institute in Solar-Terrestrial Physics was held at Boston College 19-26 August 1982. The program consisted of a two-week School followed by the first theory conference in the field. This book is based upon the lectures presented at the School.

Several years ago there was a convergence of efforts to promote the role of theory in space plasma physics. Reports from the National Academy of Sciences and NASA advisory committees documented the disciplinary maturity of solar-terrestrial physics and recommended that theorists play a greater role in the continued development of the field. The so-called theory program in solar-terrestrial physics was established by NASA in 1979 and implemented in accordance with the guidelines set forth by a panel of scientists, primarily theorists, in the field. The same panel motivated the Boston College program. Published proceedings of the school would provide curricular materials for the training of graduate students in solar-terrestrial physics.

J.M. Forbes, T.E. Holzer, A.J. Hundhausen, A.D. Richmond, and G.L. Siscoe were the principal architects of the curriculum of the School, and I am grateful for their contributions. Each also lectured at the School. The chapters in this book were prepared by the authors themselves with one exception. The chapters by Parker are edited reproductions of his lectures. Unfortunately, it is our loss that the lectures of Holzer and Hundhausen are not included in the book.

NASA officials who deserve special acknowledgement are H. Glaser, S. Tilford and T. Birmingham. Glaser got the theory program established while he was director of the Solar Terrestrial Division of the Office of Space Science. Tilford is the capable custodian of the program as director of the Environmental Observations Division (which now includes the solar-terrestrial programs) of the Office of Space Science and Applications, following NASA's reorganization in 1981. Birmingham helped in many ways to improve our Institute and diminish costs.

The Institute was held at Boston College under the most cordial and accommodating circumstances. F.A. Campanella, Executive Vice President, urged that I hold the Institute on campus and provided ample resources and characteristic hospitality. Maura E. Hagan flawlessly handled all Institute arrangements including memorable social events. Campus offices including housing, athletics, parking, special affairs, and food services were responsive and thoughtful of our needs. Special gratitutde is expressed to my assistant, Shirley Lynch, and departmental secretaries, Joan Feeney, Yavalux Sopchockchai, and Ellen Duggan for their diligence, cheerfulness and assistance at all times.

ROBERT L. CAROVILLANO

February 28, 1983

THE SOLAR-TERRESTRIAL SYSTEM

R. L. Carovillano
Office of Naval Research - London Branch
223 Old Marylebone Road
London NW1 5TH

In solar-terrestrial physics, the primary objective is to understand the processes that involve both the sun and the earth. In all areas of science, as understanding increases the boundaries of the discipline become more and more encompassing. Thus, the more we learn of energetic processes in the sun, the more we realize how these affect the complex solar atmosphere, the characteristics of interplanetary space, and the generation of geophysical processes. The sun, the interplanetary medium, and the geophysical environment are the *basic components* of solar-terrestrial system, and each of these components is, of course, worthy of independent study. The ultimate and demanding requirement of solar-terrestrial physics, however, is to understand the coupling processes among these components. For example, a geomagnetic storm may be thought of purely as a disturbance in the geophysical environment wherein altered particle populations and a ring current "explain" the phenomenon. But this description would ignore the basic question of how energy and momentum are transmitted to the magnetosphere by the interplanetary medium (or solar wind) and, in turn, what solar processes are responsible for the temporal characteristics of the disturbed solar wind.

Until very recently it was believed that interplanetary space was unimportant and consisted of a perfect vacuum traversed primarily by electromagnetic radiation and an occasional cosmic ray. Only about three decades ago were astronomical observations of the corona and comet tails used to deduce that the solar atmosphere extended deep into space, far beyond the inner planets. Sydney Chapman offered the first theory of the extended corona with a static description. Parker gave the definitive description, however, allowing for a radial flow that he called the *solar wind*. Among the many results predicted by Parker were that the solar wind is a transonic flow, a mass flux permanently escapes the sun, the flow is heated efficiently to great distances from the sun, and the solar magnetic field convects into space with the solar wind and has a spiral configuration because of the solar rotation.

R. L. Carovillano and J. M. Forbes (eds.), Solar-Terrestrial Physics, 1–9.

The interaction of the solar wind with the geomagnetic field gives rise to the so-called bow shock wave and magnetopause, which represents the boundary of the magnetosphere (cf., Figure 1). The essential characteristics giving rise to a bow shock and magnetosphere are not unique to Earth, and we now more broadly talk of planetary (and even stellar) magnetospheres. The physical environment of each of the planets is different, and accordingly each planetary environment interacts differently with the solar wind. Studying these interactions collectively - the subject of comparative magnetospheres - increases nonlinearly the amount to be learned and the quality of our understanding. It is quite the same as studying, say, the Sun in many wave length ranges of the electromagnetic spectrum rather than with just visible light. Because the broad approach has proven to be both useful and necessary, perhaps the term *solar-terrestrial physics* should be replaced with *solar-planetary physics* but the former term is far more traditional if not accurate. Retaining the term may be a remnant of the tattered notion that Earth plays a central role in the universal scheme of things.

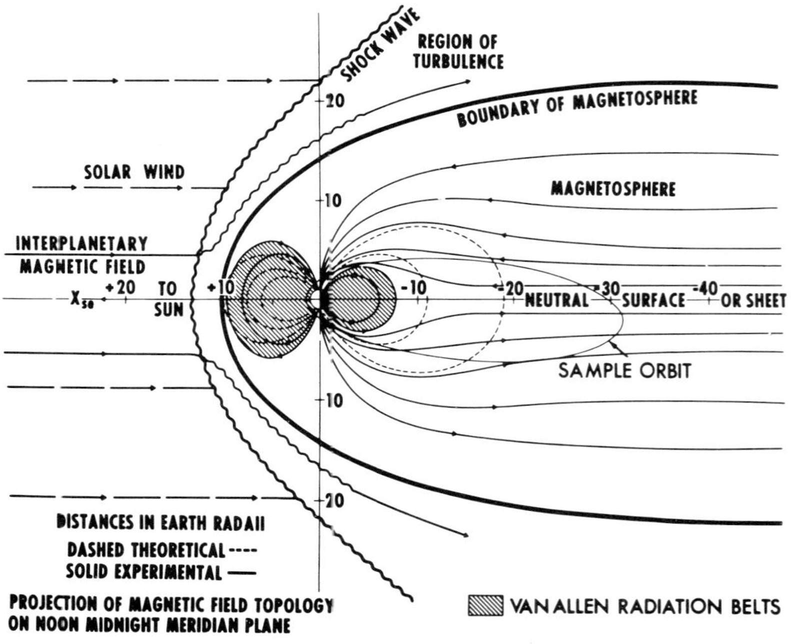

Figure 1. Schematic diagram of the earth's bow shock and magnetosphere in the noon-midnight plane. (Courtesy of NASA.)

The essential task of solar-terrestrial physics is to understand the solar system and to apply this knowledge to the cosmos beyond. In our treatment, the sun and the earth will be given primary emphasis, but considerations of the other planets, comets and other occupants of the solar system, and other stars are not excluded or avoided. Questions easily posed on the interactions between the sun and earth preclude a narrow approach. Thus, our considerations include solar and heliospheric physics, magnetospheric and thermospheric physics, coupling processes and mechanisms, and basic plasma physics.

The importance of solar-terrestrial physics is easy to perceive. The sun gravitationally maintains the location of the earth and sustains here the only known life in the universe. We have come to realize how dependent life is on environmental and political conditions. It is possible today to destroy the human race, advertently or inadvertently, by a variety of means. Security, quality of life, economic well-being, and carrying on the great human adventure to discover and explore the unknown all require an ever improving understanding of the earth's space environment. In the past we spoke of *outer space* as everything a few miles above the surface of the earth, and we felt that we live rather non interactively with outer space. Today we speak of *earth-space* that extends at least out to the bow shock, and we are quite aware that we live interactively with this domain and the region beyond.

Although plasma is overwhelmingly the most commonly occurring state of matter in the universe, it is not common at all on earth. Laboratory plasma is difficult to create, difficult to confine to a localized volume of space, difficult to sustain in time, and difficult to study diagnostically. One very special virtue of the magnetosphere is that it presents us with one of nature's own plasma laboratories in which we can make measurements, develop and apply concepts, perform diagnostic tests, and develop valid theories of low density, essentially collisionless, plasma processes. The capability of accessing the magnetosphere to conduct space plasma studies has been absolutely necessary for the progress made to date. Very few discoveries in space were anticipated by theoretical or other means prior to *in situ* observations and measurements. Other components of the solar-terrestrial system, such as the solar wind, similarly constitute natural space plasma laboratories that can be accessed, studied, and understood.

Components of the solar-terrestrial system include:

- The sun and its atmosphere. By *solar atmosphere* is meant the regions immediately above the photosphere out, say, to the peak coronal temperature.

- The heliosphere which in general terms extends out to interstellar space. The heliosphere contains the solar wind and extends essentially to the outer limits of the flow.

- The magnetosphere of the planets, especially the earth.

- The ionospheres and/or upper atmospheres of the planets.

- Boundaries or interfaces between components of the system.

Methods of studying the solar-terrestrial system include:

- Ground based observations and measurements.

- Balloon and rocket experiments.

- Satellite observations and *in situ* measurements.

- Data analysis.

- Theory.

In this delineation of methods, data analysis is elevated to a separate technique distinct from experimentation or theory. This is done to emphasize the unique role of data analysis in solar-terrestrial physics. Perhaps only in this field and in elementary particle physics does data analysis achieve this special status.

In simple terms, the "pure" data analyst does not design or perform experiments and does not develop theoretical descriptions. The work of the data analyst, as the name suggests, is simply to analyze the data. This entails, to greater or lesser degree, converting raw data into physical variables, organizing the data into coherent descriptions, and interpreting the data. These tasks require great familiarity with the details of the experimental operations and measurements and with the contents and predictions of prevailing theories. The main reason for the rapid evolution of this new *kind* of professional in our field is the variety and volume of data that is often available and necessary to describe a physical event or process. Some events have been studied using the collective results from dozens of experiments and the collective expertise of dozens of analysts. Today, data analysis is done by experimentalists, theorists, *and* data analysts.

One of the major problems and tasks in developing satisfactory space missions and observational campaigns is development of the data management system required to handle the vast amount of data to be assembled and to render this data available in suitable form for analysis. A recent development that may contribute to the design of future data management systems is the so-called CDAW (Coordinated Data Analysis Workshop). In a CDAW a vast amount of data relating to a physical event or process is collected and made available for comparative analysis in common format. The data derives from all available experiments and observations relating to the CDAW and may include satellite, rocket, balloon, or ground-based measurements. A CDAW requires a large central computer, elaborate software capability eventually including remote access, and substantial financial resources - not to mention the time and sustained effort of many scientists. Progress has been gradual, but the CDAW concept should prove to be quite valuable if properly developed.

With the exclusion of data analysis, theoretical methods essentially fall into two (related) categories:

- analytical methods

- computer simulation methods

The analytical approach is the traditional one of the theorist. Simulation methods are only recently playing a role because of their dependence on the large, super computer.

The analytical approach is the most desirable method to use since its conclusions are precise and may be tested. If the analytical deductions are made directly from the governing force laws and the laws of motion, then the conclusions are theoretical predictions unlikely to be refuted. More commonly the analytical approach is used on a reduced set of equations or laws thought to be adequate for the applications in mind. Here, any inadequacy in the analytical conclusions reflects directly upon the inadequacy of certain physical assumptions in the theory. The theory may then be modified until satisfactory conclusions are reached. In this way, a satisfactory theory will evolve with the most important physical processes operative identified.

An evolution in the analytical approach is the development of large scale physical models. This approach was probably used first in meteorology in trying to model weather patterns on a large or global scale. It is the enormous scope of the effort, namely to describe an atmospheric flow quantitatively and predictively, that limits use of pure analytical methods. The objective in large scale modelling is to identify the primary operative physical processes of interest. The model will only approximate the actual physical system and the equations used are a reduced version of the full set.

Adeptness in using large computers is a prerequisite to large scale modelling. Models can be quite elaborate and numerical procedures very complex and expensive. An issue with large scale modellers is always the reliability of their calculated results. This is no simple matter: numerical idiosyncrasies of coupled non linear equations can be as difficult to unravel as finding analytical solutions.

Computer simulation is a relatively recent approach in theoretical physics. In a computer simulation of a physical process, the objective is to simulate completely the implications of a governing set of equations. Computer simulation is essentially the maturation of the physical modelling approach; i.e., the successful computer-based physical model is a computer simulation of the physical system. The practical advantage of a computer simulation is often in treating an idealized system that is far simpler to handle than the real one.

An example of computer simulation is to examine the rate of thermalization of N particles in a box. This is done, say, by placing 200 hard spheres in a cubical box with rigid walls. The spheres exert no interparticle forces and collide elastically with each other and the walls. Starting with a uniform initial velocity distribution, the velocity distribution is then computer calculated at subsequent times. The finding is that after a surprisingly small number of collisions, the Boltzmann distribution substantially results. The simulation may then be improved by turning on an interparticle force, increasing the temperature, etc.; in the process, one also searches for new physical results. In this way, computer simulation has been used to examine the time required for thermalization, fluctuations about the equilibrium configuration, and aspects of phase transitions.

A second fascinating example of computer simulation has been to follow the evolutionary formation of a planet from an assemblage of planetesimals bound to a star. Realistic conditions are imposed: a fixed total mass, a variety of initial conditions, Keplerian orbits, and application of Newton's laws of motion. The results are simple and striking and suggest that planetary configurations similar to the solar system - smaller, denser inner planets and larger, less dense outer planets - may be commonplace among stars.

In the solar-terrestrial context, the computer simulation methods are being applied to microscopic and macroscopic processes. This approach will remain a permanent one in the field. Because of the large number of variables, the necessity of working in three dimensions, non linear coupling, and no simple symmetry conditions, realistic computer simulation calculations are highly complex. Simulation work must cope with both numerical and physical instabilities -and sort them out. Particularly troublesome are singularities and critical points (which govern the topology and many physical features of the solution) and the proper imposition of stationary and time dependent boundary conditions.

We have attempted to meet several objectives in this book:

- Presentation of basic principles that apply to all components of the solar-terrestrial system.

- Coverage of important areas and recent results in solar, heliospheric, magnetospheric and atmospheric physics. In this effort, undoubtedly there will be some omissions in coverage and non-uniformity in emphasis.

- Identification of major questions to be addressed.

The solar system has always offered mystery and fascination to encourage scientific investigation, and excitement and amazement to the successful investigator. Examples are in the meticulous work of Kepler and the unparalleled achievements of Newton. Yet, compared to today, the solar system seemed mundane and ordinary a quarter of century ago, at the beginning of the Space Age. Since 1957, thousands of satellites have been launched into space. Spectacular accomplishments include the discovery of the Van Allen radiation belts, discovery and exploration

Figure 2. John W. Young, Apollo 16 commander, walking on the moon, April 21, 1972. (Courtesy of NASA.)

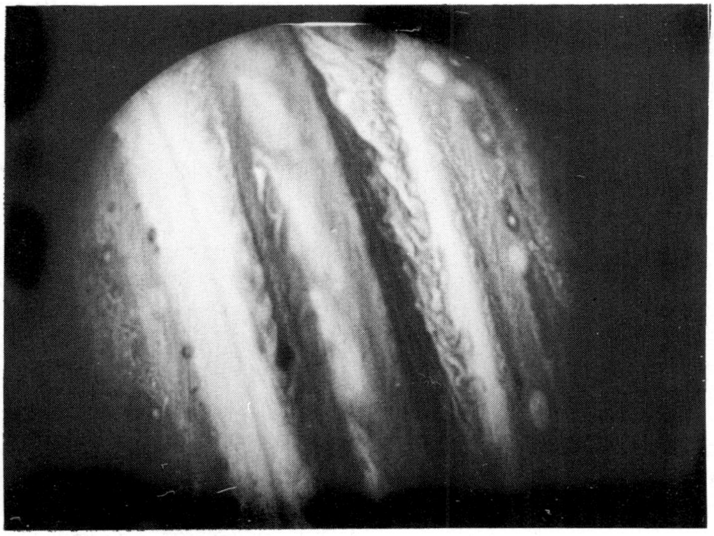

Figure 3. Voyager 1 photograph of Jupiter taken at a range of 37.5 million kilometers on January 27, 1979. (Courtesy of NASA.)

Figures 4. Voyager 1 photograph of the underside of Saturn's rings taken at a range of 740,000 kilometers on November 12, 1980. (Courtesy of NASA.)

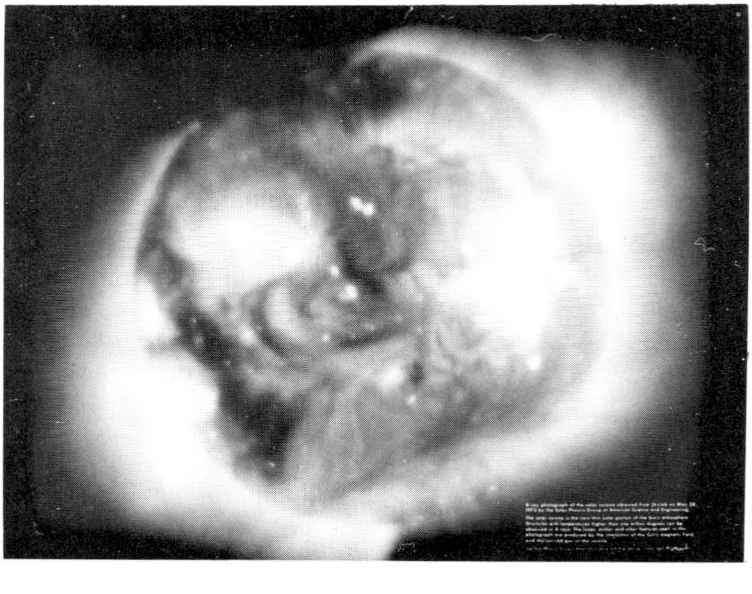

Figure 5. A view of the x-ray corona of the sun from Skylab. (Courtesy of NASA).

of the earth's magnetosphere, manned missions and sample returns from the Moon (Figure 2), exploration of the planets out to Saturn (Figures 3 and 4), discovery of the UV and x-ray emission spectra of the sun (Figure 5), the finding of coronal and ionospheric holes, discovery of new planetary satellites, sampling the amazing outer environments of Jupiter, Saturn, and Venus, witnessing a sandstorm on Mars, and much else. The freshman non science major who takes a one semester course today in solar system physics has more knowledge of the earth, sun, and each of the planets than the research scientist of only two decades ago. But the Space Age is only in its infancy. The view of the solar system and the cosmos a century from now may differ more from ours than ours does from the primitive, pre Copernican view of five centuries ago.

Although these will not be recounted, equally spectacular progress has been made in astronomy during the Space Age (for example, pulsars and quasars were hitherto unknown!). And when the Space Telescope flies, astronomy text books may require monthly revisions. Solar-terrestrial physics will always be a fundamental part of astronomy. Because of its proximity and intrinsic importance, the sun has been and will continue to be studied far more quantitatively than any other star. Knowledge of the sun will profoundly influence astronomical investigations and interpretations. More and more it is becoming apparent that solar characterics and processes are common features of stars. Examples are flares, the corona, x-ray emission, and the overriding importance of magnetic fields. The detailed quantitative knowledge developed in solar-terrestrial physics -a field that has become sophisticated and mature - will be a major asset to developments in astronomy, particularly through contraining interpretations and guiding investigations. It will be somewhat like how the success of the quantum theory in atomic and molecular physics has affected developments in wide areas of chemistry and biology.

SOLAR SYSTEM MAGNETOHYDRODYNAMICS

George L. Siscoe
Department of Atmospheric Sciences, University of California
Los Angeles

I. THE MACROSCOPIC EQUATIONS OF A PLASMA

I.1. From Particles to Fluids

Continuum mechanics is that branch of physics that treats the motions of infinitely deformable matter. It embraces hydrodynamics, aerodynamics, magnetohydrodynamics (MHD), and magnetogasdynamics. The first two differ in that the former is incompressible and the latter compressible fluid dynamics. The prefix, magneto, signifies the addition of the ponderomotive force(colloquially called the J-cross-B force) to the usual pressure gradient, gravitational and viscous forces of fluid dynamics. Magnetofluid mechanics applies to fluids that can carry electrical currents, such as liquid metals and plasmas. Our interest in Solar System MHD is confined to the latter.

In contrast to particle mechanics and rigid body mechanics, the mass on which the forces act in continuum mechanics is distributed throughout some volume of space, and any portion of the distributed mass can in principle move in an arbitrary direction or manner relative to any other portion. This has as one of its consequences the necessity of including the coordinates of space among the independent variables; and thus the equations of continuum mechanics are partial differential equations in space and time rather than ordinary differential equations in time.

The dependent variables of hydro- and aerodynamics are velocity (a vector), mass density (a scalar) and pressure (usually a scalar but sometimes a tensor). In magnetohydro- and magnetogasdynamics, one adds the magnetic field, and in magnetogasdynamics the occasions when the pressure must be treated as a tensor are more common.

For those who are concerned with applying the equations of continuum mechanics to describe the behavior of the oceans or the atmosphere, there are instruments that measure directly the required dependent variables, such as wind and pressure. The winds of space generally

R. L. Carovillano and J. M. Forbes (eds.), Solar-Terrestrial Physics, 11–100.

greatly exceed those which occur in the troposphere. The solar wind traverses in one second a three hour track of the hurricane wind. Yet the plasmas of space are so tenuous that the pressures they exert even with the full force of their winds are immeasurable by direct pressure sensing devices. The instruments for measuring the properties of plasmas in space take advantage of the fact that the individual particles that comprise the "continuum" there are electrically charged. These instruments either detect the charges of the particles individually or collectively as an electrical current. There is then a gap between the dependent variables that enter into the equations of continuum mechanics and the measured variables. The gap must be filled by mathematical processing of the data, in order to go from information about the individual particles to information about the fluid that they constitute in their bulk. There is a formalism that treats this problem specifically. This formalism begins with the phase space density.

I.2 Phase Space Density

It is necessary to specify the physical state of our collection of individual particles precisely. The physical state of a particle is given by its mass and charge, which are known constants (we treat only the non-relativistic case), its position $\vec{x}$ and its velocity $\vec{v}$. Therefore six independent numbers are needed to fix the physical state of a particle, $(x_1, x_2, x_3, v_1, v_2, v_3)$, where the subscripts denote the three independent orthogonal axes of configuration and velocity space. We may think of the six numbers as being the coordinates of the particle in a six-dimensional space, called phase space.

If we were so to locate each particle of our plasma, we would build up a non-homogeneous distribution of points occupying some portion of phase space. In virtually all problems that are concerned with the behavior of space plasmas on a macroscopic scale, the number of particles is truly enormous. Thus, it is meaningful to speak of a density of points in phase space, $f(x,v)$, which is a function of the six independent variables, and when multiplied by the six-dimensional volume element d^3xd^3v gives the number of particles with coordinates that lies in the range $(x_1, x_2, x_3, v_1, v_2, v_3)$ to $(x_1 + dx_1, x_2 + dx_2, x_3 + dx_3, v_1 + dv_1, v_2 + dv_1, v_3 + dv_3)$. Since we must allow for the change of the distribution with time, the phase space density is actually a function of seven independent variables, $f(\vec{x}, \vec{v}, t)$. The quantity f as we have defined it is also referred to as the single particle distribution function.

I.3 The Continuity Equation in Phase Space

The number density in six-dimensional phase space is $f(\vec{x}, \vec{v})$, to which we can formally assign a current, viz. $\dot{\vec{x}} f + \dot{\vec{v}} f$, analogous to the particle current or flux given by $n\vec{v}$ in usual language. Applying the generalized conservation principal

$$- \frac{\partial Q}{\partial t} = \nabla_n \cdot (Q\vec{v}_n) \tag{I.1}$$

where Q is any density, ∇_n is the divergence in n-space and $\vec{V}_n$ is n-space velocity, to f we find

$$- \frac{\partial f}{\partial t} = \frac{\partial}{\partial x_i} (\dot{x}_i f) + \frac{\partial}{\partial v_i} (\dot{v}_i f) \qquad (I.2)$$

in which the summation convention is used (i.e. $a_i b_i = a_1 b_1 + a_2 b_2 + a_3 b_3$). Since $\dot{x}_i = v_i$ and $\dot{v}_i = F_i/m$, (I.2) can be written

$$- \frac{\partial f}{\partial t} = \nabla \cdot (\vec{v} f) + \nabla_v \cdot (\frac{\vec{F}}{m} f) \qquad (I.3)$$

where $\nabla_v = \frac{\partial}{\partial v_1} \hat{e}_1 + \frac{\partial}{\partial v_2} \hat{e}_2 + \frac{\partial}{\partial v_3} \hat{e}_3$, and $\hat{e}_i$ is the unit vector corresponding to the ith coordinate.

The only forces to which our individual particles are subject are the gravitational and Lorentz forces

$$\vec{F} = m\vec{g} + q \vec{E} + q \vec{v} \times \vec{B} \qquad (I.4)$$

where q is the electrical charge on the particle. (The only other forces at this fundamental, single particle level are the weak and strong forces of nuclear interaction, with which we are not concerned.)

We now proceed to reduce (I.3) to a more specific form. Since x and v are independent variables, we may set $\nabla \cdot (\vec{v} f) = \vec{v} \cdot \nabla f$. Also it is evident that $\nabla_v \cdot (m \vec{g}) = \nabla_v \cdot (q \vec{E}) = 0$ since none of m, g, q, or E depends on v. Finally

$$\nabla_v \cdot q(\vec{v} \times \vec{B}) = q \vec{B} \cdot (\nabla_v \times \vec{v}) - q \vec{v} \cdot (\nabla_v \times \vec{B})$$

But both curls vanish, therefore

$$\nabla_v \cdot (\frac{\vec{F}}{m} f) = \frac{\vec{F}}{m} \cdot \nabla_v f + f \nabla_v \cdot \frac{\vec{F}}{m}$$

and

$$- \frac{\partial f}{\partial t} = \vec{v} \cdot \nabla f + \frac{\vec{F}}{m} \cdot \nabla_v f \qquad (I.5)$$

I.4 The Boltzmann Equation

We have until now ignored the effects of collisions. We can include them by noting that the effect of a collision is to relocate a particle along the velocity axes of phase space "instantaneously". Therefore we may represent the effect of collision by adding a separate collision term to (I.5) as follows:

$$\frac{\partial f}{\partial t} + \vec{V} \cdot \nabla f + \frac{\vec{F}}{m} \cdot \nabla_V f = (\frac{\partial f}{\partial t})_{coll} \qquad (I.6)$$

This form of the conservation equation is referred to as the Boltzmann equation.

I.5 Liouville's theorem:

In the absence of collisions, (I.6) becomes

$$\frac{\partial f}{\partial t} + \vec{V}\cdot\nabla f + \frac{\vec{F}}{m}\cdot\nabla_V f \equiv \frac{df}{dt} = 0 \qquad (I.7)$$

with the meaning that f is constant as it is convected with the particles.

I.6 Forming the Macroscopic Variables

We consider a two component gas composed of positive and negative ions, which we take to protons and electrons in practice. The charge, q, is then either + e or -e, where e is the value of the electronic charge. The macroscopic or bulk quantities are then formed from the phase space density as follows:

<u>mass density</u> $\rho = \sum_a m_a \int f_a \, d^3 v$ (I.8)

<u>charge density</u> $\rho_c = \sum_a e_a \int f_a \, d^3 v$ (I.9)

<u>bulk velocity</u> $\vec{V} = \frac{1}{\rho} \sum_a m_a \int \vec{v} f_a \, d^3 v$ (I.10)

<u>current density</u> $\vec{J} = \sum_a e_a \int \vec{v} f_a \, d^3 v$ (I.11)

<u>pressure tensor</u> $P_{ij} = \sum_a m_a \int (v_i - V_i)(v_j - V_j) f_a \, d^3 v$ (I.12)

<u>internal energy</u> $u = \frac{1}{2} \text{Trace}(P) = \frac{1}{2} \sum_a m_a \int (\vec{v} - \vec{V})^2 f_a \, d^3 v$ (I.13)

<u>heat flux vector</u> $\vec{q} = \frac{1}{2} \sum_a m_a \int (\vec{v} - \vec{V})^2 (\vec{v} - \vec{V}) f_a \, d^3 v$ (I.14)

where a is an index with values 1 and 2 for the two components.

One can see that the macroscopic variables are constructed out of appropriate <u>velocity moments</u> of f. Mass and charge density result from the zeroth velocity moment, bulk velocity and electrical current density from the first moment, pressure and internal energy are related to the second moment, and the heat flux vector to the third moment.

I.7 Derivation of the Macroscopic Equations

If we were given continuous measurements of f for all regions of space, or if f were computed from the Boltzmann equation and a comprehensive set of initial measurements, the preceding relations could be used to determine any of the macroscopic quantities of interest at any time and place of interest. Of course, neither procedure is practical, and even if it were, it would be wasteful, since f contains information on the distribution of the particles in velocity space, which is completely ignored in the macroscopic variables. Phase space density is a function of seven variables, whereas the macroscopic quantities depend on only four. There is a great advantage then, if one is interested only in the macroscopic quantities, to find equations to predict subsequent values of the macroscopic variables from some initial set which is specified by a model or obtained from a definite set of measurements of f. What we require is the macroscopic analog of the Boltzmann equation for f. Since in the macroscopic description there are a number of dependent variables, it is evident that more than one such analog is needed. We have noted that the macroscopic variables are obtained from f through the operation of taking velocity moments. It is natural then to seek the desired macroscopic equations by taking velocity moments of the Boltzmann equation.

With the foregoing motivation, we may proceed in a strictly formal manner, and define the n-th moment operator M_n operating on any quantity or expression designed by the symbol $[\cdots]$ by

$$M_n[\cdots] = \sum_a \frac{m_a}{n!} \int v_i^n [\cdots]_a d^3v, \quad n = 0,1,2 \cdots \tag{I.15}$$

where the summation convention is implied, i.e. $(v_i)^2 = v^2$, $(v_i)^3 = v^2 v_i$, etc. We apply this operator to the Boltzmann equation in the form which is most convenient for our purpose

$$\frac{\partial f}{\partial t} + \nabla \cdot (f \vec{v}) + \nabla_v \cdot (\frac{\vec{F}}{m} f) = (\frac{\partial f}{\partial t})_{coll} \tag{I.16}$$

Since the number of particles, the momentum, and the kinetic energy are conserved in collisions, we have as a direct consequence

$$M_n [(\frac{\partial f_a}{\partial t})_{coll}] = 0 \qquad n = 0,1,2 \text{ only} \tag{I.17}$$

<u>Mass</u> <u>n = 0</u>: Applying (I.15) to (I.16) with n = 0 and with definitions (I.8) and (I.10) there results immediately the <u>continuity equation</u>.

$$\frac{\partial \rho}{\partial t} + \nabla \cdot (\rho \vec{V}) = 0 \tag{I.18}$$

where it has been assumed that $\frac{\vec{F}}{m}$ f vanishes sufficiently rapidly at infinity in velocity space. Equation (I.18) is the expression for the conservation of mass as can be seen by integrating over an arbitrary volume element

$$\int \frac{\partial \rho}{\partial t} d^3x = \frac{d}{dt} \int \rho \, d^3x = \frac{dM}{dt} = - \oint_s \rho \vec{V} \cdot \hat{n} \, d^2 x \tag{I.19}$$

where s is the surface of the volume. The integrand is the mass flux per unit area. Thus, the rate of change of mass in the volume dM/dt is the net rate at which it is entering or leaving the volume.

The continuity equation is often written in terms of the convective derivative (also known as the total derivative and the substantial derivative) defined by

$$\frac{d}{dt} \equiv \frac{\partial}{\partial t} + \vec{V}\cdot\nabla \qquad (I.20)$$

It represents the time rate of change in a frame of reference moving with the fluid. Then (I.19) becomes

$$\frac{d\rho}{dt} + \rho\ \nabla\cdot\vec{V} = 0 \qquad (I.21)$$

For future reference we note that (I.21) is sometimes used as an equation for replacing $\nabla\cdot\vec{V}$ by an expression involving only ρ, namely

$$\nabla\cdot\vec{V} = -\frac{1}{\rho}\frac{d\rho}{dt} = -\frac{d\ell n\,\rho}{dt} = \frac{d}{dt}\ell n\left(\frac{1}{\rho}\right) = \rho\frac{d}{dt}\left(\frac{1}{\rho}\right) \qquad (I.22)$$

Momentum $n = 1$: After several steps of fairly straightforward algebra, the first moment of the Boltzmann equation can be cast in the form of a momentum equation also known as an Euler Equation which written in component notation becomes

$$\rho\frac{dV_i}{dt} = -\frac{\partial P_{ij}}{\partial x_j} + \rho_c E_i + (\vec{J}\times\vec{B})_i + \rho g_i \qquad (I.23)$$

in which the second term is the divergence of the pressure tensor defined by eq. (I.12).

The right hand side of the Euler equation enumerates the various forces that can accelerate an element of plasma, namely imbalances in the pressure forces acting on its surface, the electrostatic force, the ponderomotive force and the gravitational force. In the absence of interactions with neutral particles, there are no other forces. The viscous force is contained in the divergence of the pressure tensor.

Viscosity arises when a fluid is in non-uniform motion. Momentum of relative motion is exchanged between adjacent fluid elements through the cross migration of the individual constituent particles in the course of their thermal wandering. The viscous force therefore must depend explicitly on velocity gradients and on parameters that characterize the degree of cross migration. The part of the pressure tensor that contains the effect of viscosity is called the viscous stress tensor. In the highly collisional domain the viscous stress tensor has the form

$$S_{ij} = -\eta\left(\frac{\partial V_i}{\partial x_j} + \frac{\partial V_j}{\partial x_i} - \frac{2}{3}\nabla\cdot\vec{V}\,\delta_{ij}\right) - \zeta\,\nabla\cdot\vec{V}\,\delta_{ij} \qquad (I.24)$$

where η and ζ are the coefficients of viscosity, both of which are positive numbers.

In the collisional domain for which (I.24) is an appropriate representation of S_{ij}, the thermal motions of the particles are isotropic and the non-viscosity, purely pressure part of the pressure tensor is also isotropic. The entire pressure then has the form

$$P_{ij} = p\,\delta_{ij} + S_{ij} \tag{I.25}$$

where p is the scalar pressure. The force that results from taking the divergence of (I.25) can be written as explicit pressure-force and viscous-force terms to the right hand side of the Euler equation. These are

$$\nabla\cdot\overleftrightarrow{P} = \nabla p + \eta\nabla^2\vec{V} + (\zeta + \frac{1}{3}\eta)\,\nabla(\nabla\cdot\vec{V}) \tag{I.26}$$

where η and ζ have been treated as constants, as is normally done for mathematical convenience, although they are functions of pressure and temperature. In ordinary fluid dynamics the terms in (I.26) alone constitute the right hand side of the Euler equation, in which form it is known as the Navier-Stokes equation.

In a large number of solar system applications, the plasma must be treated as basically collisionless. Then the magnetic field greatly inhibits cross migration in the direction perpendicular to itself, but not in the parallel direction. The viscous stress tensor then has an influence mainly on velocity gradients parallel to the field. (See Rossi and Olbert, 1970, for a discussion of the general viscous stress tensor in a magnetized plasma). As a general observation, viscosity has not yet played a major role in discussions of solar system plasmas, although there are occasional exceptions to this statement in solar wind theory and in the theory of the solar wind coupling to planetary magnetospheres and ionospheres. The viscous stress tensor and viscous forces therefore will be omitted in the remainder of the chapter.

When collisions are sufficiently rare that charged particles gyrate around the magnetic field many times before colliding, the pressure tensor tends to become anisotropic. With reference to a local cartesian coordinate system which has its z-axis parallel to the field, the gyromotion acts to isotropy the pressure in the xy-plane. The component that relates to the z-axis is decoupled to a degree that depends on the actual extent of collisions. Two scallars are therefore necessary to represent the two partially or completely decoupled motions. The pressure tensor then has the form

$$P_{ij} = (p_{\parallel} - p_{\perp})\,b_i b_j + p_{\perp}\,\delta_{ij} \tag{I.27}$$

or in vector notation

$$\overleftrightarrow{P} = (p_{\parallel} - p_{\perp})\,\hat{b}\,\hat{b} + p_{\perp}\,\overleftrightarrow{I} \tag{I.28}$$

where $\hat{b}$ is a unit vector parallel to the magnetic field, $\vec{B}$, and $\overleftrightarrow{I}$ is the unit, diagonal tensor. To verify that (I.27) and (I.28) express the properties attributed to the pressure tensor in this case, scalar multiply (I.28) with unit vectors that are perpendicular and parallel to $\vec{B}$. Thus, if we once again let $\hat{b} = \hat{z}$,

$$\hat{x} \cdot \overleftrightarrow{P} \cdot \hat{x} = \hat{y} \cdot \overleftrightarrow{P} \cdot \hat{y} = p_{\perp} \tag{I.29}$$

$$\hat{z} \cdot \overleftrightarrow{P} \cdot \hat{z} = p_{\parallel} \tag{I.30}$$

Equations (I.29) and (I.30) define the scalars $p_{\perp}$ and $p_{\parallel}$.

Energy n = 2: The second moment of the Boltzmann equation leads after several intermediate but straightforward steps to an expression relating energy densities in the plasma

$$\frac{\partial}{\partial t}\left(\frac{1}{2}\rho V^2 + u\right) + \frac{\partial}{\partial x_j}\left[\left(\frac{1}{2}\rho V^2 + u\right) V_j + P_{jk} V_k + q_j\right] = J_j E_j + \rho V_j g_j \tag{I.31}$$

Inspection of equations (I.18), (I.23), and (I.31) shows that they are prognostic equations for mass density, ρ, material momentum density, $\rho\vec{V}$, and the total kinetic energy density, $\frac{1}{2}\rho V^2 + u$, that is, they are equations that specify the time rates of change of these quantities. Since we know that mass momentum and energy are subject to conservation laws, it should be possible to recast these expressions into the general conservation form given by equation (I.1). To achieve the desired restructuring we need the field equations for the electromagnetic and gravitational fields.

I.8 The Field Equations

The electric, magnetic and gravitational fields are related to the macroscopic quantities defined in I.6 by Maxwell's equations and Newton's equations.

$$\nabla \cdot \vec{E} = \rho_c/\varepsilon_o \tag{I.32}$$

$$\nabla \cdot \vec{B} = 0 \tag{I.33}$$

$$\nabla \times \vec{E} = -\partial\vec{B}/\partial t \tag{I.34}$$

$$\nabla \times \vec{B} = \mu_o \vec{J} + \mu_o \varepsilon_o \partial\vec{E}/\partial t \tag{I.35}$$

and

$$\nabla \cdot \vec{g} = -4\pi G \rho \tag{I.36}$$

$$\nabla \times \vec{g} = 0 \tag{I.37}$$

$\varepsilon_o = 8.854\text{x}10^{-12}$ Farad/m, $\mu_o = 4\pi\text{x}10^{-7}$ Henry/m

$1/\varepsilon_o\mu_o = c^2$, $c = 2.998\text{x}10^8$ m/sec.

$G = 6.670 \times 10^{-11}$ m^3/kg-sec.

The fields exert stresses and possess momentum and energy. The stresses are given by the Maxwell stress tensor

$$T_{ij} \equiv \varepsilon_o E_i E_j + \frac{1}{\mu_o} B_i B_j - (\frac{\varepsilon_o}{2} E^2 + \frac{1}{2\mu_o} B^2) \delta_{ij} \qquad (I.38)$$

and the gravitational stress tensor

$$\Gamma_{ij} = - \frac{1}{4\pi G} (g_i g_j - \frac{1}{2} g^2 \delta_{ij}) \qquad (I.39)$$

The divergences of the stress tensors can be seen to correspond to the body forces on the plasma that appear in equation (I.23)

$$\frac{\partial T_{ij}}{\partial x_j} = \varepsilon_o\mu_o \frac{\partial S_i}{\partial t} + \rho_c E_i + (\vec{J} \times \vec{B})_i \qquad (I.40)$$

$$\frac{\partial \Gamma_{ij}}{\partial x_j} = \rho g_i \qquad (I.41)$$

where

$$\vec{S} \equiv \frac{\vec{E} \times \vec{B}}{\mu_o} \qquad (I.42)$$

is the Poynting vector.

The energy densities of the fields are given by $\varepsilon_o E^2/2$, $B^2/2\mu_o$, and $\rho\phi$ where ϕ is the gravitational potential defined by

$$\vec{g} = - \nabla \phi \qquad (I.43)$$

The flux of energy in the electromagnetic field is the Poynting vector, as can be identified from Poynting's theorem

$$\nabla \cdot \vec{S} = \nabla \cdot \frac{\vec{E} \times \vec{B}}{\mu_o} = \frac{1}{\mu_o} [\vec{B} \cdot (\nabla \times \vec{E}) - \vec{E} \cdot (\nabla \times \vec{B})]$$

which becomes upon substitutions from Maxwell's equations

$$\frac{\partial}{\partial t} (\frac{\varepsilon_o}{2} E^2 + \frac{1}{2\mu_o} B^2) + \nabla \cdot \vec{S} = - \vec{E} \cdot \vec{J} \qquad (I.44)$$

The term on the right hand side of this expression, $-\vec{E}\cdot\vec{J}$, is the electromechanical energy conversion term. It represents a source or a sink of electro-magnetic energy, depending on the sign of $\vec{E}\cdot\vec{J}$.

The flux of gravitational energy can be constructed in the usual way, by multiplying the gravitational energy density, $\rho\phi$, by the velocity, $\vec{V}$. The choice is verified by taking the divergence of the trial expression and showing that it satisfies the appropriate conservation equation

$$\nabla \cdot (\rho \vec{V} \phi) = -\phi \frac{\partial \rho}{\partial t} - \rho \vec{V} \cdot \vec{g}$$

or

$$\frac{\partial}{\partial t}(\rho \phi) + \nabla \cdot (\rho\phi\vec{V}) = -\rho\vec{V}\cdot\vec{g} + \rho\frac{\partial \phi}{\partial t} \qquad \text{(I.45)}$$

On the right hand side the combination $-\rho\,\vec{V}\cdot\vec{g}$ is recognizable as the gravitational-mechanical energy conversion term. The second term, while formally present, in practice is absent since the gravitational fields are assumed to be constant.

I.9 The Conservation Equations

The continuity equation (I.18) is already in the form of a conservation equation for mass. It remains to find similar expressions for momentum and energy. Combination of equations (I.23) (I.40) and (I.41) produces the desired result for momentum

$$\frac{\partial}{\partial t}(\rho V_i + \varepsilon_o \mu_o S_i) + \frac{\partial}{\partial x_j}(A_{ij}) = 0 \qquad \text{(I.46)}$$

where

$$A_{ij} \equiv P_{ij} + \rho V_i V_j - T_{ij} - \Gamma_{ij} \qquad \text{(I.47)}$$

is the grand momentum stress tensor. It contains the pressure, Maxwell and gravitational stress tensor, which have already been introduced, and a new term, $\rho V_i V_j$ which is called the Reynold's stress tensor. It is the macroscopic analog of the viscous stress tensor (which at this level of exposition is still latent in the pressure stress tensor). It represents momentum in the i-direction, ρV_i, that transferred by the velocity component, V_j, across a plane which has its normal in the j-direction. In turbulence theory, the analog between the transport of momentum by random eddys on the macroscale and by the random motions of molecules on the microscale is particularly appropriate, and it is where the Reynolds stress tensor as an explicit concept was developed.

Equation (I.46) shows that the combination $\varepsilon_o\mu_o\vec{S} = \vec{S}/c^2$ is the electromagnetic momentum flux density. In virtually all solar system applications $\vec{S}/c^2$ is completely negligible compared to $\rho\vec{V}$, and it can be ignored in the conservation equation with impunity. The validity of this statement will be demonstrated in Section II.2.

The expression for energy conservation results from combining equations (I.31), (I.44) and (I.45).

$$\frac{\partial}{\partial t}\left(\frac{1}{2}\rho V^2 + u + \frac{\varepsilon_o}{2} E^2 + \frac{1}{2\mu_o} B^2 + \rho\phi\right) +$$

$$\nabla \cdot \left[\left(\frac{1}{2}\rho V^2 + u + \overleftrightarrow{P}\right) \cdot \vec{V} + \vec{q} + \vec{S} + \rho\vec{V}\phi\right] = \rho \frac{\partial\phi}{\partial t} \tag{I.48}$$

We recognize the terms acted on by the time derivative to be the various forms of energy densities. The expression shows our list of energy densities to be complete. The terms acted on by the divergence operator represent processes by which energy can enter and leave a volume through its surface. It is clear that $(\frac{1}{2}\rho V^2 + u + \rho\,\phi)\,\vec{V}$ is the flux of the total kinetic energy density $(\frac{1}{2}\rho V^2 + u)$ plus the gravitational energy density through the surface. The combination $\overleftrightarrow{P} \cdot \vec{V}$ is the work done on the volume by the pressure stress as a result of the motion $\vec{V}$ across the surface. In the case of a scalar pressure, the combination u + p which enters the conservation equation as an effective convected energy density is referred to as the enthalpy of the flow.

The heat flux vector, $\vec{q}$, is seen to have the character expected of it, namely the flow of energy across a surface, or more precisely, it is an energy flux density. Note that we may have the transfer of kinetic energy in the absence of convection if $\vec{q} \neq 0$ and $\vec{V} = 0$. The transfer in this case results completely from directional asymmetries in the single particle distribution function, that is, the heat flux represents a microscopic transport process. The macroscopic medium itself is not moving, but energy is flowing in or through it.

The flow of electromagnetic energy is represented solely by the Poynting vector, as claimed earlier. In contrast to the electromagnetic momentum flux density, the electromagnetic energy flux density is often comparable to or greater than the largest of the other terms in the equation. It must be retained in solar systems applications of MHD.

Charge To complete the discussion of conservation equations, we derive the equation for the conservation of charge. For this we define a new operator

$$Q_n\,[\cdots] \equiv \sum_a \frac{e_a}{n!} \int d^3v\; v_i^n\,[\cdots]_a \tag{I.49}$$

Comparison with equation (I.15) shows that we can distinguish between the two operators by calling the first the "mass-velocity moment operator" and the second the "charge-velocity moment operator." For the zeroth charge-moment of the Boltzmann equation we find

$$\frac{\partial \rho_c}{\partial t} + \nabla \cdot \vec{J} = 0 \tag{I.50}$$

The collision term has again been set equal to zero since charge is conserved in collisions. This result is consistent with Maxwell's equations from which it can also be derived. The divergence of (I.35) gives

$$0 = \mu_o \nabla \cdot \vec{J} + \mu_o \varepsilon_o \frac{\partial}{\partial t} \nabla \cdot \vec{E}$$

which with (I.32) is seen to be the same as (I.50).

The first charge-moment of the Boltzmann equation will be discussed in Section II.1

I.10 Inclusion of Neutral Particle Interactions

Ionization of neutrals, charge exchange between ions and neutrals and the production of neutrals from ions by recombination are important processes in solar system plasma. Examples of interacting bodies of plasmas and neutrals include the solar wind and the interstellar medium, the solar wind and comets, the solar wind and planetary atmospheres, the ionized and neutral components of satellite torii in the magnetospheres of Jupiter and Saturn, and planetary ionospheres and the neutral atmospheres in which they occur. The effects of neutrals can be included in the MHD equations as source and loss terms on the right hand sides of the conservation equations just derived.

Designate the source and loss terms corresponding to the equations for the conservation of mass momentum and energy by Q_M, Q_P, and Q_E respectively. Symbolically Q_M can be written as

$$Q_M = \sum_k \frac{(m_n)_k n_k}{(\tau_i)_k} - \frac{\rho}{\tau_r} + \sum_{k,\ell} \frac{[(m_n)_k - (m_i)_\ell] n_k}{(\tau_{ex})_{k\ell}} \tag{I.51}$$

in which n_k is the number density of the k-th species of neutral particle which gives rise to an ion of mass $(m_n)_k$ in a characteristic ionization time $(\tau_i)_k$. In general ionization can result from photoionization, with time scale τ_{ph}, and electron impact ionization with time scale $\tau_{e\ell}$. Thus

$$\frac{1}{(\tau_i)_k} = \frac{1}{(\tau_{ph})_k} + \frac{1}{(\tau_{e\ell})_k} \tag{1.52}$$

The second term in (I.51) represents loss by recombination, with time scale τ_r. The time scale itself depends on the plasma density ρ

since the number of electron encounters per second a given ion experiences increases with the density of the plasma. The third term gives the contribution to the spontaneous increase or decrease in mass density by charge exchange between neutral species k and ionized species ℓ. Since in a charge exchange interaction, the plasma gains the mass of the former neutral and losses the mass of the former ion, the net gain is the difference in the two masses, which can be negative. If an ion species charge exchanges with its own neutral species, there is no net change in mass density. In this expression the time scale $(\tau_{ex})_{k\ell}$ is referenced to the neutral species and therefore implicitly contains the density of the ion species.

Corresponding expressions can now readily be written for Q_p and Q_E. The composite momentum of the newly created ions adds to Q_p and the composite momentum of the newly lost ions subtracts from it. Similarly the energy of the new ions adds to Q_E and the energy of the lost ions subtracts from it. It is important in this case also to include the energies of the electrons that are either gained or lost in the process. In connection with Q_E it should be noted that at least one plasma in the solar system, the Io torus, exhibits a substantial loss of energy by electromagnetic radiation which must be included in Q_E.

The purpose of this section is to indicate the existence of an important class of phenomena in solar system plasmas involving interactions with neutrals. To treat these situations the equations of ideal MHD need to be modified in the way outlined here.

I.11 The Prognostic Equation for Scalar Pressure

The equations for mass, momentum, and energy can be combined to produce a simple prognostic equation for pressure. We will treat the case of a scalar pressure first. Then there is one such equation. Later we retain the tensor character of pressure that is appropriate to a magnetized plasma. Two equations are then required. The pressure equation can replace one of the original three, and it is usual to exchange it for the most complex of these, the energy equation.

In the case of scalar pressure (I.25 without S_{ij}) the internal energy (I.13) is given by

$$u = \frac{3}{2} p \qquad (I.53)$$

The scalar product of the Euler equation (I.23 with scalar pressure) with $\vec{V}$ can be manipulated into the following form with the use of the continuity equation (I.18)

$$\frac{d}{dt}\left(\frac{1}{2}\rho V^2\right) + \frac{1}{2}\rho V^2\, \nabla\cdot\vec{V} + \vec{V}\cdot\nabla p = \rho_c\, \vec{V}\cdot\vec{E} + \vec{V}\cdot(\vec{J}\times\vec{B}) + \rho V\cdot g \qquad (I.54)$$

With the substitution (I.53), the energy equation (I.31) can be manipulated into a similar form

$$\frac{d}{dt}\left(\frac{1}{2}\rho V^2\right) + \frac{1}{2}\rho V^2\, \nabla\cdot\vec{V} + \frac{\partial}{\partial t}\left(\frac{3}{2}\,p\right) + \vec{V}\cdot\nabla\left(\frac{5}{2}\,p\right) + \frac{5}{2}\,p\,\nabla\cdot\vec{V}$$

$$= -\nabla\cdot\vec{q} + \vec{J}\cdot\vec{E} + \rho\vec{V}\cdot\vec{g} \qquad (I.55)$$

Subtracting (I.54) from (I.55), and eliminating $\nabla\cdot\vec{V}$ by use of (I.22), we find after some rearranging of the right hand side

$$\frac{d}{dt}\left(\frac{3}{2}\,p\right) - \frac{5}{2}\frac{p}{\rho}\frac{d\rho}{dt} = -\nabla\cdot\vec{c} + (\vec{J}-\rho_c\vec{V})\cdot(\vec{E} + \vec{V} \times \vec{B}) \qquad (I.56)$$

Equation (I.56) can now be rewritten in a more revealing form

$$\frac{d}{dt}\left(\frac{p}{\rho^{5/3}}\right) = \frac{-\nabla\cdot\vec{q} + \vec{j}\cdot\vec{E}^*}{\frac{3}{2}\,\rho^{5/3}} \qquad (I.57)$$

in which the symbols $\vec{j}$ and $\vec{E}^*$ are defined by

$$\vec{j} \equiv \vec{J} - \rho_c\,\vec{V} \qquad (I.58)$$

$$\vec{E}^* = \vec{E} + \vec{V} \times \vec{B} \qquad (I.59)$$

Before describing the interpretation and use of the primary result, equation (I.57), we digress briefly to note the meaning of these newly defined current and electric field variables. The current density $\vec{j}$ is the current that arises from the relative motion of positive and negative charges in equal numbers, and can be referred to as the conduction current density. The conduction current density is to be distinguished from the convection current density, $\rho_c\vec{V}$, which is simply the transport with the fluid of any excess positive or negative charge in the plasma. The total current density, $\vec{J}$, is then the sum of the conduction and convection current densities.

From electromagnetic theory, we know that in the non-relativistic limit and in the presence of a magnetic field, $\vec{B}$, the electric field, $\vec{E}^*$ in a frame of reference moving with velocity $\vec{V}$ relative to a frame in which the electric field is $\vec{E}$ is given precisely by equation (I.59). Thus, $\vec{E}^*$ is the electric field that exists in the frame of reference moving with the plasma.

Equation (I.57) is most meaningfully interpreted in the context of thermodynamics where it is seen that the time derivative is operating on a term that is related to the specific entropy of the gas. The right hand side then displays the sources and sinks of specific entropy. In order to convert to a description in these terms, we need first to discuss one of the most fundamental quantities of thermodynamics, temperature.

I.12 Temperature and Related Concepts

Temperature is not a necessary variable for our description of the macroscopic behavior of a gas. This is because temperature is uniquely related to pressure and density, which we have already included. However, it is often convenient to use temperature in place of pressure or density and also to express results or concepts in the language of thermal physics. Since our approach has been to progress from the microscale to the macroscale, we introduce temperature from the point of view of the kinetic theory of gases, where it is a measure of the total internal energy of a gas.

$$u_{total} = \frac{f}{2} nkT \tag{I.60}$$

Here n is the number density of gas particles, f denotes the <u>number of degrees of freedom</u> of motion that the gas particles possess and k is <u>Boltzmann's constant</u> ($k = 1.38 \times 10^{-23}$ Joule/deg Kelvin).

$$f = f_{translational} + f_{rotational} + f_{vibrational} \tag{I.61}$$

The number of <u>translational degrees of freedom</u> can be either 1, 2 or 3 depending on whether the thermal motion of the gas particles is constrained to one-dimension, two dimensions, or is unconstrained. Examples of motion constrained to one-dimension are beads on a wire, and, more relevant to us, a gas of charged particles constrained to move parallel to a magnetic field. Similarly a gas of charged particles constrained to move perpendicular to a magnetical is an example of a case for which $f_{translational} = 2$.

The number of <u>rotational degrees of freedom</u> can be three for a gas of non-colinear molecules. A non-colinear molecule must contain three or more atoms. Of course, if the rotation is constrained to one or two axes, the number of degrees of rotational freedom is correspondingly reduced. Diatomic molecules have at most two degrees of rotational freedom, since the moment of inertia about the common axis is too small to allow it to carry rotational energy. A monatomic gas, such as are virtually all space plasmas, has zero rotational degrees of freedom.

Diatomic and polyatomic molecules can have <u>vibrational degrees of freedom</u> There are two degrees for every vibrational mode, because an oscillator possesses both kinetic and potential energy. Vibrational modes are usually excited at temperatures well above room temperature. For example, the effective number of vibrational degrees of freedom of air in the atmosphere is zero.

The internal energy defined by equation (I.13)results purely from translational motion, but for space plasmas the other kinds of motions do not apply. To be strictly correct, however, equation (I.51) for scalar pressures should be written

$$u_{translational} = \frac{3}{2} p$$

Comparison with (I.60) with f = 3 shows that

$$p = n k T \tag{I.62}$$

This is one of the most important relations of the kinetic theory of gases. To convert it fully to our macroscopic variables defined in I 6, we need only use the mean molecular weight of the gas molecules,

$$\bar{m} \equiv \frac{\rho}{n} \tag{I.63}$$

In a two-component electron-ion plasma, (I.63) becomes (subscripts denote ion and electron quantities)

$$\bar{m} = \frac{m_i n_i + m_e n_e}{n_i + n_e} \simeq \frac{m_i}{2} \tag{I.64}$$

in which we have utilized $m_i >> m_e$ and $n_i \simeq n_e$. Equation (I.62) is then

$$p = \left(\frac{k}{\bar{m}}\right) \rho T \tag{I.65}$$

This may also be written in the usual form given in thermodynamics

$$pv^* = RT \tag{I.66}$$

Here $R \equiv k/m^*$ is the gas constant and $v^* \equiv 1/\rho$ is the specific volume, that is the volume occupied by a kilogram of gas. It is usual in thermodynamics to express densities in units of per-unit-mass rather than per-unit-volume and to designate such quantities as "specific" densities. We will indicate specific densities by a superscript asterisk. Thus the specific internal energy density, u^*_{total} is from (I.60) and the definition for R

$$u^*_{total} = \frac{f}{2} RT \tag{I.67}$$

Equation (I.67) completes the set of relations between the thermodynamic and macroscopic variables that is needed in the following. We turn now to the subject of the heat capacity of the gas, by which is meant the amount of heat that must be added to raise the temperature by one degree. Two thermodynamic variables must be specified to determine the state of a gas, i.e., either p and v^*, p and T or v^* and T, since equation (I.66) gives the third variable once two are known. Clearly in connection with determining heat capacity, T should be one of the chosen variables. There are then two specific heat capacities, c_v and c_p corresponding to whether v^* or p, respectively, is chosen as the second variable. In determining the amount of heat required to change the temperature by one degree, the second variable is held fixed. Thus

$$c_v \equiv \left(\frac{\partial Q^*}{\partial T}\right)_{v^*} \tag{I.68}$$

and

$$c_p \equiv \left(\frac{\partial Q^*}{\partial T}\right)_p \tag{I.69}$$

where the specific heat density Q^* is related to the other macroscopic variables by the first law of thermodynamics

$$du^*_{total} = dQ^* - pdv^* \tag{I.70}$$

The equation says that the internal energy of a fixed amount of gas can be changed by the addition or subtraction of heat to that gas or by work done on or by the gas. To convert (I.70) into a form that can be used when p is chosen as the second variable, substitute for pdv^* from the differential form of equation (I.66)

$$du^*_{total} = dQ^* + v^*dp - RdT \tag{I.71}$$

Then (I.67), (I.68), and (I.70) taken together yield

$$c_v = \frac{f}{2} R \tag{I.72}$$

and similarly (I.67), (I.69), and (I.71) yield

$$c_p = \frac{f}{2} R + R = c_v + R \tag{I.73}$$

The ratio of specific heats, c_p/c_v, is a quantity that appears so frequently it is given its own symbol

$$\gamma \equiv c_p/c_v \tag{I.74}$$

From (I.72) and (I.73), it is readily seen that

$$\gamma \equiv 1 + \frac{2}{f} \tag{I.75}$$

The following is a list of examples of values of γ.

f	γ	Example
1	3	motion constrained ∥ to B
2	2	motion constrained ⊥ to B
3	$\frac{5}{3}$	isotropic space plasma
5	$\frac{7}{5}$	air (Earth)
∞	1	many modes of thermal motion

The last thermodynamic quantity with which we desire to make contact is the specific entropy, defined by

$$ds^* \equiv \frac{dQ^*}{T} \tag{I.76}$$

It proves to be most profitable to express (I.76) in terms of the variables p and v^*. For this first convert to the variable T and v^* through (I.67) and (I.70), with (I.72) and (I.66)

$$ds^* = c_v \frac{dT}{T} + R \frac{dv^*}{v^*} \tag{I.77}$$

Then eliminate $\frac{dT}{T}$ with the logarithmic differential form of (I.66)

$$\frac{dp}{p} + \frac{dv^*}{v^*} = \frac{dT}{T} \tag{I.78}$$

which when substituted into (I.77) and combined with (I.73) yields

$$ds^* = c_v \frac{dp}{p} + c_p \frac{dv^*}{v^*} \tag{I.79}$$

This can be rewritten as

$$ds^* = c_v \, d \, \ell n \frac{p}{\rho^\gamma} \tag{I.80}$$

where we have used the definitions of v^* and γ.

I.13 Return to the Prognostic Equation for Scalar Pressure

As promised in I.11, we will now interpret equation (I.57) as a statement regarding the behavior of specific entropy. Since for our gas $\gamma = 5/3$, the combination p/ρ^γ can be eliminated between (I.57) and (I.80) resulting in

$$\frac{ds^*}{dt} = (\frac{k}{m}) \frac{- \nabla \cdot \vec{q} + \vec{j} \cdot \vec{E}^*}{p} \tag{I.81}$$

where we have used (I.72) and the definition of R. Equation (I.81) makes it clear that (I.57) is a prognostic equation for specific entropy. If $\nabla \cdot \vec{q} = 0$ and $\vec{j} \cdot \vec{E}^* = 0$, s^* is a constant of the motion. Such a situation is referred to as an <u>adiabatic flow</u>, or <u>constant entropy flow</u>. The specific entropy will change as one moves with the flow only if there is a non-zero divergence of the heat flux or if there is <u>Joule dissipation</u> in the frame of reference proving with the plasma, i.e. if $\vec{j} \cdot \vec{E}^* \neq 0$. Note that $\frac{ds^*}{dt}$ can be zero even if $\vec{J} \cdot \vec{E} \neq 0$.

I.14 Adiabatic, Isentropic and Polytropic Flows

In an adiabatic flow $ds^*/dt = 0$, and from equation (I.80), this implies

$$p = \alpha\rho^{\gamma} \tag{I.82}$$

where α is a constant of the motion. Note that it is not necessary for two different <u>fluid parcels</u> or <u>fluid elements</u> to have the same value of α, but it is necessary for any given fluid parcel to retain the same value of α as the parcel moves with the flow. To appreciate the special status of α, consider (I.82) to be its defining equation. Then (I.80) shows that α is a function of only the specific entropy. Consequently, for an adiabatic flow $d\alpha/dt = 0$. Consider the case of a steady state, adiabatic flow. The equation for α becomes

$$\vec{V}\cdot\nabla\alpha = 0 \tag{I.83}$$

This is the equation for a <u>streamline constant</u>. Thus α may vary from one streamline to the next, but it has the same value on any given streamline. This means that if α is specified on a two-dimensional surface through which all of the streamlines pass, it is specified throughout the flow.

If it is true that α is the same for all of the fluid elements comprising the fluid, then the flow is said to be <u>isentropic</u>. In this case α is a constant in space and time. An isentropic flow is fully determined by the continuity equation (I.18), the momentum equation (I.23) and the isentropic relation (I.82). An adiabatic flow is also described by these equations, but an additional prescription must be given to specify α.

The enormous simplification that results when the energy equation (I.31) is replaced by (I.82) has lead to the generalization of (I.82) in the form

$$p = \alpha\rho^{n} \tag{I.84}$$

in which α and n are specified constants. The exponent, n, is called the <u>polytropic index</u>.

The polytropic assumption is an artifice to render the fluid equations more tractable. The value of n is chosen to produce the best simulation of the thermal condition of the gas possible within the polytropic assumption. For example, if the heat conductivity is very high, heat conduction will keep the gas at a nearly uniform temperature, even though that common temperature may change in time. One can see by comparing equations (I.65) and (I.84) that in order for T to remain uniform while p and ρ are allowed to vary in space, we must have $n = 1$. The following is a list of commonly used values of n

n	simulation
0	isobaric - constant pressure
1	isothermal - constant temperature - high heat conduction
γ	isentropic
∞	isometric - constant density

There is a more basic way of formulating the simulation that the polytropic assumption performs. The polytropic equation (I.84) is exact for steady state problems (i.e. $\frac{\partial}{\partial t} = 0$) in which the heat flux is proportional to the convected flux of internal energy, $u\vec{V}$, that is

$$\vec{q} = \kappa u\vec{V} \tag{I.85}$$

where κ is a constant, and $\vec{j}\cdot\vec{E}^* = 0$. Then n and κ are related by

$$n = (\frac{5}{3} + \kappa)/(1 + \kappa) \tag{I.86}$$

Note that if $\kappa = 0$, i.e. $\vec{q} = 0$, $\gamma = 5/3$ and if $\kappa \to \infty$, $\gamma \to 1$. Thus the adiabatic and isothermal limits are recovered properly.

I.15 The Bernoulli Equation

A special class of solutions to the momentum equation (I.23) exists in the case of steady state, polytropic flows. Rewrite the equation for the case of a scalar pressure and ignore the electrostatic term. We have retained this term for completeness but it is virtually always negligible compared to the other terms, because a plasma tends to maintain itself electrically neutral to a high degree of precision. (§II.2). Then

$$\frac{\partial\vec{V}}{\partial t} + (\vec{V}\cdot\nabla)\,\vec{V} + \frac{\nabla p}{\rho} = \frac{\vec{J}\times\vec{B}}{\rho} - \nabla\phi \tag{I.87}$$

The gravitational force has been replaced by $-\nabla\phi$ and the convective derivative of the velocity has been written out in terms of separate intrinsic and convective terms. With the polytropic relation (I.86), we may write

$$\frac{\nabla p}{\rho} = \nabla\, h^* \tag{I.88}$$

$$\text{where } h^* = \frac{n}{n-1}\,\frac{p}{\rho} \tag{I.89}$$

It is readily verified by the definitions and relations given in I.12 that h^* is the <u>specific enthalpy</u>, that is, $(u+p)/\rho$ where u is defined as if the polytropic index n were the actual ratio of specific heats. Expand $(\vec{V}\cdot\nabla/\vec{V})$ by use of the vector identity

$$(\vec{A}\cdot\nabla)\vec{A} = \nabla\,\frac{A^2}{2} - \vec{A}\times(\nabla\times\vec{A}) \tag{I.90}$$

where $\vec{A}$ is an arbitrary vector. The scalar product between (I.87) and the velocity vector $\vec{V}$ then becomes

$$\vec{V}\cdot\nabla\left(\frac{V^2}{2} + h^* + \phi\right) = \frac{\vec{J}\cdot(\vec{B}x\vec{V})}{\rho} \quad (I.91)$$

where the vector identity $(\vec{J}x\vec{B})\cdot\vec{V} = \vec{J}\cdot(\vec{B}x\vec{V})$ has been used on the right hand side.

In non-magnetized flows (e.g. in ordinary fluid dynamics) and in flows for which $\vec{B}x\vec{V} = 0$ everywhere, the right hand side of (I.91) is identically zero. That MHD situations exist for which $\vec{B}||\vec{V}$ everywhere will be demonstrated in the next section. With the right hand side of (I.91) set equal to zero the equation expresses the constancy of the quantity $V^2/_2 + h^* + \phi$ along streamlines of the flow. Thus we arrive at the result referred to as the Bernoulli equation

$$\frac{V^2}{2} + h^* + \phi = W \quad (I.92)$$

where W is a streamline constant.

I.16 Divergence of the Anisotropic Pressure Tensor

In situations where it is desired to retain in the Euler equation the anisotropic form of the pressure tensor as given by (I.27), it is useful to reduce the divergence of the tensor to vector operations on scalars and vectors. For this write the pressure tensor in component form as

$$P_{ij} = (p_{\shortparallel} - p_{\perp})\frac{B_iB_j}{B^2} + p_{\perp}\,\delta_{ij} \quad (I.93)$$

The divergence of P_{ij} is a vector the i-th component of which is

$$\frac{\partial P_{ij}}{\partial x_j} = \frac{B_iB_j}{B^2}\frac{\partial}{\partial x_j}(p_{\shortparallel} - p_{\perp}) + \frac{\partial p_{\perp}}{\partial x_i} + (p_{\shortparallel}-p_{\perp})\frac{\partial}{\partial x_j}\left(\frac{B_iB_j}{B^2}\right) \quad (I.94)$$

The component of this vector parallel to $\vec{B}$ is found from

$$B_i\frac{\partial P_{ij}}{\partial x_j} = B_j\frac{\partial p_{\shortparallel}}{\partial x_j} + (p_{\shortparallel} - p_{\perp})\,B_i\frac{\partial}{\partial x_j}\left(\frac{B_iB_j}{B^2}\right) \quad (I.95)$$

It is readily verified that

$$B_i\frac{\partial}{\partial x_j}\left(\frac{B_iB_j}{B^2}\right) = -\,B_i\frac{B_j}{B^2}\frac{\partial B_i}{\partial x_j} \quad (I.96)$$

and hence

$$B_i \frac{\partial P_{ij}}{\partial x_j} = B_i \left[\frac{\partial p_{\shortparallel}}{\partial x_i} - (p_{\shortparallel} - p_{\perp}) \frac{B_j}{B^2} \frac{\partial B_i}{\partial x_j} \right] \tag{I.97}$$

In vector notation, this is

$$(\nabla \cdot \overleftrightarrow{P})_{\shortparallel} = \nabla_{\shortparallel} p_{\shortparallel} - (p_{\shortparallel} - p_{\perp}) \left[\frac{(\vec{B} \cdot \nabla)\vec{B}}{B^2} \right] \tag{I.98}$$

in which the $\shortparallel$ signs on the vectors denote the component parallel to $\vec{B}$.

To determine the components of (I.94) in the plane perpendicular to $\vec{B}$, Let $\vec{A}$ be any vector in that plane. Then

$$A_i \frac{\partial P_{ij}}{\partial x_j} = A_i \frac{\partial p_{\perp}}{\partial x_i} + (p_{\shortparallel} - p_{\perp}) A_i \frac{\partial}{\partial x_j} \left(\frac{B_i B_j}{B^2} \right) \tag{I.99}$$

Equation (I.96) holds also when the initial B_i on each side is replaced by A_i. Thus (I.99) becomes

$$A_i \frac{\partial P_{ij}}{\partial x_j} = A_i \left[\frac{\partial p_{\perp}}{\partial x_i} + (p_{\shortparallel} - p_{\perp}) \frac{B_j}{B^2} \frac{\partial B_i}{\partial x_j} \right] \tag{I.100}$$

or in vector notation

$$(\nabla \cdot \overleftrightarrow{P})_{\perp} = \nabla_{\perp} p_{\perp} + (p_{\shortparallel} - p_{\perp}) \left[\frac{-(\vec{B} \cdot \nabla)\vec{B}}{B^2} \right]_{\perp} \tag{I.101}$$

where the $\perp$ signs on the vectors denote the componenent perpendicular to $\vec{B}$.

I.17 Single Particle Drifts and the Euler Equation

It is usual to begin a course on plasma physics with a derivation of the motion that a point particle of mass m and charge q executes in certain simple electric and magnetic field configurations. The force governing the motion is the Lorentz force

$$\vec{F}_L = q\,\vec{E} + q\,\vec{v} \times \vec{B} \tag{I.102}$$

The second term on the right hand side, the magnetic Lorentz force, always acts perpendicular to the velocity vector. In the absence of an electric field and in a uniform magnetic field, this force causes the particle to move in a circular loop in a plane perpendicular to the magnetic field. The radius r_g of the circle (the gyroradius, r_g) and the angular frequency at which the particle goes around the loop (the gyrofrequency, $\vec{\omega}_g$) can be be found by balancing the centrifugal force of the motion against the magnetic Lorentz force. Let $\vec{v}_\perp$ be the component of the velocity in the plane perpendicular to $\vec{B}$ (the gyrovelocity). Then the balance of forces is

$$m\vec{\omega}_g \times \vec{v}_\perp = q\vec{v}_\perp \times \vec{B} \tag{I.103}$$

from which we see that

$$\vec{\omega}_g = -\frac{q\vec{B}}{m} \tag{I.104}$$

and from $v_\perp = r_g \omega_g$, we find

$$r_g = \frac{mv_\perp}{qB} \tag{I.105}$$

A charged particle in circular motion generates a magnetic dipole, the magnetic moment of which is the product of the current that the circulating charge produces and the area of the circle

$$\vec{\mu} = (\pi r_g^2)(q\frac{\vec{\omega}_g}{2\pi}) = -\frac{\frac{1}{2}mv_\perp^2}{B}\frac{\vec{B}}{B} = -\mu\hat{b} \tag{I.106}$$

where

$$\mu \equiv \frac{\frac{1}{2}mv_\perp^2}{B} \tag{I.107}$$

is the magnitude of the dipole moment.

After the treatment of the motion of a particle in a uniform magnetic field, one considers the case of motion in a non-uniform magnetic field and in the presence of an electric field. One of the first non-trivial results of plasma physics is the demonstration that if the spatial and temporal scales of the gyromotion that a particle would execute around the local magnetic field are much less than those of the magnetic and electric fields themselves, μ is a constant of the motion. In this context μ is sometimes called the first adiabatic invariant.

The constancy of μ can be thought of as an example of Lenz's law of electromagnetism, which states that electrical circuits change their currents to counteract externally caused changes in linked magnetic fluxes. In our case, μ is directly proportional to the magnetic flux Φ_g linked by the gyrocircle. Explicitly

$$\mu = \frac{q^2}{2\pi m}\Phi_g \tag{I.108}$$

The constancy of μ is a result of the constancy of Φ_g. Let EMF be the electromotive force per unit charge around the gyrocircle induced by either an intrinsic change in $\vec{B}$ or by a movement of the gyrocircle through a spatially varying magnetic field. Then the particle changes its energy at the rate

$$\frac{d}{dt}\left(\frac{1}{2} m v_\perp^2\right) = q \text{ EMF } \frac{\omega_g}{2\pi} \tag{I.109}$$

But

$$\text{EMF} = \pi r_g^2 \frac{dB}{dt} \tag{I.110}$$

Substitution of (I.110) into (I.109) results after a few steps in

$$\frac{d}{dt} \ell n \left(\frac{\frac{1}{2} m v_\perp^2}{B}\right) = 0 \tag{I.111}$$

The fact that the gyromotion produces a magnetic dipole allows the magnetic Lorentz force acting on the gyro-component of velocity to be replaced by the magnetic force acting on a magnetic dipole. The magnetic force on a field aligned dipole is given in general by

$$\vec{F}_\mu = - \mu \nabla B \tag{I.112}$$

The minus sign occurs because $\vec{\mu}$ is antiparallel to $\vec{B}$ for gyrating charged particles. The original problem of determining the motion of a charged particle with velocity components $\vec{v}_\perp$ perpendicular to $\vec{B}$ and $v_\parallel$ parallel to $\vec{B}$ becomes replaced by the problem of determining the motion of a current loop of mass m, charge q and magnetic moment $\vec{\mu}$ which has velocity components $\vec{V}_D$ perpendicular to $\vec{B}$ and $v_\parallel$ parallel to $\vec{B}$.

The motion of the current loop perpendicular to the field as given by its drift velocity, $\vec{V}_D$, can be different for ions and electrons and, therefore, can result in electrical currents, called drift currents. The purpose of this section is to show that when all of the drift currents are combined, one recovers the Euler equation.

The drift velocity is obtained by considering the forces acting on the charged loop in the plane perpendicular to $\vec{B}$

$$\vec{F}_\perp = q\vec{E}_\perp + q\,(\vec{V}_D \times \vec{B}) - \mu \nabla_\perp B - m v_\parallel^2 \left[\left(\frac{\vec{B}}{B}\cdot\nabla\right)\frac{\vec{B}}{B}\right]_\perp + m\vec{g}_\perp \tag{I.113}$$
$$= m \frac{d\vec{V}_D}{dt}$$

The fourth term on the right hand side is the centrifugal force that results from the motion of the loop along a curved line of force. The term in brackets is the curvature of the field line. The other terms require no further explanation. Solve (I.113) for $\vec{V}_D$ to find

$$\vec{V}_D = \frac{\vec{E}\times\vec{B}}{B^2} + \frac{\mu}{qB}\frac{\vec{B}}{B} \times \nabla B + \frac{mv_{\parallel}^2}{qB}\frac{\vec{B}}{B} \times \left[\left(\frac{\vec{B}}{B}\cdot\nabla\right)\frac{\vec{B}}{B}\right]$$

$$+ \frac{m}{qB}\frac{\vec{B}}{B} \times \frac{d\vec{V}_D}{dt} + \frac{m}{qB}\vec{g} \times \frac{\vec{B}}{B} \qquad \text{(I.114)}$$

The subscript $\perp$ is now superfluous and has been dropped.

The first term on the right hand side of (I.114) is the <u>electric field drift</u>, the second term is the <u>gradient drift</u>, the third term is the <u>curvature drift</u>, the fourth term is the <u>inertial drift</u> and the last term is the <u>gravity drift</u>. The inertial drift is sometimes written in terms of the electric field, in which form it is called the polarization drift. The replacement is possible because in drift motion calculations, all quantities except $\vec{V}_D$ and $\vec{E}$ are assumed to be constant. Then by differentiating (I.114) with respect to time and dropping the second time derivative of $\vec{V}_D$, we find

$$\frac{d\vec{V}_D}{dt} = \frac{\partial\vec{E}}{\partial t} \times \frac{\vec{B}}{B^2} \qquad \text{(I.115)}$$

In this way the inertial drift term becomes the expression for the <u>polarization drift</u>

$$\frac{m}{qB}\frac{\vec{B}}{B} \times \frac{d\vec{V}_D}{dt} = \frac{m}{qB^2}\frac{\partial\vec{E}}{\partial t} \qquad \text{(I.116)}$$

If now the drift velocities of the ions and electrons are combined to form an expression for the drift current

$$\vec{J}_D = e[n_i(\vec{V}_D)_i - n_e(\vec{V}_D)_e] \qquad \text{(I.117)}$$

one finds by inspection of (I.114) (with $q_i = e$ and $q_e = -e$)

$$\vec{J}_D = \rho_c \frac{\vec{E}\times\vec{B}}{B^2} + \frac{p_\perp}{B^2}\frac{\vec{B}}{B} \times \nabla B$$

$$+ \frac{p_{\parallel}}{B}\frac{\vec{B}}{B} \times \left[\left(\frac{\vec{B}}{B}\cdot\nabla\right)\frac{\vec{B}}{B}\right] + \frac{\rho}{B}\frac{\vec{B}}{B} \times \frac{d\vec{V}_D}{dt}$$

$$+ \frac{\rho}{B}\vec{g} \times \frac{\vec{B}}{B} \qquad \text{(I.118)}$$

In this result the values of $p_\perp$ and $p_{\parallel}$ have been constructed out of the single particle parameters in a manner that is consistent with their definitions in (1.12).

$$p_\perp = \frac{1}{2} n_i m_i (v_\perp^2)_i + \frac{1}{2} n_e m_e (v_\perp^2)_e \tag{I.119}$$

$$p_\parallel = n_i m_i (v_\parallel^2)_i + n_e m_i (v_\parallel^2)_e \tag{I.120}$$

The factor ½ appears in (I.119) and not in (I.120) because $v_\perp^2 = v_x^2 + v_y^2$ embodies two components of motion.

To obtain the total current flowing in the plasma that has now been created out of the bringing together of separate ion and electron components, the magnetization current $\vec{J}_M$ must be added, which arises whenever there is an inhomogeneous distribution of magnetic dipoles.

$$\vec{J} \equiv \vec{J}_{Total} = \vec{J}_D + \vec{J}_M \tag{I.121}$$

where $\vec{J}_M$ is given explicitly in terms of dipole moment distribution by

$$\vec{J}_M = \nabla \times (n_i \vec{\mu}_i + n_e \vec{\mu}_e) \tag{I.122}$$

This reduces immediately to

$$\vec{J}_M = - \nabla \times (\frac{p_\perp}{B} \frac{\vec{B}}{B}) \tag{I.123}$$

$$= \frac{\vec{B}}{B^2} \times \nabla p_\perp - 2 \frac{p_\perp}{B^2} \frac{\vec{B}}{B} \times \nabla B - \frac{p_\perp}{B^2} \nabla \times \vec{B} \tag{I.124}$$

The addition of (I.118) and (I.124) gives

$$\vec{J} = \rho_c \frac{\vec{E} \times \vec{B}}{B^2} + \frac{\vec{B}}{B^2} \times \nabla p_\perp + \frac{p_\parallel}{B} \frac{\vec{B}}{B} \times [(\frac{\vec{B}}{B} \cdot \nabla) \frac{\vec{B}}{B}]$$

$$- p_\perp (\frac{\vec{B} \times \nabla B}{B^3} + \frac{\nabla \times \vec{B}}{B^2}) + \frac{\rho}{B} \frac{\vec{B}}{B} \times \frac{d\vec{V}_D}{dt} + \frac{\rho}{B} \vec{g} \times \frac{\vec{B}}{B} \tag{I.125}$$

Two further algebraic reductions are needed to reach the desired form. These are the two vector identities

$$\frac{\vec{B}}{B} \times [(\frac{\vec{B}}{B} \cdot \nabla) \frac{\vec{B}}{B}] = \frac{B}{B} \times [\frac{(\vec{B} \cdot \nabla) \vec{B}}{B^2}] \tag{I.126}$$

and

$$\frac{\vec{B} \times \nabla B}{B^3} + \frac{\nabla \times \vec{B}}{B^2} = \frac{\vec{B}}{B^2} \times [\frac{(\vec{B} \cdot \nabla) \vec{B}}{B^2}] \tag{I.127}$$

(The validity of the second equation requires $\vec{J} \perp \vec{B}$, which is appropriate to this discussion). The identities allow (I.125) to be rewritten as

$$\vec{J} = \frac{\vec{B}}{B^2} \times \left[\rho \frac{d\vec{V}_D}{dt} + \nabla p_{\perp} + (p_{\parallel} - p_{\perp}) \frac{(\vec{B}\cdot\nabla)\vec{B}}{B^2} - \rho_c \vec{E} - \rho\vec{g} \right] \quad (I.128)$$

Equation (I.128) can now be inverted to solve for $\rho \frac{d\vec{V}_D}{dt}$, with the understanding that the result applies in the plane perpendicular to $\vec{B}$

$$\rho \frac{d\vec{V}_D}{dt} = -\nabla p_{\perp} - (p_{\parallel} - p_{\perp}) \frac{(\vec{B}\cdot\nabla)\vec{B}}{B^2} + \rho_c \vec{E} + \vec{J}\times\vec{B} + \rho\vec{g} \quad (I.129)$$

The final expression, equation (I.129), is seen to be identical with the perpendicular component of the Euler equation (I.23) in which $\vec{V}_D$ is identified with $\vec{V}$ and the form of the divergence of the pressure tensor is given by (I.101). It should be noted that while the result demonstrates that the microscopic and macroscopic descriptions are formally identical, the use of particle drift theory is restricted to situations in which the magnetic field is presumed known. In the MHD description, the magnetic field is one of the dependent variables, and thus situations can be treated in which the magnetic field is in part or completely determined by the plasma.

I.18 Limitations to the Use of the Macroscopic Equations

As they now stand the macroscopic equations are not a closed set. To obtain a fully complete macroscopic description of a plasma, it is in principle necessary to compute all of the moments of the Boltzmann equation. The truncation at $n = 2$ necessarily leaves more dependent variables than equations. There is no equation for the highest order dependent variable, the heat flux vector, $\vec{q}$. The problems that can be treated either set $\vec{q}$ equal to zero (adiabatic flows), or use the artifice of a polytropic index to simulate the effect of heat flux (polytropic flows), or introduce an explicit equation for $\vec{q}$, such as a thermal conduction equation. More elaborate forms of equations for $\vec{q}$ are being evolved in connection with the theory of the solar wind.

The components of the pressure tensor also are not completely determined within the derived equations. It was noted that the specification of the viscous-like components required results from kinetic theory to obtain the coefficients of viscosity. The anisotropic form of the pure-pressure terms depends on the validity of the approximation that the pressure is isotropic in the plane perpendicular to the magnetic field. When gradients in the plasma or the fields are comparable to the gyroradius of the ions, this is not a valid approximation.

There is an even more fundamental limitation to the equations at the stage of the development we have now reached. Consider a steady state ($\partial/\partial t = 0$) problem which has only the minimum number of dependent variables to still qualify as an MHD problem, namely $\vec{V}$, ρ, p and $\vec{B}$. The current density $\vec{J}$ can be expressed in terms of $\vec{B}$ through the Maxwell equation (I.35). There are a total of eight unknowns, if the vector components are taken into account. However, the continuity equation, the Euler equation, the adiabatic relation and the remaining Maxwell equation involving $\vec{B}$ (I.33) total only six, including the three vector components of the Euler equation.

The missing equations are supplied through the computation of the first charge-moment of the Boltzmann equation (I.49 with n = 1). This results in an equation called the generalized Ohm's law that relates the electric field vector $\vec{E}$ to the other dependent variables. Then the Maxwell equation (I.34) provides three additional equations. (At the same time Maxwell equation (I.33) becomes redundant because it follows from (I.34) with the prescription that $\nabla\cdot\vec{B} = 0$ at some initial instant.)

The role that the generalized Ohm's law and its approximate form the hydromagnetic approximation play in MHD is sufficiently important to be considered in a separate major section.

II. THE HYDROMAGNETIC APPROXIMATION AND ITS CONSEQUENCES

II.1 The Generalized Ohm's Law

The first charge moment of the Boltzmann equation (I.49) leads after a fairly lengthy series of intermediate steps (e.g. Rossi and Olbert, 1970) to the following expression in which no approximations have been made other than replacing the collision integral by an effective collision time.

$$\frac{\partial J_i}{\partial t} + \frac{e}{m_a}\frac{\partial}{\partial x_j}(P_{ij})_a - \frac{e}{m_b}\frac{\partial}{\partial x_j}(P_{ij})_b + \frac{\partial}{\partial x_j}(J_iV_j + J_jV_i - \rho_c V_iV_j)$$
$$= \alpha\{\vec{E} + [\vec{V} + \frac{e}{\alpha}(\frac{1}{m_a} - \frac{1}{m_b})(\vec{J}-\rho_c\vec{V})] \times \vec{B}\}_i + \rho_c\vec{g} - (\vec{J}-\rho_c\vec{V})/\tau \qquad \text{(II.1)}$$

in which subscripts a and b signify ions and electrons, respectively. The ions are assumed to be singly charged and all to have the same mass, m_a. The parameter α is defined by

$$\alpha \equiv \frac{e^2}{m_a m_b}\rho + e(\frac{1}{m_a} - \frac{1}{m_b})\rho_c \qquad \text{(II.2)}$$

The quantity τ that enters into (II.1) is the time scale for momentum exchange by means of collisions between the ion gas and electron gas that together make up the plasma. The time scale for coulomb collisions is given by Spitzer (1956).

To transform (II.1) into a more useful expression, we make two highly accurate approximations, the first of which is obvious, and the second will be justified in Section II.2.

$$\frac{1}{m_a} << \frac{1}{m_b} \tag{II.3}$$

$$\rho_c \vec{V} << \vec{J} \tag{II.4}$$

All terms containing the left hand sides of (II.3) and (II.4) will be dropped in (II.1). At this point, we revert to vector notation and to the standard nomenclature in which subscripts i and e denote the ion and electron components of the plasma. Also to simplify nomenclature, let

$$\eta \equiv \frac{1}{\alpha\tau} = \frac{m_e}{e^2 n \tau} \tag{II.5}$$

Then if (II.1) is solved for $\vec{E}$, there results

$$\vec{E} = - \vec{V} \times \vec{B} + \eta\vec{J} - \frac{1}{ne} \nabla\cdot\overleftrightarrow{P}_e + \frac{1}{ne} \vec{J} \times \vec{B} + \frac{m_e}{e^2 n} \left[\frac{\partial \vec{J}}{\partial t} + \nabla\cdot(\vec{J}\vec{V} + \vec{V}\vec{J}) \right] \tag{II.6}$$

in which an obvious diadic notion has been used.

It can be seen from equation (I.59) that the sum of all of the terms on the right hand side of (II.6) exclusive of the first is precisely $\vec{E}^*$, the electric field in the co-moving frame of reference. Thus (II.6) can be written in the more revealing form

$$E = - \vec{V} \times \vec{B} + \vec{E}^* \tag{II.7}$$

The electric field is the sum of a motional part owing to translation of the plasma in the presence of a magnetic field (or better said, as will be seen later, owing to translation of the plasma carrying a magnetic field) and a static or material part which is the electric field in the plasma itself. $\vec{E}^*$ is given explicitly by

$$\vec{E}^* = \eta\vec{J} - \frac{1}{ne} \nabla\cdot\overleftrightarrow{P}_e + \frac{1}{ne} \vec{J} \times \vec{B} + \frac{m_e}{e^2 n} \left[\frac{\partial \vec{J}}{\partial t} + \nabla\cdot(\vec{J}\vec{V} + \vec{V}\vec{J}) \right] \tag{II.8}$$

The first contribution to E^* is the usual ohmic resistance term. The second term gives the ambipolar electric field that is needed to restrain the electrons when they are driven by a pressure gradient and thereby to maintain the plasma neutral. The third term expresses the Hall effect, the meaning of which will be given below. The final term in brackets represents the effect of electron inertia.

In any intended application of (II.6), some terms will be much smaller than the largest terms and they can therefore be neglected. To determine which are the negligible terms, we perform a scale analysis by comparing the magnitudes of all terms with that of some reference term. Take the first member of the right hand side to be the reference term and proceed to define dimensionless ratios with the remaining terms

$$R_1 \equiv \frac{VB}{\eta J} = \frac{VB}{\frac{m_e}{e^2 nT}\frac{B}{\mu_o L}} = nVLT\,\frac{\mu_o e^2}{m_e} \tag{II.9}$$

$$R_2 \equiv \frac{VB}{\frac{1}{ne}\frac{P_e}{L}} = \frac{VB}{\frac{k}{ne}\frac{nT}{L}} = \frac{e}{k}\frac{VBL}{T} \tag{II.10}$$

$$R_3 = \frac{VB}{\frac{1}{en}\frac{B^2}{\mu_o L}} = \frac{VLn}{B}\,e\mu_o \tag{II.11}$$

$$R_4 = \frac{VB}{\frac{m_e}{e^2 n}\frac{J}{T}} = L^2 n\,\frac{\mu_o e^2}{m_e} \tag{II.12}$$

In these expressions, L and T represent characteristic length and time scales of the flow.

As an example of a typical set of space parameters to evaluate the sizes of these ratios, take $V = 10^5\,ms^{-1}$, $L = 10^6 m$, $n = 10^7 m^{-3}$, $B = 10^{-8} T$, $T_e = 10^5\ {}^oK$, and note that $\mu_o e^2/m_e \simeq 3x10^{-12}$, $e/k \simeq 10^4$, $e\mu_o \simeq 2x10^{-25}$ in MKS units. Then $R_1 = 3x10^4 T >> 1$, $R_2 = 10^2 >> 1$, $R_3 = 10 > 1$, $R_4 = 3x10^5 >> 1$.

The example shows that in a typical space situation, $-\vec{V}x\vec{B}$ is comfortably larger than all of the other terms on the right hand side, except possibley $\vec{J}x\vec{B}/ne$. That is, generally

$$|\vec{E}^*| << |\vec{V}x\vec{B}| \tag{II.13}$$

Thus in most space applications, the generalized Ohm's law is replaced by the hydromagnetic equation, also known as the hydromagnetic approximation or the hydromagnetic limit of the generalized Ohm's law.

$$\vec{E} = -\vec{V}x\vec{B} \tag{II.14}$$

That is, to a good approximation, the electric field may be regarded as a purely motional field. It is important to be aware of the possibility that other terms in (II.6) could dominate in restricted regions of space.

The subject of magnetic merging, to which we return briefly later, is an example where (II.14) is assumed to be violated.

Even in normal space conditions there is a somewhat better approximation for $\vec{E}$ than that given by (II.14). As was seen in the numerical example, the term which tends most seriously to invalidate (II.14) is $\vec{J} \times \vec{B}/en$. If this term is retained, then the resulting expression for $\vec{E}$ will be improved. In our numerical example, the remaining corrections would be one percent or less. Thus a more accurate expression is

$$\vec{E} = -(\vec{V} - \frac{\vec{J}}{en}) \times \vec{B} \qquad \text{(II.15)}$$

If we consider the plasma to be composed of an ion gas component moving with velocity $\vec{V}_i$ and an electron gas component moving with velocity $\vec{V}_e$, then with the same two approximations given by (II.3) and (II.4) we may write

$$\rho\vec{V} = m_i n\vec{V}_i \qquad \text{(II.16)}$$

$$\vec{J} = en(\vec{V}_i - \vec{V}_e) \qquad \text{(II.17)}$$

where n is the common number density of the two gases (cf. II.4). With these expressions for $\vec{V}$ and $\vec{J}$, (II.15) becomes

$$\vec{E} = -\vec{V}_e \times \vec{B} \qquad \text{(II.18)}$$

Equation (II.18) shows that in the more accurate version of the hydromagnetic approximation in which the Hall effect term is retained, the electric field is the motional field of the electron gas component of the plasma.

II.2 Charge Neutrality and Related Approximations

The expression (II.6) for the electric field permits a verification of the basic condition of charge neutrality that has already been invoked several times. The relative difference in the number densities of ions and electrons can be parameterized by the dimensionless ratio ρ_c/en, where n is the common (average) number density of the two charge particles. The space charge density ρ_c is related to $\vec{E}$ through the Maxwell equation (I.32). As in the previous section, we perform a scale analysis of this equation with the foreknowledge that the term $-\vec{V}\times\vec{B}$ makes the biggest contribution to $\vec{E}$.

$$\frac{\rho_c}{en} \approx \frac{\varepsilon_o VB}{eL} \qquad \text{(II.19)}$$

Next make the replacements $B/L = \mu_o J = \mu_o en|\vec{V}_i - \vec{V}_e|$, $V = V_i$, and $\varepsilon_o \mu_o = 1/c^2$. Then (II.19) becomes

$$\frac{\rho_c}{en} \sim \frac{V_i|\vec{V}_i - \vec{V}_e|}{c^2} \tag{II.20}$$

In the non-relativistic plasmas that populate the solar system, charge neutrality is seen to be strongly obeyed.

If instead of VB, we use the largest term in $\vec{E}^*$, namely $\frac{1}{ne}\vec{J}\times\vec{B}$, to represent $\vec{E}$, the dimensionless ratio of number densities becomes

$$\frac{\rho_c}{en} \sim \frac{(\vec{V}_i - \vec{V}_e)^2}{c^2} \tag{II.21}$$

A similar analysis can be done for the electrostatic force in the Euler equation (I.23) which was stated earlier to be negligible compared to the ponderomotive force. This claim can be verified directly by a scale analysis

$$\frac{\rho_c E}{JxB} \sim \frac{\varepsilon_o E^2}{LJB} \tag{II.22}$$

Make the substitutions $E = VB$, $J = B/\mu_o L$, $\varepsilon_o \mu_o = c^2$ to find

$$\frac{\rho_c E}{JxB} \sim \frac{V^2}{c^2} \tag{II.23}$$

This ratio is of the same order of smallness as the space charge density-number density ratio. Scale analysis shows the electrostatic energy density, which enters into Poynting's theorem (I.44) and the equation for the conservation of energy, is smaller than the magnetic energy density by the same velocity ratio

$$\frac{\frac{\varepsilon_o}{2}E^2}{B^2/2\mu_o} \sim \frac{\varepsilon_o \mu_o V^2 B^2}{B^2} = \frac{V^2}{c^2} \tag{II.24}$$

Finally, the displacement current $\varepsilon_o \mu_o \partial\vec{E}/\partial t$ that appears in the Maxwell equation (I.35) is also negligible compared to $\vec{\nabla} \times \vec{B}$.

$$\frac{\varepsilon_o \mu_o \partial\vec{E}/\partial t}{\nabla x\vec{B}} \sim \frac{\varepsilon_o \mu_o VB/T}{B/L} \tag{II.25}$$

With the substitutions $\varepsilon_o \mu_o = {}^1/_{c^2}$ and ${}^L/_T = V$, this becomes

$$\frac{\varepsilon_o \mu_o \ \partial\vec{E}/\partial t}{\nabla \times \vec{B}} \approx \frac{V^2}{c^2} \tag{II.26}$$

II.3 Poynting's Theorem in the Hydromagnetic Limit

In this and subsequent sections the consequences that follow from the hydromagnetic approximation (II.14) are developed. Poynting's theorem (I.44) becomes especially simple under the hydromagnetic approximation and has an easily understood meaning where Equation (II.19) is used to replace the electric field, and the electric energy density term is dropped in accordance with (II.24). There results

$$\frac{\partial}{\partial t}\left(\frac{B^2}{2\mu_o}\right) + \nabla\cdot\left(\frac{B^2}{\mu_o}\vec{V}_\perp\right) = -\vec{V}_\perp\cdot(\vec{J}\times\vec{B}) \tag{II.27}$$

in which the Poynting vector appears as

$$\vec{S} = \frac{-(\vec{V}\times\vec{B})\times\vec{B}}{\mu_o} = \frac{B^2}{\mu_o}\left[\vec{V}-(\hat{b}\cdot\vec{V})\hat{b}\right]$$

$$= \frac{B^2}{\mu_o}\vec{V}_\perp \tag{II.28}$$

where $\vec{V}_\perp$ is the component of $\vec{V}$ perpendicular to $\vec{B}$. $\vec{S}$ has the appearance of a flux of magnetic energy density, except that magnetic energy density is $B^2/2\mu_o$. Poynting's vector, $\vec{S}$, is properly interpreted as the magnetic enthalpy flux density. In strict analogy with the mechanical enthalpy flux density $(u+p)\vec{V}$, $\vec{S}$ contains both the energy flux density of the field $(B^2/2\mu_o)\vec{V}$ and the work required to move that energy against the pressure $B^2/2\mu_o$.

The electromechanical energy conversion term $-\vec{E}\cdot\vec{J}$ becomes after an intermediate step $-\vec{V}\cdot(\vec{J}\times\vec{B})$. Since $\vec{V}_{\parallel}\cdot(\vec{J}\times\vec{B}) = 0$, where $\vec{V}_{\parallel} = (\hat{b}\cdot\vec{V})\hat{b}$ is the component of $\vec{V}$ parallel to $\vec{B}$, the expression $\vec{V}\cdot(\vec{J}\times\vec{B})$ can be written as $\vec{V}_\perp\cdot(\vec{J}\times\vec{B})$ without approximation. The term is seen to be the power resulting from applying the ponderomotive force $(\vec{J}\times\vec{B})$ against the flow velocity $\vec{V}_\perp$. Thus if $(\vec{J}\times\vec{B})$ opposes the motion, the flow is slowed and energy is transformed from mechanical form to magnetic form, and vice versa.

II.4 Equipotential Fieldlines and Streamlines in Steady State Hydromagnetic Flows

An immediately deduced and far reaching property of <u>steady state</u>

hydromagnetic flows results from the Maxwell equation (I.34, also known as Faraday's induction law). In steady state, (I.34) becomes

$$\nabla \times \vec{E} = 0 \tag{II.29}$$

The electric field is therefore given by the gradient of an electrical potential Φ_E

$$\vec{E} = -\nabla\Phi_E \tag{II.30}$$

The vector $\vec{E}$ is everywhere normal to surfaces of constant Φ_E (equipotential surfaces). In the hydromagnetic limit (II.14) both the velocity vector $\vec{V}$ and the magnetic field vector $\vec{B}$ are everywhere orthogonal to $\vec{E}$, and therefore lie in equipotential surfaces. It follows that streamlines of the flow and magnetic field lines are confined to equipotential surfaces and are therefore equipotential lines. This result can be obtained formally by replacing $\vec{E}$ by $-\nabla\Phi_E$ in (II.14) and scalar multiplying the resulting relation first with $\vec{V}$ and then with $\vec{B}$.

$$\vec{V}\cdot\nabla\Phi_E = 0 \tag{II.31}$$

$$\vec{B}\cdot\nabla\Phi_E = 0 \tag{II.32}$$

It should be kept in mind that some time dependent flows can be converted to steady state flows by an appropriate Galilean transformation. For example it is sometimes useful in problems involving propagating plane waves or planar discontinuities to transform to the frame of reference moving with the wave or the discontinuity to take advantage of (II.31) and (II.32). Solar system examples exist in the form of structures in the solar wind that corotate with the sun and possibly in the form of corotating structures in the magnetospheres of Jupiter and Saturn. The structures would appear stationary in the corresponding corotating frame of reference.

In the case of the sun a further simplification applies that illustrates a special but important class of steady flows. Assume constancy prevails in the frame of reference corotating with the sun. This may be considered an idealization of a circumstance in which solar surface conditions change slowly on a relevant solar wind flow time. Then (II.31) and (II.32) may be used as valid approximations to the actual situation. In the photosphere the solar wind velocity is essentially zero and the electrical resistivity is low. Thus in the photosphere the main contributors to $\vec{E}$ in the generalized Ohm's law can be taken to be negligibly small. This results in the characterization of the photosphere as an equipotential surface in the corotating frame. (It would not be so characterized in the inertial frame since the rotational motion of the photospheric plasma gives a $-\vec{V}_{rotation} \times \vec{B}_{photosphere}$ contribution to $\vec{E}$ in that frame.) Now in the corotating reference frame all of space filled by the solar wind is linked to the photosphere by equipotential field (and flow) lines.

It follows that since the photosphere is an equipotential surface, Φ_E = constant in the region of space filled by the solar wind. Equation (II.30) then gives $\vec{E} = 0$ everywhere in the solar wind in the corotating reference frame. The hydromagnetic equation (II.14) allows the further deduction that $\vec{V} || \vec{B}$ everywhere in the solar wind in the corotating reference frame. It will be recalled from Section (I.15) that this condition permits the construction of a Bernoulli integral for the flow in the corotating reference frame.

The situation just described for the solar wind can be generalized to a statement about a special class of steady flows. If the flow streamlines or the magnetic field lines comprising a continuous flow or field domain pass through an equipotential surface anywhere within the domain, the entire domain is an equipotential volume, the electric field within it is zero and the streamlines and flowlines within it coincide.

A further useful relation between the magnitudes of the velocity and magnetic field can be derived for these equipotential domain flows. Since it is usually necessary to make a Galilean transformation from the reference frame in which the velocity $\vec{V}$ is given and the reference frame of the equipotential domain, let $\vec{V}^*$ be the required transformation velocity. The flow velocity in the reference frame of the equipotential domain is then $(\vec{V}-\vec{V}^*)$. In the solar wind problem in which $\vec{V}$ is given in the inertial reference frame, $\vec{V}^*$ is the corotation velocity $\vec{\Omega} \times \vec{r}$, where $\vec{\Omega}$ is the solar angular velocity vector and $\vec{r}$ is the radius vector in a heliocentric spherical polar coordinate system. The condition $\vec{B} || (\vec{V}-\vec{V}^*)$ can be expressed as

$$\vec{B} = \kappa\rho(\vec{V}-\vec{V}^*) \qquad \text{(II.33)}$$

where κ is an as yet unknown scalar function of space and ρ is mass density. The reason for including ρ explicitly as part of the coefficient to $(\vec{V}-\vec{V}^*)$ is to take advantage of the continuity equation (I.18), which in steady state is

$$\nabla\cdot[\rho(\vec{V}-\vec{V}^*)] = 0 \qquad \text{(II.34)}$$

The divergence of (II.33) is zero by Maxwell's equation (I.33). Thus with (II.34) there results

$$\rho(\vec{V}-\vec{V}^*)\cdot\nabla\kappa = 0 \qquad \text{(II.35)}$$

This is the equation of a streamline constant of the flow in the reference frame of the equipotential domain, that is, κ is constant on the streamlines of $(\vec{V}-\vec{V}^*)$. If κ is constant across any surface linked by all of the streamlines in the equipotential domain, then κ is a constant throughout the domain.

II.5 Freezing Laws

In 1942, Alfvén showed that if a fluid moves such that its velocity is related to the electric and magnetic fields in accordance with equation (II.14), then the Faraday induction law (eq. I.34) imposes a constraint on the motion that can be described as a freezing of the fluid to the magnetic field, or in Alfvén's words "the matter of the liquid is fastened to the lines of force" (Alfvén, 1942). If we substitute for the electric field from eq. (II.14) into (I.34), we obtain

$$\frac{\partial \vec{B}}{\partial t} = \nabla \times (\vec{V} \times \vec{B}) \qquad \text{(II.36)}$$

The freezing law, which follows simply as a kinematical consequence of (II.36), states that the magnetic flux through a closed loop that moves with the fluid is constant in time. The demonstration of this result proceeds as follows.

The magnetic flux through a closed loop, ℓ, is defined by

$$F \equiv \int \vec{B} \cdot \hat{n} \, da \qquad \text{(II.37)}$$

where da is an element of area on any surface which has ℓ as its perimeter. Gauss' theorem and the divergence-free condition on $\vec{B}$ (eq. 1.33) guarantee that the magnetic flux is the same through all surfaces sharing a common perimeter. The freezing law can be expressed mathematically by

$$\frac{dF}{dt} = 0 \qquad \text{(II.38)}$$

The symbol for the total time derivative is used to indicate that F is to be evaluated in reference to a linked set of fluid elements that move with the fluid. Equation (I.20) for the total derivative is not appropriate, however, because F is not a locally defined quantity. We must evaluate (II.38) using the integral form of F, eq. (II.37) explicitly.

Refer to Figure (II.1) which shows a closed loop of fluid elements, ℓ, at two successive instants, t and $t + \Delta t$. An enclosed volume is formed by the two surfaces S_1 and S_2, that have $\ell(t)$ and $\ell(t + \Delta t)$ as their perimeters, and the generalized cylinder, S_3 generated by the motion of ℓ. Let F be the magnetic flux enclosed by ℓ, and denote by subscripts 1, 2 and 3 the fluxes through the surfaces, S_1, S_2 and S_3. Then if the normal vectors to S_1 and S_2 are chosen to lie on the same side of each surface relative to the flow, as indicated,

$$\frac{dF}{dt} = \frac{F_2(t + \Delta t) - F_1(t)}{\Delta t} \qquad \text{(II.39)}$$

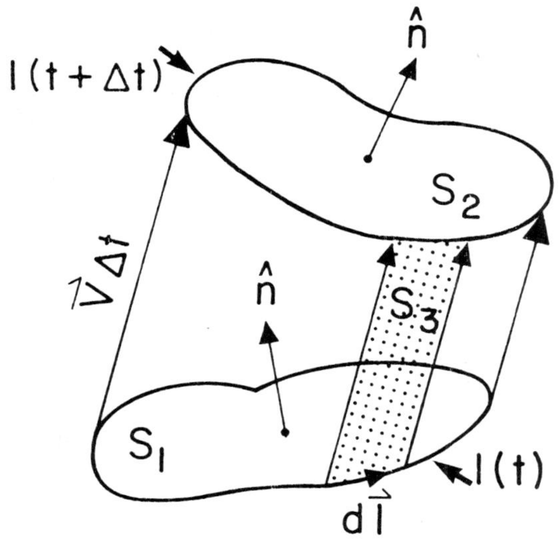

Figure II.1 Generalized cyclinder formed by the motion of closed line "frozen" to the fluid.

where the limit $\Delta t \to 0$ is understood. The divergence-free condition on $\vec{B}$ requires a zero net flux through the three surfaces at any time. In particular

$$-F_1(t + \Delta t) + F_2(t + \Delta t) + F_3 = 0 \qquad \text{(II.40)}$$

In this equation it is recognized that the outward pointing normal to S_1 is needed to utilize Gauss' theorem, and hence, the change in the sign on F_1 relative to F_2 and F_3.

Now eliminate $F_2(t + \Delta t)$ between eqs. (II.39) and (II.40) and replace the fluxes by their integral forms, (I.27)

$$\frac{dF}{dt} = \frac{1}{\Delta t} [\int_{S_1} [\vec{B}(t + \Delta t) - B(t)]\cdot\hat{n}da - \int_{S_3} \vec{B}\cdot\hat{n}da] \qquad \text{(II.41)}$$

The first integral is evidently $\int_{S_1} \frac{\partial B}{\partial t}\cdot\hat{n}da$. The second term on the right hand side can also be converted into an integral over the surface S_1 with the following identities

$$\int_{S_3} \vec{B}\cdot\hat{n}da = \int_{\ell(t)} \vec{B}\cdot(d\vec{\ell} \times \vec{V}\Delta t) = \int_{\ell(t)} (\vec{V} \times \vec{B})\cdot d\vec{\ell}\Delta t$$

$$= \int_{S_1} [\nabla \times (\vec{V} \times \vec{B})]\cdot\hat{n}da\Delta t \qquad \text{(II.42)}$$

The last equality is an application of Stokes' theorem. Thus (II.42) becomes

$$\frac{dF}{dt} = \int_{S_1} \left[\frac{\partial \vec{B}}{\partial t} - \nabla \times (\vec{V} \times \vec{B}) \right] \cdot \hat{n} da \qquad \text{(II.43)}$$

The frozen-in flux condition, eq. (II.38), then follows immediately from the hydromagnetic equation in the form of eq. (II.36).

The flux-preserving character of a plasma in the hydromagnetic limit imposes important constraints on possible fluid motions, which we will now enumerate. First define a magnetic flux tube to be a surface generated by moving any closed loop parallel to the magnetic field lines it intersects at a given instant. One creates this way a generalized cylindrical surface which encloses a constant amount of magnetic flux. By the definition it is clear that any patch of the surface of a flux tube encloses zero magnetic flux. Thus as a consequence of flux preservation, the fluid elements that form a flux tube at any instant, form a flux tube at all instants.

Flux preserving plasmas are also line preserving in the following sense. Imagine two fluid elements labeled A and B to be linked at time t by a magnetic field line. Now a field line can always be defined as the intersection of two flux tubes, and let us so define the field line linking A and B at time t. At this time the two fluid elements both belong to the surfaces of the two defining flux tubes. According to our previous corollary, they therefore must share the surfaces of two flux tubes in common at all times. That is, A and B must always lie at the intersection of two flux tubes, from which we may conclude that if two fluid elements are linked by a field line at any instant, they are always so linked. (For a fuller study see Stern 1966).

From the first corollary it is also easy to see that if a fluid element lies inside of a flux tube at one time, it always lies inside of it; and conversely if a fluid element lies outside of a flux tube at one time, it always lies outside of it.

II.6 Thawing of Magnetic Flux

If the generalized Ohm's law (II.6) in its abbreviated form (II.7) is solved in terms of $\vec{V} \times \vec{B}$ and the result substituted into the general expression for dF/dt (eq. II.43), we find with the aid of Faraday's induction law (I.35)

$$\frac{dF}{dt} = - \int_{S_1} (\nabla \times \vec{E}^*) \cdot \hat{n} da = - \oint_{\ell(t)} \vec{E}^* \cdot d\vec{\ell} = -EMF^* \qquad \text{(II.44)}$$

where EMF^* is the electromotive force per unit charge around $\ell(t)$ that the intrinsic electric field $\vec{E}^*$ produces. This result is not very surprising since it is formally the standard relation between time rate of change of magnetic flux and EMF. However, it is much more useful than a merely formal relation, because $\vec{E}^*$ is given explicitly in terms of the macroscopic plasma parameters by (II.8). Recall that the only approximations that entered into the derivation of (II.8) were $m_e \ll m_i$ and $\rho_c V \ll J$, both of which are well obeyed. Hence, if the freezing of the magnetic flux to the flow and its logical consequences are to be violated at any time or any place, the term or terms responsible for the thawing are contained in (II.8).

In the numerical example which was meant to typify solar system plasmas, the largest contributor to $\vec{E}^*$ was the Hall effect term. However as we saw in eq. (II.18) this term only moves the condition of freezing of the flux from the plasma as a whole to the electron gas component. That is, if we define F_e for the electron gas in the manner analogous to the definition of F for the plasma as a whole, namely the magnetic flux through a closed perimeter moving with the electron gas, then we can write in an obvious notation

$$\frac{dF_e}{dt} = -\int_{S_1} (\nabla \times E_e^*) \cdot \hat{n} da = -\int_{\ell_e(t)} \vec{E}_e^* \cdot d\vec{\ell}_e = -(\text{EMF})_e^* \qquad \text{(II.45)}$$

in which

$$\vec{E}_e^* = \eta \vec{J} - \frac{1}{ne} \nabla \cdot \overleftrightarrow{P}_e + \frac{m_e}{e^2 n} \left[\frac{\partial \vec{J}}{\partial t} + \nabla \cdot (\vec{J}\vec{V} + \vec{J}\vec{V}) \right] \qquad \text{(II.46)}$$

is the intrinsic electric field in the frame of reference moving with the electron gas component of the plasma. If thawing of the magnetic flux from the electron gas component is to occur at any time or any place, the responsible terms are contained in (II.46).

The ohmic term is ordinarly thought of in connection with magnetic thawing, especially in collisonal plasmas. However, in the highly collisionless plasmas of space, the other two terms need to be considered, including the off-diagonal parts of the electron pressure tensor. These matters have been pursued in studies of magnetic merging,which entails strong but spatially localized violation of freezing. (For a review of this subject see Vasyliunas, 1975).

In the case of an isotropic electron pressure and a polytropic electron gas the second term in (II.46) becomes a pure gradient term and contributes nothing to $(\text{EMF})_e^*$

The notion of freezing of a vector flux to the plasma flow can be usefully reformulated at this level of retention of terms in the Ohm's law if certain flow parameters are incorporated in the frozen quantity, as will be shown in the next section.

II.7 The Generalized Vorticity Theorem

One of the most important properties of ordinary fluids is their tendency to preserve vorticity in the same sense that MHD fluids tend to preserve magnetic flux. In this section we determine whether MHD fluids have a vorticity conserving property in addition to their flux conserving property. First the definition of vorticity, $\vec{\omega}$, is needed

$$\vec{\omega} \equiv \nabla \times \vec{V} \qquad \text{(II.47)}$$

The vorticity theorem in ordinary hydrodynamics ($\vec{E} = \vec{B} = 0$) is derived for a non-viscous, polytropic fluid. Then the Euler equation (I.87) is simply

$$\frac{\partial \vec{V}}{\partial t} + (\vec{V}\cdot\nabla)\,\vec{V} + \nabla h^* = -\nabla\phi \qquad \text{(II.48)}$$

where ϕ is again the gravitational potential and the quantity h^* is the specific enthalpy defined by (I.88), which we repeat here

$$\nabla h^* \equiv \frac{\nabla p}{\rho} \qquad \text{(II.49)}$$

More generally, one may say that a function, h^*, that satisfies (II.49) exists if and only if p is a function of ρ. Such a fluid is called <u>barotropic</u>. In a barotropic fluid, surfaces of constant pressure coincide with surfaces of constant density. In general constant pressure surfaces and constant density surfaces need not coincide, and in such a case the fluid is termed <u>baroclinic</u>. In a baroclinic fluid, a function, h^*, satisfying (II.49) does not exist, and the preservation of vorticity, which as we shall see depends on its existance, does not obtain.

Take the curl of (II.48) after substituting from the vector identity (I.90) to find

$$\frac{\partial \vec{\omega}}{\partial t} = \nabla \times (\vec{V} \times \vec{\omega}) \qquad \text{(II.50)}$$

Equation (II.50) is formally identical to (II.36) with $\vec{B}$ replaced by $\vec{\omega}$. All of the consequences derived from it pertaining to $\vec{B}$ then apply without change to $\vec{\omega}$. In particular, in a non-MHD, barotropic, nonviscous fluid, the flux of vorticity (called the <u>circulation</u>, Γ) is preserved.

$$\Gamma \equiv \int \vec{\omega}\cdot\hat{n}\,da \qquad \text{(II.51)}$$

$$\frac{d\Gamma}{dt} = 0 \qquad \text{(II.52)}$$

in strict analogy with equations (II.37) and (II.38).

Let us now try to extend the result to the MHD situation by retaining the $\vec{J} \times \vec{B}$ force in the momentum equation. At first appearance, the attempt would seem certain to fail, because the derivation of (II.50) required all non-inertial terms in the momentum equation to vanish under the curl operator. There is no reason for $\nabla \times (\vec{J} \times \vec{B})$ to be zero. However, a generalized vorticity theorem can be obtained if we utilize a more exact form of the generalized Ohm's law. We retain the three largest terms according to the dimensional scale analysis given earlier.

$$\vec{E} = -\vec{V} \times \vec{B} - \frac{1}{ne} \nabla p_e + \frac{1}{ne} \vec{J} \times \vec{B} \qquad \text{(II.53)}$$

in which only the case of a scalar pressure is considered. Now by breaking the pressure explicitly into its two components, $p = p_i + p_e$, the Euler equation (I.87) can be written as

$$\frac{m_i}{e} \left[\frac{d\vec{V}}{dt} + \nabla h_i^* + \nabla \phi \right] = \frac{1}{ne} \left[-\nabla p_e + \vec{J} \times \vec{B} \right] \qquad \text{(II.54)}$$

in which we have used $\rho = m_i n$, with n the common electron and ion number density, and

$$\nabla h_i^* \equiv \frac{\nabla p_i}{\rho} \qquad \text{(II.55)}$$

Substitution of (II.54) into (II.53) gives

$$\vec{E} = -\vec{V} \times \vec{B} + \frac{m_i}{e} \left[\frac{d\vec{V}}{dt} + \nabla(\phi + h_i^*) \right] \qquad \text{(II.56)}$$

and thus

$$\frac{\partial \vec{B}}{\partial t} = \nabla \times (\vec{V} \times \vec{B}) - \frac{m_i}{e} \left[\frac{\partial \vec{\omega}}{\partial t} - \nabla \times (V \times \vec{\omega}) \right] \qquad \text{(II.57)}$$

Recombining then yields

$$\frac{\partial \vec{\Omega}}{\partial t} = \nabla \times (\vec{V} \times \vec{\Omega}) \qquad \text{(II.58)}$$

where

$$\Omega \equiv \vec{\omega} + \frac{e\vec{B}}{m_i} \qquad \text{(II.59)}$$

Note that $\dfrac{e\vec{B}}{m_i} = \vec{\omega}_i$ (II.60)

is the gyrofrequency of the ions. Hence

$$\vec{\Omega} = \vec{\omega} + \vec{\omega}_i \tag{II.61}$$

In most applications $\omega_i >> \omega$, that is, the vorticity of the fluid motion is much less than the ion gyrofrequency. In this limit we recover the flux-freezing law given by equation (II.36). If the opposite limit should ever occur, equation (II.50) reduces to the ordinary vorticity theorem.

In conclusion, one has either freezing of the magnetic field or of vorticity or of their properly weighted sum, but not of both separately.

II.8 The MHD Helmholtz Equation:

We have seen that in the MHD limit the amount of magnetic field inside of a closed loop of fluid elements is a constant of the motion. Also it is evident that the amount of mass inside of a closed volume of fluid elements is a constant of the motion. If there is no compression or stretching of the fluid in the direction parallel to $\vec{B}$, it must follow that the ratio of field density (i.e. B) and mass density is a constant. This kinetmatic relationship between the magnetic field and mass density is expressed formally and more generally in the MHD Helmholz equation, the derivation of which proceeds as follows.
Write the continuity equation in the form (I.22)

$$\nabla \cdot \vec{V} = \rho \frac{d}{dt} \left(\frac{1}{\rho}\right) \tag{II.62}$$

Expand the curl of the vector cross product in equation (II.36) to arrive at (note $d\vec{B}/dt = \partial\vec{B}/\partial t + (\vec{V}\cdot\nabla)\vec{B}$)

$$\frac{d\vec{B}}{dt} + \vec{B} (\nabla\cdot\vec{V}) - (\vec{B}\cdot\nabla) \vec{V} = 0 \tag{II.63}$$

Divide (II.63) through by ρ, substitute in for $\nabla\cdot\vec{V}$ from (II.62) and recombine to find

$$\frac{d}{dt} \left(\frac{\vec{B}}{\rho}\right) - \left(\frac{\vec{B}}{\rho}\cdot\nabla\right) \vec{V} = 0 \tag{II.64}$$

This result with $\vec{B}$ replaced by the vorticity, $\vec{\omega}$, is known in hydrodynamics as the Helmholtz equation. Its formal similarity to the continuity equation is evident. To confirm the observation made at the beginning of this section, note that if $\vec{V}$ does not change in the direction of $\vec{B}$, then $\vec{B}/\rho$ is a constant of the motion.

Equation (II.64) has an important application to steady flows in which a stagnation point occurs (i.e. a point where $\vec{V} = 0$), for example the stagnation point in the solar wind at the magnetopause of planetary magnetospheres. In steady state, the equation may be rewritten as

$$(\vec{V}\cdot\nabla)\frac{\vec{B}}{\rho} = (\frac{\vec{B}}{\rho}\cdot\nabla)\vec{V} \qquad \text{(II.65)}$$

At an ordinary stagnation point, the right hand side is not zero because $\vec{V}$ is changing from one direction to the opposite direction across the stagnation point in the direction of $\vec{B}$. But the left hand side ought to be zero since $\vec{V} = 0$ there. To maintain the equality expressed by (II.65) it is necessary for ρ to vanish at the stagnation point. This is a characteristic of MHD flows which is not shared by ordinary fluids.

An alternative resolution of the dilemma has been suggested (see Sonnerup, 1980). Instead of forming a stagnation point, MHD flows may form stagnation lines such that $\vec{V} = 0$ along a finite stretch in the direction parallel to B. The stagnation line would terminate at both ends at a neutral point in the magnetic field (i.e. a point where $\vec{B} = 0$). Equation (II.65) is satisfied at a point which is both a stagnation point and a neutral point. Note that in parallel flows ($\vec{V}||\vec{B}$), this condition is met automatically and an ordinary neutral point can occur in the flow.

II.9 The Double Adiabatic Invariants

It is possible in the hydromagnetic limit to derive prognostic equations for the scalars $p_{\parallel}$ and $p_{\perp}$ of the anisotropic pressure tensor (I.27) in the manner in which equation (I.57) was derived for the scalar pressure p. The strictly analogous procedure entails algebra too lengthy for inclusion in this chapter. However, a kinetic theory argument will be given here that results in the correct forms of adiabatic invariants for $p_{\perp}$ and $p_{\parallel}$ which correspond to the adiabatic form $p/\rho^{5/3}$ for the scalar pressure.

Consider a particle enclosed in a container, one wall of which is a piston, which can be moved in and out in order to change the volume of the container at will. The container will be used to simulate adiabatic changes in the three equal components of the isotropic pressure tensor of a collision dominated gas and the two perpendicular components and the one parallel component of the anisotropic tensor of a collisionless magnetized plasma. In the first instance the energy gained (or lost) by the particle in colliding with the moving wall will be shared by the three components equally to preserve isotropy. In the second instance the energy change will be shared equally by the two components of $p_{\perp}$. In the third instance the one component of $p_{\parallel}$ will retain the entire change.

The adiabatic condition is imposed by moving the wall slowly compared to the speed of the particle in the container, and the collision between the particle and all of the walls is assumed to be perfectly elastic.

Let the velocity $\vec{v}$ of the particle be decomposed into cartesian (x,y,z) components with the x-axis parallel to the motion of the piston. Let u be the speed of the piston measured positive when the piston moves in the direction to decrease the volume of the container (compression). Then the change in v_x after one collision with the moving wall is

$$v_x(\text{after}) = v_x(\text{before}) + 2u \tag{II.66}$$

The corresponding <u>first order</u> energy change Δw_x is then

$$\Delta w_x = \frac{1}{2} m \left[v_x^2(\text{after}) - v_y^2(\text{before})\right] = 2muv_x \tag{II.67}$$

Let ν be the number of components that share the energy acquired in one collision before a second collision with the moving wall occurs. Thus $\nu = 3$ for the isotropic gas, $\nu = 2$ for $p_\perp$ and $\nu = 1$ for $p_\parallel$. After the sharing of the energy takes place, the net change in Δw_x is

$$\Delta w_x = \frac{2}{\nu} muv_x \tag{II.68}$$

Now the number of collisions the particle has with the moving wall each second f_{coll} is given by

$$f_{coll} = \frac{v_x}{2L_x} \tag{II.69}$$

where L_x is the separation between the face of the piston and the stationary wall opposite from it. Thus

$$\frac{dw_x}{dt} = \frac{1}{\nu}\frac{u}{L_x} mv_x^2 = \frac{2}{\nu}\frac{u}{L_x} w_x \tag{II.70}$$

But by the definitions of u and L

$$u = -\frac{dL_x}{dt} \tag{II.71}$$

Substitution of (II.71) into (II.70) and subsequent integration give

$$\frac{d}{dt} \ln (w_x L_x^{2/\nu}) = 0 \tag{II.72}$$

The length L_x is related to the mass density ρ by the expression

$$\rho = \frac{mN}{AL_x} \tag{II.73}$$

in which the total number of particles in the container, N, the area of the container in the yz-plane, $A = L_y L_z$, and the mass of the particles, m, remain constant during the motion of the piston.

The energy in the x-component of motion is related to the temperature of the gas and consequently to the pressure by

$$w_x = \frac{1}{2} k T_x = \frac{1}{2} \frac{m p_x}{\rho} \tag{II.74}$$

where (I.62) and (I.63) have been used.

Consider first the isotropic case. Then combining (II.71) with $\nu = 3$ and (II.72) and (II.74) gives

$$\frac{d}{dt} \ln \left(\frac{p}{\rho^{5/3}}\right) = 0 \tag{II.75}$$

Thus we recover the correct result for adiabatic changes in an ideal monatomic isotropic gas (cf. eq. I.57).

To treat the case of the magnetized collisionless plasma, it is necessary to recognize that two of the walls of our container form a magnetic flux tube. Thus for $p_\perp$, we must put $\nu = 2$ in (II.72) but also the equation for L_x becomes

$$F = B L_x L_y \tag{II.76}$$

where F is the magnetic flux enclosed by the container, which must remain constant as the piston moves, by the freezing laws described in Section II.5. The distance L_y is also constant by design. Hence, after substituting into (II.72) from (II.76) and dropping derivatives of quantities that remain constant as the piston moves, there results

$$\frac{d}{dt} \ln \left(\frac{w_x}{B}\right) = 0 \tag{II.77}$$

This is seen to be the expression for the constancy of the first adiabatic invariant (I.111), derived here by a statistical mechanics argument. Equation (II.74) relating W_x, pressure and mass density can now be used to convert (II.77) into the expression for the first of the two adiabatic invariants of magnetohydrodynamics

$$\frac{d}{dt} \ln \left(\frac{p_\perp}{\rho B}\right) = 0 \tag{II.78}$$

The expression for $p_\parallel$ is found by setting $\nu = 1$ in (II.72). Also since the motion of the piston in this case is parallel to the flux tube, the equation for L_x is again (II.73), but with $A = F/B$, where F is constant. The energy is expressed in terms of pressure by (II.74). Making the indicated substitutions and dropping derivatives of constant quantities result in

$$\frac{d}{dt} \ln \left(\frac{p_{\parallel} B^2}{\rho^3}\right) = 0 \qquad \text{(II.79)}$$

The quantities

$$\alpha_{\perp} \equiv \frac{p_{\perp}}{\rho B} \qquad \text{(II.80)}$$

$$\alpha_{\parallel} \equiv \frac{p_{\parallel} B^2}{\rho^3} \qquad \text{(II.81)}$$

are the two adiabatic invariants of collisionless MHD. In adiabatic MHD flows, $\alpha_{\perp}$ and $\alpha_{\parallel}$ are constants of the motion. The discussion relating to the quantity α defined in (I.82) applies to $\alpha_{\perp}$ and $\alpha_{\parallel}$ as well.

Note that B can be eliminated from (II.80) and (II.81) by combining them into a hybrid adiabatic invariant

$$(\alpha_{\perp}{}^2 \alpha_{\parallel})^{1/3} = \frac{(p_{\perp}{}^2 p_{\parallel})^{1/3}}{\rho^{5/3}} \qquad \text{(II.82)}$$

In the case $p_{\perp} = p_{\parallel}$, (II.82) reverts to the expression for the adiabatic invariant for an isotropic pressure (II.75).

A more important hybrid combination of the (II.80) and (II.81) is the expression for the pressure ratio

$$\frac{p_{\perp}}{p_{\parallel}} = \frac{\alpha_{\perp}}{\alpha_{\parallel}} \frac{B^3}{\rho^2} \qquad \text{(II.83)}$$

The ratio B^3/ρ^2 is not constant for any known flow in solar system MHD and is perhaps never constant in natural plasmas. It is therefore a fundamental property of collisionless MHD flows to become anisotropic. The sense of the anisotropy ($p_{\perp} > p_{\parallel}$ or $p_{\parallel} > p_{\perp}$) depends on how the ratio B^3/ρ^2 changes as the flow progresses. It will be seen in Section IV that a plasma becomes unstable if either sense of anisotropy gets too large. Thus, there is a tendency for collisionless MHD flows to evolve toward instabilities driven by the pressure anisotropy.

III. MHD WAVES AND DISCONTINUITIES

III.1 Linearized Plane Waves in an Isotropic Magnetized Plasma

A significant portion of the phenomenology of MHD plasmas concerns the presence of waves and discontinuities. We look first at the kinds of small amplitude isentropic waves that can propagate in a homogeneous, uniform magnetized plasma with isotropic pressure. Gravity will be ignored, which excludes the possibility of coupled gravity-MHD waves.

The relevant equations are

Continuity eq.
$$\frac{\partial \rho}{\partial t} + (\vec{V}\cdot\nabla)\ \rho + \rho\nabla\cdot\vec{V} = 0 \qquad \text{(III.1)}$$

Euler eq.
$$\rho[\frac{\partial \vec{V}}{\partial t} + (\vec{V}\cdot\nabla)\vec{V}] + \nabla p = \vec{J} \times \vec{B} \qquad \text{(III.2)}$$

Isentropic eq.
$$p = \alpha\rho^{\gamma} \qquad \text{(III.3)}$$

Maxwell's eqs.
$$\nabla \times \vec{B} = \mu_o \vec{J} \qquad \text{(III.4)}$$

$$\frac{\partial \vec{B}}{\partial t} = \nabla\times(\vec{V} \times \vec{B}) \qquad \text{(III.5)}$$

The electric field has been dropped where appropriate by the approximations given in Section II.2 and eliminated from Faraday's induction law by use of the hydromagnetic approximation. The pressure p and current density $\vec{J}$ can be eliminated by direct substitution into the Euler equation.

Euler eq.
$$\rho[\frac{\partial \vec{V}}{\partial t} + (\vec{V}\cdot\nabla)\vec{V}] + \alpha\nabla\rho^{\gamma} = -\frac{1}{\mu_o}\vec{B}\times(\nabla\times\vec{B}) \qquad \text{(III.6)}$$

Equations (III.1, 5 and 6) are a complete, deterministic set for the variables ρ, $\vec{V}$ and $\vec{B}$, from which p and $\vec{J}$ can be considered derived quantities.

These equations will now be linearized and subjected to an usual plane wave expansion. Denote zero-order quantities by subscripted zeros and first-order quantities by a prefixed δ. Then

$$\rho = \rho_o + \delta\rho \qquad \text{(III.7)}$$

$$\vec{V} = \vec{V}_o + \delta\vec{V} \qquad \text{(III.8)}$$

$$\vec{B} = B_o + \delta\vec{B} \qquad \text{(III.9)}$$

Zero-order quantities are constant in space and time. Quadratic and higher order terms in first-order quantities will be dropped. The condition that first-order quantities propagate as plane waves is imposed by the relation

$$\delta Q \rightarrow \delta Q \quad e^{i(\omega' t - \vec{k}\cdot\vec{r})} \tag{III.10}$$

where δQ denotes an arbitrary first-order quantity, and its value on the right hand side is now regarded as a constant amplitude to the plane wave. The prime on ω denotes the Doppler frequency. Since we have allowed the possibility of a zero-order velocity of the medium, the frequency of the waves will be Doppler shifted in our frame of reference.

It will be useful to carry out the calculation in a cartesian coordinate system with the z-axis defined to be parallel to $\vec{B}_o$ and the xz-plane defined to contain the propagation vector $\vec{k}$ (Fig. III.1). No loss of generality is incurred by this choice.

$$\vec{B}_o = B_o \hat{z} \tag{III.11}$$

$$\vec{k} = k_x \hat{x} + k_z \hat{z} \tag{III.12}$$

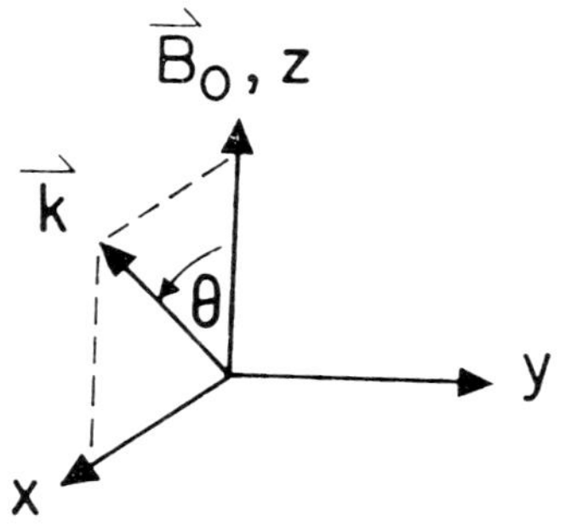

Figure III.1 Coordinate system for describing MHD plane waves

Substitution of (III.8 through 10) into (III 1, 5 and 6) and making the plane wave replacements, $\partial/\partial t \rightarrow i\omega'$, $\nabla Q = -i\vec{k}Q$, $\nabla\cdot\vec{Q} = -i\vec{k}\cdot\vec{Q}$ and $\nabla \times \vec{Q} = -i\vec{k} \times Q$, result in

$$(\omega' - \vec{k}\cdot\vec{V}_o) \frac{\delta\rho}{\rho_o} = \vec{k}\cdot\delta\vec{V} \tag{III.13}$$

$$(\omega' - \vec{k}\cdot\vec{V}_o)\, \delta\vec{V} = C_S^2 \frac{\delta\rho}{\rho_o} \vec{k} + C_A^2\, \hat{z} \times (\vec{k} \times \frac{\delta\vec{B}}{B_o}) \tag{III.14}$$

$$(\omega' - \vec{k}\cdot\vec{V}_o) \frac{\delta\vec{B}}{B_o} = (\vec{k} \times \hat{y})\, \delta V_x - k_z \delta V_y\, \hat{y} \qquad \text{(III.15)}$$

Where the parameters C_S and C_A, are characteristic velocities of the medium and are defined by

$$C_S^2 \equiv \gamma\alpha\rho^{\gamma-1} = \gamma \frac{p}{\rho} \qquad \text{(III.16)}$$

$$C_A^2 \equiv \frac{B_o^2}{\mu_o \rho_o} \qquad \text{(III.17)}$$

Equation (III.16) is the usual definition of the speed of sound. The meaning of (III.17) is given later. To arrive at (III.15) it is necessary to expand the curl of the cross product with the vector identity

$$\nabla\times(\vec{V} \times \vec{B}) = (\vec{B}\cdot\nabla)\,\vec{V} - (\vec{V}\cdot\nabla)\,\vec{B} + \vec{V}\,(\nabla\cdot\vec{B}) - \vec{B}(\nabla\cdot\vec{V}) \qquad \text{(III.18)}$$

and use the condition $\nabla\cdot\vec{B} = 0$.

The coefficients on the left hand sides of (III.13-15) show that the Doppler frequency ω' is related to the frequency in the plasma rest frame ω by

$$\omega' = \omega + \vec{k}\cdot\vec{V}_o \qquad \text{(III.19)}$$

In the remainder of the section, we will use the plasma rest frame frequency ω.

The variables $\delta\rho$ and $\delta\vec{B}$ can be eliminated by substitution to produce a single equation for $\delta\vec{V}$

$$\omega^2\delta\vec{V} = C_S^2(\vec{k}\cdot\delta\vec{V})\vec{k} + C_A^2\,(k^2\delta V_x\hat{x} + k_z^2\delta V_y\,\hat{y}) \qquad \text{(III.20)}$$

The three components of (III.20) can now be written to provide three equations for the three components of $\delta\vec{V}$. Since the equations are homogeneous, non-trivial solution exist only when the determinant of the coefficient matrix vanishes. This then provides the dispersion equation for the waves.

The resulting dispersion equation will be cubic in ω^2, implying the existence of three wave modes, unless some are degenerate. One mode can be isolated already at this point, leaving only two modes to be obtained by the formal approach. The y-component of (III.20) involves only δV_y, and is therefore a pure mode, the dispersion equation for which can be seen to be

$$\omega_i^2 = C_A^2 k_z^2 \tag{III.21}$$

where the subscript i on ω in this instance designates the intermediate mode to distinguish this one from the two others that are described below. The phase velocity of the wave which in general is given by

$$\vec{V}_{ph} = \frac{\omega}{k} \hat{k} \tag{III.22}$$

in the case of (III.21) becomes

$$(\vec{V}_{ph})_i = C_A \cos\theta \; \hat{k} \tag{III.23}$$

where θ is the angle between $\vec{k}$ and $\vec{B}_o$ (see Figure III.1).

This very simple mode has a number of interesting properties. In terms of the velocity perturbation, it is a purely transverse mode, both to $\vec{k}$ and to $\vec{B}_o$, since δV_y is orthogonal to both of these vectors. From (III.15) it is seen that $(\delta\vec{B})_i = \delta B_y \hat{y}$ is also transverse to $\vec{k}$ and $\vec{B}_o$. The mode does not entail a perturbation in density, pressure or magnetic field strength. That is

$$(\delta\rho)_i = (\delta B)_i = 0 \tag{III.24}$$

The fact that $(\delta\rho) = 0$ follows from (III.13) with $\vec{k}\cdot(\delta\vec{V})_i = 0$. It then follows from (III.3) that there is then no perturbation in pressure. The perturbation in field strength is given in general by

$$\frac{\delta B}{B_o} = \frac{\delta B_z}{B_o} \tag{III.25}$$

as can be seen by expanding $(\vec{B}_o + \delta\vec{B})^2/B_o^2$ to first order and comparing the result with $(B_o + \delta B)^2/B_o^2$. Since $(\delta B_z)_i = 0$ for this mode, $(\delta B)_i = 0$ also.

The energy carried by the mode propagates strictly parallel to $\vec{B}_o$ at the characteristic speed C_A. The velocity of energy propagation is the group velocity, which is defined in general by

$$\vec{V}_g = \frac{d\omega}{d\vec{k}} = \frac{\partial\omega}{\partial k_x}\hat{x} + \frac{\partial\omega}{\partial k_y}\hat{y} + \frac{\partial\omega}{\partial k_z}\hat{z} \tag{III.26}$$

From (III.21) there results

$$(\vec{V}_g)_i = \pm C_A \hat{z} \tag{III.27}$$

The relationship between $(\delta\vec{B})_i$ and $(\delta\vec{V})_i$ is useful in data interpretation applications to determine the direction of propagation of the wave. From (III.15)

$$\omega_i \frac{(\delta B_y)_i}{B_o} = - k_z (\delta V_y)_i \qquad \text{(III.28)}$$

or

$$k_z = - \frac{\omega_i}{B_o} \frac{(\delta B_y)_i}{(\delta V_y)_i} \qquad \text{(III.29)}$$

Thus, if the magnetic field and velocity perturbations are in the same direction (i.e. both in the +y or both in the -y direction at a given point and time), the wave energy is propagating antiparallel to $\vec{B}_o$. If they are in opposite directions, the wave energy is propagating parallel to $\vec{B}_o$

The mode we have just described was discovered by Alfven who recognized that its basic properties result from a balance between the inertial response of the mass of the plasma and the magnetic tension resulting from stretching a field line. The characteristic speed C_A is called the Alfvén speed. As already noted, in the context of the three wave modes of MHD, this mode is called the intermediate wave and C_A is the intermediate wave speed.

The dispersion relation for the other two modes is found by writing out the x and z-components of equation (III.20), which gives after some rearranging

$$(\omega^2 - C_S^2 k_x^2 - C_A^2 k^2)\ \delta V_x - C_S^2 k_x k_z \delta V_z = 0 \qquad \text{(III.30)}$$

$$- C_S^2 k_y k_z \delta V_x + (\omega^2 - C_S^2 k_z^2)\ \delta V_z = 0 \qquad \text{(III.31)}$$

Setting the determinant of the coefficient matrix to zero and solving for $\omega^2/k^2 = V_{ph}^2$, we find

$$(V_{ph}^2)_{f,S} = \frac{1}{2}\left[(C_S^2+C_A^2) \pm \sqrt{(C_S^2+C_A^2)^2 - 4C_S^2\ C_A^2 \cos^2\theta}\right] \qquad \text{(III.32)}$$

The two modes are differentiated by their phase speeds and accordingly are referred to as the fast and slow modes. It is instructive to consider the propagation of these waves parallel to $\vec{B}_o$ $(\theta = 0)$ and perpendicular to $\vec{B}_o$ $(\theta = \pi/2)$ as special cases.

$$\vec{k} \| \vec{B}_o : (V_{ph}^2)_{f,s} = \{ \begin{matrix} C_S^2 \\ C_A^2 \end{matrix} \tag{III.33}$$

$$\vec{k} \perp \vec{B}_o : (V_{ph}^2)_{f,s} = \{ \begin{matrix} C_S^2 + C_A^2 \\ 0 \end{matrix} \tag{III.34}$$

A wave that propagates parallel to $\vec{B}_O$ is seen to be either a pure sound wave (which is a longitudinal wave $\delta\vec{V} \| \vec{k}$) or a pure Alfven wave (which is a transverse wave $\delta\vec{V} \perp \vec{k}$). The phase speed of a fast mode wave for which $\vec{k} \perp \vec{B}_O$ is the Pythagorean sum of the sound speed and the Alfven speed. The slow mode does not propagate perpendicular to $\vec{B}_O$, which is also true of the intermediate mode, as can be seen from (III.23).

To demonstrate that the intermediate mode phase speed is indeed intermediate between the fast and slow phase speeds, define the two velocity ratios

$$R_{\pm} \equiv \frac{2C_A^2 \cos^2\theta}{(C_S^2+C_A^2) \pm \sqrt{C_S^2+C_A^2 - 4C_A^2\cos^2\theta}} \equiv \frac{A}{S \pm \sqrt{S^2-D}} \tag{III.35}$$

where the symbols A, S and D are introduced for brevity and are defined by their positions in context. We want to show that unity lies between R_+ and R_-, that is, one is greater than and one is less than unity. The possibility that either ratio may also be unity exists in the limits $\theta = 0$ and $\theta = \pi/2$. Unity lies between R_+ and R_- if and only if

$$(R_+ - 1)(R_- - 1) \leq 0 \tag{III.36}$$

In terms of A, S and D (III.36) is

$$\left(\frac{A-S-\sqrt{S^2-D}}{S+\sqrt{S^2-D}}\right)\left(\frac{A-S+\sqrt{S^2-D}}{S-\sqrt{S^2-D}}\right) = \frac{A^2-2AS+D}{D} \leq 0 \tag{III.37}$$

Substituting back the original variables, we find

$$4C_A^4 (\cos^2\theta-1) \cos^2\theta \leq \theta \tag{III.38}$$

That (III.38) is an obviously true statement verifies that the ordering of phase velocities implied by the names of the three MHD wave modes is correct.

Figure III.2 shows four examples of the relative phase speeds of the three MHD modes and how they depend on the angle between $\vec{k}$ and $\vec{B}_o$.

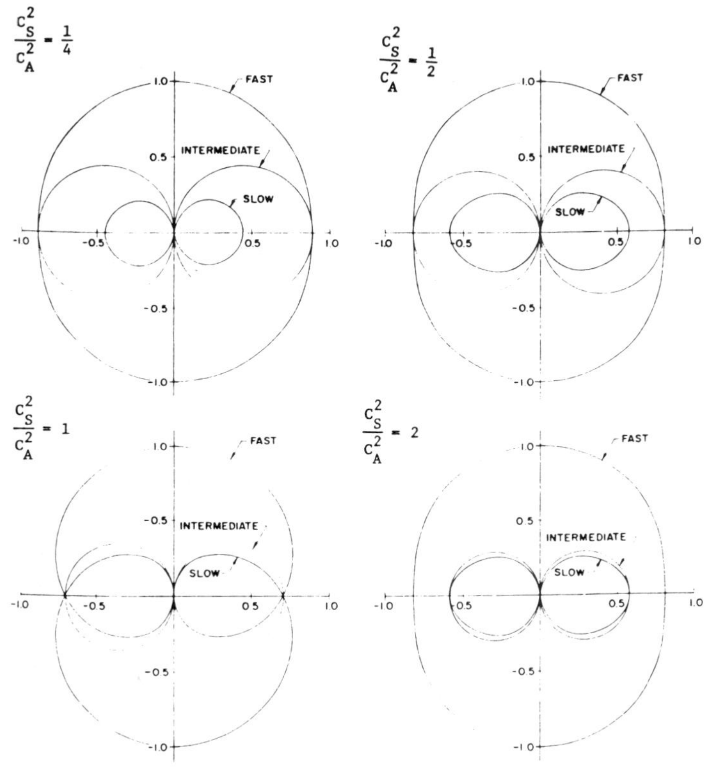

Figure III.2 Friedrichs Diagram. Polar plots showing the dependence of the propagation speeds of the three linear wave modes on the angle between the wave normal $\hat{k}$ and the magnetic field $\vec{B}_o$ for several values of the ratio of sound speed C_S to Alfvén speed C_A. In these plots B_θ is parallel to the horizontal axis. Speeds have been normalized with respect to $\sqrt{C_S^2 + C_A^2}$ (From Kantrowitz and Petschek, 1964).

In contrast to the intermediate mode, both the fast and slow modes are compressive. That is, they entail changes in density (and hence pressure) and changes in the field strength. Equations (III.13, 15 and 31) can be combined to give the dependence of $(\delta\rho)_{f,s}$ and $(\delta B)_{f,s}$ on the velocity perturbations.

$$C_S^2 \frac{(\delta\rho)_{f,s}}{\rho_c} = \frac{\omega_{f,S}}{k_z}(\delta V_z)_{f,s} \qquad \text{(III.39)}$$

$$\omega_{f,S} \frac{(\delta B)_{f,s}}{B_o} = k_x(\delta V_x)_{f,s} \qquad \text{(III.40)}$$

Mass compression (and rarefaction) is seen to result from velocity perturbations parallel to $\vec{B}_o$ and field compression (and rarefaction) to result from velocity perturbations perpendicular to $\vec{B}_o$

As a last characteristic of fast and slow modes, we find with the use of (III.31) a single expression relating the density and field strength perturbations.

$$\frac{(\delta\rho)_{f,s}}{\rho_o} = \frac{\omega_{f,s}^2}{\omega_{f,s}^2 - C_S^2 k_z^2} \frac{(\delta B)_{f,s}}{B_o} \qquad \text{(III.41)}$$

The combination $\omega^2 - C_S^2 k_z^2$ is positive for fast mode waves and negative for slow mode waves, the demonstration of which is as follows

$$(\omega_{f,s}^2 - C_S^2 k_z^2) \times \frac{2}{k^2} = (V_{ph}^2) - 2C_S^2\cos^2\theta$$

$$= C_S^2 + C_A^2 - 2C_S^2\cos^2\theta \pm \sqrt{(C_S^2+C_A^2)^2 - 4C_S^2C_A^2\cos^2\theta}$$

But

$$C_S^2 + C_A^2 - 2C_S^2\cos^2\theta = \sqrt{(C_S^2+C_A^2)^2 - 4C_S^2C_A^2\cos^2\theta - 4C_S^2\cos^2\sin^2\theta}$$

$$\leq \sqrt{(C_S^2+C_A^2)^2 - 4C_S^2C_A^2\cos^2\theta}$$

An important distinction between the fast and slow modes in thereby revealed. Density and field strength perturbations are in phase for the fast mode but out of phase for the slow mode. Correlations between density and field strength oscillations can be used as a diagnostic to identify the mode type.

Figure (III.3) indicates the geometrical basis for the variation in field strength. The top figure depicts the field configuration of an intermediate mode. The propagation vector makes an arbitrary angle in the xz-plane, perpendicular to the paper, such that the wave fronts intersect the plane of the paper in lines parallel to the y-axis. It is evident from the equidistant spacing of the field lines that this mode is non-compressive. The bottom figure shows a wave propagating in

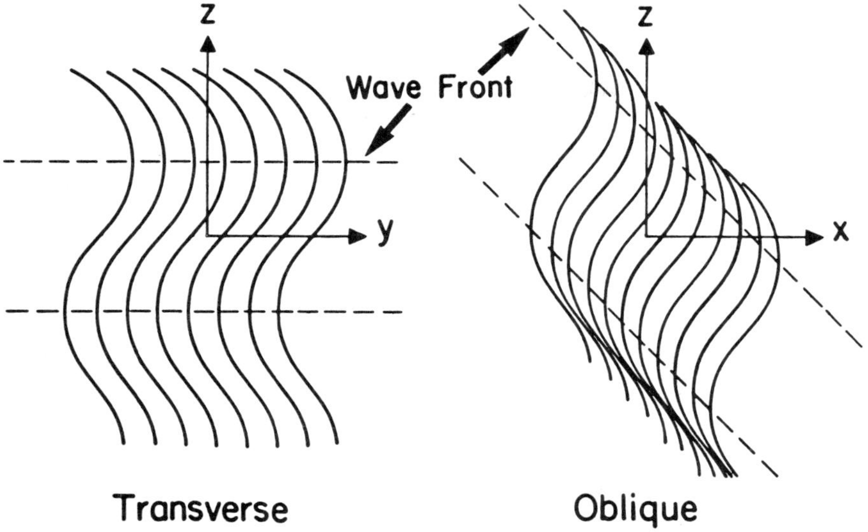

Figure III.3 The compression and rarefaction of magnetic field strength is determined geometrically by the alignment of wave crests and troughs along horizontal or oblique wave fronts.

the same plane as the field perturbation, such as is true for both the fast and slow modes. The fact that the wave crests lie along the oblique lines that represent the wave fronts geometrically causes the field lines to lie close together in one part of the wave and to be relatively separated in another part. In this picture the fast and slow modes are distinguished by whether the density compressions coincide with the strips where the field lines are close together or with the strips where they are farther apart.

An important type of problem arises in solar system plasmas when magnetic field lines connect plasmas that are in motion relative to each other. Examples of such situations are found in the motion of the solar wind relative to the sun and the motions of the plasmas of a planetary magnetosphere relative to the planetary ionosphere. The magnetic field plays the role of an elastic medium in MHD and a ponderomotive force is generated which acts to reduce the relative motion. In order that the ponderomotive force act equally and oppositely in the two plasmas, the associated electrical current must link both bodies. The current is therefore required to flow back and forth parallel to the magnetic field that connects the two regions. The currents are carried by MHD waves, which in a steady state situation will be standing waves but otherwise they will be propagating waves attached to one or the other or both of the coupled plasmas. It is readily shown that the only MHD wave capable of carrying an electrical current parallel to the magnetic field is the obliquely propagating intermediate mode waves. Thus, the oblique intermediate wave is responsible for the mechanical coupling of magnetically linked plasmas. The verification of this statement follows from MHD wave form of Faraday's induction law (III.15) which because of its importance to the present demonstration is re-written here

$$\omega \frac{\delta \vec{B}}{B_o} = (\vec{k} \times \hat{y})\ (\delta V_x)_{f,s} - k_z (\delta V_y)_i \hat{y} \qquad \text{(III.42)}$$

Now from the perturbation form of (II.4) the current carried by the wave is

$$\delta \vec{J} = - \frac{i}{\mu_o} \vec{k} \times \delta \vec{B} \qquad \text{(III.43)}$$

Substitution of (III.42) into (III.43) yields

$$\delta \vec{J} = \frac{iB_o}{\mu_o \omega} k^2 (\delta V_x)_{f,s}\ \hat{y} + \frac{iB_o}{\mu_o \omega} k_z (\delta V_y)_i\ (-k_z \hat{x} + k_x \hat{z}) \qquad \text{(III.44)}$$

This first term of the right hand side is the current associated with fast and slow waves, and the second term is the current associated with the intermediate wave. One sees that only the second term has a vector component parallel to $\vec{B}_o$, i.e. parallel to $\hat{z}$. The amplitude of the parallel component is

$$(\delta \vec{J})_{\parallel} \equiv \hat{z} \cdot \delta \vec{J} = \frac{iB_o}{\mu_o \omega} k_x k_z (\delta V_y)_i = \frac{iB_o}{\mu_o \omega} k^2 (\delta V_y)_i \sin\theta \cos\theta \qquad \text{(III.45)}$$

or in terms of the perturbation magnetic field (III.28)

$$(\delta \vec{J})_{\parallel} = - \frac{i}{\mu_o} k (\delta B_y)_i \sin\theta \qquad \text{(III.46)}$$

The factor i in the coefficient means that in the context of sinusoidally oscillating plane waves, $(\delta \vec{J})_{\parallel}$ is ninety degrees out of phase with $(\delta V_y)_i$ and $(\delta B_y)_i$. Note that $(\delta \vec{J})_{\parallel} = 0$ when $\theta = 0$ that is in the case of parallel propagation. Thus, only obliquely propagating waves carry parallel current.

The properties of the MHD waves become significantly modified when the pressure is anisotropic. The dispersion relation for an anisotropic plasma is given in Section IV.1 on the firehose and mirror instabilities.

III.2 MHD Discontinuities

In the previous section the differential MHD equations were transformed into a set of algebraic relations by first linearizing them and then making the assumption that they had plane wave solutions. The procedure lead to a description of the three MHD wave modes. A different set of algebraic relations can be found to describe the opposite limit, that is discontinuous rather than small amplitude variations.

Assume that a sudden change in the MHD parameters occurs across a surface in space, the local radius of curvature R_D of which is much greater than the local thickness δ_D. In effect, this is a statement of what is meant by a discontinuity. Then there exists a scale regime intermediate between δ_D and R_D in which the MHD parameters may be considered to be uniform on either side of the surface but to change discontinuously across it.

If the discontinuity is moving with (non-relativistic) velocity $\vec{V}_S$ relative to a frame of reference in which the MHD parameters are specified and in particular in which the flow velocity is $\vec{V}$, it is always possible to make a Galilean transformation into a frame of reference in which the discontinuity is at rest. In the rest frame of the discontinuity, the flow velocity, $\vec{U}$, is

$$\vec{U} = \vec{V} - \vec{V}_S \qquad \text{(III.47)}$$

The conservation form of the MHD equations are particularly well suited for the investigation of changes across a discontinuity. In the rest frame of the discontinuity, we may assume time independence. Then the conservation equations for mass, momentum and energy (I.21, 46 and 48) become

$$\nabla\cdot(\rho\vec{U}) = 0 \qquad \text{(III.48)}$$

$$\nabla\cdot(\rho\vec{U}\vec{U} + \overleftrightarrow{P} - \overleftrightarrow{T}) = 0 \qquad \text{(III.49)}$$

$$\nabla\cdot[(\tfrac{1}{2}\rho U^2 + u + \overleftrightarrow{P})\cdot\vec{U} + \vec{S}] = 0 \qquad \text{(III.50)}$$

The gravitational terms have been dropped, as in the previous study of MHD waves. The heat flux vector has also been dropped in the energy equation, although in the case of compressive, collisionless MHD shocks, it could represent an important sink of energy. The electric field terms will be dropped from the Maxwell stress tensor, in accordance with the approximations discussed in Section II.2. The anisotropic form of the pressure tensor (eq. I.27) will be retained, but examples will be given also for the case of a scalar pressure. The internal energy corresponding to the anisotropic pressure is

$$u = p_\perp + \frac{1}{2}p_\parallel \qquad \text{(III.51)}$$

The differential equations (III.48-50) are converted to algebraic equations by integrating them over a volume in the shape of a thin cylinder with unit area faces aligned parallel to the surface of the discontinuity and bracketing it, such that the discontinuity cuts completely through the side of the cylinder, dividing it into equal parts. The cylinder is designed such that the area of the side is negligible compared to unity. Gauss' law is then used to replace the volume integrals by an integral over the surface of the cylinder. The contribution from the side is negligible by design, and since the unit area faces lie in the regions of constant parameters on the two sides of the discontinuities, the integrals over them are the integrals themselves. The final result is then

$$\hat{n}_i \cdot \vec{Q}_i + \hat{n}_2 \cdot \vec{Q}_2 = 0 \tag{III.52}$$

in which subscripts 1 and 2 distinguish between quantities evaluated on the two sides of the discontinuity, $\hat{n}$ is the outward pointing unit normal to the cylinder faces, and $\vec{Q}$ designates any of the three composite quantities on which the divergence operator operates in eqs. (III.48-50). In the case of (III.49), Q is a tensor.

Since by construction $\hat{n}_1 = -\hat{n}_2$, let

$$\hat{n} \equiv \hat{n}_2 = -\hat{n}_1 \tag{III.53}$$

replace $\hat{n}_1$ and $\hat{n}_2$ in (III.52), and define the difference operator [[Q]] by

$$[[Q]] \equiv Q_1 - Q_2 \tag{III.54}$$

Then the equation for the conservation of mass, momentum and energy take the form

$$[[\rho U_n]] = 0 \tag{III.55}$$

$$[[\rho U_n \vec{U} + (\overleftrightarrow{P} - \overleftrightarrow{T}) \cdot n]] = 0 \tag{III.56}$$

$$[[(\frac{1}{2} \rho U^2 + 2p_\perp + \frac{1}{2} p_\parallel + (p_\parallel - p_\perp) \frac{B_n^2}{B^2}) U_n + S_n]] = 0 \tag{III.57}$$

where a subscripted n denotes components parallel to $\hat{n}$. From the magnetohydrodynamic expression for the Poynting vector (II.28) we can write S_n in the form

$$S_n = \frac{B_t}{\mu_o} (B_t U_n - B_n U_t) \tag{III.58}$$

To this set of equations we must add the difference equations that result from integrating the two Maxwell equations $\nabla\cdot\vec{B} = 0$ and $\nabla\times\vec{E} = 0$ (in steady state) across the discontinuity in a manner analogous to the treatment of the conservation equations. In the case of $\nabla\times\vec{E} = 0$ the domain of integration is the area of a thin strip with unit length sides parallel to and bracketing the discontinuity, and Stoke's theorem is used. The result is

$$[[B_n]] = 0 \tag{III.59}$$

$$[[E_t]] = 0 \tag{III.60}$$

where $E_t = \vec{E}\cdot\hat{t}$ and $\hat{t}$ is a arbitrary unit vector tangent to the surface of the discontinuity ($\hat{t}\cdot\hat{n} = 0$). Replacing the electric field by $-\vec{V}\times\vec{B}$ gives in place of (III.60)

$$[[U_n B_t - U_t B_n]] = 0 \tag{III.61}$$

It is useful to decompose the vector equation (III.56) into components parallel and perpendicular to $\hat{n}$. Scalar multiplication of (III.56) with $\hat{n}$ and $\hat{t}$ results in

$$[[\rho U_n^2 + (p_{\shortparallel} - p_{\perp}) \frac{B_n^2}{B^2} + p_{\perp} + \frac{B_t^2}{2\mu_o}]] = 0 \tag{III.62}$$

$$[[\rho U_n U_t - \xi \frac{B_n B_t}{\mu_o}]] = 0 \tag{III.63}$$

in which the combination

$$\xi \equiv 1- \frac{p_{\shortparallel} - p_{\perp}}{B^2/\mu_o} \tag{III.64}$$

appears so frequently, it is given its own symbol, and we have used the relation

$$[[B^2]] = [[B_n^2 + B_t^2]] = [[B_t^2]] \tag{III.65}$$

that follows from (III.59).

The six equations (III.55, 57, 59, 61, 62 and 63) form the complete set of continuity relations available from the conservation equations and Maxwell's equations. If the problem of interest is to determine the parameters on one side of the discontinuity given the parameters on the other side, there are seven unknowns ρ, U_n, U_t, $p_\perp$, $p_\parallel$, B_t, and B_t. If the pressure is isotropic so that it contributes one scalar to the list of unknowns instead of two, the set is completely deterministic. Otherwise an additional relation must be imposed by theory or measurement to close the set.

There exists a classification scheme for MHD discontinuities that divides them according to whether or not the plasma flows through the discontinuity and if it does not flow through it, whether or not the magnetic field penetrates it. The scheme is

Contact Discontinuity	$U_n = 0$,	$B_n \neq 0$
Tangential Discontinuity	$U_n = 0$,	$B_n = 0$
Shock Waves	$U_n \neq 0$	___
Parallel Shocks	"	$B_t = 0$
Perpendicular Shocks	"	$B_n = 0$
Oblique Shocks (Fast, Slow, Intermediate)		$B_t \neq 0$, $B_n \neq 0$.

Contact Discontinuities: If the no-flow condition $U_n = 0$ is imposed on the six continuity relations, one quickly finds that all of the listed parameters are continuous across the discontinuity except density, ρ, which is left unspecified. Thus the density may change across a contact discontinuity, but since the pressures ($p_\parallel$ and $p_\perp$, or p in the case of isotropic pressure) are continuous, the temperature must change also to maintain the pressure constant. Since a discontinuity in temperature should rapidly be dispersed by heat flux parallel to the magnetic field (recall $B_n \neq 0$) such a discontinuity is not expected to occur in solar system plasmas except possibly for short intervals of time.

Tangential Discontinuities: The dual imposition of the conditions of no cross flow (U_n=0) and no field penetration (B_n=0) leads to one nontrivial continuity relation

$$[[p_\perp + \frac{B_t^2}{2\mu_o}]] = 0 \qquad \text{(III.66)}$$

This equation expresses the condition of static pressure balance in the direction normal to the discontinuity. Thus $p_\perp$ may change across the discontinuity, but $B_t^2/2\mu_o$ must change to maintain constant total static pressure. This type of discontinuity appears to be relatively common in solar system plasmas.

Shock Waves 1: Ordinary (non-MHD) Shock Waves:

It is helpful to begin a discussion of MHD shock waves with a review of ordinary gas dynamic shocks. Since the pressure in this case is a scalar, we may make the treatment more general by leaving the ratio of specific heats unspecified.

The continuity relations reduce to

$$[[\rho U_n]] = 0 \quad \text{(III.67)}$$

$$[[\rho U_n^2 + p]] = 0 \quad \text{(III.68)}$$

$$[[\rho U_n U_t]] = 0 \quad \text{(III.69)}$$

$$[[(\tfrac{1}{2}\rho U^2 + \tfrac{\gamma}{\gamma-1} p)U_n]] = 0 \quad \text{(III.70)}$$

Where the enthalpy $u+p = \gamma p/(\gamma-1)$. Combining (III.67) and (III.69) shows that $[[U_t]] = 0$, that is, the tangential component of the flow is continuous in gas dynamic shocks. We may therefore transform away the common U_t by a motion along the shock plane. This transformation does not affect any of the other quantities, but the index n on U_n may now be suppressed, since in the new frame $U_t = 0$.

Equations (III.67, 68 and 70) form a complete set for the downstream variables ρ_2, U_2 and p_2 if the upstream variables ρ_1, U_1, and p_1 are regarded as known. A single equation for U_2 is obtained by eliminating ρ_2 and p_2 between the three equations. This can be written in the form

$$U_2^2 - \frac{2\gamma}{\gamma+1}\left(1+\frac{1}{\gamma M_1^2}\right)U_1U_2 + \frac{\gamma-1}{\gamma+1}\left[1+\frac{2}{(\gamma-1)M_1^2}\right]U_1^2 = 0 \quad \text{(III.71)}$$

in which the upstream sonic Mach number, M_1, is defined as the ratio of the flow speed to the sound speed. It is given in general in terms of the other variables by

$$M^2 = \frac{\rho U^2}{\gamma p} \quad \text{(III.72)}$$

Now since the derivation of the continuity relations did not presume the existence of a discontinuity, they must also be consistent with the absence of a discontinuity. That is equation (III.71) must have $(U_2-U_1) = 0$ as one of its roots. Equation (III.71) can be factored into

$$(U_2-U_1)\left[U_2 - \frac{\gamma-1}{\gamma+1}\left(1 + \frac{2}{(\gamma-1)M_1^2}\right) U_1\right] = 0 \qquad \text{(III.73)}$$

The shock solution is obviously the second factor, which we write in terms of the downstream-to-upstream velocity ratio

$$\frac{U_2}{U_1} = \frac{\gamma-1}{\gamma+1} + \frac{2}{(\gamma+1)M_1^2} \xrightarrow[M_1^2 \to \infty]{} \frac{\gamma-1}{\gamma+1} \qquad \text{(III.74)}$$

The density and pressure can now be found by substituting (III.74) back into the original shock equations

$$\frac{\rho_2}{\rho_1} = \frac{U_1}{U_2} = \frac{\gamma+1}{\gamma-1 + \frac{2}{M_1^2}} \xrightarrow[M_1^2 \to \infty]{} \frac{\gamma+1}{\gamma-1} \qquad \text{(III.75)}$$

$$\frac{p_2}{p_1} = \frac{2\gamma M_1^2 - (\gamma-1)}{\gamma + 1} \xrightarrow[M_1^2 \to \infty]{} \frac{2\gamma}{\gamma+1} M_1^2 \qquad \text{(III.76)}$$

Note that in the limit of weak shocks ($M_1^2 \to 1$), the parameters become continuous across the shock. That is, the shock disappears.

The shock waves that occur in solar system plasmas are in many instances strong shocks that satisfy the condition of the hypersonic limit, $M_1^2 >> 1$. The bow shocks of planetary magnetospheres and flare-driven solar blast waves are examples of strong shock waves that occur in the solar wind. The shock solutions have the interesting property of predicting an asymptotic limit on the degree of compression that a strong shock can produce. The hypersonic limits to the density and velocity ratios are given explicitly in (III.74, 75). The hypersonic limit of the post-shock pressure is given in (III.76). If the physically correct value of γ (γ = 5/3) is used, the asymptotic limits to the ratios are $\rho_2/\rho_1 = 4$ and $U_2/U_1 = 1/4$. The gas can be compressed and slowed down by a factor of no more than four. The shock wave also heats the gas. The shock relations do not predict an upper limit to the degree of heating as the Mach number increases

$$\frac{k}{m} T_s = \frac{p_2}{\rho_2} \xrightarrow[M_1^2 \to \infty]{} \frac{\gamma-1}{(\gamma+1)^2} M_1^2 U_1^2 \qquad \text{(III.77)}$$

The temperature of the solar wind gas heated by bow shocks and blast waves can be more than an order of magnitude greater than the pre-shock temperature.

In gas dynamics shock waves play the role of converting supersonic ($M>1$) flows to subsonic ($M<1$) in situations where the $M<1$ condition is needed to communicate to the fluid by means of pressure waves information about externally imposed constraints so that the fluid can adjust to them. The flow of a supersonic gas around a blunt body (e.g. the solar wind around a planetary magnetosphere), requires the interposition of a standing shock wave (usually detached from the body) to allow the gas to flow around the body. We can demonstrate directly from the shock solutions that the downstream sonic Mach numer is less than or equal to unity.

$$M_2^2 = \frac{\rho_2 V_2^2}{\gamma p_2} = \frac{(\gamma-1)M_1^2 + 2}{2\gamma M_1^2 - (\gamma-1)} \xrightarrow[M_1^2 \to \infty]{} \frac{\gamma-1}{2\gamma} \qquad \text{(III.78)}$$

Note that in the limit of weak shocks $M_1^2 = 1$, (III.78) gives $M_2^2 = 1$, that is, there is no change, as expected. Differentiation of (III.78) shows that dM_2^2/dM_1^2 is a negative definite quantity

$$\frac{dM_2^2}{dM_1^2} = -\left[\frac{\gamma+1}{2\gamma M_1^2-(\gamma-1)}\right]^2 < 0 \qquad \text{(III.79)}$$

Hence $M_2^2 < 1$ for $M_1^2 > 1$. In the hypersonic extreme, M_2^2 approaches a limiting value, which for $\gamma = 5/3$ is $M_2^2 \to 1/5$.

A fundamental property of shock waves is their adiabatic character. They dissipate some of the flow energy and convert it to heat, thereby raising the specific entropy of the gas. This statement can be verified directly by calculating the change in the adiabatic constant α of eq. (I.82).

$$\frac{\alpha_2}{\alpha_1} = \frac{p_2}{p_1}\left(\frac{\rho_1}{\rho_2}\right)^{\gamma} = \left(\frac{1}{\gamma+1}\right)^{\gamma+1} \left[2\gamma M_1^2 - (\gamma-1)\right] \left(\gamma-1 + \frac{2}{M_1^2}\right)^{\gamma} \qquad \text{(III.80)}$$

The relation reduces to $\alpha_2 = \alpha_1$ when $M_1^2 = 1$. The trend in the change in α_2/α_1 for higher Mach numbers is given by

$$\frac{d}{dM_1^2}\left(\frac{\alpha_2}{\alpha_1}\right) = \left(\frac{1}{\gamma+1}\right)^{\gamma+1} \frac{2\gamma(\gamma-1)\left(\gamma-1 + \frac{2}{M_1^2}\right)^{\gamma-1}}{M_1^4} (M_1^2-1)^2 \qquad \text{(III.81)}$$

Because of the $(M_1^2-1)^2$ term, α_2/α_1 changes very slowly in the low Mach number range ($M_1^2 \gtrsim 1$). For $M_1^2 \neq 1$ the gradient is always positive, implying that $\alpha_2 > \alpha_1$ when $M_1^2 > 1$.

Shock Waves 2: Parallel Shocks We return now to the discussion of shock waves in magnetized plasmas. The dissipative nature of the shock tends to change the components of the anisotropic pressure in ways that are not predicted by the continuity relations (recall that there is one more unknown than there are equations if $p_{\parallel}$ and $p_{\perp}$ are retained). Without additional information, it is not possible to learn how $p_{\parallel}$ and $p_{\perp}$ separately change across a shock wave. Therefore, we must sacrifice this level of detail, and represent the pressure again by its isotropic form. In this case we may also leave γ arbitrary.

A parallel MHD shock wave is characterized by the condition $\vec{B}||\hat{n}$, or $B_t = 0$. When this condition is imposed on the MHD continuity relations one finds that they reduce identically to the ordinary gas dynamic shock relations that were reviewed under the previous heading (that is, eqs. III.67 through 70). The magnetic field strength disappears as an explicit variable except in Equation (III.59) which since $B_n = B$ becomes in this instance

$$[[B]] = 0 \qquad \text{(III.82)}$$

The field strength is continuous across the shock.

The treatment for the non-magnetic variables is identical to that developed for the gas dynamic case and all of the solutions given there apply in this situation also. In particular we were allowed to transform to a frame of reference in which $U_t = 0$. In this frame then $\vec{U}||\vec{B}$ and we have prescribed the condition necessary for the existence of an equipotential domain (see Section II.4). Thus there exists a streamline constant κ such that

$$\vec{B} = \kappa\rho\vec{U} \qquad \text{(III.83)}$$

It follows that since both B and ρU are conserved across the shock in this frame of reference

$$[[\kappa]] = 0 \qquad \text{(III.84)}$$

This result emphasizes the inert response of the magnetic field to its passage through a parallel shock.

Shock Waves 3: Perpendicular Shocks. Consider next as the opposite limit to the parallel shock, the perpendicular shock in which $\vec{B}\perp\hat{n}$, that is $B_n = 0$ and $B_t = B$. We note at the outset that the continuity relation for tangential momentum (III.63) again shows that U_t is continuous across the shock, and hence as in the previous examples may be set equal to zero.

The continuity relations for the perpendicular shock can be written as

$$[[\rho U]] = 0 \tag{III.85}$$

$$[[\rho U^2 + p + \frac{B^2}{2\mu_o}]] = 0 \tag{III.86}$$

$$[[(\frac{1}{2}\rho U^2 + \frac{\gamma}{\gamma-1}p + \frac{B^2}{\mu_o})U]] = 0 \tag{III.87}$$

$$[[UB]] = 0 \tag{III.88}$$

The first and last of these can be combined to show that the field strength must change across the shock in the same proportion as does the density. That is

$$[[B/\rho]] = 0 \tag{III.89}$$

As in the procedure adopted for treating the continuity relations of ordinary gas dynamic shocks, we assume the upstream parameters to be given, in this case including the field strength B_1, and solve for the downstream parameters. The elimination of ρ_2, p_2 and B_2 results in a cubic equation for U_2. The trivial solution $U_2 - U_1 = 0$ can be factored out leaving the quadratic equation

$$(\frac{U_2}{U_1})^2 - [\frac{\gamma-1}{\gamma+1} + \frac{2\gamma}{\gamma+1}(\frac{1}{\gamma S_1^2} + \frac{1}{2A_1^2})](\frac{U_2}{U_1}) - \frac{2-\gamma}{\gamma+1}\frac{1}{A_1^2} = 0 \tag{III.90}$$

In (III.90) the sonic Mach number is designated by the symbol S and A is the Alfvén Mach number

$$S^2 \equiv M^2 = \frac{\rho V^2}{\gamma p} \tag{III.91}$$

$$A^2 \equiv \frac{\rho V^2}{B^2/\mu_o} \tag{III.92}$$

One of the two solutions of the quadratic equation is non-physical in that it gives a negative value for U_2. The other solution gives the velocity ratio for a perpendicular MHD shock of arbitrary sonic and Alfvén Mach number. In the limit $B \to 0$, i.e. $A \to \infty$, equation (III.90) reduces to the corresponding equation for ordinary gas dynamic shocks (III.74). Therefore, in the dual hypersonic limit, $A^2 \to \infty$ and $S^2 \to \infty$, we recover the same limiting expression for U_2/U_1, ρ_2/ρ_1 and p_2/p_1. In particular in the dual hypersonic limit

$$\frac{\rho_2}{\rho_1} = \frac{B_2}{B_1} = \frac{U_1}{U_2} = \frac{\gamma+1}{\gamma-1} \tag{III.93}$$

Thus in the case of a strong perpendicular shock such as a planetary bow shock or a solar blast wave (for both of which $\gamma = 5/3$), the density and the field strength jump by a factor of four, and the velocity (in the reference frame of the shock) drops by the same factor.

Shock Waves 4: Oblique Intermediate Mode Shocks

For the case of oblique shocks ($B_{\parallel} \neq 0$, $B_t \neq 0$), there exist three types of MHD shocks that correspond to the three modes of small amplitude MHD waves. The two modes corresponding to the fast and slow mode waves are compressive. The mode corresponding to the intermediate mode wave is non-compressive, but if the pressure is anisotropic and $p_{\parallel}$ and $p_{\perp}$ change across the shock, the density will also change. (Recall that in the case of an anisotropic plasma, the continuity relations do not specify the change in $p_{\parallel}$ and $p_{\perp}$). We take up here the intermediate mode shock and retain the anisotropic form of the pressure tensor.

The distinction between intermediate shocks and the compressive shocks is revealed immediately by multiplying the continuity relation for tangential momentum (III.63) by the continuous variable $(B_n/\rho U_n)$ and adding to the result the continuity relation for tangential electric field (III.61). One obtains this way

$$[[(1- \xi \frac{B_n^2}{\mu_o \rho U_n^2}) B_t U_n]] = C \tag{III.94}$$

When this condition is satisfied by the vanishing of the term in parentheses, the shock described is the intermediate mode. The equation shows that when $(1- \xi B_n^2/\mu_o \rho U_n^2) = 0$ on one side, it is zero on the other side as well. (The hybrid possibility that $B_t = 0$ on the other side, referred to as a switch-off shock, will not be considered. Such structures, if they exist, have not been demonstrated to play a significant role in solar system plasmas.) The term in parentheses is not zero in the compressive mode cases. Thus

$$1- \xi \frac{B_n^2}{\mu_o \rho U_n^2} = 0 \text{ (intermediate mode shocks)} \tag{III.95}$$

in which $\xi B_n^2/\mu_o \rho U_n^2$ may be evaluated on either side of the shock and thus this factor is continuous across an intermediate mode shock. Note that since B_n and ρU_n are continuous, it follows further that

$$[[\xi\rho]] = 0 \tag{III.96}$$

Since the change in ξ can not be specified and ξ can in principle be different on the two sides, (III.96) shows that the density can in principle change across an intermediate mode shock in an anisotropic plasma. From this one sees that the same statement can be made concerning the propagation velocity of the shock, U_n, since ρU_n is continuous. The special but important case of an intermediate mode shock in an isotropic plasma should be noted separately

$$[[U_n]] = [[\rho]] = 0 \quad (\text{isotropic pressure, } \xi = 1) \tag{III.97}$$

By use of the intermediate mode propagation equation (III.95), the change in the velocity vector can be found from (III.61 and 63). The result can be written in the form (Hudson, 1970)

$$[[\vec{U}]] = \left(\frac{\xi\rho}{\mu_o}\right)^{1/2} [[\frac{\vec{B}}{\rho}]] \tag{III.98}$$

Equation (III.95) together with the continuity of B_n reduces the continuity relation for the normal component of momentum (III.62) to

$$[[p_\perp + \frac{B_t^2}{2\mu_o}]] = 0 \quad (\xi \text{ arbitrary}) \tag{III.99}$$

It is worth noting that in the isotropic case (ξ=1) the continuity relation for energy for this mode simplifies to

$$[[p]] = 0 \qquad (\xi = 1) \tag{III.100}$$

From which with (III.99) and $[[B_n]] = 0$ it follows that

$$[[B_t^2]] = [[B]] = 0 \quad (\xi = 1) \tag{III.101}$$

Thus an intermediate mode shock wave propagating in an isotropic plasma is non-compressive and non-dissipative. It merely changes the directions of the magnetic field and the flow, while perserving their magnitudes.

Shock Waves 5: Oblique Fast and Slow Mode Shocks

The two compressive MHD shock modes have the following characteristics in common with the small amplitude fast and low modes waves.

The density increases across both of them, but in the fast mode shock the field strength increases while in the slow mode shock it decreases. Both types of shock can result from the non-linear evolution of finite amplitude fast and slow mode MHD waves. Thus, the relationship between the shocks and the waves is a direct heritage. We will consider only the case of isotropic pressure in the discussion of the compressive shocks.

Both types of compressive MHD shocks possess the notable property that the magnetic fields on the two sides on the shock and the normal to the shock are coplanar (that is $(\vec{B}_1 x \vec{B}_2) \cdot \hat{n} = 0$). This result, which is referred to as the coplanarity theorem, follows directly from (III.94) in which $\xi = 1$ and the factor in parentheses is not zero. Choose the tangential vector $\hat{t}$ to be perpendicular to $(\vec{B}_t)_1 \equiv \vec{B}_1 - B_n \hat{n}$. That is $\vec{B}_1 \cdot \hat{t} = 0$ for this choice of $\hat{t}$. But by (III.94) $\vec{B}_2 \cdot \hat{t}$ is then zero also. Thus we have shown there is a choice of $\hat{t}$ which is perpendicular to $\vec{B}_1$ and $\vec{B}_2$, and by definition it is also perpendicular to $\hat{n}$. Hence $\vec{B}_1$, $\vec{B}_2$ and $\hat{n}$ are coplanar.

The coplanarity theorem has found use in the analysis of data on shock waves obtained from measurements in space. The orientation of the shock surface as given by its normal $\hat{n}$ can be obtained from a measurement of the magnetic fields on the two sides of the shock by the following procedure. Since B_n is continuous, $\vec{B}_1 - \vec{B}_2$ contains no normal component. Thus $\vec{B}_1 - \vec{B}_2$ lies in the plane of the shock. By the coplanarity theorem, $\vec{B}_1 x \vec{B}_2$ also lies in the plane of the shock. Hence

$$\hat{n} \,||\, (B_1 - B_2) \times (B_1 x B_2) \tag{III.102}$$

It is not useful in the case of compressive oblique MHD shocks to proceed as in the previous examples of eliminating all but one of the downstream variables to arrive at a single equation for, say, $(U_n)_2$. The resulting equation is fifth order, and become fourth order after the trivial solution $(U_n)_2 - (U_n)_1 = 0$ has been factored out. However, an important ordering of the post-shock normal flow velocities can be stated on physical grounds. In order to avoid the non-physical possibility of an intermediate mode shock catching up to a preceding fast mode shock or a slow mode shock catching up to a preceeding intermediate mode shock, the post-shock speeds must be ordered according to

$$(\rho U_n^{\,2})_2 \text{ (fast)} > \frac{B_n^{\,2}}{\mu_o} > (\rho U_n^{\,2})_1 \text{ (slow)} \tag{III.103}$$

in which $B_n 2/\mu_o \rho$ is the propagation speed of an intermediate mode shock in an isotropic gas (eq. III.95). This ordering determines whether the field strength increases or decreases across the shock, as can be seen by multiplying (III.94) by the continuous variable ρU_n

$$[[(\rho U_n^2 - \frac{B_n^2}{\mu_o})B_t]] = 0 \qquad \text{(III.104)}$$

and solve for $(B_t)_2$ in terms of $(B_t)_1$

$$(B_t)_2 = \frac{(\rho U_n^2)_1 - \frac{B_n^2}{\mu_o}}{(\rho U_n^2)_2 - \frac{B_n^2}{\mu_o}} (B_t)_1 \qquad \text{(III.105)}$$

Consider first the case of a fast mode shock. Then both $(\rho U_n^2)_1 > B_n^2/\mu_o$ and $(\rho U_n^2)_2 > B_n^2/\mu_o$, the first because the upstream flow speed must be super-Alfvenic in order to have a fast shock at all, and the second by (III.103). But $(\rho U_n^2)_1 > (\rho U_n^2)_2$ since by factoring out the continuous term ρU_n, this becomes $(U_n)_1 > (U_n)_2$, which is guaranteed by the compressive nature of the shock. In summary, then both the numerator and the dencminator in (III.104) are positive, and the numerator is larger than the denominator. Hence

$$(B_t)_2 > (B_t)_1 \quad \text{(fast mode shock)} \qquad \text{(III.106)}$$

and since $(B_n)_2 = (B_n)_1$, it also follows that

$$B_2 > B_1 \quad \text{(fast mode shock)} \qquad \text{(III.107)}$$

In the case of a slow mode shock, both the numerator and denominator are negative, the first because of (III.103) and the second because $(U_n)_2$ must be even smaller than $(U_n)_1$ by the compressive nature of the shock. The latter condition also means that the magnitude of the denominator is greater than the magnitude of the numerator. Hence

$$(B_t)_2 < (B_t)_1 \quad \text{(slow mode shock)} \qquad \text{(III.108)}$$

and again since $(B_n)_2 = (B_n)_1$

$$B_2 < B_1 \quad \text{(slow mode shock)} \qquad \text{(III.109)}$$

Figure III.4 shows the difference in the magnetic signatures of fast and slow mode shocks. On the downstream side of a fast mode shock the field bends away from the shock normal, giving rise to a compression of flux density, which is the same as an increase in field strength. On the downstream side of a slow mode shock, the field bends toward the shock normal resulting in a reduction of field strength.

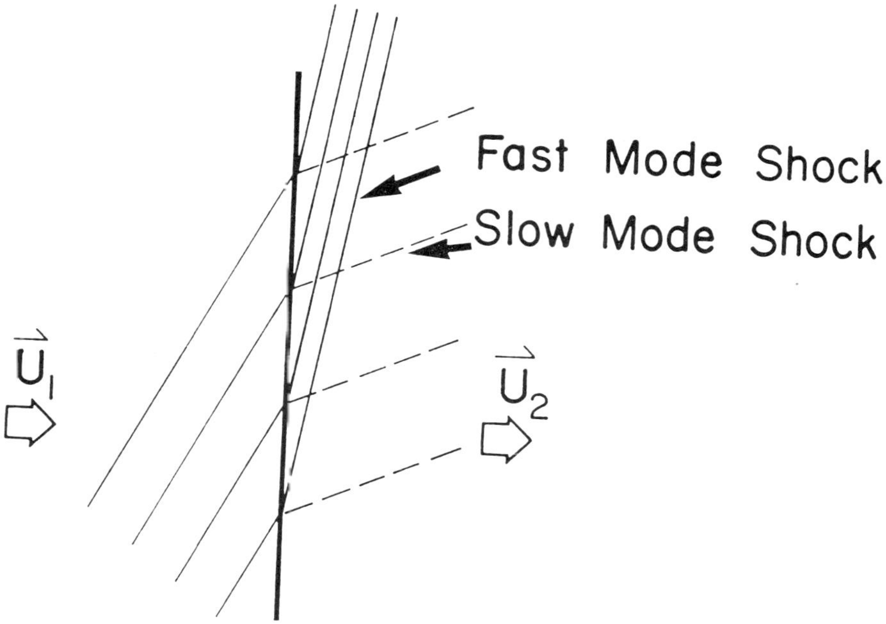

Figure III.4 Magnetic signatures of fast and slow mode shock waves

All three MHD shock modes have been observed in solar system plasmas, especially in the solar wind. Slow mode shocks play an expecially important role in some models of magnetic merging (Vasyliunas, 1975).

IV. MHD INSTABILITIES

This section reviews the four basic instabilities to which MHD fluids are subject. Two of the four are common to all fluids, namely the Kelvin-Helmholtz instability, colloqually known as the "wind over water" instability, and the Rayleigh-Taylor instability, which is called the flute instability in plasma physics and the interchange instability in magnetospheric physics. The other two, the firehose instability and the mirror instability, are caused by differences between $p_{\parallel}$ and $p_{\perp}$ in an anisotropic magnetized plasma, and are therefore peculiar to collisionless MHD fluids. We begin with instabilities driven by pressure anisotropy.

IV.1 The Firehose and Mirror Instabilities

In the theory of small amplitude MHD waves in an anisotropic plasma, these instabilities present themselves in the form of non-propagating, purely exponentially growing waves. The firehose instability is an exponential growth of the intermediate mode, and the mirror instability as an exponential growth of the slow mode. The mirror instability materializes first and grows fastest when the normal to the wave front $\hat{k}$ is oriented nearly perpendicular to the magnetic field.

To confirm the existence of these instabilities, to determine the conditions under which they occur, and to establish their properties, it is necessary to derive the dispersion relation for small amplitude MHD waves in an anisotropic plasma. The analysis proceeds analogously to that carried out in Section III.1 for small amplitude MHD wave in an isotropic plasma, except that in the Euler equation (III.2) the gradient of the scalar pressure is replaced by the expressions for the divergence of the anisotropic pressure tensor given in Section I.16, and the scalar adiabatic relation (III.3) is replaced by the double adiabatic relations (II.80 and 81). Explicitly, this procedure leads to the following form for the plane wave representation of the pressure perturbation

$$\nabla\cdot\overleftrightarrow{P} \longrightarrow -\left[\frac{\delta\rho}{\rho_o}(p_\perp)_o + \frac{\delta B_z}{B_o}(p_\parallel)_o\right]k_x\hat{x}$$
$$-\left[\frac{\delta B_z}{B_o}(p_\perp)_o + 3\left(\frac{\delta\rho}{\rho_o} - \frac{\delta B_z}{B_o}\right)(p_\parallel)_o\right]k_z\hat{z}$$
$$+\ \frac{(p_\parallel - p_\perp)_o}{B_o^2/\mu_o}\ \vec{J}\times\vec{B} \qquad \text{(IV.1)}$$

in which the unit imaginary number i has been suppressed since it ultimately multiplies all terms, and $\vec{J}\times\vec{B}$ has not been replaced by its wave perturbation form, because this reduction has already been given in (III.14). The equation is referenced to the coordinate system which was used previously and which is shown in Figure III.1.

The wave perturbation forms of the continuity equation and Faraday's induction law (eq's III.13 and 15) are now used to eliminate the density and field perturbations, leading to a single equation for the velocity perturbation, corresponding to eq. (III.20) for the isotropic pressure case.

$$\omega^2\delta\vec{V} = [C_\perp^2 k_x(\vec{k}\cdot\delta\vec{V}) - C_\parallel^2 k_x^2\delta V_x]\,\hat{x}$$
$$+\ [C_\perp^2 k_x k_z\delta V_x + 3C_\parallel^2 k_z^2\delta V_z]\,\hat{z}$$
$$+\ \xi_o C_A^2[k^2\delta V_x\hat{x} + k_z^2\delta V_y\hat{y}] \qquad \text{(IV.2)}$$

in which we have defined for notational convenience the (pseudo) anisotropic sound speeds

$$C_{\perp}^2 = \frac{(p_{\perp})_o}{\rho_o} , \quad C_{\parallel} = \frac{(p_{\parallel})_o}{\rho_o} \tag{IV.3}$$

The symbols C_A^2 and ξ have been defined previously (eq's III.17 and 64).

As before the procedure from this point is to write out explicitly the three vector components of (IV.2), identify the coefficient matrix of the dependent variables δV_x, δV_y and δV_z, and set the determinant of that matrix to zero to obtain the dispersion equation for the waves. Again however the y-component contains only the independent variable δV_y, and therefore is itself a pure mode. One obtains the anisotropic form of the dispersion equation for the intermediate mode wave directly from the y-component,

$$\omega_i^2 = \xi_o \ C_A^2 \ k_z^2 \tag{IV.4}$$

Comparison with the isotropic form of this equation (III.21) shows that the effect of anisotropy is the introduction of the multiplicative factor ξ_o. Whereas the isotropic form of this equation is positive definite, ξ_o may be negative in which circumstance the intermediate mode will exhibit non-propagating, pure exponential growth. This is the firehose instability.

The threshold for the onset of the instability is $\xi_o = 0$. In general we may classify the behavior of the mode according to whether ξ_o is positive, zero or negative.

$\xi_o > 0$, Propagating intermediate mode

$\xi_o = 0$, Non-propagating, non-growing perfectly inelastic perturbations

$\xi_o < 0$, Firehose instability (non-propagating, pure exponential growth)

In the propagating wave regime ($\xi_o > 0$), the phase speed (and the group velocity) can be greater or less than its value for the isotropic case (eq. III.23) depending on whether ξ_o is greater than or less than unity.

$$(V_{ph})_i \text{ (anisotropic)} = \sqrt{\xi_o}\ (V_{ph})_i \text{ (isotropic)} \tag{IV.5}$$

To understand the physical reason for the dependence of ω_i on ξ_o in (IV.4), it is useful to rewrite the expression for ξ_o in the form

$$\xi_o - 1 = \frac{(p_{\perp} - p_{\parallel})_o}{B_o^2/\mu_o} \tag{IV.6}$$

The right hand side of (IV.6), which contains the full effect of pressure anisotropy on this mode, shows that one must consider the contributions that all three pressures, $p_\perp$, $p_\parallel$ and $B_o^2/2\mu_o$, make to the frequency of the wave. Recall that in the isotropic case, the frequency is fixed by balancing the inertial force exerted by a volume of plasma that is oscillating transversely to the magnetic field against the magnetic tension the motion engenders in stretching the field. Since an increase in the restoring force increases the frequency, and the right hand side of (IV.6) measures the change in frequency resulting from pressure anisotropy, it is evident that $p_\perp$ acts to increase the restoring force and $p_\parallel$ acts to decrease it. When the two pressures are equal (i.e. isotropy) their effects cancel. It can be seen qualitatively from Figure IV.1 that both effects can be described in terms of a centrifugal force. The centrifugal force exerted by the thermal motions of particles moving parallel to the bent flux tube acts against the magnetic tension, which is attempting to straighten the tube. The centrifugal force exerted by the thermal motions of particles gyrating perpendicular to the flux tube acts against the tension on the oustide but with the tension on the inside. However, the bend increases the density of particles gyrating on the inside and decreases it on the outside, thus there is a net force tending to straighten the tube.

When $p_\perp > p_\parallel$ the combined pressure effect and magnetic tension increase the restoring force over that of pure isotropy, the frequency and wave speed therefore increase. When $p_\parallel > p_\perp$, the reverse occurs, and if the imbalance $p_\parallel - p_\perp$ should exceed the magnetic tension B^2/μ_o, the net restoring force becomes negative, and the bend grows under the force the bend itself produces.

Consider next the dispersion equation for the two compressive modes in an anisotropic plasma. Setting to zero the determinant of the coefficient matrix obtained from the x and z-components of (IV.2) gives

$$(\omega^2_{f,s} - 3C_\parallel^2 k_z^2)\,[\omega^2_{f,s} - (2C_\perp^2 + C_\parallel^2)\,k_x^2 - \xi_o\, C_A^2\, k_z^2] - C_\perp^4\, k_x^2\, k_z^2 = 0 \qquad \text{(IV.7)}$$

Before isolating the mirror instability, it is instructive to see how parallel propagation and perpendicular propagation are modified by the anisotropy. As in the isotropic case, when $k_x = 0$, equation (IV.7) has two solutions which can be written in terms of the phase speed.

$$\vec{k}\,||\,\vec{B}_o \; : \; (V^2_{ph})_{f,s} = \begin{cases} 3C_\parallel^2 \\ \xi_o C_A^2 \end{cases} \qquad \text{(IV.8)}$$

The first of these corresponds to a sound wave propagating parallel to the field in a gas for which $\gamma = 3$, which is the value appropriate to one degree of thermal freedom. The second is the solution for the intermediate mode wave (eq. IV.5) with $\theta = 0$. When $k_z = 0$, equation (IV.1) again has two solutions

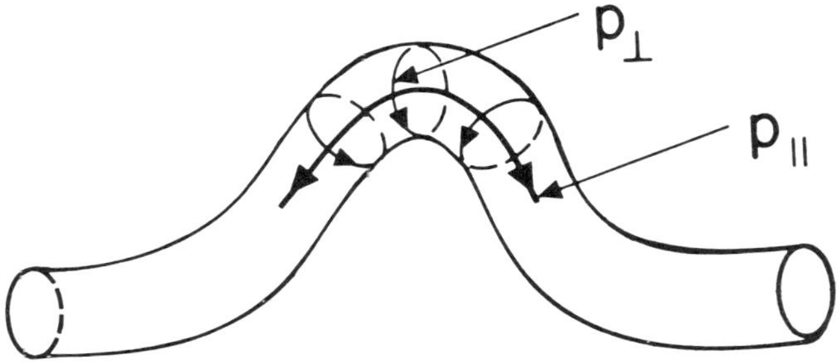

Figure IV.1 Magnetic flux tube in "firehose" configuration.

$$\vec{k} \perp \vec{B}_o : \; (V_{ph})^2_{f,s} = \begin{cases} 2C_\perp^2 + C_A^2 \\ 0 \end{cases} \qquad \text{(IV.9)}$$

The first corresponds to the perpendicular fast mode wave, which has a phase speed equal to the Pythagorean sum of the Alfvén speed and the sound speed of a gas with two degrees of freedom. The second solution corresponds to the slow mode wave, which does not propagate perpendicular to the magnetic field.

That a slow mode wave with its wave normal oriented nearly perpendicular to the magnetic field can be unstable is seen by utilizing the approximations that apply to this mode and this orientation. These are $\omega_s \approx 0$ (and therefore the ω_s^4 term can be dropped compared to the ω_s^2 term) and $k_z^2/k_x^2 << 1$. Then the dispersion equation, can be solved for ω_s^2/k^2

$$\frac{\omega_s^2}{k^2} = \frac{3C_\parallel^2(2C_\perp^2 + C_A^2) - C_\perp^4}{2C_\perp^2 + C_A^2} \qquad \text{(IV.10)}$$

The right hand side is negative, and therefore the wave is unstable, if

$$\frac{C_\perp^2}{C_\parallel^2} > 6\left(1 + \frac{C_A^2}{2C_\perp^2}\right) \qquad \text{(IV.11)}$$

Rewriting the instability in terms of pressure ratios, (IV.11) becomes

$$\left(\frac{p_\perp}{p_\parallel}\right)_o > 6\left(1 + \frac{1}{\beta_\perp}\right) \quad \text{(MHD)} \qquad \text{(IV.12)}$$

in which the symbol β is defined in general in plasma physics as the ratio of the particle pressure to the magnetic field pressure. In this case it is

$$\beta_{\perp} \equiv \frac{(p_{\perp})_o}{B_o^2/2\mu_o} \qquad \text{(IV.13)}$$

The criterion for the existence of the mirror instability given by eq. (IV.12) was derived with the use of the double adiabatic invariants of MHD. The qualification (MHD) has accordingly been affixed to the result. A treatment of this instability by the use of plasma kinetic theory, which takes into account diabatic heating by parallel heat flux, arrives at a similar result, but the factor of six in this case is replaced by unity (e.g. Krall and Trivelpiece, 1973)

$$\left(\frac{p_{\perp}}{p_{\parallel}}\right)_o > 1 + \frac{1}{\beta_{\perp}} \qquad \text{(Kinetic Theory)} \qquad \text{(IV.14)}$$

A sketch of mirror geometry in the magnetic field produced by an obliquely propagating compressive mode wave is shown in Figure IV.2. In the case of a slow mode wave, the particle pressure is strongest where the magnetic pressure is weakest, namely, in the middle of the magnetic bottles formed by the periodic constriction of each flux tube. If the mirror instability criterion is met, the increase in the destabilizing component of the pressure, $p_{\perp}$, that attends an oblique slow mode perturbation exceeds the increase in the restraining tensions in the field and in $p_{\parallel}$, and the perturbation grows.

IV.2. The Kelvin-Helmholtz Instability

The pressure perturbations that arise in a fluid when it is forced to flow over a wavy wall are such that if the wall were non-rigid the amplitude of the wave in the wall would tend to grow. The example commonly given to illustrate the preceding statement is the waves that form on the surface of open water on a windy day. There are two equivalent ways of understanding the origin of the destabilizing perturbations in this situation. In one we are to notice that the centrifgual force generated in the fluid as it moves along the wall in a serpentine motion to conform to waves in the wall are such as to apply extra force on the wall in the troughs and to diminish the force on the wall in the crests. The troughs are thereby impelled to deepen and the crests to heighten.

An alternative (but, it should be emphasized, equivalent) way to understand the phenomenon focuses on the change in the pressure in an element of fluid next to the wall as it descends into a trough or surmounts a crest of the wave. In the frame of reference of the wall, these pressure changes are of course time stationary and are characteristic of troughs and crests generally. The essential physical principle

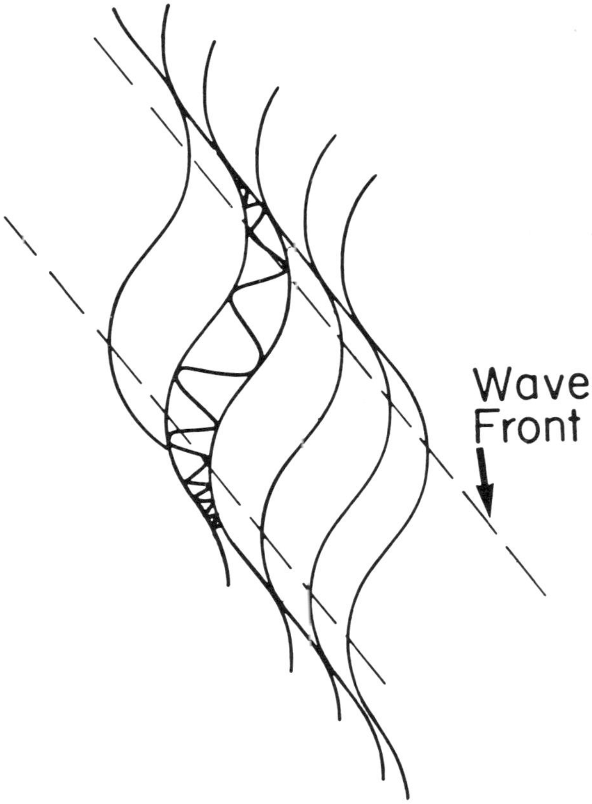

Figure IV.2 Oblique slow mode wave showing the "mirror" geometry prerequisite for the mirror instability. The path of a particle bouncing between two mirror points is indicated.

is most clearly seen if we consider the case of an incompressible fluid, that is one for which the density is constant in space and time. The distinguishing equation for an incompressible fluid is

$$\nabla . \vec{V} = 0 \qquad \text{(IV.15)}$$

which follows from the constant density condition and the continuity equation (I.22). In ordinary (non-magnetic) fluid dynamics in which gravity and viscosity are ignored, an incompressible fluid satisfies the following Bernoulli equation (from I.89 and 92 with the isometric polytropic index $n = \infty$)

$$\frac{1}{2} \rho V^2 + p = \rho W \qquad \text{(IV.16)}$$

The right hand side of (IV.16) is a constant. This equation expresses the well-known Venturi effect: when along a given streamline the velocity of the fluid increases the pressure drops, and vice versa.

The Venturi effect explains why the pressure distribution in a fluid moving along a wavy wall acts in the sense to amplify the wave. As we shall see shortly, the perturbation caused by the wavy wall decreases exponentially away from the wall with a scale length equal to $\lambda/2\pi$, where λ is the wave length of the wave in the wall. For many purposes a problem involving a perturbation that decreases exponentially with a scale length ℓ can be replaced by a problem with a constant perturbation extending over the same distance ℓ, but with zero perturbation beyond this distance. Adopting for a moment the second description, one can say that the zero-order area through which the wave-affected part of the fluid flows is given by some constant width in the direction parallel to the wave troughs and crests multiplied by a height equal to the exponential scale length $\lambda/2\pi$. Consider then how the perturbation in the area caused by the wavy wall effects the flow velocity of the fluid. In a steady incompressible flow the total quantity of fluid passing over a trough must be equal to the total quantity of fluid passing over a crest in the same interval of time. But a trough increases the area available to the flow and a crest decreases it. The velocity must then change to compensate for the change in area to keep the total flow continuous. Thus, the velocity over a trough must be less than the velocity over a crest. By the Venturi effect, the pressure is therefore increased over a trough and decreased over a crest. As in the case of the explanation in terms of the centrifugal force, the differential force on the wall exerted by the pressure acts to amplify the wave. The actual force that the fluid exerts on the wall is the pressure force. A demonstration of the equivalence between the explanation in terms of centrifugal force and pressure force follows upon deriving an explicit expression for the pressure perturbation and comparing it with the centrifugal force associated with the fluid motion over the wavy wall.

It is convenient for the purpose of fixing ideas with a concrete example to begin a treatment of the Kelvin-Helmholtz instability with a discussion of the flow of fluid past a wavy wall. In retrospect one can see that the essential effect does not depend on the presence of a physical wall, but rather it depends only on the existence of a shear in the flow velocity of the fluid. If there exists a frame of reference in which one layer of the fluid is at rest (representing the wall) while an adjacent layer is moving, any naturally occurring deviation from smoothness in the interface between the two layers will grow under the resulting pressure perturbation. Manifestations of the instability therefore should be nearly as plentiful as appearances of sheared flows. However in many situations fluids exhibit some degree of elasticity and resist the growth of the initial perturbation. Elastic resistance to the Kelvin-Helmholtz instability results from surface tension in the case of wind-over-water waves, stable stratification of the atmosphere in the case of atmospheric gravity waves, and magnetic tension in the case of surface MHD waves. In each of these instances the shear in velocity must exceed some threshold value before the onset of unstable growth occurs.

To illustrate the general nature of the treatment of the Kelvin-Helmholtz instability in a MHD context, we consider here the problem of a sheared flow between two incompressible, isotropic, inviscid MHD fluids separated by a tangential discontinuity. Gravity will be ignored. The example has application to the magnetopauses of planetary magnetospheres, the quasi-equatorial current sheet in the solar wind and the myriad accidental tangential discontinuities lacing the solar wind.

On both sides of the discontinuity, the continuity equation for an incompressible fluid given by (IV.15) applies as well as the Euler and hydromagnetic equations, which we repeat here for convenience

$$\rho \frac{d\vec{V}}{dt} + \nabla p = - \frac{\vec{B} \times \nabla \times \vec{B}}{\mu_o} \tag{IV.17}$$

$$\frac{\partial \vec{B}}{\partial t} = \nabla \times (\vec{V} \times \vec{B}) \tag{IV.18}$$

As in the treatment of the MHD plane waves, the variable parameters, $\vec{V}$, p, and $\vec{B}$, are decomposed into zero-order and perturbation parts.

$$\vec{V} = \vec{V}_o + \delta\vec{V} \tag{IV.19}$$

$$p = p_o + \delta p \tag{IV.20}$$

$$\vec{B} = \vec{B}_o + \delta\vec{B} \tag{IV.21}$$

Let (x,y,z) be a Cartesian coordinate system in which the z-axis is normal to the plane of the discontinuity. Then $\vec{V}_o$ and $\vec{B}_o$ lie in the (x,y) plane. Assume the perturbations have the form of propagating plane waves along the surface of the discontinuity, but decay in strength away from the discontinuity. This is the cannonical form of linearized plane surface waves. That is the perturbation of any quantity Q has the explicit space and time dependence given by

$$\text{Perturbation (Q)} = \delta Q \, e^{i(\omega' t - \vec{k}_t \cdot \vec{r}) - k_z z} \tag{IV.22}$$

where δQ is the (possibly complex) amplitude of the perturbation, ω' is the Doppler shifted frequency of the wave (that is the frequency in our frame of reference), $\vec{k}_t \equiv k_x\hat{x} + k_y\hat{y}$ is the component of the propagation vector that lies in the plane of the discontinuity, and k_z is the reciprocal of the scale length for the exponential decay of the strength of the perturbation in the direction normal to the surface of the discontinuity. Thus, we must have $k_z>0$ for $z>0$ and $k_z<0$ for $z<0$.

Substitution of the perturbation forms of the dependent variables with the plane surface wave assumption into the continuity, Euler and hydromagnetic equations gives respectively

$$\vec{\kappa}_{\pm} \cdot \delta\vec{V} = 0 \tag{IV.23}$$

$$\omega\rho\delta\vec{V} - \vec{\kappa}_{\pm}\,\delta p = -\,(\vec{B}_o \cdot \vec{k})\,\frac{\delta\vec{B}}{\mu_o} + \left(\frac{\vec{B}_o}{\mu_o} \cdot \delta\vec{B}\right)\vec{\kappa}_{\pm} \tag{IV.24}$$

$$\omega\delta\vec{B} = -\,(\vec{B}_o \cdot \vec{k})\,\delta\vec{V} \tag{IV.25}$$

where $\omega = \omega' - \vec{V}_o\cdot\vec{k}$ is the frequency in the plasma rest frame (cf. eq. III.19), and the complex wave number $\kappa_{\pm}$ is defined by

$$\vec{\kappa}_{\pm} \equiv k_x\hat{x} + k_y\hat{y} \pm ik_z\hat{z} \tag{IV.26}$$

in which the + sign applies for $z>0$ and the - sign for $z<0$. By this definition of $\vec{\kappa}_{\pm}$, we have arranged for k_z to be everywhere positive. In the combinations $\vec{V}_o\cdot\vec{k}$ and $\vec{B}_o\cdot\vec{k}$, the subscript t on $\vec{k}$ is superfluous.

The special properties of the boundary waves are determined by the applicable continuity relations. The total pressure is continuous across a tangential discontinuity (eq. III.66) and the displacement of the two fluids in the z-direction must be continuous in order to avoid separation or interpenetration of the fluids. Let P_T designate the total pressure (i.e. $P_T = p + (B^2/2\mu_o)$). Then the two continuity relations can be expressed as

$$[[P_T]] = 0 \tag{IV.27}$$

$$[[\delta z]] = 0 \tag{IV.28}$$

Now

$$\delta P_T = \delta p + \frac{1}{\mu_o}\vec{B}_o \cdot \delta\vec{B} \tag{IV.29}$$

The quantities δp and $\delta\vec{B}$ can be eliminated in favor of $\delta\vec{V}$ through the use of (IV.24 and 25) resulting in

$$\delta P_T\vec{\kappa}_{\pm} = [\omega^2-(\vec{V}_A\cdot\vec{k})^2]\,\frac{\rho\delta\vec{V}}{\omega} \tag{IV.30}$$

where $\vec{V}_A = \vec{B}_o/\sqrt{\mu_o\rho}$ is the Alfvén velocity. The scalar product of (IV.30) with $\vec{\kappa}_{\pm}$ together with (IV.23) show that

$$\delta P_T\,\kappa_{\pm}^2 = 0 \tag{IV.31}$$

Thus either $\delta P_T = 0$ or $\kappa_{\pm}^2 = 0$. The first option leads directly to the usual relations for an intermediate mode MHD wave, as can be seen from eq. (IV.30), which in this case gives $\omega^2 = (\vec{V}_A \cdot \vec{k})^2$. The second option is specific to surface waves, and yields the important relation

$$k_t^2 = k_z^2 \tag{IV.32}$$

This expresses the condition stated earlier that the decay length for the strength of the perturbation away from the discontinuity (or wall) is $k_z^{-1} = k_t^{-1} = \lambda/2\pi$, where λ is the wave length of the surface wave.

The two continuity relations (IV.27 and 28) will now be used together with the expression for the perturbation in the total pressure (IV.30) to obtain the dispersion equation for the surface waves. The mathematical procedure has the following structure. The quantity $\delta\vec{V}$ in (IV.30) will be eliminated in favor of δz to arrive at a relation that has the algebraic form

$$\delta P_T = A\ \delta z \tag{IV.33}$$

where the quantity A will be determined below. Then the condition $[[P_T]] = 0$ gives $[[A\delta z]] = 0$. But since δz is also continuous it can be factored out, resulting in

$$[[A]] = 0 \tag{IV.34}$$

Equation (IV.34) with A made explicit is the dispersion equation.

To find the expression represented by A, scalar multiply equation (IV.30) by $\hat{z}$ and solve for δP_T

$$\delta P_T = \frac{1}{\pm ik_z}\,[\omega^2 - (\vec{V}_A \cdot \vec{k})^2]\ \frac{\rho\ \delta\vec{V}\cdot\hat{z}}{\omega} \tag{IV.35}$$

But

$$\delta\vec{V}\cdot\hat{z} = \frac{d\delta z}{dt} = \frac{\partial\delta z}{\partial t} + \vec{V}_o \cdot \nabla\delta z = i(\omega' - \vec{V}_o \cdot \vec{k})\ \delta z = i\omega\delta z \tag{IV.36}$$

Thus

$$A = \mp\ \rho[\omega^2 - (\vec{V}_A \cdot \vec{k})^2] \tag{IV.37}$$

The common factor k_z has been dropped from the final expression since it would cancel out in the continuity relation (IV.34).

If we denote quantities that refer to the plasma above the discontinuity ($z>0$ corresponding to the + sign) by the subscript 1 and quantities that refer to the plasma below the discontinuity by the subscript 2, the discontinuity relation (IV.34) becomes

$$\rho_1[\omega_1^2 - (\vec{V}_{A1}\cdot\vec{k})^2] + \rho_2[\omega_2^2 - (\vec{V}_{A2}\cdot\vec{k})^2] = 0 \qquad \text{(IV.38)}$$

This can be rewritten as a quadratic equation for the frequency ω' in the initially chosen frame of reference. This is the observed frequency and it must be the same for both sides of the discontinuity. The solution of the quadratic equation for ω' is

$$\omega' = \frac{\rho_1\vec{V}_1\cdot\vec{k} + \rho_2\vec{V}_2\cdot\vec{k}}{\rho_1 + \rho_2} \pm \frac{1}{(\rho_1 + \rho_2)} \{(\rho_1 + \rho_2)[\rho_1(\vec{V}_{A1}\cdot\vec{k})^2 + \rho_2(\vec{V}_{A2}\cdot\vec{k})^2] - \rho_1\rho_2[(\vec{V}_1 - \vec{V}_2)\cdot\vec{k}]^2\}^{\frac{1}{2}} \qquad \text{(IV.39)}$$

in which the subscripted zeros on $\vec{V}_1$ and $\vec{V}_2$ have been dropped.

It is evident that ω' will have an imaginary part corresponding to the functioning of the Kelvin-Helmholtz instability if the argument of the radical is negative. It can be quickly verified from this equation that the condition for the operation of the instability can be expressed in terms of a threshold condition on the velocity shear $\Delta\vec{V} \equiv \vec{V}_1 - \vec{V}_2$, namely

$$(\Delta\vec{V}\cdot\vec{k})^2 > \frac{1}{\mu_o}\left(\frac{1}{\rho_1} + \frac{1}{\rho_2}\right)[(\vec{B}_1\cdot\vec{k})^2 + (\vec{B}_2\cdot\vec{k})^2] \quad \text{(Kelvin-Helmholtz instability)} \qquad \text{(IV.40)}$$

in which $\vec{B}_1$ and $\vec{B}_2$ are the zero-order fields on the two sides of the discontinuity.

In order to make the left hand side of (IV.40) as large as possible for a given $\Delta\vec{V}$, choose $\vec{k}$ to be parallel (or antiparallel) to $\Delta\vec{V}$. Then it is apparent the value which $\Delta\vec{V}$ must exceed in order for the instability to operate depends on the size of components of $\vec{B}_1$ and $\vec{B}_2$ parallel (or antiparallel) to $\Delta\vec{V}$. This is readily understood to be a consequence of the stretching by the wave of that component of $\vec{B}$ (on either side) that lies parallel to $\vec{k}$. The stretching occurs because if there is a component of $\vec{B}$ parallel to $\vec{k}$, the field lines cut across the troughs and crests of the wave and are stretched in length, according to their obliquity relative to k, in proportion to the ratio that the area of a wavy surface makes to a smooth one. Since the tension inherent in a magnetic field resists any force acting to stretch the field, the velocity shear is required to exceed a certain value given by (IV.40) in order to overcome this resistance. Note that if $\Delta\vec{V}$ is perpendicular to $\vec{B}$ on both sides of the discontinuity, the right hand side of (IV.40) is zero for a wave propagating parallel to $\Delta\vec{V}$. In this case the wave does not stretch the field lines and the surface is unstable for arbitrarily small values of ΔV.

As a final observation concerning the properties of MHD surface waves, consider the wavelength dependence of the instability. In the expression for the threshold criterion (IV.40), the wave vector $\vec{k}$ multiplies all terms. The expression would be unaffected numerically therefore if $\vec{k}$ were replaced by the unit vector pointing in the propagation direction, $\hat{k}$. The threshold criterion is thereby seen to be independent of the wavelengths of the surface wave. That is, if the surface is unstable for one wavelength, it is unstable for all wavelengths. On the other hand, the dispersion equation (IV.39) shows that when the instability criterion is met, the imaginary part of the frequency is directly proportional to k. Thus the growth rate of the instability is greatest for small wavelength waves. However, we have treated only the linear problem. The result concerning the growth rate therefore only implies that the short wavelength waves reach their nonlinear form more quickly than the long wavelength waves. The linear treatment can not predict which wavelength waves will have the largest amplitude after they have evolved into the nonlinear domain.

IV.3 The Magnetospheric Interchange Instability

The magnetospheric interchange instability is a particular type of Rayleigh-Taylor instability in a MHD fluid. The Rayleigh-Taylor instability occurs whenever an adiabatic interchange of fluid parcels results in a reduction of stored energy, whether the energy be stored as potential energy or as kinetic energy of compression. The kinetic energy associated with the interchange motion can then be supplied by the release of stored energy, the motion becomes self-propelled, and the initial arrangement of fluid parcels proves to be unstable.

In the most commonly cited example of a Rayleigh-Taylor instability, one incompressible fluid overlies another which has a smaller mass density, or more informally expressed, a heavy fluid overlies a light one. Then an interchange of a parcel of the heavy fluid and an equal volume of light fluid results in lowering their common center of gravity, thereby releasing stored gravitational energy. Such interchanges therefore will occur spontaneously, overturning the unstable configuration until the light fluid completely overlies the heavy fluid. In contrast to the initial state, the final state is stably stratified.

For this simple example it is a trivial matter to write down a mathematical criterion which must be satisfied in order for the instability to occur

$$\vec{g} \cdot \nabla \rho < 0 \quad \text{(unstable)} \qquad \text{(IV.41)}$$

In (IV.41), $\vec{g}$ is the force of gravity, but it could as well represent any inertial force such as the centrifugal force. The combination of gravitational and centrifugal forces, which is called the geopotential force, is used to describe the "effective" gravitational force in the frame of reference corotating with a planet. Later in this subsection we will need to use the geopotential force and we will denote it by $\vec{g}^*$. Then $\vec{g}$ in (IV.41) is replaced by $\vec{g}^*$.

The Rayleigh-Taylor instability manisfests itself in stellar and planetary atmospheres as well, but here the compressibility of the fluid (in this case a gas) must be taken into account. To guarantee stability, it is not sufficient that a less dense gas overlie more dense gas since in a vertical interchange of gas parcels, a descending parcel is compressed adiabatically and becomes denser as it moves through increasing atmospheric pressure. Conversely, an ascending parcel expands adiabatically and becomes less dense as it moves through decreasing atmospheric pressure. Thus, instead of comparing mass densities at two different levels, it is necessary to compare the masses in volumes that increase with height according to the adiabatic relation

$$V = c\,(p_{st})^{-\frac{1}{\gamma}} \qquad \text{(IV.42)}$$

which follows from eq. (I.82) and the definition of a fluid parcel, which entails a volume of fixed total mass and thus which obeys ρV=const. The pressure p in (IV.42) is meant to be the actual pressure of the atmosphere. To make this designation explicit, we have used the subscript st, which denotes "structual". For a given, fixed value of the parameter c, atmospheric parcels with volumes given by (IV.42) can be interchanged vertically with no change in the volume of the surrounding gas. Thus no work in the form of compression attends such interchanges, and the only change in energy can result from a change in the gravitational potential. It is now evident by direct analogy with (IV.41) the instability criterion in this case is

$$\vec{g} \cdot \nabla (\rho_{st} V) < 0 \quad \text{(unstable)} \qquad \text{(IV.43)}$$

in which $\rho_{st}V$ is the mass of volume-equivalent fluid parcels. Again $\vec{g}$ can be replaced by $\vec{g}^*$, which represents the generalized inertial force.

It is convenient to cast (IV.43) in terms of more readily available observables. To do this first note that

$$\rho_{st} V = c\,\rho_{st} (p_{st})^{-1/\gamma} \qquad \text{(IV.44)}$$

and recall that the specific entropy s is proportional to $\ell n(p/\rho^{\gamma})$. In terms of s_{st}, the instability criterion becomes (upon multiplying (IV.43) by $-\gamma/\rho_{st}V$)

$$\vec{g} \cdot \nabla s_{st} > 0 \quad \text{(unstable)} \qquad \text{(IV.44)}$$

Equation (IV.44) states that an atmosphere is unstable if the specific entropy decreases with height. Of course, the specific entropy is also not a readily available observable, but equation (IV.44) reveals that the condition of isentropy (s_{st} = const.) divides stable from unstable atmospheres. This suggests that the instability criterion should be expressed by reference to a convenient variable in an adiabatically stratified atmosphere. Since the variation of pressure with height in an atmosphere changes in time much less than the vertical variation of temperature does, the vertical pressure profile is usually regarded as fixed and known. Then the family of temperature profiles can be used for reference purposes that satisfys the adiabatic relation with respect to the given structural pressure profile (see eq's I.62 and 82)

$$T_{ad}(p_{st})^{-\frac{\gamma-1}{\gamma}} = \text{const.} \tag{IV.45}$$

where the constant on the right hand side is the family parameter. To arrive at an instability criterion in terms of a comparison between the vertical profiles of the structural temperature and the adiabatic temperature given by (IV.45), eliminate ρ_{st} on the right hand side of (IV.44) by use of the ideal gas law (I.62) to find

$$\frac{1}{\rho_{st}V} = c_1\, T_{st}(p_{st})^{-\frac{\gamma-1}{\gamma}} = c_2\, T_{st}(T_{ad})^{-1} \tag{IV.46}$$

in which the second equality follows from (IV.45), and c_1 and c_2 are constants. Equation (IV.46) is substituted into (IV.43) and evaluated with T_{ad} initiallized to T_{st}. Then the instability criterion becomes

$$\vec{g} \cdot (\nabla T_{st} - \nabla T_{ad}) > 0 \quad \text{(unstable)} \tag{IV.47}$$

In atmospheric application, instead of referencing directions of change to $\vec{g}$, it is usual to use height, z. Then noting that $\vec{g} = -g\hat{z}$, we may write (IV.47) in the most common form for the instability criterion

$$\frac{dT_{st}}{dz} < \frac{dT_{ad}}{dz} \quad \text{(unstable)} \tag{IV.48}$$

Since temperature normally decreases with height, equation (IV.48) states that an atmosphere is unstable if the temperature decreases in it more rapidly with height than in an adiabatic atmosphere.

With the incompressible and compressible forms of the gravitational Rayleigh-Taylor instability treated as preliminary examples, we turn next to consider the algebraically somewhat more complicated case of the magnetospheric interchange instability. We are here concerned with a planetary magnetosphere or portion of magnetosphere for which for mathematical convenience the following idealizations and approximations may be made. The magnetic field is a pure dipole field. The dipole axis is parallel to the rotation axis of the planet. The kinetic energy density of the plasma in the magnetosphere is small compared to the magnetic energy density. This last condition is to ensure that the magnetic field retains its dipole geometry, that is that the distortions caused by magnetospheric currents may be ignored. As noted by Gold (1959), the magnetic field can nevertheless undergo interchange motions as a result of forces generated by the plasma. This type of motion merely involves the interchange of entire magnetic flux tubes, all of which enclose the same quantity of magnetic flux. Such interchange motions can occur without causing any change in the magnetic field configuration, and thus entail no change in magnetic energy. It is clear that interchange motions take the form of circulations, since each flux tube that moves to fill the place of another must have its place filled in turn. An example of an interchange circulation pattern is shown in Figure IV.3.

It should be noted explicitly that the notion of interchanging flux tubes applies to a plasma for which the hydromagnetic approximation is valid. Then by the freezing law, the flux tube plays the role of a fluid parcel. It retains a constant quantity of plasma within it as it moves in interchanging circulations. As we shall see, an important difference between this and the previous compressible case we studied is that the volume of a flux tube is fixed by the quantity of magnetic flux it contains and its position in the magnetosphere. By our "low β" assumption, the volume of the flux tube is independent of the pressure of the gas it contains.

If an interchange motion such as the one depicted in the figure results in reducing the amount of energy stored in the plasma which is enclosed by the participating flux tubes, it is reasonable to assume that the motion will occur spontaneously, driven by the released energy. By analogy with the previous examples it is easy to anticipate the general structure that the resulting instability criterion will take. If the interchange affected only the geopotential energy, as previously, the criterion would have the general form given by eq. (IV.43) with $\vec{g}$ replaced by $\vec{g}^*$ and V taken to be the volume of equi-flux flux tubes. (The apparent inconsistency between the local nature of the equation and the global nature of V is resolved by specifying the point of application to be the equatorial plane.) However, in this case the interchange can also result in a net change in the kinetic energy of compression. This was not true in the previous example because there each interchange involved equal and opposite changes in the volumes of the interchanging parcels as they passed in opposite directions through the identical pressure variation in the surrounding atmosphere. Thus, the work done on a descending parcel was identically cancelled by the work done by an ascending parcel, and there was no net change in energy of compression in the system of interchanging gas parcels. In the present situation, the volume is governed by the geometry of the dipole field. The pressure within each flux tube is arbitrary, in principle, as long as it satisfies the low β requirement. Thus, while the changes in the volumes of interchanging flux tubes are equal and opposite, the pressure attending the volume changes can be different for rising and sinking flux tubes, or, to switch to magnetospheric parlance, for outward

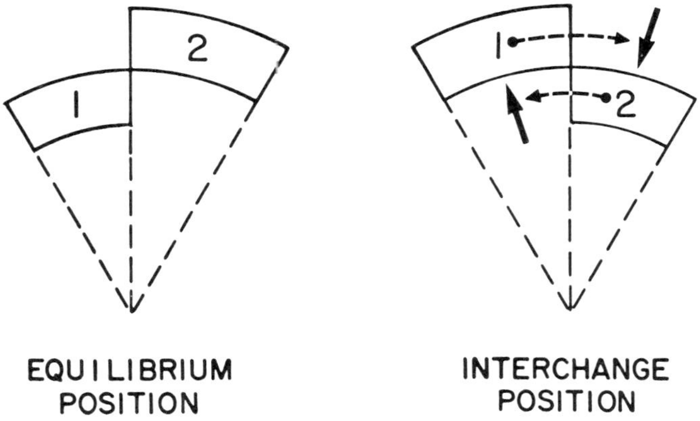

Figure IV.3 Sketch of interchange motion in the equatorial plane of a dipolar magnetosphere. The interchange involves two radially moving elements labeled 1 and 2. The radial displacements are indicated by solid arrows. The positions vacated by elements 1 and 2 are filled by azimuthal motions of the elements which formerly occupied the positions into which 1 and 2 moved, as indicated by the dashed arrows.

and inward moving flux tubes.Since inward moving flux tubes absorb energy of compression and outward moving flux tubes release energy of compression, it is evident that a magnetosphere that is radially stratified such that the compressed energy per unit magnetic flux decreases outward is unstable to interchange motions, as far as the criterion based on stored energy of compression is concerned (Gold, 1959). That is, in this situation more energy will be released by outward moving flux tubes than is absorbed by the ones moving inward to replace them.

A criterion for the interchange instability based on the energy principle has been given by Sonnerup and Laird (1963) in which both geopotential and compressional energies are included (see also Melrose 1967). The approach to be adopted here is based on the method used in discussions of the flute instability in plasma physics. It serves thereby to demonstrate the equivalence between the flute instability of plasma physics and the MHD interchange instability.

Figure IV.3 represents two situations in the equatorial plane of a magnetosphere. The equilibrium state is assumed to be static and to consist of radially stratified flux shells, that is, there are no motions initially and no variations in plasma parameters in the azimuthal direction. The radial profile of plasma parameters is assumed to be known and, as before, will be designated by a subscript st. The equilibrium is maintained by azimuthally flowing electrical currents that provide the ponderomotive force to balance the pressure gradient and geopotential forces. The requisite current is found by solving the momentum balance equation for $\vec{J}$,

$$\vec{J}_{\perp} = \frac{\vec{B}}{B^2} \times (\nabla p - \rho \vec{g}^*) \tag{IV.49}$$

in which the subscript $\perp$ on $\vec{J}$ merely makes explicit what is implicit in the right hand side of the equation, that the ponderomotive current flows perpendicular to the magnetic field, and has no parallel component. The pressure p in (IV.49) is taken to be a scalar to simplify the discussion.

After the interchange has occurred the radial profile of plasma parameters becomes locally adiabatic, and will be designated by a subscript ad. The current required to provide force balance as given by (IV.49) will now be discontinuous across the azimuthal interfaces between the interchanged and the ambient plasmas. In the discussion of the flute instability, the discontinuity in current builds up electric space charges on the two azimuthal walls of a radially displaced interchange element. The space charges will have opposite algebraic signs on the two sides of the element by the symmetry of the problem. An azimuthal electric fields is thereby generated. In the presence of the magnetic field the electric field will produce an $\vec{E} \times \vec{B}$ drift which is either radially out or in. If $\vec{E} \times \vec{B}$ is in the direction of the initial displacement that gave rise to the $\vec{E}$ field, the initial stratification is unstable, and all such perturbations will continue to grow. If $\vec{E} \times \vec{B}$ is in the direction opposite to the initial displacement, the plasma is stably stratified.

The only change to the above discussion that is called for because of the magnetospheric setting of the problem is the replacement of space charge build up by parallel currents. The space charge is discharged through the conducting ionosphere by flowing down the magnetic field lines. However, in this process also an electric field is generated across the flux tube as a result of the discharge current crossing the finite electrical resistance of the ionosphere. This electric field is in the same direction as the field that the space charge would have created. Thus, the instability criterion is the same in both cases.

In the absence of space charge build up, the divergence of the total current is zero. The equation for the parallel current is then

$$\frac{\partial J_{\parallel}}{\partial z} = - \nabla_{\perp} \cdot \vec{J}_{\perp} = - \nabla_{\perp} \cdot [\frac{\vec{B}}{B^2} \times (\nabla p - \rho \vec{g}^*)] \tag{IV.50}$$

in which the z direction is defined by $B = B\hat{z}$. The subscript $\perp$ on the del operator denotes a two dimensional divergence in the equatorial plane. If we adopt a cylindrical polar coordinate system (r,ϕ,z), the two dimensional divergence of the cross product of any two vectors $\vec{G}$ and $\vec{H}$ can be written explicitly as

$$\nabla_{\perp}\cdot(\vec{G}\times\vec{H}) = \vec{H}\cdot\nabla\times\vec{G} - \vec{G}\cdot\nabla\times\vec{H} + G_\phi \frac{\partial H_r}{\partial z} - G_r \frac{\partial H_\phi}{\partial z} - H_\phi \frac{\partial G_r}{\partial z} + H_r \frac{\partial G_\phi}{\partial z} \tag{IV.51}$$

By use of (IV.51), equation (IV.50) can be evaluated with the aid of the following conditions which are appropriate to the equatorial plane and our initial assumptions: $\nabla \times \vec{B} = 0$, and therefore $\partial B_r/\partial z = \partial B_z/\partial r$, $B = B_z$, $\partial B/\partial r = -3B/r$ (dipole field), $B_r = B_\phi = g^*_\phi = 0$, $g^* = g^*(r)\hat{r}$ and $\nabla \times \vec{g}^* = 0$ ($\vec{g}^*$ is given by the gradient of a potential). Eq. (IV.50) then reduces after several intermediate steps to

$$\frac{\partial J_{\parallel}}{\partial z} = \frac{1}{rB}\,\frac{\partial}{r\partial\phi}\,(3p + r\rho g^*) \tag{IV.52}$$

Thus as stated above, an azimuthal contrast in p or ρ brought about by a radial interchange motion gives rise to field-aligned currents. It should perhaps be noted that equation (IV.52) for $J_{\parallel}$ is general (for an isotropic plasma in the equatorial plane of a rotation-aligned dipole field) and does not depend on how the azimuthal variations in p and ρ are created.

The instability criterion can now be deduced from (IV.52). By our definition of the z-direction ($\vec{B} = B\hat{z}$) and the fact that (r,ϕ,z) forms a right handed coordinate system, we see that a positive E_ϕ produces an outward $\vec{E} \times \vec{B}$ drift. Thus, the magnetospheric stratification is unstable if an outward adiabatic interchange displacement produces a positive E_ϕ. The sign of E_ϕ across the interchange element will be the same as the sign of $\partial J_{\parallel}/\partial z$ on the clockwise (smaller ϕ) side of the element, since a positive $\partial J_{\parallel}/\partial z$ would correspond to a build up of positive charge in the absence of a field aligned current. From this we concluded that the stratification is unstable if $(3p + r\rho g^*)$ is greater on the adiabatic side of the azimuthal interface created by an outward displacement than it is on the ambient side. Mathematically the criterion can be expressed by

$$\frac{d}{dr}\,(3p + r\rho g^*)_{st} < \frac{d}{dr}\,(3p + r\rho g^*)_{ad} \quad \text{(unstable)} \tag{IV.53}$$

In the case of an isotropic plasma that uniformly fills the flux tubes, the adiabatic gradients can be given explicitly, since the volume of a dipolar flux tube is to a good approximation proportional to r^4. In an adiabatic displacement pV^γ = constant and ρV = constant (where V is the volume of the flux tube). The adiabatic gradient therefore can be written as

$$\frac{d}{dr}(3p + r\rho g^*)_{ad} = -4\gamma\frac{p}{r} - 4\frac{r\rho g^*}{r} + \rho\frac{drg^*}{dr} \qquad \text{(IV.54)}$$

If the radial profiles of the structural pressure and density are expressed also as power-law variations

$$p_{st} \propto r^{\lambda} \quad , \quad \rho_{st} \propto r^{\mu} \qquad \text{(IV.55)}$$

then the instability criterion becomes

$$3(\lambda+4\gamma)p + (\mu+4)r\rho g^* < 0 \quad \text{(unstable)} \qquad \text{(IV.56)}$$

(Since p_{ad} and ρ_{ad} are initiallized to the values of p_{st} and ρ_{st} at the same value of r, the subscripts are unnecessary in (IV.54 and 56). The term $\rho\, {}^{drg^*}/dr$ is common to both sides of (IV.53), and therefore cancels out in IV.56). It remains only to give an explicit expression for g^*. In the equatorial plane g^*, the combination of the gravitational and centrifugal accelerations is given by

$$g^* = -g\frac{R_p^2}{r^2} + \Omega_p^2 r \qquad \text{(IV.57)}$$

where g is the gravitational acceleration at the surface of the planet, R_p is the radius of the planet and Ω_p is the angular velocity associated with the rotation of the planet.

The criterion (IV.56) can be modified readily to make it applicable to situations in which the plasma does not fill the flux tube completely, such as in the case of anisotropic pressure or when the plasma is confined to the equatorial plane by the centrifugal force. The plasma formation in Jupiter's magnetosphere which is composed of matter originating on Io exhibits equatorial confinement. In this case the volume occupied by the plasma varies with distance more nearly as r^3. The factor four that appears in (IV.56) should then be replaced by the factor three. In the Jovian case there is also a background of energetic particles for which the factor four is appropriate. The terms in (IV.56) must be evaluated by combining both populations to determine whether or not the stratification of the Jovian magnetosphere is stable.

References:

Alfvén, H.: 1942, "On the existence of Electromagnetic-hydrodynamic waves", Nature, 150, 405.

Gold, T.: 1959, "Motions in the magnetosphere of the earth", J. Geophys. Res., 64, pp. 1219-1224.

Hudson, P. D.: 1970, "Discontinuities in an anisotropic plasma and their identification in the solar wind", Planet. Space Sci., 18, pp. 1611-1622.

Kantrowitz, A. F., and Petschek, H. E.: 1964, "MHD characteristics and shock waves", AVCO-Everett Research Report 185.

Krall, N. A., and Trivelpiece, A. W.: 1973, Principles of Plasma Physics, McGraw-Hill Book Company.

Melrose, D. B.: 1967, "Rotational effects on the distribution of thermal plasma in the magnetosphere of Jupiter", Planet. Space Sci., 15, pp. 381-393.

Rossi, B., and Olbert, S.: 1970, Introduction to Space Physics, McGraw-Hill Book Company, New York.

Sonnerup, B. U. Ö., and Laird, M. J.: 1963, "On the magnetospheric interchange instability", J. Geophys. Res., 68, pp. 131-139.

Sonnerup, B. U. Ö.: 1979, "Transport mechanisms at the magnetopause", in Dynamics of the Magnetosphere, edited by S.-I. Akasofu, D. Reidel Publishing Co., Dordrecht-Holland, pp. 77-100.

Spitzer, L., Jr.: 1956, Physics of Fully Ionized Cases, Interscience Publishers, Inc., New York.

Stern, D. P.: 1966, "The motion of magnetic field lines", Space Sci. Rev., 6, pp. 147-173.

Vasyliunas, V. M.: 1975, "Theoretical models of magnetic field line merging, 1", Rev. Geophys. Space Phys., 13, pp. 303-336.

GENERATION OF SOLAR MAGNETIC FIELDS I

Eugene N. Parker
Department of Physics
University of Chicago
Chicago, Illinois 60637

1. Introduction

Although the stated topic of the next two lectures is the generation of solar magnetic fields, they will be elementary lectures on magnetic fields in general because the sun is by no means an isolated body. In these lectures I will talk about the properties of magnetic fields that allow them to destroy themselves remarkably rapidly, producing solar activity, stellar activity, geomagnetic activity, and so forth. It might be called the birth and death of magnetic fields.

The universe poses a problem which is still very mysterious, namely, that nearly every object seems to have a magnetic field associated with it, whether you talk about planets, stars, or galaxies. Wherever you have magnetic fields, they produce activity, usually in the form of superheated gases. We know about the million degree corona that is the result of solar activity in some way. One million degrees is superheated, certainly hotter than obtained from black body radiative considerations.

It is the products of magnetic fields, the flares, plages, eruptions and streamers, that we see. It is a complicated process, but somehow the magnetic fields are the agitators that produce the activity. The same is true for the magnetospheres of the earth and the other planets. The same is true for most other stars; in fact, one infers the existence of magnetic fields in other stars from the x-rays which we can see coming from them. X-rays would not come from a star based on classical models where you have a photospheric temperature and everything gets cooler from there on out. The same is true for galaxies, but even less is known there.

2. Theories of Magnetic Field Origins

Many astrophysical objects possess a magnetic field. You could ask whether the magnetic fields we see today might not be primordial. Although primordial is a catch-all term, we mean something that was

R. L. Carovillano and J. M. Forbes (eds.), Solar-Terrestrial Physics, 101–128.

swept into the body when it was formed by collapsing gases, and then held there for ever after, or at least 10^{10} years, by the high electrical conductivity in the volume.

Such a possibility exists in the sun, as was first pointed out by Cowling (1953) some 30 years ago. The decay time for a magnetic field trapped in the core of the sun is approximately 5×10^9 years, because of the large dimensions of the core and the high electrical conductivity. If there were an initial magnetic field, it may have decayed by a factor of e since the beginning, but this poses no problem. Similar arguments have been offered for the galaxies.

The reason we are inclined to discount such arguments, at least for the sun and the galaxies, is the dynamic state of observed fields. Also, magnetic fields are buoyant. It is obvious that the convection zone of the sun is turbulent; the fields tend to mix rapidly, so that the decay time which you compute for the core of the sun may not be applicable to its outer part. In any case, the turbulent diffusion time for the sun tends to be 10 or 20 years, and for the galaxies perhaps as much as 10^8 years, but certainly not 10^{10} years.

The final point is that the sun's magnetic field is observed to reverse every 10 years or so. A number of people have thought very hard about alternative ways to dress up primordial fields to produce what is observed. For instance, the oscillating east-west field of the sun might somehow be produced by torsional oscillations from a primordial field trapped in the core. You can imagine that the outer part of the sun, oscillating relative to the inner part, can shear the field and produce a number of effects. Perhaps so, but the observations do not show large changes in the rate of rotation near the surface of the sun to account for it. Ways of salvaging primordial fields should be considered because it is difficult to understand why there should not be some remnant of the original field; nonetheless, it has not been a very fruitful line of thought so far. We are left with the idea that most objects, planets, stars, and galaxies have magnetic fields because the fields are being generated now. They are not left over from some earlier epoch of the universe.

If that is the case, we are left with the problem of why they have magnetic fields. I am going to base my discussion on the earth because it involves fewer complications. In some respects it is a simpler structure than the sun. The principles we arrive at are applicable to the sun and some of the solutions are periodic like the solar field.

There are a number of points we might touch on before plunging into the question of where the magnetic field comes from. A number of ideas have been proposed over the years. We cannot try to review them all, although some of them are very interesting. The existence of magnetic fields, as opposed to the rather rare occurrence of electrostatic fields, in the universe is a consequence of the electrical nature of matter. As you well know, the universe is full of free charges which short circuit electric fields.

So far there is no evidence of magnetic monopoles. Perhaps, the Stanford event is a monopole, but in any case there is a definite limit on the flux of magnetic monopoles that one could have, given the wide occurrence of magnetic fields. [With the present concept of magnetic monopoles there are living organisms which are smaller than the fundamental particle of exchange (10^{-8} gm).] If you do the arithmetic, you find that, assuming velocities on the order of 200 km/sec, the upper limit of the monopole flux is about $10^{-15} cm^{-2} sr^{-1} sec^{-1}$. If there were more than that it would be hard to account for the existence of a galactic magnetic field, which would be short circuited by the freely moving monopoles. So if the Stanford event is a monopole, the observing period implies a flux of about a million times that which the galactic magnetic field can tolerate (Turner, Parker, and Bogdan, 1982).

We have three possible explanations for the apparent discrepency. The Stanford event could be a glitch in the superconducting properties of matter. This appears to be the most likely explanation. Or, the monopoles, if they exist are an entirely local phenomenon, perhaps trapped in orbit in the solar system. [Finally, we could be just plain crazy and don't know anything about the universe, an idea which should not be entirely discounted.] In any case, there are a lot of interesting implications of magnetic field's which we don't have time to discuss here.

Returning to the origins of magnetic fields, we ask the question, why should the earth and some other planets, the sun, a star, or a galaxy have magnetic fields? Thermal effects can be appealed to. They behave much like a battery as a way of balancing thermal forces and gravity in a rotating star, and produce very slight magnetic fields. There are thermoelectric effects, proposed for the earth some time ago which keep rearing their heads, but apparently none are anywhere near large enough. Of course, the earth has a habit of flipping its field occasionally, which certainly does not look like a thermal effect. The field flips in a matter of a couple thousand years, and it is hard to believe that the internal thermal constitution of the earth reverses in that period of time.

The idea of rotating electrostatic charges is a recurring idea. It is an absurd idea, but somehow it grabs people who fail to plug in the basic numbers. [It comes up about every 15 years with every new generation of graduate students.] The problem is that it produces magnetic fields of the order of V/C relative to the electric field, and electric fields are difficult to produce because we live in an electrically conducting universe. So, such ideas are ill conceived. [Nonetheless they come up rather regularly. I've been trying to correlate this 15 year cycle with the 11 year sunspot cycle, but so far there has been no matching.]

We can consider an idea that was bandied about for a long time. The idea was that fluid motions could generate induced magnetic fields just by flowing across magnetic fields and producing currents which lead to additional magnetic fields. It was put on a firm foundation by Elsasser (1945) in connection with the magnetic field of the earth. He wrote down the induction equations, put in reasonable fluid velocities, and gave us our present understanding of the generation of magnetic fields, as far as it goes. A lot of people at the time had other ideas, mostly in opposition to Elsasser, but most of them some years later rediscovered what he already had done and thought they discovered it themselves.

Let me remind you about the history of our knowledge of the earth's magnetic field. The earth's magnetic field has been known for a long time. It was discovered by the Chinese who used it for practical matters, like navigation and surveying, for a thousand years. Not too long after that the compass appeared in Europe, probably from the Orient. The study of the earth's magnetic field began in a global sense around 1600 with the work of Gilbert who pointed out that the earth was like a lodestone, or a large bar magnet.

The study of the sun's magnetic field has a spottier history. Hale discovered the sun's field by observing sunspots in 1908. He reported fields of a few thousand Gauss in the sunspots and a 50 G dipole field across the disk of the sun. It is a scandal, and a severe criticism of the state of mind of most people in astronomy, that nobody, thought to reproduce Hale's observations until 40 years later when the Babcocks built their magnetometer and studied solar and stellar magnetism. Nobody cared; it was not interesting; it did not pose a challenge. [Older books occasionally mentioned solar magnetic fields but they attached it to nothing. It was an isolated fact. They talked about other things like sunspots and flares, but the general field of the sun was of no interest.] When the Babcocks started looking at the sun they found 2 G, not 50 G.

3. Induction Equations

With that preamble let us return to the main subject. To derive the hydromagnetic equations we start with Maxwell's equations and neglect all terms that are small, of order $V^2/C^2 \ll 1$. We are dealing with very nonrelativistic motions and we also neglect the displacement current. The current density is

$$j = \sigma E' = \frac{\sigma(\underline{E} + \underline{V}/C \times \underline{B})}{(1 - V^2/C^2)^{1/2}} \simeq \sigma(\underline{E} + \underline{V}/C \times \underline{B}) \qquad (1)$$

The primed electric field is the field in the frame of reference of the material; that is the important field. The Lorentz transformation to a stationary reference frame and the requirement that $V^2/C^2 \ll 1$ give the final result.

After we substitute (1) into Maxwell's equations we get the familiar hydromagnetic equation.

$$\frac{\partial B}{\partial t} = \underline{\nabla} \times (\underline{V} \times \underline{B}) - \underline{\nabla} \times \eta(\underline{\nabla} \times \underline{B}) \tag{2}$$

The first term on the right hand side of equation (2) arises because the fluid moves across the magnetic field and that field is carried bodily with the fluid. The second term is a diffusion term where in egs units $\eta = C^2/4\pi\sigma$ is the diffusion coefficient with dimensions cm^2/sec. It is usually assumed that the conductivity is uniform so that equation (2) may be rewritten.

$$\frac{\partial B}{\partial t} = \underline{\nabla} \times (\underline{V} \times \underline{B}) + \eta\nabla^2 B \tag{3}$$

If there were no fluid motion equation (3) would be a heat flow equation for each Cartesian component of the magnetic field. Equation (3) can be integrated once, in a manner of speaking, by expressing the magnetic field in terms of a vector potential $\underline{A}$.

$$\underline{\nabla} \times \frac{\partial A}{\partial t} = \underline{\nabla} \times (\underline{V} \times \underline{\nabla} \times \underline{A}) - \eta\underline{\nabla} \times [\underline{\nabla} \times (\underline{\nabla} \times \underline{A})] \tag{4}$$

If we then "uncurl" equation (4) we get

$$\frac{\partial A}{\partial t} = \underline{V} \times (\underline{\nabla} \times \underline{A}) - \eta\underline{\nabla} \times (\underline{\nabla} \times \underline{A}) - \nabla\varphi \tag{5}$$

where φ is an arbitrary scalar function.

Equation (5) is the induction equation, and what Elsasser did was to put in reasonable fluid velocities. It should be emphasized here that we will be discussing kinematic dynamo equations; that is, we are going to take the velocities for granted. By making an intelligent choice we can establish the principles by which magnetic fields can be regenerated. Later we will consider the much more difficult part of the problem; namely, what happens when we have to solve the dynamical equations of the fluid velocity.

Rather than getting bogged down in details, a brief sketch of these ideas will aid our understanding. Suppose we have a magnetic field and let it decay. There are no fluid motions and therefore we have only the diffusion term in equation (3). Dimensionally, the left hand side is essentially B divided by a characteristic decay time. The Laplacian term on the right hand side can be expressed as B divided by the square of some characteristic scale, Λ. Thus equation (3) with $V=0$ may be approximated by

$$\frac{B}{t} = \eta \frac{B}{\Lambda^2}$$

From this we get

$$t = \frac{\Lambda^2}{4\eta} \tag{6}$$

where the 4 inserted in the denominator turns out to be better in most circumstances. In books on geophysics or geology we find that the earth has a molten core with a radius of 3400 km, which we can approximate to 3×10^8 cm for Λ. The core being molten iron or nickel, is nonmagnetic and is an electric conductor of no particular quality. Letting $\sigma = 10^{16} sec^{-1}$ (cgs units), we obtain a diffusion coefficient, η of 10^4 cm^2/sec. Equation (6) yields a diffusion time, t, of about 10^5 years. With all the factors of π included a better value is about 1/2 x 10^5 years. Any field in the core, whether azimuthal or north-south, would decay in a matter of 50,000 years. We will come back to this point later when we discuss convection.

Returning to Elsasser's problem, what fluid motions do we expect in the core? At this stage of our ignorance we should look at the observations and not expect anything. There is evidence for fluid motions in the earth's core based on secular variations of the magnetic field. From this we conclude that the field is locked into the molten core for periods of thousands of years, and if the core moves relative to the crust we walk on (i.e. the mantle) the field would appear to change. Indeed, there are large scale inhomogenerties which move around. These grow and decay over periods of several thousand years based on maps compiled since Gauss' time. The inhomogenerties tend to drift westward at the slow rate of about 0.3 mm/sec. There are also upwellings and sinking motions, again at a very slow rate.

If we have a rapidly rotating body in which there is convection, the Coriolis force is significant and we could expect some sort of nonuniform rotation in the earth, possibly leading to a westward drift. In the most elementary way, a westward drift suggests that the core does not rotate quite as fast as the mantle. Conservation of angular momentum suggests the same thing if there is an upwelling of material.

Elsasser first supposed azimuthal, east-west, flow in the core of the earth which is a function of distance from the center of the earth and polar angle θ from the axis of rotation. He showed that a simple axisymmetric, azimuthal fluid velocity causes a growth in the azimuthal magnetic field which arises from the shearing of the dipole (Elsasser, 1950, 1956 a,b).

Briefly, shearing of the dipole field by a nonuniform rotation can be described by the following equation.

$$\left[\frac{\partial}{\partial t} - \eta\left(\nabla^2 - \frac{1}{\omega^2}\right)\right]B_\varphi = \frac{1}{\gamma}\frac{\partial(\Psi,\Omega)}{\partial(\theta,r)} \tag{7}$$

The coordinates are cylindrical with $\omega = (x^2+y^2)^{1/2}$ and φ the azimuthal angle. R and θ are the usual polar coordinates. Ω is the nonuniform angular velocity in the core and Ψ is a kind of vector potential for B.

$$B_r = \frac{1}{r^2 \sin\theta}\frac{\partial\Psi}{\partial\theta} \qquad B_\theta = -\frac{1}{r\sin\theta}\frac{\partial\Psi}{\partial r} \tag{8}$$

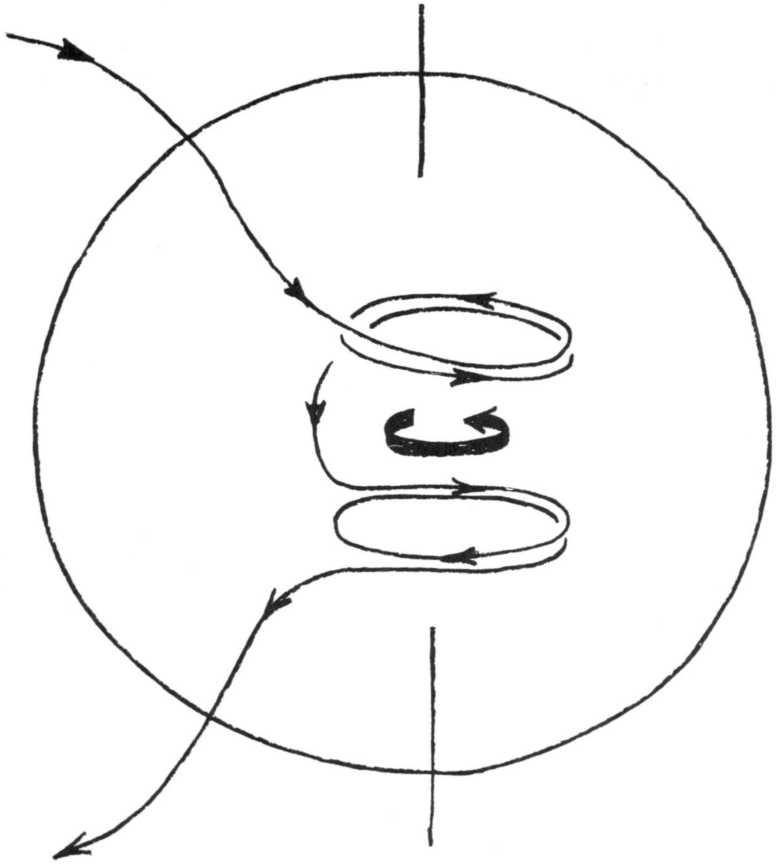

Figure 1. A sketch of the formation of the toroidal (azimuthal) magnetic field in the core of Earth as a consequence of the faster rotation of the inner part of the core.

The physically obvious results can be computed exactly and the field in the core looks like Figure 1. The inner part of the core rotates more rapidly than the outer part, which tends to carry the lines of force around where they are closest to the axis. We get an east-west field that is stretched out one way in the northern hemisphere and the opposite way in the southern hemisphere. That field can be very strong; it may be rather stronger than the dipole field. The dipole field extrapolates down to the core to about 5 G at the poles. The core field is probably at least 20 G, and could be 200 G as far as anyone knows. It is a very difficult problem and others are working very hard on the dynamics of the earth with much more complicated convection patterns than described here. Nevertheless, this part of earth's main field is forever hidden in the core. For the field to leave the core a current would have to be in the mantle, by Stoke's Theorem. But the mantle is essentially a nonconductor and the field remains trapped inside.

The next question is, how do we avoid Cowling's Theorem? In 1918 Larmor incorrectly interpreted the filamentary structure around sunspots as evidence for cyclonic motions, or swirls, on the surface of the sun. It is a perfectly obvious interpretation, but it just happens to be wrong. He suggested that somehow the swirling motion of the fluid generated a magnetic field. Had he written down some calculations on the back of an envelope, he would have seen that there was no way it could work because of the V/C factor mentioned earlier. Nonetheless he was an important scientific figure and did not have to do such things, and everybody accepted it until Cowling attacked the problem from an entirely different point of view.

Cowling (1934) showed that there is no way that an axisymmetric field can be maintained by magnetic induction. In a meridional cross section through an axisymmetric field there must be at least one point around which all the lines of force circulate. It is a general topological fact in two dimensions that if lines do not extend to infinity they must circle around and connect together. (That is not true in three dimensions, by the way, in spite of what many textbooks say.) The magnetic field circulates around the neutral point, so there must be an electric current through the neutral point perpendicular to the meridional planes and extending all the way around the axis. But an electric current cannot be induced or maintained at the neutral point because there is no magnetic field in the cross product with the flow velocity that induces the emf to drive the current. In other words, Cowling showed that the idea of induction broke down at some point in an axisymmetric field. Elsasser, Bullard, and others were aware of Cowling's Theorem, so they introduced nonsymmetric expansions in Legendre polynomials. They found nonconvergent solutions. This gave rise to the conjecture that there was a super Cowling's Theorem which said that there were no solutions to the stationary induction equation. If we start out by setting the left hand side of equation (3) equal to zero it becomes an eigenvalue problem because the ratio of the velocity to the resistivity must have just the right value so that we have neither growing nor decaying fields. It was later discovered that there are many stationary dynamos and the convergence failures have recently been handled more successfully (see discussion and references in Parker, 1979, pp 532-559).

4. Short-Sudden Approximation

We will now discuss the basic dynamo effects using a time dependent dynamo. We can add a nonaxisymmetric part of the field to get around Cowling's Theorem and see what kind of magnetic fields we can generate. Remember that there are convective cells in the earth, each one producing local inhomogeneities in the earth's field. Depending on how finely structured a map is, we can identify 10 or 15 of these cells around the earth. This suggests large rising and falling convective cells with periods of 500 to 1000 years.

Let us consider an eddy, or convection cell, as the basic element of nonaxisymmetric velocity. Imagine that we have an upwelling of fluid, and, because of the rapid rotation of the earth, there is a strong Coriolis force which tends to make it rotate in the same direction as the earth. If we tackle equation (3) we find that there is no way to solve it analytically for such eddies.

What we must do is resort to subterfuge (Parker, 1955, 1970) by taking the terms in equation (3) one at a time. Suppose we start at time t = 0 with a large scale field. In the earth's case this will be the dipole field and the strong azimuthal field in the core caused by the nonuniform rotation. Then for a short period of time (short compared to the diffusion time) we turn on the fluid motions. This is called the "short-sudden" approximation or idealization. We can make the fluid move as fast and as far as we like during that period, but to simplify the mathematics suppose that the scale of the fluid motion ℓ is small compared to the large diffusion scale Λ used earlier. When the fluid velocity is switched on, the fields can wrap up quite a bit, so that after this brief period of violence we are left with some very complicated fields. Then the fluid is clamped motionless so that only the diffusion term in equation (3) is involved. This is easy to solve, and after a time which is long compared to the characteristic diffusion time of the small scale fluctuations, but short compared to the diffusion time of the large scale fields, the local fluctuations are smoothed out. What remains is some large scale field but not necessarily the dipole and azimuthal fields we started with. The cycle can then be started over again.

This method keeps the mathematics simple. There are other ways of going about it, of course. Sometimes the quasilinear approximation is used where the conductivity of the fluid is sufficiently low. Unfortunately, while that produces the basic dynamo effect, it does not describe other effects like flux ejection and negative diffusivity, which will come up later. It is also much more complicated mathematically than what I propose to cover.

Therefore, we will continue with this idealization, our goal being to demonstrate dynamo principles. Certainly, the eddies in the earth's core do not have the remarkable property of suddenly moving around and then sitting still while their fields smooth out. In any case, we can do the mathematics and not lose any physical effects.

Substituting $\underline{\nabla} \times \underline{A} = \underline{B}$ we have the following result from equation (3).

$$\frac{\partial A}{\partial t} = \underline{V} \times (\underline{\nabla} \times \underline{A}) - \eta \underline{\nabla} \times (\underline{\nabla} \times \underline{A}) - \nabla\varphi \qquad (9)$$

where φ is an arbitrary scalar function and $\nabla \times \nabla\phi \equiv 0$ ensures that equation (3) is recovered when the curl of equation (9) is taken. Since we are free to choose a so-called gauge condition, it is convenient to set $\varphi = \underline{V} \cdot \underline{A}$. This reduces (9) to

$$\frac{\partial A_i}{\partial t} + V_j \frac{\partial A_i}{\partial x_j} = - A_j \frac{\partial V_j}{\partial x_i} \tag{10}$$

where, as usual, summation is carried out over repeated indices. Equation (10) has the solution

$$A_j(x_k,t) = A_j(X_k,\ 0) \frac{\partial X_j}{\partial x_i} \tag{11}$$

We have taken the Lagranian point of view where x_k is the position of a fluid element at time t that was initially at position X_k at time t = 0. Then we must specify the velocity field or Lagrangian displacements in terms of initial positions, which is not always easy. We can write the final position as

$$x_k = X_k + \xi(X_k,t) \tag{12}$$

Suppose that at time t = 0 we start with a magnetic field described by the vector potential $A(X_k,0)$. We then switch on the velocity field, very quickly so that there is no diffusion, and we have a formal solution after a short period τ_1

$$A_j(x_k,\tau_1) = A_j(X_k,0)\frac{\partial X_j}{\partial x_i} \tag{13}$$

This gives a formal prescription for the field, but in terms of the field at a position X_k at some other point. To handle this, remember that while we may have large displacements during the period of violent turbulence, they are not large compared to the large scale dimensions of the mean field. An eddy can turn many times but it is localized so that a fluid element is not far from where it started compared to the <u>large scale</u> field. Thus we can expand $A_j(X_k,0)$ in a rapidly converging series.

$$A_j(X_k,0) = A_j(x_k,0) - \frac{\partial A_j(x_k,0)}{\partial X_n} \xi_n \tag{14}$$

$$+ \frac{1}{2!}\frac{\partial^2 A_j(x_k,0)}{\partial x_m \partial x_n} \xi_m \xi_n + \ldots$$

For the other factor in equation (13)

$$\frac{\partial X_k}{\partial x_j} = \delta_{ij} - \frac{\partial \xi_k}{\partial x_j} \tag{15}$$

Remember that this is a kinematical calculation which ignores the equations of motion. It is an arbitrary but general fluid motion which leads to a change in the magnetic field. We are not doing the whole job at once, but when all the pieces are understood they can be assembled into a complete story. That is what a number of people are working on now.

Continuing on to the final result, we can average over an ensemble of systems, where the mean displacement is zero in a large volume of of systems, where the mean displacement is zero in a large volume of fluid, and arrive at the change in the field after the short period of deformation by turbulence.

$$\Delta A_i(x_k,\tau_1) = \langle \xi_k \frac{\partial x_j}{\partial x_i} \rangle \frac{\partial A_j}{\partial x_k}$$
$$+ \frac{1}{2!} (\delta_{ij}\langle \xi_k \xi_l\rangle - \langle \xi_k \xi_\ell \frac{\partial \xi_j}{\partial x_i} \rangle) \frac{\partial^2 A_j}{\partial x_k \partial x_l} \qquad (16)$$
$$+ \frac{1}{3!} \langle \xi_k \xi_l \xi_m \frac{\partial \xi_j}{\partial x_i} \frac{\partial^3 A_j}{\partial x_k \partial x_\ell \partial x_m}$$

Then the fluid is held motionless to complete the cycle, and all the small scale ripples die away over a long time $t > L^2/\eta$. Only the mean field, with large scale L, remains. During this waiting period there are slow, large scale velocities which deform the fields steadily, corresponding to the nonuniform rotation in the earth's core. Letting time τ be the total period of the cycle consisting of violent deformation by turbulence and waiting for diffusion, a final result in terms of Lagrangian displacements can be calculated (Parker, 1970, 1979, pp 567-570).

$$\frac{\partial A_i}{\partial t} + \eta \frac{\partial}{\partial x_i} \frac{\partial A_n}{\partial x_n} = - A_j \frac{\partial V_i}{\partial x_i} - V_j \frac{\partial A_i}{\partial x_j} + \frac{1}{\tau} \langle \xi_k \frac{\partial \xi_j}{\partial x_i} \rangle \frac{\partial A_j}{\partial x_k}$$
$$+ [\eta \delta_{ij} \delta_{kl} \frac{1}{2\tau} (\delta_{ij}\langle \xi_k \xi_l\rangle - \langle \xi_k \xi_l \frac{\partial \xi_j}{\partial x_i} \rangle)] \frac{\partial^2 A_j}{\partial \xi_k \partial \xi_l}$$
$$+ \ldots \qquad (17)$$

This equation tells us, in open form, the complete result of a Lagrangian displacement. It contains all the known dynamo effects as well as negative turbulent diffusion. Simple examples exploiting equation (17) will be considered in the next lecture.

References:

Cowling, T.G.: 1934, Mon. Not. Roy. Astron. Soc., 94, 39.

Cowling, T.G.: 1953, Solar electrodynamics, in The Sun. (ed. G.P. Kuiper) (Chicago, University of Chicago Press).

Elsasser, W.M.: 1945, Phys. Rev., 69, 106.

Elsasser, W.M.: 1950, Rev. Mod. Phys., 22, 1.

Elsasser, W.M.: 1956a, Rev. Mod. Phys., 28, 135.

Elsasser, W.M.: 1956b, Amer. J. Phys., 24, 85.

Parker, E.N.: 1955, Astrophys. J., 122, 293.

Parker, E.N.: 1970, Astrophys. J., 160, 383; 162, 665.

Parker, E.N.: 1979, Cosmical Magnetic Fields (Oxford, Clarendon Press).

Turner, M.S., Parker, E.N. and Bogdan, T.J.: 1982, Phys. Rev. D, 26

GENERATION OF SOLAR MAGNETIC FIELDS II

Eugene N. Parker
Department of Physics
University of Chicago
Chicago, Illinois 60637

1. Introduction

In this lecture we continue our discussion concerning the generation of magnetic fields from the motion of conducting fluids. We proposed the "short sudden" idealization, which led to equation (17) in the preceding lecture. V_j in that equation is the large scale, steady fluid velocity that continually shears the field. In addition we introduce quick bursts of turbulence during which any amount of rotation or twisting of the field can be accomplished. Following these quick bursts of motion the fluid is held motionless so that small scale irregularities die off and we are left with a smooth, average, large scale state. This cycle is repeated at time intervals τ, producing the dynamo equations for the mean vector potential A_i (x_k, t).

2. Cyclonic Convection

We are now at liberty to imagine any Lagrangian displacement we wish, for the intermittent turbulence subject to the condition that their scales be small compared to the large dimensions of the field. Naturally, there are infinitely many possibilities so we must choose our way rather carefully. It is beneficial to explore one particularly simple case because it has a great deal of physics in it. Imagine that the eddies are cylindrical in form and composed of an upwelling surrounded by a downdraft (Figure 1). This is a common form observed in nature and on the sun, in particular (Parker, 1955, 1970, 1979 pp. 573-583).

Take the z axis in Figure 1 as the axis of symmetry. We will work in rectangular coordinates since there is no point at the moment for complicating things by folding the eddies into spherical shapes. Let us also add some rotation into the rising column only, just to keep things simple. This is typical in cyclonic convection because the converging flow at the base tends to accelerate the rising column and give it more angular velocity than the descending column.

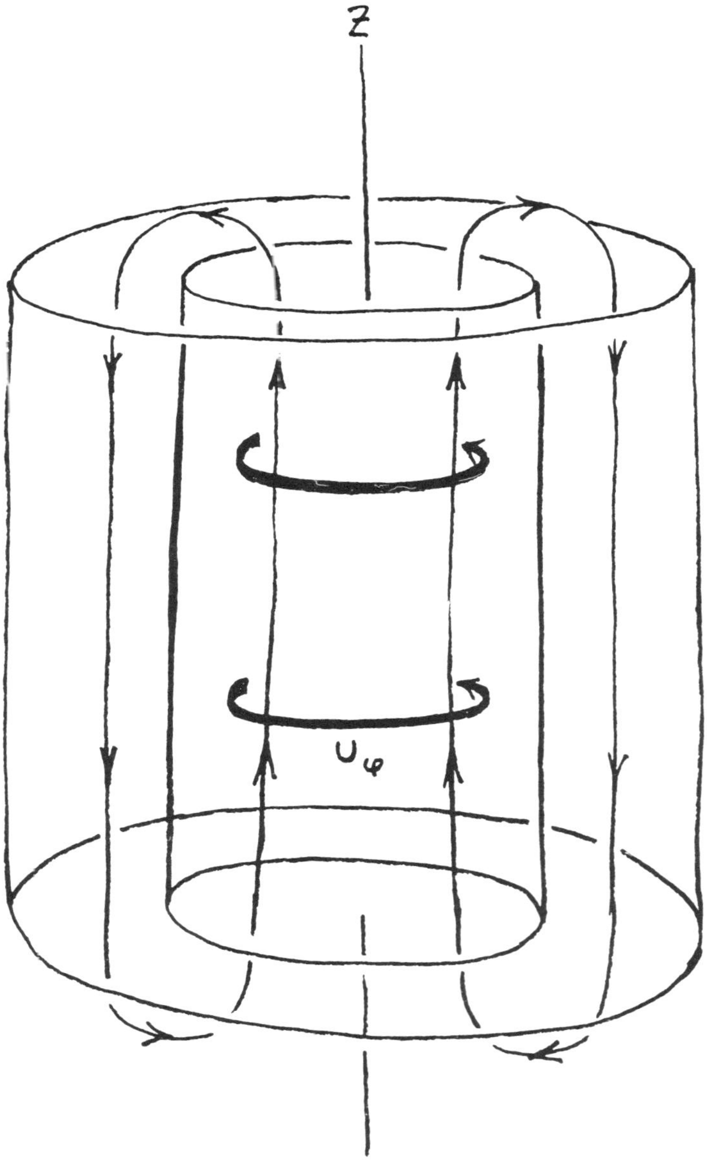

Figure 1. A sketch of the idealized cylindrical convective cell used to calculate the dynamo coefficients $\langle \xi_i(\partial \xi_j/\partial x_k)\rangle$. The inner, rising column is subjected to a rotational velocity v_φ. The streamlines indicate the closed circulation of the fluid.

With this model we can compute the following coefficients which appear in equation (17) in the first lecture.

$$\langle \xi_x \frac{\partial \xi_x}{\partial x} \rangle = - \langle \xi_z \frac{\partial \xi_x}{\partial x} \rangle$$

$$= \langle \xi_y \frac{\partial \xi_z}{\partial y} \rangle = - \langle \xi_z \frac{\partial \xi_y}{\partial y} \rangle \equiv - Q_T \qquad (1)$$

$$\langle \xi_x \frac{\partial \xi_z}{\partial y} \rangle = - \langle \xi_z \frac{\partial \xi_x}{\partial y} \rangle$$

$$= - \langle \xi_y \frac{\partial \xi_z}{\partial x} \rangle = \langle \xi_z \frac{\partial \xi_y}{\partial x} \rangle \equiv \Gamma_T \qquad (2)$$

The remaining coefficients (with 3 indices there are 27 in all) are zero, fortunately. The 8 nonzero coefficients divide into 2 classes, those with 2 repeated indices, (equation (1)) and those with no repetitions (equation (2)) They are not very interesting in themselves; all we want to know is their magnitude. If the eddies are close packed, the average of Γ is essentially the rotational velocity U_Φ. Very roughly, $\Gamma \sim U \sin \Phi$, where Φ is the angle of rotation. Q is of the same magnitude but goes like $U_\Phi \sin^2 1/2\, \Phi$. Note that Γ maximizes at $\Phi = \pi/2$ and then decreases while Q maximizes at $\Phi = \pi$, so the two coefficients are rather distinct effects.

The first term in row 2 of equation (17) in the previous lecture is a sort of glorified diffusion coefficient. It involves the ordinary diffusion coefficient, and the Kronecker deltas reduce it to a simple Laplacian describing diffusion. However, in the presence of cyclonic turbulence we have second order coefficients which are not as simple as the scalar coefficient When worked out they involve the vertical displacement and the angle of rotation. Under some conditions the negative term dominates and we actually get negative diffusion.

This effect was first discovered by Kraichnan (1976) on the basis of physical arguments and justified by some calculations and finally with numerical simulations. Precisely and literally it is negative diffusion. It is just like ordinary diffusion with time running backwards. Of course, in that case the solutions steepen up and become singular, so that the assumption that the scale of the field is large compared to the scale of the turbulence breaks down. It is not a dynamo effect, in the sense that it conserves flux; it does not generate new flux. We will return to this effect later.

We should mention that the word "dynamo" is used in a variety of ways in the literature, which is unfortunate because it leads to confusion. In magnetospheric physics the common practice is to refer to any inductive effect as a dynamo effect. In other fields one tends to refer to a dynamo effect as a self-sustaining field amplification, something that would increase the total flux. That is the sense in which we are using the word.

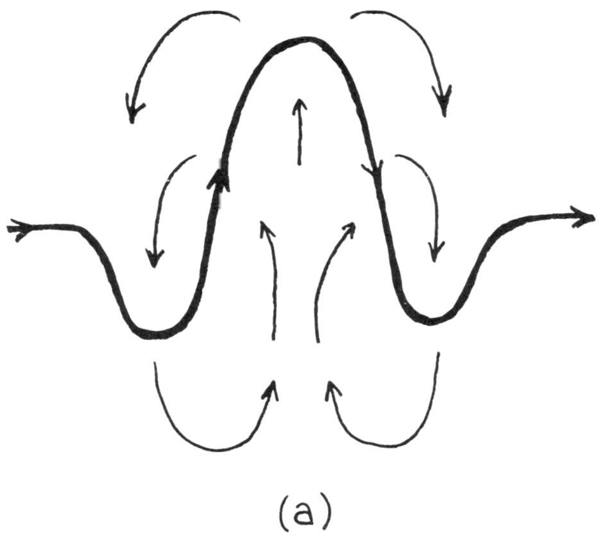

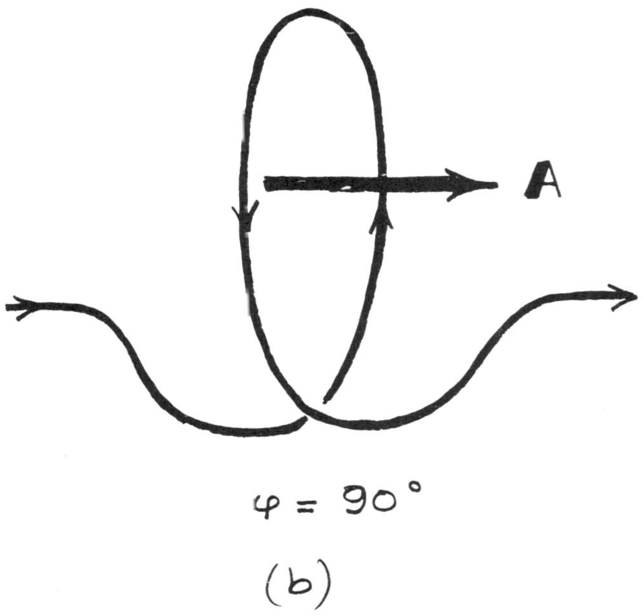

Figure 2. (a) A sketch of an initial horizontal line of magnetic force as it is deformed by a simple convective cell.
(b) The effect if the rising column is subject to a rotation of $\varphi = 90^{\circ}$.

With that preamble, let us discuss what each of the effects means. Taking the Γ coefficient first, suppose that there is a rising column of fluid surrounded by descending fluid. Remember that a line of force is carried bodily with the fluid, so that, the deformed field line looks something like Figure 2a. That accomplishes no dynamo effect because as much flux is pushed up as down. But if a 90° rotation is added, a single line of force would be a twisted loop like Figure 2b. We now have a loop which is perpendicular to to the original field. This is called the α effect, the one that seems to be the most prevalent dynamo effect in nature. An equivalent vector potential is produced in the direction of the original magnetic field. Without going through the mathematical details, we find that large scale magnetic field can generate a vector potential given by

$$\left(\frac{\partial}{\partial t} - \eta \nabla^2 \right) A_y = \Gamma B_y \tag{3}$$

Most likely this is the dominant dynamo effect in the sun and the earth, although there are also other effects.

If we write down all of the coefficients in equation (17) in the previous talk, we have quite a mess. But it can be simplified a great deal by taking a lesson from nature, which is that under most conditions, by far the strongest dynamo effect is the large scale shear produced by nonuniform rotation. It just grabs the field and stretches it out, usually in the azimuthal direction because most objects have azimuthal symmetry. For simplicity we can replace the rotational shear with a velocity shear in the y direction only, so that V_y is a function of z. Any field in the z direction is stretched in the y direction while preserving its form in the z direction. The hydromagnetic equation becomes

$$\left(\frac{\partial}{\partial t} - \eta \nabla^2 \right) B_y = \frac{dV_y}{dz} B_z = \frac{dV_y}{dz} \frac{\partial A_y}{\partial x} \tag{4}$$

Substituting B_y from equation (3) into equation (4) we have

$$\left(\frac{\partial}{\partial t} - \eta \nabla^2 \right)^2 A_y = \Gamma \frac{dV_y}{dz} \frac{\partial A_y}{\partial x} \tag{5}$$

The strong shear makes $B_y > B_x$ and B_z, and the equations for the other components can be neglected because they involve the interaction of the turbulence with weak components of the field. With a steady shear, dV_y/d_z = constant, solutions to equation (5) are of the form (Parker, 1955, 1957, 1979 pp. 625 -640).

$$A_y = C \exp \left(\frac{t}{\tau} + ik_x x + ik_z z \right) \tag{6}$$

where

$$\left[\frac{1}{\tau} + \eta(k_x^2 + k_z^2)\right]^2 = \Gamma \frac{dV_y}{dz} (ik_x)$$

Hence,

$$\frac{1}{\tau} = - \eta(k_x^2 + k_z^2) \mp (\frac{\Gamma}{2} \frac{dV_y}{dz} k_x)^{1/2} (1 + i) \qquad (7)$$

Since we are interested in growing fields we should look for solutions with positive τ, that is, where the real part of $1/\tau$ is positive. This is possible only with the positive sign in equation (7). Thus, a dynamo effect occurs if the product of Γ, the strength the eddies, and the shear is large enough. Also notice that k_z, the vertical wave number, does not enter into the generation in any essential way; it is k_x which is most important. So, we can simplify things for the sake of discussion by letting $k_z = 0$. We do not get quite the right result for the dissipation, but nonetheless it's neglect does not leave out any of the essential physics.

Suppose that the real part of $1/\tau$ is exactly zero (no growth), that is, growth exactly balances dissipation. From equation (7) we get

$$k_x^3 = \frac{\Gamma}{2\eta^2} \frac{dV_y}{dz} \quad (k_z = 0) \qquad (8)$$

The stronger the field is generated, the more rapidly it dissipates, (to maintain zero growth) for which purpose the wave number becomes larger. Notice that the imaginary part of $1/\tau$ is not zero:

$$\omega = \mathrm{Im}(\frac{1}{\tau}) = (\frac{k_x}{2} \Gamma \frac{dV_y}{dz})^{1/2} = - \eta(k_x^2 + k_z^2) \qquad (9)$$

We have an oscillatory field with a frequency that is determined by the characteristic diffusion time.

To understand the origin of the oscillatory field, the sketch in Figure 3 may be helpful. Imagine that there are alternating bands of field perpendicular to the figure and the cyclonic turbulence forms loops with the sense shown in the figure. The small loops diffuse and cancel where they abut their neighbors. We are left with a large-scale circulation of the magnetic field. There is a vertical field between the two circulations, and the shear, which has been running all the time, stretches the field and reinforces the right hand edge of the band and destroys the left hand edge. The net result is that the whole pattern is shifted over a bit, leading to a migrating, os-scillating field, with a phase velocity $V = \omega/k_x$.

One might ask how a migrating field is related to the earth's quasi-stationary field. Some sort of boundary is necessary to block the migration so that instead of real wave numbers, which we have assumed, we let $ik_x = K_x$, which is real, and obtain

$$\frac{1}{\tau} = \eta K_x^2 + (K_x \Gamma \frac{dV_y}{dx}) \quad (10)$$

Thus, we can have a growing field which is stationary because it has come up against a boundary, and it is found that when the solutions are put into the confines of a sphere, such as the core of the earth, they can be blocked so that we are left with a stationary mode.

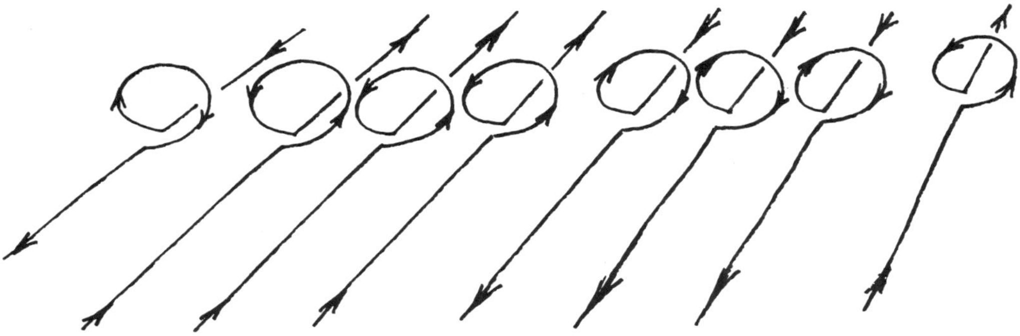

Figure 3. A sketch of the loops of flux in the perpendicular plane produced by identical convective cells interacting with alternating bands of field.

3. Induction in the Earth

Figure 4 shows how induction works in spherical geometry. In Figure 4a the dipole field has been stretched to the east in the northern hemisphere and to the west in the southern hemisphere by the nonuniform rotation of the core. We discussed this effect in the previous lecture and it is by far the largest effect, producing fields of 100 or more Gauss in the core. If a rotating eddy rises through the core, this field becomes twisted as shown in Figure 4b for the northern hemisphere. Converging flow leads to cyclones with one sense of rotation in the northern hemisphere and the opposite sense in the southern hemisphere. This happens to be the sense that leads to field amplification; with the opposite sense the field can be destroyed very quickly. Figure 4b shows only one twisted loop; there are others extending all the way around the earth as sketched in Fig. 4c. Then, if we imagine that the loops are held still, the small scale field diffuses away and we are left with a large scale circulation in the same sense as the original dipole field. The outer side of the loop diffuses through the core and some new lines of force have been added to the dipole field.

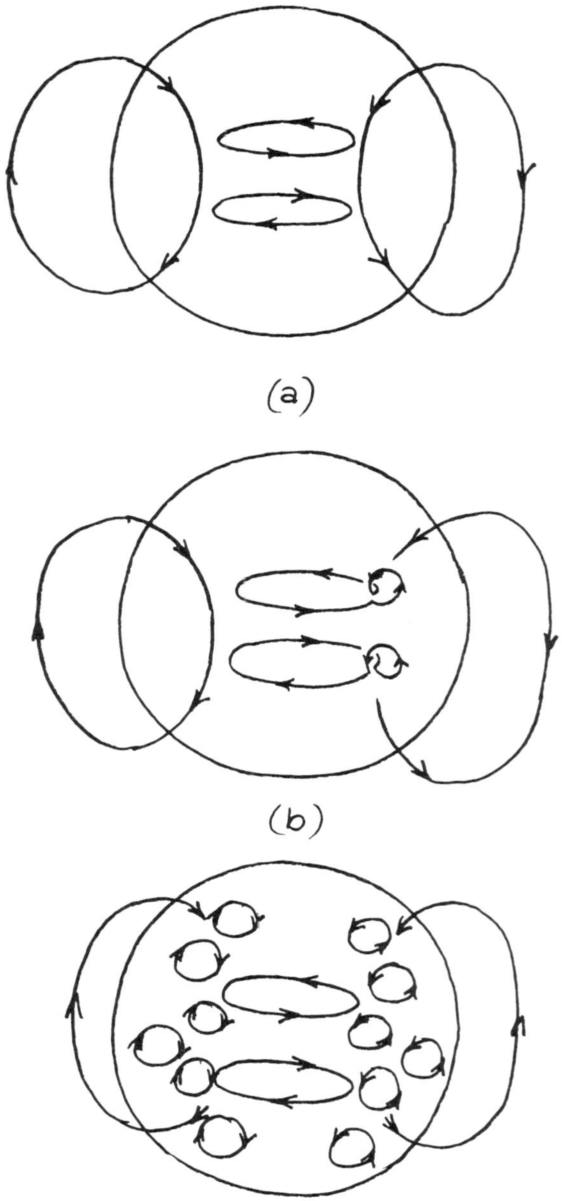

Figure 4. A sketch (a) of the initial toroidal (azimuthal) and poloidal (meridional) magnetic fields in the dynamo cycle proposed for Earth; (b) shows the convective formation of small loops in the meridional planes; (c) shows many small meridional loops, whose coalescence leads to a large-scale circulation of flux with the sense shown.

If the dynamo is more vigorous the field slips into an oscillating mode. And, of course, even in the present quasi-stationary mode the modest number of convective cells provides continual fluctuation of the strength of the dipole. It so happened that a few thousand years ago the earth's field was increasing with a characteristic time of about 2,000 years. About 1,000 years ago it went through a maximum and now it is tapering off, again with a characteristic time of about 2,000 years. There are always people, who may be called "professional extrapolators," who extrapolate curves and say that during the time the earth's field is declining the field will eventually reach zero. Thus, we continually hear the statement that the earth's field is reversing. What would they have said 2,000 years ago when the field was increasing?

The earth's field fluctuates because of the irregularities in the convection in the core, or less balances out. If the dynamo runs faster a number of things happen, but it is simplest to think about it in terms of the ordinary modes or their characteristic wave numbers. The stationary dipole seems to be the lowest mode. With more vigorous convection in the core the next mode is a periodic quadrupole. The period of oscillation is about 2,000 years depending on the wave numbers used. These fluctuating modes could reverse the earth's field.

Another possibility would be to shift the cyclonic distribution by 20° in latitude. Calculations show that in that case the field can be actively destroyed by the dynamo in 1,000 years or so, and then reconstitute itself with either sign being equally probable.

Consider, then, the dimensionless dynamo number,

$$N_D \equiv \frac{\frac{\Gamma}{2}\frac{dV_y}{dx}k_x}{\eta^2(k_x{}^2 + k_z{}^2)^2} = \frac{\frac{\Gamma}{2}\frac{dV_y}{dx}}{\eta^2 k_x{}^3} \quad (k_z = 0) \qquad (11)$$

It is a characteristic number which describes whether the rate of generation exceeds the dissipation. If $N_D > 1$ the field grows and if $N_D < 1$ the dynamo cannot keep up with the dissipation and the field decays. It has been pointed out that N_D is the product of two magnetic Reynold's numbers, $N_\Gamma \equiv \Gamma/2\eta(k_x{}^2 + k_z{}^2)^{1/2}$, which describes the eddies, and

$$N_G \equiv \frac{\frac{dV_y}{dx}k_x}{\eta(k_x{}^2 + k_z{}^2)^{3/2}} ,$$

which is the effective Reynold's number for the shear. Either N_Γ or N_G can be small as long as the other is large enough to sustain the dynamo.

Based on dimensional considerations and observational inferences of the cyclonic motions, Coriolis force, and nonuniform rotation of the core, among other things, it is found that both N_Γ and N_G are rather large. In fact, their product is sufficiently large that the earth's field should be in a relatively high order oscillatory state, whereas it actually is in its lowest state. The only thing one can say is that this is a kinematical dynamo where the chosen fluid motions, namely the cylindrical cyclonic eddies, are probably the most efficient form of motion. Most likely, the east-west azimuthal field becomes so strong that it supresses some of those motions, so that we have grossly overestimated the cyclonic aspect of the convection. The point is that nature is overly generous with the ability to generate magnetic fields. The same proves to be true in the sun, although there are more complications in that case.

This concludes the discussion of the basic dynamo effect stemming from the Γ coefficient, which is related to 90° rotations or some fraction.

4. Other Dynamo Effects

We now go on to the Q coefficient, which is a bit different in its func-tion. It does not produce a dynamo effect in an infinite space, while the Γ coefficient, which can be called the zero order dynamo effect, produces net flux with almost any boundary conditions. The Q coefficient is the dominant dynamo effect when the eddies rotate through 180°. We do not know where or if it takes place in nature , but we are studying the physics of dynamo effects to see what we can get out of them (Parker, 1979, pp. 618-623).

A rising eddy rotating by 180° would twist a field line in the manner shown in Figure 5a. If we draw a lot of them with square sides, for clarity, we can see why there is no net dynamo effect. The upper end of each tier of eddies has reverse field because of the 180° twist. We see that across the bottom of each tier there is a double strand of field, but immediately above there is a reverse field and the sum is just what we started with, a single line of force. However, in a finite space with a surface somewhere (for instance at the top of Figure 5b) the reverse flux at the top can be lost by diffusion. If we count lines we realize that there is one set of lines pointing to the right that has not been canceled by the reverse line that was lost. We end up with more lines pointing to the right than we did before. The net effect of this is a downward transport of flux through the surface and a creation of new flux through the mechanism of ejection of reverse flux (Parker, 1982 a,b,c).

Losing reverse field is the same as bringing in field in the original direction. This is a flux ejection dynamo which, in effect, brings in flux from the top surface until it stops at the bottom of the convection zone, where it piles up. Any component, B_x or B_y, can be locally amplified without involving any other component. We could imagine that in some circumstances a star has local patches of field

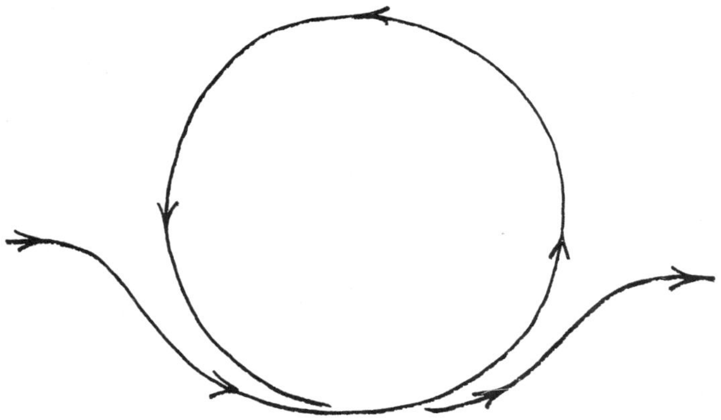

(a)

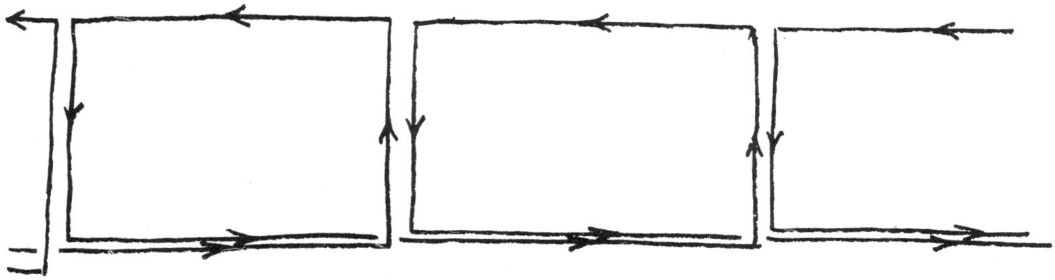

(b)

Figure 5. (a) The loop of flux produced by a convective cell that undergoes 180° of rotation,
(b) A sequence of such loops shown in rectangular form so that the cancellation of adjacent sides is obvious, with the double band of field across the base.

amplified by the flux ejection dynamo effect. These would float around like islands, more or less independent of anything else going on. We do not know whether that actually happens; it is merely a conjecture.

Where might we expect to find cyclonic rotation on the order of 180° so that the flux ejection mechanism could occur? Obviously it must take place where the Coriolis force is strong, where the rate of rotation of the parent body, the earth, the sun, or some star, is rapid compared to the lifetime of the individual eddy or convective cell. This does not occur on the surface of the sun. Granules and super granules live no longer than a day or so, while the sun rotates every 25 days in an inertial reference frame. But deeper in the convective zone where scales are 50,000 -100,000 km the turnover time for convection is rather long, possibly longer than a month. Therefore, at least in our present state of ignorance, there exists at least the possibility of eddies that rotate by as much as 180° inside the sun.

The earth is an example where one might also look for this effect. The convective cells have periods of 500 -1,000 years, and the earth spins around many times during their lifetimes. In fact, because of the earth's uniform density, the powerful Coriolis force apparently re-orients the convective motions into cylindrical forms, so the model used here may not be applicable. Therefore, we don't know whether this flux ejection mechanism occurs in nature, but if it does, it is quite different from the familiar α effect where the rotation is smaller.

Another effect when eddies rotate by 180° is negative turbulent diffusion, which was mentioned at the beginning of this lecture. It is one of those curious effects where it would be interesting to know whether it really exists. Although negative turbulent diffusion works in its purest form with loops twisted 180°, it does not have to be exactly that amount. One clear distinction between the situation where loops are twisted by 180° and the α-effect, which has loops twisted by 90°, is that for the α-effect the helicity of the fluid has a non-vanishing average or

$$\langle \underline{V} \cdot \nabla \times \underline{V} \rangle \neq 0$$

It is a corkscrew effect. If we have an equal number of eddies with opposite rotations we have nothing, so that the α-effect is clearly a result of the helicity of cyclonic convection.

One might be tempted to say that the flux ejection effect also depends on helicity, but the fact is that it does not. Whether the loop rotates to the right or to the left the effect is the same. Even though rotated in opposite directions, the loops are identical magnetically, and the effect does not require net helicity. It is sometimes stated that hydromagnetic dynamos are based on the helicity of fluid motion but this is too simple a statement. The nicest example of all was in a report by G. O. Roberts some years ago. He wrote down steady fluid velocities for which the helicity is identically zero at all points and he got a dynamo effect out of it.

The same applies to negative turbulent diffusion; it does not require a net helicity. Figure 6 shows how it works. Imagine that we have a field of varying strength with overlapping loops formed by cyclonic motion. The first impression would be that the loops twisted 180° in Figure 6 cancel each other because they are in opposite directions. In fact, they do not because one of them is coming from a weak field region and has fewer lines entrained in it. That is the physical basis for the effect (see Parker, 1979 pp. 585-592).

In Figure 6 the general field is in the +x direction and has a positive gradient $\partial B_x/\partial_z > 0$. For a loop of radius L twisted up from the weak field region, at position z = L and z = 0, we have $B_x = -B_x(-L) = \nabla \times A \cong \delta\Delta/L$, or $\delta A = LB_x(-L) \cong L[B_x(o) - L\ \partial B_x/\partial_z]$. Similarly, for a loop coming down from the strong field region, at z = +L to z = 0 we have $B_x = -B_x(L) \simeq \delta A/L$, or $\delta A = -B_x(-L) \simeq -L\ \lfloor B(0) + L(\partial B_x/\partial z)\rfloor$. Adding the two contributions, we get for a total change in the vector potential $\delta A = -2L^2\ \partial B_x/\partial_z$. And since $B_x \simeq \partial\delta A/\partial z$, we have

$$\frac{\partial B_x}{\partial t} \simeq \frac{(\frac{\partial \delta A}{\partial z})}{\tau} = -\frac{2L2}{\tau}\frac{\partial 2B_x}{\partial z^2} \qquad (12)$$

This peculiar addition of reverse loops builds the field up where it is already strongest. It grows where it is convex and decays where it is concave, opposite to what we are accustomed to seeing in ordinary "positive" diffusion diffusion.

⌊This lecture is something like Tom Lehrer's song about the elements. At the end he says that there may be many others, but they have not been discovered.] There may be many other dynamo effects buried in the calculations that people have made, but they have not been pulled out explicitly and demonstrated in simple diagrams. However the ones discussed here seem to be the ones available to us when we consider the origins of magnetic fields in the earth and the sun.

5. Magnetic Field of the Sun

Let us turn briefly to the sun's magnetic field. The first thing we discover when we apply these dynamo effects to the sun is that they are absurd; they don't work because the scales seem to be completely wrong. For an ionized hydrogen gas $\sigma \simeq 2 \times 10^7\ T^{3/2} = 2 \times 10^{13}\ sec^{-1}$ where T = 10,000 K at the sun's surface. Hence, $\eta = c^2/4\eta\sigma \cong 10^7$(cm/-sec) and with a characteristic solar dimension Ro the diffusion time becomes $t \sim Ro^2/4\eta = 1/2 \times 10^{22}/4 \times 10^7 = 10^{14}$ sec = 3 x 10^7 years. That is simply too long to wait compared to the 22 year cycle. It has been pointed out that turbulent diffusion is involved (Leighton, 1969) . An estimate of the turbulent diffusion is $\eta_T \sim .1\ V\ell$, but there is not a good theory for it at present. [Curiously enough, some people turn around and say, that therefore it does not exist. This is very interesting because there is no theory for the human brain, either.]

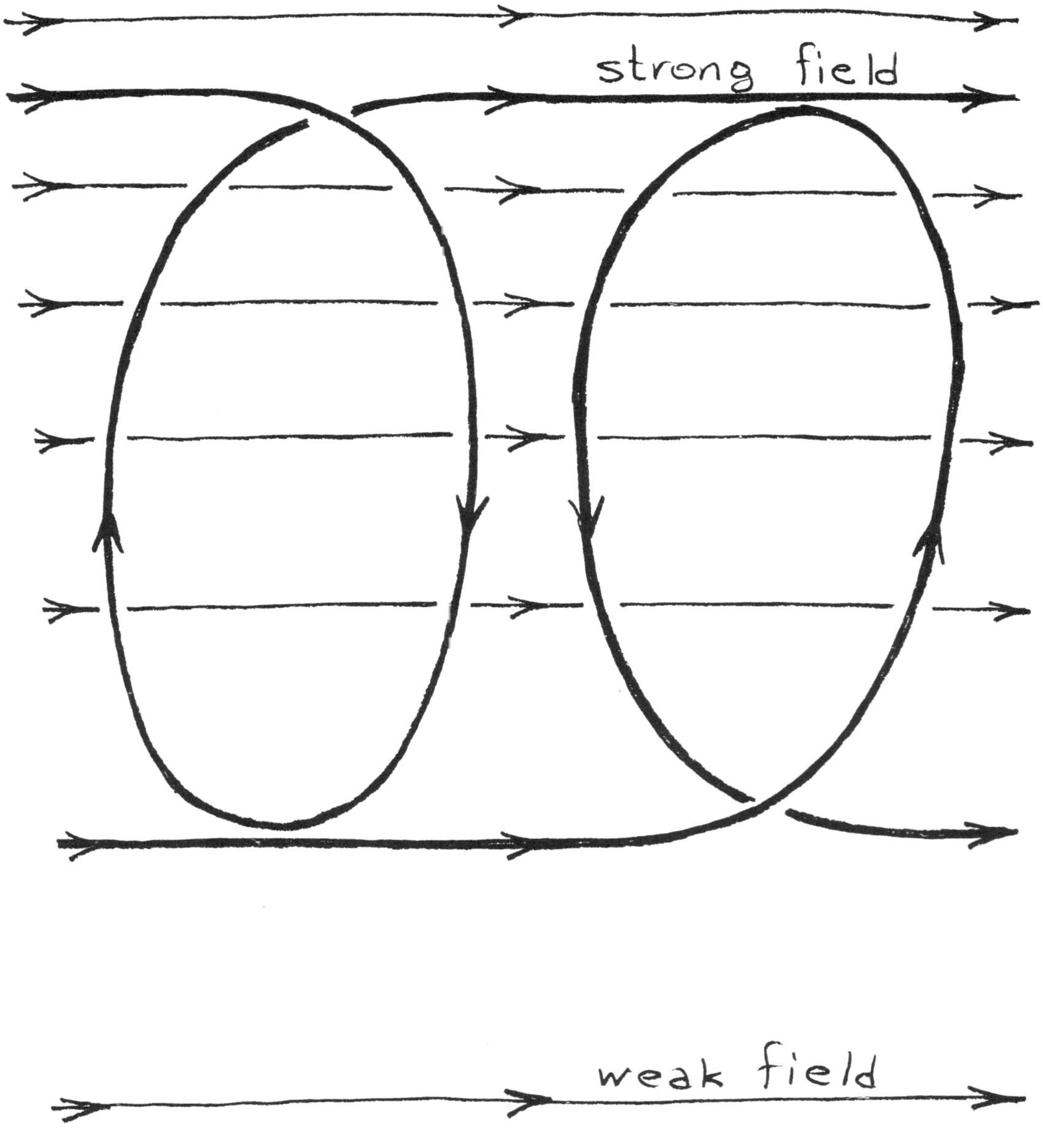

Figure 6. A sketch of two loops of field, each rotated 180°, with one pushed downward from the region of strong field at the top and the other pushed up from the region of weak field at the bottom.

In any case, we can substitute numbers to obtain an estimate. If we consider granules, the equivalent velocity is about 1 km/sec and the scale is about 1,000 km, leading to $\eta_T \sim 10^{12}(cm^2/sec)$. This is the value at the surface of the sun; other values are possible at different depths. [Since we have plenty of free parameters we can play all kinds of heuristic games.] Computing the diffusion time for turbulent diffusion we get about 10^9 sec which is about 30 years. This is still a bit too long, but the numbers are very rough. The supergranules have velocities of 0.3 km/sec and dimensions of 3×10^4 km, providing $\eta_T \sim 10^{13}$ cm^2/sec, with an effective diffusion time of 3 years. It seems that the only reason that anything can be done with the sun's field is because of turbulence. It mixes the field and allows it to slip through the fluid, although it is much more complicated than this, as we shall see in the next lecture. Nevertheless, when the field is moved around it is stretched in the turbulent fluid and on this basis we have dynamo waves. If the angular velocity of the sun increases downward through the convective zone, the equations predict, in a very simple way, that the east-west field migrates toward the equator, producing a magnetic oscillation with a period easily adjusted to the observed 22 years. Three or four free parameters can do remarkable things like reproducing "butterfly" diagrams for sunspots (see review and references in Parker, 1979 pp. 746-764).

Thus, it appears that nonuniform rotation with depth may account for the sun's magnetic field. That, together with cyclonic convection in the convective zone of the sun seems quite adequate to explain the 22 year magnetic cycle. Nobody, however, has been able to look below the surface of the sun to prove that. One should keep an open mind for alternatives since we are always coming up against the fact that nature is more imaginative than we are. There might be flux ejection dynamo effects working at the sun's polar regions; we simply do not know.

The sun's field is remarkably irregular. Even though we talk about a sunspot cycle, a migration of azimuthal fields toward the equator, and a dipole field, they are all buried in a lot of noise. Sometimes the dipole moment points at right angles to the rotation axis, although it is not true on the average. Leighton has pointed out that there are times when the sun appears as if it were in a quadrupole moment with a strong asymmetry between the two hemispheres. In fact, Leighton and others have run numerical models that show this characteristic.

Finally, let us conclude with the fact that there is still no good theory for turbulent diffusion of vector fields. With scalar fields such as smoke or some chemical the diffusion is straightforward. But when dealing with vector fields we have many unanswered questions, such as what happens when a small scale field is built up. One answer is that they may destroy themselves in ways that we shall discuss in the next lecture. This will lead us into questions of field dissipation and eventually the heating of the solar corona.

References

Kraichnan, R.H.: 1976 J. Fluid Mech. 75, 657; 77, 753.

Leighton, R.B.: 1969 Astrophys. J. 156, 1.

Parker, E.N.: 1955 Astrophys. J. 122, 293.

Parker, E.N.: 1957, Proc. Nat. Acad. Sci. 43, 8.

Parker, E.N.: 1970, Astrophys. J. 162, 665.

Parker, E.N.: 1979, Cosmical Magnetic Fields (Oxford, Clarendon Press).

Parker, E.N.: 1982a, Geophys. Astrophys. Fluid Dynamics 20, 165.

Parker, E.N.: 1982b, Astrophys. Space Sci. 85, 167.

Parker, E.N.: 1982c, Astrophys. Space Sci. 85, 183.

HEATING OF THE OUTER SOLAR ATMOSPHERE I

Eugene N. Parker
Department of Physics
University of Chicago
Chicago, Illinois 60637

In the previous lecture we discussed dynamo effects which seem to provide the origin of magnetic fields beneath the surface of the sun. The observations, of course, are limited to the surface, but through the surface there continually emerges fresh magnetic fields. This leads to the idea that there must be a source somewhere below. We can start by considering present day ideas about the sun's internal structure. The sun has a radius of approximately 700,000 km, of which the outer 100,000 km or so is the convective zone, according to mixing-length models. In the convective zone the material is so opaque that it does not let heat pass easily so that the temperature gradient becomes steeper than the adiabatic gradient, which leads to convective, rather than radiative, transport of heat.

The transition from radiative to convective transport deep in the sun is continuous. At the bottom of the convective zone densities are about .1 gm/cm^2, whereas at the top (the photosphere) it is about 2 x 10^{-7} gm/cm^3. This large difference makes the dynamics very difficult to handle.

The Solar Magnetic Field

The dynamo is believed to operate in the convective zone, across which there may be a 5 -10% variation in the angular velocity. Convective velocity estimates vary according to whose mixing-length theory is applied, but velocities of 0.1 -1 km/sec appear reasonable deep into the convective zone. There are the stretched east-west fields similar to the ones in the earth's core which we discussed earlier. Associated with these are poloidal fields which contribute to a net dipole moment of the sun and are generated by a dynamo that operates like the one described earlier in Cartesian coordinates. The direction of migration is toward the equator with a period of 22 years, as judged from the surface fields.

The question of why the fields come to the surface is related to the phenomenon of magnetic buoyancy (Parker, 1955) . A magnetic field

R. L. Carovillano and J. M. Forbes (eds.), Solar-Terrestrial Physics, 129–154.

contributes a pressure $B^2/8\pi$ to the material in which it is embedded, while it contributes essentially no density. An element of fluid which has a magnetic field permeating it will have a lower density at a given total pressure since the temperature is the same as its surroundings, and therefore it tends to be buoyant. Another way of looking at it is that the fields are not in equilibrium.

If we equate the buoyant force to the hydromagnetic drag we find that, to an order of magnitude, a flux tube would rise at the Alfvén speed computed in that flux tube. That is very fast, which is one of the difficulties. A field of 1000 G at the bottom of the convective zone would rise out of the convective zone in a matter of six months. There are effects which could hold it down, but nonetheless it is a complication in constructing a satisfactory dynamo theory for the sun.

It is not entirely clear how long the delay is between the formation of the field and its buoyant arrival at the surface. It may be a few months or a few years. There are many difficulties but most people are inclined to take the simplest point of view, which may be wrong, and interpret the fields bulging through the surface as a direct indication of what is being formed below. From the arching magnetic fields over active regions on the surface of the sun we infer that east-west fields are buried beneath the surface.

The magnetic field coming up through the surface of the sun is responsible for the solar activity that heats the outer solar atmosphere. The field behaves in ways that we do not expect and do not understand to a large degree. The sun has fields extending out from the poles and a general disordered field over the middle latitudes of 1-5 G. Then there are active regions with stronger fields of 10-100 G. Years ago it became clear that there was something inconsistent in the magnetic field measurements. Different values of the field came from straightforward interpretation of observations of Zeeman broadening at the same height, depending on the atomic line used. It was finally shown to be a consequence of the fibril nature of the magnetic field.

It was a surprise to find (Beckers and Schröter, 1969; Stenflo, 1973) that the field of the sun has a fibril structure (everywhere outside of sunspots) made up of separate, compressed flux tubes. Very roughly they have fields of 1,000-2,000 G and diameters of 100-500 km. Compared to an average 5 G on the surface, that represents an enormous increase in energy. Some force with considerable muscle must work on these fields, split them up into individual tubes, and then compress them to an extreme degree. Figure 1 shows how such flux tubes might look in a vertical section. What lies below the surface is conjecture, but immediately above the surface the field expands to fill the available volume, there being nothing to prevent it from doing so. This structure is an indication that a considerable force in the sun compresses the field to over 1,000 G. Near the surface the gas pressure is hardly more than the field pressure so that $\beta \sim 1$.

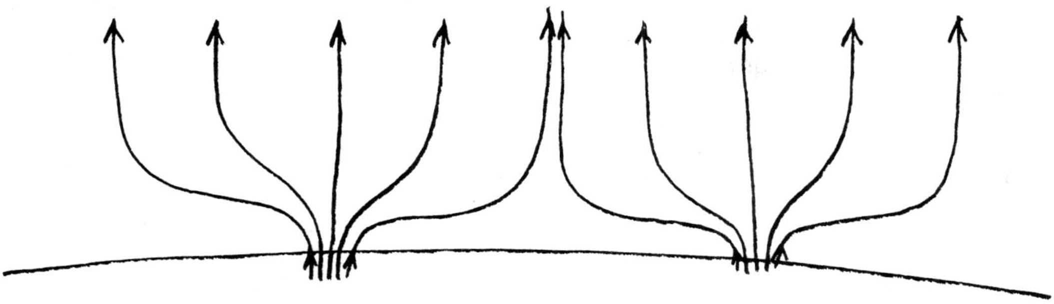

Figure 1. A sketch of the concentrated flux tubes extending vertically through the photosphere, with rapid divergence immediately above.

It is not entirely clear why the field at the surface has this remarkable property. However, powerful downdrafts of 1/2 km/sec are observed in the flux tubes. These downdrafts may evacuate 1/2 or 2/3 of the gas from the interior, which would lessen the gas pressure and account for compression by the external gas pressure (Parker, 1978, 1979 pp 260-271, Spruit, 1979; Spruit and Zweibel, 1979). Why they are stable is unknown. Why there is a downdraft is also not understood, except for some hand-waving arguments.

Some people speculate that the flux tubes might extend all the way down through the convective zone. Since we have never been there we do not know, but, if that is the case it makes questions about turbulent diffusion much easier to handle theoretically. The curious fact is that those who argue that the field is fibril all the way to the bottom of the convective zone also argue there is no theory of turbulent diffusion of magnetic fields.

In any case we do not know what is happening below the surface of the sun. The fibril state does not invalidate any of the things discussed in the previous lecture. The fibrils are carried along with the fluid, where the material is sufficiently dense, so that there is hardly any slippage. With only minor modifications the previous discussion remains valid.

Heating of the Corona

The subject of the heating of the solar corona is an old one. When mysterious spectral lines were observed in the solar corona, the hypothetical element "coronium" was invented but became increasingly difficult to justify as chemists and physicists filled in the periodic table. Theory finally suggested that the spectral pattern indicated highly ionized elements like iron, calcium, and silicon heated to a million degrees. This was immediately contradicted by the high priests because it was absurd to think that a 5600 K photosphere could heat the corona to a million degrees.

Then it was pointed out that the convective zone is a heat engine that produces mechanical work. Mechanical work can be used in many different ways. For instance, we can generate electricity and the temperature of an electric arc is in no way limited by the boiler temperature back at the powerhouse. A number of people, Biermann, Schwartzschield, and Alfven, pointed out that various kinds of mechanical motions could be produced in the convective zone. These motions could propagate up into the tenuous atmosphere and then dissipate through friction and very easily produce higher temperatures than in the convective zone. Another way of explaining it is that a boy scout with a body temperature of 37°C is capable of rubbing sticks together and producing a fire. The conversion of mechanical energy to heat is in no way limited by the Second Law of Thermodynamics.

Ever since the 1940's people have searched for the mechanical work that is dissipated in the corona. So far the solution has been very elusive. Acoustical waves were popular but it is now known that they dissipate too quickly to heat the corona. They could heat the chromosphere but they just do not get into the corona. Progress began when x-ray observations became possible, because then we could look, in a quantitative way, directly at the super hot gas in the corona. From the detailed locations of x-ray emissions we could see where heat was coming from and where it was deposited. Quantitative models could be proposed based on thermal conduction, density, and barometric effects.

In the x-ray images of coronal loops and arches fine striations are often seen, apparently extending along the lines of force. Typical temperatures are 2×10^6 K, and one of the first conclusions that came from the model building was that it was not sufficient to heat the loops only at their footpoints in the lower corona. There must be a distributed heat source. The apex is often the hottest part, and that suggests that the heat is deposited more or less uniformly because the gas is least dense at the apex and least able to radiate and cool itself.

Model for Magnetic Merging

A paper by Rosner, Tucker, and Vaiana (1978) pulled together the collected evidence and pointed out that the only interpretation for heating the visible corona was a direct transfer of energy from the magnetic field to the gas. This did not rule out magnetohydromagnetic waves, but they could be ruled out on other grounds. In particular, both compressional and transverse Alfvén waves tend to propagate without much dissipation unless their amplitude is extremely large. Their long dissipation lengths ($\sim 10^5$ km) contradict the fact that very large regions, like active loops, and ephemeral magnetic regions ten times smaller, can have the same brightness.

Therefore, it seems that somehow the magnetic field is dissipating its energy into heat without some intermediate form of wave motion. For two decades it has been known that the high conductivity in the corona poses a serious difficulty for dissipation. With a temperature of 1×10^6 K the electrical conductivity $\sigma \approx 2 \times 10^{16}$ sec^{-1} and $\eta = c^2/4\pi\sigma \approx 10^4$ cm^2/sec. Assuming a scale on the size of a granule ($L = 1000$ km $= 10^8$ cm), we get a diffusion time $t = L^2/4\eta = 10^{16}/4 \times 10^4 = 2.5 \times 10^{11}$ sec or about 10^4 years. This is much too long; the heat must be deposited in a few hours.

However, under some conditions there can be enhanced resistivity. If conduction electrons are driven up to the ion sound speed, ion cyclotron and ion acoustic waves are generated and they can scatter electrons. This is called anomalous resistivity and it is caused by plasma turbulence. It was believed that it might help, but it was found recently from numerical simulations that in regions of strong field this effect would only double or quadruple the resistivity. This still leaves a diffusion time of 2500 years and therefore it does not solve the problem. It turns out that in strong fields we have to drive the conduction velocity all the way up to the electron thermal velocity. The thermal velocity is very high, and in order to drive the conduction velocity that high we must have currents concentrated in remarkably thin sheets. If we go through the calculations we find that the currents in a 1000 km diameter flux tube must be concentrated in a few cm, which is impossible. The debye length and cyclotron radius are much larger than that. In any case, if currents were concentrated that much we would hardly need any more help; the scale size l is much smaller and the dissipation becomes much more rapid in the current.

Rosner, Tucker, and Vaiana attacked the problem by noting that a concentration of electric current must be produced in some way. Or, from Ampere's Law, the shear in the magnetic field must be concentrated into thin layers in order to have the rapid dissipation which observations and theoretical interpretations of coronal loops seem to demand.

Normally, we describe the equilibrium of a magnetic field in a conducting fluid with pressure p by the following equations.

$$\rho \frac{d\underline{v}}{dt} = 0 = -\underline{\nabla}\, p + \frac{(\nabla \times \underline{B}) \times \underline{B}}{4\pi} = -\nabla \left(p + \frac{B^2}{8\pi}\right) + \frac{(B.\nabla)B}{4\pi} \qquad (1)$$

$$\rho \frac{dv_i}{dt} = 0 = -\frac{\partial p}{\partial x_i} + \frac{\partial M_{ij}}{\partial x_j}$$

M_{ij}, the Maxwell stress tensor, is the sum an isotropic magnetic pressure, $B^2/8\pi$, and a tension, $B^2/4\pi$, in the direction of the magnetic field. In the equilibrium state $\underline{B}\cdot\underline{\nabla}p = 0$, so that the pressure is constant along the field. Alternatively, the Lorentz force, $(\nabla x\underline{B})x\underline{B} = \underline{j}x\underline{B}$, has no component along the field and the gas is free to slide back and forth along the field. In this way each flux tube is like a pipe, but when the gas tries to move across the field it must take the field with it. The field is elastic and tends to resist motion across it.

The final result of this discussion will be that essentially no magnetic field configuration has an equilibrium. In an arbitrary magnetic field topology there is active dissipation taking place for reasons which gradually will become apparent through the lecture . We can start by illustrating one nonequilibrium situation. Imagine that we have two highly conducting plates in which there is anchored a magnetic field (Figure 2). Suppose that the fields are anti-parallel as shown and that the plates are pressed together, squeezing the fields. Along a vertical line perpendicular to the interface equilibrium requires that the total pressure, fluid plus magnetic, must be constant. Also, since the magnetic field goes from +B to −B in the middle there must be a neutral line perpendicular to the plane of the figure where B=0.

Along the vertical line of symmetry in Figure 2 $(\underline{B}\cdot\underline{\nabla})\underline{B}=0$, and from equation (1), $\underline{\nabla}(p + B^2/8\pi) = 0$ or $p + B^2/8\pi$ = constant. Thus, the fluid pressure p is highest in the middle where B=0, and equilibrium can be easily maintained by pumping in gas from the sides. But that is a special requirement. Generally speaking there would be merely a sea of gas all at the same pressure, with the result that the gas is squeezed out both ends. The pressure variation from the central area to a distance L at each end is essentially $B^2/8\pi$, and this pressure difference is approximately equal to $1/2\ \rho V^2$ where V is the velocity of the gas squeezed out at both ends. Thus, $V = B/\sqrt{4\pi\rho} \equiv V_A$ which is the Alfven speed computed in the field.

As the plates are held together the gas flows out the ends. There is always a neutral line between the fields where the gas pressure is forced to be higher. Eventually the fields come together so close at some small distance l that no matter how high the initial conductivity, diffusion begins to occur. Remember, the characteristic diffusion time τ is approximately l^2/n, which becomes very short at small l. In fact, the diffusion reaches a quasi-steady situation where the rate at which the opposite fields move together is $u \cong l/\tau = \eta/l$. Equating the amount of material entering the middle region at velocity u to the amount leaving at velocity V_A we have $uL \cong V_A l$. Hence, $u = V_A/(V_A L/\eta)^{1/2}$, where $(V_A L/\eta) \equiv N$ is the magnetic Reynold's number. N is generally very large in this situation because of the broad dimension L. Thus, there is an enhancement of the field dissipation because of the peculiar nonequilibrium. Regardless how high the conductivity is, the gradient between the fields becomes so steep that diffusion sets in and a sort of steady state is reached with the field being eaten away at the midplane at a speed u (Parker, 1957).

One might object that u is small because N and $N^{1/2}$ are very large. Although this is true for the simple case described above, Petsheck pointed out that it is not over the whole breadth L where the fields press together but in a smaller region in the middle. A com-

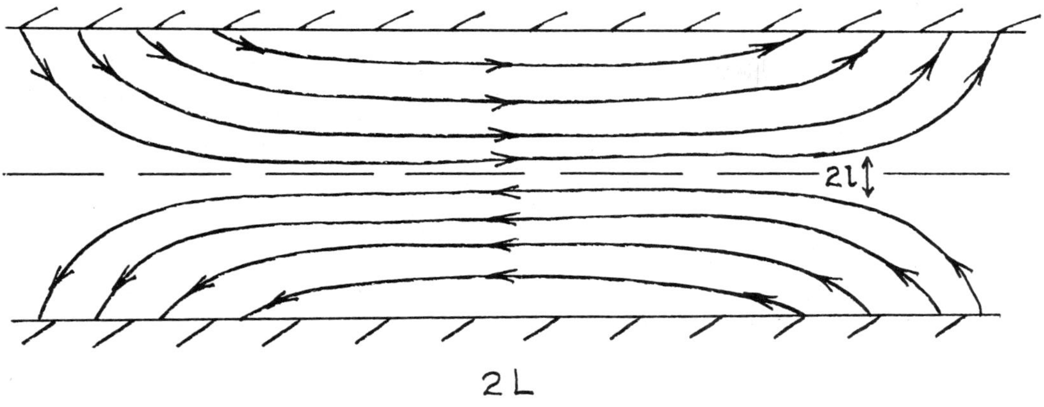

Figure 2. A sketch of two magnetic fields of opposite orientation anchored in their respective footplanes and pressed firmly together.

plete analysis shows that the merging rate u may be as large as (V_A/lnN) (Petschek, 1964). Even if N is a large number, its logarithm is not. The rate u could be as large as (V_A/20), and that is interesting. Which of the two limits for u is applicable depends on the boundary conditions in rather subtle ways (see discussion and references in Parker, 1979 pp 392-437; Priest, 1981).

Thus, there is an indication that in spite of high conductivity there are ways that magnetic fields can destroy themselves relatively quickly. Now the question is where does this occur in nature. As we shall see, it appears to happen all the time. There seems to be no way of avoiding it and it seems the normal fate of magnetic fields which are pushed up through the surface of the sun.

To conclude this lecture we will set up an idealized problem and take it up in the last lecture. Obviously, we cannot write down the general equilibrium and dynamic equations for magnetic fields and solve them; they are far too nonlinear for that. We have to use our intelligence instead. Consider a magnetic field extending between two planes (Figure 3). Initially the field is uniform, but suppose that we are at liberty to move the footpoints arbitrarily in the lower plane, while the field is fixed in the highly conducting upper plane. This simulates the motion of granules on the surface of the sun which shuffles the footpoints about. We can wrap the field around in any way. We can twist one flux tube one way and its neighbor the other way. We can imagine many complicated wrappings. We also suppose that the distance between the planes is much larger than the dimensions at the ends of the flux tubes, or $L \gg \lambda$. Since we do not want boundaries interfering with this discussion, let the arrangement extend to infinity on all sides.

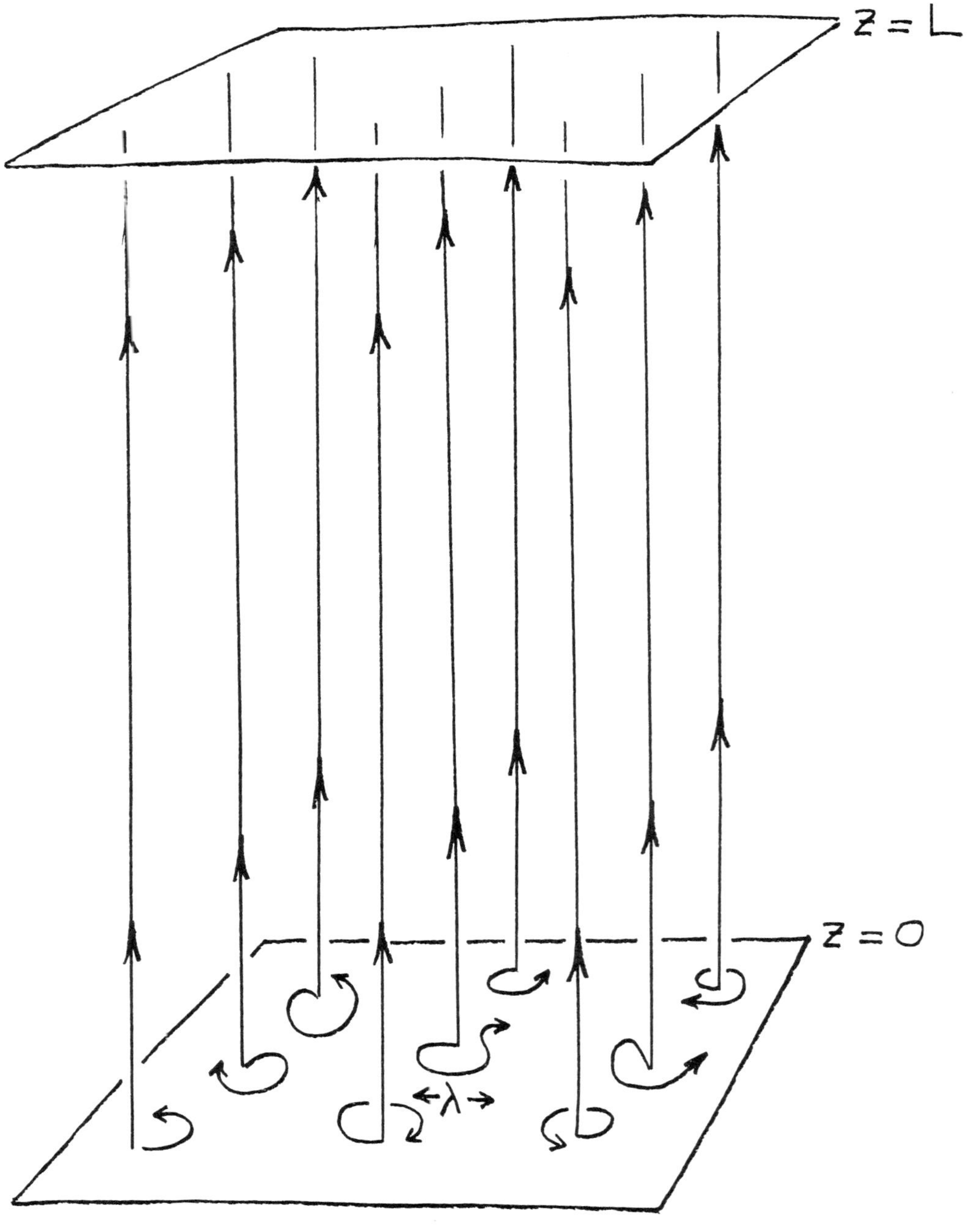

Figure 3. A sketch of the lines of force of the initially uniform field B_0 anchored in the planes z=0, L and with the footpoints at z=0 subjected to random shuffling on a scale $\lambda(\ll L)$ as indicated by the arrows.

We will discover that except in a small measure of special cases is there no possibility of equilibrium. Suppose we are careful to twist the flux tubes uniformly throughout their entire length. Then we find that there are solutions to the equations 1, but when we examine them we find that they have symmetries that would exist only in the mind of mathematician and not in nature. Considerations of individual cases indicate that the neutral point conversion described above occurs in all of them. Apparently, when magnetic fields are manipulated in the way shown in Figure 3, they dissipate quickly in spite of the high conductivity, and any work done on the footpoints twists the field and is dissipated in fluid motions and heating. That proves to be a universal property of magnetic fields which cannot be avoided, and it probably is responsible for a major part of the heating of the sun's outer atmosphere. We will go into details in the next lecture.

References

Beckers, J.M. and Schröter, E.H.: 1968, Solar Phys. 4, 142, 165.

Parker, E.N.: 1955, Astrophys. J., 122, 293.

Parker, E.N.: 1957, J. Geophys. Res. 62, 509.

Petschek, H.E.: 1964, The physics of solar flares, AAS-NASA Symposium, NASA SP-50 (ed. W.N. Hess) p. 425.

Priest, E.R.: 1981, Solar Flare Magnetohydrodynamics (New York, Gordon and Breach Science Pub.) pp. 139-213.

Rosner, R., Tucker, W.H., and Vaiana, G.S.: 1978, Astrophys. J., 220, 643.

Spruit, H.C.: 1979, Solar Phys., 61, 363.

Spruit, H.C. and Zweibel, E.G: 1979, Solar Phys., 62, 15.

Stenflo, J.O.: 1973, Solar Phys., 32, 41.

HEATING OF THE OUTER SOLAR ATMOSPHERE II

Eugene N. Parker
Department of Physics
University of Chicago
Chicago, Illinois 60637

1. Introduction

Before we continue the discussion which was begun in the previous lecture, remember that we are talking about active regions on the sun. The fields are closed; they come out at one point and reenter at another (Figure 1). The foot points of the lines of force are continually being shuffled around at random, twisting the fields above. At the end of the previous lecture we considered an idealized problem. It consists of an initially uniform field extending between two parallel planes at z = 0 and z = L. The field at z = L is fixed while the footpoints at z = 0 are shuffled about, which twists the field lines into complicated patterns, and simulates the situation shown in Figure 1. The equilibrium equation is

$$0 = -\nabla p + \frac{(\nabla \times B) \times B}{4\pi} \qquad (1)$$

From the above we readily deduce that the pressure is constant along a line of force, or $\underline{B} \cdot \nabla p = 0$. We are neglecting gravity; barometric effects could be included but they play no essential role.

In the previous lecture we also described a particular kind of dynamical nonequilibrium where two oppositely directed fields are squeezed together. They move together slowly and dissipate in the middle. Fluid is ejected out the sides unless there is external pressure to oppose the flow. The fluid is ejected at about the Alfvén speed and carries away most of the energy from the dissipated fields.

2. Idealized Problem

Returning to the idealized situation of the twisted field between two planes, first we imagine that L is very large so that the spiralling of the field is extremely small. We can express the field as

$$\underline{B} = \hat{e}_z B_o + \xi \underline{b}(x,y,z)$$

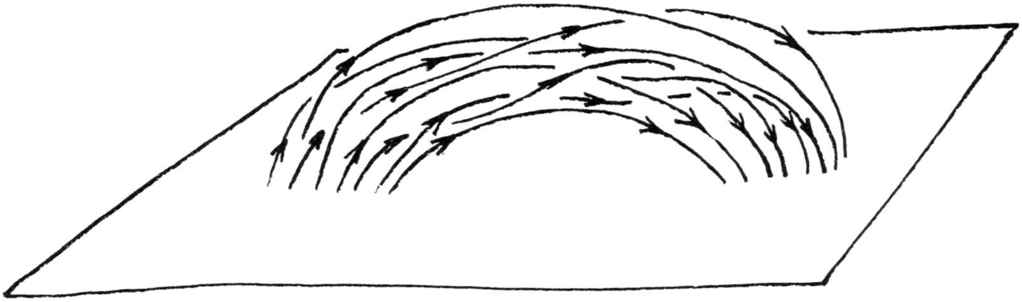

Figure 1. A sketch of the mutual wrapping and shuffling of the lines of force of a bipolar magnetic field above the photosphere.

It is the sum of the uniform field B_o plus the product of a small number ξ and the complicated winding field. To maintain equilibrium it would be necessary to perturb the pressure by ξp in equation (1). The zero order terms are obviously satisfied for a uniform pressure in a uniform field. The first order terms yield the following equations.

$$\frac{\partial}{\partial x}\left(p + \frac{B_o b_z}{4\pi}\right) = \frac{B_o}{4\pi}\frac{\partial b_x}{\partial z} \tag{2a}$$

$$\frac{\partial}{\partial y}\left(p + \frac{B_o b_z}{4\pi}\right) = \frac{B_o}{4\pi}\frac{\partial b_y}{\partial z} \tag{2b}$$

$$\frac{\partial}{\partial z}\left(p + \frac{B_o b_z}{\partial z}\right) = \frac{B_o \partial b_z}{4\pi} \tag{2c}$$

These equations describe the nature of the equilibrium and in particular the relation between the perturbation pressure p and b_z. Since $\underline{\nabla} \cdot \underline{b} = 0$, the equations yield

$$\nabla^2\left(p + \frac{B_o b_z}{4\pi}\right) = 0 \tag{3}$$

which is simply Laplace's equation. The boundary conditions have the planes separated by a distance L which is large compared to the scale λ over which the footpoints are moved. The fields extend to infinity in the transverse directions.

Thus, we have solutions to Laplace's equation in the midst of an arbitrarily large space. There are surface perturbations which extend into the region from $Z = \mp L$ and die off over distance as $e^{-(z/\lambda)}$, so that they are negligible throughout the interior of the "wilderness" for large L. The only bounded solution to Laplace's equation extending throughout the interior of the space is a constant,

$$p + (B_o b_z/4\pi) = \text{constant} \tag{4}$$

Combining equation (4) with equations (2a-c) we find that $\partial b/\partial z = 0$, that is, the perturbation field does not vary in the z direction. In other words, unless the winding pattern is constant along the field direction, there is no equilibrium (Parker, 1972).

This sounds mysterious when we first consider it, but let us see what goes wrong when the winding pattern changes. Consider two isobaric surfaces where one surface is at pressure p and the other is at pressure $p + \delta p$ (Figure 2). The lines of force lie on the two surfaces ($\underline{B} \cdot \underline{\nabla} p = 0$). Let us pick two corresponding field lines on each surface separated by δW and let the planes be separated by δh. For equilibrium the force caused by the pressure difference must be balanced by the Lorentz force.

$$\frac{\delta p}{\delta h} = \frac{j_\perp B}{c} \tag{5}$$

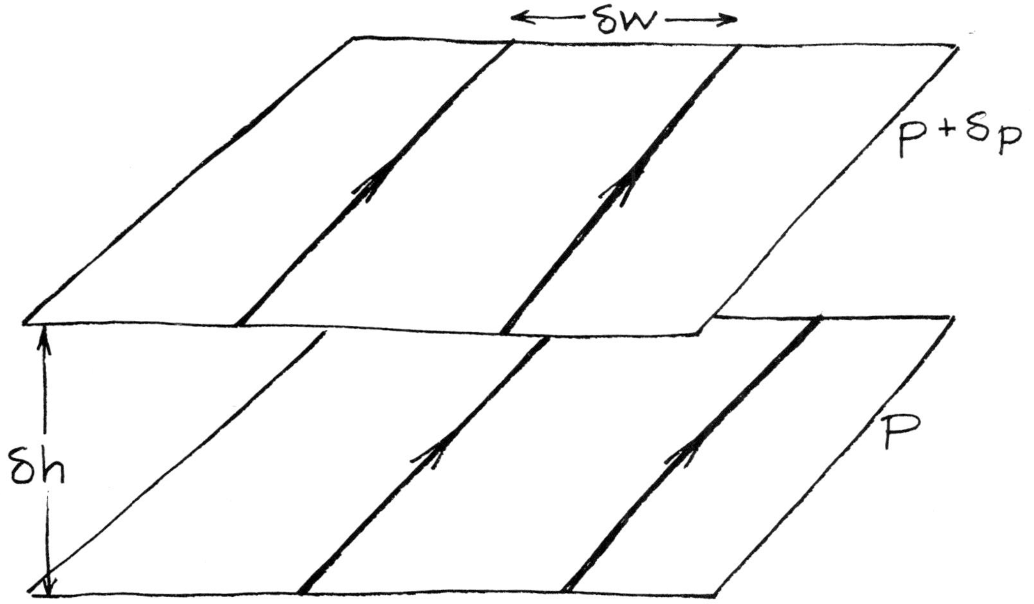

Figure 2. A sketch of the lines of force separated by δW and δh in the planes p = constant and $p + \delta p$ = constant.

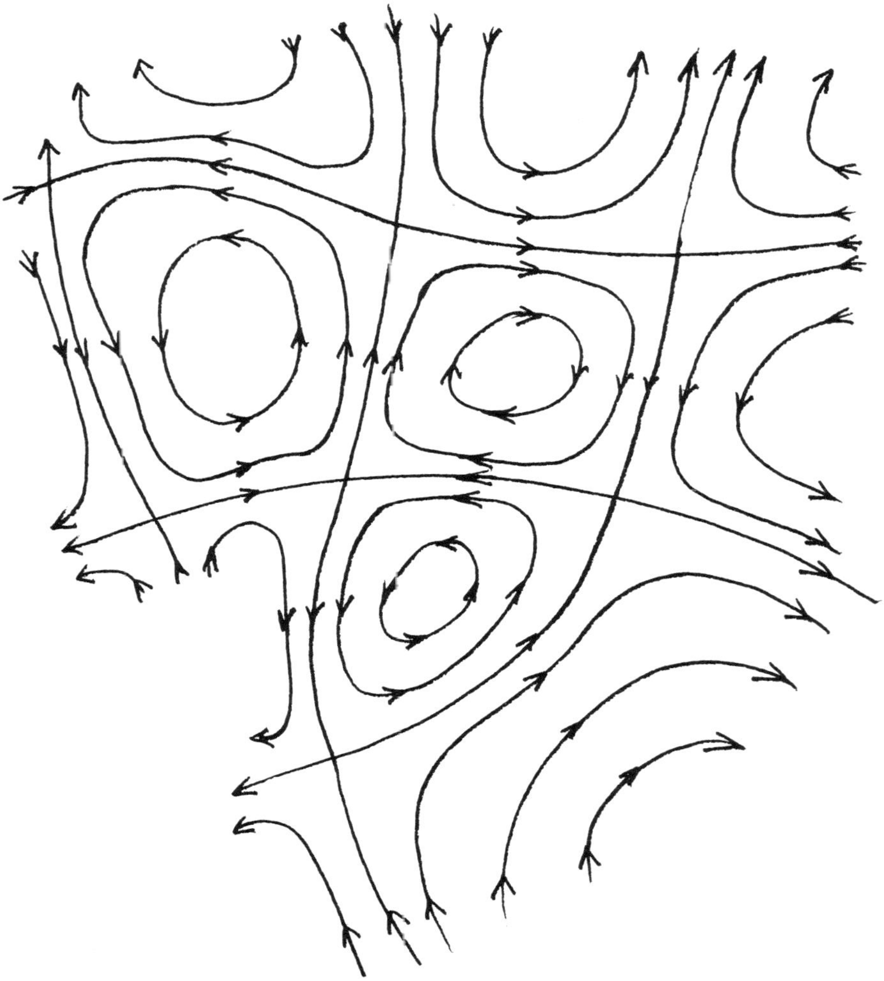

Figure 3. A sketch of the cross section (z = constant) through a bundle of close packed flux tubes.

Since the parallelogram in Figure 1 describes the boundary of a flux tube, $B(\delta W)(\delta h)$ = constant along the lines of force, from which it follows that

$$\frac{\delta p}{\delta} = \frac{j_\perp}{c} \frac{\text{constant}}{\delta W \delta h} \Rightarrow j_\perp \propto \delta w \qquad (6)$$

Now consider what happens if we change the winding pattern. To do this some lines of force which were initially very close together must separate and become increasingly distant from each other. Thus, the separation δW would grow without bound. The current density $j\perp$ would also grow, and to conserve flux, δh would be required to decrease, causing the pressure gradient to grow without bound. Thus equilibrium in the face of a change in the winding pattern implies a singularity in the field, which is unphysical (Yu, 1973).

There is another way of expressing the problem along similar lines. Imagine that we take a section through the field in the lower part of Figure 1. Various kinds of twisted flux tubes are packed together (Figure 3). An equilibrium solution could be found for this cross section and there would be a fixed pressure on any line of force bounding the flux tubes. But if this pattern is split up somewhere else along the field we are no longer at liberty to adjust the pressure which is determined by the fact that p is mapped along the lines of force. The pressure distribution of one pattern does not map properly to produce equilibrium in another pattern. This is the basic difficulty of changing the winding pattern. If we went into detail we would find that in every case neutral point nonequilibrium situations would develop (Parker, 1972, 1979 pp. 378-391).

3. Application to the Sun

If the granules on the surface of the sun shuffle about and change the winding pattern, at least part of the pattern will be burned out fairly soon by neutral point reconnection. However, the situation is even worse than that. Suppose we have some well behaved granules where all the varying parts of the winding pattern have already been burned out by neutral point reconnection. We are left with parallel cylinders twisted one way or the other, so that the field is essentially independent of z now and we can write down the equilibrium solutions. Since everything is independent of the z direction we can write $B_x = \partial A_z/\partial y$ and $B_y = -\partial A_z/\partial x$, which means that in the x-y plane $\underline{B} \cdot \underline{\nabla} A_z = 0$ or A_z = const. along the lines of force projected onto the x-y plane. When these expressions for B_x and B_y are put into equation (1) we find that B_z is a function of A, or $B_z = B_z(A)$. We also find that $p + B_z{}^2/8\pi = F(A)$ and hence is const. along a line of force, and that

$$\nabla^2 A + 4\pi F'(A) = 0 \tag{7}$$

where F is an arbitrary function of A and the prime denotes differentiation with respect to A. Because there are infinitely many functions F(A), there appears to be infinitely many equilibrium solutions.

That was the happy state of mind we enjoyed until some difficulties recently arose. The equilibrium we have established depends on packing together some well ordered flux tubes, wound by the circulating granules at their feet into constant, invariant, winding patterns. In general, these flux tubes would be of different strengths and diameters, but they are packed together as described by the equilibrium equations. If we had separate flux tubes with fluid between them they would have circular cross sections, because the field lines that wrap a flux tube are under tension and tend to compress it. Although this kind of equilibrium can be described it is not the situation that we are dealing with here. We are talking about close packed flux tubes.

We are interested in a configuration where the flux tubes are compressed together as shown in Figure 4. When flux tubes, which are elastic bodies, are pushed together they press harder in the middle of each face than at the vertices. The difficulty arises because the boundary between individual cells is a line of force, and therefore a line on which the pressure is constant. The fluid squeezes out from the boundary, which increases the field strength $B_\perp^2 = B_x^2 + B_y^2$ and maintains equilibrium. But suppose opposite fields are pressed together. The fluid squeezes out as before and the field on each side may increase, but the field has to pass through zero from one side to the other. This sets up neutral point reconnection. Thus, the lesson is clear; any equilibrium pattern must avoid opposite fields meeting anywhere. That is, any equilibrium pattern must avoid all contact between flux tubes with the same sense of twist.

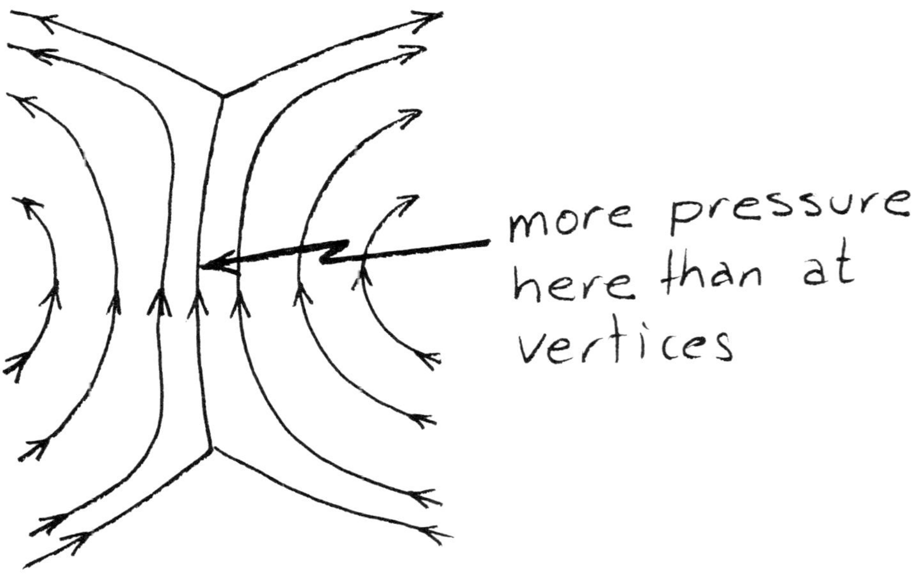

Figure 4. A sketch of the transverse field in two flux tubes squashed together at a common boundary causing higher contact pressure in the middle of the boundary.

If we examine equilibrium solutions from equation (7) we find that the sense of twist must have the property described above. To take an example, suppose that $4\pi F(A) = 1/2\, k^2A^2$; then we have $\nabla^2 A + k^2 A = 0$ which is just the wave equation. A solution on a square grid (Figure 5) is $A_z = \sin\lfloor (k/\sqrt{2})x\rfloor \sin\lfloor (k/\sqrt{2})y\rfloor$. $B_\perp$ is in the same direction at every boundary and there is no problem with equilibrium.

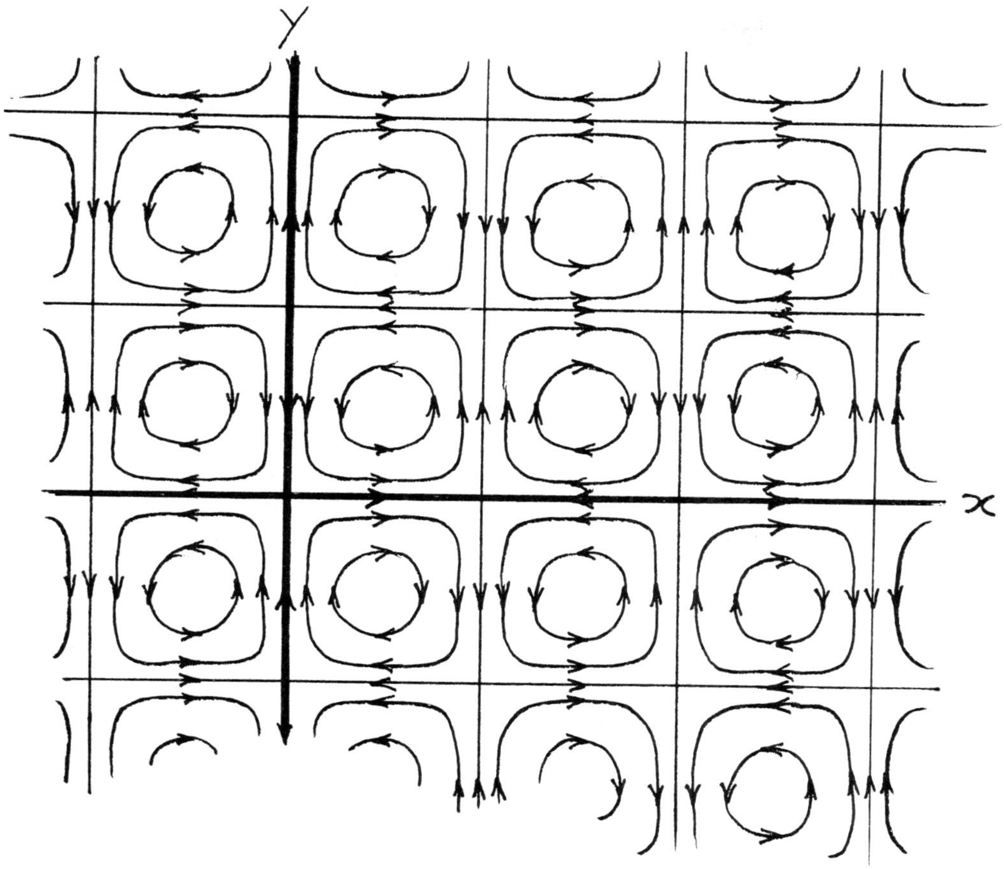

Figure 5. A sketch of the lines of force in the regular solution $A = \sin(kx/\sqrt{2}) \sin(ky/\sqrt{2})$.

Even if all the flux tubes have the same radius and internal structure the condition of oppositely circulating neighbors requires that we must avoid all 3 way vertices (Figure 6). Figure 5 has 4 way vertices which are tolerated for equilibrium, but in Figure 6 two of the circulations can be chosen to give equilibrium while the third cannot. This can be generalized to the statement that only even order vertices maintain equilibrium. The next step is to ask whether we can avoid odd vertices, and the answer is no, in general. A pair 3-way vertices could be pushed together to merge into a single 4-way vertex (Figure 7), but nature is seldom so obliging when flux tubes are packed together. One may object that the square configuration in Figure 5 is a solution. It certainly is a mathematical possibility, but although it is an equilibrium, it is also unstable. When elastic bodies are packed together the minimum energy state is hexagonal close packing, like a honeycomb. And since the preferred state has all 3-way vertices, the quest for equilibrium appears hopeless.

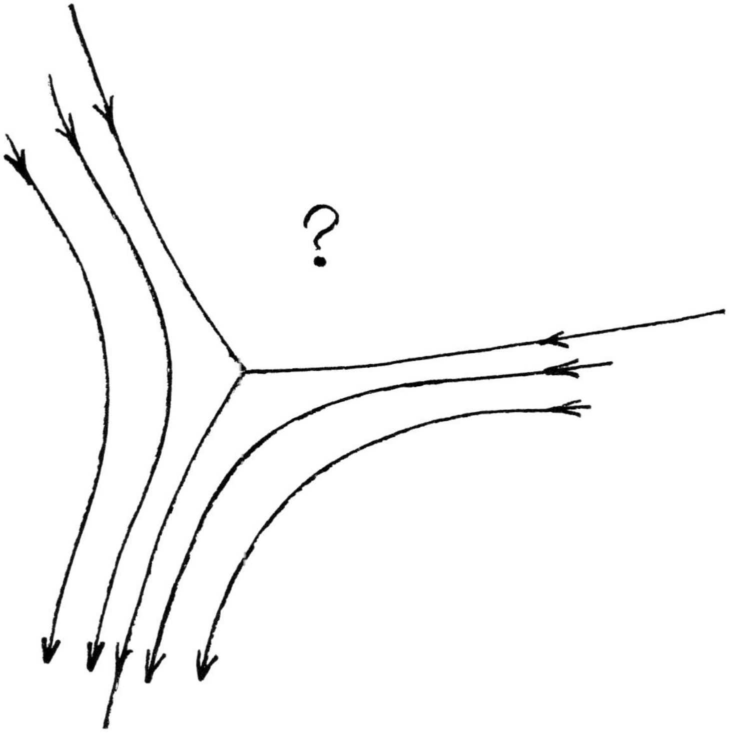

Figure 6. A sketch of the topology of the fields in the cells contiguous to a three-fold vertex. The field in the sector marked with a question mark is necessarily opposite to one of the two others.

The point of all this is that even if the winding patterns of the flux tubes are well behaved, the field is still not capable of equilibrium (Parker, 1982a,b). There are mathematical solutions but most have little or nothing to do with nature. Figure 8 depicts the development of a single cell from two adjacent cells with the same circulations once neutral point reconnection has begun. Plasma physicists call this phenomenon "coalescence of islands." The tension tends to open up the neck and pull the two cells together. Thus, the winding pattern evolves toward larger and larger cells until there are only two left with opposite twist, which can coexist, and the process ceases. For an infinitely broad region essentially all of the transverse energy is dissipated by the time the field is reduced to two oppositely twisted tubes.

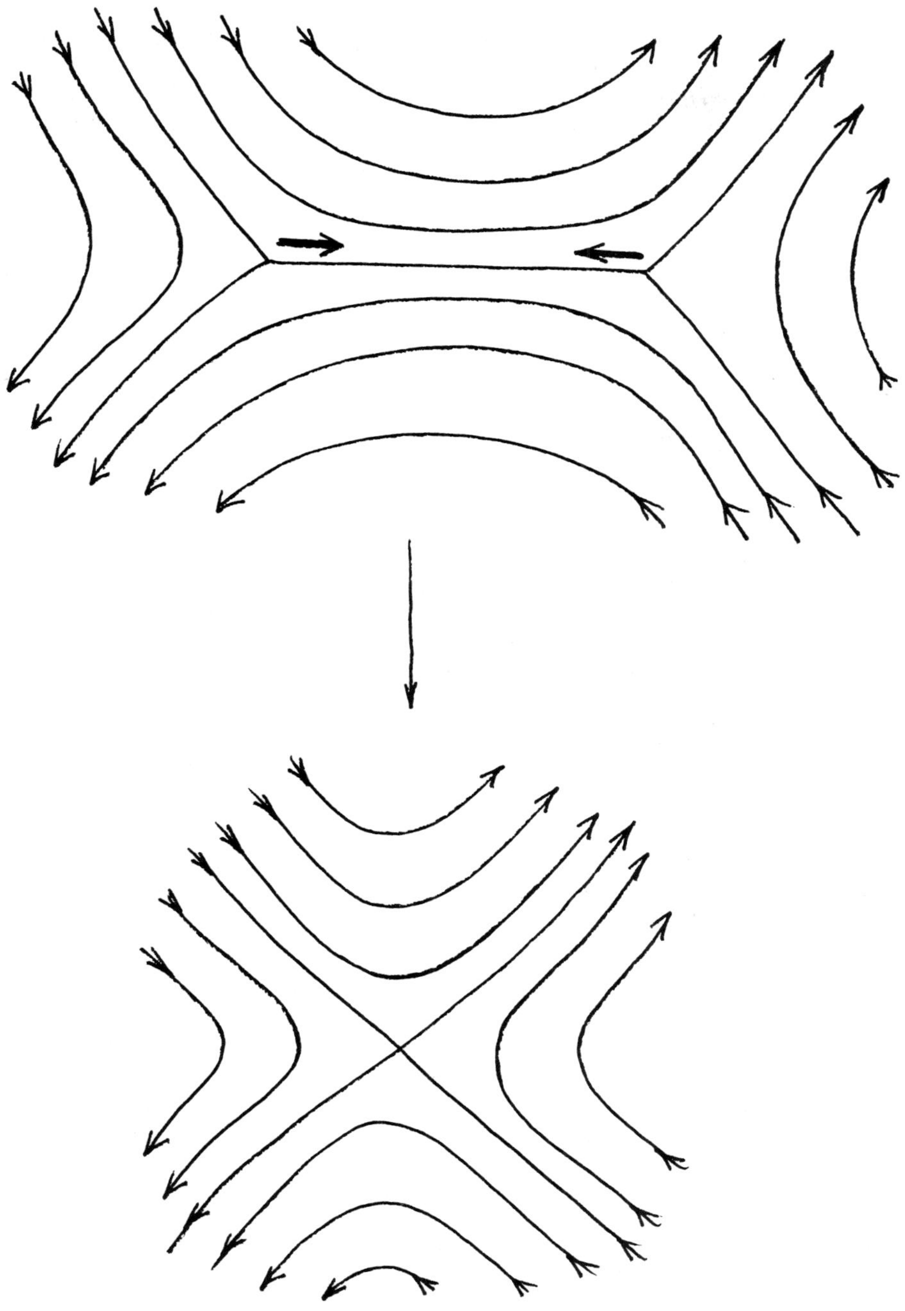

Figure 7. A sketch of the process of forcing two three-fold vertices together to make a single four-fold vertex.

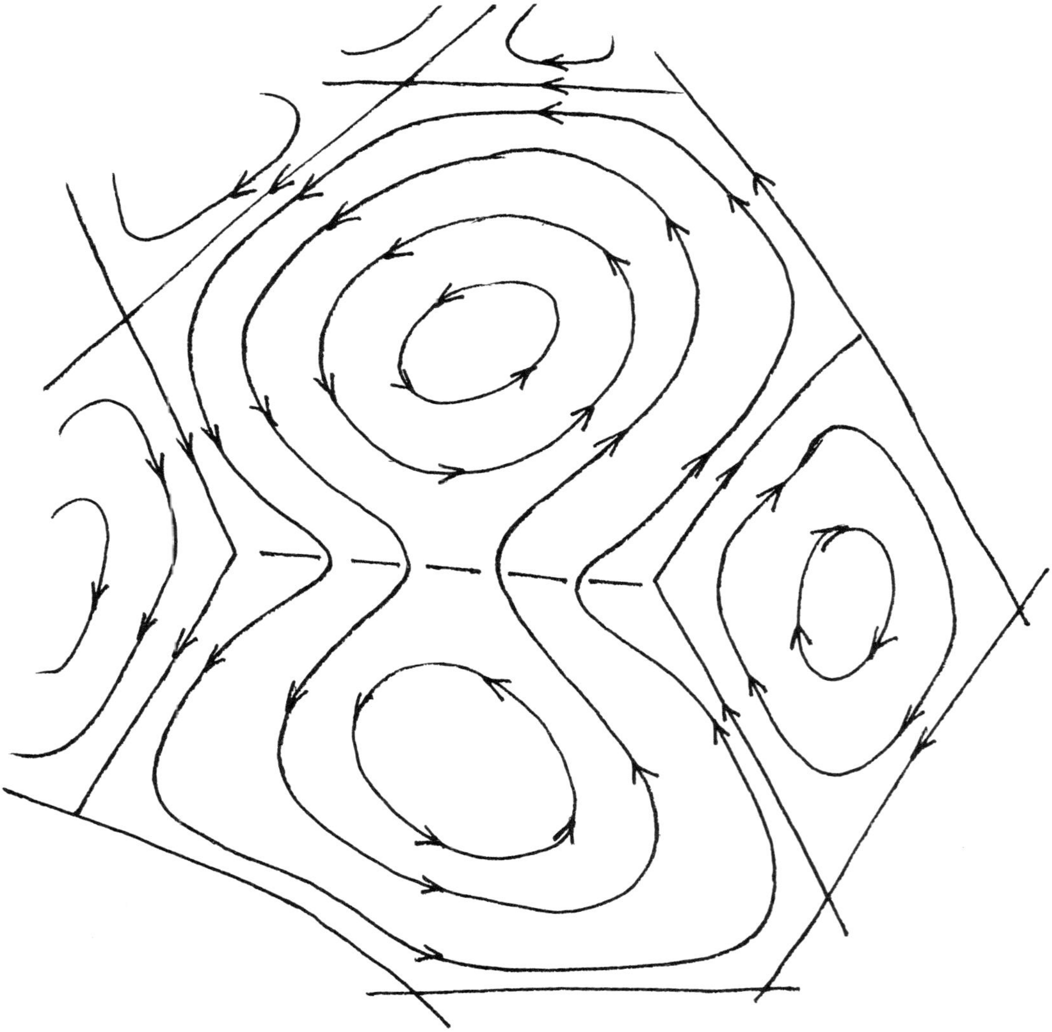

Figure 8. A sketch of the field topology resulting from reconnection of the transverse field between two flux tubes with the same sense of twist.

A flow does not aid equilibrium, because addition of a Reynold's stress, $-\rho v_i v_j$, to the Maxwell stress tensor acts as a compression along the lines of force. This only makes the situation worse. There seems to be no way of avoiding the reconnection process with flow solutions.

Formally, we can write down general solutions to equation (7) by parameterizing F(A). The equation is then solved by perturbation techniques for small values of the expansion parameter. Any specific random array of flux tubes can be fitted only with a special choice of the expansion coefficients. Since the equation for F(A) also gives the pressure it means that the pressure supplied by nature must be just right, which is not very likely. This is another way of showing, in general, the incompatability of equilibrium conditions with arbitrary flux tube cross sections. But the physics of the real situation demands 3-way vertices, which are forbidden in equilibrium.

Now we can apply the results of these discussions to the sun, our main topic. The situation on the sun is different because the flux tubes are curved, not straight, but that is not an essential part of the physics. The granules are constantly shuffling the feet of the field lines on a fairly small scale, roughly 1,000 km. This motion extends up into the winding pattern and can be thought as a very long wavelength Alfven wave. The magnetic field in the corona over an active region may be 100 G. The gas density would be about 10^9 cm^{-3} in the x-ray regions where the temperature is 2 x 10^6K. Under these conditions the Alfven speed turns out to be 500 -1,000 km/sec, indicative of a strong field, low β gas.

We can only guess to what degree the fields are twisted by the granular motion. The question is how much work does the fluid motion accomplish in twisting the fields. This is the basic calculation in the problem of coronal heating. The power input in ergs/cm^2-sec is roughly equal to the characteristic velocity of the foot points times the transverse Maxwell stress, $p \cong VB_z B_\perp/4\pi$. Some representative numbers are $v \cong 1/2$ km/sec = 5 x 10^4 cm/sec, $B_z \simeq 100$ G, and $B_\perp \simeq 20$ G, yielding a power of about 10^7 ergs/cm^2-sec. This value applies to active regions; over quiet regions it may be as low as 10^5 ergs/cm^2-sec. Incidentally, in coronal holes, to which this mechanism probably does not apply, the energy consumption is almost as high as it is in the active regions because of the large solar wind expansion of the coronal hole. Thus, it seems that the idea of heating the corona through work done by the granules is correct, based on simple energy considerations. The fields are wound up like clock springs, which then leads to a non-equilibrium situation if the winding is not uniform. There may be hydromagnetic waves, also; this discussion in no way excludes them. But whether it is waves or a quasi-static process that winds up the fields, the work gets done in some way.

It must be emphasized that it is not possible to calculate a unique merging rate. As we pointed out before, the nonequilibrium could conceivably go as slowly as $V_A/N^{1/2}$, N being the magnetic Reynold's number, or under some conditions as rapidly as $V_A/\ln N$. Just exactly where the true rate falls is a matter of detailed calculation which cannot be done now.

To obtain a transverse field of 20 G the twisting must accumulate for a period on the order of 10-15 hours. If the neutral point annihilation proceeds very rapidly, the perpendicular Alfvén speed leads to a cutting rate of about 50 km/sec. This rate slices across a 1000 km diameter flux tube in 20 sec, which is not enough time to build up much twisting. In essence, the springs cannot be wound up tightly and the amount of energy going into the corona is negligible. On the other hand, if we pick the very slow rate, the flux tubes can sit quietly side by side, and the twisting can accumulate to enormous levels, so that the energy input can reach 10^9 erg/km^2 -sec. Interestingly enough, the merging rate consistent with the observed power inputs proves to be the harmonic mean of the extreme rates. We do not attach any significance to that fact, but merely point out the coincidence.

Let us return now to some details of the merging process which we passed over when we were interested in establishing its existence. The energy derived from the magnetic field is converted mostly into kinetic energy of the expelled fluid. Remember, merging involves a slow inflow until the lines are cut and reconnected, after which an outgoing speed of the order of the Alfvén speed is produced by the field tension and extra pressure in the central region (Figure 9). It is interesting to follow the dissipation of the ejected high speed jet by calculating the Reynold's number. For an ionized hydrogen gas the viscosity is $\mu \simeq 1.2 \times 10^{16} T^{5/2}$ gm/cm-sec, and with $T = 2 \times 10^6$K, $\mu \simeq 1$ gm/cm-sec, which is fairly viscous. With an Alfvén speed of about 500 km/sec , a scale size $L \simeq 1000$ km $= 10^8$ cm, and a mass density $\rho \simeq 1.66 \times 10^{-15}$ gm/cm^3, the Reynold's number $LV_A\rho/\mu \simeq 10$, which is a moderately viscous flow.

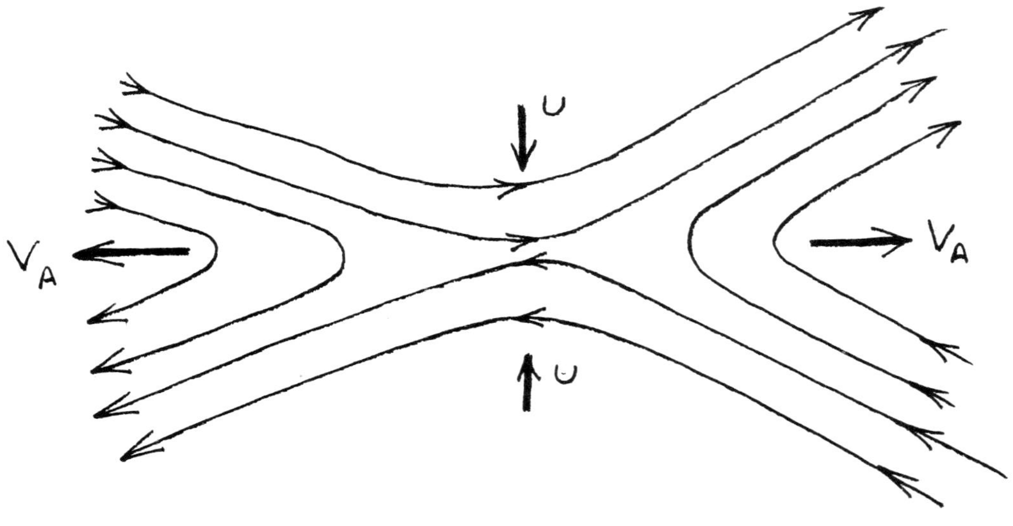

Figure 9. A sketch of the rapid neutral point reconnection between opposite field components, indicating the ejection of fluid from the region at the Alfvén speed V_A.

However, there is a longitudinal field in the direction of the flow which does not interfere with reconnection but enormously inhibits the viscosity. The ions, which are principally responsible for viscosity are forced to move in cyclotron orbits with a radius $R \simeq 20$ cm for a 2×10^6 K gas in a 100 G field. The mean free path λ tends to be of the order of 1 km or more, so there is a large inhibition which is proportional to $(\lambda/R)^2$. Hence the Reynolds number is not of order unity but more like $10^8 - 10^{10}$ for the transverse motion.

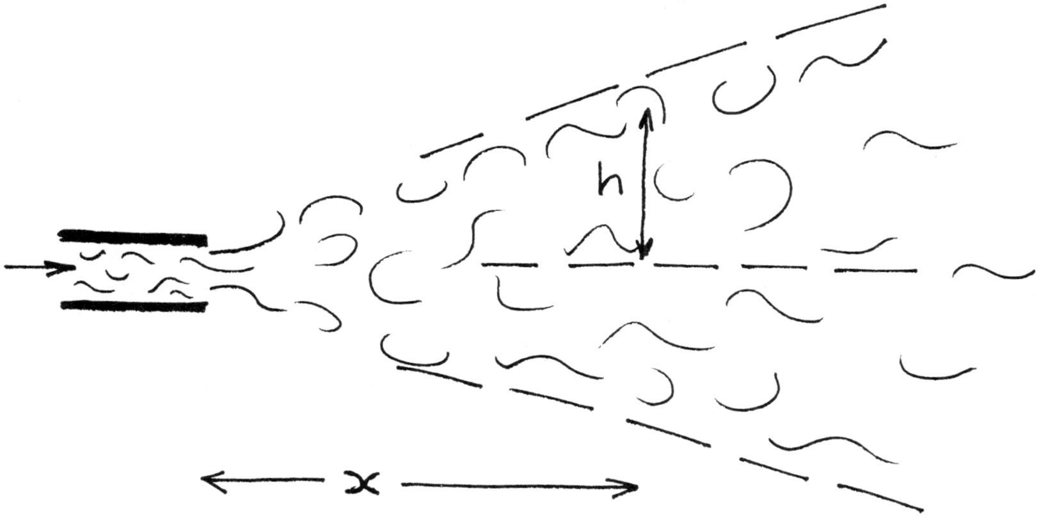

Figure 10. A sketch of a turbulent jet of fluid of half angle $\theta \simeq 20^o$ and half width $h \simeq x \tan \theta$.

Fortunately, flows with high Reynold's number have been studied at great lengths by aerodynamicists and they have some rather interesting universal properties. Consider a turbulent jet of fluid emerging from a pipe so that the stream does not start out broad (Figure 10). It is found that such flows generally diverge with half angle θ of about 20°. This involves a very rapid development of turbulence which entrains the surrounding fluid. Momentum flux is conserved $((vh)(\rho v) \sim \rho v^2(\theta x) = \text{constant})$, so that $V \ (1/x)^{1/2}$ downstream in the jet. The kinetic energy flux, however, is equal to $1/2\ \rho v^2(vh) \cong 1/2\ \rho\ \theta x/x^{3/2}$ $1/x^{1/2}$, indicating that the kinetic energy declines asymptotically to zero down the jet. The energy has degraded into eddies which survive only a fraction of a rotation and rapidly cascade to small scales.

The scale of the fluid jets in the corona is of the order of a few kilometers, according to rough estimates. Initial velocities can be 200 km/sec but they degrade very rapidly. An important question is whether these features have been observed, or indeed whether they are observable at all. Their detection seems unlikely because the material is extremely transparent and even a few hundred kilometers does not offer much optical depth. However, there is a recent paper which describes observations of small scale coronal motions with velocities of 50 km/sec. They are spectacular measurements of up and down motions, but because the typical scales are about 1,000 km they probably are not the phenomenon that we are discussing here. Possibly they are connected with the mysterious spicule features which were described briefly in the previous lecture.

4. Alternative Mechanisms

Now let us return to some auxiliary topics which were mentioned briefly before, double layers and anomalous resistivity. When electron conduction velocities reach the ion thermal velocity, plasma turbulence is excited, and even more so when the conduction velocity approaches the electron thermal velocity. The ion thermal velocity in the corona is typically 200 km/sec and the electron thermal velocity is approximately 10,000 km/sec. The current density required for a velocity at the ion thermal velocity is easily calculated. Substituting typical quantities at the base of the corona, we find that $j = Nev = (10^9\ cm^{-3})(4.8 \times 10^{-10}\ esu)(2 \times 10^7\ cm/sec) \simeq 10^7\ esu/cm^2$-sec. The scale length of the magnetic shear accompanying this current can be estimated from Ampere's Law, $j = c/4\pi\ \nabla \times B \simeq c/4\pi\ \Delta B/\ell$. For a magnetic field in this case we can use the 100G longitudinal field. We then get $\ell = c/4\pi\ \Delta B/j \simeq (3 \times 10^{10})(10^2)/(12)10^7 \simeq 1/4 \times 10^5$ cm, which means that the transition from 0 to 100 G must take place over a distance of 200 m, a very sharp shear. If we want very strong anomalous resistivity under such conditions where the electron cyclotron frequency exceeds the plasma frequency, the conduction velocity must increase by a factor of 50, producing a correspondingly sharper shear over about 4 m. This is an extremely concentrated current density, and long before this point is reached other effects such as resistive diffusion will come into play.

We raise this question because it is not clear under what circumstances we can or cannot invoke either anomalous resistivity or double layers. Unfortunately, some of the more enthusiastic proponents of these ideas refuse to make the calculations. Or, if they make the calculation they remark that the scale lengths are not unreasonable. But as we just noted, other effects can handle the dissipation before such extremes are produced. Anomalous resistivity and double layers may very well play a crucial role in heating the corona, but at this point it is not clear how to establish or reject them.

5. Conclusion

We conclude this series of lectures with some general remarks summarizing our present state of knowledge. Most other stars exhibit active phenomena like the sun. This was not known definitely until x-ra. and EUV observations were possible. Now we know that essentially all classes of stars have activity, and even those that do not are suspect This means that activity goes on in stars under quite different conditions. Temperatures vary and mass loss from some stars occurs much more easily than from the sun, which alters the situation enormously. This suggests that all stars have magnetic fields of some kind. There may be ways of producing activity without magnetic fields but if so they have not been worked out.

If stars have magnetic fields one is obliged to think about the origin of the fields. Some stars may have primordial fields. We did not dwell on this in the lecture on dynamos, but the so-called magnetic A stars, the first magnetic stars discovered by the Babcocks, are fairly young stars. They are massive and luminous. We use the term young in its ordinary sense, not in the strange way that astronomers use it. To them, a young star is in a particular class of stars which are short lived, but which may, in fact, be in the final stage of senility. That is not what young means to most people.

Type A stars may be in an early stage of their normal life span and therefore may be exhibiting magnetic fields captured when these stars were formed by collapsing gas clouds.

The reason for this comment is that some of the hotter stars and type A stars have strong magnetic fields, and since, by present day thinking, these stars do not have convective zones which can work as dynamos, their fields may be primordial. The field varies with time simply because the star is rotating and the field is not aligned with the rotation axis. There does not seem to be any obvious case where the field varies in intensity or changes sign for an observer sitting on the surface of the star.

Our final point is that many of the things which we presently find puzzling in the sun are probably examples of very general effects that occur throughout the universe. This makes it very important that we get the details straight before we extrapolate them to other objects. You should be left with the impression that, while we have many ideas about the sun, and we can do some hard physics, nevertheless the applications are difficult because we cannot see everything that we would like to see in the sun. We cannot see below the surface and we cannot resolve the structure that we discussed in the corona. We still have a long way to go and, as usual, nature will continue to surprise us and demolish our favorite ideas. This is one of the things that makes the subject fun. If matters were as well established as some of our colleagues think, then we should look for other fields of work.

References

Parker, E.N.: 1972, Astrophys. J., 174, 499.

Parker, E.N.: 1979, Cosmical Magnetic Fields (Oxford, Clarendon Press).

Parker, E.N.: 1983a, Geophys. Astrophysical Fluid Dynamics (in press).

Parker, E.N.: 1983b, Astrophys. J., 181, (in press).

Yu, G.: Astrophys. J., 181, 1003.

HYDROMAGNETIC WAVES, TURBULENCE, AND COLLISIONLESS PROCESSES IN THE INTERPLANETARY MEDIUM

Aaron Barnes
Theoretical and Planetary Studies Branch, NASA Ames Research Center, Moffett Field, California 94035

I. HYDROMAGNETIC WAVES IN THE INTERPLANETARY MEDIUM

Introduction

The solar wind does not flow quietly. It seethes and undulates, fluctuating on time scales that range from the solar rotation period down to fractions of milliseconds. Most of the power in interplanetary waves and turbulence lies at hydromagnetic scales. These fluctuations are normally of large amplitude, containing enough energy to affect solar and galactic cosmic rays, and may be the remnants of a coronal turbulence field powerful enough to play a major role in accelerating the solar wind itself. The origin and evolution of interplanetary hydromagnetic waves and turbulence, and their influence on the large-scale dynamics of the solar wind are among the most fundamental questions of solar-terrestrial physics. However, the motivation for studying interplanetary fluctuations goes beyond that. The solar wind is at present the best laboratory we have for studying hydromagnetic waves and turbulence, because the fluctuations are abundant and for the most part not complicated by nearby boundaries. This field of study thus provides an opportunity to advance the understanding of one of the most fundamental processes of astrophysical plasmas.

Fluctuations in a plasma are said to be of hydromagnetic scale if they satisfy two criteria: (1) their time scale (as measured in the local plasma rest frame) must be much longer than the local proton gyroperiod, and (2) their length scale must be much longer than the local mean proton and electron Larmor radii. In the solar wind at the orbit of Earth the proton gyroperiod is ~10 s and the mean proton gyroradius is ~100 km. At 1 astronomical unit (AU), hydromagnetic fluctuations propagate at ~40 km/s relative to the plasma, much slower than the typical ~400-km/s flow speed of the plasma itself. Thus a propagating structure 1 gyroradius thick is swept past a stationary observer (like a spacecraft) in a fraction of a second. In practice, all interplanetary fluctuations whose time scale is observed to be >1 s can be regarded as hydromagnetic fluctuations.

R. L. Carovillano and J. M. Forbes (eds.), Solar-Terrestrial Physics, 155–199.

On the other hand, the length and time scales for interplanetary hydromagnetic fluctuations are very much smaller than the corresponding Coulomb collision scales of the plasma ions and electrons. Therefore, strictly speaking, the theory of hydromagnetic fluctuations should be a collisionless theory. However, it is much simpler, and in certain respects rather satisfactory, to model the interplanetary variations as fluctuations in a magnetohydrodynamic fluid. This model will be used uncritically in the present lecture, but the reader should be warned that collisionless phenomena can be of deep and qualitative significance! Some of these effects will be discussed in a subsequent lecture.

It must also be pointed out that the interplanetary hydromagnetic fluctuations are sometimes described as waves, sometimes as turbulence. These terms are not mutually exclusive in all instances, but our experience with water and air suggests that often one or the other mode of description is much more useful. The truth is that we really do not know which language works best for the interplanetary medium; the problem is compounded by the fact that the theory of hydromagnetic turbulence is still in a rather primitive state. The best that can be done for now is to try out both the wave and turbulence descriptions, and see how well they work in various observed situations. This topic will be discussed further in my next lecture. The present lecture will be restricted to a discussion of the theory of magnetohydrodynamic waves.

Equations of Magnetohydrodynamics

Consider an inviscid ideal gas of infinite electrical conductivity, containing a magnetic field. The fluid is characterized by density ρ, pressure P, flow velocity $\underset{\sim}{v}$, heat flow vector $\underset{\sim}{q}$, magnetic field $\underset{\sim}{B}$, and electric field $\underset{\sim}{E}$. The dynamical equations are the continuity equation,

$$\partial\rho/\partial t + \underset{\sim}{\nabla} \cdot (\rho\underset{\sim}{v}) = 0 \; ; \tag{1a}$$

the momentum equation (neglecting gravity),

$$\rho\left(\frac{\partial}{\partial t} + \underset{\sim}{v} \cdot \underset{\sim}{\nabla}\right)\underset{\sim}{v} + \underset{\sim}{\nabla}P = \frac{1}{4\pi}\,(\mathrm{curl}\ \underset{\sim}{B}) \times \underset{\sim}{B}$$

$$= \frac{1}{4\pi}\,\underset{\sim}{B} \cdot \underset{\sim}{\nabla}\underset{\sim}{B} - \frac{1}{8\pi}\,\underset{\sim}{\nabla}B^2 \; ; \tag{1b}$$

the first law of thermodynamics for an ideal gas,

$$\frac{P}{\gamma - 1}\left(\frac{\partial}{\partial t} + \underset{\sim}{v} \cdot \underset{\sim}{\nabla}\right)\ln(P\rho^{-\gamma}) = -\underset{\sim}{\nabla} \cdot \underset{\sim}{q} \; ; \tag{1c}$$

the condition of infinite electrical conductivity,

$$\underset{\sim}{E} + \underset{\sim}{v} \times \underset{\sim}{B}/c = 0 \; ; \tag{1d}$$

Faraday's law

$$\frac{\partial \mathbf{B}}{\partial t} = -c \text{ curl } \mathbf{E} \; ; \tag{1e}$$

and, effectively as an initial condition on Eq. (1e), the solenoidal condition

$$\nabla \cdot \mathbf{B} = 0 \; . \tag{1f}$$

The electric field is eliminated between Eqs. (1d) and (1e) to give

$$\frac{\partial \mathbf{B}}{\partial t} = \text{curl}(\mathbf{v} \times \mathbf{B}) = -\mathbf{v} \cdot \nabla\mathbf{B} - \mathbf{B}\, \nabla\cdot\mathbf{v} + \mathbf{B} \cdot \nabla\mathbf{v} \tag{1e'}$$

If we take the scalar products of $\mathbf{v}$ with Eq. (1b) and $\mathbf{B}$ with (1e'), the resulting equations and (1a) and (1c) can be manipulated to give the energy conservation equation,

$$\frac{\partial \mathscr{E}}{\partial t} + \nabla \cdot \boldsymbol{\mathscr{F}} = 0 \; ,$$

where

$$\mathscr{E} = (1/2)\rho v^2 + P/(\gamma - 1) + B^2/8\pi$$

and

$$\boldsymbol{\mathscr{F}} = \mathbf{v}[(1/2)\rho v^2 + \gamma P/(\gamma - 1)] + \mathbf{q} + [\mathbf{v}B^2 - \mathbf{B}\mathbf{B} \cdot \mathbf{v}]/4\pi \; . \tag{1c'}$$

The last two terms in $\boldsymbol{\mathscr{F}}$ are the Poynting vector.

An Exact Solution — Alfvénic Fluctuations

Alfvén (1942) showed that incompressible magnetohydrodynamics admits an exact nonlinear propagating solution. This solution is readily generalized to compressible media and, in fact, holds for quite general magnetic field geometry (Goldstein et al., 1974), even for randomly tangled field lines. Although the plasma is compressible under arbitrary fluctuations, this particular solution yields constant density (and pressure and magnetic field strength). A particularly simple derivation of this solution, due to L. Davis, Jr. (private communication), starts with the assumption that there is some frame of reference S in which the plasma flows steadily, and that the electric field vanishes so that $\mathbf{v}$ and $\mathbf{B}$ are parallel. In addition, the flow is nondissipative, so that $\mathbf{q} = 0$. Thus Eqs. (1c) and (1e) are automatically satisfied. Because $\tilde{\rho}$ is constant, we must have

$$\nabla \cdot \mathbf{v}_S = \nabla \cdot \mathbf{B} = 0 \; , \tag{2a}$$

where $\underset{\sim}{v}_S$ is the velocity as measured in S. Then the momentum equation (1b) with constant $|\underset{\sim}{B}|$ requires that

$$\underset{\sim}{v}_S \cdot \underset{\sim}{\nabla}\underset{\sim}{v}_S - \underset{\sim}{B} \cdot \underset{\sim}{\nabla}\underset{\sim}{B}/(4\pi\rho) = 0 \ . \tag{2b}$$

Equations (2a)-(2b) are satisfied if $\underset{\sim}{B}$ is solenoidal and

$$\underset{\sim}{v}_S = \mp \underset{\sim}{B}/(4\pi\rho)^{1/2} \ . \tag{2c}$$

Note in particular that the (Eulerian) mean flow velocity is

$$\langle \underset{\sim}{v}_S \rangle = \mp \langle \underset{\sim}{B} \rangle/(4\pi\rho)^{1/2} \ , \tag{2c'}$$

so that the flow speed is nonzero in S. It follows immediately that for an arbitrary frame of reference

$$\underset{\sim}{v} = \langle \underset{\sim}{v} \rangle \mp (\underset{\sim}{B} - \langle \underset{\sim}{B} \rangle)/(4\pi\rho)^{1/2} \ , \tag{3}$$

and $\underset{\sim}{v}$ and $\underset{\sim}{B}$ are functions of $\underset{\sim}{x} - [\langle \underset{\sim}{v} \rangle \pm \langle \underset{\sim}{B} \rangle/(4\pi\rho)^{1/2}]t$; $\underset{\sim}{B}$ can be any solenoidal field that satisfies $\underset{\sim}{B}^2$ = constant. Note that this last constraint makes the Alfvénic solution truly nonlinear; in particular, the linear superposition of two Alfvénic solutions is generally not a solution (Alfvénic or otherwise) of the magnetohydrodynamic equations.

It is instructive to consider the special case of the plane Alfvén wave propagating in the direction, say, of unit vector $\underset{\sim}{n}$. In this case the fluctuating part of the magnetic field is normal to $\underset{\sim}{n}$, that is,

$$\underset{\sim}{n} \cdot (\underset{\sim}{B} - \langle \underset{\sim}{B} \rangle) = 0 \ , \tag{4}$$

so that $\underset{\sim}{n}$ can be described as the direction of zero magnetic variance. The magnetic field now depends on the single independent variable

$$\underset{\sim}{n} \cdot \underset{\sim}{x} - \underset{\sim}{n} \cdot [\langle \underset{\sim}{v} \rangle \pm \langle \underset{\sim}{B} \rangle/(4\pi\rho)^{1/2}]t \ , \tag{4a}$$

so that the upper sign of Eq. (3) corresponds to phase propagation more nearly parallel than antiparallel to the mean magnetic field, and the lower sign corresponds to phase propagation in the opposite sense. Note that this correlation between the sign in Eq. (3) and propagation sense persists even in the most general nonplanar solution. This fact is a powerful diagnostic of observed Alfvénic fluctuations, because it provides information about the propagation direction even when (as is normally the case) observations from only one spacecraft are available.

It might be thought that the Alfvénic solution described above is very special, perhaps requiring boundary conditions not likely to be found in natural plasmas. It is a remarkable fact that this solution turns out to be a very good approximate representation of an important class of fluctuations that is commonly found in the solar wind; an example is shown in Fig. I.1. Alfvénic fluctuations are identifiable, perhaps dominant, in ~50% of the observed interplanetary microstructure;

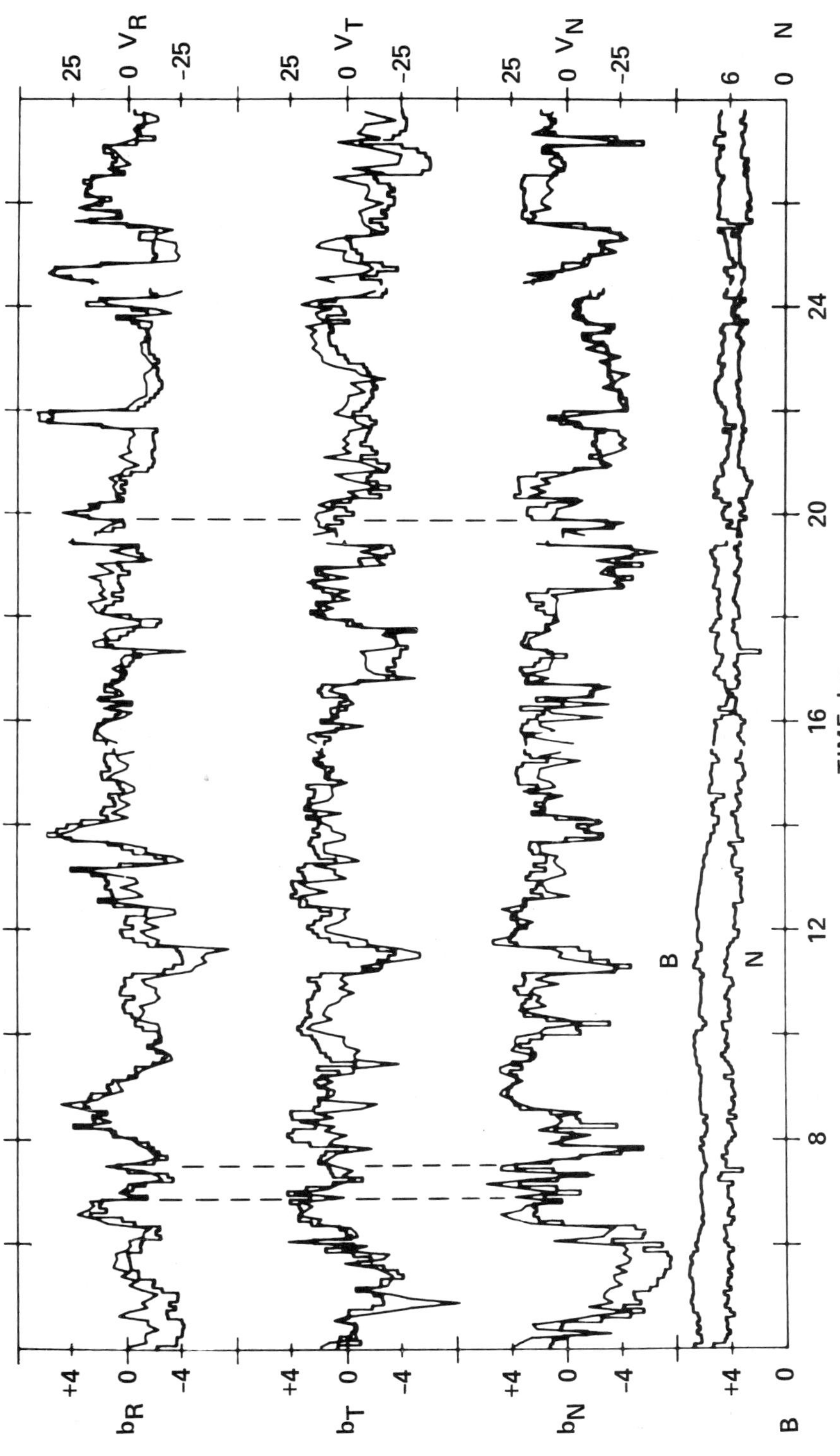

Figure I.1. Twenty-four hours of magnetic field and plasma data, showing the presence of nearly pure Alfvén waves (taken from Belcher and Davis, 1971). The upper six curves are 5-min averages of the fluctuations in components of $\underset{\sim}{v}$ and $\underset{\sim}{B}$; the lower two curves are magnetic field strength and proton number density.

the purest examples of Alfvénic fluctuations are found in high-velocity solar wind streams and in their trailing edges (Belcher and Davis, 1971). It is found that the cross-correlation of velocity and magnetic fluctuations in Alfvénic fluctuations is essentially always of the sign [cf. Eq. (3)] corresponding to propagation away from the Sun. This observation, coupled with the fact that the solar-wind flow speed is much larger than the local Alfvén speed beyond a few tens of solar radii from the Sun, provides a strong argument for the solar origin of the interplanetary Alfvénic component (Belcher and Davis, 1971). If, on the contrary, the waves had been generated in the region of super-Alfvénic flow, then either they must have been produced by some mechanism that strongly favors outward propagation, or else inward propagating waves were selectively dissipated, or scattered into the forward direction. These possibilities cannot be completely rejected, but no convincing mechanism has yet been put forward.

The energetics of Alfvénic fluctuations can also be expressed in exact terms. Referring to Eq. (1c'), we see that for pure Alfvénic fluctuations the total (Eulerian) average energy density is

$$\langle \mathscr{E} \rangle = \mathscr{E}_B + \mathscr{E}_A , \tag{5}$$

where the background contribution is

$$\mathscr{E}_B = (1/2)\rho\langle \underset{\sim}{v}\rangle^2 + P/[\gamma - 1] + \langle \underset{\sim}{B}\rangle^2/8\pi , \tag{5a}$$

and the Alfvénic contribution is

$$\mathscr{E}_A = \langle(\delta \underset{\sim}{B})^2\rangle/4\pi \tag{5b}$$

($\delta \underset{\sim}{B} \equiv \underset{\sim}{B} - \langle \underset{\sim}{B}\rangle$). The calculation of the energy flux is somewhat more complicated. What is probably the simplest approach begins with the observation that [Eq. (3)]

$$\underset{\sim}{\alpha} \equiv \underset{\sim}{v} \pm \underset{\sim}{B}/(4\pi\rho)^{1/2} = \langle \underset{\sim}{v}\rangle \pm \langle \underset{\sim}{B}\rangle/(4\pi\rho)^{1/2} = \text{constant} .$$

Note also that $\langle B^2\rangle = \langle \underset{\sim}{B}\rangle^2 + \langle(\delta \underset{\sim}{B})^2\rangle$. Expressing $\underset{\sim}{v}$ in terms of $\underset{\sim}{\alpha}$, $\underset{\sim}{B}$, and ρ, we have

$$\begin{aligned}(1/\rho)\underset{\sim}{\mathscr{F}}' &\equiv (1/\rho)\underset{\sim}{\mathscr{F}} - \gamma/(\gamma - 1)\underset{\sim}{v}P/\rho \\ &= \underset{\sim}{\alpha}[(1/2)\alpha^2 \mp \underset{\sim}{\alpha} \cdot \underset{\sim}{B}/(4\pi\rho)^{1/2} + 3B^2/8\pi\rho] \\ &\quad \mp (1/2)(\alpha^2 + B^2/4\pi\rho)\underset{\sim}{B}/(4\pi\rho)^{1/2} .\end{aligned}$$

Then, because $\underset{\sim}{\alpha}$ and $\langle B^2\rangle$ are constant,

$$\begin{aligned}(1/\rho)\langle \underset{\sim}{\mathscr{F}}'\rangle &= \underset{\sim}{\alpha}[(1/2)\alpha^2 \mp \underset{\sim}{\alpha} \cdot \langle \underset{\sim}{B}\rangle/(4\pi\rho)^{1/2} + 3(\langle \underset{\sim}{B}\rangle^2 + \langle(\delta \underset{\sim}{B})^2\rangle)/8\pi\rho] \\ &\quad \mp (1/2)\{\alpha^2 + [\langle \underset{\sim}{B}\rangle^2 + \langle(\delta \underset{\sim}{B})^2\rangle]/4\pi\rho\}\langle \underset{\sim}{B}\rangle/(4\pi\rho)^{1/2} .\end{aligned}$$

If we now express $\underset{\sim}{\alpha}$ in terms of $\langle \underset{\sim}{v} \rangle$ and $\langle \underset{\sim}{B} \rangle$ we obtain

$$\langle \underset{\sim}{\mathscr{F}} \rangle = \underset{\sim}{\mathscr{F}}_B + \underset{\sim}{\mathscr{F}}_A \, , \tag{6}$$

where

$$\underset{\sim}{\mathscr{F}}_B = \langle \underset{\sim}{v} \rangle \left\{ \frac{1}{2} \rho \langle \underset{\sim}{v} \rangle^2 + \frac{\gamma}{\gamma - 1} P + \frac{\langle \underset{\sim}{B} \rangle^2}{4\pi} \right\} - \frac{\langle \underset{\sim}{B} \rangle \langle \underset{\sim}{B} \rangle \cdot \langle \underset{\sim}{v} \rangle}{4\pi} \tag{6a}$$

is the background energy flux and

$$\underset{\sim}{\mathscr{F}}_A = \left\{ \frac{3}{2} \langle \underset{\sim}{v} \rangle \pm \langle \underset{\sim}{B} \rangle / (4\pi\rho)^{1/2} \right\} \langle (\delta \underset{\sim}{B})^2 \rangle / 4\pi \tag{6b}$$

is the energy flux due to Alfvénic fluctuations. In the (average) plasma rest frame $\underset{\sim}{\mathscr{F}}_A$ is parallel (or antiparallel) to $\langle \underset{\sim}{B} \rangle$. It should be noted explicitly that in a general frame of reference $\underset{\sim}{\mathscr{F}}_A \neq [\langle \underset{\sim}{v} \rangle \pm \langle \underset{\sim}{B} \rangle / (4\pi\rho)^{1/2}] \mathscr{E}_A$. The reader is also reminded that Eqs. (5b) and (6b) are valid for large amplitudes and general magnetic field geometry, and in particular are not restricted to plane waves.

This entire discussion of Alfvénic fluctuations is based on fluid magnetohydrodynamics, and therefore is somewhat limited as a description of interplanetary Alfvénic fluctuations. For example, all the equations in this section must be modified for pressure anisotropy. Moreover, real interplanetary fluctuations are normally not purely Alfvénic; for example, small but nonzero fluctuations in the magnitude of $\underset{\sim}{B}$ are typically present (Burlaga and Turner, 1976). Nevertheless, the magnetohydrodynamic theory has provided a useful if approximate description of the interplanetary Alfvénic fluctuations, and its relative simplicity facilitates the study of physical processes related to these fluctuations.

Classification of Wave Modes — Simple Waves

The preceding theory of Alfvénic variations is simple, but very general. No such general theory is available for other kinds of magnetohydrodynamic fluctuations. However, a complete classification of one-dimensional MHD plane waves exists. The reader is probably already acquainted with the simplest version of this classification scheme, the small-amplitude (linearized) theory of waves in a uniform background plasma. However, linearization is not necessary, and in fact the fully nonlinear theory of MHD simple waves both expands and clarifies the classification. This conceptual advantage, as well as the fact that fluctuations observed in the solar wind are generally of large amplitude, recommends the simple-wave theory as a starting point for the study of plane waves.

The method of characteristics provides a powerful and elegant tool for the analysis of one-dimensional waves (and two-dimensional flows) in gasdynamics and magnetohydrodynamics (Courant and Friedrichs, 1948;

Jeffrey and Taniuti, 1964). However, our present goal is limited to finding the MHD simple waves, and may be attained without the full generality of the characteristics theory (Barnes and Hollweg, 1974). Let (ξ,η,ζ) be the axes of a Cartesian coordinate system, and suppose a plane simple wave is propagating in the ζ direction. Then all dependent variables must be functions of a single phase $\psi(\zeta,t)$. Then, for example, $P = P[\psi(\zeta,t)]$, and

$$\partial P/\partial t = \psi_t P' \equiv P'\partial\psi/\partial t$$

$$\partial P/\partial\zeta = \psi_\zeta P' \equiv P'\partial\psi/\partial\zeta ,$$

where $P' = dP/d\zeta$. The solenoidal condition (1f) gives $B_\zeta' = 0$, so that

$$B_\zeta \equiv B_{\zeta o} = \text{const} . \tag{7}$$

[Note that $B_{\zeta o}$ is the component of magnetic field in the propagation direction and is not in general the average (or "background") magnetic field]. Then define

$$\begin{aligned} \underset{\sim}{B}_\perp &\equiv \underset{\sim}{B} - \underset{\sim}{B}_{\zeta o} \\ \underset{\sim}{v}_\perp &\equiv \underset{\sim}{v} - v_\zeta \underset{\sim}{B}_{\zeta o}/|\underset{\sim}{B}_{\zeta o}| . \end{aligned} \tag{8}$$

For $\underset{\sim}{q} = 0$, Eqs. (1a)-(1c) and (1e') can be written as

$$(\psi_t + v_\zeta\psi_\zeta)\rho' + \psi_\zeta\rho v_\zeta' = 0 , \tag{9a}$$

$$(\psi_t + v_\zeta\psi_\zeta)\rho\underset{\sim}{v}_\perp' - \psi_\zeta B_{\zeta o}\underset{\sim}{B}_\perp'/4\pi = 0 , \tag{9b}$$

$$(\psi_t + v_\zeta\psi_\zeta)\rho v_\zeta' + \psi_\zeta[P' + (\underset{\sim}{B}^2)'/8\pi] = 0 , \tag{9c}$$

$$(\psi_t + v_\zeta\psi_\zeta)\underset{\sim}{B}_\perp' + \underset{\sim}{B}_\perp\psi_\zeta v_\zeta' - B_{\zeta o}\psi_\zeta\underset{\sim}{v}_\perp' = 0 , \tag{9d}$$

$$(\psi_t + v_\zeta\psi_\zeta)(P\rho^{-\gamma})' = 0 \tag{9e}$$

Note that the ζ component of Eq. (1e') is an identity because of (7).

Our task is to find all self-consistent solutions of Eqs. (9). First, consider nonpropagating structures ($\psi_t + v_\zeta\psi_\zeta = 0$). Equation (9e) is automatically satisfied, and (9a) immediately gives $v_\zeta = v_{\zeta o} = \text{const}$. It is easily shown that the remaining equations admit two solutions. The first, the tangential pressure balance, is characterized by $B_{\zeta o} = 0$ and conservation of total pressure $P + B^2/8\pi$. The second solution is the entropy "wave," characterized by constant $\underset{\sim}{B}$, $\underset{\sim}{v}$, and P, but with variable density (and temperature) admitted. These solutions are summarized in Table 1.

For propagating waves $\psi_t + v_\zeta\psi_\zeta \neq 0$, so that $P\rho^{-\gamma} = \text{const}$. It is clear that one solution must be the plane Alfvén wave, as may readily be checked; the self-consistency condition of ψ turns out to be

$\psi_t + [v_{\zeta o} \pm B_{\zeta o}/(4\pi\rho)^{1/2}]\psi_\zeta = 0$, whose solution is (except for notation) given by Eq. (4a). The other solutions turn out to be the magnetoacoustic waves, which are compressive and characterized by

$$\psi_t + (v_\zeta \pm \mathcal{V}_f)\psi_\zeta = 0$$

or

$$\psi_t + (v_\zeta \pm \mathcal{V}_s)\psi_\zeta = 0 \ , \tag{10}$$

where

$$2\mathcal{V}^2_{(f,s)} = \frac{B^2}{4\pi\rho} + \frac{\gamma P}{\rho} \pm \left[\left(\frac{B^2}{4\pi\rho} + \frac{\gamma P}{\rho}\right)^2 - \frac{\gamma P B^2_{\zeta o}}{\pi\rho^2}\right]^{1/2} . \tag{10a}$$

Details of the calculation are given by Barnes and Hollweg (1974). The properties of all these modes are summarized in Table 1.

Table 1. Summary of simple waves

Name	Description	Phase speed	Nonlinear steepening?
Tangential pressure balance	$P + B^2/8\pi$, $v_{\zeta o}$ constant, $B_{\zeta o} = 0$, $\underset{\sim}{B}_\perp, \underset{\sim}{v}_\perp$ variable	0	No
Entropy "wave"	$P, \underset{\sim}{v}, \underset{\sim}{B}$ constant, ρ variable	0	No
Alfvén wave	$\rho, \lvert\underset{\sim}{B}\rvert, P$ constant	$\pm B_{\zeta o}/(4\pi\rho)^{1/2}$	No
Magnetoacoustic waves	Linear polarization	$\pm\mathcal{V}_f$, $\pm\mathcal{V}_s$	Yes

For magnetoacoustic waves, the correlations between the fluctuations in $\underset{\sim}{B}$, $\underset{\sim}{v}$, etc., may be obtained from Eqs. (9) and (10); $\psi(\zeta,t)$ = const. on the characteristic lines in the $\zeta - t$ plane:

$$\frac{d\zeta}{dt} = v_\zeta(\psi) \pm \mathcal{V}_{(f,s)}(\psi) \ . \tag{11}$$

Because ψ is constant, the velocity $v_\zeta \pm \mathcal{V}_{(f,s)}$ is constant along characteristics and (11) is the equation of a straight line. If all quantities are specified as functions of ζ at $t = 0$, then the characteristics $\zeta(t)$ may be labeled by their values $\lambda = \zeta(0)$ taken at $t = 0$. Then the solution of (11) is

$$\zeta(\lambda,t) = \lambda + [v_\zeta(\lambda) \pm \mathcal{V}_{(f,s)}(\lambda)]t \ . \tag{12}$$

The spatial derivative of any dependent variable (ρ_ζ, for example) is given by an equation of the form

$$\partial\rho/\partial\zeta = (d\rho/d\lambda)(\partial\lambda/\partial\zeta) = (d\rho/d\lambda)/[1 + t\, d(v_\zeta \pm \mathcal{V}_{(f,s)})/d\lambda] \ . \qquad (13)$$

If $d(v \pm \mathcal{V})/d\lambda$ is negative, the waveform will have infinite slope after the time

$$T(\lambda) = -\left[\frac{\partial}{\partial\lambda}(v_z \pm \mathcal{V}_{(f,s)})\right]^{-1} \ . \qquad (14)$$

For given initial conditions, this time will be a minimum T_S for some value of λ; thus, a magnetoacoustic wave steepens into a shock that begins to form at T_S. This steepening is analogous to that of a large-amplitude sound wave, occurring because regions of high compression propagate faster than those of lower compression and eventually overtake them.

Steepening and rarefaction do not occur for the Alfvén wave or for the nonpropagating structures. Nevertheless, it is possible to form thin Alfvén waves, entropy waves, and tangential pressure balances; in the limit of zero thickness (from the point of view of MHD) these structures are called, respectively, rotational discontinuities, contact discontinuities, and tangential discontinuities.

In the limit of small-amplitude, the propagation speed of all modes becomes independent of amplitude, and the results of the linearized wave theory are recovered. The waves are nondispersive (i.e., their phase speeds are independent of frequency), but the phase speeds depend on the direction of propagation relative to the background magnetic field (illustrated in Fig. I.2). According to the full nonlinear theory, the magnetic fluctuations in magnetoacoustic waves are linearly polarized, whereas the Alfvén mode is characterized by $|\underset{\sim}{B}|$ = const. Thus, in a plane Alfvén wave the tip of the magnetic vector moves along a segment of a circle (this is true circular polarization only if the vector moves around the entire circle), and therefore can never be linearly polarized. It is thus something of a paradox that the polarization of Alfvén waves is linear in the small-amplitude limit. This arises because the requirement that B^2 be constant means that $\underset{\sim}{B} \cdot \delta\underset{\sim}{B}$ is of the order $(\delta\underset{\sim}{B})^2$, and therefore vanishes in the linear limit.

Geometrical Hydromagnetics

The preceding discussion of hydromagnetic waves has been restricted to plasmas that are spatially uniform, at least in an average sense. It is frequently important to consider the variation of hydromagnetic waves in a nonuniform medium. The analysis is relatively simple for the case of small-amplitude waves that propagate and are convected in a plasma whose background properties are steady in time, with spatial gradients whose length scales are much longer than a typical wavelength. In this case, the method of calculation is the approximation of geometrical

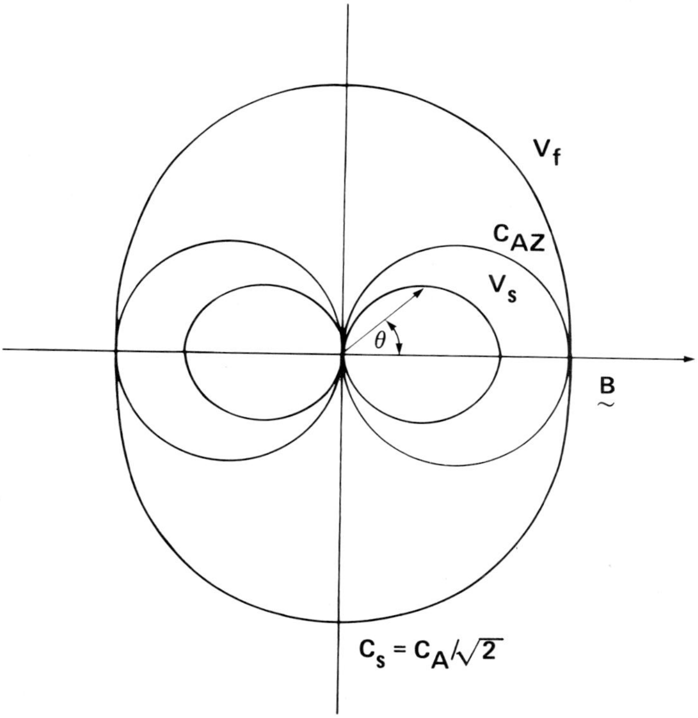

Figure I.2. Polar plot of phase-speeds of small-amplitude, fast, Alfvén, and slow waves for $C_A^2 = B^2/4\pi\rho > C_S^2 = \gamma P/\rho$.

hydromagnetics, sometimes called the eikonal method, or the WKB method (Weinberg, 1962; Bazer and Hurley, 1963; Jeffrey and Taniuti, 1964). Although the analysis is conceptually simple, the detailed computations are cumbersome. Therefore, we outline the method schematically, giving detailed results only for the case of Alfvén waves.

By way of introduction, consider again the linearized theory of waves in a uniform background. Separate background and fluctuation by writing $\underset{\sim}{B} = \underset{\sim}{B}_o + \delta\underset{\sim}{B}$, etc., and assume that the fluctuations are plane waves,

$$\delta\underset{\sim}{B} \propto \exp i(\underset{\sim}{k} \cdot \underset{\sim}{x} - \omega t) \ .$$

For isentropic fluctuations

$$\delta P = \mathscr{C}_o^2 \delta\rho \ ,$$

where $\mathscr{C}_o^2 = \gamma P_o/\rho_o$ is the square of the background sound speed. If we use this relationship to eliminate δP from Eq. (1b), then multiply Eq. (1a) by $\mathscr{C}_o^2/\rho_o$, Eq. (1b) by 1, and Eq. (1e') by $1/4\pi$, the linearized form of the resulting equations is

$$iA(\underset{\sim}{k},\omega)\delta R = 0 \ , \tag{15}$$

where δR is the seven-component column vector whose transpose is

$$(\delta R)^T = (\delta\rho,\ \delta v_x,\ \delta v_y,\ \delta v_z,\ \delta B_x,\ \delta B_y,\ \delta B_z) \equiv (\delta\rho,\ \delta\underset{\sim}{v},\ \delta\underset{\sim}{B}) \ , \tag{16}$$

and $A(\underset{\sim}{k},\omega)$ is a 7×7 matrix that depends on the frequency ω, wave vector $\underset{\sim}{k}$, and the background parameters. The explicit expression for A is not required for the present discussion, and is left as an exercise to the reader. One important fact, which will be useful later, is that A is real and symmetric [this was the reason for the choice of the multiplicative factors of Eqs. (1a), (1b), and (1e')]. Therefore, the transpose of (15) is just

$$i(\delta R)^T A = 0 \ . \tag{17}$$

Equation (15) is self-consistent if and only if $\det A = 0$; this condition then gives a functional relation between ω and $\underset{\sim}{k}$, which defines the wave modes of the plasma. These modes are, of course, just the linearized versions of the propagating modes summarized in Table 1.

Now suppose that the background plasma is nonuniform (but temporally steady), and that the gradients of the background quantities have a length scale much longer than $1/k$. The eikonal approximation is based on the assumption that the fluctuating variables may be expanded in the form

$$\delta R = [R_1(\underset{\sim}{x}) + R_2(\underset{\sim}{x}) + \ .\ .\ .\ ]\exp i[S(\underset{\sim}{x}) - \omega t] \ . \tag{18}$$

The rapid spatial variations are contained in the phase $S(x)$, and the spatial variations of R_1, R_2, and $\underset{\sim}{k}(\underset{\sim}{x}) = \underset{\sim}{\nabla} S$ are much slower, on the same scale as the background gradients. In addition, we assume that $R_2/R_1 \sim O(1/kL)$, where L is the length scale of the background gradients. Then the dynamical equations (1) may be linearized and systematically expanded as a perturbation series in powers of $1/kL$. For example, the first two equations in the expansion of the continuity equation are

$$i(\underset{\sim}{k} \cdot v_0 - \omega)\rho_1 + i\rho_0 \underset{\sim}{k} \cdot \underset{\sim}{v}_1 = 0$$

and

$$i(\underset{\sim}{k} \cdot \underset{\sim}{v}_0 - \omega)\rho_2 + i\rho_0 \underset{\sim}{k} \cdot \underset{\sim}{v}_2 = -\underset{\sim}{\nabla} \cdot (\rho_0 \underset{\sim}{v}_1 + \rho_1 \underset{\sim}{v}_0) .$$

Carrying out the linearization in analogy to the derivation of Eq. (15) (including the multiplicative factors), the eikonal perturbation scheme yields, in first and second order,

$$iA(\underset{\sim}{\nabla} S, \omega, \underset{\sim}{x})R_1 = 0 \tag{19a}$$

and

$$iA(\underset{\sim}{\nabla} S, \omega, \underset{\sim}{x})R_2 = L(R_1) , \tag{19b}$$

where A is the same symmetric matrix as in (15), with k replaced by $\underset{\sim}{\nabla} S$ and an additional slow spatial dependence due to the variation of the background. The column vector $L(R_1)$ is linear in R_1 and its derivatives. The components of L are

$$L_1 = -\frac{\mathscr{C}_0^2}{\rho_0} \underset{\sim}{\nabla} \cdot (\rho_0 \underset{\sim}{v}_1 + \rho_1 \underset{\sim}{v}_0) ,$$

$$\begin{aligned} L_{2,3,4} = {} & -(\rho_0 \underset{\sim}{v}_1 + \rho_1 \underset{\sim}{v}_0) \cdot \underset{\sim}{\nabla} \underset{\sim}{v}_0 - \rho_0 \underset{\sim}{v}_0 \cdot \underset{\sim}{\nabla} \underset{\sim}{v}_1 - \underset{\sim}{\nabla}(\mathscr{C}_0^2 \rho_1) \\ & - \frac{1}{4\pi} \underset{\sim}{\nabla}(\underset{\sim}{B}_0 \cdot \underset{\sim}{B}_1) + \frac{1}{4\pi} \underset{\sim}{B}_1 \cdot \underset{\sim}{\nabla} \underset{\sim}{B}_0 + \frac{1}{4\pi} \underset{\sim}{B}_0 \cdot \underset{\sim}{\nabla} \underset{\sim}{B}_1 , \end{aligned} \tag{20}$$

and

$$\begin{aligned} L_{5,6,7} = {} & -\frac{1}{4\pi} \{\underset{\sim}{v}_1 \cdot \underset{\sim}{\nabla} \underset{\sim}{B}_0 + \underset{\sim}{v}_0 \cdot \underset{\sim}{\nabla} \underset{\sim}{B}_1 + \underset{\sim}{B}_0 \underset{\sim}{\nabla} \cdot \underset{\sim}{v}_1 + \underset{\sim}{B}_1 \underset{\sim}{\nabla} \cdot \underset{\sim}{v}_0 \\ & - \underset{\sim}{B}_0 \cdot \underset{\sim}{\nabla} \underset{\sim}{v}_1 - \underset{\sim}{B}_1 \cdot \underset{\sim}{\nabla} \underset{\sim}{v}_0\} . \end{aligned}$$

First of all, note that Eq. (19a) is identical to (15). This means that the relation between $\underset{\sim}{k}$ and ω that characterized the linear modes in a homogeneous background is preserved in the eikonal theory; in addition, (19a) guarantees that the cross-correlations of the fluctuations are exactly as they are for a homogeneous background. For Alfvén waves, for example,

$$\omega = \{\mathbf{v}_0 \pm \mathbf{B}_0/(4\pi\rho_0)^{1/2}\} \cdot \mathbf{k}(\mathbf{x}) \tag{21a}$$

and

$$\mathbf{v}_1 = \mp\mathbf{B}_1/(4\pi\rho_0)^{1/2} , \tag{21b}$$

where $\mathbf{B}_0$, etc., are evaluated locally at $\mathbf{x}$. Equation (21) and its counterparts for magnetoacoustic waves are of the form

$$\omega = \omega(\nabla S, \mathbf{x}) = \text{const.} \tag{22}$$

that arises in Hamilton-Jacobi theory. Hence we may formally treat $\mathbf{k} = \nabla S$ as the momentum of a quantum whose Hamiltonian is $\omega(\mathbf{k},\mathbf{x})$. The Hamiltonian equations of motions are thus

$$\frac{d\mathbf{x}}{dt} = \frac{\partial\omega}{\partial\mathbf{k}} \equiv \mathbf{v}_g \tag{23a}$$

and

$$\frac{d\mathbf{k}}{dt} = -\frac{\partial\omega}{\partial\mathbf{x}} . \tag{23b}$$

In the language of optics, Eqs. (23) are the ray equations (parameterized by t). The operator d/dt is closely related to the spatial derivative along a ray:

$$\frac{d}{dt} = \mathbf{v}_g \cdot \nabla . \tag{24}$$

If the background geometry is given, Eq. (23) can be integrated (at least in principle) to give the ray along which any wave must propagate. The Alfvén rays are given by

$$\begin{aligned} d\mathbf{x}/dt &= \mathbf{v}_g = \mathbf{v}_0 \pm \mathbf{B}_0/(4\pi\rho_0)^{1/2} , \\ d\mathbf{k}/dt &= -k_j\nabla[v_{0j} \pm B_{0j}/(4\pi\rho_0)^{1/2}] \end{aligned} \tag{25}$$

(repeated indices imply summation).

A case of special interest in solar-wind studies occurs for $|\mathbf{v}_0|$ much larger than the propagation speed relative to the plasma. In this case, $\omega \approx \mathbf{k} \cdot \mathbf{v}_0$. For radial flow at constant speed $\mathbf{v}_0 = v_0\mathbf{e}_r$, and Eqs. (23) give

$$\begin{aligned} d\mathbf{x}/dt &\approx v_0\mathbf{e}_r \\ d\mathbf{k}/dt &= -(v_0/r)(\mathbf{k} - k_r\mathbf{e}_r) . \end{aligned} \tag{26}$$

The rays are radial, k_r is constant, and we readily show that

$$\frac{d}{dt}(\underset{\sim}{k} - k_r \underset{\sim}{e}_r)^2 \cong -\frac{2v_0}{r}(\underset{\sim}{k} - k_r \underset{\sim}{e}_r)^2 ,$$

which can be integrated immediately to give

$$k^2 - k_r^2 = \text{const}/r^2 . \tag{27}$$

This result has important implications for steady radial flow models of the solar wind. Waves of near-solar origin, whatever their initial propagation direction, will have their wave normals oriented radially from the Sun by the time they have propagated into the distant wind. For this reason, an initial superposition of various propagation directions would eventually evolve into a spherical wave, which locally would look like a plane wave. The relation of this prediction to observation will be discussed in the next lecture.

The second approximation of geometrical hydromagnetics, Eq. (19b), can be used to evaluate the evolution of the wave amplitude along the rays. Because A is symmetric, the transpose of (19a) gives

$$R_1^T A = 0 ,$$

so that multiplication of (19b) on the left by R_1^T yields

$$R_1^T L(R_1) = 0 , \tag{28}$$

a bilinear equation in R_1 and its derivatives. Equation (28) gives the evolution of the amplitude along the rays.

For the case of Alfvén waves, Eq. (28) can be cast into a simple form. In the linear limit $\underset{\sim}{B}_0 \cdot \underset{\sim}{B}_1 = \rho_1 = 0$, and $\underset{\sim}{v}_1$ and $\underset{\sim}{B}_1$ are related by (21b). Thus Eq. (20) is simplified, and Eq. (28) becomes, after straightforward manipulation,

$$(d/dt)\, ln(B_1^2/\rho_0^{1/2}) = -\underset{\sim}{\nabla} \cdot \underset{\sim}{v}_0 \tag{29}$$

Hence, in a static medium $|\underset{\sim}{B}_1|$ varies $\propto \rho_0^{1/4}$ along every ray.

The above discussion of geometrical hydromagnetics requires linearization of the dynamical equations. However, it can be shown that the fundamental equations (25) and (29) for the Alfvén mode remain valid in the nonlinear regime, with $\underset{\sim}{B}_0 \to \langle \underset{\sim}{B} \rangle$, etc. (Barnes and Hollweg, 1974).

Alfvén Wave Pressure

Consider a plasma in which the background electric field vanishes, so that $\underset{\sim}{v}_0$ and $\underset{\sim}{B}_0$ are parallel (this is a reasonable idealization of the solar wind as viewed from a frame of reference that corotates with the Sun). The velocity streamlines, magnetic field lines, and Alfvén rays coincide [see Eq. (25)]. It follows from conservation of mass and magnetic flux that $\rho_0 v_0/B_0$ is conserved along each ray. The Alfvénic Mach number M_A is defined by the relation $M_A^2 = 4\pi\rho_0 v_0^2/B_0^2$; clearly $M_A\rho_0^{1/2}$ is conserved along each ray. Also,

$$\underset{\sim}{\nabla} \cdot \underset{\sim}{v}_0 = -\underset{\sim}{v}_0 \cdot \underset{\sim}{\nabla} \ln\rho_0 = 2\underset{\sim}{v}_0 \cdot \underset{\sim}{\nabla} \ln M_A ,$$

so that Eq. (29) can be written as

$$\frac{d}{dt} \ln(B_1^2 M_A) = -2\underset{\sim}{v}_0 \cdot \underset{\sim}{\nabla} \ln M_A , \tag{30}$$

so that on each ray

$$(M_A \pm 1) d\ln(B_1^2 M_A) = -2M_A \, d\ln M_A , \tag{31}$$

and

$$B_1^2 = \text{const.}\ M_A^{-1}(M_A \pm 1)^{-2} . \tag{32}$$

In the limit $M_A^2 \ll 1$, $B_1^2 \propto 1/M_A \propto \rho_0^{1/2}$. For $M_A \gg 1$, as would be expected far from the Sun, $B_1^2 \propto 1/M_A^3 \propto \rho_0^{3/2}$; since asymptotically $\rho_0 \propto r^{-2}$, $B_1^2 \propto r^{-3}$. Thus, the Alfvén wave energy is not conserved, for $\mathscr{F}_A$ does not vary as r^{-2} [cf. Eq. (6b)]. Evidently there is an exchange of energy between waves and the background medium, that is, the waves perform work by exerting a force on the medium.

To analyze this process in greater generality, take the divergence of (6b), and apply Eq. (29) and the continuity equation $\underset{\sim}{\nabla} \cdot (\rho_0 \underset{\sim}{v}_0) = 0$. Straightforward calculation leads to

$$\underset{\sim}{\nabla} \cdot \underset{\sim}{\mathscr{F}}_A = \underset{\sim}{v}_0 \cdot \underset{\sim}{\nabla}(B_1^2/8\pi) . \tag{33}$$

Because the flow is steady, $\underset{\sim}{\nabla} \cdot (\underset{\sim}{\mathscr{F}}_B + \underset{\sim}{\mathscr{F}}_A) = 0$; thus the work done on the background by the waves is given by

$$\underset{\sim}{\nabla} \cdot \underset{\sim}{\mathscr{F}}_B = \underset{\sim}{f}_A \cdot \underset{\sim}{v}_0 ,$$

where

$$\underset{\sim}{f}_A = -\underset{\sim}{\nabla}(B_1^2/8\pi) . \tag{34}$$

This force can be derived directly. Because $\rho_1 = P_1 = 0$ for Alfvén waves, the contribution of the waves to the momentum equation, taken to second order in wave amplitude and averaged over phase, is just

$$\rho_0 \underset{\sim}{v}_1 \cdot \nabla \underset{\sim}{v}_1 - \frac{1}{4\pi} \underset{\sim}{B}_1 \cdot \nabla \underset{\sim}{B}_1 + \frac{1}{8\pi} \underset{\sim}{\nabla} B_1^2 = \frac{1}{8\pi} \underset{\sim}{\nabla} B_1^2 ,$$

the first two terms canceling by Eq. (21b). Thus, Eq. (34) indicates that Alfvén waves exert a sort of radiation pressure on the plasma in which they propagate. This wave pressure may be an important, perhaps dominant, process in the acceleration of the solar wind (Alazraki and Couturier, 1971; Belcher, 1971; Hollweg, 1973).

How large is this wave pressure force in the solar wind? First, consider the region of sub-Alfvénic flow ($M << 1$). Then

$$-\nabla(B_1^2/8\pi) \sim -(B_1^2/16\pi\rho)\nabla\rho \ .$$

Compare this to the ordinary gas pressure force $-\nabla P$, which in order of magnitude is $-\mathscr{C}^2\nabla\rho$, where $\mathscr{C}^2$ is the speed of sound. The ratio of the two forces is of the order

$$\frac{\nabla P}{\nabla(B_1^2/8\pi)} \sim \frac{\mathscr{E}_B}{\mathscr{E}_A} \ .$$

However, in the sub-Alfvénic region $M_A << 1$, so that even if $\mathscr{F}_A$ is as large as $\mathscr{F}_B$, the wave pressure force is $O(M_A)$ in comparison with the gas pressure force. A similar calculation for super-Alfvénic flow shows that the ratio of the two forces is typically of the order $\mathscr{F}_A/\mathscr{F}_B$. Thus, Alfvén waves could appreciably accelerate the solar wind if they carry a significant fraction of the total energy of the wind, but this acceleration would be most effective in regions where the flow speed is greater than or comparable to the Alfvén speed.

The Alfvénic fluctuations observed near the orbit of Earth typically have an amplitude of a few times 10^{-5} gauss; according to Eq. (6b), this Alfvénic flux would carry less than 1% of the energy of the wind. Extrapolating back to the Sun by Eq. (31), we find an Alfvénic flux of less than 10% of the wind's energy flux. Thus, the geometrical theory indicates that the observed interplanetary Alfvénic fluctuations imply a near-solar flux that is too small to be a major acceleration mechanism for the solar wind. On the other hand, it has been suggested (Hollweg, 1973) that in fact Alfvén waves should undergo nonlinear dissipation, and, if so, the near-solar Alfvén flux could be much larger. The observational evidence to decide this issue is not yet at hand.

Analogous pressures are associated with the magnetoacoustic modes. However, the calculations are more complicated; in addition, a conceptual quandary arises. For a steady flow, for example, the continuity equation is

$$\nabla \cdot (\langle \rho \rangle \langle \mathbf{v} \rangle) = -\nabla \cdot \langle \delta\rho\delta\mathbf{v} \rangle ,$$

where, as usual, the brackets denote Eulerian averages. In the Alfvénic case, there are no density fluctuations, so that $\langle \rho \rangle$ and $\langle \mathbf{v} \rangle$ are well-defined "background" quantities. In the magnetoacoustic case, however, the waves transport mass (from the Eulerian viewpoint), and the Eulerian definition of the background quantities is not completely satisfactory. A powerful and elegant way to circumvent this problem involves the use of Lagrangian averaging, which yields a compact theory of wave energy and the associated wave pressure (Dewar, 1970). This approach has been used in solar wind models (Jacques, 1977); a detailed discussion of this advanced topic is beyond the scope of these lectures, however.

References

Alazraki, G. and Couturier, P.: 1971, Astron. Astrophys., 13, p. 380.
Alfvén, H.: 1942, Ark. för Mat., Astron., o. Fys., 29B, p. 1.
Barnes, A. and Hollweg, J. V.: 1974, J. Geophys. Res., 79, p. 2302.
Bazer, J. and Hurley, J.: 1963, J. Geophys. Res., 68, p. 147.
Belcher, J. W.: 1971, Astrophys. J., 168, p. 509.
Belcher, J. W. and Davis, L., Jr.: 1971, J. Geophys. Res., 76, p. 3534.
Burlaga, L. F. and Turner, J. B.: 1976, J. Geophys. Res., 81, p. 73.
Courant, R. and Friedrichs, K. O.: 1948, Supersonic Flow and Shock Waves, Interscience, New York.
Dewar, R. L.: 1970, Phys. Fluids, 13, p. 2710.
Goldstein, M. L., Klimas, A. J., and Barish, F. D.: 1974, "On the Theory of Large-Amplitude Alfvén Waves" in Solar Wind Three (ed. by C. T. Russell), UCLA Press, p. 385.
Hollweg, J. V.: 1973, Astrophys. J., 181, p. 547.
Jacques, S. A.: 1977, Astrophys. J., 215, p. 942.
Jeffrey, A. and Taniuti, T.: 1964, Nonlinear Wave Propagation, Academic, New York.
Weinberg, S.: 1962, Phys. Rev., 126, p. 1899.

II. HYDROMAGNETIC TURBULENCE IN THE INTERPLANETARY MEDIUM

Introduction

Imagine a mountain stream swollen with water from the spring snowmelt. It flows quietly, through swiftly, through a valley; here the channel is deep and fairly level. In a higher, steeper place the channel is strewn with boulders, and the flow is violent and chaotic. This portion of the stream seems always to be changing. A physicist hiking by is struck with the beauty of the scene and muses on the complicated patterns made by the water. He then thinks about the problem of trying to explain these patterns and feels glad that he is not there on business. Those motions are just too complicated, and no pattern reproduces itself in an obvious way. Yet, as time passes, he realizes that there are constant and relatively simple aspects to the flow. On the time scale of his visit the water stays in the stream bed, with the same average depth at a given point. If a twig is thrown into the water, it is carried away by the current, tracing out a tortuous pattern. If another twig is thrown in at the same point, it describes a pattern that differs in detail from the first. Still, on the average, the paths of the first and second twigs are very similar. Dropping a third twig into the water near the shore, just upstream of a boulder that diverts the current, he finds to his surprise that the twig heads upstream at first, then moves out from shore, starts downstream, then turns in again toward shore. In fact, it is trapped in a vortex, and repeats its cyclic pattern for a long time. The path of the stick is not smooth, and does not repeat itself exactly, but the cyclic pattern persists nonetheless; this vortex is a stable and relatively permanent feature of the flow. In the middle of the stream several submerged boulders lie hidden; the frothing water marks their presence by surging up and breaking. Our physicist then notices that these surges in the flow, though variable, on the average are fixed in the same place. He has discovered another relatively permanent feature, standing waves. He further notices that the waves break on the upstream side; this, he decides, happens because the waves are propagating upstream relative to the water, and he is observing breaking due to nonlinear steepening. He begins to realize that even extremely chaotic, turbulent flows can be at least partly described in relatively simple terms, though of course that description can only be a statistical one.

In a system as complicated as the one just described, there is no hope of predicting the motion of any one fluid element, no matter how precisely initial conditions are known. The best that can be hoped for is to find a means of describing the flow statistically, treating the dynamical variables as random quantities. Flow with this level of complexity is described as "turbulent."

Likewise, patterns of, say, magnetic field in the interplanetary medium almost never repeat themselves in precise detail. From this point of view it is sensible to describe the interplanetary medium as turbulent. On the other hand, interplanetary fluctuations have also been described as "waves." What is the difference, if any? We often think of waves as propagating sinusoidal oscillations. However, small-amplitude sinusoidal waves can be superposed in an arbitrary way, resulting in an infinite variety of wave forms. If the source of a wave field fluctuates in a random way, the propagating waves will exhibit the same randomness. Should we define such a stochastic wave field as a form of turbulence? Some plasma physicists do so, and in fact there is a whole subfield called "weak turbulence theory" that studies the interaction of small-amplitude waves with themselves and with the plasma that contains them.

On the other hand, the phenomena in water and air that we commonly think of as turbulent are not as simple as a superposition of waves. In fact, an essential feature of the theory of turbulence for ordinary fluids is the nonlinear character of the turbulent flow. Magnetohydrodynamics is the generalization of fluid mechanics to electrically conducting fluids. It is to be anticipated that under some conditions plasmas will exhibit turbulent behavior; this hydromagnetic turbulence will in some sense be analogous to the turbulence of ordinary fluids, though doubtless more complicated. It is reasonable to suppose that the large-amplitude stochastic interplanetary fluctuations are, at least in part, a manifestation of hydromagnetic turbulence. The question of waves versus turbulence in the interplanetary flow will be discussed further. First, however, it is necessary to introduce the reader to some concepts of the theory of fluid turbulence.

Some Comments on the Theory of Fluid Turbulence

The theory of turbulence is still far from a complete state, despite decades of effort by many of the best physicists. Nevertheless, considerable progress has been made, and useful descriptions of some turbulent phenomena have been developed. A general overview of the theory of turbulence is far beyond the scope of this lecture; the discussion will be aimed at introducing the reader to the language and to some basic concepts.

The first point to be stressed is that sound waves have little to do with fluid turbulence (this point should be obvious, because the most familiar examples of turbulence occur in incompressible fluids). In a sense, the fundamental element of fluid turbulence is the "eddy," a fluid element whose vorticity curl $\underset{\sim}{v}$ does not vanish. The distinction between sound wave and eddy is clearest for fluid variations whose gradients are restricted to one spatial direction (although we would not describe such a simple situation as turbulent). To make the point even simpler, consider a uniform ideal gas with small-amplitude fluctuations; that is, we write the density as $\rho = \rho_0 + \delta\rho$, the pressure as $P = P_0 + \delta P$, and the flow velocity as $\underset{\sim}{v} = \underset{\sim}{v}_0 + \delta\underset{\sim}{v}$. The background quantities P_0, etc., are treated as uniform, and the perturbations are assumed to vary sinusoidally as $\exp i(\underset{\sim}{k} \cdot \underset{\sim}{x} - \omega t)$. If the flow is adiabatic, the continuity and momentum equations and the first law of thermodynamics give

$$(\underset{\sim}{k} \cdot \underset{\sim}{v}_0 - \omega)\delta\rho + \rho_0\underset{\sim}{k} \cdot \delta\underset{\sim}{v} = 0 \ , \tag{1a}$$

$$(\underset{\sim}{k} \cdot \underset{\sim}{v}_0 - \omega)\rho_0\delta\underset{\sim}{v} + \underset{\sim}{k}\delta P = 0 \ , \tag{1b}$$

$$(\underset{\sim}{k} \cdot \underset{\sim}{v}_0 - \omega)\delta(P\rho^{-\gamma}) = 0 \ . \tag{1c}$$

The self-consistent solutions of these equations are sound waves and entropy waves, whose properties are summarized in Table 1. The sound waves propagate relative to the fluid; entropy waves do not. The sound waves are compressive; the entropy waves are incompressible. The sound waves are irrotational; the entropy waves have nonzero vorticity. The entropy wave, unlike the sound wave, is an example of a (very simple) "turbulent eddy."

Table 1. One-dimensional fluctuations

	Sound wave	Entropy "wave"
Dispersion relation	$\omega = \underset{\sim}{k} \cdot \underset{\sim}{v}_0 \pm k(\gamma P_0/\rho_0)^{1/2}$	$\omega = \underset{\sim}{k} \cdot \underset{\sim}{v}_0$
Compressible?	Yes, $\underset{\sim}{k} \cdot \delta\underset{\sim}{v} = \pm k(\gamma P_0/\rho_0)^{1/2}\delta\rho/\rho_0$	No, $\underset{\sim}{k} \cdot \delta\underset{\sim}{v} = 0$
Vorticity	$\underset{\sim}{k} \times \delta\underset{\sim}{v} = 0$	$\underset{\sim}{k} \times \delta\underset{\sim}{v} \neq 0$
Other properties	$\delta P = \gamma(P_0/\rho_0)\delta\rho$	$\delta\rho \neq 0,\ \delta P = 0$

We note that the entropy wave is incompressible in the sense that the density of any given fluid element is preserved, even though the density is not uniform throughout the flow. Mathematically, the condition is

$$\left(\frac{\partial}{\partial t} + \underset{\sim}{v} \cdot \underset{\sim}{\nabla}\right)\rho = -\rho\underset{\sim}{\nabla} \cdot \underset{\sim}{v} = 0 . \tag{2}$$

In fact, the flow of ordinary gases is essentially incompressible under many common conditions. This may be verified by considering the momentum equation for inviscid flow.

$$\rho\left(\frac{\partial}{\partial t} + \underset{\sim}{v} \cdot \underset{\sim}{\nabla}\right)\underset{\sim}{v} + \underset{\sim}{\nabla}P = 0 . \tag{3}$$

In general the density is not strictly constant and, in fact, pressure gradients are associated with density gradients; in order of magnitude,

$$\delta P \sim \mathscr{C}^2\delta\rho ,$$

where $\mathscr{C}$ is the speed of sound. Let the flow in the region under consideration be characterized by gradients of length scale L or temporal variations of time scale T or both. Then by the momentum equation, $\mathscr{C}^2\delta\rho/L\rho$ is comparable to the larger of $\delta v/\tau$ or $v\delta v/L$. By the continuity equation $\text{div}\ \underset{\sim}{v} \sim$ the larger of $\delta\rho/\rho\tau$ or $v\delta\rho/\rho L$. Clearly $|\underset{\sim}{\nabla} \cdot \underset{\sim}{v}| \ll \delta v/L$ if both $|\underset{\sim}{v}| \ll \mathscr{C}$ and $L \ll \mathscr{C}\tau$. Therefore, to a good approximation, a fluid flow is incompressible if both (1) the fluid velocity is everywhere much smaller than the speed of sound and (2) the time scale for significant evolution of the flow is long in comparison with the time required for a sound wave to cross the region under consideration. In particular, these conditions are satisfied for the turbulent flow of gases in many situations of experimental or practical interest, so that the theory of incompressible turbulence can be relevant both to gases and liquids. The assumption of incompressibility simplifies the mathematics considerably, and is made in most treatments of turbulence theory.

The motion of an individual fluid element in turbulent flow is too complicated to analyze, and even if it could be analyzed the result would be so complex as to be useless, at least to human minds. Therefore, turbulence theory is statistical in nature. It is usual to think of a given turbulent flow as a realization of the flow in a large ensemble of systems whose average properties are identical. Denote the emsemble average of any variable A by $\langle A \rangle$. A useful statistical concept is the cross-correlation function of two dynamical variables A and B,

$$R_{AB}(\underset{\sim}{x},\underset{\sim}{x}',t) = \langle [A(\underset{\sim}{x},t) - \langle A \rangle][B(\underset{\sim}{x}',t) - \langle B \rangle] \rangle . \tag{4}$$

If the turbulence is spatially homogeneous (in the average sense) then the correlation function depends only on $\underset{\sim}{x} - \underset{\sim}{x}'$ and t. The correlation function may be expressed in terms of its spatial Fourier transform $S_{AB}(\underset{\sim}{k},t)$,

$$R_{AB}(\underset{\sim}{x} - \underset{\sim}{x}',t) = R_{BA}(\underset{\sim}{x}' - \underset{\sim}{x},t) = \int d^3k S_{AB}(\underset{\sim}{k},t) \exp i\underset{\sim}{k} \cdot (x - \underset{\sim}{x}') . \tag{5}$$

The S_{AB} term is called the cross-correlation spectrum. In the special case that $A = B$, R_{AA} is the autocorrelation function of A and S_{AA} is the power spectrum.

An especially important correlation function is the energy correlation tensor (we choose a frame of reference in which $\langle \underset{\sim}{v} \rangle = 0$),

$$W_{ij}(\underset{\sim}{x} - \underset{\sim}{x}',t) = \langle v_i(\underset{\sim}{x},t) v_j(\underset{\sim}{x}',t) \rangle /2 , \tag{6}$$

and its associated spectral tensor $E_{ij}(\underset{\sim}{k},t)$. The trace of (6) is

$$W(\underset{\sim}{x} - \underset{\sim}{x}',t) = \langle \underset{\sim}{v}(x,t) \cdot \underset{\sim}{v}(x',t) \rangle /2 = \int d^3k E(\underset{\sim}{k},t) \exp i\underset{\sim}{k} \cdot (x - \underset{\sim}{x}') \tag{7}$$

where $E = \mathrm{Tr}(E_{ij})$. The mean energy is

$$W_0(t) = W(0,t) = \langle v(\underset{\sim}{x},t) \cdot \underset{\sim}{v}(x,t) \rangle /2 = \int d^3k E(\underset{\sim}{k},t) . \tag{8}$$

The velocity fluctuations can also be expressed in terms of their Fourier transforms,

$$v_i(\underset{\sim}{x},t) = \int d^3k V_i(\underset{\sim}{k},t) \exp i(\underset{\sim}{k} \cdot \underset{\sim}{x}) . \tag{9}$$

Since $\underset{\sim}{v}(\underset{\sim}{x},t)$ is a random variable, so is $\underset{\sim}{V}(\underset{\sim}{k},t)$. It can be shown that

$$\langle V_i(\underset{\sim}{k},t) V_j(\underset{\sim}{k}',t) \rangle /2 = E_{ij}(\underset{\sim}{k},t) \delta(\underset{\sim}{k} + \underset{\sim}{k}') , \tag{10}$$

where δ is the Dirac delta function. Equation (10) follows from the definitions above by a straightforward application of the theory of generalized functions (Lighthill, 1960).

It is our experience that turbulent systems fluctuate on a wide range of length and time scales. The turbulence is generally stirred up on a fairly large scale (consider an eggbeater in a bowl of liquid or a boulder in a stream), but is manifested on much smaller scales as well. If the energy source that produces the turbulence is taken away, the turbulence will die out, converting its energy into heat by viscous dissipation. What seems to happen is that the energy is introduced into large eddies, and then cascades through eddies of shorter and shorter scale (larger values of k) until it reaches a scale at which viscous dissipation is important. To view this process from a more formal viewpoint, consider an incompressible fluid described by the Navier-Stokes equation

$$\frac{\partial \underset{\sim}{v}}{\partial t} + \underset{\sim}{v} \cdot \underset{\sim}{\nabla}\underset{\sim}{v} + \frac{1}{\rho}\underset{\sim}{\nabla} P - \nu \nabla^2 \underset{\sim}{v} = 0 \tag{11}$$

(here ν is the kinematic viscosity, taken to be constant). The spatial Fourier transform of (11) is (for constant ρ, ν),

$$[(\partial/\partial t) + \nu k^2]\underset{\sim}{V}(\underset{\sim}{k},t) + i\underset{\sim}{k}P(\underset{\sim}{k},t)/\rho + i\int d^3k''\underset{\sim}{k}'' \cdot \underset{\sim}{V}(\underset{\sim}{k} - \underset{\sim}{k}'',t)\underset{\sim}{V}(\underset{\sim}{k}'',t) = 0 \ . \tag{12}$$

If we take the scalar product of $V(k',t)$ with (12), and take the ensemble average of the result [noting that $\underset{\sim}{k} \cdot \underset{\sim}{V}(k,t) = 0$], we obtain

$$2[(\partial/\partial t) + \nu k^2]E(k,t)\delta(k + k') + i\int d^3k''\underset{\sim}{k}'' \cdot \langle \underset{\sim}{V}(\underset{\sim}{k} - \underset{\sim}{k}'')\underset{\sim}{V}(\underset{\sim}{k}') \cdot \underset{\sim}{V}(\underset{\sim}{k}'')\rangle = 0 \ . \tag{13}$$

We have made use of Eq. (10) in the derivation of (13). This result states that the mean energy at a given wave number can be dissipated by viscosity or transferred to or from other wave numbers by means of the Reynolds stress $\underset{\sim}{v} \cdot$ grad $\underset{\sim}{v}$. It should be noted that this wave-number transfer, or cascade, is essentially a nonlinear process.

The energy cascade process can be analyzed further (though with less rigor) by the approach of Kolmogorov and Obukhov. It is assumed that the turbulence is (statistically) steady in time and that the length scale associated with the energy input is much larger than that associated with viscous dissipation. It is also assumed that at scales considerably smaller than the energy-input length all direct effects of the boundaries are negligible. This condition implies that there is no preferred spatial direction (recall we have chosen a frame with $\langle \underset{\sim}{v} \rangle = 0$), so that the turbulence is isotropic, and in particular

$$E(\underset{\sim}{k})d^3k = U(|\underset{\sim}{k}|)d|\underset{\sim}{k}| \ . \tag{14}$$

Furthermore, the only physical quantities that can affect the spectrum are the energy input rate (per unit mass) Q and the viscosity ν. The only quantity with the dimension of length that can be formed from Q and ν is the dissipation length

$$\lambda = (\nu^3/Q)^{1/4} .$$

The only quantity with the dimension of U (velocity^2 $\times$ length) that can be formed from Q and ν is $Q^{1/4}\nu^{5/4}$, so that U must be of the form

$$U(|\underset{\sim}{k}|) = \alpha Q^{1/4}\nu^{5/4}F(|\underset{\sim}{k}|\lambda) , \tag{15}$$

where α is a dimensionless scale factor. If λ is very much smaller than the scale L of the driving eddies, it is plausible that there will be a range of wave numbers $1/L << k_1 < |\underset{\sim}{k}| < k_2 << 1/\lambda$ in which dissipation is negligible. In that case, $U(|\underset{\sim}{k}|)$ must be independent of ν, so that (15) requires that

$$U(|\underset{\sim}{k}|) = \alpha Q^{2/3}|\underset{\sim}{k}|^{-5/3} \qquad (k_1 < k < k_2) . \tag{16}$$

This power-law energy spectrum, called the Kolmogorov-Obukhov spectrum, is valid for the wave-number range k_1 to k_2, usually called the inertial range. There is no external energy input into the inertial range, nor is there dissipation; the energy simply cascades from smaller to greater wave number.

In practice the measurements of correlation functions are usually made only along one direction, say the x-direction. Thus, if the turbulence is homogeneous, what is actually measured is a reduced correlation function such as

$$R^*(x - x',t) = R(x - x',0,0,t) ,$$

and the associated reduced spectrum

$$S^*(k_x,t) = \int S(\underset{\sim}{k},t)dk_y\ dk_z .$$

If the energy spectrum is an isotropic power law such that $E(\underset{\sim}{k})d^3k = \text{const}|\underset{\sim}{k}|^{-q}\ d|\underset{\sim}{k}|$, the reduced spectrum is also a power law, with the index q,

$$E^*(k_x,t) = \text{const}|k_x|^{-q} .$$

The above remarks about fluid turbulence are introductory and intended to provide groundwork for our subsequent discussion of interplanetary hydromagnetic turbulence. The reader who wishes to pursue the theory of fluid turbulence in greater depth is referred to the excellent monograph of Batchelor (1960).

Rugged Invariants of Hydromagnetic Turbulence

We have seen that the energy correlation tensor W_{ij} plays a central role in the theory of fluid turbulence. This is not surprising, for its trace W_0 [Eqs. (7), (8)] is the energy per unit mass. If we take the scalar product of $\underset{\sim}{v}$ with the Navier-Stokes equation (11), and recall that $\underset{\sim}{\nabla} \cdot \underset{\sim}{v} = 0$ for incompressible flow, the result is

$$\frac{\partial}{\partial t}\left(\frac{1}{2}\underset{\sim}{v}^2\right) + \underset{\sim}{\nabla} \cdot \left(\underset{\sim}{v}\left[\frac{1}{2}v^2 + \frac{P}{\rho}\right]\right) = \nu\underset{\sim}{v} \cdot \nabla^2\underset{\sim}{v} . \tag{17}$$

For suitable boundary conditions (periodic, or $\underset{\sim}{v}$ tending to 0 at large distances), a volume integration of the divergence term vanishes. Carrying out the volume integration on (17), and defining the total energy as

$$\mathscr{U}(t) = \int d^3x\langle \underset{\sim}{v}^2\rangle/2 ,$$

we obtain

$$\frac{d\mathscr{U}}{dt} = \nu \int d^3x\underset{\sim}{v} \cdot \nabla^2\underset{\sim}{v} . \tag{18}$$

Thus $\mathscr{U}(t)$ is a conserved quantity in the absence of dissipation; furthermore, in the case of homogeneous incompressible turbulence it is expressible as a bilinear form in the Fourier components of a dynamical variable [Eqs. (7), (10)]. Any such quantity is unaffected by nonlinear processes, and is called a rugged invariant. Invariants of this general kind play a central role in the vocabulary of turbulence theory. It may be shown that a second rugged invariant of incompressible turbulence is the helicity

$$H_V = \int d^3x(\text{curl }\underset{\sim}{v}) \cdot \underset{\sim}{v} . \tag{19}$$

Let us now consider the rugged invariants of magnetohydrodynamics. Keeping in mind that any such invariant will be destroyed by dissipation, we restrict our attention to ideal dissipationless MHD, and look for conservation laws. Equation (1c') in Section I gives energy conservation, so that for appropriate boundary conditions

$$d/dt \int d^3x\{(1/2)\rho v^2 + P/(\gamma - 1) + B^2/8\pi\} = 0 . \tag{20}$$

Note that if the density is constant, the fluctuating part of the energy density is a bilinear form analogous to W_0. The energy equation was derived (in part) by taking the scalar product of $\underset{\sim}{v}$ into the momentum equation, and of $\underset{\sim}{B}$ into Faraday's law. Another invariant may be found by taking the scalar product of $\underset{\sim}{B}$ into the momentum equation and $\underset{\sim}{v}$ into Faraday's law, and adding the results. We obtain

$$(\partial/\partial t)(\mathbf{v} \cdot \mathbf{B}) + \nabla \cdot \{\mathbf{v}\mathbf{v} \cdot \mathbf{B} - \mathbf{B}v^2/2\} = -(\mathbf{B}/\rho) \cdot \nabla P \;.$$

This equation is not yet in conservation form (unless ρ is uniform). If the conducting fluid is barytropic, that is, if $\rho = \rho(P)$, we can rewrite our result as

$$(\partial/\partial t)(\mathbf{v} \cdot \mathbf{B}) + \nabla \cdot (\mathbf{v}\mathbf{v} \cdot \mathbf{B} - \mathbf{B}\{(1/2)v^2 - \int dP/\rho(P)\}) = 0 \;. \tag{21}$$

Integrating over volume as before, and assuming suitable boundary conditions, we have a second conserved quantity, the cross-helicity

$$H_c = \int d^3x \mathbf{v} \cdot \mathbf{B} \;, \tag{22}$$

where $dH_c/dt = 0$ in the absence of dissipation. A third rugged invariant involves the vector potential $\mathbf{A}$. The electric field can be written as

$$-\mathbf{E}c = \mathbf{v} \times \text{curl}\, \mathbf{A} = v_j \nabla A_j - \mathbf{v} \cdot \nabla\mathbf{A} = \partial\mathbf{A}/\partial t + c \,\text{grad}\, \phi \;.$$

This equation can be rewritten as

$$[(\partial/\partial t) + \mathbf{v} \cdot \nabla]\mathbf{A} - v_j \nabla A_j + c\nabla\phi = 0 \;.$$

Take the scalar product of $\mathbf{B}$ into this last equation, and of $\mathbf{A}$ into Faraday's law, and add the two results. We obtain

$$(\partial/\partial t)(\mathbf{A} \cdot \mathbf{B}) + \nabla \cdot (\mathbf{v}\mathbf{A} \cdot \mathbf{B} + c\mathbf{B}\phi - \mathbf{B}\mathbf{A} \cdot \mathbf{v}) = 0 \;. \tag{23}$$

Because no special gauge was assumed in the derivation of (23), this result must be gauge-invariant. Performing the required integrations, and invoking suitable boundary conditions, we have conservation of magnetic helicity,

$$H_M = \int d^3x \mathbf{A} \cdot \mathbf{B} \;. \tag{24}$$

This invariant was first noted by Woltjer (1958).

The physical meanings of the energy and cross-helicity are fairly obvious, but that of the magnetic helicity is somewhat more subtle. It is a measure of the linkage or "knottedness" of the magnetic field lines, as the following argument, due to Moffatt (1978), illustrates. Suppose the magnetic field $\mathbf{B}$ vanishes everywhere except in two flux tubes 1 and 2 of infinitesimal cross section, following the curves C_1 and C_2 (Fig II.1). Let the fluxes in the two tubes be F_1 and F_2. Let $\mathbf{t}$ be the unit vector tangent to the curves, and ds the element of length on each curve. On the curve C_k we have $\mathbf{B}\, d^3x = F_k \mathbf{t}\, dl$, so that, for example, the magnetic helicity for the tube 1 is

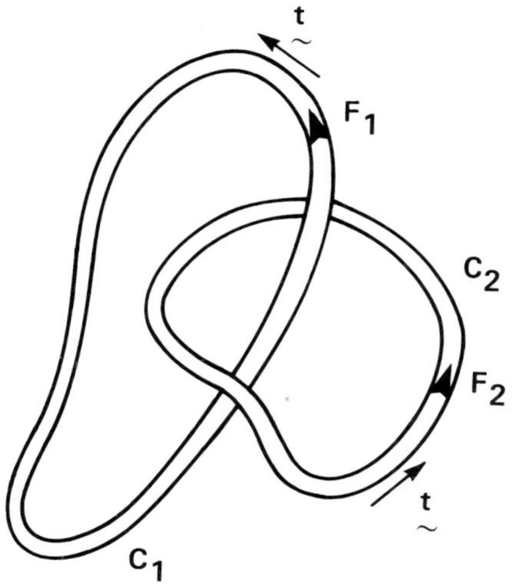

Figure II.1. Two linked flux tubes showing positive magnetic helicity (after Moffatt, 1978).

$$H_{M1} = F_1 \oint_{C_1} \underset{\sim}{A} \cdot \underset{\sim}{t}\, dl = F_1 F_2 ,$$

and by a similar argument we find that $H_{M2} = F_2 F_1 = H_{M1}$. If the field direction of (say) tube 2 is reversed, so that the "handedness" of the linkage is reversed, the magnetic helicity reverses sign. If the magnetic loops are shifted so that they are no longer linked, the magnetic helicity vanishes. If we have many linked loops, the magnetic helicity is the sum of the helicities of the various individual linkages; if the linkages are predominantly right-handed the magnetic helicity is positive, and if right- and left-handed linkages are equally mixed, the magnetic helicity vanishes. Similarly, right-circularly polarized waves have positive magnetic helicity and left-circularly polarized waves have negative magnetic helicity.

The same topological interpretation of the helicity of ordinary fluid dynamics [Eq. (19)] can be made, replacing $\underset{\sim}{B}$ by the vorticity curl $\underset{\sim}{v}$, and $\underset{\sim}{A}$ by $\underset{\sim}{v}$.

Up to the present, theoretical studies of hydromagnetic turbulence have used the assumption of incompressibility. In most cases, it has also been assumed that the mean magnetic field vanishes. Relaxing this assumption unleashes a host of theoretical problems, which are only just beginning to be addressed. For example, nonzero $\langle \underset{\sim}{B} \rangle$ automatically introduces a preferred direction, and therefore asymmetry of the turbulence (Montgomery and Turner, 1981); this problem does not arise in fluid turbulence, because we can always choose a reference frame in which $\langle \underset{\sim}{v} \rangle = 0$. Moreover, for nonzero $\langle \underset{\sim}{B} \rangle$ the fluctuations will propagate at the Alfvén speed, thereby calling into question the assumption of incompressibility. Other subtleties arise, such as the necessity to revise the magnetic-helicity invariant (Matthaeus and Goldstein, 1982).

The first attempt to measure and characterize the spectral properties of the rugged invariants in interplanetary turbulence has been reported by Matthaeus and Goldstein (1982). They find that the energy spectra are consistent with a power-law behavior over ~2 decades in wave number, and that the length scale for maximum magnetic helicity is larger than or comparable to that for maximum energy. This latter result is consistent with theoretical suggestions that magnetic helicity (in contrast to energy) may cascade downward in wave number. Extension and refinement of studies of this general kind will provide a test of the utility of the language of turbulence theory as a description of interplanetary fluctuations.

Observed Alfvénic Fluctuations — Waves or Turbulence?

We showed in Section I that Alfvénic fluctuations are an exact solution of the MHD equations, and that their geometry can be quite general. In particular, the field lines can be tangled and otherwise disorderly. Such a field is turbulent, at least from a kinematic viewpoint. Alfvénic fluctuations have the additional desirable property that density (and, for that matter, pressure and magnetic field strength) are strictly constant. We reiterate that this property of constant density, etc., is not imposed at the outset, but rather is simply one aspect of the exact solution.

Let us now try to address the question of whether the observed interplanetary Alfvénic fluctuations are waves or turbulence. Let us first test the hypothesis that they are plane waves. According to Eq. (4) in Section I, plane waves have a direction of zero magnetic variance (because $\operatorname{div} \underset{\sim}{B} = 0$ and all gradients are restricted to a single direction in space). In addition, Alfvén waves cannot be linearly polarized (because $|\underset{\sim}{B}|$ is constant), so that the zero-variance direction is unique (except for sign). Empirical studies of Alfvénic fluctuations show that in most cases a well-defined direction of minimum variance does exist (Daily, 1973, and many later corroborations). This result, then, is consistent (to a good approximation) with the interpretation of the fluctuations as plane Alfvén waves.

We have good evidence that these fluctuations are of solar origin. If they are simply plane waves, the simplest picture of their evolution is that they simply propagate out from the Sun, and that this propagation can be described by geometrical hydromagnetics. As discussed in Section I, a radially flowing wind will refract the wave vectors (and hence the minimum variance directions) into the antisunward direction far from the Sun. This model gives a clear prediction that the minimum variance directions observed by spacecraft should cluster about the radial direction. The various empirical studies of minimum variance directions have tested this hypothesis. It turns out that the minimum variance directions do have a strongly preferred direction, but it is not radial! The minimum variance directions cluster strongly around the average magnetic field direction (at 1 AU this direction is about 45° from the radial). Somehow the model of plane waves propagating in a radially flowing wind has run into trouble.

Refinements of the theory to allow for nonsteady nonradial flow or violation of the ordering required for the eikonal approach have not materially improved the agreement with observation. These difficulties of the wave interpretation suggest that perhaps these fluctuations are better described as turbulence. The behavior of the minimum variance direction may simply be a consequence of randomness of the fluctuations. If the magnetic fluctuations are modeled by the random walk of a vector of constant length, there is a well-defined minimum variance direction that tends to be aligned with the mean field direction (Barnes, 1981). In addition, observed Alfvénic fluctuations exhibit power-law spectra over several decades in wave number, suggesting a fully developed turbulent cascade.

However, there are also unsatisfactory features to the interpretation of pure Alfvénic fluctuations as turbulence. Recall that in ordinary turbulence the cascade of energy is due to the Reynolds stress

$\mathbf{v} \cdot \nabla \mathbf{v}$ [cf. Eq. (13)]. In magnetohydrodynamics the analogous cascade is associated with the sum of the Reynolds and Maxwell stresses, $\mathbf{v} \cdot \nabla \mathbf{v} - \mathbf{B} \cdot \nabla \mathbf{B}/4\pi\rho$. But, as was shown in Section I, in pure Alfvénic fluctuations, the fluctuations of velocity are related to those of magnetic field by

$$\Delta \mathbf{v} = \mp \Delta \mathbf{B}/(4\pi\rho)^{1/2} , \tag{25}$$

so that the sum of magnetic and Reynolds stress must vanish. Therefore, pure Alfvénic fluctuations do not admit an energy cascade!

Why, then, to the observed interplanetary Alfvénic fluctuations exhibit power-law spectra? One plausible explanation is that the energy cascade is not occurring locally, but evolved much nearer the Sun, where there was an appreciable non-Alfvénic compponent to the turbulence. From this viewpoint, the observed interplanetary Alfvénic spectra are fossils of a more complicated field of turbulence near the Sun. This interpretation is supported by the consistency of the observed sign of Eq. (25) (or in the language of turbulence, the consistency of the sign of cross-helicity) with outward propagation from the Sun. However, if the "fossil spectrum" idea is correct, we have to account for the disappearance of the non-Alfvénic component. A straightforward explanation arises from the fact that any non-Alfvénic component of the fluctuations will be associated with compression of the magnetic field (i.e., $|\mathbf{B}|$ will vary); compressive fluctuations are likely to result in significant collisionless dissipation, as will be discussed in Section III. Thus, the compressive component may simply have disappeared by damping, leaving only the Alfvénic component. An alternative mechanism suggested by recent preliminary theoretical studies is that nonlinear processes make a turbulent system evolve toward the maximal cross-helicity [i.e., Eq. (25)] characteristic of Alfvénic fluctuations. In addition, of course, real interplanetary Alfvénic fluctuations are never quite pure, so that in fact some cascade of energy may be going on locally. However, a local origin of the spectrum is difficult to reconcile with the consistency of outward propagation and the velocity-magnetic field correlation.

References

Barnes, A.: 1981, J. Geophys. Res., 86, p. 7498.
Batchelor, G. K.: 1960, The Theory of Homogeneous Turbulence, Cambridge.
Daily, W. D.: 1973, J. Geophys. Res., 78, p. 2043.
Lighthill, M. J.: 1960, Introduction to Fourier Analysis and Generalized Functions, Cambridge.
Matthaeus, W. H. and Goldstein, M. L.: 1982, J. Geophys. Res., 87, p. 6011.
Moffatt, H. K.: 1978, Magnetic Field Generation in Electrically Conducting Fluids, Cambridge.
Montgomery, D. C. and Turner, L.: 1981, Phys. Fluids, 24, p. 825.
Woltjer, L.: 1958, Proc. Nat. Acad. Sci. U.S.A., 44, p. 833.

III. COLLISIONLESS PROCESSES IN THE INTERPLANETARY MEDIUM

Introduction

Magnetohydrodynamics, strictly speaking, is the theory of the dynamics of electrically conducting fluids in which interparticle collisions occur on length and time scales very short in comparison with the scales of interest. This ordering of scales is opposite to that in many space plasmas, including the interplanetary medium. Nevertheless, experience shows that magnetohydrodynamics often provides a useful model of phenomena that occur in collisionless plasmas. A necessary condition for the application of magnetohydrodynamics to a collisionless medium is that the phenomena of interest must be of hydromagnetic scale (length longer than mean particle gyroradii, time longer than particle gyroperiods). This condition is not sufficient, however, and it sometimes turns out that the full collisionless theory yields results that cannot be anticipated in magnetohydrodynamics.

General Remarks on the Collisionless Kinetic Theory

Consider a plasma whose various charge species (ions and electrons) are labeled by α. Let $f_\alpha(\underset{\sim}{v},\underset{\sim}{x},t)$ be the velocity distribution of the αth species, that is,

$$f_\alpha(v,x,t)\,d^3v\ d^3x$$

is the number of particles of type α in the volume element $d^3v\ d^3x$ of configuration space containing the point $\underset{\sim}{v},\underset{\sim}{x}$. If Coulomb collisions are negligibly infrequent, the kinetic equation for f_α is the Vlasov equation (neglecting gravity),

$$\left(\frac{\partial}{\partial t} + v \cdot \nabla\right)f_\alpha + \frac{q_\alpha}{m_\alpha}\left(E + \underset{\sim}{v} \times \frac{\underset{\sim}{B}}{c}\right) \cdot \frac{\partial f_\alpha}{\partial \underset{\sim}{v}} = 0 \ , \qquad (1)$$

where $\underset{\sim}{E}(\underset{\sim}{x},t)$ and $\underset{\sim}{B}(\underset{\sim}{x},t)$ are the electric and magnetic fields, q_α and m_α are the charge and mass for charge species α, and c is the speed of light. The fields are related to the f_α by Ampere's law and Coulomb's law,

$$\text{curl}\ \underset{\sim}{B} - \frac{1}{c}\frac{\partial \underset{\sim}{E}}{\partial t} = \frac{4\pi}{c}\sum_\alpha q_\alpha \int \underset{\sim}{v} f_\alpha\, d^3v \ , \qquad (2a)$$

and

$$\underset{\sim}{\nabla} \cdot \underset{\sim}{E} = 4\pi \sum_\alpha q_\alpha \int f_\alpha\, d^3v \ . \qquad (2b)$$

In addition, of course, $\underset{\sim}{B}$ must be solenoidal and Faraday's law must be satisfied. A rigorous, physically correct description of the dynamics of a collisionless plasma requires the self-consistent solution of the Vlasov-Maxwell equations.

It is often convenient to describe a plasma in terms of the first few velocity moments of f_α. The number density is given by

$$n^\alpha = \int f_\alpha \, d^3v \,, \tag{3a}$$

the flow velocity V is given by

$$n^\alpha \underset{\sim}{V}^\alpha = \int f_\alpha \underset{\sim}{v} \, d^3v \,, \tag{3b}$$

the stress (or pressure) tensor is

$$\underset{\approx}{P}^\alpha = m_\alpha \int f_\alpha (\underset{\sim}{v} - \underset{\sim}{V}^\alpha)(\underset{\sim}{v} - \underset{\sim}{V}^\alpha) d^3v \,, \tag{3c}$$

and the (third rank) heat flow tensor is

$$\underset{\approx}{Q}^\alpha = (m_\alpha/2) \int f_\alpha (\underset{\sim}{v} - \underset{\sim}{V}^\alpha)(\underset{\sim}{v} - \underset{\sim}{V}^\alpha)(\underset{\sim}{v} - \underset{\sim}{V}^\alpha) d^3v \,. \tag{3d}$$

The tensors $\underset{\approx}{P}$, $\underset{\approx}{Q}$ are symmetric with respect to interchange of any indices. The heat-flow vector may be obtained from Q by contraction,

$$q_i^\alpha = Q_{ijj}^\alpha \,. \tag{3e}$$

From these moments one derives other macroscopic quantities of interest, such as the mass density

$$\rho = \sum_\alpha m_\alpha n^\alpha \,,$$

electric current density

$$\underset{\sim}{J} = \sum q_\alpha n^\alpha \underset{\sim}{V}^\alpha \,,$$

total plasma stress tensor

$$\underset{\approx}{P} = \sum_\alpha \underset{\approx}{P}^\alpha \,,$$

total heat flow vector

$$\underset{\sim}{q} = \sum_{\alpha} \underset{\sim}{q}^{\alpha} ,$$

etc.

Fluid equations for these moments are obtained by multiplying Eq. (1) by 1, $\underset{\sim}{v}$, $\underset{\sim}{v}\underset{\sim}{v}$, etc., and integrating over velocity. The zeroth, first, and second moment equations are the equation of continuity

$$\partial n^{\alpha}/\partial t + \mathrm{div}(n^{\alpha}\underset{\sim}{V}^{\alpha}) = 0 , \tag{4a}$$

the momentum equation

$$m_{\alpha}n_{\alpha}\{\partial/\partial t + \underset{\sim}{V}^{\alpha} \cdot \underset{\sim}{\nabla}\}\underset{\sim}{V}^{\alpha} + \underset{\sim}{\nabla} \cdot \underset{\approx}{P}^{\alpha} - q_{\alpha}n^{\alpha}\{\underset{\sim}{E} + \underset{\sim}{V}^{\alpha} \times \underset{\sim}{B}/c\} = 0 , \tag{4b}$$

and the (tensor) heat flow equation

$$\begin{aligned}(1/2)[(\partial/\partial t) + \underset{\sim}{V}^{\alpha} \cdot \underset{\sim}{\nabla}]P^{\alpha}_{ij} + \partial Q_{ijk}/\partial x_k + (q_{\alpha}/2m_{\alpha}c)(\varepsilon_{kim}P^{\alpha}_{jk} \\ + \varepsilon_{kjm}P^{\alpha}_{ik})B_m + (1/2)P^{\alpha}_{ij}\underset{\sim}{\nabla} \cdot \underset{\sim}{V}^{\alpha} + (1/2)(P^{\alpha}_{jk}\partial V^{\alpha}_{i}/\partial x_k \\ + P^{\alpha}_{ik}\partial V^{\alpha}_{j}/\partial x_k) = 0 .\end{aligned} \tag{4c}$$

Equation (4a) was used in the derivation of (4b), and both (4a) and (4b) were used to obtain (4c). Higher-order moment equations can also be derived. The problem, of course, is that at any given level we have more dependent variables than equations, and the system cannot be closed at any finite level. Therefore, in order to solve the momentum equations we must find an external closure condition, either assumed, or (in the best of worlds) derived from solution of the kinetic equation (1). In practice one usually works with Eqs. (4a) and (4b) and the contraction of (4c),

$$(1/2)(\partial/\partial t)T_r\underset{\approx}{P}^{\alpha} + (1/2)\underset{\sim}{\nabla} \cdot (\underset{\sim}{V}^{\alpha}T_r\underset{\approx}{P}^{\alpha}) + \underset{\sim}{\nabla} \cdot \underset{\sim}{q}^{\alpha} + P^{\alpha}_{jk}\, V^{\alpha}_{j}/\partial x_k = 0 . \tag{4c'}$$

Equations (4a), (4b), and (4c') are the generalization of Eqs. (1a)-(1c) in Section I. Manipulations analogous to those that led to Eq. (1c') in Section I yield an alternative form of the energy equation

$$\begin{aligned}(\partial/\partial t)\{(1/2)m_{\alpha}n^{\alpha}(\underset{\sim}{V}^{\alpha})^2 + (1/2)T_r\underset{\approx}{P}^{\alpha}\} + \underset{\sim}{\nabla} \cdot \{\underset{\sim}{V}^{\alpha}[(1/2)m_{\alpha}n^{\alpha}(\underset{\sim}{V}^{\alpha})^2 \\ + (1/2)T_r\underset{\approx}{P}^{\alpha}] + \underset{\sim}{q}^{\alpha} + \underset{\sim}{V}^{\alpha} \cdot \underset{\approx}{P}^{\alpha}\} = n^{\alpha}q_{\alpha}\underset{\sim}{V}^{\alpha} \cdot \underset{\sim}{E} .\end{aligned} \tag{4c''}$$

One other preliminary point to be made is that fluctuations in a plasma can, in a sense, act as scattering centers and produce transport somewhat analogous to transport associated with interparticle collisions.

If we write $f_\alpha = \langle f_\alpha \rangle + \partial f_\alpha$, where $\langle\ \rangle$ denotes an Eulerian ensemble average, the average of Eq. (1) is

$$[(\partial/\partial t) + \mathbf{v} \cdot \nabla]\langle f_\alpha\rangle + (q_\alpha/m_\alpha)(\langle \mathbf{E}\rangle + \mathbf{v} \times \langle \mathbf{B}\rangle/c) \cdot \partial\langle f_\alpha\rangle/\partial \mathbf{v}$$
$$= -(q_\alpha/m_\alpha)\langle(\delta\mathbf{E} + \mathbf{v} \times \delta\mathbf{B}/c) \cdot \partial(\delta f_\alpha)/\partial\mathbf{v}\rangle . \quad (5)$$

The right-hand side of Eq. (5), due to fluctuations, thus plays the formal role of a collision operator in the kinetic equation for $\langle f\rangle$.

So far the discussion of the collisionless kinetic theory has been quite general. The hydromagnetic limit of the collisionless theory is of special interest. Let ε be a smallness parameter, of the order of the largest of the ratios r_G/L, τ/T, where L and T are the macroscopic length and time scales, and r_G and τ are the smallest gyroradius and gyroperiod for any relevant particle. If $\varepsilon \ll 1$, dimensional analysis shows that the first two terms of (1) are $O(\varepsilon)$ in comparison with the terms containing the electromagnetic fields. Hence the Vlasov equation can be satisfied for all $\mathbf{v}$ only if

$$\{c\mathbf{E} + (\mathbf{v} \times \mathbf{B})\}\partial f_\alpha/\partial\mathbf{v} = O(\varepsilon)Bf_\alpha ,$$

and, similarly, by the momentum equation (46) we have $\mathbf{E} + \mathbf{V}^\alpha \times \mathbf{B}/c = O(\varepsilon)VB/c$. Then,

$$f_\alpha = G_\alpha(v_\parallel, v_\perp, \mathbf{x}, t)\{1 + O(\varepsilon)\} , \quad (6)$$

where $v_\parallel = (\mathbf{v} - \mathbf{V}^\alpha) \cdot \mathbf{B}/|\mathbf{B}|$ and $v_\perp^2 = (\mathbf{v} - c\mathbf{E} \times \mathbf{B}/B^2)^2 - v_\parallel^2$, and

$$\mathbf{E} + \mathbf{V}^\alpha \times \mathbf{B}/c = O(\varepsilon)V^\alpha B/c \quad (7)$$

for all α. Equation (6) states that in the first approximation the plasma is gyrotropic, that is, velocity distributions are symmetric (in velocity space) under rotations about the local magnetic field direction. Physically, this is so because each particle encounters only small variations in field during each gyration. One consequence of this gyrotropy is that the stress tensor is diagonal, in fact,

$$\mathbf{P}^\alpha[1 + O(\varepsilon)] = P_\parallel^\alpha \mathbf{BB}/B^2 + P_\perp^\alpha\{1 - \mathbf{BB}/B^2\} . \quad (8)$$

An analogous symmetry exists for heat flow. Equation (7) states that in the first approximation all charge species flow with a velocity whose components transverse to $\mathbf{B}$ are given by the familiar $\mathbf{E} \times \mathbf{B}$ guiding-center drift.

Thus, Eq. (7) is the generalization of (1d) from Section I. Note, however, that Eq. (7) is only a first approximation. In particular, a component of $\mathbf{E}$ parallel to $\mathbf{B}$ is allowed,

$$\mathbf{E} \cdot \mathbf{B} \sim O(\varepsilon)|\mathbf{E}||\mathbf{B}| .$$

It is clear from (4b) that this parallel component can be a significant term in the momentum equation of each charge species. When the momentum equation is summed over species, the net contribution due to $E_{\parallel}$ disappears, because the plasma must be electrically neutral (to very high accuracy). However, $E_{\parallel}$ can have major effects on the details of the velocity distributions, and in particular can result in strong dissipation. This parallel component of $\underset{\sim}{E}$ is thus one important effect about which classical MHD provides no information.

It may also be anticipated from Eq. (7) that each species has components of velocity transverse to the field that differ from the $\underset{\sim}{E} \times \underset{\sim}{B}$ drift in $O(\varepsilon)$. This correction will in general be different for different species, and thus contributes to the currents that flow in the plasma.

Consider a two-component plasma (one species of ion and electrons). In most situations of interest $\underset{\sim}{V}^i \cdot \underset{\sim}{B} = \underset{\sim}{V}^e \cdot \underset{\sim}{B}\{1 + O(\varepsilon)\}$. We also note that in nonrelativistic problems, displacement current can be neglected in the hydromagnetic limit. Then summing Eqs. (4a), (4b), and (4c") over species and applying Maxwell's equations and Eq. (7) gives, to first order in ε,

$$\frac{\partial \rho}{\partial t} + \underset{\sim}{\nabla} \cdot (\rho \underset{\sim}{V}) = 0 \, , \tag{9a}$$

$$\rho\left(\frac{\partial}{\partial t} + \underset{\sim}{V} \cdot \underset{\sim}{\nabla}\right)\underset{\sim}{V} + \underset{\sim}{\nabla}\left(P_{\perp} + \frac{B^2}{8\pi}\right) - \frac{1}{4\pi}\,\underset{\sim}{B} \cdot \underset{\sim}{\nabla}\left\{\left(1 + \frac{4\pi(P_{\perp} - P_{\parallel})}{B^2}\right)\underset{\sim}{B}\right\} = 0 \, , \tag{9b}$$

and

$$\frac{\partial}{\partial t}\left(\frac{1}{2}\rho V^2 + P_{\perp} + \frac{1}{2}P_{\parallel} + \frac{B^2}{8\pi}\right) + \underset{\sim}{\nabla} \cdot \left[\underset{\sim}{V}\left(\frac{1}{2}\rho V^2 + 2P_{\perp} + \frac{1}{2}P_{\parallel} + \frac{B^2}{4\pi}\right) + \underset{\sim}{q} - \frac{\underset{\sim}{B}\underset{\sim}{B} \cdot \underset{\sim}{V}}{4\pi}\left\{1 + \frac{4\pi(P_{\perp} - P_{\parallel})}{B^2}\right\}\right] = 0 \, . \tag{9c}$$

The utility of classical MHD for describing collisionless plasmas is remarkable, though not, as some would have it, a deep mystery requiring a subtle explanation. The reason, of course, is that the collisionless Eqs. (9) are nearly the same as the classical MHD equations, being statements of the conservation of mass and energy, Newton's laws, and Maxwell's equations. Statements to the effect that a collisionless plasma behaves like a fluid because "the particles are tied to magnetic field lines" or because "particles collide with fluctuations in the field" are beside the point. Even in a collisionless electrically neutral gas, for example, a gradient in pressure is associated with flow (consider a finite sphere of collisionless gas placed in a vacuum). Laminar solutions of the collisionless equations are typically quite similar to results in classical MHD. It is certainly true that there are some important differences between the classical and collisionless theories, some of which will be discussed below, but these are primarily concerned with dissipation, transport, and stability.

Confusion has sometimes arisen about how the collisionless fluid equations (9) are related to the guiding-center drifts. For example, the well-known mirror force $-\mu\nabla|B|$ is not manifest in (9b). This has led, on occasion, to the erroneous conclusion that it has somehow been left out, and needs to be added as an extra term. In fact, the mirror force and all other effects of the guiding-center theory are automatically included in Eqs. (9). That this is not obvious reflects the special character of the guiding-center theory; it follows the motions of guiding centers, not particles. For example, the electric current cannot be obtained simply by summing the guiding-center drifts. In contrast, the Vlasov equation in effect follows the detailed motion of particles moving under the Lorentz force, and leads in a straightforward way to a fluid theory by way of its moment equations. Informative discussions of the relation between fluid and guiding-center approaches can be found in the books of Spitzer (1962) and Northrop (1963), and in the lectures by Siscoe appearing in this volume.

Evolution of Velocity Distributions in Solar Wind Flow

The density of the interplanetary medium near the orbit of Earth is typically of the order of ~5 protons/cm^3, the proton temperature of the order of ~5 × 10^4 K, and the electron temperature of the order of ~2 × 10^5 K. Hence the Coulomb collision time is of the order of days for electrons, and somewhat longer for protons. The mean time of flow from the Sun to 1 AU is about 4 days. Hence the flow of protons is essentially collisionless, and the electrons are marginally collisionless. On the other hand, the coronal plasma at the source of the solar wind is much denser, and near the Sun the flow must be collisional for both protons and electrons. We do not know exactly where the transition from collisional to collisionless flow occurs, but typical estimates are ~10 solar radii, possibly much nearer for protons.

It can be instructive to model the solar wind as a purely collisionless flow. The simplest model is steady, laminar radial flow, with the magnetic field directed radially outward. The inertial and gravitational terms can be neglected in the electron momentum equation, so that we have

$$(\nabla \cdot \mathsf{P}^e)_r + neE_r = 0 \qquad (10)$$

($-e$ is the electron charge). Equation (10) shows clearly that a component of electric field along the magnetic field is required. Because the electron pressure decreases outward, $E_r > 0$; hence this polarization field pulls back on the electrons and pushes out on the ions of the plasma. The field may be written as the gradient of a potential,

$$E_r = -dU/dr \ .$$

If we set the boundary condition that U tends to 0 for large r, $U(r) > 0$.

If the flow is collisionless, Eq. (1) applies (with $\partial f/\partial t = 0$). Because the Vlasov equation is derived from the Liouville equation, it is essentially a restatement of the equations of motion. Therefore, in particular, any function of the constants of particle motion is a solution of the Vlasov equation. As an illustration, consider the total energy

$$\mathscr{E}(\underset{\sim}{x},\underset{\sim}{v}) = (1/2)m\underset{\sim}{v}^2/2 + qU(x) ,$$

as a function of $\underset{\sim}{x}$ and $\underset{\sim}{v}$ (in the case of ions the gravitational potential energy should be added to this expression). Clearly

$$\underset{\sim}{v} \cdot \underset{\sim}{\nabla}\mathscr{E} + \frac{q}{m}\left(\underset{\sim}{E} + \underset{\sim}{v} \times \frac{\underset{\sim}{B}}{c}\right) \cdot \frac{\partial \mathscr{E}}{\partial \underset{\sim}{v}} = 0 ,$$

so that any function of $\mathscr{E}$ is a solution. Because the magnetic moment $\mu = mv_\perp^2/2B$ is an adiabatic invariant for gradients of hydromagnetic scale, it follows that functions of μ are also good approximate solutions to (1) (the validity of this approximation can be formally justified, but the calculations will not be given here). In our special geometry, the local flow speed $V(r)$ is radially oriented (as is the magnetic field), so that elementary calculations give

$$v_\perp^2 = 2\mu B/m$$

and (11)

$$v_\parallel = -V(r) \pm [2\{\mathscr{E} - (qU + \mu B)\}/m]^{1/2} .$$

Now consider a plasma with base conditions given at some radius r_o, where the velocity distributions are $f_o(v_\parallel ,v_\perp^2)$. We can express f_o in terms of μ and $\mathscr{E}$,

$$f_o(v_\parallel ,v_\perp) = f_o(-V_o \pm [2\{\mathscr{E} - (qU_o + \mu B_o)\}/m]^{1/2}, 2\mu B_o/m) ,$$

where the subscript o denotes conditions at r_o. Then at any r we have

$$f(r,v_\parallel ,v_\perp) = f_o(-V_o \pm \{v_\parallel^2 + 2v_\parallel V(r) + V^2(r) + v_\perp^2(1 - B_o/B) + 2q(U - U_o)/m\}^{1/2}, v_\perp^2 B_o/B) . \tag{12}$$

Thus, for example, and initially isotropic velocity distribution will not be isotropic at all r. Physically, this occurs because the simultaneous conservation of both energy and magnetic moment requires exchange of energy between motion transverse to $\underset{\sim}{B}$ and parallel to $\underset{\sim}{B}$. The initial velocity distribution will be distorted in other ways as well; an initially Maxwellian velocity distribution will not stay Maxwellian.

Detailed measurements of the velocity distributions of solar wind protons and electrons have been made. In general these distributions are far from Maxwellian. Moreover, it is well established in plasma physics that sufficiently distorted velocity distributions may be unstable. Thus, it is quite probable that the velocity distributions of solar wind ions and electrons evolve into unstable states during the course of flow from the Sun. A detailed analysis of the kinds of instabilities that might develop in the ion flow suggests that the solar wind may be characterized by three different regions, namely, the collision dominated outer corona, a spherical shell of laminar exospheric expansion, and an outer region in which the state of the plasma is determined by the nonlinear evolution of plasma instabilities (Eviatar and Schulz, 1970).

The situation for the electrons is much more complicated. In the first place, Coulomb collisions may not be entirely negligible. However, even if the electrons are collisionless, an additional complication arises. Note that for electrons, $v_{\parallel}$ is given by

$$mv_{\parallel}^2/2 = mv_{\parallel o}^2/2 + \mu B_o(1 - B/B_o) + e(U - U_o) \ . \tag{13}$$

An electron cannot escape from r_o to infinity (where $B = U = 0$) if

$$mv_{\parallel o}^2/2 + \mu B_o - eU_o < 0 \ . \tag{14}$$

On the other hand, let $r_c < r_o$ be the radial distance at which the transition between collisional and collisionless flow occurs. If

$$mv_{\parallel o}^2/2 + \mu(B_o - B_c) + e(U_c - U_o) < 0 \ , \tag{15}$$

as must be the case for sufficiently large μ, an electron will not have access to r_c from r_o. Electrons that satisfy both (14) and (15) will thus be trapped in a collisionless region; such an electron initially spiraling outward will be turned around by the polarization electric field, and then spiral inward until it is mirrored back by the interplanetary magnetic field.

These considerations indicate that the solar-wind electrons may comprise several different populations. Some propagate out with the wind, others are not energetic enough to escape, and some, as described above, may be trapped in the interplanetary medium beyond the corona. The associated velocity distributions may well be unstable, thereby modifying the electron component and its associated heat transport (Perkins, 1973; Hollweg, 1976).

The discussion so far has ignored the fact that the Coulomb cross section, and hence collision time, are strongly energy dependent, so that in fact the most energetic electrons are the ones least affected by collisions. Hence, a plausible refinement of the collisionless electron model described above might be a velocity distribution consisting of a low-energy collisional component and a higher-energy collisionless component. One might expect a fairly abrupt transition between the two components (Fig. III.1); the high-energy tail would consist of outward-propagating electrons that have made the transition to the point of observation without Coulomb collision.

As might be expected, the real situation turns out to be more complicated. Observations of the interplanetary electron distributions show them to be in a subtle intermediate state between collisional and collisionless flow. Typically, there are in fact two components, separated at a fairly well-defined break at ~60 eV. In high-speed solar wind streams the high-energy electrons may be strongly beamed (Rosenbauer et al., 1976), consistent with the model represented by Fig. III.1. However, in slower solar winds it is common to find that the high-energy component can be represented by a modified Maxwellian distribution, and, in particular, that there are abundant high-energy electrons moving back toward the Sun (Feldman et al., 1975). In either case, the electron heat flux is carried by the high-energy component.

The existence of sunward-propagating particles among the higher-energy (or "halo") component of the electrons cannot be accounted for by purely laminar collisionless expansion. One possibility is that electron distributions become unstable as they evolve, generating fluctuations that scatter the energetic electrons into the backward direction. An alternative is that backward streaming electrons may be produced by Coulomb collisions in the outer heliosphere; the collision mean free path is long, but the outward-moving electrons must spiral a very long way along a tightly wound interplanetary magnetic field. In this picture, local properties of interplanetary electrons are determined by the global properties of the heliosphere. Both kinds of theory have been discussed at length in the literature, and it is not at present clear which is more likely to be correct. For more detailed accounts of the interplanetary electron distributions, the reader is referred to Feldman (1979) and Scudder and Olbert (1979).

Collisionless Waves, Damping, and Instability

Let us now consider what new elements collisionless phenomena introduce to the theory of hydromagnetic structures. Some are virtually unchanged. A case in point is the tangential pressure balance, which is as described in MHD theory, except that the pressure P is replaced by $P_\perp$. Alfvén waves, too, are virtually the same in collisionless and collisional theory, except that the propagation speed and correlation between magnetic and velocity variations are modified by pressure anisotropy. In contrast, the entropy wave in effect does not exist; if we try to set one up, the particles stream freely along the magnetic field lines and wipe it out (diffusion has an analogous effect on entropy waves in a collisional fluid).

The linearized theory of small-amplitude waves in a uniform collisionless plasma is solved, at least in principle. The starting point,

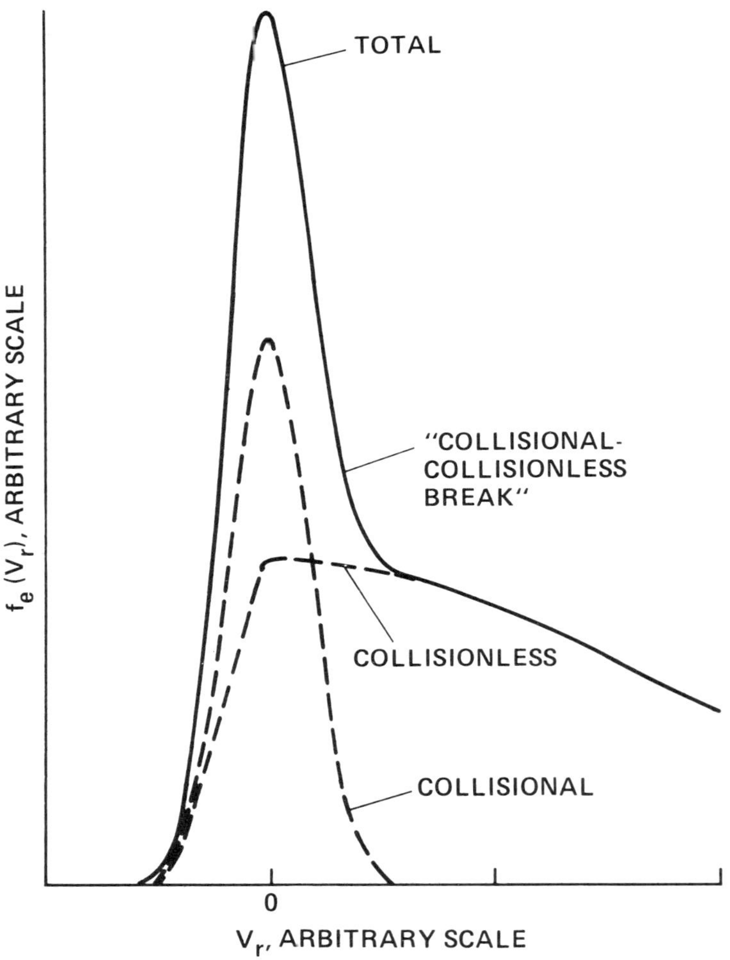

Figure III.1. Schematic representation of electron radial-velocity distribution for ideal superposition of low-energy collisional and high-energy collisionless electrons. Real interplanetary electron distributions are typically much more complex (see text).

of course, is to assume that $f = f_o + \delta f$, etc., and to linearize the Vlasov-Maxwell set. The problem may be solved by the approach of Landau (1946). The linearized equations are Fourier-analyzed in space and Laplace-analyzed in time. The result is that long after the initial disturbance, the Fourier components of the perturbed fields and moments of the perturbed velocity distributions oscillate in proportion to $\exp - (i\omega(\underset{\sim}{k})t)$, where the complex frequency $\omega(\underset{\sim}{k})$ is related to the wave vector $\underset{\sim}{k}$ by a dispersion relation,

$$D\{\underset{\sim}{k},\omega(\underset{\sim}{k})\} = 0 \ . \qquad (16)$$

The details of the method are discussed, for example, in Montgomery and Tidman (1964).

The solution of the linearized Vlasov equation, using Faraday's law to eliminate magnetic fluctuations, gives a linear relation between the Fourier components of the fluctuating electric fields and plasma (flow) velocities,

$$\underset{\sim}{V}^{\alpha}(\underset{\sim}{k},\omega(\underset{\sim}{k})) = (c/B_o)\underset{\approx}{M}^{\alpha}(\underset{\sim}{k},\omega\{\underset{\sim}{k}\}) \cdot E(\underset{\sim}{k},\omega\{\underset{\sim}{k}\}) \ , \qquad (17)$$

where $\underset{\approx}{M}^{\alpha}$ is called the effective mobility tensor, and depends on the properties of the unperturbed plasma and magnetic field. If $V_o^{\alpha} = 0$, the current density is

$$\underset{\sim}{J}(\underset{\sim}{k},\omega(\underset{\sim}{k})) = \sum_{\alpha} \frac{q_{\alpha}n_o^{\alpha}c}{B_o} \underset{\approx}{M}^{\alpha} \cdot \underset{\sim}{E}(\underset{\sim}{k},\omega(\underset{\sim}{k}))$$

$$\equiv \frac{\omega(\underset{\sim}{k})}{4\pi i} [\underset{\approx}{K}(\underset{\sim}{k},\omega(\underset{\sim}{k})) - \underset{\approx}{1}] \cdot \underset{\sim}{E}(\underset{\sim}{k},\omega(\underset{\sim}{k})) \ , \qquad (18)$$

where $\underset{\approx}{K}$ is the effective dielectric tensor. Substitution of Eq. (18) in Ampere's law leads to

$$\left(\frac{c}{\omega(\underset{\sim}{k})}\right)^2 \{k^2\underset{\approx}{1} - \underset{\sim}{k}\underset{\sim}{k}\}\underset{\sim}{E}(\underset{\sim}{k},\omega(\underset{\sim}{k})) = \underset{\approx}{K}(\underset{\sim}{k},\omega(\underset{\sim}{k})) \cdot \underset{\sim}{E}(\underset{\sim}{k},\omega(\underset{\sim}{k})) \ , \qquad (19)$$

which has nontrivial solutions if and only if

$$\det\left[\left(\frac{c}{\omega(\underset{\sim}{k})}\right)^2 (\underset{\sim}{k}\underset{\sim}{k} - k^2\underset{\approx}{1}) + \underset{\approx}{K}(\underset{\sim}{k},\omega(\underset{\sim}{k}))\right] = 0 \ . \qquad (20)$$

Equations (19) and (20) can be derived for arbitrary frequency and wave number (see Montgomery and Tidman, 1964, for details). The expressions are very complicated and numerical solution is generally required, even in the hydromagnetic limit. One exception is the Alfvén wave, which shows up with the simple dispersion relation

$$(\omega/k_{\parallel} C_A)^2 = 1 + (4\pi/B_o^2)(P_{\perp} - P_{\parallel}) \ , \qquad (21)$$

where $k_{\parallel} = \underset{\sim}{k} \cdot \underset{\sim}{B}_0/B_0$ and $C_A^2 = B_0^2/4\pi\rho_0$. The magnetoacoustic waves are governed by a much more complicated dispersion relation (Barnes, 1966), and have to be solved by computer. It turns out that the phase speeds have magnitudes similar to those of their MHD counterparts, and the anisotropy of phase speeds is similar. The striking and very important difference between collisional and collisionless theories is that the collisionless dispersion relation forces an imaginary part to the frequency. In an isotropic Maxwellian plasma $\text{Im}\omega < 0$, so that the waves damp, generally quite rapidly. For sufficiently distorted velocity distributions $\text{Im}\omega$ can be positive, so that an instability develops.

For $\underset{\sim}{V}^{\alpha} = 0$, the dispersion relation for magnetoacoustic waves in a collisionless hydrogen plasma depends on four parameters, $\beta_{\perp}^{+}$, $\beta_{\perp}^{-}$, $\beta_{\parallel}^{+}$, and $\beta_{\parallel}^{-}$, defined as $\beta_{\perp}^{+} \equiv 8\pi P_{\perp 0}^{+}/B_0^2$, etc. A plot of damping rates versus θ (the angle between $\underset{\sim}{k}$ and $\underset{\sim}{B}_0$) for two isotropic Maxwellian hydrogen plasmas is given in Fig. III.2. In both cases the damping rate is substantial. The sound-like mode is so strongly damped that it essentially does not exist as a propagating wave. Damping of the fast magnetoacoustic mode is more moderate, but far from negligible; it is characterized by two peaks in the $\text{Im}\omega - \theta$ plots, one at $\theta \sim 25°$ corresponding to resonant heating of protons, the second at $\theta \sim 87°$ corresponding to resonant heating of electrons. The proton-peak damping rate depends sensitively on $\beta_{\parallel}^{+}$, but is fairly strong for values of $\beta_{\parallel}^{+} > 0.3$, as is commonly found in the interplanetary medium. According to these calculations, under typical interplanetary conditions, compressive waves should be damped out in 10-100 wave periods. Therefore, any magnetoacoustic waves of (spacecraft frame) period < 1 hr would be expected to be of local origin.

Clearly, this damping cannot be anticipated from classical MHD theory. Moreover, any approximation that is a straightforward generalization of the fluid theory, such as relating $P_{\parallel}$ and $P_{\perp}$ to ρ and B by assuming double-adiabatic invariance, will fail to predict Landau damping. This dissipation process can be found only by analyzing the microscopic details of the field-particle dynamics.

The collisionless damping of hydromagnetic waves is associated with the Landau resonance [i.e., it occurs when the resonant velocity $\text{Re}(\omega)/k_{\parallel}$ is comparable to the electron or ion thermal speed] and is thus an example of Landau damping. However, Landau damping requires not only a substantial number of particles moving along the field lines at the resonant velocity, but also the existence of a suitable accelerating force. In the case of hydromagnetic waves, the necessary acceleration is best understood from the guiding-center viewpoint. A particle's guiding center is accelerated along the magnetic field lines by the mirror force and parallel component of electric field,

$$m(dv_{\parallel}/dt) = qE_{\parallel} - \mu\partial|\underset{\sim}{B}|/\partial s \quad , \tag{22}$$

where $\partial B/\partial s$ is the gradient along the local field direction. The electric field component $E_{\parallel}$ is nonzero because the mirror force, acting

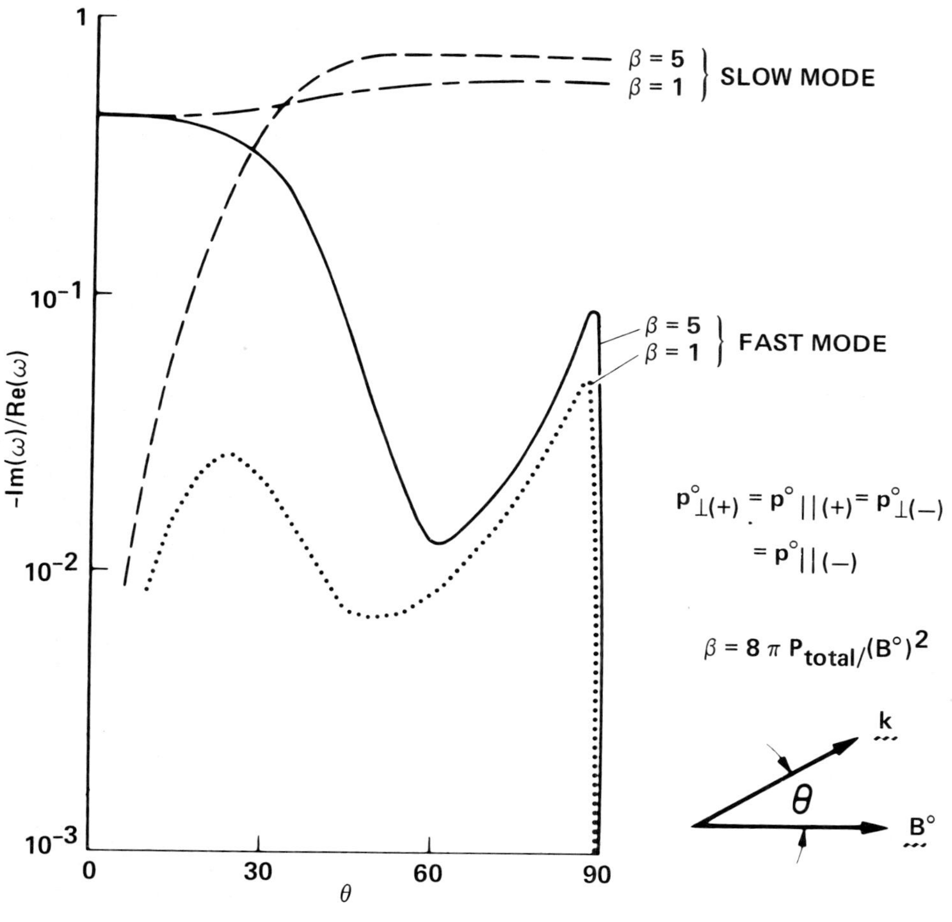

Figure III.2. Landau-damping rate versus propagation direction for collisionless magnetoacoustic waves (after Barnes, 1966).

differently on protons and electrons, tends to produce a net charge density that has to be opposed by a polarization field. On the average, this acceleration tends to bring the velocities of resonant particles toward $\mathrm{Re}(\omega)/k_{\parallel}$, and results in a net exchange of energy between the wave and the resonant particles. The sign of the energy exchange depends on whether more particles gain energy than lose energy (or vice versa), and hence depends on the details of the velocity distributions. For Maxwellian plasmas, the particles show a net gain of energy, so that the waves must damp.

It is immediately clear why the Alfvén wave is exempt from this Landau damping. The magnitude of the field strength is constant in an Alfvén wave, so that the mirror force vanishes (which in turn means that $E_{\parallel}$ vanishes). Hence, even though there are resonant particles, the resonant force (22) does not occur, and the Alfvén wave is undamped.

For a more detailed summary of the theory and the resonant acceleration viewpoint, the reader is referred to Barnes (1979).

If magnetoacoustic waves are strongly Landau-damped, they should be observed only near source regions (except for the special case of the fast-mode propagating transverse to the field, which is undamped). It seems likely, moreover, that the resonant acceleration will occur in compressive fluctuations under rather general (even turbulent) conditions. This viewpoint is consistent with the widespread observation of relatively pure Alfvénic fluctuations in the solar wind, and may account for the disappearance of the non-Alfvénic component of turbulence that appears to be necessary for the development of a turbulent cascade. The strongest compressive fluctuations are normally found in the interaction regions between fast and slow solar wind streams, and presumably are locally generated. In addition, compressive fluctuations of local origin are common in the shocked flow sheaths behind planetary bow shocks.

Various collisionless instabilities can occur for sufficiently distorted velocity distributions. A simple example may be found from the Alfvén-wave dispersion relation (21), for the frequency becomes imaginary if

$$P_{\parallel o} > P_{\perp o} + B_o^2/4\pi \ . \qquad (23)$$

This nonresonant instability is called the fire-hose instability. A second nonresonant instability, the mirror instability, occurs when

$$1 + \beta_{\perp}\left(1 - \frac{\beta_{\perp}}{\beta_{\parallel}}\right) - \frac{1}{2\beta_{\parallel}} \frac{(\beta_{\parallel +}\beta_{\perp -} - \beta_{\parallel -}\beta_{\perp +})^2}{\beta_{\parallel -}\beta_{\parallel +}} < 0 \ . \qquad (24)$$

It has occasionally been suggested that these instabilities regulate the anisotropy in the solar wind; however, higher-frequency resonant instabilities sometimes give more stringent restrictions on stable anisotropy.

References

Barnes, A.: 1966, Phys. Fluids, 9, p. 1483.
Barnes, A.: 1979, Hydromagnetic Waves and Turbulence in the Solar Wind, in Solar System Plasma Physics, Vol. 1 (ed. by C. F. Kennel, L. J. Lanzerotti, and E. N. Parker), North-Holland, p. 249.
Eviatar, A. and Schulz, M.: 1970, Planet. Space Sci., 18, p. 321.
Feldman, W. C.: 1979, Kinetic Processes in the Solar Wind, in Solar System Plasma Physics, Vol. 1 (ed. by C. F. Kennel, L. J. Lanzerotti and E. N. Parker), North Holland, p. 321.
Feldman, W. C., Asbridge, J. R., Bame, S. J., Montgomery, M. D., and Gary, S. P.: 1975, J. Geophys. Res., 80, p. 4181.
Hollweg, J. V.: 1976, J. Geophys. Res., 81, p. 1649.
Landau, L.: 1946, J. Phys. USSR, 10, p. 25.
Montgomery, D. C. and Tidman, D. A.: 1964, Plasma Kinetic Theory, McGraw-Hill, New York.
Northrop, T. G.: 1963, The Adiabatic Motion of Charged Particles, Interscience, New York.
Perkins, F.: 1973, Astrophys. J., 179, p. 637.
Rosenbauer, H., Miggenrieder, H., Montgomery, M., and Schwenn, R.: 1976, Preliminary Results of the Helios Plasma Measurements, in Physics of Solar-Planetary Environments (ed. by D. J. Williams), AGU, Washington, D.C., p. 319.
Scudder, J. D. and Olbert, S.: 1979, J. Geophys. Res., 84, p. 2755.
Spitzer, L., Jr.: 1962, Physics of Fully Ionized Gases, Interscience, New York.

SOLAR COSMIC RAYS - THEIR INJECTION, ACCELERATION, AND PROPAGATION

L. A. Fisk
Department of Physics
University of New Hampshire
Durham, NH 03824

Selected aspects of the injection, acceleration, and propagation of solar cosmic rays are discussed. In particular, an injection mechanism which determines the composition of the accelerated particles is reviewed; the limitations on our current understanding of the flare acceleration process are described briefly; and the theory governing particle diffusion in the solar wind and its apparent inadequacy is discussed.

1. Introduction

When a flare explodes in the atmosphere of the Sun, it produces energetic particles--the so-called solar cosmic rays--which propagate outward through the solar wind. Solar cosmic rays contain electrons, protons, and heavier nuclei. They can range in energy from just above the solar wind to several GeV.

We will be concerned in this section with selected aspects of the injection of the particles into the acceleration process, their subsequent acceleration, and their propagation. We will concentrate in particular on the injection and propagation because there is considerable theoretical work on these problems, with some well-documented successes and challenges for current theories. The acceleration process is clearly important and indeed the central theme; however, the several theories for the acceleration are relatively primitive and current observations cannot distinguish between them.

We begin by reviewing some of the inherent limitations in flare observations for understanding particle acceleration. We then discuss the injection of particles into the acceleration process with particular emphasis on the role of the injection process in altering the composition of the accelerated particles. We mention briefly various acceleration theories. Finally we discuss the propagation of solar cosmic rays with particular emphasis on how the particles are expected to diffuse among the irregularities in magnetic field of the solar wind, but how at present we are unable to calculate correctly the relevant diffusion parameters.

R. L. Carovillano and J. M. Forbes (eds.), Solar-Terrestrial Physics, 201–215.

2. Solar flare observations

Energetic particles produced by flares produce electromagnetic radiation when they interact with the matter and magnetic fields in the solar atmosphere. This radiation generally occurs in two phases. The first phase is characterized by impulsive bursts of hard X-rays and microwaves and by Type III radio emissions. In the second phase there are type II bursts, microwave and metric Type IV emission, flare continuum radio emission and gradual hard X-rays. Gamma-ray emission occurs from particles present in both phases. (see Forman et al.(1982) for a summary of observations of flare radiation produced by energetic particles.)

It would seem that we could learn about the acceleration of the solar cosmic rays seen in the solar wind by studying the observations of this electromagnetic radiation. Unfortunately, there is mounting evidence that this may not be the case. The connection between the particles producing the electromagnetic radiation and the ones escaping into the solar wind may be loose.

For example, gamma-ray lines are produced by energetic ions interacting with the solar atmosphere and thus can serve as a measure of the number of ions accelerated at the flare site. The correlation, however, between the number of ions accelerated and the number seen in the solar wind is not strong (von Rosenvinge et al., 1981). Similarly, the charge states of flare energetic ions seen in the solar wind are generally characteristic of those expected in the average $1\text{-}2 \cdot 10^6$ °K corona. Yet the temperatures in flares can be many times hotter than this, with an ionization state for the particles much higher than the average corona (Ma Sung et al., 1981).

It may be that the particles that produce the electromagnetic radiation tend to remain trapped near the flare site. The solar cosmic rays we observe in the solar wind may be produced in the more tranquil surrounding regions of the corona, perhaps by shocks or turbulence generated by the flare. To study this aspect of the flare acceleration process, which is our primary concern here, we may have to rely primarily on the information in the spectra and composition which is carried with the energetic particles themselves.

3. The injection

In most acceleration processes, e.g. in statistical or shock acceleration, some injection mechanism is required to raise the particles from thermal energies to above some threshold, either for the acceleration process itself or beyond where energy losses are important. However, until recently this aspect of the flare process was not given much thought. The injection and the acceleration were considered as part of the same process, and neither was well understood.

In recent years, however, it has been discovered that there are wide variations in the composition of solar cosmic-ray ions, particularly in small flare events, and that a likely cause for this variation is the injection process. Conversely, by studying the composition in flares we can learn about the injection process, or equivalently about the plasma conditions that must prevail at the acceleration site for this injection mechanism to work.

The most dramatic example of a compositional variation in a solar flare event is the so-called ^3He-rich solar flare in which the $^3He/^4He$ ratio at energies ~1 MeV/nucleon can exceed unity. In contrast the $^3He/^4He$ ratio in the solar wind or in solar prominences is only $\sim 4 \cdot 10^{-4}$. We will review the plasma injection mechanism of Fisk (1978) for yielding ^3He enhancements. This mechanism is a good example of an injection mechanism and one for which there is considerable observational support of its occurrence. Indeed, as we will see, this mechanism may be a common occurrence in the solar corona and account for other compositional enhancements in flare events besides ^3He.

Consider first the basic properties of ^3He-rich flares: The first property of these events is of course that the $^3He/^4He$ ratio is high. We shall use here, as an operational definition of ^3He-rich flares, that the $^3He/^4He$ ratio at energies ~a few MeV/nucleon is $>10\%$. Flares with smaller $^3He/^4He$ ratios are of course observed. However, it may be that the ^3He content in these events can be explained by more conventional means involving straightforward nuclear reactions between higher solar cosmic rays and ambient coronal material.

The second property of ^3He-rich events is that only upper limits on deuterium or tritium are observed. In the events with a large $^3He/^4He$ ratio--on the order of unity--the limits on the deuterium and tritium are fairly severe, e.g., $d/^3He <$ a few percent. For events with smaller $^3He/^4He$ ratios of ~10%, the observational limits are less severe, e.g., $d/^3He < 10\%$. These less severe limits of course do not imply a higher deuterium or tritium content in these events, but merely reflect observational difficulties in observing the deuterium and tritium.

The observed restriction on the deuterium and tritium content in ^3He-rich flares places severe constraints on the possibility that the ^3He is produced by nuclear reactions. Nuclear reactions tend to produce deuterium and tritium as roughly equal by-products with ^3He.

The third property of ^3He-rich flares is that they are the result of quite small flare events on the Sun. The $^3He/^4He$ ratio increases with decreasing event size, and the largest ratios, $^3He/^4He \sim 1$, are observed in very small flares.

The fourth property of these events is that there is a correlation, not always one to one, between large $^3He/^4He$ ratios and enhancements in the heavier elements, particularly iron. Indeed, the enhancements in iron over normal coronal abundances can be comparable in magnitude to the enhancements in ^3He.

The final property of ^{3}He-rich flares is that the ^{4}He content tends to be high. The events with the largest ^{3}He/^{4}He ratios tend to have small H/^{4}He ratios, or equivalently a large ^{4}He content.

Consider the plasma conditions that we might reasonably expect to occur at the site of a ^{3}He-rich flare, which we suppose lies in the low corona. First, we should expect that the β of the plasma, the ratio of thermal-to-magnetic pressure, is small, as is characteristic of conditions in the low corona. Indeed, the β might be quite small. We recall that ^{3}He-rich flares tend to be small events, which in many cases are not readily visible on the Sun. We expect, then, that the thermal energy or equivalently the thermal pressure at the flare site is not large. However, particles are to be accelerated here so that some reasonable magnetic-field strength might be required. We thus assume that $\beta > 10^{-3}$, which is the β of a plasma at a few million degrees (i.e., just above average coronal temperatures), with a density of $\sim 10^9$ cm^{-3}, in a field of 75 Gauss. we further assume, consistent with our requirement that ^{3}He-rich flares are weak events, that the electron and ion temperatures are not particularly different at the flare site, i.e., $T_e/T_i < 10$, and that the currents in the plasma are relatively weak.

In such a plasma a likely instability to occur is the excitation of electrostatic ion cyclotron waves. These waves are excited when the plasma electrons drift relative to the ions at a speed small compared with the electron thermal speed, but large compared with the ion thermal speed (Drummond and Rosenbluth, 1962). This instability is one of the most likely to occur when the β of the plasma is as low as specified here, and the T_e/T_i ratio is not large. It is more likely to occur in these conditions than, e.g., the excitation of ion acoustic waves (Kindel and Kennel, 1971).

In a plasma that contains only protons and electrons, the electrostatic ion cyclotron waves occur in a narrow frequency range just above the proton cyclotron frequency. We recall, however, that ^{3}He-rich flares can be overabundant in ^{4}He. We suppose that the reason for this overabundance is that the ^{4}He content in the ambient material at the flare site is unusually high. After all, the solar wind, as it expands, tends to leave ^{4}He behind in the corona, and thermal diffusion between the corona and the chromosphere tends to yield a relatively high concentration of ^{4}He in the low corona. In the case where the ^{4}He content in the plasma is larger than normal, electrostatic ion cyclotron waves are also excited in a narrow frequency range just above the ^{4}He cyclotron frequency, which, since the ^{4}He is fully ionized, is at half the proton cyclotron frequency.

The electrostatic ion cyclotron waves that are excited by the drifting electrons will be damped in part by the ions in the plasma, or equivalently the ions will be heated by the waves. This heating will be particularly effective for any ion species whose cyclotron frequency lies near the frequency of the unstable waves, since then the bulk of the ions can resonate with the waves at the first harmonic of their cyclotron frequency. There is, however, only one ion species--$^3He^{+2}$-- that has its cyclotron frequency between the ^4He and proton cyclotron frequencies, where the waves are excited. Any fully stripped ion with an equal number of protons and neutrons has the same cyclotron frequency as $^4He^{+2}$. Any partially stripped ion has a lower cyclotron frequency than $^4He^{+2}$. Also the isotopes expected in the corona, with the exception of ^3He, should all be neutron rich; i.e., they have more neutrons than protons, or a cyclotron frequency lower than that of $^4He^{+2}$.

Thus, if we take reasonable conditions for the plasma at the site of a ^3He-rich flare, we find that we can excite waves that will preferentially interact with and heat ambient ^3He.

The plasma physics for this mechanism can be worked out in some detail, as has been reported in Fisk (1978). This instability has been considered for the ionosphere, and these latter analyses can be applied directly to the solar problem. A linear analysis shows that the mechanism works well provided that the $^4He/H$ ratio in the ambient plasma is >20%. It is found also that the mechanism works best when $T_e/T_i \sim 5$, since then the frequency of the unstable waves lies very close to the $^3He^{+2}$ cyclotron frequency. Neither of these requirements is unreasonable. For example, the electron current which is required to drive the

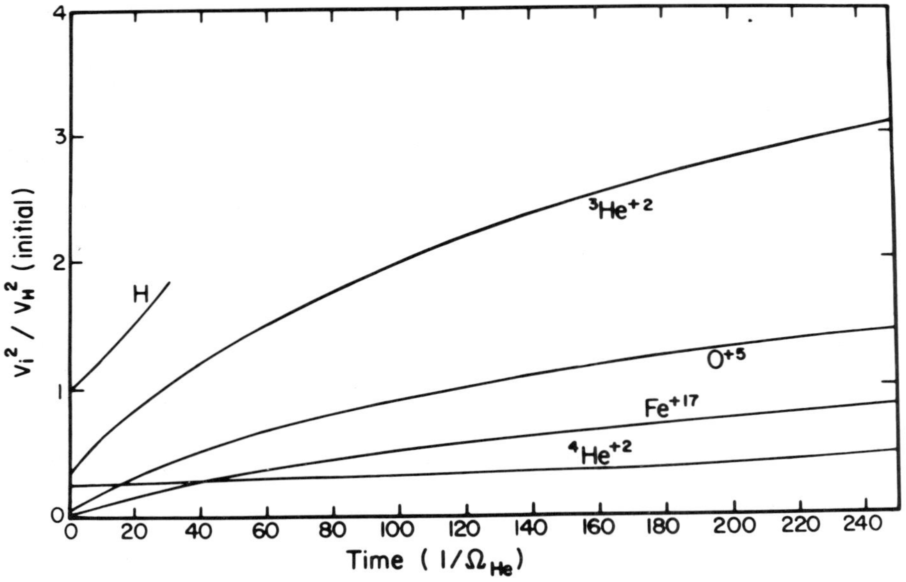

Figure 1. The increases in the square of the various ion thermal speeds, with time (in units of hydrogen cyclotron period), that results from the injection mechanism of Fisk (1978). The ion thermal speeds are compared to the initial proton thermal speed v_H (initial). The ions are assumed to be at equal temperatures initially.

instability is expected to raise the electron temperature relative to the ion temperature by Joule heating alone.

In Fisk (1978) a nonlinear analysis similar to that of Palmadesso et al. (1974) is also performed, from which it is possible to estimate the heating rates for various ion species. The results of this analysis are shown in Figure 1, where the square of the thermal speed of various ion species is plotted versus time, as the instability proceeds.

As can be seen in Figure 1, the instability also heats, by nonlinear processes, the protons and ^{4}He in the plasma. In fact, this is the mechanism by which the instability is expected to saturate. Increasing the temperature of the protons and ^{4}He raises the threshold for the instability and shuts it off; i.e., increasing the temperature increases the critical drift speed beyond which the electrons must be drifting for the instability to proceed. With the parameters chosen here, i.e., a relative concentration of ^{4}He to H of 30%, $T_e/T_i \sim 5$, and an electron drift speed of 15% of their thermal speed, the ^{4}He and proton temperatures are increased by a factor $\sim$ 1.8. When the protons have been heated by this factor, which occurs more rapidly than for ^{4}He, the instability at frequencies above the proton cyclotron frequency saturates. When ^{4}He is heated by this factor, the waves located between the ^{4}He and proton cyclotron frequencies saturate.

However, while waves are present between the ^{4}He and proton cyclotron frequencies, the $^3He^{+2}$ is very effectively heated since its cyclotron frequency lies near the frequency of these unstable waves. As can be seen in Figure 1, its temperature is increased by a factor $\sim$ 9.

The flare acceleration process presumably takes particles from the tail of the Maxwellian distribution of the thermal plasma and accelerates them to high energies. This acceleration should act only on particles above some threshold speed, which must be located well into the tail of the distribution since only a small fraction of the available particles are actually accelerated to high energies. Suppose that this threshold speed for ^{4}He is such that one part in 10^4 of the ^{4}He is accelerated; such a threshold is compatible with flare observations. Suppose also that ^{3}He has approximately the same threshold speed. However, following the heating described in Figure 1, the ^{3}He has a much broader thermal distribution than ^{4}He, and essentially all of the ^{3}He will have speeds above the threshold speed and will be accelerated. If we start then with a ^{3}He/^{4}He ratio of $\sim 10^{-4}$ in the ambient plasma, and we inject one part in 10^4 of the ^{4}He and essentially all of the ^{3}He, we will get a ^{3}He/^{4}He ratio on the order of unity in the accelerated particles. Thus, the mechanism described here can in principle yield the observed ^{3}He/^{4}He ratio on the order of unity.

Although $^3He^{+2}$ is the only ion that has its cyclotron frequency near the frequencies of the unstable waves, certain partially stripped heavier ions have the second harmonics of their cyclotron frequencies in this range. As is discussed in Fisk (1978) these certain ions will also interact with the unstable waves and can be heated by them. Thus, they too may be preferentially injected into the flare acceleration process, or equivalently the mechanism described here may be able to account for other compositional anomalies seen in flare events. Also shown in Figure 1 is the heating that is expected for $^{16}O^{+5}$ and $^{56}Fe^{+17}$, which are two ions with the second harmonics of their cyclotron frequencies near the frequencies of the unstable waves.

The enhancements in the partially stripped heavy ions provide a good test of this mechanism since neither O^{+5} nor Fe^{+17} occur commonly in flare events. In fact, Ma Sung et al. (1981) have found that, in a large ^3He-rich flare for which the charge states of the energetic ions could be measured, that O^{+5} and Fe^{+17} appear to be present. These observations have been extended recently to other ^3He-rich flares (Gloeckler, private communication) and again O^{+5} and Fe^{+17} was observed.

By varying the plasma parameters in this mechanism, a variety of compositional enhancements can be generated. For example, it is possible to yield enhancements in the partially stripped heavier ions without an enhancement of ^3He. In fact, Mason et al. (1980) have shown that this plasma mechanism can account for all of the compositional enhancements observed in a four-year study of small flare events.

4. The acceleration

There are three principal mechanisms considered for accelerating energetic particles in flare events: statistical acceleration in turbulence generated by the flare, shock acceleration in a flare-generated shock wave, and acceleration in collapsing magnetic fields at the flare site. The theories for the acceleration are relatively primitive and offer little opportunity at present for distinguishing which mechanism is operative. We will mention here only the current limitations with these theories; for more details the reader is referred to Forman et al. (1982).

In statistical acceleration particles interact randomly with turbulence in a plasma and diffuse upward in momentum. (See the chapter on Acceleration of Energetic Particles in the Solar Wind for a more detailed discussion of the theory of statistical acceleration.) In the case of solar flares, however, the nature of the turbulence at or near the flare site is essentially unknown, and thus, the time-scale for the acceleration and detailed spectra cannot be determined.

In shock acceleration, particles gain energy by being compressed between the magnetic irregularities upstream from the shock and the shock front and/or the downstream irregularities. (See the chapter on Acceleration of Energetic Particles in the Solar Wind for a more detailed discussion of shock acceleration.) In the case of solar flares, however, the geometry of the medium through which the shock is passing is very complex; e.g., it can contain complicated magnetic loop structures, a diverging solar wind flow, etc. Thus, the ability of the particles to remain near the shock front and be accelerated is difficult to calculate, as is the spectra of the accelerated particles.

It should be noted that in simple calculations which have been done, statistical acceleration and shock acceleration in flare events each yield very similar spectra (Forman et al., 1982). In statistical acceleration particles gain and lose momentum randomly, with a small fraction diffusing to high energies. In shock acceleration particles gain energy at the shock front but lose energy in the expanding plasma ahead of and behind the shock, with a small fraction remaining near the shock for sufficient duration to reach high energies. The statistical aspect of the acceleration in both cases yields similar spectra.

In acceleration in collapsing magnetic fields, particles gain energy in the electric fields generated by magnetic reconnection. Again, the geometries assumed in calculations of this acceleration process are substantially simpler than the complex magnetic configurations which occur in real flares. Moreover, this acceleration process may be more applicable to the acceleration of particles that remain trapped near the flare site and produce the observed electromagnetic radiation. As we discussed in section 2, the solar cosmic rays seen in the solar wind may be produced in the corona surrounding the flare site, where magnetic reconnection is not an important process.

5. The propagation

Following their acceleration in solar flares the energetic particles can leak from the lower corona and propagate outward into the solar wind. Here the particles are channeled along the Archimedes spiral pattern of the large-scale heliospheric magnetic field and are scattered in pitch angle by small-scale magnetic irregularities. These irregularities control the rate at which the particles leak outward through the solar wind; they determine the variation of the particle intensity with time and the spectra which are observed.

The interactions of the particles with the magnetic irregularities in the solar wind is normally considered to be a diffusive process which is governed by a diffusion coefficient $\kappa_{\parallel}$ that describes the rate at which particles diffuse parallel to the mean magnetic field B_0 due to pitch-angle scattering. The coefficient $\kappa_{\parallel}$ can be determined in two ways: (i) in a heuristic way from direct observations of solar cosmic-ray behavior during flare events, or (ii) directly from observations of the magnetic field irregularities. In principle these two determinations should agree. In practice they are in wide disagreement. It is one of the major unsolved theoretical problems in solar cosmic-ray physics to determine why.

Consider now the quasi-linear derivation for the rate of change of a particle's mean square pitch angle due to scattering from magnetic irregularities, which we will shortly relate to $\kappa_{\parallel}$. We will follow closely the derivation presented in Fisk (1979).

The energetic particles are assumed to interact only with the electromagnetic fields in the solar wind, and not with each other. These fields, in turn, are assumed to be unaffected by the particles since the particle energy density is much less than that of the field. The fields are also taken to be known quantities, which can be specified at least in a statistical sense. The only equation that is required here is then the Vlasov equation, which governs the behavior of the particle one-dimensional distribution function $f(\underset{\sim}{r}, \underset{\sim}{p}, t)$:

$$\left(\frac{\partial}{\partial t} + \frac{\underset{\sim}{p}}{\gamma m} \cdot \nabla + \underset{\sim}{F} \cdot \frac{\partial}{\partial \underset{\sim}{p}}\right) f = 0 \qquad (1)$$

Here, r is position and p, particle momentum. The particle rest mass is m and the Lorentz factor is $\gamma = (1 + p^2/m^2c^2)^{1/2}$, where c is the speed of light. The electromagnetic force is

$$\underset{\sim}{F} = q(\underset{\sim}{E} + \underset{\sim}{p} \times \underset{\sim}{B}/\gamma mc) \qquad (2)$$

where $\underset{\sim}{E}$ and $\underset{\sim}{B}$ are the electric and magnetic fields respectively, and q is charge.

We are not interested, of course, in the detailed behavior of f, but rather in its average behavior on time-scales long compared to the time during which the particles are scattered by interactions with the field. This coarse-grained behavior is what we require for our diffusion equation. To obtain this behavior we invoke the usual ergodic hypothesis in which averages of f over time are equated to averages of f over ensembles of $\underset{\sim}{F}$.

By denoting ensemble averages by brackets <>, and by assuming that f and F have an average and fluctuating part (e.g., $f = \langle f\rangle + \delta f$) we can rewrite (1) as two equations:

$$\left(\frac{\partial}{\partial t} + \frac{\underset{\sim}{p}}{\gamma m} \cdot \nabla + \langle \underset{\sim}{F}\rangle \cdot \frac{\partial}{\partial \underset{\sim}{p}}\right) \langle f\rangle = -\frac{\partial}{\partial p} \cdot \langle \delta \underset{\sim}{F}\, \delta f\rangle \qquad (3)$$

and

$$\left(\frac{\partial}{\partial t} + \frac{\underset{\sim}{p}}{\gamma m} \cdot \nabla + \langle \underset{\sim}{F}\rangle \cdot \frac{\partial}{\partial p}\right) \delta f + \delta \underset{\sim}{F} \cdot \frac{\partial}{\partial p}\, \delta f - \langle \delta \underset{\sim}{F} \cdot \frac{\partial}{\partial p}\, \delta f\rangle$$

$$= -\delta \underset{\sim}{F} \cdot \frac{\partial \langle f\rangle}{\partial \underset{\sim}{p}} \qquad (4)$$

We then approximate (4) in such a manner that we can solve it exactly for δf in terms of its initial value $\delta f(r, p, t_o)$, and $\langle f\rangle$. Upon substituting this result into (3) we obtain a single equation for the required ensemble averaged $\langle f\rangle$.

The most basic of the approximations to (4) is the quasi-linear approximation, in which terms that are second order in the fluctuating components δf and δF are ignored, or (4) becomes

$$\left(\frac{\partial}{\partial t} + \frac{\underset{\sim}{p}}{\gamma m} \cdot \nabla + \langle \underset{\sim}{F} \rangle \cdot \frac{\partial}{\partial \underset{\sim}{p}}\right) \delta f = - \delta \underset{\sim}{F} \cdot \frac{\partial \langle f \rangle}{\partial \underset{\sim}{p}} \qquad (5)$$

This equation can be readily solved for δf, and the result substituted into (3) to yield

$$\left(\frac{\partial}{\partial t} + \frac{\underset{\sim}{p}}{\gamma m} \cdot \nabla + \langle \underset{\sim}{F} \rangle \cdot \frac{\partial}{\partial \underset{\sim}{p}}\right) \langle f \rangle = - \frac{\partial}{\partial \underset{\sim}{p}} \cdot \langle \delta F \; U_o(t,t_o) \delta f(\underset{\sim}{r},\underset{\sim}{p},t_o)) \rangle$$

$$+ \frac{\partial}{\partial \underset{\sim}{p}} \cdot \langle \delta F \int_{to}^{t} d\tau U_o \; (t,\tau) \delta F(\underset{\sim}{r},\underset{\sim}{p},\underset{\sim}{t}) \rangle \cdot \frac{\partial}{\partial \underset{\sim}{p}}$$

$$\langle f \rangle \; (r,p,\tau) \qquad (6)$$

Here, $U_o(t,t')$ is an operator which propagates the $\underset{\sim}{r}$ and $\underset{\sim}{p}$ dependence of all functions to the right of it backwards in time from t to t' along trajectories in the force field $\langle F \rangle$. Since there are no mean electric fields in interplanetary space ($\langle E \rangle = 0$), this operator then causes the functions to the right of it to be evaluated along the trajectory of a particle as it moves backwards in time, spiralling about the mean magnetic field $\langle \underset{\sim}{B} \rangle \equiv \underset{\sim}{B}_o$.

Equation (6), however, is not yet in the form of a diffusion equation, and some additional approximations must be made. The first and probably most reasonable is that the initial value term in (6) can be ignored. We assume that correlations between δF and $U_o(t,t_o)\delta f(\underset{\sim}{r},\underset{\sim}{p},t_o)$ decrease sufficiently rapidly so that after a reasonable time the ensemble average of these quantities is zero.

We also assume for simplicity that $\langle f \rangle$ is gyrotropic about $\underset{\sim}{B}_o$. Time variations in $\langle f \rangle$ on the scale of the gyroperiod are thus not considered here. This assumption, however, precludes consideration cross-field diffusion.

We assume further that a change in a particle's orbit (i.e., a scattering of a particle) occurs on time-scales that are long compared with the time over which the particle makes a correlated interaction with the fluctuating fields, i.e., long compared with the time $(t-t_o)$ that is required for the integral in (6) to saturate and to obtain essentially the full value that it will reach as $t-t_o \to \infty$. This latter assumption is of course consistent with the quasi-linear approximation. We assumed in deriving (6) that the interaction of the particles with the fields could be calculated by knowing only the particle's unperturbed trajectory along the mean magnetic field. As a consequence of this and the previous assumption, however, we can assume that $\langle f \rangle$ does not evolve on the time-scales required for the integral in (6) to saturate, i.e., that $U_o(t,\tau)$ $\langle f\rangle(\underset{\sim}{r},p,\tau)\simeq\langle f\rangle(\underset{\sim}{r},\underset{\sim}{p},t)$ and that $\langle f\rangle$ can be removed from under the integral. This last approximation is known as the adiabatic approximation.

Clearly, both the quasi-linear the adiabatic approximation will have difficulties in dealing with particles that have pitch angles near 90°. With a small change in pitch angle such particles can reverse direction and thus hardly maintain the unperturbed orbit that is assumed in quasi-linear theory. Moreover, these particles move slowly along the mean field, and thus for them there is no clear separation of the scattering time scale and the time required for the integral in (6) to saturate.

With the quasi-linear and adiabatic approximations, however, and with some manipulation, (6) can be written in the form of a Fokker-Planck equation, or as

$$\left(\frac{\partial}{\partial t} + \frac{\underset{\sim}{p}}{\gamma m} \cdot \nabla + \langle \underset{\sim}{F} \rangle \cdot \frac{\partial}{\partial \underset{\sim}{p}}\right)\langle f\rangle = \frac{\partial}{\partial \underset{\sim}{p}} \cdot \underline{\underline{D}} \cdot \frac{\partial \langle f\rangle}{\partial \underset{\sim}{p}} \tag{7}$$

where the diffusion tensor $\underline{\underline{D}}$ has components

$$D_{ij} = \hat{e}_i \cdot \underline{\underline{D}} \cdot \hat{e}_j = \hat{e}_i \; Y \int_0^\infty d\tau \delta F(\underset{\sim}{r},\underset{\sim}{p},t) U_o(t,t-\tau) \delta F(\underset{\sim}{r},\underset{\sim}{p},t-\underset{\sim}{\tau})\rangle \cdot \hat{e}_j \tag{8}$$

Here, $\hat{e}_i$ is a unit vector which projects $\underset{\sim}{p}$ onto the axes of an orthogonal coordinate system. We have considered only diffusion in momentum space under the assumption that to lowest order the spatial gradients in $\langle f\rangle$ can be ignored in the calculation of particle interactions with the field. Note also that U_o acts on $\hat{e}_j$ and that the limits of the integral have been extended from $(0,t-t_o)$ to $(0,\infty)$. This last simplification is possible because we are dealing by assumption with times $(t-t_o)$ long compared with the time required for the integral to saturate.

For the propagation problem we are concerned with directional changes in $\underset{\sim}{p}$ as opposed to magnitude changes. Moreover, we are concerned here only with pitch-angle changes since cross-field diffusion was explicitly eliminated from consideration by our assumption of gryotropy. Consider, then, that we express $\underset{\sim}{p}$ in a spherical coordinate system with the polar angle θ measured relative to the mean magnetic field $\underset{\sim}{B}_o$ (i.e. θ is the pitch angle). Equation (7), then, for propagation parallel to $\underset{\sim}{B}_o$ only, reduces to

$$\left(\frac{\partial}{\partial t} + \frac{\underset{\sim}{p}}{\gamma m} \cdot \nabla + \langle F \rangle \cdot \frac{\partial}{\partial p}\right)\langle f\rangle = \frac{\partial}{\partial \mu} \; D_{\mu\mu} \; \frac{\partial \langle f\rangle}{\partial \mu} \tag{9}$$

where we have followed the usual convention and described the pitch-angle scattering in terms of $\mu=\cos\theta$. Here

$$D_{\mu\mu} = (1-\mu^2)D_{\theta\theta}/p^2 \tag{10}$$

where $D_{\theta\theta}$ is determined from (8).

In evaluating (10) we must of course specify the fluctuating force $\delta\underset{\sim}{F}$ or equivalently the fluctuating magnetic ($\delta\underset{\sim}{B}$) and electric ($\delta\underset{\sim}{E}$) field. For the propagation problem we need concern ourselves only with the $\delta\underset{\sim}{B}$. In the solar wind, the fluctuations which affect energetic particles should arise from waves that are hydromagnetic in nature. For these waves the ratio of the force that is due to the electric field to that from the magnetic field is $\sim\omega/kv$, where ω/k is the phase speed of the waves and v, the particle speed. Since ω/k is the Alfvén speed which is typically ~50 km/sec and thus small compared wtih v, the fluctuating magnetic field dominates in producing the scattering. Our problem then reduces in effect to a magnetostatic problem.

The knowledge of δB that we require for (8) and (10) is the behavior of its two-point correlation tensor $\underline{\underline{C}}(\underset{\sim}{r}') = \langle\delta\underset{\sim}{B}(r)\delta\underset{\sim}{B}(\underset{\sim}{r}+\underset{\sim}{r}')\rangle$, where it is assumed here that the turbulence is homogeneous. Our present knowledge of this tensor, however, is incomplete. Spacecraft measure the interplanetary magnetic field as it is convected past by the solar wind. The correlations then can be measured only a along a line which lies in the heliocentric radial direction, a procedure that is not sufficient to specify the full tensor without additional assumptions (e.g., that the tensor is isotropic).

The measured correlations of the field are often expressed in terms of their Fourier transforms, i.e. as one-dimensional power spectra. As the field is convected past the spacecraft, what is measured, of course, is a time series, which can be converted into a power spectrum as a function of observed frequency ω'. This power spectrum, in turn can be expressed in terms of k_r, the component of the wave vector k in the heliocentric radial direction. The wave speeds are much smaller than the solar wind speed, with the result that $\omega' \simeq k_r V$.

Suppose then that we take the Fourier transform or $\underline{\underline{C}}(r')$ which for homogeneous turbulence is

$$\tilde{\underline{\underline{C}}}(\underset{\sim}{k}) = \delta\tilde{\underset{\sim}{B}}(\underset{\sim}{k})\ \delta\underset{\sim}{B}(-\tilde{\underset{\sim}{k}}) \tag{11}$$

where

$$\delta\tilde{\underset{\sim}{B}}(\underset{\sim}{k}) = \int_{-\infty}^{\infty} e^{i\underset{\sim}{k}\cdot\underset{\sim}{r}}\ \delta\underset{\sim}{B}(\underset{\sim}{r})d^3r \tag{12}$$

Note that we have ignored time variations in δB under the assumption that the scattering is, in effect, a magnetostatic problem. The measured power spectra then provide us only with $\underline{\underline{P}}(k_r)$ where

$$\underline{\underline{P}}(k_r) = \int \tilde{\underline{\underline{C}}}(\underset{\sim}{k})d^2k \tag{13}$$

is the integration of $\tilde{\underline{\underline{C}}}(k)$ over the components of $\underset{\sim}{k}$ normal to k_r.

Since we cannot specify $\tilde{\underline{\underline{C}}}(\underset{\sim}{k})$ in detail, we must thus make assumptions about this tensor. The usual procedure here is to assume that the fluctuations are due only to normal mode oscillations in the solar wind, which in our case restricts us to transverse Alfvén waves, and fast and slow mode magnetosonic waves. We then construct from these waves acceptable forms of $\tilde{\underline{\underline{C}}}$.

For example, suppose that we assume that the waves are Alfven waves, which have the property that $\delta\underset{\sim}{B}$ must lie in the direction which is normal to the plane defined by $\underset{\sim}{k}$ and $\underset{\sim}{B}_o$. We thus assume that

$$\delta\underset{\sim}{\tilde{B}}(k) = \frac{A(\underset{\sim}{k})}{k_\perp B_o}\, \underset{\sim}{k} \times \underset{\sim}{B}_o \tag{14}$$

where $k_\perp$ is the component of $\underset{\sim}{k}$ normal to $\underset{\sim}{B}_o$. The weighting function in $A(\underset{\sim}{k})$ must be normalized such that

$$\int d^3k\, A(\underset{\sim}{k})\, A(-\underset{\sim}{k}) = \langle(\delta B)^2\rangle \tag{15}$$

and it is chosen such that (13) yields a power spectrum $\underline{\underline{P}}(k_r)$ which resembles the observations. If the waves are magnetosonic then

$$\delta\underset{\sim}{\tilde{B}}(k) = \frac{A(\underset{\sim}{k})}{k\, k_\perp\, B_o}\, \underset{\sim}{k} \times (\underset{\sim}{k} \times \underset{\sim}{B}_o) \tag{16}$$

We can also consider correlation tensors which are the result of a combination of Alfvén and magnetosonic waves.

Upon substituting our assumed form for the correlation tensor into (8) we can then evaluate $D_{\theta\theta}$, and in turn through (10) obtain the pitch-angle diffusion coefficient $D_{\mu\mu}$. This procedure will yield the usual result that particles respond only to waves which satisfy the resonance condition

$$\mu\, k_\parallel\, v = \pm\, n\, \Omega \tag{17}$$

where $k_\parallel$ is the component of $\underset{\sim}{k}$ parallel to $\underset{\sim}{B}_o$; Ω is the particle gyrofrequency in the mean field and n is an integer denoting the harmonic. The result in general will involve an infinite sum over these harmonics.

In most cases the infinite sum of $D_{\mu\mu}$ needs to be evaluated numerically. There are, however, two exceptions which illustrate in a simple form the scattering that is predicted by quasi-linear theory. As has been pointed out by Jokipii (1972), if we assume that the turbulence is the result of Alfvén waves which propagate only along the mean magnetic field direction, then

$$D_{\mu\mu} = \frac{2\pi(1-\mu^2)v}{r_g^{\,2} B_o^{\,2}}\, P_\perp\, (k=1/|\mu| r_g) \tag{19}$$

to a good approximation. Note the similarities between (18) and (19); essentially $D_{\mu\mu}$ in (19) is only a factor of $|\mu|$ greater than $D_{\mu\mu}$ in (18).

Measurements of particle behavior during flare events do not generally yield $D_{\mu\mu}$ directly, but rather yield $\kappa_{\parallel}$. According to Earl (1974), who has studied this aspect of the propagation problem in problem in detail, the proper averaging scheme relates $\kappa_{\parallel}$ to $D_{\mu\mu}$ according to the formula

$$\kappa_{\parallel} = \frac{v2}{4} \int_{-1}^{1} \frac{d\mu(1-\mu 2)}{D_{\mu\mu}} \tag{20}$$

Waves and turbulence in the solar wind have relatively large amplitude so that the quasi-linear approach used above to derive $D_{\mu\mu}$ and in turn $\kappa_{\parallel}$ should be suspect. However, the non-linear corrections which have been considered to date generally predict more scattering than the quasi-linear approach (e.g. Volk, 1973, 1975; Jones et al. 1973). Thus, the quasi-linear result provides an upper limit on the predicted scattering, which is sufficient, as we shall see, to demonstrate the discrepancy between theory and observations.

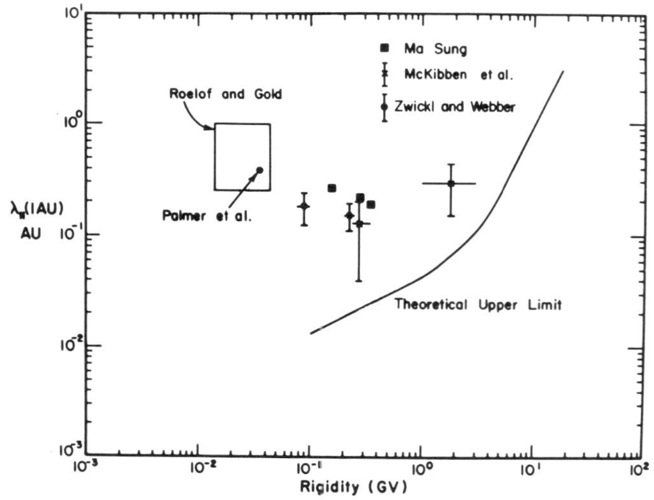

Figure 2. A comparison between the mean free path parallel to the mean magnetic field ($\lambda_{\parallel}$) as inferred from particle observations, and the quasi-linear prediction for $\lambda_{\parallel}$, which is an upper limit to current theoretical predictions. The data was compiled by Zwickl and Webber (1978).

Shown in Figure 2 are the mean free path $\lambda_{\parallel}$ ($=3\kappa_{\parallel}/v$ where v is the particle speed) for propagation parallel to $\underset{\sim}{B}_o$, as was determined from several studies of flare observations. Shown also is the quasi-linear result for $\lambda_{\parallel}$, which we consider an upper limit on current theoretical predictions. This result is determined from (18) and (20), with observed values for $P_{\perp}$. Clearly, the quasi-linear result lies systematically below the data points. There is no evidence in the theoretical curve for the flattening or increasing at low rigidities which appears to be required by the data. Indeed, at low rigidities there is roughly one order of magnitude difference between theory and observations.

Clearly, we do not understand the nature of waves and turbulence in the solar wind, and how energetic particles interact with them. This is a theoretical problem which requires attention.

REFERENCES:

Drummond, W.E., and Rosenbluth, M.N.: 1962, Phys. Fluids 5, 1507.

Fisk, L.A.: 1978, Astrophys. J., 224, 1048.

Fisk, L.A.: 1979, Solar System Plasma Physics, North-Holand Publishing Co., 1, 179.

Forman, M.A., Ramaty, R., and Zweibel, E.G.: 1982, The Physics of the Sun, in press.

Jokipii, J.R.: 1972, Astrophys. J., 172, 319.

Jones, F.C., Kaiser, T.T., and Birmingham, T.J.: 1973, Phys. Rev. Letters, 31, 485.

Kendel, J.M., and Kennel, C.Y.: 1971, J. Geophys. Res., 76, 3055.

Mason, G.M., Fisk, L.A., Hovestadt, D., and Gloeckler, G.: 1980; Astrophys. J., 239, 1070.

MaSung, L.S., Gloeckler, G., Fan, C.Y., and Hovestadt, D.: 1981, Astrophys. J., 245, L45.

Palmadesso, P.L., Coffey, T.P., Ossakow, S.L. and Papadopoulos, K.: 1974, Geophys. Res. Letters, 1, 105.

Völk, H.J.: 1973, Astrophys. Space Sci., 25, 471.

Völk, H.J.: 1975, Rev. Geophys. Space Phys., 13, 547.

von Rosenvinge, T.T., Ramaty, R., and Reames, D.V.: 1981, 17th International Cosmic Ray Conf., 3, 28.

Zwickl, R.D., and Webber, W.R.: 1977, Solar Phys., 54, 452.

SOLAR MODULATION OF GALACTIC COSMIC RAYS

L.A. Fisk
Department of Physics
University of New Hampshire
Durham, New Hampshire 03824

A brief description is provided of the basic principles of and the techniques which are used for describing the behavior of galactic cosmic rays in the solar wind. Several numerical examples which illustrate features of the cosmic-ray behavior are presented.

1. Introduction

The purpose of this section is to provide an overview of the techniques which are used to describe the behavior of galactic cosmic rays in the solar wind--the so-called solar modulation of the galactic cosmic-ray flux. This subject has been worked on extensively for more than twenty years, and there are many aspects to this problem which are discussed in the extensive literature that is available. We will concentrate here only on the basic points, leaving it to the reader to pursue the finer details (see, e.g., the review article by Fisk (1979) and references therein). We will begin by describing the basic data on the cosmic-ray behavior over the solar cycle. We will then discuss the ideas behind modulation theory and derive the equations which describe these ideas. We will review the techniques for and the assumptions made in solving the modulation equations, and then we will describe some illustrative examples of these solutions.

2. Solar-cycle variations

Shown in Fig. 1 is the galactic cosmic-ray flux as measured by the Mt. Washington neutron monitor. Cosmic rays striking the upper atmosphere produce neutrons, which with proper corrections become a direct measure of the cosmic-ray flux. You will note that the flux in Fig. 1 varies with time roughly in anti-coincidence with solar activity. During periods of low activity on the Sun, such as in recent years or in the mid-1960's, the cosmic-ray flux is high. During high activity, such as in 1970, the flux is low. This temporal variation in the cosmic-ray flux, which is induced by changing conditions in the heliosphere--the region in space over which the Sun has a major influence--is what we refer to as the solar-cycle modulation of the cosmic rays.

You will note that no attempt has been made here to plot the

R. L. Carovillano and J. M. Forbes (eds.), Solar-Terrestrial Physics, 217–230.

cosmic-ray flux vs. the mean sunspot number, which is a common correlation to make. Sunspots, in themselves, have no effect on cosmic-ray modulation. The magnetic fields of sunspots are strong and closed, whereas the cosmic rays, as we shall discuss, respond to the weaker photospheric field which is dragged out into the heliosphere by the solar wind. Of course, changes in sunspot number may correlate with changes in heliospheric conditions; but exactly how this correlation works is presently unknown.

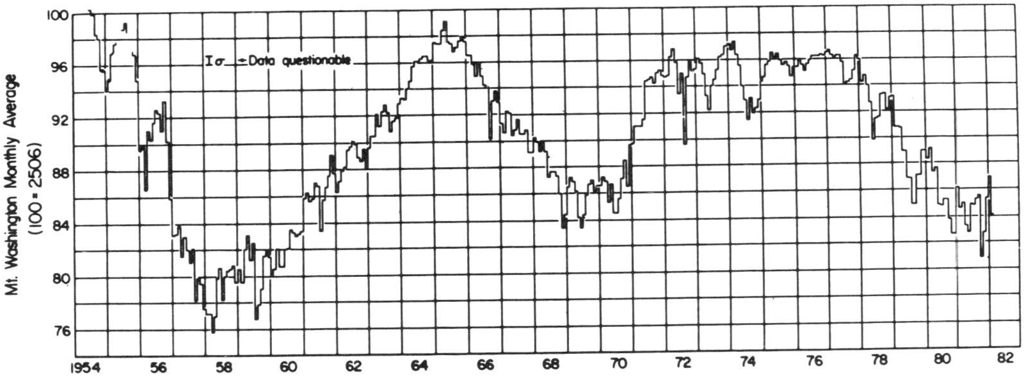

Figure 1. The monthly average counting rate of the Mt. Washington neutron monitor from 1954-1982 (courtesy of J.A. Lockwood).

You will note also that the variation in flux shown in Fig. 1 is not large. The scale is linear and the total change in flux from minimum to maximum activity conditions is only about 20%. However, neutron monitors measure the flux of relatively high energy particles; e.g., the Mt. Washington neutron monitor samples particles with energies ~5 GeV. As we move down in energies to particles that are measured by spacecraft, the modulation becomes larger.

Shown in Fig. 2 is a plot, versus energy, of the differential intensity of galactic cosmic-ray protons in space, i.e., the number of protons observed per unit time, area, steradian and energy. The data points are spacecraft measurements in different years; the curves are a crude fit to these data. As can be seen in this figure, at lower energies, e.g., ~50 MeV, the difference in the flux between 1965 or 1972, which are solar-minimum conditions, and the 1969, which is near solar maximum, is a factor of 5 -10.

3. The basic idea

The basic concepts which describe how cosmic rays behave in the heliosphere can be stated relatively simply:

The solar wind flows radially outward from the Sun in all directions and drags with it, as a passive partner, the solar magnetic field. The field, however, remains attached to the rotation Sun, and thus on the large scale executes an Archimedes spiral pattern. On the small scale, the magnetic field contains many irregularities: waves, turbulence, and discontinuities.

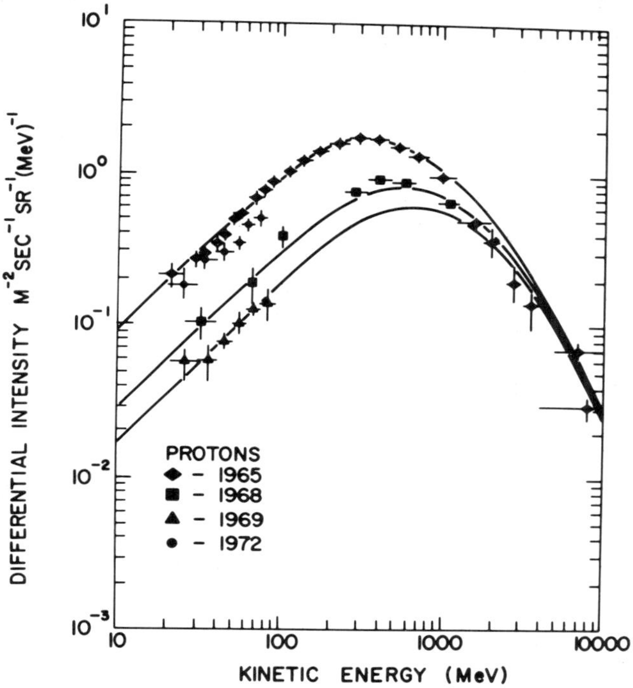

Figure 2. The differential proton intensity near earth in 1965, 1968, 1969 and 1972 (from Fisk, 1972).

The cosmic rays are charged particles and thus interact with the magnetic field in the solar wind. The motion of the particles is directed along the large-scale field. The particles are scattered in pitch angle by the small-scale irregularities.

The solar wind, then, tends to sweep galactic cosmic rays out of the heliosphere, or equivalently the cosmic rays must fight their way upstream against the outward flow of the wind. Not all the cosmic rays successfully make this trek, with the result that the cosmic-ray flux seen in the inner heliosphere is lower than it is in the interstellar medium. And if we vary conditions in the heliosphere, which presumably occurs over the solar cycle, the cosmic-ray flux will change in time.

However, within these relatively simple statements are contained some fairly complicated and, in some cases, some fairly subtle physics.

4. The modulation equations

We require, of course, a set of equations to describe the behavior of cosmic rays in the heliosphere. There are several, equivalent derivations for these equations available (e.g., Parker, 1965; Gleeson and Axford, 1967). We will follow closely here the derivation put forth by Fisk (1979). Although not completely rigorous in all regards, this latter derivation is designed to illustrate the underlying physics of the cosmic-ray behavior. The essential assumption in this derivation, as with previous ones, is that the interaction of the cosmic rays with the magnetic irregularities in the solar wind is a diffusive process.

Let U denote the differential number density of the cosmic rays, per interval of kinetic energy T. Let $\underline{S}$ denote the differential current density or streaming--the net number of particles in a given energy interval which crosses a unit area normal to $\underline{S}$ in unit time. Both U and $\underline{S}$, and also T, are measured here in the frame fixed with respect to the Sun (as opposed to, e.g., the frame moving with the solar wind). The equation of continuity of number of particles is then (Gleeson and Axford, 1967; Fisk, 1974)

$$\frac{\partial U}{\partial t} + \nabla \cdot \underline{S} = - \frac{\partial}{\partial T} (\underline{V} \cdot \nabla P) \tag{1}$$

Here t is time and V is again the solar wind speed. The differential pressure of the cosmic rays is denoted by P. Since galactic cosmic rays are observed to exhibit only small anisotropies in interplanetary space,

$$P = \frac{\alpha}{3} TU \tag{2}$$

where $\alpha = (T+2T_o)/(T+T_o)$ with T_o particle rest energy (Parker, 1966). The term on the right side of (1) results from the work done on the cosmic rays by the solar wind, as it sweeps them out of the solar cavity. The solar wind does work against the pressure gradient of the cosmic rays, or equivalently it alters the cosmic-ray energy and as a result the differential number density.

As a result of our assumption that particle propagation is a diffusive process, we consider that cosmic rays stream relative to the solar wind because there are gradients in U, and that these gradients are directly proportional to the streaming. Hence,

$$-\underline{\underline{\kappa}} \cdot \nabla U = \underline{S} - \underline{S}_o \tag{3}$$

where $\underline{\underline{\kappa}}$ is the diffusion tensor and $\underline{S}_o$ is the streaming that is seen in the frame fixed with the Sun, when the cosmic-ray distribution is isotropic in the frame moving with the solar wind (i.e., when there is no streaming relative to the solar wind.

The term $\underline{S}_o$ must be treated with some care. Consider a particle with a given energy in the frame moving with the solar wind. If the particle is moving in the direction of the solar wind, it will be seen at a higher energy in the frame fixed with respect to the Sun; conversely, if it is moving opposite to the wind it is seen at a lower energy in the fixed frame. Thus, the streaming in the fixed frame which results from an isotropic distribution in the frame of the wind depends not only on $\underline{V}$ but also on the shape of the cosmic-ray spectrum. This last effect is known as the Compton-Getting effect. It has been cast in its modern context by Gleeson and Axford (1968), and by Forman (1970), who find that

$$\underline{S}_o = \underline{V}U - \frac{\underline{V}}{3}\frac{\partial}{\partial T}(\alpha TU) \equiv C\underline{V}U \qquad (4)$$

where $C = 1-(1/3U)(\partial(\alpha TU)/\partial T)$ is the so-called Compton-Getting coefficient. It should be noted that, while the streaming in the frame moving with the solar wind differs from that in the fixed frame, the differential number density in these two frames is the same to $O(V^2/v^2)$, where v is particle speed (Jokipii and Parker, 1967).

The diffusion approximation which is used in (3) has been invoked as a valid description of cosmic-ray behavior for more than twenty years. Indeed, in cases where we can think of particles scattering off of well-defined 'scattering centers' this approximation can be rigorously justified. However, the validity of this approximation becomes less clear when we think of a cosmic-ray spiraling about a mean magnetic field and slowly changing pitch angle as it interacts, resonately or otherwise, with a random field component. In fact, there is currently no defendable proof that low-energy particles diffuse. Of course, the ultimate justification for this approximation will come from how well it can explain particle observations. In this context it appears to do quite well.

Upon substituting (2), (3) and (4) into (1), we obtain a Fokker-Planck euqtion for U (Parker, 1965)

$$\frac{\partial U}{\partial t} = \nabla\cdot(\underline{\underline{\kappa}}\nabla U) - \nabla\cdot(\underline{V}U) + \frac{\nabla\cdot\underline{V}}{3}\frac{\partial}{\partial T}(\alpha TU) \qquad (5)$$

This is the basic equation which governs cosmic-ray behavior in interplanetary space.

We will not be concerned with azimuthal variations in U because we will be dealing with times long compared with a solar rotation period (~27 days). Thus, it is convenient to express (5) in spherical coordinates with θ the polar angle relative to the rotation axis of the Sun, and r heliocentric radial distance, or as

$$\frac{\partial U}{\partial T} + \frac{1}{r^2}\frac{\partial}{\partial r}(r^2\kappa_{rr}\frac{\partial U}{\partial r} + \frac{1}{r^2\sin\theta}\frac{\partial}{\partial\theta}(\sin\theta\,\kappa_{\theta\theta}\frac{\partial U}{\partial\theta})$$
$$+ \frac{1}{r\sin\theta}\frac{\partial}{\partial\theta}(\sin\theta\kappa_{\theta r}\frac{\partial U}{\partial r}) + \frac{1}{r^2}\frac{\partial}{\partial r}(rk_{r\theta}\frac{\partial U}{\partial\theta})$$
$$-\frac{1}{r^2}\frac{\partial}{\partial r}(r^2VU) + \frac{1}{3r^2}\frac{\partial}{\partial r}(r^2V)\frac{\partial}{\partial T}(\alpha TU) = 0 \qquad (6)$$

where the solar wind has been assumed here to flow only in the heliocentric radial direction.

The radial component of the diffusion tensor κ_{rr} can also be expressed as (Parker, 1967)

$$\kappa_{rr} = k_{\parallel} \cos^2\psi + \kappa_{\perp} \sin^2\psi \tag{7}$$

where $\kappa_{\parallel}$ is the diffusion coeffient for propagation parallel to the mean field direction, and $\kappa_{\perp}$, for propagation normal to this direction, along cones of constant θ. The angle between the mean field direction and the radial direction is ψ. If the mean field lies along cones of constant θ, as it does, for example, if it executes the Archimedes spiral pattern, the $\kappa_{\theta\theta}$ results from cross-field diffusion and in most circumstances is equal to $\kappa_{\perp}$. The off-diagonal elements of the diffusion tensor, $\kappa_{\theta r}$ and $\kappa_{r\theta}$, describe the effects of particle drift motion in the large-scale magnetic field and can be expressed as (Jokippi and Parker, 1970)

$$\kappa_{\theta r} = - \kappa_{r\theta} = \frac{1}{3} v r_g \sin\psi \tag{8}$$

Here v is particle speed and r_g is the gyro-radius of the particle, which is assumed to have a positive charge. If the charge or the mean field change sign, $\kappa_{\theta r}$ (and $\kappa_{r\theta}$) also reverses sign. The effects of these drift terms have been considered in detail in the recent work of Jokipii and co-workers (e.g., Jokipii and Kopriva, 1979).

4.1 Energy Loss

It is interesting to note that in the final equations, (5) and (6), the term which describes the effects of cosmic-ray energy changes has the form

$$- \frac{\partial}{\partial T} \left(\frac{dT}{dt} U\right) \tag{9}$$

where

$$\frac{dT}{dt} = - \frac{1}{3r^2} \frac{\partial}{\partial r} (r^2 V) \alpha T \tag{10}$$

is the rate at which particles are adiabatically cooled due to the expansion of the solar wind. The expression $(1/r^2)(\partial(r^2V)/\partial r)$ is just the rate at which a volume in the solar wind expands as it moves radially outward. Indeed, Parker (1965) first derived (5) by assuming simply that the principal energy change suffered by the cosmic rays results from this adiabatic cooling.

We have here, however, a slight puzzle. We derived (5) and (6) under the assumption that the solar wind does work against the cosmic rays. The cosmic-ray intensity in the inner solar system is lower than it is in the interstellar medium, or equivalently $\underline{V}\cdot\nabla p = VdP/dr>0$. Yet in the final analysis the cosmic rays are losing energy. To understand this apparent paradox we must consider in detail who does work on what, and where.

It is instructive for this explanation of cosmic-ray energy changes to consider the following simplified model, which was discussed by Fisk (1974). Imagine a series of infinitely thin concentric spherical shells. At a given time these shells are all expanding with radial speed V, and particles are all expanding with radial speed V, and particles are trapped between any two shells. The density between successive shells increases with radial distance.

Consider now the work done by a given shell at radius r_1. Particles that strike the shell from radial distances $r>r_1$ gain energy since the particles make 'head-on' collisions with the outward moving wall. Particles striking the surface from $r<r_1$ lose energy in 'overtaking' collisions. Since the density of particles increases with r, there are more particles striking from $r>r_1$ than from $r<r_1$ and the moving shell does work on the particles.

Consider, however, the particles trapped between two successive shells. The particles gain energy by striking the inner shell (at r_1) and lose energy by striking the outer shell (at $r_2, r_2 > r_1$). However, the surface area of the outer shell exceeds that of the inner shell. Thus, overtaking collisions outnumber head-on collisions and the particles are cooled. Indeed since the differential number density is the same in the frame moving with the shells as it is in the frame when the shells are moving (to $O(V^2/v^2)$), this cooling occurs at the adiabatic cooling rate given in (10).

The cooling, however, must stop at the outermost shell which we place at r=R. Particles in r>R can only make head-on collisions; i.e., they can only gain energy.

The analogy between the solar wind and this simple model is straightforward. The solar wind does work against the cosmic-ray gradient everywhere in the interplanetary medium, as do the shells. However, the work which is done on the cosmic rays through head-on collisions at one radial distance is immediately undone, and more so, by the more numerous overtaking collisions which occur at slightly larger radial distances. Thus, cosmic rays in the solar wind, as with the particles trapped between shell, are adiabatically cooled continuously. Particles, however, which penetrate only ~ one mean free path into the solar wind and depart (strike only the outermost shell) do gain energy.

Of course, energy must be conserved in the modulation process. The energy which is carried off by particles making only ~ one collision with the solar wind must equal the work done on all the cosmic rays by the wind, throughout the solar cavity. Indeed, as has been shown by Jokipii and Parker (1967), (5) can be integrated to prove that energy is properly conserved.

For observations made deep within the solar wind (e.g., for observations made near earth) only the energy loss process is important. The particles that we see have been continuously cooled in the solar wind at the rate given in (10).

It should be noted also that the energy loss results directly from our assumption that the solar wind tends to sweep the cosmic rays out of the solar cavity. If there is convection by the solar wind, there will also be energy loss. The two processes cannot be decoupled.

5. Solving the modulation equations

Over the years numerous analytic and approximate solutions have been found to the modulation equations (e.g., Fisk and Axford, 1968). In practice, however, particularly if the aim is to fit data, the simplest and most useful approach is to solve (6) numerically. The technique for numerically solving the steady-state spherically symmetric form of (6) was developed by Fisk (1971). This technique was expanded in Fisk (1976) to include transport of cosmic rays in heliographic latitude by cross-field diffusion. Jokippi and Kopriva (1979) developed a code which includes both cross-field diffusion and drift effects. Perko and Fisk (1982) have solved numerically a time-dependent form of (6).

In solving (6), which includes all known effects, we must ask whether each of the terms is in fact important. Here we must be guided by the cosmic-ray data. In principle, we can determine the diffusion coefficients in (6) directly from observations of the irregularities in the heliospheric magnetic field. In practice, however, the theory which relates the diffusion coefficients to the field data is inadequate for this purpose (e.g., Fisk, 1979). Moreover, we have observed the magnetic field only in a narrow region near the equatorial plane of the Sun out to ~30 AU. The cosmic rays could respond to a much larger region, both in heliocentric radial distance and in latitude.

In the earlier days of modulation work(<1970), it was common to use the steady-state form of (6), in which only radial transport is allowed. This form had the advantage that it was readily solved.

In the mid-1970's, latitude variations in this modulation were argued to be important. In Figure 3, a schematic drawing of the Archimedes spiral pattern of the heliospherical magnetic field is shown, at different latitudes. In the equatorial plane, rotation effects wind the field into a tight spiral, whereas over the solar poles, the field remains nearly radial. As was pointed out by Fisk (1976) and others, this changing orientation of the field can have a strong effect on the inward radial diffusion of the particles. From (7), the radial diffusion coefficient depends strongly on Ψ, the angle between the field and the radial direction. By assuming that $\kappa_{\parallel} << \kappa_{\perp}$, as is believed to be the case, κ_{rr} in the equatorial plane at large radial distances is small, where $\cos\psi \sim 1$. Thus, large latitude gradients in the cosmic-ray flux can result, with perhaps significant effects for the cosmic-ray flux near earth due to latitude transport by cross-field diffusion.

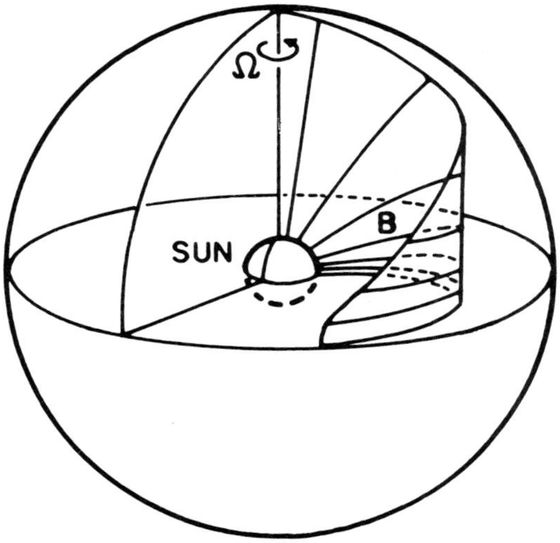

Figure 3. A schematic drawing of the pattern of the mean magnetic field in the heliosphere.

In the late 1970's, Jokipii and co-workers (e.g., Jokipii, Levy and Hubbard, 1977; Jokipii and Kopriva 1979) have argued that latitude transport by particle drifts through the large-scale Archimedes spiral field should be important. Drift effects were of course known prior to this, but their possible importance was never systematically calculated. Jokipii and his co-workers point out that the average drift speed of the particles can exceed the solar wind speed, and that there are models in which transport in latitude by drifts are the most important aspect of the modulation process.

Unfortunately, curent data suggests that latitude transport effects may not be of particular significance for the cosmic-ray flux seen in the equatorial plane. Consider, for example, the Pioneer 10/11 data of McDonald (Webber, private communication). The decline in the cosmic-ray flux as solar-maximum conditions set in 1977, occurred as a series of sharp drops, which were seen sequentially at earth, at Pioneer 11 (at ~20 AU). The drops appear to propagate outward with the average solar wind speed ~400 km/s. Moreover, the cosmic-ray radial gradient is approximately the same before and after the drop. These data are consistent, then, with the modulation resulting from some disturbance in the solar wind propagating outward beyond each spacecraft, and thereafter shielding it from the inward radial flow of the cosmic rays. There is no evidence of particles filling in behind the disturbances, which would be the case if there was significant latitude transport.

The lack of latitude transport due to cross-field diffusion is easy to understand. Evidently, the cross-field diffusion coefficient $\kappa_\perp$ is very small. Thus, significant latitude gradients of the cosmic-ray flux could occur, as is expected from the large-scale configuration of the heliospheric magnetic field. However, the transport of the cosmic rays due to these gradients could still be small.

The lack of latitude transport due to drifts is less clear. The large-scale Archimedes spiral pattern of the heliospheric magnetic field should yield a significant drift effect, as Jokipii and co-workers have argued. However, as Lee and Fisk (1981) have argued, the small-scale irregularities in the heliospheric field may mitigate the systematic drift effects, and render them unimportant.

In the remainder of this article, we will consider that only radial transport is important in (6), i.e., only $\kappa_{rr} \neq 0$.

6. Some illustrative examples

Consider the results of some numerical solutions to (6) which illustrate some of the features and consequences of the theory for solar modulation.

The first example, due to Goldstein et al. (1970), illustrates the extent of the energy loss of the cosmic rays due to the adiabatic deceleration of the solar wind. In this calculation a reasonable form for the interstellar spectrum of cosmic-ray protons is divided up into a series of essentially mono-energetic spectra. These spectra are then each subjected to the modulation predicted by the steady-state, radial form of (6), and the resulting spectra at 1 AU calculated. The parameters in (6) are chosen so as to give a reasonable level for the overall modulation in the heliosphere. Clearly, by comparing the spectra at 1 AU with the monoenergetic spectra in the interstellar medium, we can see how a particle which enters the heliosphere with a given energy is likely to behave.

The results of the calculations of Goldstein et al. are reproduced in Figure 4. The essentially monoenergetic interstellar spectra are shown in the solid curves and the resulting modulated spectra at 1 AU (each labeled with the same number as its interstellar input spectrum), in the light dashed curves. (The heavy dashed curve represents the sum of all the light dashed curves.) Note that at high energies (spectra 8 -10) the effects of the modulation are rather small. There is only a slight reduction in the intensity at 1 AU as compared with the interstellar intensity, and only a slight shift in the mean energy of the

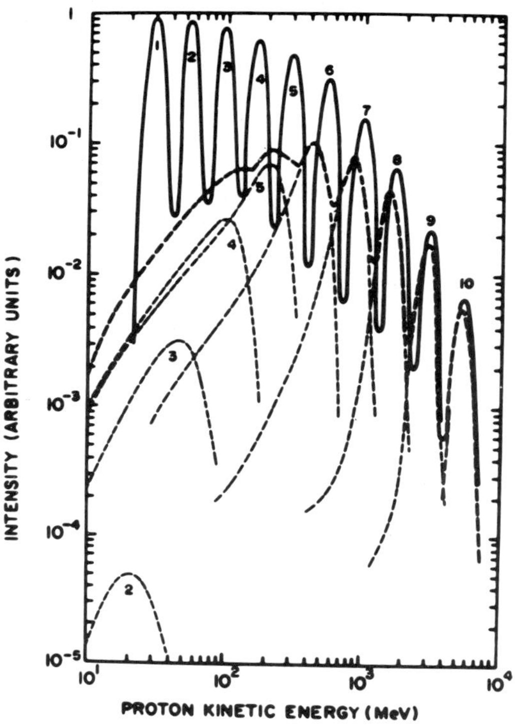

Figure 4. A series of essentially monoenergetic proton spectra in interstellar space (solid) and their resulting modulated spectra at earth (dashed) (from Goldstein et al., 1970).

particles as they are modulated. At intermediate energies (spectra 4 - 7) the reduction and shift are more pronounced. There is also a considerable spread in energy here. Since the particles diffuse, there is a distribution in the time which particles spend in the solar cavity, and thus also a distribution in the energy is extreme. In fact, the intensity of protons that enter the interplanetary medium at energies below ~100 MeV is so reduced at 1 AU that the probability of observing these particles is quite small. As can be seen in Figure 4, at energies below ~100 MeV we are more likely to see particles from spectra 4 and 5, which entered the interplanetary medium at energies above 100 MeV and were cooled down.

From observations near earth we are thus unable to sample directly cosmic rays which have low energies in the interstellar medium. For example, we cannot measure the composition of these particles and infer whatever information is contained in this composition about the nature of cosmic ray sources. Similarly, our observations near earth are insensitive to the shape of low-energy interstellar spectra. We could, for example, raise spectrum 2 in Figure 4 by two orders of magnitude and yet produce no appreciable effect on the total spectrum at 1 AU (the heavy dashed curve). Thus, we cannot infer from our observations at earth whether there is sufficient energy density in low-energy galactic cosmic rays to affect the pressure or ionization state of the interstellar medium.

The second example, due to Perko and Fisk (1982), illustrates some of the features of the solar-cycle variation in the cosmic-ray flux. As is suggested by Pioneer 10/11 observations, the solar-cycle variation is assumed to be due to a series of disturbances which are carried outward with the solar wind, sweeping the cosmic rays out of the heliosphere. These disturbances could be due to, for example, flare-generated shock waves, and their frequency would increase with increasing solar activity. The disturbances are simulated in the calculation by isolated regions of reduced diffusion coefficient, or equivalently enhanced scattering, which are convected outward with the solar wind. The frequency of the scattering regions increases from essentially none in solar-minimum conditions, to one every few AU during solar maximum. During the second half of the solar cycle, from solar-maximum to solar-minimum conditions, the number of disturbances in the heliosphere is the mirror image of the number in the first half of the cycle.

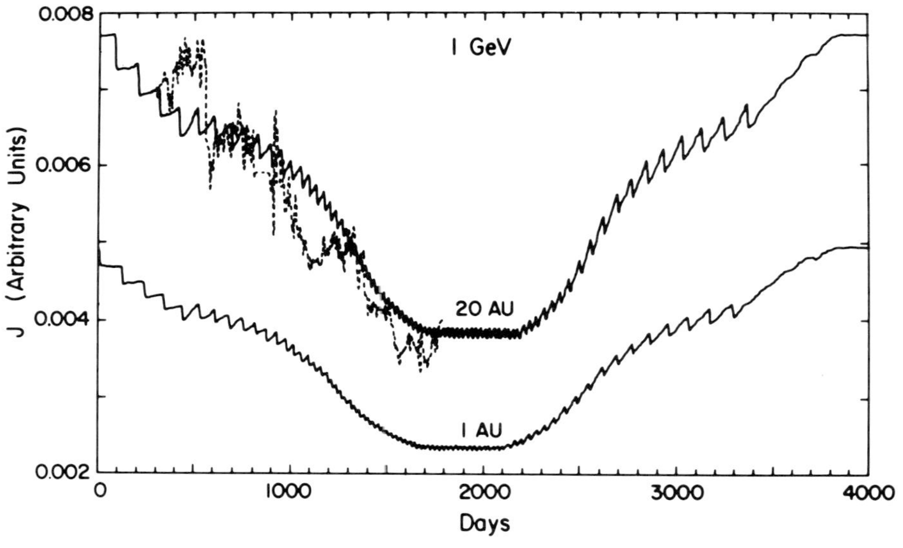

Figure 5. The differential intensity for 1-GeV protons versus time as calculated in the model for the solar cycle of Perko and Fisk (1982). The data is the proton intensity >60 MeV (which has a mean response ~1 GeV) from Pioneer 10 (courtesy of W. R. Webber).

Shown in Figure 5 is the variation of the intensity of ~1 GeV protons over the solar cycle, as calculated by Perko and Fisk (1982).

This result is from a numerical solution to the time-dependent, radial form of (6). The increasing number of disturbances in the solar wind causes the intensity to drop from solar-minimum conditions, where the calculation starts, to the mid-point of the time series which represents solar-maximum conditions. Data from Pioneer 10/11 is superimposed on the calculated curve to indicate that the reduction in intensity over the first half of the cycle is roughly the correct magnitude.

A close examination of Figure 5 reveals that the rise in the cosmic-ray intensity during the second half of the cycle is not the mirror image of the decline in the intensity during the first half, despite the fact that the number of disturbances in the heliosphere during the two halves are mirror images. A disturbance will depress the cosmic-ray intensity at earth so long as the disturbance is between the earth and the outer boundary of the modulation region, which in this model is placed at 100 AU from the sun. Since the transit time of a disturbance over this distance is ~1 year, there is then a considerable delay between when the number of disturbances in the solar wind decreases and the cosmic-ray flux undergoes a commensurate increase. This effect is known as the hysteresis in solar-cycle variation of the cosmic-ray flux. The magnitude for the hysteresis calculated by Perko and Fisk (1982) is in good agreement with the observed effect.

References:

Fisk, L. A., and Axford, W. I.: 1968, J. Geophys. Res. 74, 4973.

Fisk, L. A.: 1971, J. Geophys. Res. 76, 221.

Fisk, L. A.: 1974, High Energy Particles and Quanta in Astrophysics, MIT Press, 170.

Fisk, L. A.: 1976, J. Geophys. Res. 81, 4646.

Fisk, L. A.: 1979, Solar System Plasma Physics, North-Holland Publishing Co. 1, 1979.

Forman, M. A.: 1970, Planet. Space Sci. 18, 25.

Gleeson, L. J., and Axford, W. I.: 1967 Astrophys. J. (Letters), 149, L115.

Gleeson, L. J., and Axford, W. I.: 1968, Astrophys. J. 154, 1011.

Goldstein, M. L., Fisk, L. A., and Ramaty, R.: 1970, Phys. Rev. Letters 25, 832.

Jokipii, J. R., and Parker, E. N.: 1967, Planet. Space Sci. 15, 1375.

Jokipii, J. R., and Parker, E. N.: 1970, Astrophys. J. 160, 735.

Jokipii, J. R., Levy, E. H., and Hubbard, W. R.: 1977, Astrophys. J. 234, 384.

Jokipii, J. R., and Kopriva, D. A.: 1979, Astrophys. J. 234, 384.

Lee, M. A., and Fisk, L. A.: 1981, Astrophys. J. 248, 836.

Perko, J. A., and Fisk, L. A.: 1982, J. Geophys. Res., in press.

Parker, E. N.: 1965, Planet. Space Sci. 13, 9.

Parker, E. N.: 1966, Astrophys. J. 144, 916.

Parker, E. N.: 1967, Planet. Space Sci. 15, 1723.

THE ACCELERATION OF ENERGETIC PARTICLES IN THE SOLAR WIND

L.A. Fisk
Department of Physics
University of New Hampshire
Durham, New Hampshire 03824

The two principal mechanisms for accelerating energetic particles in the solar wind, statistical acceleration and shock acceleration, are reviewed briefly. Statistical accleration in both Alfvén and magnetosonic turbulence is considered, but it is argued that neither may be of particular importance in the solar wind. Shock acceleration is illustrated with the case of acceleration at the forward and reverse shocks which bound corotating interaction regions in the solar wind. The anomalous cosmic-ray component, which may be the most important example of acceleration in the solar wind, is also discussed briefly.

1. Introduction

There are two mechanisms which are commonly considered for accelerating energetic particles in the solar wind: statistical accleration and shock acceleration. For the former there is no clear evidence that it actually occurs in the solar wind. For the latter, there are numerous examples of its occurrence.

We will begin by describing in simple terms the theory for statistical acceleration, and discuss why it may not be important in the solar wind. We will then describe the illustrative example of shock acceleration in corotating interaction regions of the solar wind. Finally, we will discuss the so-called anomalous cosmic-ray component, which may be the most important example of acceleration in the solar wind, but for which the mechanism of accleration is unclear.

2. Statistical Acceleration

Energetic particles interacting with random waves or turbulence will be accelerated. The particles will undergo a mean square change in momentum per unit time, which we denote by D_{pp}, and will thus diffuse to higher momentum.

R. L. Carovillano and J. M. Forbes (eds.), Solar-Terrestrial Physics, 231–241.

If we describe the behavior of the energetic particles by a transport spectrum for the omni-directional particle distribution function f (number of particles per volume of phase space averaged over particle direction), the effect of the statistical acceleration is included in this equation by a term

$$\frac{1}{p^2}\frac{\partial}{\partial p}\left(D_{pp}\, p^2 \frac{\partial f}{\partial p}\right) \qquad (1)$$

If we write the transport equation in terms of the differential number density U (number of particles per unit volume and kinetic energy T), it becomes

$$\frac{\partial}{\partial T}\, D_{TT}\frac{\partial U}{\partial T} - \frac{\partial}{\partial T}\left(\frac{D_{TT}}{2T}\, U\right) \qquad (2)$$

when $D_{TT} = v^2\, D_{pp}$ and v is particle spread. Clearly $D_{TT}/2T$ is the mean change of energy per time that results from the statistical acceleration; D_{TT} is the rate of mean square change. It should be noted that in some treatments of statistical acceleration only the mean change in energy is considered, whereas in most cases the mean square change is of equal importance.

Magnetic irregularities in the solar wind that affect energetic particles are presumably the result of the superposition of Alfvén and magnetosonic waves. The former are observed to dominate, as should be the case, since the latter are readily damped.

Consider first the interactions of particles with Alfvén waves. As a result of the interaction with any wave, a nonrelativistic particle will experience a change in energy by an amount

$$\Delta T \simeq m\, u_{\parallel}\, \Delta_{\parallel} \qquad (3)$$

where m is particle mass, $u_{\parallel}$ is the phase speed of the wave parallel to the mean field $\underset{\sim}{B}_o$, and $\Delta v_{\parallel}$ is the change in v parallel to $\underset{\sim}{B}_o$ as a result of the interaction, as viewed in the frame moving with $u_{\parallel}$. For Alfvén waves, the principal interaction is a change in pitch angle, a scattering of the particle. For a full collision with a wave, in which the particle reverses its direction $\Delta v_{\parallel} \simeq 2v_{\parallel}$; $u_{\parallel} = V_A$, the Alfven speed. Thus, we can estimate that for Alfven waves D_{TT} is

$$\frac{D_{TT}}{T^2} \simeq \frac{(\Delta T)^2}{T^2}\frac{v_{\parallel}}{\lambda_{\parallel}} \simeq \frac{16}{3\sqrt{3}}\frac{V_A{}^2}{v\lambda_{\parallel}} \qquad (4)$$

where $\lambda_{\parallel}$ is the mean free path for particle scattering along $\underset{\sim}{B}_o$, and $v_{\parallel} \simeq v/\sqrt{3}$.

Clearly, from (4), for statistical acceleration by Alfvén waves to be important in the solar wind the mean free path $\lambda_{\parallel}$ needs to be small. Unfortunately, $\lambda_{\parallel}$ is observed to be ~0.1 AU (see Fisk 1979 for a summary of the observations), which with $V_A \simeq 50$ km/s makes D_{TT} in

(4) unimportant in most applications. For example, with these values of $\lambda_{\parallel}$ and V_A, the predicted acceleration rate for 1 MeV protons is substantially slower than the rate at which particles are being adiabatically cooled by the expansion of the solar wind.

For magnetosonic waves, we can calculate the acceleration rate by considering the interaction of the particles with a characteristic wave, which we assume has scale-lengths λ_z' and $\lambda_\perp'$ in a direction along and normal to $\underset{\sim}{B}_o$, respectively. These scale-lengths are assumed to be long compared with the particle gyro-radius r_g. We note first that $u_{\parallel}$ in (4) is $\sim u\, \lambda_z'/\lambda_\perp'$, where u is the phase speed of the wave. Since the Alfven and sound speeds are comparable in the solar wind, $u \sim \sqrt{2}\, V_A$. We next find $\Delta v_{\parallel}$ by noting that in the frame moving with $u_{\parallel}$ there is no electric field and thus the change in $v_{\parallel}$ is caused by $\partial B/\partial z$, the gradient in the magnitude of the magnetic field along $\underset{\sim}{B}_o$, which is produced by the characteristic wave. Hence $v_{\parallel}$ should satisfy

$$\frac{dv_{\parallel}}{dt} \simeq -\frac{v_\perp^2}{B_o}\frac{\partial B}{\partial z} \tag{5}$$

or

$$|\Delta v_{\parallel}| \simeq -\frac{v_\perp^2}{\lambda_z'}\left|\frac{\delta B}{B_o}\right| \Delta t \tag{6}$$

Here, $v_\perp$ is the component of the particle's velocity normal to $\underset{\sim}{B}_o$, $|\delta B|$ is the amplitude of the magnitude fluctuation in the field, and Δt is the time during which the particle makes a coherent interaction with the wave. If the correlation length for the turbulence along B_o is λ_z, then $\Delta t \sim \lambda_z/v_{\parallel}$. Upon substituting for $u_{\parallel}$, $\Delta v_{\parallel}$, and Δt in (4) and (6), we then estimate that the acceleration rate for magnetosonic waves is

$$\frac{D_{TT}}{T^2} \simeq \frac{32}{3\sqrt{3}}\frac{V_A^2}{v}\left(\frac{\delta B}{B_o}\right)^2 \frac{\lambda_z}{\lambda_\perp^2} \tag{7}$$

The length scales in (7) λ_z and $\lambda_\perp$ are on the order of the correlation lengths of the turbulence, which in the solar wind are ~0.01 AU, much smaller than the ~0.1 AU length for the mean free path. Thus, the statistical acceleration by magnetosonic waves can in principle be much more efficient than that by Alfven waves.

Unfortunately, the efficiency for magnetosonic acceleration is also its downfall. Because these waves provide strong acceleration, they are equivalently rapidly damped. The amplitude for magnetosonic waves in the solar wind is typically $(\delta B/B_o)^2 \sim 0.01 - 0.03$ (Smith, 1974), and the acceleration rate applied by (7) is too small to be of interest.

3. Shock accleration

There are numerous examples of acceleration of energetic particles by shocks in the solar wind. Particles can be accelerated at planetary bow shocks (e.g., the earth's or Jupiter's bow shock), at propagating interplanetary shocks generated by flares, at corotating shocks which bound stream-stream interaction regions (the so-called corotating interaction regions or CIRs), or perhaps at the termination shock of the solar wind. There are some differences in the theory of shock acceleration for each of these cases. For example, in propagating interplanetary shocks time variations can be important; at bow shocks, which have finite size, the geometry of the shock can matter. However, the basic physics of how particles are accelerated, and the general techniques for describing this acceleration, are the same in all cases. We shall illustrate this physics and the general approach with the CIR case, which is a straightforward example of shock accleration, and one in which the theory seems to be particularly successful. We will follow closely here the treatment of Fisk and Lee (1980).

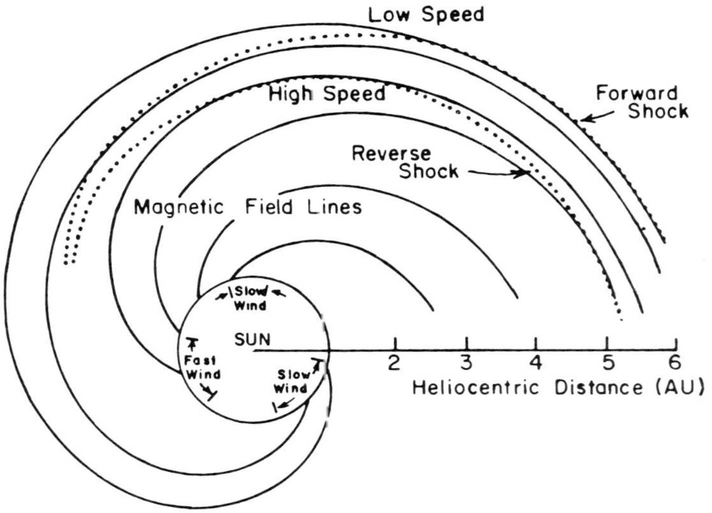

Figure 1. A schematic drawing of a corotating interaction region, as viewed from the frame corotating with the solar wind.

Shown in Figure 1 is a schematic drawing of a CIR, as viewed from the frame corotating with the Sun. High-speed solar wind from an isolated region of the Sun flows outward and collides with lower-speed wind from some adjacent region. At several AU from the Sun, a substantial interaction region, the CIR, develops between the two flows. The CIR is bounded on the front side by a forward shock, where the solar wind is accelerated discontinuously, and on the back side by a reverse shock, where it is decelerated. During solar-minimum conditions, the high-speed stream is relatively steady in time, and thus the CIR and the shocks are stationary in the frame corotating with the Sun. Note also, in this frame, the solar wind flows parallel (or antiparallel) to the mean magnetic field, and follows the streamlines shown in Figure 1.

The accelerated energetic particles are seen in two locations. A broad increase in the intensity of ~1 MeV/nucleon particles occurs in the high-speed solar wind stream. This increase extends from the reverse shock back to the orbit of earth, and is seen as far inward towards the Sun as ~0.4 AU, from Helios or Mariner 10 (Van Hollebeke et al., 1982; Christon and Simpson, 1979; Kunow et al., 1977). An increase in energetic particles is also seen in front of the forward shock. However, this increase generally involves fewer particles than the increase behind the reverse shock, has a softer spectrum, and does not appear to extend back to earth (Barnes and Simpson, 1976; Van Hollebeke et al., 1979).

The spectra of the energetic particles observed at earth (in the high-speed stream) have an exponential character. The University of Maryland/Max-Planck Institute group report that spectra in several events are well-fit if they are plotted as a distribution function f (number of particles per unit volume of phase space) that is an exponential in velocity independent of particle species (Gloeckler, Hovestadt and Fisk, 1979). This results holds over a wide energy range, from ~0.15 - 8 MeV/nucleon, and for all major species from protons through iron.

The e-folding velocity of the spectra of particles in the high-speed stream is also observed to be essentially independent of radial distance; i.e., it is roughly the same at 0.4 AU, as observed by Helios, at earth, and at 3 - 4 AU as seen from Pioneer 11. The gradient in the intensity of the particles in the high-speed stream is observed to be steepest in the inner solar system (~350%/AU from ~0.4-1 AU), and then decreases somewhat beyond earth (~100%/AU)(Van Hollebeke et al., 1982).

Particles can gain substantial energy at the forward and reverse shocks which bound the CIRs by being compressed between the shock front and the magnetic irregularities in the solar wind upstream from the shock (which are moving at different speeds from the shock), or by being compressed between the upstream irregularities and those downstream from the shock. This acceleration is most readily discussed in the frame corotating with the Sun, which is depicted in Figure 1. In this frame there is no electric field associated with bulk motion of the interplanetary magnetic field, since the solar wind flow is aligned with the field direction. Particles thus experience no energy change while crossing the shock fronts. The energy changes result from particles interacting with moving magnetic irregularities upstream and downstream from the shocks, which can be treated by using a standard diffusion approach.

The particle intensity increases upstream from the CIR are relatively steady in time, and thus their behavior should be described by the standard steady-state equation for propagation in the solar wind, which was developed in the chapter on solar modulation. In the corotating frame, and in terms of f (rather than differential number density U) in a given magnetic flux tube, or equivalently along a given streamline in Figure 1, this equation is (Parker, 1965; Fisk, Forman and Axford, 1973; Ng, 1972)

$$- \frac{1}{3r^2} \frac{\partial}{\partial r} (r^2 V) v \frac{\partial f}{\partial v} = \frac{1}{r^2} \frac{\partial}{\partial r} (r^2 \kappa \frac{\partial f}{\partial r}) - V \frac{\partial f}{\partial r} \qquad (8)$$

It is assumed here that the particles are nonrelativistic; i.e., Equation (8) is written in terms of particle speed as opposed to particle momentum. The first term on the right side describes the spatial diffusion of the particles. It is assumed that particles move only along and not across magnetic field lines, in which case $\kappa = \kappa_{\parallel} \cos^2\psi$, where $\kappa_{\parallel}$ is the diffusion coefficient for propagation along the mean magnetic field $\underset{\sim}{B}$, and ψ is the angle between $\underset{\sim}{r}$ and $\underset{\sim}{B}$. The remaining two terms describe the effects of convection and adiabatic deceleration by the solar wind; V is the speed of the solar wind in the intial frame fixed wtih respect to the Sun, i.e., it is the radial component of the solar wind in the corotating frame.

The effects of the shock can be described by means of a boundary condition on the solution to Equation (8). We note that for a given magnetic flux tube the product of the particle differential streaming S and the cross-sectional area A of the flux tube should be conserved across the shock front; i.e., in a steady state there is no buildup of particles at the shock. We note also that magnetic flux is conserved across the shock front, i.e. that BA is constant. By writing S in terms of f (Gleeson and Axford, 1967; Fisk, Forman and Axford, 1973), it can be readily shown that an appropriate boundary condition for f at a shock located at $r = r_s$ is then

$$- \frac{Vv}{3} \frac{\partial f}{\partial v} - \frac{\partial f}{\partial r} = - \beta \frac{Vv}{3} \frac{\partial f}{\partial v} \quad , \text{ at } r = r_s \qquad (9)$$

where

$$\beta = \frac{V'^2 + \Omega^2_{\odot} r_s^2)^{1/2}}{(V^2 + \Omega_{\odot}^2 r_s^2)^{1/2}} \frac{B}{B'} \qquad (10)$$

Here, V, κ, and B are, respectively, the solar wind speed, the diffusion coefficient and the magnetic field strength upstream from the shock; V' and B' refer to conditions downstream. We have assumed for simplicity in Equation (9) that there is substantial particle scattering in the CIR so that particles are transported downstream from the shock primarily by convection. We also assume that f is continuous across the shock front. If the distribution function were strictly isotropic, Liouville's theorem would require that f is continuous (e.g., Parker, 1963).

The diffusion coefficient in undisturbed solar wind, at least for particles with energies $\gtrsim$ 1MeV/nucleon, is observed to be sufficiently large so that on the right side of Equation (8) the second term tends to be small in comparison with the first (Zwickl and Webber, 1977). The leading dependence of f on v for conditions upstream from the shocks can be obtained from the approximate equation:

$$-\frac{1}{3r^2}\frac{\partial}{\partial r}(r^2V)\ v\frac{\partial f}{\partial v} \simeq \frac{1}{r^2}\frac{\partial}{\partial r}\ r^2\kappa\frac{\partial f}{\partial r} \tag{11}$$

The diffusion coefficient for low-energy particles is also observed to be of the form $\kappa = v\ g(r)$, where $g(r)$ is a function of radial distance only; i.e., the mean free path is independent of particle rigidity (Zwickl and Webber, 1977). With this form for κ it can readily be seen that Equation (11) is satisfied by various separable solutions which can be expressed as an exponential in v times a function of r. However, the solution must also satisfy the boundary condition given in Equation (9) and remain finite as $r \to 0$, conditions which in general can be satisfied by only one of the possible separable solutions.

Consider, for example, the case where upstream from the shocks, $\kappa = \kappa_o\ v\ r$ (κ_o is a constant), and where V is constant. The appropriate solution to Equation (11), which satisfies Equation (9), and remains finite as $r \to 0$, is then

$$f \propto \left(\frac{r}{r_s}\right)^{2\beta/(1-\beta)} \exp\left(-\frac{6\ \kappa_o\ \beta\ v}{V(1-\beta)^2}\right) \tag{12}$$

where r_s is the shock location.

Clearly, the solution in Equation (12) can account for many of the observed features of energetic particles upstream from the CIR's. The leading dependence of the distribution function of particle velocity is an exponential which is independent of particle species. The e-folding velocity $[V(1-\beta)^2/(6\kappa_o\beta)]$ is independent of radial distance. Further, the radial gradient of the intensity, $(1/f)(\partial f/\partial r)$, varies as $1/r$ and thus is steepest in the inner solar system.

The solution in Equation (12) can also yield a steeper spectrum at the forward shock than at the reverse, again as is observed. There are several quantities in the e-folding velocity in Equation (12) which should be roughly equal for the two shocks. For shock interactions at large radial distances β in Equation (10) is $\beta \simeq \beta'/B$; which is observed to be a similar ratio for the two shocks (Smith and Wolfe, 1977). Also κ_o could be similar for the high- and low-speed streams. However, the e-folding velocity varies directly as the upstream solar wind speed V, which for the forward shock is, of course, substantially slower.

4. The anomalous component

The most challenging example of acceleration in the solar wind may well be the so-called anomalous cosmic-ray component. The observations of this component suggest that large-scale acceleration of energetic particles is occurring in the outer heliosphere; the challenge is to understand how.

Starting in 1972, a variety of groups began to observe a new component in the energetic particle flux, the so-called anomalous component (Hovestadt et al., 1973; McDonald et al., 1974). Shown in Figure 2 are the spectra of several elements as were measured by Voyager (Webber, private communication). The dotted curves are reasonable extrapolations of the solar energetic particle flux or perhaps particles associated with corotating interaction regions. The dashed curves are the galactic cosmic-ray flux. The anomalous component occurs between these other two components, and, as is evident in the figure, consists primarily of He, N, O, and Ne.

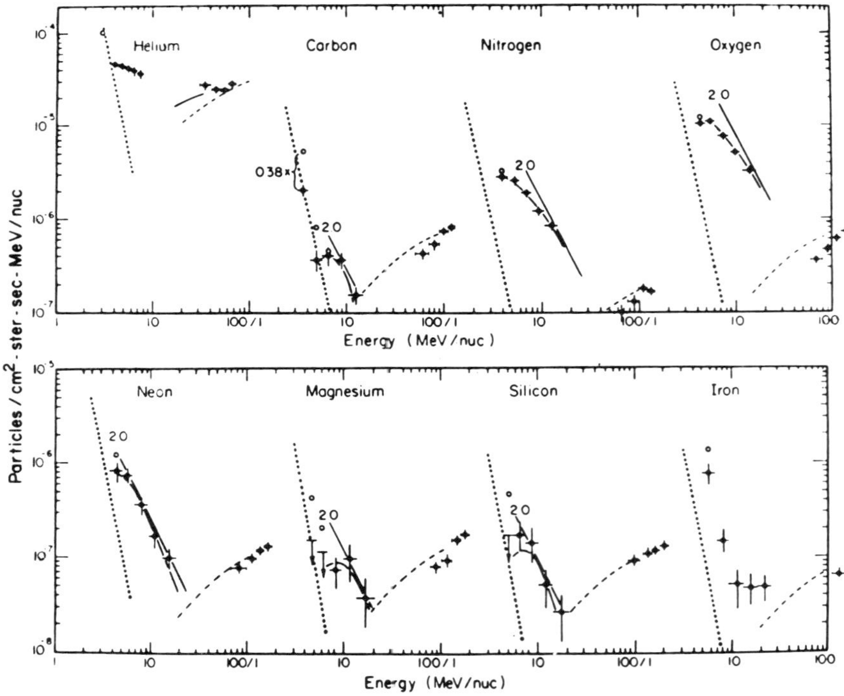

Figure 2. The spectra of the anomalous component as seen from Voyager (Webber, private communication).

The principal theory for the anomalous component is due to Fisk et al., (1974), who argue that these particles are due to interstellar neutral gas which is ionized and accelerated in the solar wind. This theory can account for the composition. Interstellar neutral gas should consist primarily of H, He, N, O, and Ne. These particles are swept into the heliosphere by the motion of the Sun through the interstellar medium. Once in the heliosphere, the neutral particles are ionized by photoionization, and by charge-exchange with the solar wind. The interstellar ions are then picked up by the solar wind and convected into the outer heliosphere. It is argued that as they are convected outward they can be accelerated up to energies to ~10 MeV/nucleon. Having obtained these higher energies, they should tend to diffuse back into the inner heliosphere. However, the interstellar ions should be singly charged; there is time to ionize them only once. Singly charged He, N, O and Ne have relatively high rigidities at ~10 MeV/nucleon, and thus they can diffuse back into the inner heliosphere easily. However, interstellar H, as with its galactic cosmic-ray counterpart, has a low rigidity at ~ 10 MeV/nucleon. The modulation process will thus exclude the interstellar H from the inner heliosphere, leaving a component with He, N, O, and Ne as is observed.

Since this theory was first proposed, a number of good theoretical arguments and experimental results have been presented which support it. For example, Fisk (1976) argued that an interpretation in which the anomalous particles were a component of the galactic cosmic-ray flux is incompatible with modulation theory. The argument was particularly strong for anomalous O. For these particles to be galactic cosmic-rays, which will be fully stripped of their electrons, a flux of O $\sim 10^{11}$ times that at earth must be present in the interstellar medium, which violates a number of known constraints on the interstellar cosmic-ray flux.

The principal prediction of Fisk et al. (1974) is that the anomalous particles should be singly charged. Direct measurements of the charge states are not possible; however, the rigidity and thus the charge state of the particles can be inferred from their spatial and temporal variations. Webber et al. (1977) argue that the spatial gradients of the anomalous O are compatible with other measured gradients only if O is singly charged. McKibben (1977) concludes that the time variation of anomalous He is compatible with the time variations of other species only if it is singly charged. A similar conclusion, based on time-variation studies, is also drawn by Klecker et al. (1981).

The acceleration of the anomalous particles is clearly a non-trivial process. When the interstellar particles are first ionized and picked up by the solar wind, they have energies ~1 keV/nucleon. When they are observed they have energies ~10 MeV/nucleon. And if we consider the number of particles that must be accelerated, particularly if a significant number of protons are accelerated, the acceleration process can be an appreciable drain on the available energy in the solar wind.

The difficulty, however, in understanding the acceleration process for the anomalous component is that it appears to occur beyond the range of Pioneer 10, the furthest spacecraft from the Sun. Thus there are no direct observations of the conditions which prevail in the acceleration region and of the acceleration mechanisms which are likely to be important.

Fisk (1976) presented a model for the acceleration of the anomalous component which was a statistical process in large-scale magnetosonic turbulence. The model yielded good fits to observed spectra (Klecker, 1977). However, whether the required turbulence exists is unknown.

Clearly, the acceleration of the anomalous component is a problem which requires more theoretical work, and hopefully, in the future, deep-space observations of the distant heliosphere.

References:

Christan, S., and Simpson, J.A.: 1979, Astrophys. J. (Letters), 227, L49.

Fisk, L.A., Forman, M.A., and Axford, W.I.: 1973, J. Geophys. Res. 78, 995.

Fisk, L.A., Kozlovsky, B., and Ramaty, R.,: 1974, Astrophys. J. (Letters), 190, L35.

Fisk, L.A.: 1976, Astrophys. J., 206, 333.

Fisk, L.A.: 1979, Solar System Plasma Physics, North Holland Publishing Co., 1, 179.

Fisk, L.A., and Lee, M.A.: 1980, Astrophys. J., 234, 620.

Gleeson, L.J., and Axford, W.I.: 1967, Astrophys. J. (Letters), 149, L115.

Gloeckler, G., Hovestadt, D., and Fisk, L.A.: 1979, Ap. J. (Letters), 230, L191.

Hovestadt, D., Vollmer, O., Gloeckler, G., and Fan, C.Y.: 1973, Phys. Rev. Let. 31, 650.

Klecker, B.: 1977, J. Geophys. Res. 82, 5287.

Klecker, B., Hovestadt, D., Gloeckler, G., and Fan, C.Y.: 1980, Geophys. Res. Letters, 7, 1033.

Kunow, H., Wibberenz, G., Green, G., Muller-Mellin, R., Witter, M., Hermpa, H., Mewaldt, R.A., Stone, E.C., and Vogt, R.E.: 1977, 15th Inter. Cosmic Ray Conf. 3, 227.

McDonald, F.B., Teegarden, B.J., Trainor, J.H., and Webber, W.R.: 1974, Astrophys. J. (Letters) 187, L105.

McKibben, R.B.: 1977, Astrophys. J. (Letters) 217, L113.

Parker, E.N.: 1963, Interplanetary Dynamical Processes, Interscience, New York.

Parker, E.N.: 1965, Planet. Space Sci. 13, 9.

Smith, E.J.: 1974, Solar Wind Three, Univ. of California Press, p. 257.

Smith, E.J., and Wolfe, J.H.: 1977, Study of Traveling Interplanetary Phenomena, D. Reidel Publishing Co.

Van Hollebeke, M.A.I., McDonald, F.B., Trainor, J.H., and von Rosenvinge, T.T.: 1981, Proc. Solar Wind Conf. 4, 497.

Webber, W.R., McDonald, F.B., and Trainor, J.H.: 1977, 15th Inter. Cosmic Ray Conf. 3, 233.

LARGE-SCALE MORPHOLOGY OF THE MAGNETOSPHERE

Vytenis M. Vasyliunas
Max-Planck-Institut für Aeronomie
D-3411 Katlenburg-Lindau 3, Federal Republic of Germany

ABSTRACT

The flow of the solar wind past the earth with its magnetic dipole field and its ionosphere gives rise to a complex configuration of plasmas and magnetic fields. This chapter surveys the principal large-scale structures associated with the interaction between the solar wind and the earth, introduces the various terms by which they are known, and sketches their physical properties and their theoretical interpretation if there is a widely accepted one.

INTRODUCTION

The large-scale configurations of the magnetic field and the plasma in and near the terrestrial magnetosphere are now known in considerable detail. A coherent survey of this topic is not an easy task; a systematic approach that might be considered suitable for a theoretically oriented book, presenting the various structures as derived from fundamental physical concepts, is hardly feasible, since some aspects of the observed configurations are reasonably well understood while others are not and their connection with the basic physics of the system can be seen only dimly if at all. Instead, this chapter is organized around a sequence of key concepts, introduced either on an empirical or a theoretical basis as appropriate in each case.

A classification of structures as related primarily to the magnetic field or to the plasma is convenient, even if somewhat artificial since magnetospheric structures generally are reflected in both. The magnetic structures include the magnetosphere, the magnetopause, and the magnetotail; the bow shock and the magnetosheath, structures external to the magnetosphere, are conveniently also discussed together with it. The plasma structures can be further classified according to whether the dominant plasma population is of interplanetary (solar-wind) or terrestrial (ionospheric) origin. The former include the plasma mantle, other boundary layers, and the plasma sheet; the latter include the

R. L. Carovillano and J. M. Forbes (eds.), Solar-Terrestrial Physics, 243–254.

plasmasphere with the plasmapause; the ring current contains important contributions from both sources. (All the structures listed here by name are described in the text; a primary aim of this chapter is to introduce the terminology, for a student unfamiliar with the subject.) A third aspect to be discussed is the projection of magnetospheric structures down along the magnetic field lines to the ionosphere and their relation to the polar aurora and other phenomena observed at low altitudes.

MAGNETIC FIELD STRUCTURES

A schematic diagram of the magnetic field lines inside the magnetosphere together with the plasma flow outside is presented in Figure 1. (See also the discussion, in a more general context, in Vasyliunas 1983b--this volume.) The *magnetosphere* is the volume of space bounded by the *magnetopause* surface. The name "magnetosphere" was introduced by Thomas Gold in 1959, but the concept itself is considerably older; under the name "geomagnetic cavity," it was discussed already by Chapman and Ferraro in 1931.

Why the magnetosphere exists is reasonably well understood in terms of basic physical concepts. Its theory exemplifies a paradox repeatedly encountered in cosmic plasma physics and elsewhere: a phenomenon may be

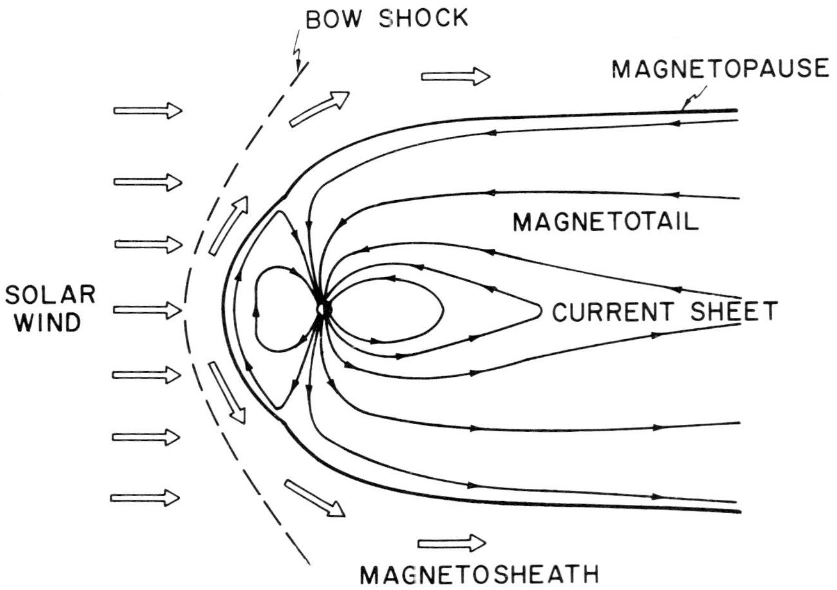

Figure 1. Schematic view of the configuration of magnetic field lines (solid lines with arrows) inside the magnetosphere and of the plasma flow (open arrows) outside, viewed in the noon-midnight meridian plane.

understandable only on the basis of some particular approximation and yet its details may depend crucially on departures from that approximation. The fundamental reason why the magnetosphere is formed as the solar wind plasma flows past the earth with its magnetic dipole field lies in the MHD (magnetohydrodynamic) approximation, which implies that plasma initially on a given magnetic flux tube remains on a single flux tube at all times (see, e.g., Siscoe 1983--this volume). With the solar wind plasma lying initially on flux tubes of the interplanetary magnetic field, the MHD approximation implies, if applied strictly, the division of space into an exterior region, occupied by plasma from the solar wind and hence threaded by magnetic field lines extending into interplanetary space, and an interior region threaded by magnetic field lines connected to the earth and identified as the magnetosphere. This concept, with the magnetopause viewed as a barrier both to plasma and to magnetic field lines, provides a good first approximation, qualitatively and quantitatively, to the observed magnetosphere; yet, as will be seen later, some of the most interesting magnetospheric phenomena would not occur if the magnetopause were strictly impenetrable.

The location and the shape of the magnetopause are determined primarily by the condition of pressure balance: the total pressure (plasma plus magnetic) just outside the magnetopause surface should equal the total pressure just inside (for a general discussion of pressure balance across discontinuities, see Siscoe 1983--this volume). The external pressure is exerted predominantly by the solar wind plasma as it is deflected and flows around the magnetosphere. Since the incident solar wind flow is supersonic with respect to sound as well as MHD or Alfvén waves, to deflect it a detached *bow shock* wave (analogous to a sonic boom in supersonic aerodynamic flow past an obstacle) is required. The region of thermalized and deflected plasma between the bow shock and the magnetopause is called the *magnetosheath* (the term "transition region" used in some earlier papers is now obsolete); the distance from the subsolar point of the magnetopause to the bow shock is often called the *standoff distance*. Plasma flow in the magnetosheath can be modeled to a good approximation by means of an aerodynamic calculation, neglecting the magnetic stresses and treating the magnetosphere as an impenetrable obstacle of known shape (for a detailed review of these models, including comparison with observations and arguments for neglect of the magnetic stresses, see, e.g., Spreiter, Alksne, and Summers, 1968). The external pressure, P, at any point on the magnetopause is obtained as part of the aerodynamic calculation; to an accuracy adequate for most purposes, it can be represented by the so-called Newtonian approximation

$$P = \xi \rho_o V_o^2 \cos^2\psi + P_o \tag{1}$$

where ρ_o, V_o, and P_o are the mass density, bulk flow speed, and thermal pressure, respectively, of the incident solar wind (upstream of the bow shock), ψ is the angle between the local normal to the magnetopause and the upstream solar wind flow direction, and $\xi \simeq 0.9$ is a constant. The last term in (1) is often neglected when dealing with the front side of the magnetosphere (where ψ is not close to 90°) since $P_o \ll \rho_o V_o^2$.

The pressure internal to the magnetopause depends largely on the magnetic field. In the idealized Chapman-Ferraro model, it is assumed that magnetic field lines do not cross the magnetopause and that electric currents within the magnetosphere can be neglected except for those on the boundary and those inside the earth (producing the magnetic dipole field). The magnetic field within the magnetosphere can then be calculated by solving a well-posed magnetostatic boundary value problem defined by the conditions (a) $\nabla \times \underline{B} = 0$ inside the magnetospheric volume, (b) the component normal to the magnetopause $B_n = 0$, and (c) $\underline{B}$ approaches a dipole field as $\underline{r} \rightarrow 0$. This calculation yields a unique field $\underline{B}$ ($\underline{r}$) for any given boundary surface; the surface appropriate for modeling the magnetopause is picked out by the pressure balance requirement

$$B^2/8\pi = P \quad (2)$$

where B is the magnetic field magnitude just inside the magnetopause and P is the external pressure given, e.g., by equation (1).

The scheme described above has been applied in a number of calculations to obtain a quantitative model for the location and shape of the magnetopause (see, e.g., Mead and Beard, 1964; review by Roederer, 1969). The results are generally in reasonable agreement with observations as far as the front side of the magnetosphere is concerned. The geocentric distance to the subsolar point of the magnetopause and its variations with changes of the solar wind dynamic pressure $\rho_o V_o{}^2$--the average distance is 10-11 R_E, but occasional extreme excursions reaching as close as 6.6 R_E or as far as 18 R_E have been observed--are well accounted for by the models. More recently, the model calculations have been refined by relaxing the current-free condition $\nabla \times \underline{B} = 0$ and including the effects of some current systems inside the magnetosphere (see reviews by Walker, 1979 and Olson, Pfitzer, and Mroz, 1979).

On the night side of the magnetosphere, on the other hand, the magnetic field structure cannot be accounted for by the simple concepts of the Chapman-Ferraro model. In contrast to the compressed configuration of field lines on the day side, a consequence of the external solar wind plasma pressure, the magnetic field lines on the night side are highly stretched out and extended in a direction aligned with solar wind flow to form the magnetospheric tail or *magnetotail* for short. The reversal of $\underline{B}$ between the extended field lines from the northern and southern hemispheres implies the existence of a *current sheet* (frequently although somewhat inaccurately called the magnetic *neutral sheet*), with electric current flowing from dawn to dusk across the center of the magnetotail. On the basis of numerous and widespread observations, an extensive and in part quantitative empirical description of the magnetotail has been developed; there is, however, as yet no adequate theoretical understanding capable of providing quantitative physical models.

The stretched-out magnetotail configuration implies a magnetic tension, whose magnitude can be estimated from the empirical models, which must be balanced by a mechanical force (in the antisolar direction) derived ultimately from the solar wind (Carovillano and Siscoe, 1973, and references therein); equivalently, one may say that the solar wind plasma exerts an effective *tangential stress* or *tangential drag* along the magnetopause in addition to a pressure normal to it (the term "viscous drag" is also used, particularly in the older literature, but is perhaps somewhat misleading since ordinary collisional viscosity is certainly insignificant). As to the mechanism of tangential drag, the most common view is that it is associated with a non-zero magnetic field component normal to the magnetopause in the magnetotail regions: the tangential drag is the mechanical stress that balances the tangential component of the $\underline{j} \times \underline{B}$ force at or near the magnetopause implied by $B_n \neq 0$. In this view, some magnetic field lines from the earth extend into the solar wind plasma and hence are pulled back by its flow to form the magnetotail. An alternative suggestion sometimes made is that it is the plasma flow in the magnetospheric boundary layers (described later in this chapter) that pulls back the geomagnetic field lines and forms the magnetotail.

The assumption of a non-zero B_n represents a departure from the strict idealized Chapman-Ferraro model. A magnetosphere where $B_n \neq 0$ and hence some magnetic field lines interconnect between the earth and the solar wind is called an *open magnetosphere*, in contradistinction to a *closed magnetosphere* where no field lines cross the magnetopause. For a discussion of the evidence for the open character of the earth's magnetosphere and its formation and properties, see Hill (1983--this volume) and Vasyliunas (1983a--this volume) and references therein. Quantitatively, the effects of the open magnetosphere are small in one sense: $B_n \ll B$, so that it is in fact difficult to establish by direct observation that $B_n \neq 0$ and most of the evidence for the open magnetosphere is indirect. In another sense, however, the effects are large: the empirically estimated amount of magnetic flux on open field lines represents a considerable fraction of the total flux that could potentially be open (estimated as the flux on field lines of an undistorted magnetic dipole that close beyond the distance ~10 R_E defined by the subsolar magnetopause).

At the interface between the day and night sides of the magnetosphere, there are two regions, one north and one south, of magnetic field lines with a funnel-shaped geometry, known as the *polar cusps*. The central field line of the cusp, with neighboring field lines bending away from it in all directions, is the topological counterpart of the field line through the poles of the dipole (although it is not the same line--it intersects the earth's surface some 10-15° of latitude sunward of the geomagnetic pole). In the case of a closed magnetosphere, it is the singular magnetic field line that intersects the magnetopause at a neutral point where $\underline{B} = 0$, with field lines then radiating from that point in all directions, umbrella-fashion, to form the magnetopause surface. An indentation of the surface at the two neutral points (north and south) is predicted by models of the magnetopause developed from the Chapman-Ferraro approach. Whether or not such an indentation is actually present is not firmly established, but the funnel geometry of the cusp is well supported by magnetic field observations (e.g., Hedgecock and Thomas, 1975). In an open magnetosphere, the magnetopause is not formed by field lines (except possibly in limited regions) and the central line of the polar cusp is not a singular line, but the general funnel configuration of the cusp region is not greatly changed as long as $B_n << B$.

PLASMA STRUCTURES

With respect to plasma, the magnetopause behaves in a similar way as with respect to magnetic field lines. To first approximation, it acts as a barrier and forces the solar wind plasma to flow around the magnetosphere; on a closer look, it is found to allow penetration of plasma by an amount that is relatively small (in the sense that it represents a small fraction of the solar wind plasma flowing through an area equal to a frontal cross-section of the magnetosphere and also the component of flow normal to the magnetopause is small compared to the tangential) but nevertheless constitutes a major source of plasma for the magnetosphere.

Exterior to the magnetosphere everywhere lies the magnetosheath. The inner boundary of magnetosheath plasma does not, however, coincide with the magnetopause but occurs somewhat earthward of it. Thus, there are several regions bounded by the magnetopause on the outside and by the boundary of magnetosheath plasma on the inside, i.e., regions of space threaded by the magnetic field of the magnetosphere but populated largely by plasma similar to that found in the magnetosheath, which collectively are called the *magnetospheric boundary layers*. They are shown, together with various other plasma regions inside the magnetosphere, schematically in Figure 2. Their existence, well established observationally (see, e.g., Haerendel et al., 1978, and references therein) is usually taken as direct evidence for penetration of plasma across the magnetopause to depths that range from under 10^3 km to several R_E and more; a firm theoretical understanding of how they are formed is, for the most part, still lacking.

The boundary layers can be classified into several distinct and variously named types (Haerendel et al., 1978, Vasyliunas, 1979, and references therein). At the magnetic polar cusps, magnetosheath plasma extends so deep into the magnetosphere that the name "boundary layer" is hardly appropriate: the entire funnel of field lines, down to the upper atmosphere, is filled with magnetosheath-like plasma. This region is referred to by some as the *polar cusp* again (or sometimes the *dayside cusp*) and by others as the *magnetospheric cleft*. Its outermost parts, near the magnetopause, have been called the *entry layer*, a term criticized as enshrining yet unproven physical assumptions in the terminology, and *interior cusp* has been suggested as a more noncommittal term ("exterior cusp" being the layer of magnetosheath plasma outside the presumably indented magnetopause). Whatever the name, the entire region is a distinct recognizable entity, with plasma densities comparable to

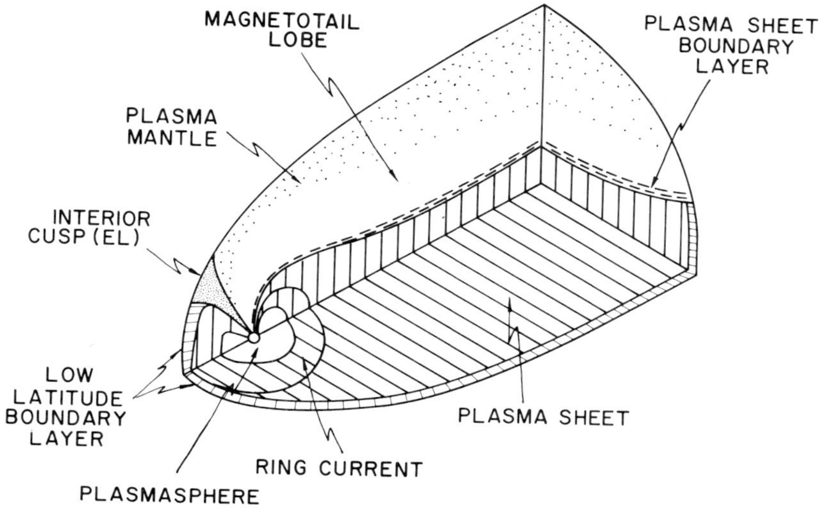

Figure 2. Schematic three-dimensional cut-away view of the principal plasma regions inside the magnetosphere. The thickness of the boundary layers is in general not to scale.

those in the magnetosheath, somewhat higher temperatures, complicated flow patterns along the magnetic field but only slow average bulk flow perpendicular to the field.

The *low-latitude boundary layer* (parts of it have also been called the front side or dayside boundary layer), on the other hand, can be as thin as 0.1 R_E near the nose of the magnetosphere and expands to perhaps 1-2 R_E at the flanks. The plasma within it has a pronounced bulk flow in a direction close to (or possibly even coincident with) that of the plasma flow in the adjacent magnetosheath. The topology of the magnetic field lines within the low-latitude boundary layer as well as, to some extent, within the cusp/cleft region--are they closed, connected to the earth in both directions, or open, crossing the magnetopause in one direction?--is a still unresolved controversial question with major implications for the physical understanding of these boundary layers.

There is little question, however, that the *plasma mantle*, adjacent to the magnetopause downstream of the polar cusps, lies on open magnetic field lines. Plasma from the magnetosheath expands essentially freely into the ever-lengthening open magnetic flux tubes as they remain attached to the earth at one end while the other is convected downstream by the solar wind flow; equivalently, ions and electrons from the magnetosheath move in, mirror in the geomagnetic field, and then move out, all the while being carried tailward by the $\underline{E} \times \underline{B}$ drift associated with the dawn-to-dusk electric field mapped (in accordance with the MHD approximation) from the solar wind along the open magnetic field lines. The resulting boundary layer exhibits a considerable decrease of density from its magnetosheath value, a strong bulk flow nearly aligned with the external magnetosheath flow and the magnetotail magnetic field direction, a small but significant inward flow component perpendicular to $\underline{B}$ and consequently a thickness that increases with increasing distance downstream. These are all properties found in the observations; the plasma mantle, unlike the other boundary layers, seems to be fairly well understood in physical terms at least qualitatively.

There are indications that, at the 60 R_E distance of the lunar orbit, the plasma mantle may become thick enough to fill a significant part of the magnetotail cross-section at least occasionally (Hardy, Hills, and Freeman, 1975). A continued expansion will in any case bring the northern and southern plasma mantles together at some point of the distant magnetotail. Also, the open magnetic field lines cannot continue to be stretched indefinitely but will eventually be cut and reconnected at a magnetic singular X line, as described in more detail in, e.g., Vasyliunas (1983a--this volume) and Hill (1983--this volume). It has been repeatedly proposed (see, e.g., Cowley, 1980, and references therein) that either one or both of these occurrences--the collision of the plasma mantles or the merging of magnetic field lines from the northern and southern halves of the magnetotail--will produce a new region of highly heated plasma, the *plasma sheet*, extending earthward through the center of the magnetotail. The plasma sheet is in any case a structure well established on the basis of direct observations, discovered (and named) independently of the theoretical suggestions. It has a planar geometry, enveloping the current sheet and stretching across the magnetotail between dawn and dusk, with a considerably variable half-thickness of the order of a few R_E. Whether the current sheet is a much thinner structure imbedded within the plasma sheet (the majority view) or whether instead to two are coextensive is a question that has not yet been conclusively settled. Both the thickness of the plasma sheet and the flow pattern within it exhibit large systematic temporal variations in association with geomagnetic and auroral activity (see, e.g., Vasyliunas, 1980, and references therein).

The plasma within the plasma sheet, compared to that in the magnetosheath, is very highly heated and expanded: while the magnetosheath typically has thermal energies of the order of several hundred eV and number densities 30-50 cm^{-3} (related to the solar wind bulk flow kinetic energy of 1 keV and density of ~10 cm^{-3}, respectively) near the subsolar region and lower values elsewhere, the plasma sheet is characterized by electron thermal energies near 1 keV, ion thermal energies of 5-10 keV and above, and number densities generally below 1 cm^{-3} (see, e.g., review by Vasyliunas, 1970). The energy density of the plasma is comparable to or larger than the energy density of the local magnetic field. The thermal energies generally decrease with increasing distance from the central plane, and adjacent to the plasma sheet above and below is a *plasma sheet boundary layer*, with thermal energies of the order of 100 eV and hence somewhat similar to those in the magnetosheath, but still with low densities. Finally, between the plasma sheet with its boundary layers and the plasma mantles lie the two *magnetotail lobes*, characterized by very low plasma densities (below 10^{-2} cm^{-3}) and by relatively steady and strong magnetic fields; the magnetic pressure within the lobes is required to balance the combined plasma-plus-magnetic pressure of the plasma sheet.

For all the plasma structures discussed so far, the supplier of the major part of the plasma population is ultimately the solar wind. It, however, is not the sole source. A smaller but significant part of the plasma sheet population is thought to come from the earth's ionosphere, as evidenced primarily by the ionic composition of the plasma, the presence of oxygen ions which are relatively abundant in the ionosphere but very rare in the solar wind (see, e.g., Johnson, 1979, and references therein). Within and near the plasma sheet boundary layers, a fairly direct indication of plasma input from the ionosphere is the observation of narrowly collimated ion beams flowing along the magnetic field lines away from the earth; there is also evidence for an ionospheric contribution to the plasma mantle (Hultqvist, 1982, and references therein).

A region populated almost entirely by plasma from the ionosphere is the *plasmasphere*, located deep within the inner magnetosphere; it extends from the ionosphere outward to a relatively sharp boundary (called the *plasmapause*) whose location is somewhat variable but typically coincides with a shell of magnetic field lines crossing the equator at distances from 4 to 6 R_E. The plasma within this region is relatively dense and cold, with number densities ranging from $\gtrsim 10^4$ cm^{-3} just above the topside ionosphere to values of the order of 10^2 cm^{-3} near the plasmapause and thermal energies of the order of a few eV or less, characteristic of the ionosphere (in contrast to plasma of ionospheric origin found in the plasma sheet, which is heated to the high thermal energies typical of that region). The plasmasphere can be viewed as just the upward extension of the ionosphere. The reason for the plasmapause, the sharply defined boundary aligned with the magnetic field, is well understood: within it the flux tubes circulate around the earth and can accumulate plasma from the ionosphere over long periods, while beyond it they are repeatedly carried out to the distant regions of the magnetosphere by the flow pattern of magnetospheric convection (Hill, 1983--this volume, Wolf, 1983--this volume, and references therein).

Finally, the complex region generally named the *ring current* forms the interface between the plasmasphere on the inside and the plasma sheet on the outside. A major source of plasma for the ring current is inward transport (with attendant compression and heating as the magnetic flux tubes contract and shrink in volume) of the plasma from the plasma sheet, including both its solar-wind and its ionospheric-source components. There are significant electric currents that can be viewed either as a consequence of the stress balance between the pressure gradient and the Lorentz force

$$\nabla P = \underline{j} \times \underline{B}/c \qquad (3)$$

or equivalently as the sum of gradient, curvature, and magnetization drifts of charged particles trapped in the magnetic field; these currents produce magnetic disturbances observable on the earth that have long been attributed to a "ring current" in space, with the average ground-level disturbance field being proportional to the total energy content of the plasma (see, e.g., Carovillano and Siscoe, 1973, and references therein). The interrelationships, within the ring current and plasmasphere regions, among currents, plasma distribution and flow, and electric fields in the ionosphere and the magnetosphere form one of the best understood chapters of magnetospheric theory, reviewed here by Wolf (1983--this volume).

LOW-ALTITUDE STRUCTURES

The detailed morphology of auroral and ionospheric phenomena and their relation to the magnetosphere is a vast topic beyond the scope of this chapter. The large-scale average distribution of these phenomena, however, sketched in Figure 3, has a recognizable correspondence to the large-scale configuration of the magnetosphere projected along

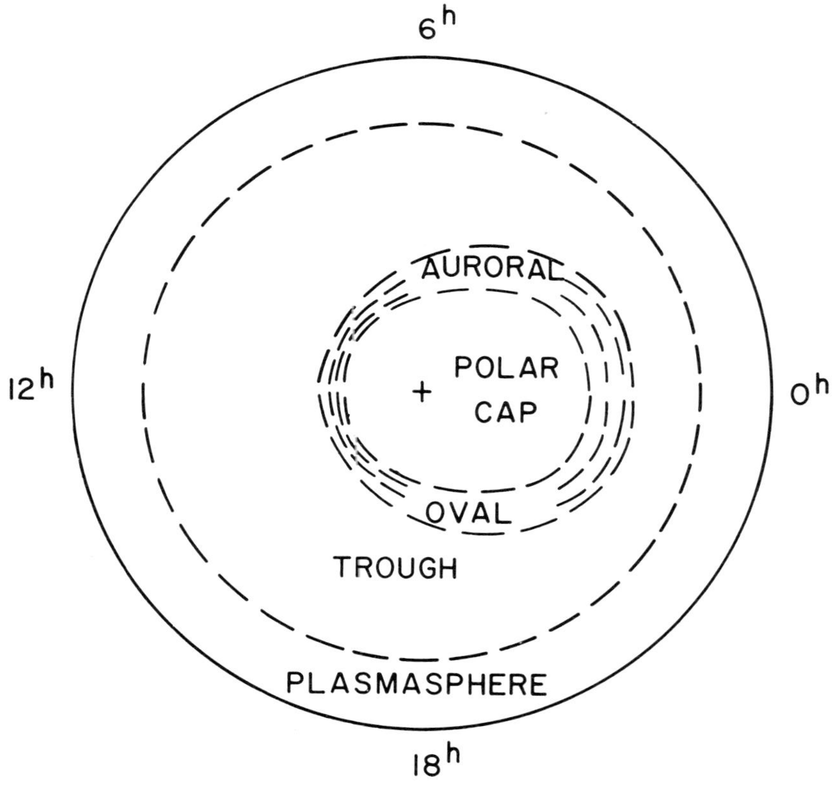

Figure 3. Sketch of the principal auroral and ionospheric regions, viewed from above the north geomagnetic pole (marked with a +). The sun is to the left and the numbers indicate the local time.

magnetic field lines. The principal feature is the *auroral oval*, defined at any given time as the region where auroral particle precipitation and associated auroral light emissions occur. The auroral oval lies close to the ionospheric projection of the demarcation between open and closed magnetic field lines and of the outer parts of the plasma sheet and/or its boundary layers. The precise correspondence is still uncertain; it is frequently assumed (but not firmly established) that its poleward boundary corresponds to the last closed field line. The dayside part of the oval, within a few horus of local noon, is probably associated with the ionospheric projection of the polar cusp/magnetospheric cleft.

Poleward of the auroral oval lies the *polar cap*. It is generally agreed that it is the ionospheric projection of the region of open field lines, or at least of the major part of them, including the field lines threading the lobes of the magnetotail.

Equatorward of the auroral oval lies a region of reduced electron density in the topside ionosphere, generally known as the *ionospheric trough*. Its equatorward boundary is the ionospheric projection of the plasmapause; its poleward boundary is the equatorward boundary of the auroral oval. Geometrically, the trough has some relation to (although probably not a precise coincidence with) the ring current, Its ionospheric-physics aspects are discussed in Schunk (1983--this volume).

ACKNOWLEDGMENTS

This chapter was written while I was a visitor at the Center for Space Research, Massachusetts Institute of Technology. I am grateful to Professor H. S. Bridge and the CSR staff for their hospitality and support.

REFERENCES

Carovillano, R. L., and Siscoe, G. L.: 1973, Rev. Geophys. Space Phys. 11, pp. 289-353.
Cowley, S. W. H.: 1980, Space Sci. Rev. 26, pp. 217-275.
Haerendel, G., Paschmann, G., Sckopke, N., Rosenbauer, H., and Hedgecock, P. C.: 1978, J. Geophys. Res. 83, pp. 3195-3216.
Hardy, D. A., Hills, H. K., and Freeman, J. W.: 1975, Geophys. Res. Lett. 2, pp. 169-172.
Hedgecock, P. E., and Thomas, B. T.: 1975, Geophys. J. R. Astr. Soc. 41, pp. 391-403.
Hill, T. W.: 1983, this volume.
Hultqvist, B.: 1982, Rev. Geophys. Space Phys. 20, pp. 589-611.
Johnson, R. G.: 1979, Rev. Geophys. Space Phys. 17, pp. 696-705.
Mead, G. D., and Beard, D. B.: 1964, J. Geophys. Res. 69, pp. 1169-1179.

Olson, W. P., Pfitzer, K. A., and Mroz, G. J.: 1979, in W. P. Olson (ed.), "Quantitative Modeling of Magnetospheric Processes," AGU Geophysical Monograph 21, Washington, D. C., pp. 77-85.
Roederer, J. G.: 1969, Rev. Geophys. Space Phys. 7, pp. 77-96.
Schunk, R. W.: 1983, this volume.
Siscoe, G. L.: 1983, this volume.
Spreiter, J. R., Alksne, A. Y., and Summers, A. L.: 1968, in R. L. Carovillano, J. F. McClay, and H. R. Radoski (ed.), "Physics of the Magnetosphere," D. Reidel, Dordrecht, The Netherlands, pp. 301-375.
Vasyliunas, V. M.: 1970, in E. R. Dyer (ed.), "Solar Terrestrial Physics/1970: Part III," D. Reidel, Dordrecht, The Netherlands, pp. 192-211.
Vasyliunas, V. M.: 1979, in B. Battrick (ed.), "Proceedings of Magtospheric Boundary Layers Conference," ESA SP-148, Noordwijk, The Netherlands, pp. 387-393.
Vasyliunas, V. M.: 1980, in C. S. Deehr and J. A. Holtet (ed.), "Exploration of the Polar Upper Atmosphere," D. Reidel, Dordrecht, The Netherlands, pp. 229-244.
Vasyliunas, V. M.: 1983a, "Magnetic Field Line Merging: Basic Concepts," this volume.
Vasyliunas, V. M.: 1983b, "Comparative Magnetosphere," this volume.
Walker, R. J.: 1979, in W. P. Olson (ed.), "Quantitative Modeling of Magnetospheric Processes," AGU Geophysical Monograph 21, Washington, D. C., pp. 9-34.
Wolf, R. A.: 1983, this volume.

MAGNETIC FIELD LINE MERGING: BASIC CONCEPTS

Vytenis M. Vasyliunas
Max-Planck-Institut für Aeronomie
D-3411 Katlenburg-Lindau 3, Federal Republic of Germany

ABSTRACT

The concept of magnetic field line merging is defined and is illustrated by reference to various possible structures of magnetic topology in the magnetosphere. A key element is the electric field tangent to the magnetic separatrix surfaces; its role and its relation to the MHD approximation can be viewed both in a global and a local context.

In many situations within solar-terrestrial physics, as well as in astrophysics and plasma physics, an important and interesting process is one variously called "magnetic field line merging" or more simply "magnetic merging", or else "magnetic field line cutting and reconnection" which is always abbreviated to "magnetic field line reconnection" or simply "magnetic reconnection" (all the names are synonymous, in actual usage). This process has been studied theoretically for decades; more recently, there have been extensive efforts to observe its manifestations within the magnetosphere as well as studies of laboratory plasma model experiments and numerical simulations. Yet the fundamentals of the process are not always easy to grasp, its applicability and description remain controversial, and certain misconceptions seem to persist among opponents as well as proponents. The purpose of this chapter is to present qualitatively some basic concepts which I consider essential for an understanding of the subject. For a detailed mathematical formulation and for specific applications to the magnetosphere, the reader is referred to Vasyliunas (1975) and to Hill (1983 - this volume), respectively, and references therein.

Magnetic field line merging can be defined as "the process whereby plasma flows across a surface that separates regions containing topologically different magnetic field lines" (Vasyliunas, 1975).

As an illustration, consider the topological configuration of the magnetic field shown in Figure 1, a particularly simple model for the

R. L. Carovillano and J. M. Forbes (eds.), Solar-Terrestrial Physics, 255–260.

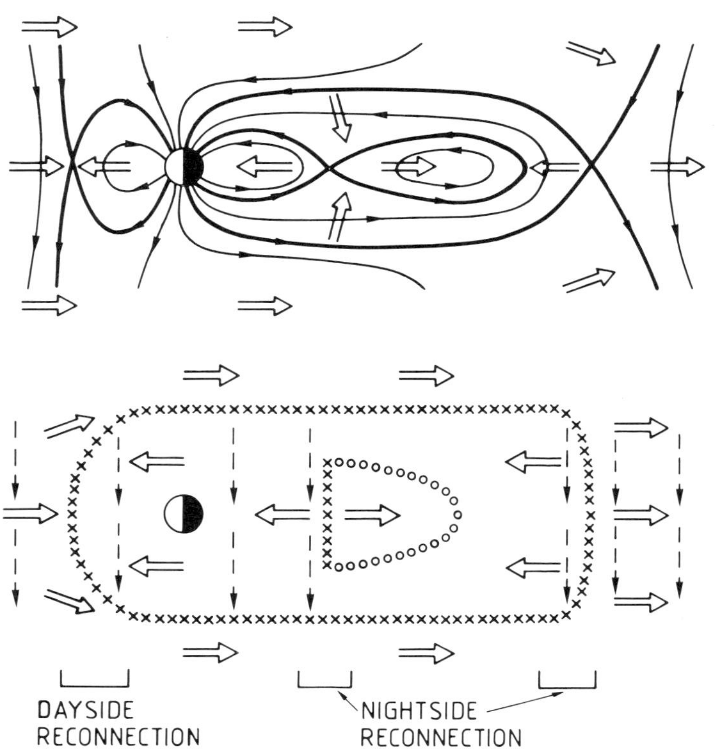

Figure 1. Sketch of a possible topology of the magnetosphere. Top: view in the moon-midnight meridian plane, showing magnetic field lines, separatrices (heavy lines), and plasma flow directions (open arrows). Bottom: projection onto a surface containing the magnetic singular lines (the equatorial plane in a case of ideal symmetry), showing the X and O lines, plasma flow (open arrows) and electric field directions (dashed arrows).

Earth's magnetosphere. The topologically different classes of magnetic field lines include (1) "closed" field lines that connect to the Earth in both directions, (2) "open" field lines that connect to the Earth at one end and to the interplanetary magnetic field at the other, and (3) interplanetary field lines that do not connect to the Earth; the bounding surface or separatrix between regions occupied by these classes of field lines, topologically a doughnut touching the inside of a cylinder, has two branches that intersect along a line, called the X line, which forms a closed ring. In addition, enclosed by the region of closed field lines there are (4) field lines that do not connect either to the Earth or to interplanetary space but form an isolated magnetic "island" or "bubble" within the magnetotail; they are bounded by a separatrix, topologically a sphere touching a ribbon, with two branches that intersect along a line segment, again an X line, which connects to an O line segment to form a closed ring. (The two topological features, the open magnetosphere and the magnetic island, are shown together for convenience, but in fact either one could exist without the other.)

Two distinct magnetic merging processes can occur in a magnetosphere with the topology shown in Figure 1. One is associated with the flow of the solar wind, which unavoidably crosses the separatrix between interplanetary and open field lines; that this magnetic merging process occurs is unquestionable, given the solar wind flow (which is indisputable) together with the open character of the magnetosphere (which is not seriously disputed any more). The other is associated with plasma flow in the magnetotail by virtue of magnetospheric convection (Hill, 1983 - this volume; Wolf, 1983 - this volume) across the separatrix of the magnetic island; whether and when this magnetic merging process occurs is still an unsettled question, since the evidence for the magnetic island, although (in my opinion) substantial, is not yet conclusive.

The general flow pattern outside and inside the magnetosphere implied by both magnetic merging processes, derived by extending the solar wind flow along magnetic field lines, in accordance with the MHD approximation (Siscoe, 1983 - this volume), and by continuity, is also sketched in Figure 1. Note that magnetic merging associated with the solar wind and the open magnetosphere involves two distinct flow regions: one in which the flow proceeds both from interplanetary and from closed to open field lines (often called the dayside reconnection region), and the other where the flow proceeds from open both to closed and to interplanetary field lines (the nightside reconnection region). On the other hand, merging associated with the magnetic island has only one flow region, the flow proceeding into the island from all sides (the term "nightside reconnection region" is also used here, the distinction from the other usage of the term often being apparent only from the context or not at all). It is obvious that magnetic merging in the case of the magnetic island is intrinsically time dependent and cannot be steady-state; for a discussion of the island's growth and further evolution, see, e.g., Vasyliunas (1980b) and references therein. The magnetic merging associated with the solar wind, on the other hand, can be steady, with flows across the dayside and the nightside reconnection regions balancing each other, at least in principle, even though time variations are certain to occur in practice.

The MHD approximation associates an electric field with plasma flow, in accordance with the equation

$$\underline{E} + \underline{V} \times \underline{B}/c = 0 \qquad (1)$$

An alternative (and equivalent) definition of the magnetic merging process is to take the existence of an electric field in the separatrix surface rather than of plasma flow across the surface as the defining element. With either definition, an unavoidable complication of the magnetic merging process is that the MHD approximation necessarily becomes inapplicable in the vicinity of the X and O lines: the zero on the RH side of equation (1) is actually an approximation to several terms that are small as long as the macroscopic gradient length scale is large compared to characteristic plasma lengths such as particle gyroradii (see, e.g., the discussion in Siscoe, 1983 - this volume), but near an X or an O line the length scale of magnetic field variations becomes comparable to the distance from the line and hence arbitrarily small. However, the breakdown of MHD occurs at a distance from the X line at most comparable, say, to a few ion gyroradii, which is very small compared to a typical dimension (such as the length or the curvature, up to tens of Earth radii) of the X line itself. Hence, from plasma flow across the separatrix one may infer the existence of an electric field in the separatrix, on the basis of the MHD approximation, down to small distances from the X line, where one can then invoke Faraday's law and the continuity of tangential components of $\underline{E}$ (without any further appeal to MHD) to conclude that a finite component of $\underline{E}$ exists along the X line itself. (A corresponding argument cannot be made for O lines since they do not lie on a separatrix or any other magnetic surface; the component of $\underline{E}$ along an O line can in fact be shown to be negligible, cf. Vasyliunas, 1980a.)

Occasionally magnetic merging is defined as the process associated with an electric field along a magnetic X line; it should be clear that such an electric field can be deduced from the more general definition given previously and need not be postulated as the defining element. Similarly, the terms "dayside reconnection region" and nightside reconnection region" are frequently restricted to denote the corresponding segments of the X line with tangential electric fields (the term "reconnection line" is also used).

From the global point of view developed so far, it is clear that magnetic field line merging is not a single specific process but a wide class of processes, namely all those involving plasma flows in topologically complex systems. (It should not be forgotten that the topology of Figure 1 is at best a first approximation - the actual topology of the Earth's magnetosphere is in all likelihood even more complex). Setting up a complete model means solving for the field and the flow throughout the system, and this obviously depends on the given conditions and specifications; there can be no question of a single, universally applicable "magnetic merging model" (any more than of a universally applicable mere flow model).

There is, however, also a local view of the magnetic merging problem, motivated largely by the unavoidable role of non-MHD effects near the X line: the component of $\underline{V} \times \underline{B}$ along the X line is zero, by the definition of the field geometry, while the component of E is non-zero if magnetic merging is occurring, and therefore the neglected terms on the RH side of equation (1) must necessarily be taken into account. The question then arises, could these non-MHD effects severely limit the value of E at the X line and hence constrain the entire magnetic merging configuration ? As an extreme possibility, if E at the X line were forced (by high conductivity or otherwise) to be negligibly small, then by reversing a previously used argument one would conclude that there could be no significant electric field anywhere in the separatrix and no flow across the separatrix, thus no magnetic merging, which would in turn imply that the topological configuration of Figure 1 was not possible - the magnetosphere would be closed and no magnetic islands could appear. To investigate such questions as well as to gain insight into the detailed flow and field configuration, one may formulate a local problem: consider a volume near the X line, large compared to the microscopic scales (gyroradii, etc.) but small compared to the macroscopic lengths, and approximate the system within the volume as a flattened two-dimensional steady configuration of nearly antiparallel magnetic fields shown in Figure 2. The occurrence of magnetic merging now manifests itself as plasma flow from both sides toward the magnetic field reversal region; the flow speed is a measure of the merging rate, usually made dimensionless by normalizing it to the Alfvén speed.

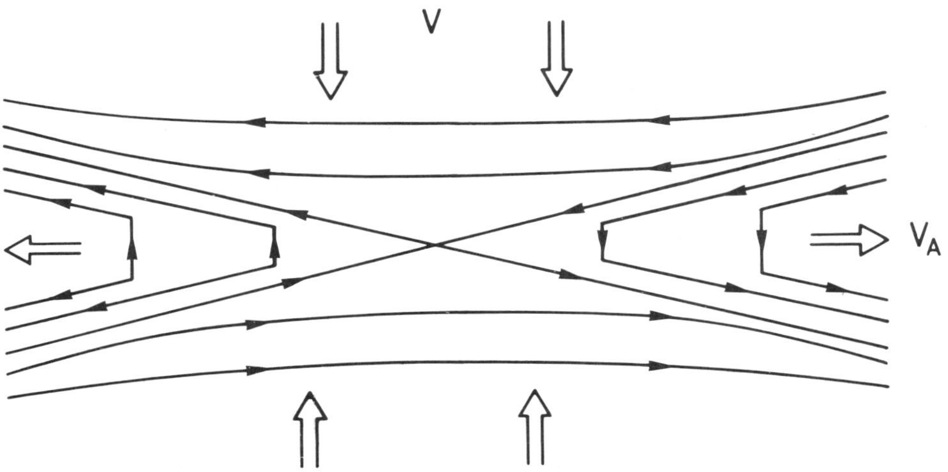

Figure 2. Idealized schematic configuration of the plasma flow and the magnetic field in the vicinity of the X line.

This local problem has been the subject of an extensive physical and mathematical development, summarized and critically reviewed by Vasyliunas (1975), whose principal conclusions include the following:

(1) After crossing the separatrix, within the field reversal region, the incoming plasma turns and flows out with speeds of the order of the Alfvén speed. These high-speed plasma jets are among the primary observational signatures used to identify reconnection regions in the magnetosphere.

(2) Non-MHD effects, such as resistivity or electron inertia, determine the microscopic structure associated with the X line but do not impose any strong limitations on the electric field along the X line.

(3) The merging rate is not determined by the local problem but can have any value, depending only on the boundary conditions. (The statement, frequently seen in the literature, that the dimensionless merging rate has an upper limit of order 0.1 applies only to a subclass of systems with particularly simple boundary conditions; rates of order 1 or higher are possible if the incoming flow is suitably forced.) Boundary conditions for the local problem are set by the global configuration. The studies of the local problem thus indicate the global view that there can be no universal magnetic merging model; the magnetic field line merging process is just an aspect of the complete dynamics of the particular system under consideration.

REFERENCES

Hill, T.W.: 1983, this volume.
Siscoe, G.L.: 1983, this volume.
Vasyliunas, V.M.: 1975, Rev. Geophys. Space Phys. 13, pp. 303-336.
Vasyliunas, V.M.: 1980a, J. Geophys. Res. 85, pp. 4616-4620.
Vasyliunas, V.M.: 1980b, in C.S. Deehr and J.A. Holtet (ed.), "Exploration of the Polar Upper Atmosphere," D. Reidel, Dordrecht, The Netherlands, pp. 229-244.
Wolf, R.A.: 1983, this volume.

SOLAR-WIND MAGNETOSPHERE COUPLING

T. W. Hill
Department of Space Physics and Astronomy
Rice University, Houston, TX 77251

ABSTRACT

The dynamical coupling between the solar wind and Earth's magnetosphere is described in general terms within the context of two antipodal coupling models: the Dungey (open) model and the Axford-Hines (closed) model. The basic element of solar-wind magnetosphere coupling is the magnetospheric convection flow driven by the solar-wind flow at their common boundary. The voltage across this flow is the fundamental measure of the coupling rate. The magnitude of this voltage, and its dependence on interplanetary magnetic-field direction, are the two basic observational data, and both data appear to favor the open model. The data, however, admit the possibility of observable closed-model coupling processes which, although they may not have global terrestrial consequences, are nevertheless of intrinsic interest.

INTRODUCTION

The solar wind is the ultimate power source for virtually all dynamical processes in Earth's magnetosphere. The magnetosphere extracts a few percent of the available solar-wind energy flux to power the aurora and other more subtle geomagnetic processes.

The magnetosphere presents an obstacle to the solar wind of cross-sectional area $A \sim \pi\ (15\ R_E)^2 \sim 3\times10^{16}\ m^2$ ($1\ R_E$ = Earth's radius = 6.4×10^3 km). The available power from a typical solar wind (mass density $\rho \sim 5\ amu/cm^3$, velocity $v \sim 400$ km/s) is thus $(1/2)\ \rho v^3 A \sim 8\times10^{12}$ W. The average total power consumed by magnetospheric phenomena is difficult to estimate, but 2×10^{11} W seems a reasonable estimate (corresponding, for example, to a current of 4 MA across a voltage drop of 50 kV); hence the estimate of a "few percent" coupling efficiency. A paramount problem of magnetospheric theory is to explain this coupling efficiency.

The classical model of solar-wind/magnetosphere interaction, i.e.,

R. L. Carovillano and J. M. Forbes (eds.), Solar-Terrestrial Physics, 261–302.

the model that would (indeed, did) follow from classical physics in the absence of geophysical and space observations, is one in which the solar wind is deflected around the magnetosphere obstacle with negligible energy loss. This "closed" model was elucidated by Johnson (1960) following the classical work of Chapman and Ferraro (1931); individual particles of the solar-wind stream are presumed to be specularly reflected by the Earth's magnetic field at their common boundary (the "magnetopause"), the net result being a potential flow of the solar wind around the magnetosphere analogous to that of a perfectly-conducting magnetized fluid around a perfectly-conducting obstacle.

Because the solar-wind flow is supersonic, such deflection would require, in fluid-dynamic terms, the existence of a shock wave standing upstream of the obstacle. The thickness of this shock wave, however, would scale classically as the collisional mean-free-path in the upstream solar wind, which is about 3×10^6 R_E ~ 100 AU (!) for the density quoted above. The actual bow shock wave was found (Ness et al., 1964) to have a thickness of no more than 1000 km ~ 6×10^{-6} AU; this factor-of-10^7 discrepancy had been anticipated (e.g., Axford, 1962; Kellogg, 1962), and it provided an early clue that the classical understanding of the solar-wind/magnetosphere interaction was likely to be in error. The essential role of the interplanetary magnetic field (IMF) was stressed by Axford (1962), who argued that the gyroradius of a typical solar-wind ion (~ 1 keV) in a typical IMF (~ 5 nT), being ~ 10^3 km, was much smaller than the collisional mean-free-path (~ 10^{10} km), and that the solar wind might thus be expected to exhibit fluid-like behavior on scales much smaller than the collisional mean-free-path. This concept has been so thoroughly verified by space observations, and thus so inconspicuously absorbed into space plasma theory, that its revolutionary nature is likely to be overlooked. The collisional interaction among charged particles in space is indeed quite negligible (to parts-per-million accuracy) compared to the large-scale Lorentz force, and yet the description of space plasmas as classical fluids is not only commonplace, it is in many applications surprisingly accurate on scales larger than the ion gyroradius.

THE SOLAR WIND — AN ANISOTROPIC FLUID

The IMF influences the flow of the solar wind around the magnetosphere in two fundamental ways. The Lorentz ($\underline{v}\times\underline{B}$) force strongly couples the two components of particle motion $\underline{v}$ in the plane perpendicular to the magnetic field $\underline{B}$, and thus produces two-dimensional fluid-like behavior in that plane. The IMF also provides a medium for the propagation of various low-frequency waves that can interact with individual particles and thus produce one-dimensional fluid-like behavior in the direction of $\pm\underline{B}$. Thus the plasma rest-frame velocity distribution function $f(\underline{v})$ is very well approximated by the "gyrotropic" form $f(v_\perp, v_\parallel)$, and likewise its second moment, the pressure tensor

$$\underline{\underline{P}} \equiv \int f(\underline{v})\, \underline{v}\,\underline{v}\, d^3v \approx P_\perp \underline{\underline{1}} + (P_\parallel - P_\perp)\, \underline{b}\,\underline{b} \tag{1}$$

where $\underline{\underline{I}}$ is the identity tensor, $\underline{b}$ is the unit vector along $\underline{B}$, and subscripts $\perp$ and $\parallel$ refer to directions perpendicular and parallel to $\underline{B}$. The pressure elements $P_\perp$ and $P_\parallel$ generally obey different equations of state, but it is mathematically convenient, and in some cases justifiable, to neglect this difference and set $P_\parallel = P_\perp$, thus reducing $\underline{\underline{P}}$ to a scalar fluid pressure obeying a single (three-dimensional) equation of state.

The dynamic effects of pressure anisotropy ($P_\parallel \neq P_\perp$) are potentially important when β, the ratio of plasma pressure to magnetic-field pressure, is of order unity. Anisotropy effects generally scale as the dimensionless ratio

$$a \equiv \mu_0(P_\parallel - P_\perp)/B^2 \tag{2}$$

which can be neglected if (but not "only if")

$$\beta \equiv \mu_0(P_\parallel + P_\perp)/B^2 \tag{3}$$

can be neglected compared with unity. For example, a plasma with $a > 1$ is subject to the firehose instability, and a plasma with $a < -1$ is subject to the mirror instability (see Siscoe, 1983 — this volume), either of which can dramatically influence the equation of state. (The dynamics can also be influenced in more subtle ways by smaller values of $|a|$. For example, a weak "pancake" anisotropy ($0 < -a \ll 1$), as is produced naturally in the ring current, can amplify gyroresonant waves ("whistlers") and the resultant pitch-angle scattering may be an important loss mechanism for ring-current particles (Kennel, 1969).)

The macroscopic flow of the solar wind around the magnetosphere exhibits two clear effects of fluid anisotropy. First, the bow shock has widely different structure and properties depending on whether its local unit-normal vector is more nearly parallel or perpendicular to the upstream magnetic field. In the former "quasi-parallel" geometry, the shock is relatively thick and turbulent; in the latter "quasi-perpendicular" geometry, it is relatively thin and laminar. (Because of the average 45° spiral angle of the IMF, the quasi-parallel geometry tends to prevail on the dawn side and the quasi-perpendicular geometry on the dusk side (Greenstadt, 1972).) Heuristically, this difference can be attributed to the fact that a parallel shock involves compression of $P_\parallel$ which, in an open magnetic-field geometry, requires a significant noise component of $\underline{B}$, whereas a perpendicular shock involves compression of $P_\perp$, which can be accomplished in principle with a quasi-steady compression of $\underline{B}$.

The second macroscopic effect involves the compression of magnetosheath (post-shock) plasma as it approaches the magnetopause. Compression in the plane perpendicular to $\underline{B}$ enhances the magnetic mirror force, leading to enhanced expansion along $\pm\underline{B}$; the "magnetic flux tube" description is quite literally applicable in this case as the plasma is "squeezed" from the two open ends of magnetosheath flux tubes by compression near the subsolar point. The result is an asymmetry of the magnetosheath flow field determined by the IMF direction (Lees, 1964), in addition to the obvious asymmetry arising from the dipole nature of the geomagnetic field.

The theory of the collisionless bow shock is a separate and rather intricate subject which will not be reviewed here; the interested reader is referred to Kennel (1981) for a brief review and introduction to the literature. The "squeezing" process has been treated in detail by Zwan and Wolf (1976). For the present purposes it suffices to recognize that the post-shock solar wind that interacts directly with the magnetosphere is likely to already have a significant and variable degree of anisotropy, and that such anisotropy may have important geometric if not dynamic effects. The inclusion of such effects has generally proved to be mathematically intractable in all but the simplest of geometrical models (e.g., Hill, 1975), but their neglect should be kept in mind whenever theoretical and observational results are compared.

The remainder of this chapter is concerned with the way in which the magnetosphere captures a small but significant fraction of the incident solar-wind energy flux to drive magnetospheric convection and related geomagnetic activity.

MAGNETOSPHERIC CONVECTION

The solar-wind magnetosphere coupling produces a system of plasma circulation or "convection" in the magnetosphere and in the magnetically-connected ionosphere. Plasma in the outermost layers of the magnetosphere, and in the magnetically-connected high-latitude ionosphere, flows away from the sun as if propelled by frictional contact with the adjacent post-shock solar wind. (Why such a frictional effect should occur in the total absence of collisions is the main subject of this chapter.) Plasma in the inner magnetosphere, and the magnetically connected lower-latitude ionosphere, flows sunward to complete the circulation pattern; this sunward return flow is a necessary consequence of the high-latitude antisunward flow (see Appendix A) and is treated elsewhere in this volume (Wolf, 1983). This magnetospheric convection system (particularly its antisunward half) is the essence of the solar-wind magnetosphere interaction.

Frozen-in-Flux

The convection system can be described either in terms of the plasma bulk flow velocity $\underline{v}$ or in terms of the electric field $\underline{E}$, the two being related by the frozen-in-flux theorem

$$\underline{E} + \underline{v}\times\underline{B} = 0 \tag{4}$$

This theorem applies whenever the following two conditions are met: (i) $\underline{E}\cdot\underline{B} = 0$; (ii) the length scale of magnetic-field gradients exceeds the largest characteristic length of the plasma (generally the ion gyroradius). Condition (ii) is generally satisfied in the magnetosphere and solar wind except within thin current sheets (e.g., the bow shock and the magnetopause). Condition (i) is known to be violated in the auroral zones, near the boundary between sunward and antisunward flow, as is discussed elsewhere in this volume (Burke, 1982). Significant values of $\underline{E}\cdot\underline{B}$ (compared to $|\underline{E}\times\underline{B}|$) are, however, generally restricted to thin velocity-shear layers corresponding to sharp gradients of the electric field, and the frozen-in-flux theorem (4) thus generally provides a good approximation except in the presence of sharp magnetic-field gradients (current layers) or sharp electric-field gradients (charge layers). It is commonly assumed that (4) provides an adequate relationship between the large-scale convective flow patterns in the ionosphere and magnetosphere. Further discussion of (4) and of the conditions for its applicability is provided by Vasyliunas (1975) and by Siscoe (1982 — this volume).

The cross-product of (4) with $\underline{B}$ gives

$$\underline{v} = \underline{E}\times\underline{B}/B^2 + v_{\parallel}\,\underline{b} \tag{5}$$

where $v_{\parallel} \equiv \underline{v}\cdot\underline{b}$; this $\underline{v}$ does not include charge-dependent drifts (e.g., magnetic gradient and curvature drifts) which are generally small by comparison whenever the two above conditions are met. In particular, if the magnetic field is time-stationary, and the electric field hence derivable from a potential ϕ, then (5) implies that the flow streamlines are confined to equipotential surfaces ($\underline{v}\cdot\nabla\phi = 0$). The term "convection velocity" generally refers to the $\underline{E}\times\underline{B}/B^2$ term of (5).

The Axford-Hines Model

The existence of a systematic convection pattern was deduced originally by Axford and Hines (1961) and by Dungey (1961) on the basis of ground-based geomagnetic and auroral observations. Subsequent ground-based and spacecraft observations have thoroughly confirmed its existence and elucidated many of its detailed characteristics. Axford and Hines introduced the concept of a frictional interaction between the solar wind and magnetosphere and proposed the circulation pattern illustrated in Figure 1. Solar-wind momentum is transmitted to the magnetosphere in a viscous boundary layer just inside the magnetopause, producing the required antisunward high-latitude flow and the implied (Appendix A) low-latitude return flow. The corresponding "convection electric field" (4) has a dawn-to-dusk orientation in the "polar cap" (region of antisunward flow in the ionosphere) and a dusk-to-dawn orientation in the lower-latitude return flow region of the ionosphere; these orientations are reversed in the magnetospheric equatorial plane (cf. Figures 1a and 1b).

Ordinary collisional viscosity fails by many orders of magnitude to account for this "viscous" boundary layer, just as in the above discussion of bow-shock thickness. Axford and Hines recognized this discrep-

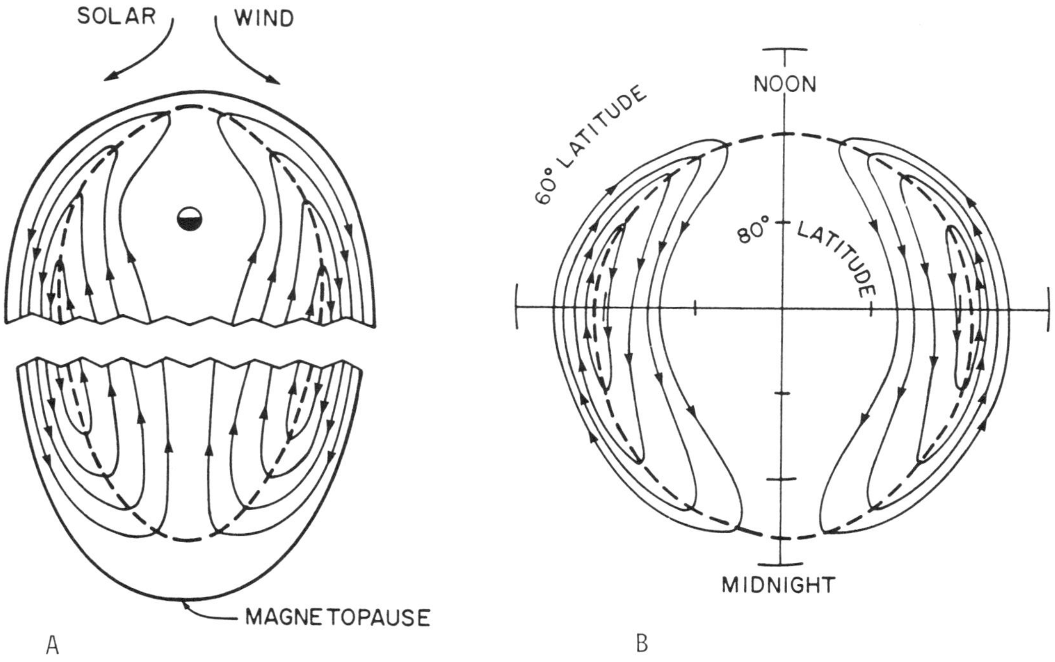

Figure 1. Streamlines of the Axford-Hines convection pattern, (a) in the magnetospheric equatorial plane, and (b) on the polar-cap ionosphere.

ancy, and suggested that wave-particle interactions might provide the required increase in effective collision frequency. Granted a sufficient effective viscosity, the Axford-Hines model provides a coherent phenomenological description of the basic solar-wind magnetosphere coupling, and it has provided an invaluable paradigm for much subsequent work. The model, however, has two as-yet unresolved difficulties: (1) it requires no interplanetary magnetic field (IMF), and hence, presumably, predicts no first-order correlation between the coupling efficiency and the IMF direction, whereas significant such correlations have been clearly observed, as described briefly below and in greater detail by Burch and Heelis (1979); and (2) a mechanism to produce sufficient effective viscosity has not yet been found.

The Dungey Model

Dungey (1961) first proposed that the required momentum transfer could be attributed to magnetic rather than viscous coupling if high-latitude geomagnetic field lines are allowed to extend into interplanetary space (i.e., "interconnect" with the IMF), as illustrated in Figure

2. In this model, magnetic tension transfers solar-wind momentum to the magnetosphere and ionosphere, and the solar-wind $-\underline{v}\times\underline{B}$ electric field (4) is correspondingly "mapped" along interconnected field lines to the magnetosphere and high-latitude ionosphere. Dungey tacitly assumed that the available magnetic stress (corresponding to the interconnection of all interplanetary magnetic flux that would intersect a cross-sectional area the size of the magnetosphere) is sufficient to drive magnetospheric convection. Subsequent analysis has verified this assumption; an "interconnection rate" (ratio of interconnected IMF flux to total unperturbed IMF flux intersecting the magnetosphere's cross section) of the order of 0.1 is evidently adequate (e.g., Hill, 1979; Reiff et al., 1981; see below).

The Dungey model (Figure 2) is depicted in the noon-midnight meridian plane with a predominantly southward IMF. (Only in this case can the magnetic field lines and plasma flow streamlines be depicted two-dimensionally.) For other IMF directions the field geometry becomes more complicated but the same topology prevails; i.e., the polar caps remain magnetically connected to interplanetary space and the solar-wind $-\underline{v}\times\underline{B}$ electric field maps (according to (4)) to a generally dawn-to-dusk polar-cap electric field (e.g., Stern, 1973). The interconnection geometry changes in two important ways as the IMF direction becomes less southward. First, the amount of interconnected flux, and hence the size of the interconnected polar caps, decreases. Secondly, any east-west component of the IMF produces an east-west component of magnetic tension which in turn produces east-west flow asymmetries in the magnetosphere

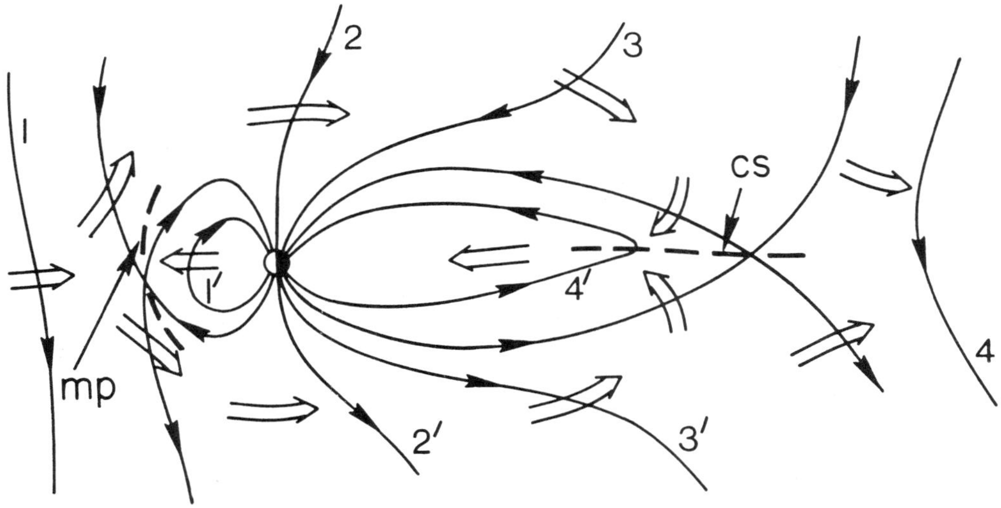

Figure 2. The Dungey magnetosphere model (noon-midnight meridian plane).

and polar-cap ionosphere, the asymmetry having opposite senses in the northern and southern polar caps. These asymmetries have been discussed by Russell and Atkinson (1973) and by Stern (1973), and can perhaps be visualized by reference to Figure 3, an attempt to depict the three-dimensional interconnection geometry.

The Dungey model thus has two immediate consequences: (1) the amount of interconnected flux, hence the strength of the solar-wind magnetosphere coupling, should be an increasing function of the southward component of the IMF, and (2) the resultant convection pattern should exhibit dawn-dusk asymmetries of opposite sense in the two polar caps, the direction being determined by the east-west IMF component (flow in the southern polar cap should be deflected toward the IMF direction and vice-versa in the northern polar cap). Both of these correlations are clearly observed (e.g., Burch and Heelis, 1979, and references therein) and their observation is largely responsible for the widespread acceptance of the Dungey model.

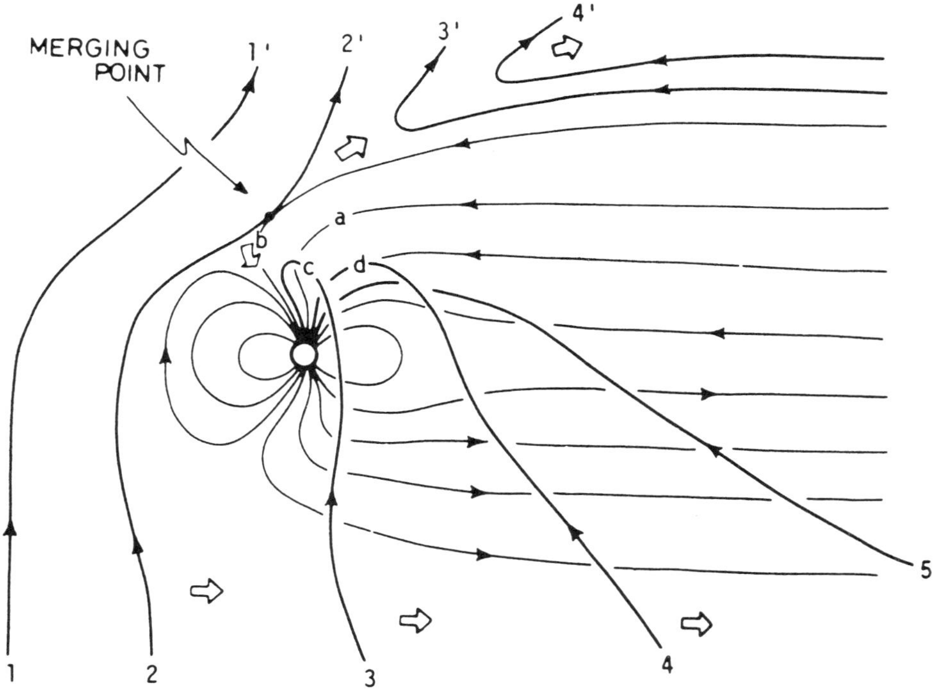

Figure 3. A perspective view from the dusk side of magnetic interconnection for a northward-tending IMF (Maezawa, 1976). An IMF with a dusk-to-dawn component connects from the dusk side through the northern cusp (as shown) and from the dawn side through the southern cusp (not shown). The reverse is true for a dawn-to-dusk IMF orientation.

Open vs. Closed

The Dungey model is usually called the "open model" because it requires the existence of open (interconnected) field lines; the Axford-Hines model is called the "closed model" because it does not. The two models are not easily distinguished solely on the basis of observed plasma flow in the high-latitude ionosphere. Both models predict the two-cell circulation topology shown in Figure 1b, the essential difference being that the antisunward flow occurs on open (interconnected) field lines in the Dungey model, and on closed geomagnetic field lines just inside the magnetopause in the Axford-Hines model. Unambiguous observational identification of open field lines has proved to be rather more difficult than might be imagined. Geomagnetically trapped (radiation-belt or ring-current) particles generally populate the sunward flow region but not the antisunward flow region, and interplanetary (solar-flare) particles generally have preferred access to the antisunward flow region (e.g., Paulikas, 1974, and references therein). These results are expected automatically in the open model, but they may also be accommodated by the closed model if the high latitude field lines are sufficiently stretched and distorted (Michel and Dessler, 1970). Moreover, the precise identification of the geomagnetic trapping boundary ("open-closed field line boundary") and of the sunward-antisunward flow boundary, and the instantaneous comparison of the locations of these two boundaries, is an intricate observational problem (see, for example, Reiff, 1979).

The IMF-dependence of the flow pattern mentioned above, however, does provide a strong indication that magnetic coupling, as in the Dungey model, is an important if not dominant factor in powering magnetospheric convection. In particular, the line integral of the electric field across the antisunward flow region (i.e., the polar-cap potential drop if a steady state prevails) is a direct measure of the strength and extent of the convection system and hence of the coupling efficiency. Recent statistical studies (Reiff, 1983 — this volume, and references therein) indicate that the potential drop responds to the IMF north-south component as expected in the Dungey model, and that, at most, a small fraction ($\lesssim$ 20%) can be attributed to non-magnetic coupling.

Birkeland Currents

Both the Dungey model and the Axford-Hines model require a magnetic transfer of antisunward momentum from the outer magnetosphere to the ionosphere, and this transfer is accomplished by an electric current circuit of the type illustrated in Figure 4. The dawn-dusk electric field drives a dawn-dusk Pedersen current j_1 in the ionosphere, producing a $\underline{j} \times \underline{B}$ force that maintains the antisunward convection against the frictional drag of the neutral atmosphere. The power dissipated by j_1 is provided by the "generator" current j_2 (note $\underline{j}_2 \cdot \underline{E} < 0$), a deceleration drift current that draws energy from the primary flow, either at the magnetopause or in the adjacent magnetosheath (Dungey model — Figure 4a) or in the viscous boundary layer (Axford-Hines model — Figure 4b).

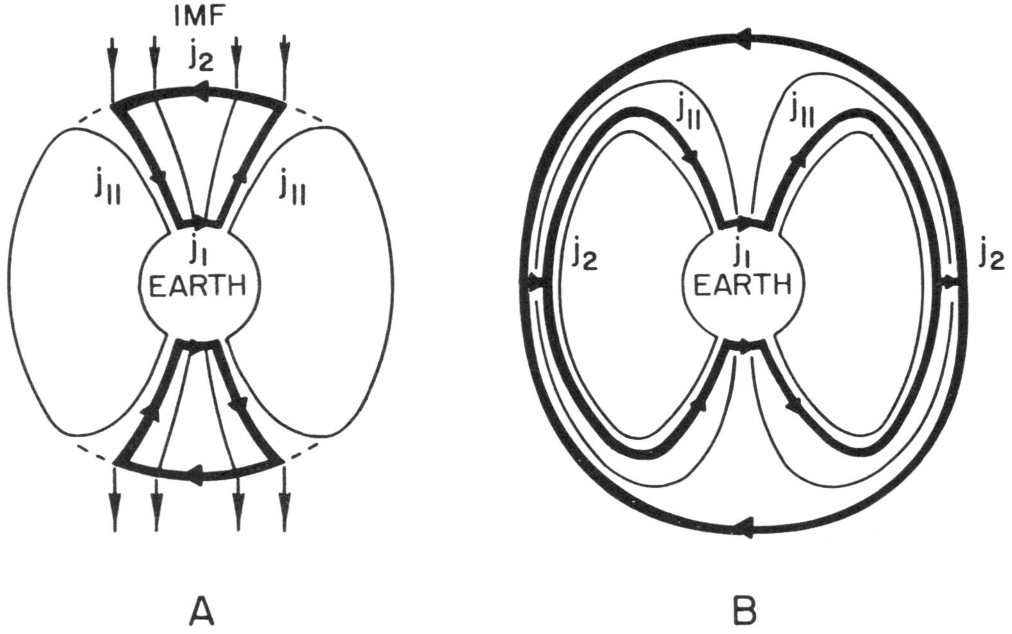

Figure 4. The "driving" current system (heavy lines) that transfers anti-sunward momentum to the polar-cap ionosphere, directly from the solar wind in the Dungey model (a), and from the low-latitude boundary layer in the Axford-Hines model (b). The view is from the Sun.

The two transverse currents j_1 and j_2 are connected by magnetic-field-aligned currents $j_\parallel$, often called Birkeland currents because their existence was originally deduced by Birkeland (1908). The Birkeland currents flow down (toward Earth) on the dawn side and up (away from Earth) on the dusk side; they are generally assumed to flow without resistance according to the assumption $\underline{E} \cdot \underline{B} \approx 0$.

This current system is part of the "dynamo" or "driving" current for magnetospheric convection; its field-aligned segment forms part of the observed "Region 1 field-aligned current system" (Iijima and Potemra, 1976a). The other part (typically the larger part) of the driving (Region 1) field aligned current connects to a lower-latitude "Region 2" Birkeland current system of opposite polarity (down at dusk, up at dawn). This system is related to the sunward return flow region of the magnetosphere and is treated elsewhere in this volume (Wolf, 1983). A third Birkeland current system ("Cusp region") is observed only near the noon meridian, slightly poleward of, and opposite in polarity to, the Region 1 current; this current system is consistent with that required to drive the east-west flow asymmetries mentioned above (Iijima and Potemra, 1976b).

The Birkeland current system is determined by the flow pattern imposed on the ionosphere and by the distribution of ionospheric conductivity (see further discussion by Vasyliunas, 1979, and by Wolf, 1983 — this volume). As in the case of the ionospheric plasma flow pattern, the mere existence of the observed Birkeland current systems does not favor either of the two (Axford-Hines vs. Dungey) models.

Substorms

The return (sunward) half of the magnetospheric convection cycle is not a smooth continuous extension of the downstream flow (as is implied, for example, by the potential flow pattern of Figure 1a). Instead, the return flow in the tail is apparently established sporadically in association with rapid changes of the tail magnetic-field geometry. These events, magnetospheric substorms, are discussed briefly below, following some further discussion of each of the two basic coupling models.

MAGNETIC COUPLING

The Requirement for Magnetic Merging

A consequence of the frozen-in-flux theorem (4) is that the plasma populations of two distinct flux tubes cannot intermix (Siscoe, 1983 — this volume). This condition is clearly violated in the open magnetosphere because solar-wind plasma coexists with magnetospheric plasma on interconnected (polar-cap) flux tubes, whereas the same solar-wind plasma did not share a flux tube with magnetospheric plasma when far upstream, nor will it do so when sufficently far downstream. The theorem (4) then must be violated at two places: once at the day side magnetopause where interplanetary and geomagnetic flux tubes become interconnected, and again in the magnetotail current sheet where they become disconnected. It is natural to expect violations of (4) at the day side magnetopause and tail current sheet, where the conditions for its validity are known to be violated, i.e., where the magnetic-field structure has a gradient scale length comparable to the ion gyroradius.

The process by which (4) is violated within a thin current sheet generally is called "magnetic merging" or "magnetic reconnection," and the theory of this process is described elsewhere in this volume (Vasyliunas, 1983). There it is shown that the merging process requires an electric-field component tangent to the current in the boundary, and it is this electric field component that allows, in the open model, a transfer of energy and momentum orders-of-magnitude larger than that provided by collisional viscosity.

It is also possible, in principle, to violate (4) by choosing a unique distribution of $\underline{E} \cdot \underline{B} \neq 0$ that exactly eliminates the tangential component of $\underline{E}$ at the magnetopause, thus decoupling totally the electric fields inside and outside the magnetosphere. This alternative is implicit in any argument (e.g., Heikkila, 1975) that would retain the open magnetic topology while neglecting (or, indeed, stoutly denying) the process of merging at the magnetopause. While this alternative is, indeed, mathematically possible, it is of little interest from a physical point of view because (1) it requires, not merely a significant magnetic-field-aligned resistivity, but a unique and complicated spatial/temporal distribution of field-aligned resistivity, with no particular mechanism in mind for producing such resistivity, and (2) in decoupling the electric fields inside and outside the magnetopause, it would sacrifice the primary reason for proposing an open magnetic-field topology in the first place, namely, to provide a viable mechanism for significant energy and momentum transfer from the solar wind to the magnetosphere.

For practical purposes, then, the open magnetosphere model requires the process of magnetic merging at the day side magnetopause and in the magnetospheric tail. Historically, this requirement has been the basis of considerable scepticism regarding the open model — the requirement for merging was recognized long before a viable theory of the merging process was generally recognized. Subsequent theoretical analysis of the merging process (e.g., Vasyliunas, 1983 — this volume; and references therein) has largely dispelled such scepticism, by showing that the merging process requires, not new physical laws, but the application of old physical laws (Maxwell's Equations and the Boltzmann equation) to a complicated geometry that, although novel, is undeniably observed.

The magnetospheric consequences of magnetic merging can be visualized if, in Figure 2, the magnetic field lines are instead viewed as loci of guiding centers of specific sets of plasma "test particles" in the various regions. These loci, by definition, move with the plasma. Except within thin current sheets (e.g., the day side magnetopause and magnetotail current sheet), the velocity of the plasma (hence of the loci) in the direction perpendicular to $\underline{B}$ is given to a very high degree of accuracy by $\underline{E} \times \underline{B}/B^2$ (equation (5)). A solar-wind locus and a magnetospheric locus (e.g., 1 and 1' in Figure 2) must therefore be broken at the dayside magnetopause and "reconnected" to form two polar-cap loci (one each in the northern and southern hemispheres — e.g., 2 and 2' in Figure 2), and two polar-cap (tail) loci (one each from northern and southern hemispheres — e.g., 3 and 3' in Figure 2) must be broken somewhere in the tail current sheet and become "reconnected" to form a downstream solar-wind locus (e.g., 4 in Figure 2) and an Earthward-flowing magnetospheric locus (e.g., 4' in Figure 2). This is a precise, if somewhat cumbersome, definition of the magnetic merging or reconnection process in the magnetospheric context. The same description is commonly given with the phrase "magnetic field lines" in place of the phrase "loci of guiding centers of specific sets of plasma test particles"; one then speaks (figuratively) of the field lines themselves as undergoing the motions described above. The description is the same, but one must bear in mind that the magnetic field lines are mathematical, not physical, entities, and their motion cannot therefore be uniquely defined (i.e., cannot be measured physically) (see, e.g., Vasyliunas, 1972). The merging concept is, in fact, still sometimes criticized (e.g., Alfvén, 1977), on the grounds that it relies on the concept of magnetic-field-line motion; the above discussion is intended to illustrate that the "motion of magnetic field lines" is a convenient mental shorthand notation, not a basic physical postulate, and that criticisms based on this notation should be taken as semantical rather than physical arguments. (The motion of magnetic field lines may not be physically measurable, but the motion of the loci of guiding centers of specific sets of plasma test particles is certainly measureable, at least in principle.)

Energy Flow in the Open Model

For the sake of illustration, we adopt the simple Dungey model in which the IMF is assumed to be predominantly southward (Figure 2, redrawn with different labels in Figure 5). The conclusions are assumed to be modified quantitatively, but not qualitatively, if the IMF has significant non-southward components (Stern, 1973). Note from Figure 5 that the open model generally requires $\underline{j} \cdot \underline{E} > 0$ at the day side magnetopause (Region 1) and in the tail current sheet (Region 3), and $\underline{j} \cdot \underline{E} < 0$

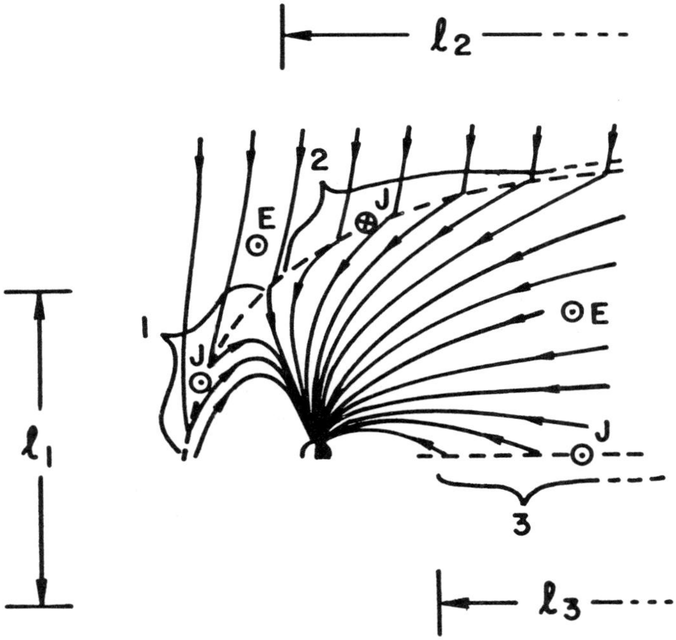

Figure 5. The Dungey model again, illustrating regions where $\underline{j} \cdot \underline{E}$ is positive (1 and 3) and negative (2). (Adapted from Hill and Reiff, 1977.)

at the tail magnetopause (Region 2). (These Regions 1 and 2 should not be confused with the "Region 1" and "Region 2" Birkeland current systems described earlier.) The dividing line on the magnetopause between the "day side" (Region 1) and the "tail" (Region 2) is here defined by the reversal of the sign of $\underline{j}\cdot\underline{E}$; it corresponds roughly to the "cusp" or "cleft" region, equatorward of which the geomagnetic field lines inside the magnetopause close across the equatorial plane on the day side, and poleward of which the field lines inside the magnetopause cross the equator only far downstream if at all. This "day-side/tail" boundary actually lies somewhat sunward of the dawn-dusk meridian plane.

The discussion of energy flow is facilitated by the use of Poynting's theorem

$$\frac{\partial}{\partial t}\left(B^2/2\mu_0 + \varepsilon_0 E^2/2\right) + \underline{j}\cdot\underline{E} + \nabla\cdot(\underline{E}\times\underline{B})/\mu_0 = 0 \tag{6}$$

which follows immediately from Maxwell's equations with the vector identity $\nabla\cdot(\underline{E}\times\underline{B}) \equiv \underline{B}\cdot(\nabla\times\underline{E}) - \underline{E}\cdot(\nabla\times\underline{B})$. The first term in (6) is the time derivative of electromagnetic energy density, the second term is the rate of conversion of electromagnetic energy to other forms of energy (basically, in this context, plasma thermal and/or kinetic energy), and the third term is the divergence of the electromagnetic energy flux ("Poynting vector")

$$\underline{S} = \underline{E}\times\underline{B}/\mu_0 \tag{7}$$

The ratio E/B is generally of the order of v even if (4) does not hold precisely; thus the ratio of electric-field energy density $\varepsilon_0 E^2/2$ to magnetic energy density $B^2/2\mu_0$ is generally of order $\mu_0\varepsilon_0 v^2 \equiv v^2/c^2$, where c is the velocity of light in vacuum. This ratio is quite small in the magnetosphere ($\lesssim 10^{-6}$ whenever v is less than the solar-wind speed), and Poynting's theorem (6) is thus very well approximated by

$$\underline{B}\cdot\frac{\partial \underline{B}}{\partial t}/\mu_0 + \underline{j}\cdot\underline{E} + \nabla\cdot(\underline{E}\times\underline{B})/\mu_0 = 0 \tag{8}$$

In a sufficiently long-term average, the explicit time derivative must vanish, and one has the familiar and convenient result that the local conversion rate of magnetic to mechanical energy is, on average, equal to the local convergence of the Poynting flux vector:

$$\langle\underline{j}\cdot\underline{E}\rangle = -\langle\nabla\cdot(\underline{E}\times\underline{B}/\mu_0)\rangle \tag{9}$$

where $\langle\rangle$ denotes a suitably long-term average.

From equation (9) derives the conventional description of energy flow in the open magnetosphere. Magnetic energy converges toward the day side magnetopause from both sides ($\nabla\cdot(\underline{E}\times\underline{B}/\mu_0) < 0$) and is converted there into plasma flow energy ($\underline{j}\cdot\underline{E} > 0$) to produce the "merging outflow," i.e., the accelerated poleward flow of solar-wind plasma around the day side magnetosphere (Region 1 in Figure 5). The accelerated outflow also has non-poleward components when the IMF has non-southward components, i.e., in general. On the tail magnetopause (Region 2 in Figure 5), part of the magnetosheath flow energy (including, but not limited to, the day side merging outflow energy) is extracted ($\underline{j} \cdot \underline{E} < 0$) to provide a flux of magnetic energy into the tail ($\nabla \cdot (\underline{E} \times \underline{B}/\mu_0) > 0$). This magnetic-energy flux is largely converted, either simultaneously or at some later time, into mechanical energy by reconnection ($\underline{j} \cdot \underline{E} > 0$) at the tail current sheet (Region 3 in Figure 5), although a fraction (see below) of the tail magnetic energy flux is diverted to provide an Earthward Poynting flux component, thus driving the Sunward return flow and the myriad geomagnetic processes therein (see, e.g., Wolf, 1983 — this volume). Of the plasma mechanical energy produced in the tail current sheet, some fraction (~ 1/2?) remains trapped on geomagnetic field lines to heat and/or accelerate the plasma sheet, and the remainder is lost to the downstream solar-wind wake, the dividing line being the tail reconnection line (x-line or neutral line — see Vasyliunas (1983 — this volume)).

In this description, magnetic energy flows in the $\underline{E} \times \underline{B}$ direction as does the plasma (Equation (5)) and as do the guiding-center loci (sometimes called "magnetic field lines"). The description is thus intuitively appealing, and it is consistent with, though not required by, the time-averaged Poynting theorem (9).

The Poynting theorem (6) specifies, not the energy-flux vector, but only its divergence. To the Poynting vector (7) we can add any divergence-free vector without contradicting the Poynting theorem (6). For example, Lai (1981) has proposed that the traditional Poynting vector (7) be replaced by

$$\underline{S}' = \underline{E} \times \underline{B}/\mu_0 + \nabla \times (\phi \underline{B})/\mu_0 \tag{10}$$

where ϕ is the electrostatic potential. The added term is clearly divergence-free (as required by (6)), and does not affect the traditional interpretation of Poynting's vector as applied to vacuum electromagnetic waves, for example, where $\phi \equiv 0$. However, in a steady-state situation with $\partial \underline{B}/\partial t = 0$, $\partial \underline{E}/\partial t = 0$, the added term has an interesting effect: the "energy flux" along $\underline{E} \times \underline{B}$ drift streamlines, as given by (7), is replaced by an "energy flux" along electric current streamlines:

$$\underline{S}' = \phi \underline{j} \quad \text{(in steady state)} \tag{11}$$

which follows from (10) with the vector identity $\nabla \times (\phi \underline{B}) \equiv \phi \nabla \times \underline{B} - \underline{B} \times \nabla \phi$ together with $\underline{E} = -\nabla \phi$ and $\mu_0 \underline{j} = \nabla \times \underline{B}$. According to (11), electromagnetic energy in a steady state flows from source to sink along current streamlines, a description that may be more intuitively appealing to some. In the magnetospheric context, this would mean that magnetic energy is transported by the magnetopause current from its source (Region 2 in Figure 5) to its two primary sinks, the dayside magnetopause (Region 1) and the tail current sheet (Region 3). Note, however, that a detailed balance between the Region 2 source and the Region 1 and 3 sinks is not to be expected, even in time average, and even allowing for entropy loss, because the day side magnetopause also draws from an additional source, namely, the compression of the IMF in the day side magnetosheath, which is associated with currents ($\underline{j}\cdot\underline{E} < 0$) in the bow shock and magnetosheath; this magnetosheath energy flux can also, of course, be described equally well by either form of Poynting's vector ((7) or (10)).

The choice between the two alternate forms of Poynting's vector, i.e., the choice between visualizing the electromagnetic energy flux as following $\underline{E}\times\underline{B}$ drift streamlines or current streamlines, is largely a matter of taste. The physically measurable quantity is the divergence of Poynting's vector, which is the same in both cases ($-\underline{j}\cdot\underline{E}$ for a steady state).

The magnitude of this energy flow can be estimated as follows. Let $\underline{j}_1$ be the day side magnetopause current density (Region 1 in Figure 5) and let $\underline{E}_1$ be the electric field at the day side magnetopause. Similarly, let $\underline{j}_2$ and $\underline{E}_2$ be the current density and electric field at the tail magnetopause (Region 2 in Figure 5), and $\underline{j}_3$ and $\underline{E}_3$ at the tail current sheet (Region 3). The rate of conversion of magnetic to mechanical energy at the day side is given by the volume integral over the day side magnetopause

$$P_1 \equiv \int_1 \underline{j}\cdot\underline{E}\ dV > 0 \tag{12}$$

Similarly, the rate of conversion of mechanical to magnetic energy at the tail magnetopause is

$$-P_2 \equiv -\int_2 \underline{j}\cdot\underline{E}\ dV > 0 \tag{13}$$

and the rate of conversion of magnetic to mechanical energy at the tail current sheet is

$$P_3 \equiv \int_3 \underline{j}\cdot\underline{E}\ dV > 0 \tag{14}$$

The term "mechanical energy" above includes both the bulk flow kinetic energy and the thermal energy of the plasma; flow energy is the principal sink of magnetic energy in simple models of the merging process (Vasyliunas — this volume, 1983).

The first integral (12) can be approximated by

$$P_1 \approx J_1 \phi_1 \ell_1 \tag{15}$$

where $\underline{J}_1$ is the integral of $\underline{j}_1$ across the magnetopause current layer, ϕ_1 is the line integral of $\underline{E}_1$ along the current layer in the direction of $\underline{J}_1$, and ℓ_1 is the extent of the current layer in the direction transverse to $\underline{J}_1$ (Figure 5). From Ampere's Law,

$$J_1 = |(\underline{B}_M - \underline{B}_s)\times\hat{\underline{n}}|/\mu_0 \sim B_M/\mu_0 \tag{16}$$

where $\hat{n}$ is the outward magnetopause unit normal, $\underline{B}_M$ is the magnetospheric magnetic field just inside the magnetopause, and $\underline{B}_s$ is the magnetosheath field just outside; the contribution of B_s to J is neglected for order-of magnitude purposes, although its magnitude and direction certainly affect the value of ϕ_1 (e.g., Hill, 1975; Vasyliunas, 1983 — this volume) which, however, will be taken as given in the present argument. If we adopt, as representative values, $B_M \approx 60$ nT, $\phi_1 \approx 50$ kV, and $\ell_1 \approx 15\ R_E$, then

$$P_1 \sim 2.3 \times 10^{11}\ \mathrm{W} \tag{17}$$

The quantity ϕ_1 is known to vary by more than an order of magnitude ($10\ \mathrm{kV} \lesssim \phi_1 \lesssim 200\ \mathrm{kV}$) in response to IMF variations (Reiff et al., 1981); thus (17) represents a typical value within a corresponding range of variation of P_1. The approximations invoked between (12) and (17) produce an error of perhaps ±50% in our estimate of P_1, an acceptable error compared to the order-of-magnitude natural variations of the quantity itself. A better estimate would require a realistic quantitative model of the open magnetosphere geometry; such a model is not presently available.

We can use (15) and (16) to estimate the energy gained by a typical solar-wind (magnetosheath) ion in the merging process. The flux of magnetosheath ions into the magnetopause is

$$f_1 \sim n_s(\phi_1/B_s)\ \ell_1 \tag{18}$$

where n_s and B_s are the magnetosheath ion density and field strength just outside the magnetopause. Dividing (15) by (18) gives the average magnetic energy converted per incident ion:

$$\langle w \rangle \approx P_1/f_1 = B_M B_s/(\mu_0 n_s) \tag{19}$$

which, in present merging models, appears largely in flow energy rather than thermal energy (Vasyliunas, 1983 — this volume); hence the emphasis on ions rather than electrons. The density n_s and field B_s are both enhanced by the bow-shock compression, but in roughly the same proportion, so we may approximate the ratio n_s/B_s by its upstream solar-wind value n_{sw}/B_{sw}. Taking $n_{sw} = 5/\mathrm{cm}^3$, $B_{sw} = 5$ nT, and $B_M = 60$ nT as above, we obtain

$$\langle w \rangle \sim 300\ \mathrm{eV} \tag{20}$$

corresponding to an ion (proton) velocity of 240 km/s. (The squeezing process (Zwan and Wolf, 1976) reduces the ratio n/B just outside the magnetopause, and would increase the estimate (20) by a similar factor.) By comparison, a typical solar-wind ion (proton) has flow energy ~ 1 keV (v ~ 400 km/s), which is largely converted to thermal and magnetic energy as ions cross the bow shock and approach the nose of the magnetosphere, then converted back into flow energy as the ions are accelerated away from the nose to rejoin the downstream magnetosheath flow. This energy conversion process occurs irrespective of whether the magnetosphere is open or closed (e.g., Spreiter and Alksne, 1969). The merging process, to the extent that it occurs, is thus a potentially significant factor, but not an overwhelmingly dominant factor (cf. Heikkila, 1975), in the acceleration of magnetosheath plasma away from the nose (stagnation) region and around the body of the magnetosphere. The merging process might produce a noticeable asymmetry in the acceleration of magnetosheath plasma away from the nose, and might tend to confine the acceleration geometrically to a narrow vicinity of the magnetopause (both of which effects are also produced by the "squeezing" process described above (Zwan and Wolf, 1976), even in the absence of merging), but it is not expected to increase dramatically the global acceleration rate, which is basically determined by the requirement that most of the incident solar wind flows around, not through, the magnetosphere.

The integrals P_2 (13) and P_3 (14) can be estimated in a similar manner, except that the dimension ℓ_2 ($\sim \ell_3$), i.e., the length of the tail, is not empirically known. We can, however, estimate the tail length by an argument originally posed (in slightly different form) by Dungey (1965). If the radius of the tail is R_T (empirically $R_T \approx 20\ R_E$) and the mean strength of the near-Earth tail magnetic field is B_T (empirically $B_T \approx 20$ nT), then the total magnetic flux entering either (northern or southern) tail lobe is $\Phi = (\pi/2)\ R_T{}^2 B_T$. In the open model, part of this magnetic flux is connected between northern and southern tail lobes through the tail current sheet and the remaining part is connected to the IMF through the tail magnetopause. We obtain a rough upper limit to the tail length by assuming that most of Φ is connected through the magnetopause, i.e.,

$$\Phi = \frac{\pi}{2} R_T{}^2 B_T \gtrsim B_n (2 R_T \ell_2) \tag{21}$$

where B_n is the mean normal component of $\underline{B}$ across the tail magnetopause. From (21),

$$B_n \ell_2 \lesssim \frac{\pi}{4} B_T R_T \tag{22}$$

The mean normal component B_n is not known empirically, but it can be related to the downstream tangential velocity v_t of the external magnetosheath plasma by (4):

$$v_t B_n \approx E_2 \tag{23}$$

Combining (22), (23), and $\phi_2 \approx 2E_2 R_T$ gives

$$\ell_2 \lesssim \left(\frac{\pi}{2}\right) B_T R_T{}^2 v_t / \phi_2 \tag{24}$$

and, with $J_2 \approx B_T/\mu_0$, we have

$$-P_2 \approx 2 J_2 \phi_2 \ell_2 \lesssim v_t \left(B_T{}^2/\mu_0\right) \left(\pi R_T{}^2\right) \tag{25}$$

which (as may have been anticipated) equals the rate of work extracted from the downstream flow (v_t) by the tail magnetic tension ($B_T{}^2/\mu_0$)

applied to the tail cross-sectional area (πR_T^2). The factor 2 in (25) accounts for equal contributions from the two (northern and southern) tail lobes. The downstream velocity v_t is very nearly equal to the solar-wind velocity $v_{SW} \sim 400$ km/s; using this value and the above values for B_T and R_T we have from (25)

$$-P_2 \lesssim 6.5 \times 10^{12} \text{ W} \tag{26}$$

The integral P_3 (14) can be estimated in a similar fashion, except that the total current-sheet current $I_3 = J_3 \ell_3$ is somewhat less than the total tail magnetopause current $I_2 = J_2 \ell_2$; part of the tail magnetopause current closes across the day side magnetopause rather than through the tail current sheet. (A small fraction (see below) also closes through the ionosphere as illustrated in Figure 4a above.) Accordingly (since $\phi_2 = \phi_3$ in a steady state or in a sufficiently long time average) P_3 is somewhat less than $|P_2|$; most, but not all, of the magnetic energy generated at the tail magnetopause is "dissipated," i.e., converted to plasma mechanical energy, at the tail current sheet. The remainder (perhaps ~ 5% — compare (17) and (26)) appears as an Earthward Poynting flux from the tail which ultimately powers the day side merging process as well as the sunward return flow and related dissipation processes, e.g., aurora (see Wolf, 1983 — this volume).

Comparison of (26) with the total solar-wind kinetic energy flux incident on the tail cross section indicates that a large fraction ($(2B_T/B_M)^2 \sim 1/2$) of the kinetic energy flux is converted to magnetic energy flux in the tail. Note, however, that a considerably smaller fraction is "captured" by the magnetosphere to drive dissipative processes in the magnetosphere and ionosphere; most is converted back into flow energy of the downstream solar wind (P_1 and the tailward half of P_3). This temporary storage of flow energy in the magnetic field is not unique to the open model. Even if the magnetosphere were closed, solar-wind flow energy would be partially converted to magnetic-field energy in the bow shock and day side magnetosheath, and later returned to the downstream flow. What is unique to the open model is that magnetic interconnection allows a fraction of the stored magnetic energy to be injected into the magnetosphere before being returned to the downstream flow.

Inserting the same numerical values in (24) gives the estimated length of the tail

$$\ell_2 \lesssim 32\ R_T \sim 640\ R_E \tag{27}$$

and (22) then gives

$$B_n \sim 0.025\ B_T \sim 0.5 \text{ nT} \tag{28}$$

A typical IMF strength is about 5 nT; thus the ratio of the interconnected flux to the total IMF flux through a magnetosphere-size area is

$$B_n / B_{IMF} \sim 0.1 \tag{29}$$

This is the "interconnection rate" referred to earlier; it is also the ratio of the average cross-polar-cap potential drop to the total potential drop across a magnetospheric diameter in the upstream solar wind. This "interconnection rate" should not be confused with the "merging rate" at the day side magnetopause, which is of a similar order of magnitude but is defined differently (Vasyliunas, 1983 — this volume).

Momentum Transfer in the Open Model

The equation of motion of the plasma, neglecting gravity and viscosity, is

$$\rho \frac{d\underline{v}}{dt} = - \nabla \cdot \underline{\underline{P}} + \rho_c \underline{E} + \underline{j} \times \underline{B} \tag{30}$$

where ρ_c is charge density. In general, the ratio of the last two terms is of order $\mu_0 \varepsilon_0 |\nabla(E^2)| / |\nabla(B^2)|$ which is very small ($\sim v^2/c^2$) unless the electric field has extraordinarily sharp gradients (e.g., within double layers); the term $\rho_c \underline{E}$ is therefore generally dropped (see Siscoe, 1982 — this volume). The explicit part of the time derivative may also be dropped in a sufficiently long time average, and the remaining terms can be combined in the form

$$\nabla \cdot [\rho \underline{v}\, \underline{v} + \underline{\underline{P}} - \underline{\underline{M}}] = 0 \tag{31}$$

where

$$\underline{\underline{M}} \equiv \underline{B}\, \underline{B}/\mu_0 - \underline{\underline{1}}\, B^2/2\mu_0 \tag{32}$$

is the Maxwell stress tensor (in the approximation $E \ll cB$) and the steady-state continuity equation $\nabla \cdot (\rho \underline{v}) = 0$ is implicit in (31). (Note $\underline{j} \times \underline{B} = \nabla \cdot \underline{\underline{M}}$.) Equation (31) has the form of a conservation law; the bracketed quantity represents the total (mechanical plus electromagnetic) momentum flux tensor. When $\underline{\underline{P}}$ is gyrotropic (Equation (1)), (31) becomes

$$\nabla \cdot [\rho \underline{v}\, \underline{v} + \underline{\underline{1}}\, (P_\perp + B^2/2\mu_0) - (\underline{B}\, \underline{B}/\mu_0)\, (1-a)] = 0 \tag{33}$$

where a is the anisotropy parameter defined in Equation (2).

The form (33) of the momentum-conservation law is particularly useful in considering momentum transfer across a sharp boundary like the magnetopause. Let $\underline{\hat{n}}$ be the local inward normal unit vector, and $\underline{\hat{t}}$ be a unit vector tangential to the magnetopause (say in the downstream direction in Region 2 of Figure 5), and assume that tangential gradients are negligible compared to normal gradients [$\nabla \approx \underline{\hat{n}}\, (\partial/\partial n)$]. The normal component of (33) integrated across the boundary then gives the familiar pressure-balance condition

$$[\rho v_n^2 + P_\perp + B^2/2\mu_0 - (B_n^2/\mu_0)(1-a)] = 0 \tag{34}$$

where [ ] denotes the change of the bracketed quantity across the boundary. The first and last terms are generally negligible at the magnetopause compared to the middle two, so that

$$[P_\perp + B^2/2\mu_0] \approx 0 \tag{35}$$

(The ratio of the neglected terms to the retained terms is of the order of the square of the dimensionless merging rate (Vasyliunas, 1983 — this volume), and is typically ~ 0.01.)

The transfer of tangential momentum, as required to drive magnetospheric convection, is described by the tangential component of (33) which, when integrated across the boundary, gives

$$[\rho v_n v_t - (B_n B_t/\mu_0)(1-a)] = 0 \tag{36}$$

The components v_n and v_t refer to the bulk flow velocity $\underline{v} = \langle \underline{u} \rangle$ where angle brackets here denote an average over the velocity distribution and $\underline{u}$ denotes individual particle velocities. Note that $v_n v_t \equiv \langle u_n \rangle \langle u_t \rangle \neq \langle u_n u_t \rangle$ if the pressure is anisotropic. Indeed, it is easy to show (e.g., Hill, 1975) that

$$\rho \langle u_n u_t \rangle = \rho v_n v_t + (P_\parallel - P_\perp) B_n B_t / B^2 \tag{37}$$

which provides a slightly more convenient form of (36):

$$[\rho \langle u_n u_t \rangle - B_n B_t/\mu_0] = 0 \tag{38}$$

in which the first term includes both bulk-flow and pressure-anisotropy contributions to the transfer of mechanical momentum.

The form (38) is convenient because the solar-wind (magnetosheath) particles interacting with the magnetopause can be divided into two classes, those that are reflected (subscript 1) and those that are transmitted across the magnetopause into the tail lobe (subscript 2). For the reflected particles, $\rho_1 \langle u_{n1} \rangle = 0$ by definition, but $\rho_1 \langle u_{n1} u_{t1} \rangle \neq 0$; the reflected particles ($u_n < 0$) have smaller tangential velocities than the incident particles ($u_n > 0$), and $[\rho_1 \langle u_{n1} u_{t1} \rangle] = \rho_1 \langle u_{n1} u_{t1} \rangle = \rho_1 \langle |u_{n1}| \rangle \langle \Delta u_{t1} \rangle$ where $\langle \Delta u_{t1} \rangle$ is the (flux-weighted) average tangential velocity decrement of reflected vs. incident particles. (The change of this quantity across the boundary is equal to its external value, its internal value being zero by definition.) The transmitted particles have $\rho_2 \langle u_{n2} \rangle$ = constant > 0, and $[\rho_2 \langle u_{n2} u_{t2} \rangle] = \rho_2 \langle u_{n2} \rangle \langle \Delta u_{t2} \rangle$. The sum of these two terms $[\rho_1 \langle |u_{n1}| \rangle \langle \Delta u_{t1} \rangle + \rho_2 \langle u_{n2} \rangle \langle \Delta u_{t2} \rangle]$ is, by definition, the first term of (38); in words, both the reflected and the transmitted particles suffer some degree of deceleration (not necessarily the same degree) during the boundary encounter, and the sum of these decelerations over all particles balances (Equation (38)) the tangential magnetic stress on the boundary, given by $(\underline{J}\times\underline{B})_t = B_n[B_t]/\mu_0$. This is the basis of the statement in the previous section to the effect that mechanical work is extracted from the flow by the magnetic tension on the tail. (Note that $\underline{v}\cdot(\underline{j}\times\underline{B}) \equiv -\underline{j}\cdot(\underline{v}\times\underline{B}) = \underline{j}\cdot\underline{E}$ when (4) is satisfied.)

It is worth distinguishing between the reflected and transmitted particle populations because only the latter population transfers anti-solar (mechanical) momentum into the magnetosphere and subsequently becomes available to populate the plasma sheet (Pilipp and Morfill, 1980) and ultimately the geomagnetic ring current, while both populations act to balance (hence "are responsible for") the stressed tail field configuration. Little is known, theoretically, about the transmission properties of the (open) tail magnetopause with respect to solar-wind particles, and the rate of mechanical momentum transfer is therefore impossible to evaluate at present. The magnetic stress on the tail, however, can be evaluated rather easily, if crudely, given our empirical knowledge of B_t (~ 20 nT) and our previous estimate of B_n (Equation (28)). The integral of this stress over the tail magnetopause area (Region 2 of Figure 5) gives the total tangential force exerted by the solar wind (both reflected and transmitted components) on the tail magnetic field (and vice-versa):

$$F_2 \simeq 2[B_nB_t/\mu_0](2R_T\ell_2) \lesssim (B_T{}^2/\mu_0)(\pi R_T{}^2) \ (= P_2/v_t) \tag{39}$$

where the initial factor of two accounts for the two (northern and southern) hemispheres. (In setting $[B_t] \approx B_T$, we have neglected the tangential component of the magnetosheath field; the resulting error in the global estimate should be small because the magnetosheath field tends to add to the stress on one side and substract on the other. The resulting asymmetry may affect the pattern of convection (e.g., Crooker, 1980) but should not seriously affect the total stress as estimated here.) The ($\lesssim$) sign, as already discussed in connection with (21), represents the fact that some, but not all, of the tail magnetic flux connects to the IMF through the magnetopause, the remainder connecting internally across the tail current sheet. Accordingly, some, but not all, of the tail magnetic tension is balanced by solar-wind deceleration at the magnetopause, the remainder being balanced at the tail current sheet by mechanical stresses in the plasma sheet, consisting of some combination of Earthward plasma pressure gradient, field-aligned pressure anisotropy, and Earthward acceleration (e.g., Rich et al., 1972; Hill, 1975). Numerically, from (39), $F_2 \lesssim 1.6\times10^7$ Nt.

This is the total force available to drive magnetospheric convection, although, as might be anticipated from the previous section, a rather small fraction of this force is actually required to balance the friction and inertia of the system. For example, the collision of ions with neutral molecules in the Pedersen-conducting layer of the ionosphere is the only significant source of friction in the system. Figure 6 illustrates a schematic model of the magnetic flux tube that connects the polar-cap ionosphere (the bottom horizontal surface labeled S_1) with the tail magnetopause (the top horizontal surface labeled S_2,

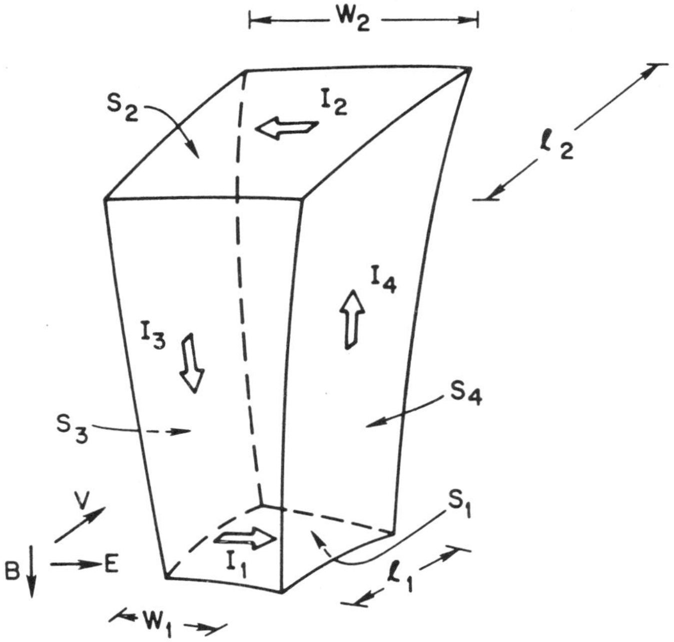

Figure 6. Schematic view of the flux tube that transmits antisunward momentum to the polar cap (as in Figure 4).

the tail magnetopause (the top horizontal surface labeled S_2, corresponding to Region 2 in Figure 5). The ionosphere conducts a dawn-to-dusk Pedersen current (open arrow labeled I_1) and the resulting $\underline{j}\times\underline{B}$ force maintains the antisunward convection against the frictional force of ion-neutral particle collisions in the ionosphere. At the magnetopause, the acceleration-drift current (open arrow labeled I_2) extracts momentum from the local antisunward flow. Part of this current is connected to the ionospheric current by the Birkeland currents I_3 (downward on the morning side) and I_4 (upward on the evening side); these are the "driving currents" of magnetospheric convection, the "Region 1 field-aligned currents" described above. The total magnitude of the current through this loop is observed to be typically one or two MA (Iijima and Potemra, 1976a).

The total $\underline{j}\times\underline{B}$ force integrated over the volume of the conducting layer of the ionosphere is

$$F_1 = I_1 B_1 w_1 \tag{40}$$

where w is the dawn-dusk width of the antisunward convection region and subscript "1" denotes the ionosphere. With $I_1 = 2$ MA, $B_1 = 5\times10^{-5}$ T, and $w_1 = 0.5\ R_E$ we require $F_1 = 3\times10^8$ Nt to balance the ion-neutral friction. This does not, however, imply that a force of similar magnitude is applied at the magnetopause. The applied force ($F_2 = I_2 B_2 w_2$) is smaller by the ratio $(B_2 w_2)/(B_1 w_1) = \ell_1/\ell_2$. (Current continuity implies $I_1 = I_2$, and the magnetic flux $Bw\ell$ is, by definition, constant along the flux tube.) Thus the solar wind, operating at a distance, enjoys a certain mechanical advantage which may be likened to that of a lever arm. To estimate the required force F_2 we assume that the field lines are equipotential (Equation (4)) so that $v_1 B_1 w_1 \approx v_2 B_2 w_2 \equiv \phi$, the total applied electrostatic potential, whence

$$F_2 = I_2 \phi / v_2 \tag{41}$$

(The same result follows from the energy relation $vF = I\phi$.) Note that I_2 in (41) is a small fraction (~ 3%) of the total tail magnetopause current $I'_2 \sim \ell_2 B_T/\mu_0$. Taking, as above, $I_2 \sim 2$ MA, $\phi \sim 50$ kV, and $v_2 \sim 400$ km/s gives the required transfer rate as $F_2 \sim 5\times10^5$ nT, where an extra factor of two has again been introduced to account for both northern and southern polar caps. (Note that $F_1/F_2 = \ell_2/\ell_1 \approx 600$; this is consistent with our earlier estimate of ℓ_2 (27) if, as expected, $\ell_1 \approx 1\ R_E$.)

This driving force is a small fraction ($= I_2/I'_2 \sim 3\%$) of the total force on the magnetospheric tail estimated above (39); this result is implicit in the discussion of the previous section where it was noted that only a small fraction of the energy delivered to the tail by the solar wind is dissipated in convection-related phenomena.

In summary, the open magnetosphere geometry implies that a large antisunward force is applied by the solar wind to the tail magnetopause, large in the sense that only a small fraction of this force is required to maintain magnetospheric convection against atmospheric friction, the remainder being available to produce a long distended magnetospheric tail. Equivalently, only a small fraction of the tail magnetopause current closes through the ionosphere, the remainder closing through the tail current sheet to maintain the tail magnetic field (apart from the fixed fraction that is required to close across the day side magnetopause).

Geometry

The popular two-dimensional geometry adopted in the above discussion becomes untenable when the IMF direction departs significantly from southward (by more than, say, 15°), that is to say, most of the time. The need for at least two qualitative modifications is immediately apparent.

First, the day side magnetopause current $\underline{J}$ acquires a north-south component, and the associated magnetic stress $(\underline{J} \times \underline{B})$ therefore acquires an east-west component. The antisunward convection streamlines in the polar cap are thus deflected away from the noon-midnight meridian, such that southern polar-cap convection tends to follow the IMF direction and vice-versa in the northern polar cap; this asymmetry is observed (Heppner, 1972) in the appropriate sense (Russell and Atkinson, 1973).

The second modification involves the orientation of the "merging line," i.e., the line on the day side magnetopause that separates the domains of inward vs. outward B_n, through which the IMF connects to the northern vs. southern polar caps, respectively. This merging line forms a symmetry axis of the merging outflow (i.e., of the tangential component of $\underline{J} \times \underline{B}$); it lies in the east-west direction for a strictly southward IMF (the case illustrated above) but must deviate from this direction as the outflow acquires east-west components in response to east-west IMF components. The amount of this deviation is, at present, an unsettled question; a plausible hypothesis (e.g., Sonnerup, 1974; Hill, 1975) is that the merging line aligns with the magnetopause current vector, although such alignment has not been demonstrated to be necessary geometrically (Cowley, 1976).

Because the merging line is, by definition, a symmetry axis of the merging outflow, it seems natural to assume that it passes through the subsolar stagnation point of the magnetosheath flow; thus assuring that the tangential $\underline{J} \times \underline{B}$ stress (and the resulting merging outflow) has everywhere a component away from, rather than toward, the subsolar stagnation point. In other words, the merging line must pass through the stagnation point in order that the merging outflow everywhere assist, rather than hinder, the magnetosheath flow around the magnetosphere. This rather vague "principle of least action" is presumably the basis of the common, if often unstated, assumption that the merging line passes through the subsolar point even for significantly non-southward IMF orientations. On the basis of this assumption, one can predict, largely on geometrical grounds, the functional dependence of the potential drop along the merging line (and hence, within the open model, of the cross-polar cap potential drop) on the IMF orientation (e.g., Sonnerup, 1974; Hill, 1975), a prediction that has met with some success (Reiff et al., 1981).

Apart from the subsolar point, the magnetosheath flow is expected to have two other stagnation points, namely, the northern and southern polar cusps, where the magnetopause is expected to be indented by virtue of the reversal of the internal tangential magnetic-field component (e.g., Spreiter and Summers, 1967). One might thus plausibly expect that merging occurs (i.e., a merging line is formed) near the cusps, rather than (or in addition to) near the subsolar point; this suggestion was made, on the basis of slightly different arguments, by Haerendel et al. (1978).

Crooker (1979, 1980) pointed out that the cusp location has a particular advantage over the subsolar point with respect to merging-line formation. The cusp is observed (Haerendel et al., 1978) to resemble a true mathematical cusp or neutral point in the sense of Chapman and Ferraro (1931); that is, the internal geomagnetic field points radially toward or away from the cusp in all directions in a spoke pattern (toward in the northern hemisphere, away in the southern hemisphere). As a result, an IMF of any given orientation will find itself exactly antiparallel to the internal field in some vicinity of the cusps, whereas such antiparallelism does not occur at the subsolar point except in the singular case of exactly southward IMF. Because the merging rate is a maximum when the opposed fields are antiparallel (e.g., Sonnerup, 1974; Hill, 1975), it is plausible to expect (Crooker, 1979) that, of the infinite number of merging lines that are geometrically permissible, the dynamically important one is the locus of points at which the instantaneous IMF and the internal geomagnetic field are locally antiparallel. This locus always intersects the cusps, but does not intersect the subsolar point except in the singular case of exactly southward IMF.

This locus of antiparallelism, and its dependence on IMF orientation, are illustrated schematically in Figure 7 (a and b, respectively), reproduced from Crooker (1979).

The consequences of this merging geometry for the magnetospheric convection pattern have been discussed qualitatively by Crooker (1979, 1980), who demonstrates an encouraging degree of consistency with observations. A quantitative model of the merging process in this intrinsically three-dimensional geometry awaits further research.

One interesting qualitative consequence of this "antiparallel merging" hypothesis is that a significant section of the day side magnetopause would be magnetically closed ($B_n = 0$), the area of the closed section increasing from zero for an exactly southward IMF to virtually the entire day side magnetopause for any northward-tending IMF (Figure 7). This geometry offers the interesting possibility that non-magnetic coupling processes (as envisioned by Axford and Hines) may have observable consequences near the low latitude day side magnetopause even if (as indicated by empirical correlation studies, e.g., Reiff et al. (1981)) such processes make a relatively unimportant contribution to the global solar-wind magnetosphere interaction.

The geometry of interconnection is undoubtedly further complicated by the spatial and temporal non-uniformity of the incident IMF and plasma flow, both of which can dramatically influence the merging rate (e.g., Vasyliunas, 1975; Hill, 1975). In particular, two types of magnetosheath flow irregularities have been suggested as responsible for "mixing" of magnetosheath and (closed) magnetospheric flux tubes, a "turbulent eddy" flow in the polar cusp indentations of the magnetopause (Haerendel et al., 1978), and an "impulsive penetration" of magnetosheath flux tubes having above-average momentum density (Lemaire, 1977). As was discussed above, any such "mixing," by definition, involves the

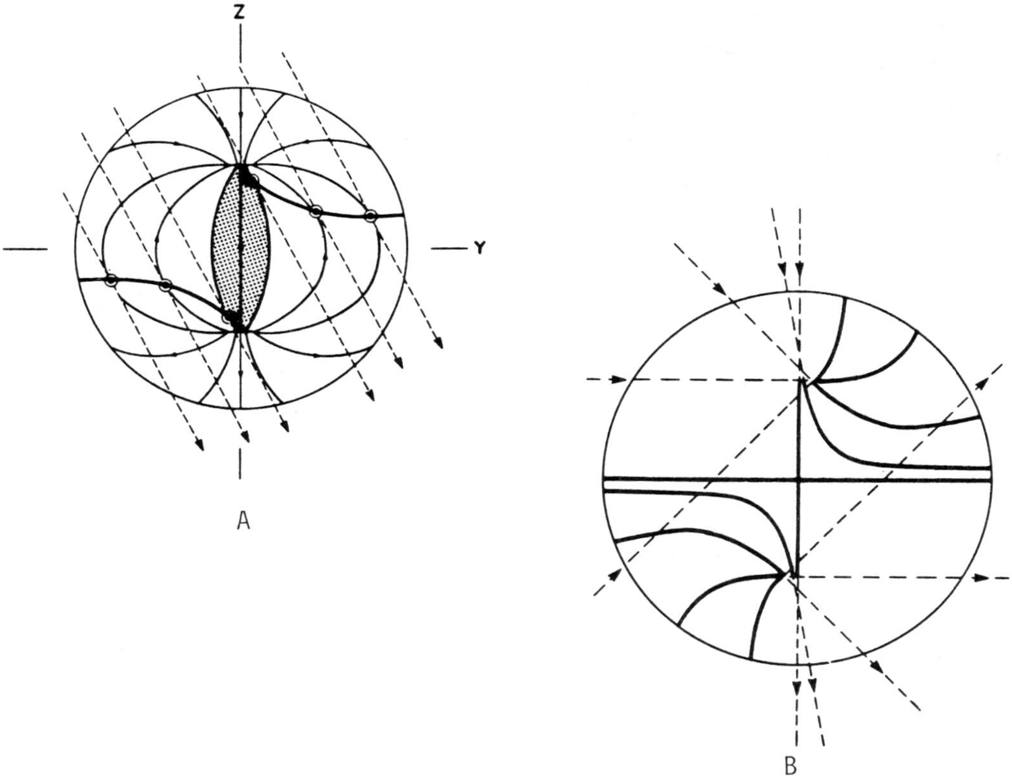

Figure 7. Illustration of the "antiparallel merging" hypothesis (Crooker, 1979) wherein the merging line (heavy lines) is the locus along which the external field (dashed lines) is antiparallel to the internal field (light solid lines in (a)). The variation of this locus with IMF orientation is illustrated in (b). Both views are from the Sun.

magnetic merging process. (Indeed, merging must occur at two different points along the same magnetosheath flux tube if its plasma content is to be transferred (in part) to a closed day side geomagnetic flux tube, the topology then being that proposed by Dungey (1963).) Discussions of such flux-tube mixing phenomena generally either mention the merging requirement only in passing (e.g., Haerendel et al., 1978), or ignore it altogether (e.g., Lemaire, 1977), but it is doubtful that such processes can be realistically assessed, even qualitatively, without explicit consideration of the implied merging geometry. These suggested processes should be viewed, not as alternatives to the merging hypothesis, but rather as bothersome, if perhaps necessary, complications to an already complicated hypothesis.

NON-MAGNETIC COUPLING

By "non-magnetic coupling" we mean any coupling process (such as envisioned by Axford and Hines) that does not rely on large-scale magnetic interconnection between the IMF and the geomagnetic field. The magnetic field invariably plays an important role even in "non-magnetic" coupling processes; the distinction is whether or not the coupling process necessarily depends on the macroscopic interconnection of interplanetary and geomagnetic fields.

In principle, momentum and energy can be transported across a closed magnetopause either by hydromagnetic waves or by plasma particles. In practice, the latter option appears to have more potential importance to the global coupling.

Wave Transfer

Discussions of wave transfer usually center around the Kelvin-Helmholtz instability (Siscoe, 1983 — this volume) of the magnetopause itself. The instability produces ripples of the magnetopause surface perpendicular to the adjacent magnetosheath flow (like wind over water); these ripples intercept the flow and thus provide a certain degree of tangential momentum transfer across the magnetopause. The fastest wave growth occurs where the flow is perpendicular to the internal magnetic field (whose tension otherwise tends to restore surface perturbations); the associated momentum transfer is therefore more likely to be important in the equatorial than the polar region.

Although the linear stability criterion is a straightforward (if tedious) problem, the nonlinear development of Kelvin-Helmholtz waves is a highly complicated and largely unsolved problem (e.g., Southwood, 1979, and references therein). It is thus impossible to assess theoretically the importance of Kelvin-Helmholtz waves as momentum and energy transfer agents — the waves must certainly become nonlinear before they can seriously affect such transfer. Empirically, such waves appear to have negligible affect on the global transfer rates, although they probably generate certain types of geomagnetic pulsations (Southwood, 1979).

Hydromagnetic waves, by definition, do not violate the frozen-in-flux theorem (4), and hence cannot inject solar-wind plasma into the magnetosphere. Kelvin-Helmholtz surface waves may accidentally resonate with magnetosheath ions and hence contribute to cross-field diffusion (Southwood, 1979), but if so the waves are no longer properly described as hydromagnetic.

Particle Transfer

Apart from magnetic merging, there are two obvious mechanisms that violate the frozen-in-flux theorem and can thereby transport magnetosheath particles onto closed geomagnetic field lines: classical magnetic (gradient and curvature) drift, and anomalous (non-collisional) cross-field diffusion. Either process can produce a low-latitude boundary layer of injected plasma flowing downstream on closed field lines at a significant fraction of the solar-wind speed (Figure 8). The sum of the potential drops across the two (dawn and dusk) boundary layers equals the polar-cap potential drop (Figure 8), and determines the magnitude of the mass, momentum, and energy transfer associated with the boundary-layer flow, as follows. The mass transfer rate (kg/s) through a given cross section of the boundary-layer flow is

$$R \sim 2nmv_t w\ell_1 = 2nm\phi\ell_1/B \tag{42}$$

where n and v_t are boundary-layer values of number density and tangential velocity, respectively (both presumably less, but not too much less, than adjacent magnetosheath values if the process is to be significant), m is the mean ion mass (1 a.m.u. will be assumed here), B is the mean boundary-layer magnetic field component perpendicular to $\underline{v}_t$, and ϕ is the potential drop across each boundary layer of thickness w and north-south extent ℓ_1 (cf. Figure 3). (The factor 2 accounts for equal contributions from dawn and dusk sides; if the contributions are unequal, then a dawn-dusk average is implied.) The quantity R increases

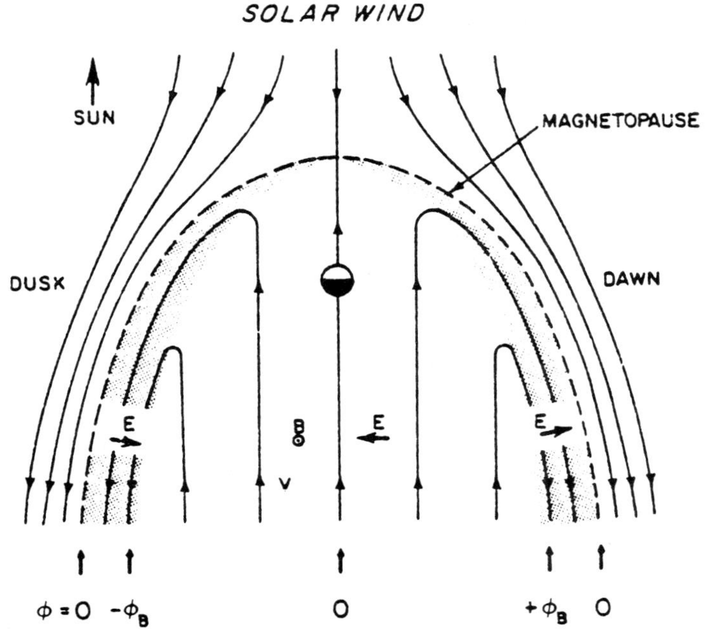

Figure 8. To the extent that the magnetosphere is closed, the sum of the voltages across the dawn and dusk boundary layers (shaded) equals the voltage across the polar cap. (View in the equatorial plane.)

with downstream distance as long as injection is occurring locally. We can obtain a rough upper limit by assigning typical magnetosheath values to n ($5/cm^3$) and v_t (400 km/s); taking $\ell_1 = 15\ R_E$ and $B \sim 20$ nT we obtain

$$R \lesssim (10^{27}\ \text{amu/s})(\phi/25\ \text{kV}) \tag{43}$$

where ϕ is (recall) the average of the individual dawn- and dusk-side potentials, not their sum. ($\phi = 25$ kV implies $w \gtrsim 1/3\ R_E$ for $v_t \lesssim 400$ km/s.)

The momentum and energy transfer rates are simply

$$F = Rv_t \lesssim (7\times10^5\ \text{Nt})(\phi/25\ \text{kV}) \tag{44}$$

and

$$P = Rv_t^{\,2}/2 \lesssim (1.3\times10^{11}\ \text{W})(\phi/25\ \text{kV}) \tag{45}$$

By reference to the preceding discussion we note that a value $\phi \sim 25$ kV is required in order for particle injection (drift or diffusion) to be competitive with magnetic coupling. The general description of energy transfer in terms of Poynting's theorem and of momentum transfer in terms of $\underline{j}\times\underline{B}$ stress are the same as in the above discussion (allowing for the obvious differences in magnetic topology and geometry), and will not be repeated here.

Magnetic gradient-drift entry was proposed by Fejer (1965) and by Wentworth (1965) and is illustrated in Figure 9. The gradient drift has a component perpendicular to the magnetopause whenever the internal field has a gradient in the plane of the boundary, the drift being generally inward for ions on the dawn side and electrons on the dusk side. The resulting charge polarization would presumably be discharged by the ionosphere, the required current system having the sense of the "Region 1" driving currents.

The normal components of magnetic gradient and curvature drift can be approximated by

$$v_n \approx kT/(qBr_B) \tag{46}$$

where r_B is a characteristic scale length of the magnetic-field geometry and kT is a characteristic particle energy. For 90° pitch-angle particles, $kT = kT_\perp = (m/2)\langle v_\perp^{\,2}\rangle$, $r_B = B/|\nabla_t B|$ where ∇_t is the component of the gradient tangent to the boundary, and v_n is pure gradient drift. For 0° pitch-angle particles, $kT = kT_\parallel = m\langle v_\parallel^{\,2}\rangle$, $r_B = |\hat{\underline{b}}\cdot\nabla_t\,\hat{\underline{b}}|^{-1}$ is the field-line radius of curvature in the tangent plane, and v_n is pure curvature drift. For a distribution of pitch angles the same expression suffices with kT and r_B representing appropriate averages.

At a downstream distance x from the first injection point, the max-

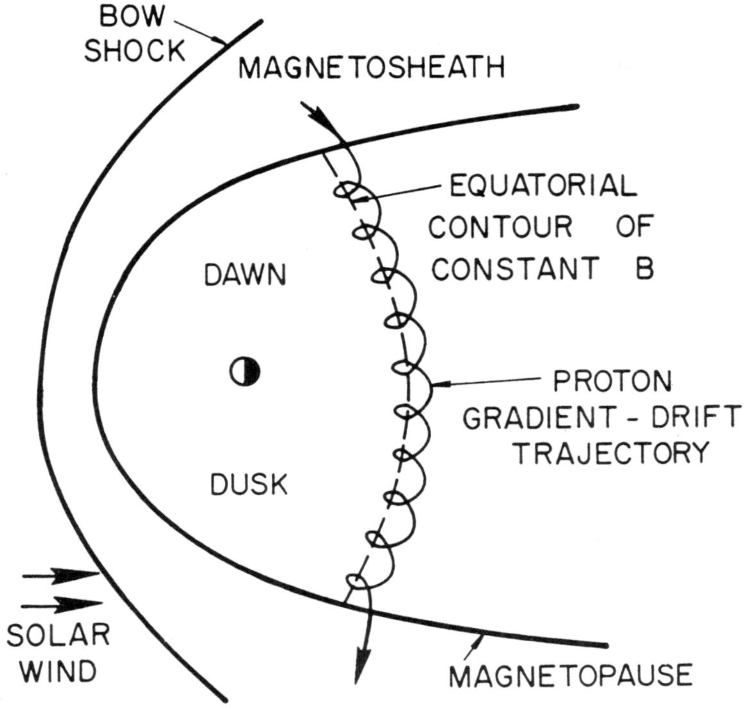

Figure 9. Schematic illustration (in the equatorial plane) of the ∇B drift entry hypothesis. This picture neglects convection ($\underline{E}\times\underline{B}$ drift) and would therefore apply only to high-energy protons (energy per charge much larger than the cross-tail potential), not to solar-wind protons.

imum particle residence time in the boundary layer is $t \sim x/v_t$ and the corresponding thickness of the boundary layer is

$$w \sim v_n t \sim [kT/(qBv_t)](x/r_B) \tag{47}$$

giving a potential drop

$$\phi = v_t Bw \sim (kT/q)(x/r_B) \tag{48}$$

The magnetic-field scale length can become quite small compared to x near the edges of the tail current sheet. However, in this geometry the bounce-averaged gradient and curvature drifts tend to cancel one another; in such a "drift-free" geometry the net result of gradient and curvature drifts along a given trajectory is much less than the instantaneous drift components (Stern, 1978). The geometry is drift-free to the same degree of approximation that B_z, the northward component of $\underline{B}$ crossing the current sheet, is constant; thus the scale length of the downstream gradient of B_z actually determines the net normal drift and hence the thickness (47) and potential drop (48) of the boundary layer. This gradient scale length cannot be much less than x, because B_z is necessarily a monotonically decreasing, positive function of x in the geometry considered. Thus $r_B \gtrsim x$ and (48) gives

$$\phi \lesssim kT/q \sim 200 \text{ V} \tag{49}$$

which is far too small to have global consequences.

The same result (49) can be deduced from energy considerations. The gradient and curvature drifts, whether adiabatic or not, transport particles "uphill" in the potential $q\phi$ only at the expense of their thermal energy kT, and the maximum potential difference that can be thus traversed is necessarily $\sim kT/q$. Acceleration drift (deceleration in the Earth's frame of reference) can add to the gradient/curvature drift (indeed it must if the boundary layer is to act as a dynamo generator) but this adds an available potential of only $mv_t^2/2q$, which is somewhat larger than (49) but still inconsequential ($\sim$ 1 kV) on the dawn side where m = proton mass, and even more inconsequential (by the electron-to-proton mass ratio) on the dawn side, since only (ions/electrons) can enter on the (dawn/dusk) side by this mechanism.

Thus, although the drift-injection mechanism is attractive in its simplicity and, perhaps, its inevitability, it apparently has a negligible role in the global coupling process. The theoretical maximum potential drop is smaller than typically observed values by at least an order of magnitude, nor is the predicted gross dawn-side asymmetry (independent of IMF orientation) observed. Both of these difficulties can, in principle, be avoided by the process of particle diffusion.

Diffusion

Compared to drift injection, the stochastic scattering of individual particles across the magnetic field by gyroresonant waves has the advantage of avoiding the gross dawn-dusk asymmetry inherent in the drift injection process. It has the additional advantage of allowing, in principle, a larger flux of mass, and hence of momentum and energy, into the boundary layer, but only insofar as the waves responsible for the scattering are nonlinear (i.e., have perturbation electric and/or magnetic fields comparable to, if not exceeding, the time-average field magnitudes). The energy constraint (49) on the boundary layer potential drop can likewise, in principle, be avoided, subject to the same requirement of nonlinearity: the generation of the boundary potential ϕ has, in addition to the plasma thermal and kinetic energy, an additional source, the energy of the waves responsible for scattering. The waves may, in principle, heat the particles in addition to scattering them, the waves in turn drawing energy from a source external to the boundary layer. The requirement for such heating and the implied nonlinearity of

TABLE 1

CANDIDATE INSTABILITIES FOR CROSS-FIELD DIFFUSION

Name of Instability	Free Energy Source	Resonant Particle Species	Reference
Two-Stream Ion Cyclotron	(Self-explanatory)	Ions	Eviatar & Wolf (1968)
Kinetic Alfvén Wave	mode-coupling from drift wave or K-H wave	Ions	Hasegawa & Mima (1978)
Lower-Hybrid Drift	microscopic density/ temperature gradients	Ions	Gary & Eastman (1979)
Electrostatic Flow Shear	(Self-explanatory)	Ions	Gary & Schwartz (1980)

the waves (not only $\delta B^2 \gtrsim B^2$, but $\delta B^2 \gtrsim \beta B^2$) is frequently overlooked.

Several micro-instabilities have been proposed to generate waves capable of scattering particles (generally ions) across the magnetopause to form a boundary layer. Table 1 provides a catalog of these instabilities, which will not otherwise be reviewed here. Generally, as indicated above, such instabilities will be potentially important only insofar as they produce grossly nonlinear gyroresonant waves ($\delta B^2 \gtrsim B^2$); an equivalent requirement is that the wave scattering must produce a cross-field diffusion coefficient (mean-square guiding-center displacement per unit time) comparable to the Bohm limit

$$D_B \equiv kT/(16\ qB) = a_c{}^2\omega_c/32 \qquad (50)$$

where ω_c and a_c are the gyrofrequency and mean-square gyroradius of the diffusing particles and kT is an ill-defined kinetic temperature, the second equality of (50) corresponding to the identification $kT = kT_\perp \equiv (m/2)\langle v_\perp{}^2\rangle$. The dimensional term $a_c{}^2\omega_c$ in (50) is dictated by the characteristic length (a_c) and frequency (ω_c) of the process, but the proposed maximum value of the dimensionless coefficient (1/32) was assigned empirically on the basis of laboratory results in the original reference (Bohm et al., 1949) and has since been commonly adopted as a theoretical limit in space plasmas. The actual maximum value of this dimensionless coefficient is probably closer to unity (i.e., 32 times larger — see below), but the point remains that local parameters (a_c and ω_c) dictate an upper limit for the diffusion rate, and hence (see below) for the boundary-layer potential drop (see Wolf, 1975; Hill, 1979).

Let us assume that one or more plasma instabilities produce, for a given particle species, an effective collision frequency ν (average frequency of large-angle scattering in the plane perpendicular to $\underline{B}$). It is then a straightforward exercise (Appendix B) to show that the mean-square displacement between collisions is

$$\langle \ell^2 \rangle = 2a_c{}^2/(1+\nu^2/\omega_c{}^2) \tag{51}$$

and the diffusion coefficient is

$$D \equiv \nu\langle \ell^2 \rangle = a_c{}^2\omega_c\ [2(\nu/\omega_c)/(1+\nu^2/\omega_c{}^2)] \tag{52}$$

The functional dependence on ν/ω_c (bracketed term in (52)) is intuitively correct: the effective collisions must "resonate" with the gyromotion to have maximum effect. The bracketed term has a maximum value of unity (32 times larger than the classical "Bohm limit") at $\nu = \omega_c$. The diffusion flux $(-D\nabla_n)$ thus has exactly the same dependence on (ν/ω_c) as does the Pedersen-current conduction flux (e.g., Hanson, 1965; Schunk, 1983 — this volume), and for the same reason: the effective collisions must be sufficiently frequent to break the first adiabatic invariant (Siscoe, 1982 — this volume), i.e., to interrupt the gyromotion around $\underline{B}$, but not so frequent as to seriously retard the net particle flux, whether that flux be due to the $-q\nabla\phi$ force (in the case of Pedersen conduction) or to the $-kT\nabla n$ "force" (in the case of diffusion).

This "coincidence" may, in fact, impose a serious limitation on the net plasma diffusion rate. If particles of one charge sign were to diffuse faster than the other (and, according to (52), this should be the rule rather than the exception), the resulting charge-separation electric field, in the plasma frame of reference, would drive a Pedersen conduction flux tending to reduce the unbalanced diffusive charge flux. The Pedersen conduction flux is a significant perturbation to the diffusion flux if the potential across the diffusion layer is comparable to the thermal energy equivalent; the ratio of the two fluxes is in fact

$$\frac{j}{qD|\nabla n|} = nq|\nabla\phi|/(2kT_\perp|\nabla n|) \approx q\phi/(2kT_\perp) \tag{53}$$

The diffusion of the plasma as a whole might then be expected to proceed at some compromise rate between the intrinsic diffusion rates of the two charge species, in a manner analogous to the ambipolar diffusion of ions and electrons along the magnetic field (see Schunk, 1983 — this volume). On the other hand, the ionosphere might supply neutralizing charge and thus allow the differential diffusion to proceed unhindered. (This hypothetical neutralizing current system, unlike that of the "Region 1" driving currents, would be symmetric between dawn and dusk.)

In any event, the existence of a firm upper limit (from (52)) for the diffusion coefficient,

$$D \leqslant D_0 \equiv a_c{}^2\omega_c = 2kT_\perp/qB \tag{54}$$

enables us to derive an upper limit for the potential drop across a diffusive boundary layer, irrespective of the details of the diffusion process. The diffusion flux is $-D\nabla n$ and the effective diffusion speed is thus

$$v_n = D|\nabla n|/n \lesssim 2kT/(qBw) \tag{55}$$

where $w \equiv n/|\nabla n|$ is the effective thickness of the diffusion layer. Comparison of (46) and (55) shows that the diffusion speed can, in principle, exceed the gradient/curvature drift speed by the large factor $2r_B/w$. The diffusion speed is, of course, self-limiting, decreasing as w increases.

The diffusion-layer thickness is generally

$$w \approx (Dt)^{1/2} \tag{56}$$

where t is the residence time of the "oldest" particles in the boundary layer. For example, if the magnetopause is approximated as a planar source boundary with density $n = n_0$, with injection beginning at $t = 0$, then the one-dimensional diffusion equation has the familiar solution

$$n/n_0 = \text{erfc}\ [y/(4Dt)^{1/2}] \tag{57}$$

where y is inward displacement normal to the boundary and erfc is the complementary error function; (57) gives $n/n_0 = 1/2$ at $y \approx 0.95(Dt)^{1/2}$. Relating t to downstream displacement as above ($t \sim x/v_t$) we have

$$\phi = v_t Bw \lesssim (2kT_\perp v_t Bx/q)^{1/2} \tag{58}$$

For representative values $kT_\perp \sim 200$ eV, $v_t \sim 200$ km/s, $B \sim 20$ nT this corresponds to

$$\phi \lesssim (15\ \text{kV})(x/20\ R_E)^{1/2} \tag{59}$$

and we find that diffusion may, in principle, become marginally competitive with magnetic coupling if diffusion were to proceed at its maximum conceivable rate (54). (One might be tempted to conclude from (59) that ϕ could increase by a significant factor at sufficiently large downstream distances (say $x \sim 100\ R_E$), but this suggestion is defeated by the decrease of B, at least as fast as $1/x$; in (58). For example, at $x \sim 100\ R_E$, a more appropriate value of B would be $\lesssim 1$ nT — recall that B in (58) is defined as the component of B transverse to v_t (essentially B_z).

Observations at $x \sim 60\ R_E$ (Sanders et al., 1980) indicate that the maximum potential drop across any diffusive boundary layer at that distance is of the order of 4 kV. This would imply, not surprisingly, that diffusion actually proceeds at considerably less than the maximum possible rate (54); the "Bohm limit" might indeed be a reasonable estimate of the actual diffusion rate.

THE TAIL

After all these steady-state arguments, applicable to "a sufficiently long-term average", one may well ask "how long is 'long-term'?"

The empirical answer is $\gtrsim$ one hour. The solar-wind magnetosphere coupling drives an antisunward plasma flow over the polar caps, and this flow requires, sooner or later, the establishment of a compensating return flow in the lower latitude magnetosphere. The establishment of the return flow evidently takes ~ 1 hr., and occurs through a characteristic sequence of events known as the magnetospheric substorm.

A fundamental observational fact (Caan et al., 1975) is that the tail magnetic field increases slowly but steadily during the initial "growth" phase of a substorm, then decreases rapidly during the substorm "expansion" phase. (I will not attempt to explain why the synonyms "growth" and "expansion" are conventionally used to describe two such opposite events.) This observation implies (through Faraday's law) the existence of an induced voltage ($\oint \underline{E} \cdot d\underline{\ell} \neq 0$) around each lobe (northern and southern) of the tail; the voltage across the anti-sunward flow (Region 2 in Figure 5) exceeds that across the return flow (Region 3 in Figure 5) during the growth phase, and vice-versa (with a vengeance) during the expansion phase. The return flow does not match the anti-sunward flow continuously, but only in a sporadic catch-up fashion.

The reason for this is not clear. A certain delay between driving perturbation and flow response is to be expected simply on the basis of wave-propagation lag, but the growth-phase time constant probably cannot be explained on this basis — the hydromagnetic wave transit time is probably no more than a few minutes (e.g., Coroniti and Kennel, 1973). Propagation delay would not, in any case, explain why a relatively smooth driving perturbation should elicit an impulsive response, as seems to be the case (e.g., McPherron, 1979).

The impulsive character of the return flow, as evidenced in magnetospheric substorms, is apparently related to the internal dynamics of the magnetospheric tail, and particularly of the plasma sheet that carries the cross-tail current. The return flow is apparently inhibited until the magnetic stress in the tail builds to some critical level. The empirically inferred nature of this process is aptly compared to the formation and separation of water drops from a leaky faucet. The driving force (gravity) gradually increases (owing to the gradual addition of mass) until it exceeds the capability of the restraining force (surface tension), whereupon the continuity equation is finally and inevitably satisfied in a dramatic event. For the magnetospheric return flow, the driving force is evidently the tension in the tail magnetic field, but the nature of the restraining force is less obvious. Two recent suggestions, reviewed by Hill (1982), are that the restraining force consists of an adverse plasma pressure gradient in the near-Earth plasma sheet (Erickson and Wolf, 1980) or the dynamic pressure of solar-wind plasma streaming into the distant plasma sheet (Hill and Reiff, 1980).

The theory of tail dynamics is in too preliminary a state to be reviewed in any comprehensive fashion here. This is perhaps fortunate — the chapter is already long enough. A simple and general argument (Appendix A) indicates that the time scale for establishing the return flow cannot exceed about three hours without catastrophic consequences; the fact that the actual time scale approaches this upper limit is food for thought.

APPENDIX A — THE REQUIREMENT FOR RETURN FLOW

Consider a cross-section of the northern lobe of the tail, sufficiently close to the Earth that most of the downstream convection streamlines intersect that cross section before turning around to form return flow streamlines. (To be safe in this regard, we could equally well consider the dawn-dusk meridian cross section of Figure 4; the argument would be the same.) We wish to show that the "amount" of return flow (in some sense) must match the "amount" of downstream flow on the average, and to find a maximum time scale for possible deviations from this average state.

The cross section is bounded by a closed curve (which need not be time stationary) resembling the top half of a "θ." The magnetic-field component in the plane of the cross section always crosses this boundary in the inward sense, regardless of whether the normal component belongs to field lines locally crossing the tail current sheet (magnetic equator) from the southern lobe, or to open field lines locally crossing the magnetopause from the IMF. Thus the component of $\underline{E} = -\underline{v} \times \underline{B}$ (4) tangent to this boundary curve is always in the counterclockwise sense (as viewed from the Sun) where the flow is sunward, and in the clockwise sense where the flow is antisunward (regardless of whether on open field lines or closed boundary-layer field lines).

Consider the clockwise line integral $\oint \underline{E} \cdot d\underline{\ell}$ around the boundary curve. Define V_1 as the total (positive) contribution to this integral from any antisunward flow region(s) and define V_2 as the magnitude of the (negative) contribution from any sunward return flow region(s). The complete line integral is then

$$\oint \underline{E} \cdot d\underline{\ell} = V_1 - V_2 = \frac{\partial \Phi}{\partial t} \qquad \text{(A1)}$$

where Φ is the total (Earthward-directed) magnetic flux through the cross-section, the second equality being Faraday's law. The total tail flux must be constant in a sufficiently long-term average, and hence $V_1 = V_2$ in the same average sense. It is precisely in this sense that the "amount" of return flow must, on average, match the "amount" of downstream flow; the two flows must, on the average, intersect the same voltage, or, equivalently, must intercept magnetic flux at the same rate. The intercepted magnetic flux is the quantity that the magnetospheric convection system must ultimately conserve. (The plasma mass flux, for example, need not be conserved on any time scale; indeed it is almost certainly not conserved — there are at least four source/sink terms (solar-wind source, solar-wind sink, ionosphere source, ionosphere sink) that operate by motion along $\underline{B}$, not to mention cross-field diffusion, gradient/curvature drift, etc.)

What if $V_1 \neq V_2$? Suppose, for example, that the downstream flow has a typical voltage $V_1 \sim 50$ kV, but that the return flow is temporarily suppressed ($V_2 << V_1$); this is a (perhaps somewhat exaggerated) description of a substorm growth phase. Then

$$\frac{\partial \Phi}{\partial t} \sim 50 \text{ kV} \tag{A2}$$

The total flux in each tail lobe is

$$\Phi = \frac{\pi}{2} B_T R_T{}^2 \sim 5\times10^8 \text{ Weber} \tag{A3}$$

(using, as above, $B_T = 20$ nT, $R_T = 20\ R_E$); this value would double, at the rate (A2), in a time

$$\tau \equiv \Phi/(\partial\Phi/\partial t) \sim 3 \text{ hr} \tag{A4}$$

The total tail flux is, however, rather narrowly constrained by the overall solar-wind pressure balance requirement; the magnetospheric configuration simply cannot tolerate factor-of-two changes in Φ. (Observed changes during substorm growth phases rarely exceed 15% (Caan et al., 1975).) For a given amount of antisunward flow (as measured by V_1), the appropriate return flow therefore must be established on a time scale of the order of one hour. We have inescapable empirical evidence that $V_2 \lesssim V_1$ during substorm growth phases, and the fact that this imbalance persists nearly as long as is theoretically possible strongly suggests that the substorm involves a large-scale configurational instability of the tail magnetic field. We may not understand the cause of substorms (i.e., the reason why $V_2 \lesssim V_1$ in the first place), but we can easily understand why substorms cannot last much longer than an hour.

APPENDIX B — MEAN-SQUARE DISPLACEMENT IN CROSS-FIELD DIFFUSION

Let ν be the average frequency of large-angle scattering (from whatever cause) for particles with charge q, mass m, and gyrospeed $v_\perp$. This means that the probability of such a scattering in any infinitesimal time interval dt is

$$dP = \nu\, dt \tag{B1}$$

As a direct consequence of this definition, the probability of surviving a time t _without_ such scattering is

$$P_1(t) = e^{-\nu t} \tag{B2}$$

and the probability of suffering the _first_ such scattering between t and t + dt is then

$$P_1(t)\, dP = \nu e^{-\nu t} dt \tag{B3}$$

(Large-angle scattering is treated here as a discrete event, although this is not a fundamental restriction — the "effective collision frequency" for any scattering process can always be defined such that (B3) holds for $t \gtrsim 1/\nu$. We do assume that the scattering is isotropic in the particle rest frame, as is implicit in the concept of diffusion.)

Following any collision, the particle will gyrate around $\underline{B}$ until the time of its next collision. This statement defines the "particle rest frame" mentioned above — all adiabatic drifts such as $\underline{E}\times\underline{B}$ drift, gradient drift, etc. are, by definition, removed in the transformation to the rest frame. After gyrating for time t without collision, the net displacement is

$$x(t) = 2a_c|\sin(\omega_c t/2)| \tag{B4}$$

where

$$\omega_c = |q|B/m \tag{B5}$$

and

$$a_c = v_\perp/\omega_c \tag{B6}$$

are the gyrofrequency and gyroradius, respectively. The result (B4) follows from simple geometric construction: $x(t)$ is the length of a chord with central angle $\omega_c t$ in a circle of radius a_c. Averaging the square of (B4) over the probability distribution (B3) gives the mean-square displacement

$$\begin{aligned}\langle \ell^2 \rangle &\equiv \int_0^\infty [x(t)]^2 P_1(t)\nu dt \\ &= 4a_c^{\,2} \int_0^\infty \sin^2(\omega_c \xi/2\nu) e^{-\xi} d\xi \\ &= 2a_c^{\,2}/(1+\nu^2/\omega_c^{\,2}) \\ &= 2a_c^{\,2}\ell_0^{\,2}/(\ell_0^{\,2}+a_c^{\,2})\end{aligned} \tag{B7}$$

where $\ell_0 \equiv v_\perp/\nu$ is the scattering mean-free-path. The last line of (B7) confirms the intuitive result that the mean-square displacement is determined by the gyroradius or by the scattering mean-free-path, whichever is the smaller.

ACKNOWLEDGMENTS

I thank P. H. Reiff and R. A. Wolf for their helpful comments. This work was supported by the National Science Foundation (Division of Atmospheric Sciences) under Grant ATM-8017316.

REFERENCES

Alfvén, H.: 1977, Rev. Geophys. Space Phys. 15, pp. 271-284.
Axford, W. I.: 1962, J. Geophys. Res. 67, pp. 3791-3796.
Axford, W. I., and Hines, C. O.: 1961, Can. J. Phys. 39, pp. 1433-1464.
Birkeland, K.: 1908, "The Norwegian Aurora Polaris Expedition 1902-3, 1, On the Cause of Magnetic Storms and the Origin of Terrestrial Magnetism," first section, H. Aschehoug and Co., Christiania.
Bohm, D., Burhop, E. H. S., and Massey, H. S. W.: 1949, in A. Guthrie and R. K. Wakerling (ed.), "Characteristics of Electrical Discharges in Magnetic Fields, McGraw-Hill, New York, pp. 13-76.
Burch, J. L., and Heelis, R. A.: 1979, in S.-I. Akasofu (ed.), "Dynamics of the Magnetosphere," D. Reidel, Boston, pp. 47-62.
Burke, W. J.: 1983, this volume.
Caan, M. N., McPherron, R. L., and Russell, C. T.: 1975, J. Geophys. Res. 80, pp. 191-194.
Chapman, S., and Ferraro, V. C. A.: 1931, Terr. Magn. & Atm. Elec. 36, pp. 171-186.
Coroniti, F. V., and Kennel, C. F.: 1973, J. Geophys. Res. 78, pp. 2837-2851.
Cowley, S. W. H.: 1976, J. Geophys. Res. 81, pp. 3455-3458.
Crooker, N. U.: 1979, J. Geophys. Res. 84, pp. 951-959.
Crooker, N. U.: 1980, J. Geophys. Res. 85, pp. 575-578.
Dungey, J. W.: 1961, Phys. Rev. Lett. 6, pp. 47-48.
Dungey, J. W.: 1963, in C. DeWitt, J. Hieblot, and A. Lebeau (ed.), "Geophysics, the Earth's Environment," Gordon and Breach, New York, pp. 526-537.
Dungey, J. W.: 1965, J. Geophys. Res. 70, pp. 1753.
Erickson, G. M., and Wolf, R. A.: 1980, Geophys. Res. Lett. 7, pp. 897-900.
Eviatar, A., and Wolf, R. A.: 1968, J. Geophys. Res. 73, pp. 5561-5576.
Fejer, J. A.: 1965, J. Geophys. Res. 70, pp. 4972-4975.
Gary, S. P., and Eastman, T. E.: 1979, J. Geophys. Res. 84, pp. 7378-7381.
Gary, S. P., and Schwartz, S. J.: 1980, J. Geophys. Res. 85, pp. 2978-2980.
Greenstadt, E. W.: 1972, J. Geophys. Res. 77, pp. 1729-1738.
Haerendel, G., Paschmann, G., Sckopke, N., Rosenbauer, H., and Hedgecock, P. C.: 1978, J. Geophys. Res. 83, pp. 3195-3216.
Hanson, W. B.: 1965, in F. S. Johnson (ed.), "Satellite Environment

Handbook," Stanford University Press, Stanford, Cal., pp. 23-49.
Hasegawa, A., and Mima, K.: 1978, J. Geophys. Res. 83, pp. 1117-1124.
Heikkila, W. J.: 1975, Geophys. Res. Lett. 2, pp. 154-157.
Heppner, J. P.: 1972, J. Geophys. Res. 77, pp. 4877-4887.
Hill, T. W.: 1975, J. Geophys. Res. 80, pp. 4689-4699.
Hill, T. W.: 1979, in B. Battrick (ed.), "Magnetospheric Boundary Layers," ESA SP-148, Noordwijk, Netherlands, pp. 325-332.
Hill, T. W.: 1982, Rev. Geophys. Space Phys. 20, pp. 671-684.
Hill, T. W., and Reiff, P. H.: 1977, J. Geophys. Res. 82, pp. 3623-3628.
Hill, T. W., and Reiff, P. H.: 1980, Geophys. Res. Lett. 7, pp. 177-180.
Iijima, T., and Potemra, T. A.: 1976a, J. Geophys. Res. 81, pp. 2165-2174.
Iijima, T., and Potemra, T. A.: 1976b, J. Geophys. Res. 81, pp. 5971-5979.
Johnson, F. S.: 1960, J. Geophys. Res. 65, pp. 3049-3052.
Kellogg, P. J.: 1962, J. Geophys. Res. 67, pp. 3805-3812.
Kennel, C. F.: 1969, Rev. Geophys. Space Phys. 7, pp. 379-419.
Kennel, C. F.: 1981, J. Geophys. Res. 86, pp. 4325-4330.
Lai, C. S.: 1981, Am. J. Phys. 49, pp. 841-842.
Lees, L.: 1964, AIAA J. 2, pp. 1576-1582.
Lemaire, J.: 1977, Planet. Space Sci. 25, pp. 887-889.
Maezawa, K.: 1976, J. Geophys. Res. 81, pp. 2289-2303.
McPherron, R. L.: 1979, Rev. Geophys. Space Phys. 17, pp. 657-680.
Michel, F. C., and Dessler, A. J.: 1970, J. Geophys. Res. 75, pp. 6061-6072.
Ness, N. F., Scearce, C. S., and Seek, J. B.: 1964, J. Geophys. Res. 69, pp. 3531-3569.
Paulikas, G. A.: 1974, Rev. Geophys. Space Phys. 12, pp. 117-128.
Pilipp, W. G., and Morfill, G.: 1978, J. Geophys. Res. 83, pp. 5670-5678.
Reiff, P. H.: 1979, in B. Battrick (ed.), "Magnetospheric Boundary Layers," ESA SP-148, Noordwijk, Netherlands, pp. 167-173.
Reiff, P. H.: 1983, this volume.
Reiff, P. H., Spiro, R. W., and Hill, T. W.: 1981, J. Geophys. Res. 86, pp. 7639-7648.
Rich, F. J., Vasyliunas, V. M., and Wolf, R. A.: 1972, J. Geophys. Res. 77, pp. 4670-4676.
Russell, C. T., and Atkinson, G.: 1973, J. Geophys. Res. 78, pp. 4001-4002.
Sanders, G. D., Maher, L. J., and Freeman, J. W.: 1980, J. Geophys. Res. 85, pp. 4607-4615.
Schunk, R. W.: 1983, this volume.
Siscoe, G. L.: 1983, this volume.
Sonnerup, B. U. O.: 1974, J. Geophys. Res. 79, pp. 1546-1549.
Southwood, D. J.: 1979, in B. Battrick (ed.), "Magnetospheric Boundary Layers," ESA SP-148, Noordwijk, Netherlands, pp. 357-364.
Spreiter, J. R., and Alksne, A. Y.: 1969, Rev. Geophys. 7, pp. 11-50.
Spreiter, J. R., and Summers, A. L.: 1967, Planet. Space Sci. 15, pp. 787-798.
Stern, D. P.: 1973, J. Geophys. Res. 78, pp. 7292-7305.
Stern, D. P.: 1978, J. Geophys. Res. 83, pp. 1079-1088.

Vasyliunas, V. M.: 1972, J. Geophys. Res. 77, pp. 6271-6274.
Vasyliunas, V. M.: 1975, Rev. Geophys. Space Phys. 13, pp. 303-336.
Vasyliunas, V. M.: 1979, in B. Battrick (ed.), "Magnetospheric Boundary Layers," ESA SP-148, Noordwijk, Netherlands, pp. 387-393.
Vasyliunas, V. M.: 1983, this volume.
Wentworth, R. C.: 1965, Phys. Rev. Lett. 14, pp. 1008-1010.
Wolf, R. A.: 1975, Space Sci. Rev. 17, pp. 537-562.
Wolf, R. A.: 1983, this volume.
Zwan, B. J., and Wolf, R. A.: 1976, J. Geophys. Res. 81, pp. 1636-1648.

THE QUASI-STATIC (SLOW-FLOW) REGION OF THE MAGNETOSPHERE

R. A. Wolf
Rice University, Houston, Texas 77251

I. INTRODUCTION

This chapter deals with the theory of the earth's inner and middle magnetosphere. With regard to the day side of the magnetosphere, we consider the region earthward and equatorward of the magnetopause and its associated boundary layers. With regard to the nightside, we consider the inner part of the plasma sheet, specifically, the part that lies on closed magnetic field lines that experience only flows that are slow compared to the fast-mode speed. Excluded from consideration are all X-lines and magnetopause boundary layers, including the polar cusp, mantle, and tail lobe. Figure 1 indicates, in a rough, qualitative way, the extent of the region being discussed.

Theoretical considerations determine the limits of the region. The boundaries are all magnetic-field aligned, because the theory treats each magnetic field line and the particles on it as a discrete entity, due to the ease with which particles and electric currents flow along $\underset{\sim}{B}$. Also, the theory to be developed here is based on the assumption that the inertial term

$$\rho \frac{d\underset{\sim}{V}}{dt} \equiv \rho \left(\frac{\partial}{\partial t} + \underset{\sim}{V} \cdot \nabla\right) \underset{\sim}{V}$$

in the MHD momentum equation can be neglected. This limits applicability to subsonic (slow-flow) regions of the magnetosphere. For much of the theory to be discussed, we also assume that the magnetic field $\underset{\sim}{B}(\underset{\sim}{x},t)$ can be estimated with adequate accuracy from observations or from other considerations that are outside the scope of the calculations. In other words, $\underset{\sim}{B}(\underset{\sim}{x},t)$ is regarded as known input. Section VII will discuss the magnetic-field modeling problem, in general.

The slow-flow condition, while only involving the neglect of one term in one of the governing equations (the momentum equation), allows a major theoretical simplification. It allows us to separate the problem of magnetospheric convection from the (difficult) problem of the pro-

R. L. Carovillano and J. M. Forbes (eds.), Solar-Terrestrial Physics, 303–368.

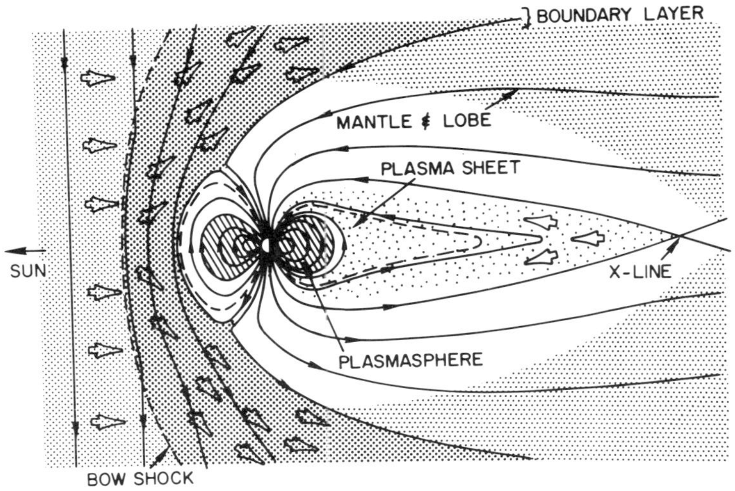

Figure 1. View of the earth's magnetosphere, in the noon-midnight meridian plane. The sun is to the left. Hollow arrows represent rapid flows. Solid curves with arrows on them are magnetic field lines. The dashed curve qualitatively indicates the boundary of the slow-flow region, the region discussed in this chapter.

pagation of MHD waves in the magnetosphere. (See chapter by W. J. Hughes.) It greatly simplifies the basic equation of magnetosphere-ionosphere electrical coupling and it reduces the instability problems involved in numerical solution of the equations.

The assumption that $\underset{\sim}{B}(\underset{\sim}{x},t)$ is known allows equally drastic additional simplifications, by separating the convection problem from the calculation of magnetic structure. The convection problem then reduces from a three-dimensional problem to two coupled two-dimensional problems.

This chapter deals with the physics of the slow-flow region of the magnetosphere, as defined above. Section II qualitatively describes the inner and middle magnetosphere, in order to provide an overview. Section III reviews the theory of adiabatic drift in the magnetosphere, including bounce-averaged drift. Section IV is devoted to some aspects of the theory of ionosphere-magnetosphere coupling; first I derive the relationship between magnetospheric particle pressure and magnetic-field-aligned current, both from the theory of particle drifts and from magnetohydrodynamics; then I consider the ionospheric end of the circuit, reviewing the theory of height-integrated ionospheric currents, and finally deriving the basic equation of global electrodynamic coupling between the earth's magnetosphere and ionosphere. Section V discusses some simple analytic models of convection in the slow-flow part of the magnetosphere. Section VI describes results of computer models of magnetospheric convection, and Section VII deals with two related theoretical problems: the theory of particle loss from the inner magnetosphere and the magnetic-field modeling problem.

Aside from the qualitative-overview section, the emphasis in this chapter is highly theoretical. Its main purpose is to derive with some care the basic physical equations governing the large-scale electric fields, electric currents, plasma distributions and magnetic fields in the region being considered. I will also illustrate how basic observed characteristics of the system can be deduced from the equations. In the interest of coherently developing one line of theoretical reasoning, while keeping the length reasonable, I have unfortunately had to make important sacrifices. Some major topics in the large-scale theory of the inner magnetosphere are not covered (most notably the theory of L-shell diffusion of particles in the Van Allen belts). Also, the reader is not given historical information on how discoveries were made or how concepts were developed. I apologize to all the researchers in the field whose contributions are not acknowledged.

II. QUALITATIVE OVERVIEW

A. Plasma Regions

Figure 2 shows an equatorial-plane view of the main plasma regions of the inner and middle magnetosphere. The boundaries between regions tend to be approximately aligned with the magnetic field. The definitions of the regions are imprecise, due partly to the fact that there are no completely sharp physical distinctions that can be drawn between the different plasma regions.

The plasma sheet extends far into the tail of the magnetosphere, and is most intense near the equatorial plane. The electron number density n_e is typically 0.1 to 1 cm^{-3}. The mean electron energy is typically ~ 1 keV, the mean ion energy ~ 5 keV. The ions are primarily H^+, but with significant admixtures of O^+, particularly near the earth. The earthward edge of the plasma-sheet electrons is often fairly well defined and lies at 8-10 R_E geocentric distance, on the average. The inner edge of plasma-sheet ions frequently is less well defined. The plasma sheet moves closer to the earth during magnetospheric substorms. Inner-plasma-sheet particles do not circle the earth on trapped orbits. Rather, they seem to come from the tail and flow past the earth to the dayside magnetopause. However, plasma-sheet flows in the magnetotail are observed to be quite irregular, with the formation of eddies. Thus systematic sunward flow is frequently not obvious, and strong tailward flows also occur, especially in substorms.

The trapped-radiation belts (the "Van Allen belts") consist of particles that circle the earth on trapped orbits with geocentric radii $\lesssim 8\ R_E$. They have energies ranging from keV to hundreds of MeV. (The

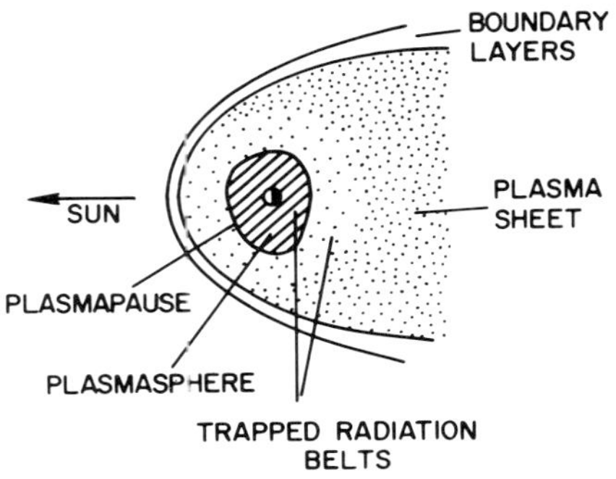

Figure 2. Major plasma regimes of the inner and middle magnetosphere. The more densely dotted region is the plasma sheet, which consists of plasma in the kilovolt range and is separated by an indistinct boundary from the trapped radiation belts (sparsely dotted). The lined region is the plasmasphere, made up of cold ionospheric plasma.

lower energy limit is arbitrary and there is no universally accepted value.) For the higher-energy ions, H^+ is the dominant species, but other components are present. For the lower energy ions, O^+ (apparently of ionospheric origin) is important and frequently dominates the H^+. Among the higher-energy particles especially, the orbits are nearly circular and quite stable, and individual particles can remain trapped for long periods, years in some cases. However, time-varying electric and magnetic fields gradually diffuse the trapped particles radially, and in pitch angle and energy. The theory of this diffusion will not be covered here, but is discussed in detail in books by Schulz and Lanzerotti (1974) and by Roederer (1970).

The plasmasphere is a region of relatively dense ($n_e \gtrsim 10\ cm^{-3}$), cold (~ eV) plasma in the inner magnetosphere. The plasma appears to be of ionospheric origin, and it rotates with the earth. The outer boundary of the plasmasphere, called the "plasmapause," is frequently quite sharp, normally involving a density drop of more than one order of magnitude. The plasmapause tends to be closer to the earth in active times than in quiet times. The region occupied by the plasmasphere also contains trapped-radiation-belt particles. The distinction between the two (both indistinct and arbitrary) is one of particle energy. The same region contains a high density of low-energy plasmaspheric particles that corotate with the earth, and also a much lower density of energetic Van-Allen belt particles, which mainly gradient/curvature drift around the earth.

The trapped-radiation belts and plasmasphere lie in the "inner magnetosphere." The slow-flow part of the plasma sheet will be referred to as "middle magnetosphere."

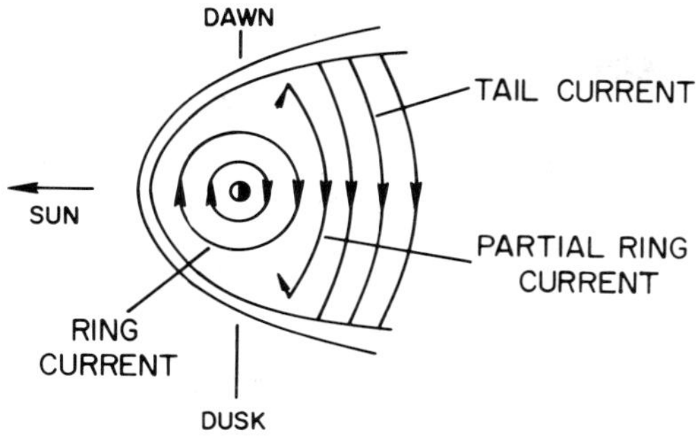

Figure 3. Electric currents in the inner and middle magnetosphere. The partial ring currents connect to Birkeland currents (magnetic-field-aligned currents), which flow to and from the ionosphere. These partial ring currents can exist in both the ring-current and tail-current regions.

B. Electric Currents

Figure 3 shows the major electric current systems in the slow-flow region of the magnetosphere.

The cross-tail current flows across the tail from the dawn magnetopause to the dusk magnetopause. The current is carried by plasma-sheet particles. The total current carried at geocentric distances from 10 to 60 R_E is typically $\sim 6 \times 10^6$ amperes. The cross-tail current is often called the "neutral sheet current," because the magnetic field in its center is much weaker than the magnetic field outside it. However, the field in the center of the sheet is seldom zero and has a systematic northward component in the inner, slow-flow region of the magnetosphere. (Regions with zero or southward equatorial magnetic field exist, but are defined here to be in the outer magnetosphere. The theoretical methods to be discussed in this chapter do not apply to the rapid flows that typically occur near such regions. See chapter by V. M. Vasyliunas.)

The ring current flows in rings that encircle the earth at 2-10 R_E geocentric distance. The total magnitude of the ring current is $\sim 10^6$ A in quiet times, perhaps ten times larger in intense geomagnetic storms. The current is carried mainly by particles of the trapped radiation belts. The "ring-current particles," i.e., the particles that carry most of the ring current, are mainly positive ions in the range 3-300 keV.

The term "partial ring current" is used to describe inner- and middle-magnetospheric currents that neither flow in closed loops near the equatorial plane nor flow into or out of the magnetopause. Rather, they connect to magnetic-field-aligned (Birkeland) currents that flow up from or down to the conducting ionosphere. These partial ring currents can exist in the same regions of the magnetosphere as the ring currents and cross-tail currents. They in fact represent a non-zero divergence of current due to gradient/curvature drift.

The Birkeland currents flow along magnetic field lines, connecting the magnetosphere to the ionosphere. As discussed elsewhere in this volume (e.g., Figure 4 of the chapter by T. W. Hill), there are two major regions of Birkeland current. Region-1 currents, the more poleward set, reach far out in the magnetosphere and connect to the "generator" of the convection circuit. The region-2 currents, the more equatorward of the currents, apparently connect mainly to the near-earth part of the plasma sheet. They flow from the dawnside ionosphere out into the magnetosphere, and from the duskside magnetosphere down to the ionosphere. It is not clear yet to what extent, if any, region-1 Birkeland currents connect to the slow-flow region of the magnetosphere.

C. Dynamics of the Inner Magnetosphere

The important types of large-scale magnetospheric plasma motion are the following: 1. gradient/curvature drift, which is dominant for the trapped radiation belts; 2. corotation with the earth, which dominates in the plasmasphere; 3. sunward convection, which is dominant in the inner plasma sheet and which affects all regions because of its variability; 4. turbulent motion, which is a major effect throughout the plasma sheet. The rotation periods for gradient/curvature drift motions range from hours (for typical ring-current particles) to minutes for the high-energy particles of the radiation belts. The convection electric field is typically ~ 10^{-3} V m^{-1} in the strong-flow regions of the magnetosphere, but is typically weaker earthward of the inner edge of the plasma sheet. An electric field of 10^{-3} V m^{-1} corresponds to a flow velocity of the order of 30 km s^{-1} at 10 R_E geocentric distance, ~ 7 km s^{-1} at 6 R_E. The time required for a typical flux tube to convect through the middle magnetosphere, from tail to dayside magnetopause, is of the order of an hour or two. Further out in the plasma sheet, convection velocities are larger, and the flow is quite turbulent. Near the inner edge of the plasma sheet, the velocities involved in MHD waves with periods of the order of minutes are typically somewhat larger than the convection velocities, although they presumably average to zero over a convection time scale. These MHD waves are discussed elsewhere in this volume (chapter by W. J. Hughes).

Particle transfer between ionosphere and magnetosphere occurs along magnetic field lines. Plasma-sheet electrons precipitate into the ionosphere and cause aurora. Electron precipitation lifetimes tend to be comparable to convection times. Thus a substantial fraction of plasma sheet electrons can be lost from a given flux tube as it convects through the middle magnetosphere. Because ions move more slowly, they tend to have considerably longer precipitation lifetimes. Beams of ions are also observed moving upward along auroral-zone magnetic field lines from the ionosphere to the magnetosphere. These upward-streaming ions, which are apparently of ionospheric origin, have typically been accelerated to energies of a few hundred electron volts. However, energies as high as 5-10 keV are sometimes observed.

Although the central source of the magnetospheric-substorm phenomenon is in the magnetotail, outside the slow-flow region of the magnetosphere, substorm effects are observed deep in the magnetosphere. Convection tends to increase, and the plasma sheet tends to move earthward during substorm activity. Near the onset time of the substorm, plasma-sheet magnetic field lines near local midnight become less tail-like, more dipolar, and fresh plasma is injected into the region near 7 R_E geocentric distance. This plasma then drifts around the earth, and some of it becomes trapped on closed, circular orbits. Dramatic increases are observed in ionospheric currents, particularly in the westward electrojet, which flows across the nightside auroral ionosphere. (For detailed descriptions, see the book by Akasofu (1979)).

The main phase of a magnetic storm is a succession of strong magnetospheric substorms, resulting in injection of many particles onto trapped orbits. The total energy contained in the trapped particles at 3-7 R_E greatly increases, causing a proportionate increase in ring current strength, and a depression in the northward magnetic field observed at the earth's surface at low latitudes. The high radiation-belt population levels gradually subside in the recovery phase of the storm, which lasts a few days. A magnetic storm often (but not always) begins with a "sudden commencement," which is a sudden compression of the magnetosphere due to increased solar-wind ram pressure, ρV^2; the sudden commencement is observed as a sudden increase in the northward magnetic field at the earth's surface, at low latitude. The storm main phase typically occurs after the sudden commencement, sometimes with a delay of several hours.

III. ADIABATIC INVARIANTS AND ADIABATIC DRIFTS

A. Adiabatic Invariants

A charged particle moving through the magnetosphere executes three motions: (i) gyration about the magnetic field line, (ii) bounce motion along the field line, and (iii) slow but systematic drift perpendicular to $\underset{\sim}{B}$. For particles in the 100 eV-100 keV range, which is what we shall consider, these motions occur on three different time scales. The gyration is the fastest, occurring at the gyro-frequency (also called the Larmor frequency or cyclotron frequency)

$$\omega_g = \frac{qB}{m} \approx (9.6\ \text{sec}^{-1})\ (\frac{q}{e})\ \frac{(B/100\gamma)}{(m/m_H)} \tag{1}$$

where e = magnititude of the electron charge, m_H = mass of hydrogen atom, and $1\gamma \equiv 1$ nanotesla = 10^{-9} tesla = 10^{-5} gauss. ($B \equiv 31{,}000\ \gamma$ at the earth's surface at the equator, about 31γ at 10 R_E if the field is dipolar). So $\omega_g \sim$ kilocycles for ions at low altitudes, megacycles for the lighter electrons. At ≈ 10 R_E, roughly the outer limit of the "inner magnetosphere," $\omega_g \sim 3\ \text{sec}^{-1}$ for hydrogen, $\sim 0.2\ \text{sec}^{-1}$ for O^+, $\sim 5000\ \text{sec}^{-1}$ for electrons. The gyro-radius is given by

$$r_g = \frac{mv_\perp}{|q|B} \approx (45\ \text{km})\ \frac{(m/m_H)^{1/2}}{|q|/|e|}\ \frac{\left(\frac{E_\perp}{1\ \text{keV}}\right)^{1/2}}{\left(\frac{B}{100\gamma}\right)} \tag{2}$$

where $v_\perp$ = velocity of gyrational motion perpendicular to $\underset{\sim}{B}$. The gyro-radius is typically several orders of magnitude smaller than magnetospheric dimensions, and is smaller for electrons than for ions of the same energy.

The particle's pitch angle α is defined to be the angle between $\underset{\sim}{v}$ and $\underset{\sim}{B}$, namely

$$\alpha = \tan^{-1}\left(\frac{v_\perp}{v_\parallel}\right) \tag{3}$$

The particle's motion parallel to $\underset{\sim}{B}$ consists of bouncing between mirror points, where $B = E/\mu$ and $\alpha = \pi/2$. (See Figure 4.) The period τ_B of the bounce motion is given by

$$\tau_B = 2\int \frac{ds}{v_\parallel} = \sqrt{2}\,\ell_b\, m^{1/2}\langle E_\parallel\rangle^{-1/2} \simeq (290\text{s})\left(\frac{\ell_b}{10\ R_E}\right)\left(\frac{m}{m_H}\right)^{1/2}\left\langle\frac{1\ \text{keV}}{E_\parallel}\right\rangle^{1/2} \tag{4}$$

where $v_\parallel$ is the particle velocity along $\underset{\sim}{B}$, ℓ_b is the distance along the field line between mirror points, and the brackets indicate an average over the length of the field line. Note that, from (1)-(3),

$$\omega_g\tau_B \sim \frac{2\ell_b}{r_g}\left(\frac{v_\perp}{v_\parallel}\right) \gg 1. \tag{5}$$

The third type of motion that is involved here is drift perpendicular to $\underset{\sim}{B}$. Elsewhere in this volume (chapter by G. L. Siscoe), the formula for the $\underset{\sim}{E} \times \underset{\sim}{B}$, gradient, and curvature drift of a charged particle was derived. It can be written in the form

$$\underset{\sim}{v}_D = \frac{\underset{\sim}{E} \times \underset{\sim}{B}}{B^2} + \frac{mv_\perp^2\ \underset{\sim}{B} \times \nabla B}{2qB^3} + \frac{mv_\parallel^2\ \hat{R}_c \times \underset{\sim}{B}}{qR_cB^2}, \tag{6}$$

where R_c = radius of curvature of the magnetic field line, and $\hat{R}_c$ = unit vector outward from the center of curvature. (In general, I will use a

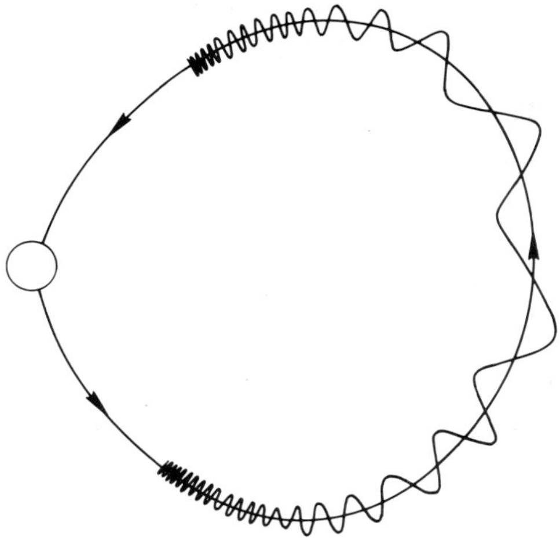

Figure 4. Qualitative sketch of the bounce motion of a charged particle on an inner-magnetospheric field line.

symbol ^ to designate a unit vector.) The time scale associated with gradient/curvature drift is given by

$$\tau_D^{-1} \sim \frac{v_D}{r} \sim \frac{mv^2}{rqRB}, \tag{7}$$

where r = geocentric distance and R = some scale length associated with the magnetic field ($B/|\nabla B|$ or R_c). Combining (5) and (7) gives

$$\frac{\tau_B}{\tau_D} \sim \left(\frac{2\ell_b}{v}\right) \frac{mv^2}{qBrR} = r_g \left(\frac{2\ell_b}{rR}\right) \sim \frac{r_g}{r} \tag{8a}$$

the last order-of-magnitude inequality being based on the assumption that, typically, $\ell_b \sim r \sim R$. Comparing (5) and (8a), we see that the ratio of bounce time to cyclotron period is about the same as the ratio of drift time to bounce time. There is a hierarchy of motions, with typically 2-3 orders of magnitude separating their time scales.

Note that τ_D in (8) refers to the time associated with gradient/curvature drift, not $\underset{\sim}{E} \times \underset{\sim}{B}$ drift. It is clearly the appropriate time scale for Van-Allen-belt particles, whose gradient/curvature drift exceeds their $\underset{\sim}{E} \times \underset{\sim}{B}$ drift. It is correct in order of magnitude for most inner-plasma-sheet particles, for which the two drifts tend to be comparable. In general, the essential condition

$$\tau_B << \tau_{E\times B \text{ drift}} \tag{8b}$$

holds if the drift velocity is small compared to the thermal velocity. In the slow-flow region, where the drift velocity is defined to be small compared to the fast-mode speed ~ $[B^2/(\mu_0\rho) + \gamma p/\rho]^{1/2}$, (8b) then holds for any particle whose thermal velocity is greater than, or comparable to, the fast-mode speed. In other words, (8b) holds for practically all electrons and for the kinds of ions that make important contributions to the total density and the total pressure $p + B^2/2\mu_0$.

The usual wide separation of the gyro, bounce, and drift time scales allows the convenient use of simple adiabatic invariants. Specifically, since the bounce and drift motions are so slow compared to the gyrational motion, the adiabatic invariant associated with the gyrational motion is conserved to high accuracy, if there are no other important forces involved other than that due to a steady $\underset{\sim}{B}$ field. The theory of adiabatic invariants implies that, for a position variable Q for which the motion is cyclic, the integral $\oint P\,dQ$ over a cycle is an adiabatic invariant. (Here $P \equiv \partial L/\partial\dot{Q}$ = momentum conjugate to Q. The general theory of adiabatic invariants is given, for example, by Landau and Lifshitz (1960) or Rossi and Olbert (1970).) To derive the adiabatic invariant for simple cyclotron motion for a uniform $\underset{\sim}{B} \parallel \hat{z}$, consider motion that is cyclic in x and y. Let $\underset{\sim}{A} = xB\hat{y}$ and consider $\oint P_x dx$. The Lagrangian is given by (e.g., Goldstein, 1959)

$$L = \frac{1}{2} m\dot{\underset{\sim}{x}}^2 + q\,\dot{\underset{\sim}{x}} \cdot \underset{\sim}{A}$$

$$= \frac{1}{2} m(\dot{x}^2 + \dot{y}^2 + \dot{z}^2) + q\dot{y}xB$$

and thus

$$P_x = \partial L/\partial\dot{x} = m\dot{x}$$

We thus find, for the adiabatic invariant

$$\oint P_x dx = m \int \left(\frac{dx}{dt}\right)^2 dt = m\,\frac{(\omega_g r_g)^2}{2} \times \frac{2\pi}{\omega_g} = \frac{\pi m^2 v_\perp^{\,2}}{eB} \tag{9}$$

(The other invariant $\oint P_y\,dy$ turns out to be zero, providing no new information.) Invariance of $\oint P_x\,dx$ thus implies that the ratio $v_\perp^{\,2}/B$ is an adiabatic invariant, as is the magnetic moment

$$\mu = \frac{mv_\perp^{\,2}}{2B} \tag{10}$$

This "first adiabatic invariant" was derived elsewhere in this volume (chapter by G. L. Siscoe) by a different set of arguments. Note that the perpendicular energy increases as B increases. This is the phenomenon of cyclotron acceleration.

Because the drift motion is so slow compared to the bouncing, it is also useful to consider the adiabatic invariant associated with simple one-dimensional bounce motion. The kinetic energy of the particle, namely

$$\frac{1}{2} mv_{\perp}^2 + \frac{1}{2} mv_{\parallel}^2 = \mu B + \frac{1}{2} mv_{\parallel}^2 = E$$

is a constant, in the absence of an electric field. This corresponds to a Lagrangian

$$L(v_{\parallel}, s) = \frac{1}{2} mv_{\parallel}^2 - \mu B(s) \tag{11a}$$

and

$$P_{\parallel} = \partial L/\partial v_{\parallel} = mv_{\parallel} \tag{11b}$$

where s = distance along the field line and $v_{\parallel} \equiv \dot{s}$. The magnetic field, whose strength varies with position along the field line, acts like a potential for simple one-dimensional particle motion. The adiabatic invariant associated with parallel motion is

$$J = \oint P_{\parallel} ds = \sqrt{2m} \oint \left(E - \mu B(s)\right)^{1/2} ds \tag{12}$$

or

$$J = (8m\mu)^{1/2} I \tag{13}$$

where the geometrical integral

$$I = \int_{s_{m1}}^{s_{m2}} \sqrt{B_m - B(s)}\, ds \tag{14}$$

does not depend on any properties of the particle except the location of the mirror points s_{m1} and s_{m2}. Here $B_m \equiv E/\mu$ = mirror magnetic field, and J is called the second adiabatic invariant. If the length of the bounce path decreases, the average parallel energy increases, which is the basic element of "Fermi acceleration."

For further discussions and derivations of the first and second adiabatic invariants, see the books by Northrup (1963) and Roederer (1970). They also derive a third adiabatic invariant which is applicable if conditions change on a time scale long compared to a drift time. We will not discuss the third invariant here.

B. Bounce-Averaged Drifts

I will primarily be discussing the adiabatic drifts of particles through the inner magnetosphere. The standard formula derived earlier (equation (6)) contains all the essential information, but we need to obtain an expression for the bounce-averaged drift velocity — an appropriately weighted average of equation (6) over the bounce path of the particle. It is convenient to represent this bounce-averaged drift velocity in terms of the motion of the intersection of the bounce path with the magnetic equatorial plane (or some other similar surface). For this discussion, we will, for simplicity, use the magnetic equatorial plane, and pretend that the earth's magnetic dipole is perpendicular to the solar-wind flow direction, as shown in Figure 1. The resulting particle-drift formulae are easily adapted to more complex geometries, but displays and bookkeeping are more difficult.

Consider first the $\tilde{E} \times \tilde{B}$ drift. This case does not require a complicated average over the bounce length. Consider two particles that are initially on the same field line, one that mirrors in the equatorial plane (a minimum in B for our assumed simple geometry), and one that mirrors at some arbitrarily chosen point along the field line. The perfect-conductor assumption,

$$\tilde{E} + \tilde{v} \times \tilde{B} = 0, \tag{15}$$

which is equivalent to the two conditions

$$\tilde{E} \cdot \tilde{B} = 0 \tag{16}$$

and

$$\tilde{v}_{\perp} = \frac{\tilde{E} \times \tilde{B}}{B^2} \tag{17}$$

then implies that any two particles that move at velocity $\tilde{v}(\tilde{x},t)$ and start out on the same field line will forever remain on the same field line. (See chapter by G. L. Siscoe.) That is the essence of the concept that a perfectly conducting fluid is frozen to the magnetic field. Returning now to our two particles, one mirroring in the equatorial plane and $\tilde{E} \times \tilde{B}$ drifting there, and one bouncing up and down between hemispheres while $\tilde{E} \times \tilde{B}$ drifting, it is clear that they should forever share the same field line if they initially shared a field line. An equatorially mirroring particle essentially $\tilde{E} \times \tilde{B}$ drifts the same as a particle mirroring anywhere else on the field line.

The same argument does not hold for gradient and curvature drifts, since these drifts represent violations of the perfect-conductor assumption (15). In the inner magnetosphere, gradient/curvature drift is often comparable to $\tilde{E} \times \tilde{B}$ drift, and may exceed it. We can derive convenient formulas for bounce-averaged gradient/curvature drifts, using conservation of energy and some algebraic manipulations. For a given magnetic-field configuration, the kinetic energy E_K of a particle of charge q is known if the equatorial crossing point $\underset{\sim}{x}_e$ and the adiabatic invariants μ and J are known:

$$E_K = E_K(\underset{\sim}{x}_e, \mu , J). \tag{18}$$

If particles with given μ and J gradient/curvature drift in a static electric and magnetic field configuration, then their total energy (kinetic plus potential) must remain constant, and

$$(\underset{\sim}{v} \cdot \nabla) E = (\underset{\sim}{v}_{EBe} + \underset{\sim}{v}_{GCe}) \cdot \nabla_e \left[qV(\underset{\sim}{x}_e) + E_K(\underset{\sim}{x}_e, \mu , J)\right] = 0 \tag{19}$$

where $V(\underset{\sim}{x}_e)$ = electrostatic potential at $\underset{\sim}{x}_e$, ∇_e = two-dimensional gradient operator in the equatorial plane; the equatorial $\underset{\sim}{E} \times \underset{\sim}{B}$ drift velocity is given by

$$\underset{\sim}{v}_{EBe} = (-\nabla_e V_e) \times \underset{\sim}{B}_e / B_e^2 \tag{20}$$

and $\underset{\sim}{v}_{GCe}$ = bounce-averaged gradient/curvature drift velocity, expressed in terms of the rate of motion of the particle's equatorial crossing point. Since (19) must hold for any V, including $V(\underset{\sim}{x}_e) = 0$, we can apply (20) to (19) and obtain

$$\underset{\sim}{v}_{GCe} \cdot \nabla_e E_K(\underset{\sim}{x}_e,\mu,J) = 0 \tag{21}$$

or

$$\underset{\sim}{v}_{GCe} = C\hat{z} \times \nabla_e E_K(\underset{\sim}{x}_e,\mu,J) \tag{22}$$

where C is an as-yet-unknown scalar and $\hat{z}$ = unit vector normal to the equatorial plane. Subtracting (21) from (19) and using (20) gives

$$\underset{\sim}{v}_{EBe} \cdot \nabla_e E_K(\underset{\sim}{x}_e,\mu,J) + \underset{\sim}{v}_{GCe} \cdot \nabla_e\left(qV(\underset{\sim}{x}_e)\right) = 0$$

Using (20) and (22), and the fact that $\underset{\sim}{B}(\underset{\sim}{x}_e) = |\underset{\sim}{B}(\underset{\sim}{x}_e)|\hat{z}$, we obtain

$$C\hat{z} \times \nabla_e E_K(\underset{\sim}{x}_e,\mu,J) \cdot \nabla_e V(\underset{\sim}{x}_e) = \frac{1}{q|\underset{\sim}{B}(\underset{\sim}{x}_e)|} \hat{z} \times \nabla_e E_K(\underset{\sim}{x}_e,\mu,J) \cdot \nabla_e V(\underset{\sim}{x}_e) \tag{23}$$

Since this must hold for arbitrary $\nabla_e V(\underset{\sim}{x}_e)$, it must be that $C = \left[qB(\underset{\sim}{x}_e)\right]^{-1}$ and, from (22),

$$\underset{\sim}{v}_{GCe} = \frac{\hat{z} \times \nabla_e E_K(\underset{\sim}{x}_e,\mu,J)}{qB(\underset{\sim}{x}_e)} \tag{24}$$

(This formula is also derived in the books by Northrup (1963) and Roederer (1970).) The quantity $E_K(\underset{\sim}{x}_e,\mu,J)/q$ acts like an effective potential from the viewpoint of bounce-averaged gradient/curvature drift. If only gradient/curvature drift is operating (no $\underset{\sim}{E} \times \underset{\sim}{B}$ drift), then the particle drifts along curves of constant kinetic energy E_K. The only difficulty with this useful and general formula lies in evaluating the function $E_K(\underset{\sim}{x}_e,\mu,J)$ for a complicated magnetic-field configuration. One simple example is available, however, namely the case of $J = 0$. In that case $E_K = \mu B$, and (24) becomes

$$\underset{\sim}{v}_{GCe}\ (J = 0) = \frac{\mu\ \hat{z} \times \nabla_e B(\underset{\sim}{x}_e)}{qB(\underset{\sim}{x}_e)}$$

This agrees exactly with the gradient/curvature part of (6) for the case where $v_\parallel = 0$ (which must hold if $J = 0$). For particles with $J = 0$, which simply mirror in the equatorial plane, the bounce-averaged formula (24) reduces to the local formula (6).

The total bounce-averaged drift velocity of particles in a static electric and magnetic-field configuration is obtained by adding (20) and (24)

$$\underset{\sim}{v}_e = \frac{\hat{z} \times \nabla_e \left[E_K(\underset{\sim}{x}_e,\mu,J) + q\ V(\underset{\sim}{x}_e)\right]}{qB(\underset{\sim}{x}_e)} \tag{25}$$

The particles follow contours of constant total energy (kinetic + potential).

Figure 5 sketches drift paths through the magnetosphere for two kinds of particles - one with $\mu = J = 0$, corresponding to simple $\underset{\sim}{E} \times \underset{\sim}{B}$ drift in a simple dawn-to-dusk electric field, and one with high μ and J. The contours of constant energy avoid the earth in the latter case, because the kinetic energy $E_K(\underset{\sim}{x}_e, \mu, J)$ gets large for $\underset{\sim}{x}_e$ near the earth. Effects of the earth's rotation have been neglected here. Trapped radiation-belt particles follow the closed loop trajectories, while plasma-sheet particles generally follow open trajectories. Formulas (24) and (25) have been extensively applied to studies of radiation-belt particles, as discussed by Roederer (1970).

We now demonstrate a sort of frozen-in-flux theorem for particles that gradient/curvature drift as well as $\underset{\sim}{E} \times \underset{\sim}{B}$ drift. Specifically, we wish to show that particles that drift according to a law of the form

$$\underset{\sim}{v}_e = \frac{\underset{\sim}{E}_e \times \underset{\sim}{B}_e}{B_e^{\,2}} + \frac{\nabla_e Y \times \underset{\sim}{B}_e}{B_e^{\,2}} \tag{26}$$

conserve the parameter η, the number of particles per unit magnetic flux. To show this, we write the law of conservation of particles, mapped to the equatorial plane, in the form

$$\frac{\partial(\eta B_e)}{\partial t} + \nabla_e \cdot (\eta B_e \underset{\sim}{v}_e) = 0 \tag{27}$$

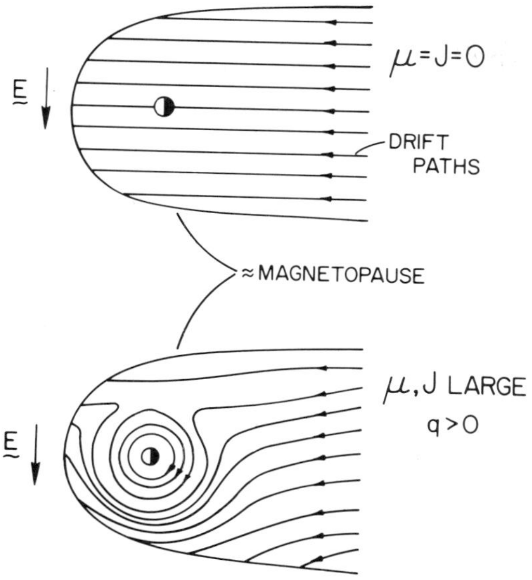

Figure 5. Qualitative sketch of particle drift paths through a static magnetosphere. The assumed electric field is from dawn to dusk (downward in diagram). Effects of the earth's rotation are neglected. The top diagram pertains to electrons or ions with zero thermal energy ($\mu = J = 0$). The bottom diagram pertains to positive ions with ($\mu > 0$, $J > 0$).

Note that ηB_e is the number of particles per unit area in the equatorial plane. Note also that we have assumed that no particles have been lost. Rearranging (27) and using (26) gives

$$B_e \left(\frac{\partial}{\partial t} + \underset{\sim}{v}_e \cdot \nabla_e\right)\eta + \eta \frac{\partial B_e}{\partial t} + \eta\nabla_e \cdot \left[\underset{\sim}{E}_e \times \hat{z} + \nabla_e Y \times \hat{z}\right] = 0$$

Faraday's law requires that the $\eta\partial B_e/\partial t$ and $\eta\nabla_e \cdot (\underset{\sim}{E}_e \times \hat{z})$ terms cancel, and we obtain, finally

$$\left(\frac{\partial}{\partial t} + \underset{\sim}{v}_e \cdot \nabla_e\right)\eta = 0 \tag{28}$$

That is, η, the number of particles per unit magnetic flux, remains constant along the drift path of a particle.

We have so far discussed gradient and curvature drift for the case where the adiabatic invariants μ and J are conserved. There is another convenient approximation that is often used, namely the approximation of isotropic pitch angles. It is a very elegant approximation, because it makes the distribution function describing particles on a given field line a function only of energy and not of either pitch angle or location on the field lines. (See chapter by G. L. Siscoe for a discussion of the distribution function and Liouville's theorem.) We thus assume that wave-particle interactions elastically scatter particle pitch angles, on a time scale short compared to a drift time, to keep the particle distribution isotropic. This approximation, though an over-simplification, is reasonable for the plasma sheet, since observed particle distributions usually are approximately isotropic there (Stiles et al., 1978). It is a less justifiable approximation for trapped radiation-belt particles. We also neglect the loss of heat or particles from the flux tube. The particles behave like a monatomic, isotropic, perfect gas that is confined by the magnetic field. Its response to adiabatic compression is described by $\gamma = 5/3$, and its kinetic energy per particle is given by

$$E_K(\underset{\sim}{x}_e, \lambda) = \lambda \left[\Xi(\underset{\sim}{x}_e)\right]^{-2/3} \tag{29}$$

where λ is constant along a drift path, and $\Xi(\underset{\sim}{x}_e)$ is an effective confining volume; specifically, consider a group of monoenergetic particles populating a certain flux tube, with equatorial crossing point $\underset{\sim}{x}_{eo}$ at $t = 0$; let the initial volume occupied by the particles be $\Xi(\underset{\sim}{x}_{eo})$; if they then drift to a flux tube with equatorial crossing point $\underset{\sim}{x}_e$, they occupy volume $\Xi(\underset{\sim}{x}_e)$. Of course, we now know the relationship between gradient/curvature drift and E_K. By the same argument that led to (24), we have

$$\underset{\sim}{v}_{GCe} = \frac{\hat{z} \times \nabla_e E_K(\underset{\sim}{x}_e, \lambda)}{q B_e(\underset{\sim}{x}_e)} \tag{30}$$

Since the right side of (30) has exactly the form of the second term on the right side of (26), we conclude that η is conserved for these isotropic particles. Since η, the number of particles per unit of magnetic flux, is conserved along a drift path, then if particles are conserved, the magnetic flux contained in the confining volume must also be conserved along a trajectory. If we define our initial confining volume to correspond to one unit of magnetic flux, then the effective confining volume $\Xi(\underset{\sim}{x}_e)$ is simply the flux tube volume per unit magnetic flux, namely $\int ds/B$, and we have

$$E_K(\underset{\sim}{x}_e, \lambda) = \lambda \left[\int ds/B\right]^{-2/3} \tag{31}$$

where the integral extends over the length of the flux tube, from the southern ionosphere to the northern. The coefficient λ is the energy invariant for these particles that suffer strong pitch-angle scattering but no change in energy. The formula for gradient/curvature drift of these isotropic-pitch-angle particles is then

$$\underset{\sim}{v}_{GCe} = \frac{\lambda \hat{z} \times \nabla_e [\int ds/B]^{-2/3}}{qB(\underset{\sim}{x}_e)} \tag{32}$$

We should remark that the formulas for bounce-averaged gradient/curvature drift, namely (24) and (32), are valid for a slowly varying magnetic-field configuration, as well as time-independent magnetic fields. Although the conservation-of-energy argument used in deriving these formulae utilized the static equation $\underset{\sim}{E} = -\nabla_e V$, that is not a condition on the validity of (24) and (32). These formulae can also (with considerable labor) be derived by integrating the gradient/curvature terms of (6) appropriately over a flux tube. Viewed in this way, it is clear that the correct formula for $\underset{\sim}{v}_{GCe}$ cannot depend on the rate of change of $\underset{\sim}{B}$ but only on the instantaneous magnetic field configuration.

IV. IONOSPHERE-MAGNETOSPHERE COUPLING

This section will be devoted to the theory of the coupling between the inner magnetosphere and the ionosphere by means of Birkeland currents (i.e., currents flowing along the magnetic field lines). First, we derive the relationship between the Birkeland current and the pressure distribution in the magnetospheric plasma, from the theory of bounce-averaged adiabatic drift developed in the previous section. Then we derive the same relationship from magnetohydrodynamics. Finally, we derive the relationship between Birkeland currents and ionospheric electric fields, thereby arriving at the basic equation that governs the coupling between ionosphere and magnetosphere.

A. Pressure/Birkeland-Current Relation (From Drift Theory)

What we basically wish to do here is to use the adiabatic-drift results derived in Section III to calculate the total current flowing across field lines in the magnetosphere, and to equate the divergence of this total current to the Birkeland current.

The general expression for currents due to adiabatic drifts consists of the following terms: $\underset{\sim}{E} \times \underset{\sim}{B}$ drift, polarization (or acceleration) drift, gradient/curvature drift, and external-force drift. The term due to $\underset{\sim}{E} \times \underset{\sim}{B}$-drift is essentially a relativistic correction, and is negligible for conditions in the earth's magnetosphere. (See chapter by G. L. Siscoe.) The polarization-drift term is neglected because of the slow-flow approximation. The external-force drift term is negligible, since the gravitational force is normally at least two orders of magnitude smaller than the centrifugal and magnetic-field-gradient forces, for particles above 100 eV energy. To the remaining gradient/curvature-drift current, we must add magnetization currents $\nabla \times \underset{\sim}{M}$ in order to get total current, but the magnetization current does not contribute to the Birkeland current, because it is divergenceless. Thus for calculating Birkeland currents connecting to the slow-flow part of the magnetosphere, we need only consider gradient- and curvature-drift currents.

In line with our procedure of keeping track of equatorial crossing points of drifting particles, we visualize the gradient/curvature-drift currents as flowing across field lines in the equatorial plane, though in reality they flow across field lines at various distances from that plane.

Applying (32) to various particle species s, we find that the gradient/curvature-drift current per unit length in the equatorial plane is

$$\underset{\sim}{j}_{GCe}(x_e) = \sum_s \eta_s(\underset{\sim}{x}_e)\, B_e(x_e)\, q_s \left[\frac{\lambda_s\, \hat{z} \times \nabla_e \{\int ds/B\}^{-2/3}}{q_s\, B_e(\underset{\sim}{x}_e)}\right]$$

$$= \sum_s \eta_s(\underset{\sim}{x}_e)\lambda_s\, \hat{z} \times \nabla_e \{\int ds/B\}^{-2/3} \qquad (33)$$

where $\eta_s(\underset{\sim}{x}_e)$ = number of particles of species s per unit magnetic flux at $\underset{\sim}{x}_e$, so that $\eta_s(\underset{\sim}{x}_e)\, B_e(\underset{\sim}{x}_e)$ = number of of particles per unit area in the equatorial plane. The Birkeland current per unit area away from the magnetosphere toward the ionosphere, but expressed as a northward current per unit area, out of the equatorial plane, is as follows:

$$J_{\| e} = -\frac{1}{2} \nabla_e \cdot \underset{\sim}{j}_{GCe}(\underset{\sim}{x}_e)$$

$$= -\frac{1}{2} \nabla_e \left[\sum_s \eta_s(\underset{\sim}{x}_e)\, \lambda_s\right] \cdot \hat{z} \times \nabla_e \{\int ds/B\}^{-2/3}$$

$$= \frac{1}{2} \hat{z} \cdot \nabla_e \left[\sum_s \eta_s(\underset{\sim}{x}_e)\, \lambda_s\right] \times \nabla_e \left[\int ds/B\right]^{-2/3} \qquad (34)$$

The factor of 1/2 accounts for the fact that the total Birkeland current generated in the magnetosphere is divided between northern and southern hemispheres. For simplicity, we are assuming an even split between hemispheres, but the theory can be generalized to account for unequal conductivities in the two hemispheres, for example.

Equation (34) has one simple and remarkable implication. Recall that we showed in Section III that η_s is constant along the drift path of a particle of species s, and λ_s is a constant. Thus $\sum_s \eta_s(\underset{\sim}{x}_e)\lambda_s$ is the same for any two flux tubes that have equivalent plasma populations — differing from each other only by adiabatic expansion or compression. Equation (34) thus implies that no Birkeland currents flow into or out of a region of the magnetosphere where all the flux tubes have equivalent plasma distributions.

We can also rewrite (34) in another form that has a more magnetohydrodynamic appearance. The particle pressure in the flux tube with equatorial crossing point $\underset{\sim}{x}_e$ is given by

$$p = \frac{2}{3} \sum_s n_s E_{Ks} \tag{35}$$

where n_s and E_{Ks} are number density and average energy for species s. We thus have

$$p(\underset{\sim}{x}_e) = \frac{2}{3} \sum_s \left[\frac{\eta_s}{\int ds/B}\right] \left[\lambda_s \left(\int ds/B\right)^{-2/3}\right] \tag{36}$$

and (34) becomes

$$J_{\| e} = -\frac{1}{2} \hat{z} \cdot \nabla_e p(\underset{\sim}{x}_e) \times \nabla_e \left[\int ds/B\right]. \tag{37}$$

Mapping the Birkeland current density $J_{\| e}$ down along field lines to the ionosphere gives, for the current per unit area perpendicular to $\underset{\sim}{B}$, down into the northern ionosphere,

$$J_{\| i}(\underset{\sim}{x}_i) = J_{\| e}(\underset{\sim}{x}_e) \left[(B_i(\underset{\sim}{x}_i))/(B_e(\underset{\sim}{x}_e))\right] \tag{38}$$

where $\underset{\sim}{x}_i$ and $\underset{\sim}{x}_e$ refer to the ionospheric and equatorial crossing points of the same field line and $B_i(\underset{\sim}{x}_i)$ is the magnetic-field strength at $\underset{\sim}{x}_i$.

B. Pressure/Birkeland-Current Relation from MHD

We now present an alternative derivation of equation (37), which is the fundamental relationship between particle pressure in the inner magnetosphere and Birkeland current. This time, instead of starting from a formula for bounce-averaged drift, we will start from the isotropic-pressure form of the MHD momentum equation:

$$\rho \frac{d\underset{\sim}{V}}{dt} = -\nabla p + \underset{\sim}{J} \times \underset{\sim}{B} \tag{39}$$

Crossing (39) with $\underset{\sim}{B}$, we can solve for the component of the current-density vector perpendicular to $\underset{\sim}{B}$:

$$\underset{\sim}{J}_\perp = \frac{\underset{\sim}{B} \times \nabla p}{B^2} + \frac{\underset{\sim}{B} \times \left(\rho \frac{d\underset{\sim}{V}}{dt}\right)}{B^2} \tag{40}$$

Interpreted in terms of particle drifts, the first term in (40) represents the sum of magnetization and gradient/curvature-drift current. The last term in (40) represents acceleration drift or polarization drift.

Conservation of current implies that

$$B \frac{\partial (J_{\parallel}/B)}{\partial s} + \nabla \cdot \underset{\sim}{J}_{\perp} = 0 \tag{41}$$

where $J_{\parallel} \equiv \underset{\sim}{J} \cdot \hat{b}$, and $\hat{b}$ is the unit vector along $\underset{\sim}{B}$. Integrating over a flux tube with unit magnetic flux from southern ionosphere (s) to northern ionosphere (n), we find that

$$\left(\frac{J_{\parallel}}{B_i}\right)_n - \left(\frac{J_{\parallel}}{B_i}\right)_s = -\int \frac{ds}{B} \nabla \cdot \underset{\sim}{J}_{\perp}$$

$$= -\int \frac{ds}{B} \nabla \cdot \left(\frac{\underset{\sim}{B} \times \nabla p}{B^2}\right) - \int \frac{ds}{B} \nabla \cdot \left[\frac{\underset{\sim}{B} \times \rho \frac{d\underset{\sim}{V}}{dt}}{B^2}\right] \tag{42}$$

Assuming as before that equal amounts of current go to each hemisphere, the left side can be written $2J_{\parallel i}/B_i$ where now $J_{\parallel i}$ refers to the northern hemisphere. Within the slow-flow region, where we can neglect the inertial ($\rho\, d\underset{\sim}{V}/dt$) term in (39), we can drop the last term in (42). Also, according to (39), or Liouville's theorem (chapter by G. L. Siscoe), we can take p to be a constant along a magnetic field line. From this p = constant condition and the assumption that the magnetic field B_i at the ionospheric end of the field line is very large compared to B near the equatorial plane, one can, with some effort, prove from simple vector calculus that

$$\int \frac{ds}{B} \nabla \cdot \left(\frac{\underset{\sim}{B} \times \nabla p}{B^2}\right) = \frac{1}{B} \hat{b} \cdot \nabla p \times \nabla\left[\int ds/B\right] \tag{43}$$

The right side of the equation can be evaluated at any point on the field line (p and $\int ds/B$ are, of course, constants along the field line). Expressed in the equatorial plane, (42) takes the form

$$\frac{J_{\parallel i}}{B_i} = \frac{J_{\parallel e}}{B_e} = -\frac{1}{2B_e} \hat{z} \cdot \nabla_e p \times \nabla_e \left[\int ds/B\right] \tag{44}$$

Equation (44), which was derived from the MHD momentum equation, is identical to (37), which was derived from the formula for bounce-averaged gradient/curvature drift.

If the flux tubes all had the same number of particles per unit magnetic flux and the same entropy per particle, then we could write

$$p = K \left[\int ds/B\right]^{-5/3} \tag{45}$$

with K = constant. In general, K will be different for different flux tubes. Using (45), we can rewrite (44) in the form

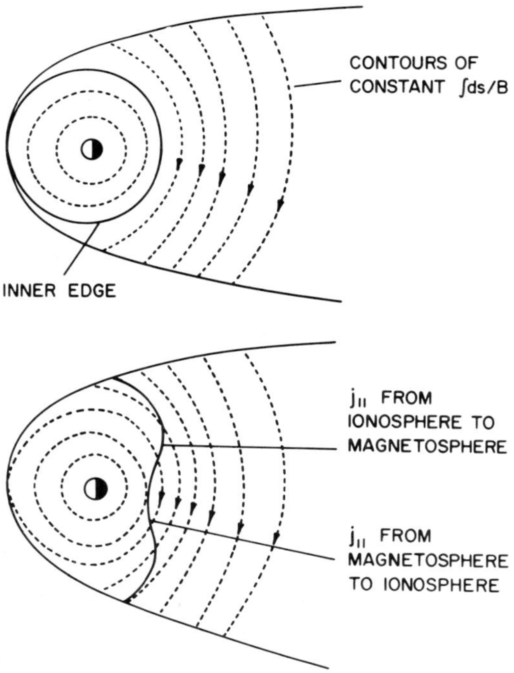

Figure 6. Illustration of the relationship between Birkeland currents and magnetospheric plasma distributions. Dashed curves are contours of constant $\int ds/B$. The top diagram illustrates the case where the plasma-sheet inner edge (assumed sharp) lines up exactly with contours of constant flux-tube volume, so that no Birkeland current is generated. The lower diagram shows a case where the inner edge is closer to earth near midnight, resulting in partial rings of gradient/curvature-drift current and Birkeland current.

$$\frac{J_{\| e}}{B_e} = - \frac{1}{2B_e} \left[\int ds/B\right]^{-5/3} \hat{z} \cdot \nabla_e K \times \nabla_e \left[\int ds/B\right]. \tag{46}$$

Wherever K = constant, there is no Birkeland current.

Figure 6 illustrates qualitatively how this pressure/Birkeland-current relation [which I've written variously as (34), (37), (44) and (46)] operates in the magnetosphere. Normally, in the magnetosphere, the particle pressure tends to increase slowly as one approaches the earth, not as fast as $[\int ds/B]^{-5/3}$. If p increases slowly, or decreases, then K tends to decrease as one approaches the earth. For the purpose of qualitative illustration in Figure 6, we pretend that K is uniform antiearthward of an inner edge. Earthward of the inner edge, K is equal to zero.

The top diagram in Figure 6 illustrates the case where there is no Birkeland current. Gradient/curvature-drift current flows along these contours of constant $\int ds/B$, and the inner edge lines up exactly with one of these contours. The gradient/curvature-drift currents flow in continuous rings that extend to the magnetopause boundary layer — outside the slow-flow region being considered. In terms of equations (34), (37), (44) and (46), $\nabla_e p$, $\nabla_e K$, and $\nabla_e \Sigma_s \eta_s \lambda_s$ are all parallel to $\nabla_e \int ds/B$, making the cross-products zero. In the lower diagram, the plasma sheet intrudes deeper into the magnetosphere near midnight, resulting in short westward partial rings of current past midnight, which connect to appropriately directed Birkeland currents. In terms of equation (46), for example, $\nabla_e K \times \nabla_e \int ds/B$ is in the $\hat{z}$-direction (-z direction) before (after) midnight. Thus $J_{\parallel e}$ is negative (positive), indicating current up from (down to) the ionosphere.

C. Ionospheric Conductivity: Conductivity $\perp \underset{\sim}{B}$ in a Partially Ionized Plasma

General Remarks. In a fully collisionless plasma with scale lengths large compared to a gyro-radius, and time scale large compared to a gyro-period, the currents are not mainly conduction currents. If one applies an $\underset{\sim}{E}$ perpendicular to $\underset{\sim}{B}$, the charged particles all drift at velocity $\underset{\sim}{E} \times \underset{\sim}{B}/B^2$. Since they all drift at the same velocity, no electric current results, assuming that the plasma is neutral overall. This often strikes students learning plasma physics as strange, since collisionless plasmas have a reputation as excellent conductors. Sometimes we idealize them as perfect conductors. How can that be if they don't carry a steady current when a steady electric field is applied across them? The answer is that they choose to preserve their reputation as perfect conductors, not by carrying any current when $\underset{\sim}{E}$ is applied, but rather by transforming themselves into a rest frame where there is no electric field. That is just the frame moving at the $\underset{\sim}{E} \times \underset{\sim}{B}$ drift velocity. When the particles gradient/curvature drift in addition to $\underset{\sim}{E} \times \underset{\sim}{B}$ drift, that causes a current but it is a violation of the perfect-conductor approximation. Incidentally, although we have been discussing a collisionless plasma, adding collisions really does not greatly affect this picture of cross-field conductivity, as long as the collisions are among charged particles. The eventual effect of putting an electric field on the plasma is still to get all the particles $\underset{\sim}{E} \times \underset{\sim}{B}$ drifting.

Conductivity Parallel to $\underset{\sim}{B}$. The effect of installing an electric field parallel to $\underset{\sim}{B}$ corresponds much better to the popular image of a plasma. If the plasma is collisionless and homogeneous, the current in fact increases continuously with time as electrons and ions are accelerated oppositely by the $\underset{\sim}{E}$ field, until some more complex collective plasma process takes over to limit the current. (See the chapter by W. J. Burke in this volume.) The effect of electron-ion collisions is also to act to limit the current, and to give a linear relation of the form $j_{\parallel} = \sigma_{\parallel} E_{\parallel}$. More complex plasma processes may or may not be representable by such a linear relation.

Conductivity in a Partially Ionized Plasma Ionosphere. The current-carrying layers of the ionosphere are not at all fully ionized. For example, in the E-region at 120 km altitude, the neutral number density is typically about 5×10^{11} cm^{-3}, whereas the electron density rarely exceeds 10^6 cm^{-3}. (See chapter by A. D. Richmond.) Collisions with neutrals are obviously important, and this profoundly changes the system's response to an electric field applied across $\underset{\sim}{B}$. Now when the charged particles try to $\underset{\sim}{E} \times \underset{\sim}{B}$ drift, they have to drag through the neutrals. If the neutrals felt no forces other than rubbing against charged particles, they would eventually accelerate to the $\underset{\sim}{E} \times \underset{\sim}{B}$ drift velocity. In the ionosphere, this happens to some extent, but the neutrals generally do not simply $\underset{\sim}{E} \times \underset{\sim}{B}$ drift with the plasma for three reasons: 1. they interact strongly with the lower atmosphere, which tends to move with the solid earth; 2. they respond to other forces (pressure gradients induced by solar heating, for example); and 3. the substantial mass of the neutrals anyhow prevents them from reacting quickly to changes in local $\underset{\sim}{E} \times \underset{\sim}{B}$ drifts. Because the neutrals do not move at velocity $\underset{\sim}{E} \times \underset{\sim}{B}/B^2$, the conduction currents flow across magnetic field lines in the ionosphere; the currents result from the friction between the neutrals, which want to do their own thing, and charged particles, which want to $\underset{\sim}{E} \times \underset{\sim}{B}$ drift.

Standard Conductivity Formulas. Formulas for ionospheric conductivity can easily be derived by considering force balance, including the friction due to collisions with neutrals. Expressed in the rest frame of the neutrals, the analysis leads to the following simple linear expression

$$\underset{\sim}{j} = \sigma_0 \, E_{\parallel} \, \hat{b} + \sigma_1 \, \underset{\sim}{E}_{\perp} + \sigma_2 \, \hat{b} \times \underset{\sim}{E}_{\perp} \tag{47}$$

Formula (47), and the expressions for the conductivity coefficients σ_0, σ_1, and σ_2 are derived in the chapter by A. D. Richmond. Typical height-profiles for σ_0, σ_1, and σ_2 are also given. The coefficient σ_0 represents direct conductivity along the magnetic field. It goes to infinity as the electron collision frequency goes to zero. The coefficient σ_1 is the Pedersen conductivity, which represents the capacity of the partially ionized plasma to conduct current in the direction of $\underset{\sim}{E}$, when an $\underset{\sim}{E}$ is applied perpendicular to $\underset{\sim}{B}$. The Pedersen conductivity goes to zero when the collision frequencies go to zero, and is generally much smaller than σ_0 as long as the electron collision frequency is small compared to the electron gyrofrequency, which is normally the case throughout the main current-carrying layers of the ionosphere. The Hall conductivity results from the fact that the electrons, being more tightly bound to the magnetic field lines than the ions, come closer to $\underset{\sim}{E} \times \underset{\sim}{B}$-drifting than the ions do; this differential drift in the $\underset{\sim}{E} \times \underset{\sim}{B}$ direction, with the negative particles drifting faster, causes a current in the $-\underset{\sim}{E} \times \underset{\sim}{B}$ direction.

Height-Integrated Ionospheric Conductivities. Suppose we construct a sphere around the earth, just above the F-region of the ionosphere. I am going to feed electrical currents down through this sphere and draw them up through it, and I wish to know what electric fields will then have to exist on the surface of the sphere. If I had complete three-dimensional information on σ_0, σ_1 and σ_2, I could solve a three-dimensional potential problem to obtain the needed information on the ionosphere's electrical response. However, solving the current-conservation equation in three dimensions would be a sizable project, and a physically reasonable simplification can be carried out. Generally, above the bottom of the E-layer of the ionosphere, we have, as noted earlier,

$$\sigma_0 \gg \sigma_1 \text{ and } \sigma_0 \gg \sigma_2 \tag{48}$$

Thus for structures whose horizontal dimensions are larger than or comparable to vertical ionospheric scale heights, we can visualize magnetic field lines as perfect conductors.

The idea that the magnetic field lines are perfect conductors leads to the concept of height-integrated conductivity. Consider first the thin top slab shown in Figure 7. (Magnetic field lines are taken to be vertical, for simplicity.) In the presence of a horizontal electric field $\underset{\sim}{E}$, horizontal current density is, according to the Ohm's law equation (47), $\sigma_{1a} \underset{\sim}{E} + \sigma_{2a} \hat{b} \times \underset{\sim}{E}$. This is current per unit area, the area being measured perpendicular to the current. Taking $\underset{\sim}{E}$, σ_{1a}, and σ_{2a} to be independent of altitude within the thin slab, then the total horizontal current carried by the whole thickness of the slab is given by

$$\underset{\sim}{j}_1 = (\sigma_{1a} \underset{\sim}{E} + \sigma_{2a} \hat{b} \times \underset{\sim}{E}) \Delta h_a$$

This is now horizontal current per unit horizontal length, the horizontal length being measured perpendicular to $\underset{\sim}{j}$. A second ionospheric layer, like the lower one shown in Figure 7, will feel the same electric field as the first layer, because the field lines are taken to be equipotentials. Then the second layer will carry an additional current

$$\underset{\sim}{j}_2 = (\sigma_{1b} \underset{\sim}{E} + \sigma_{2b} \hat{b} \times \underset{\sim}{E}) \Delta h_b$$

For an arbitrary number of layers, the height-integrated horizontal cur-

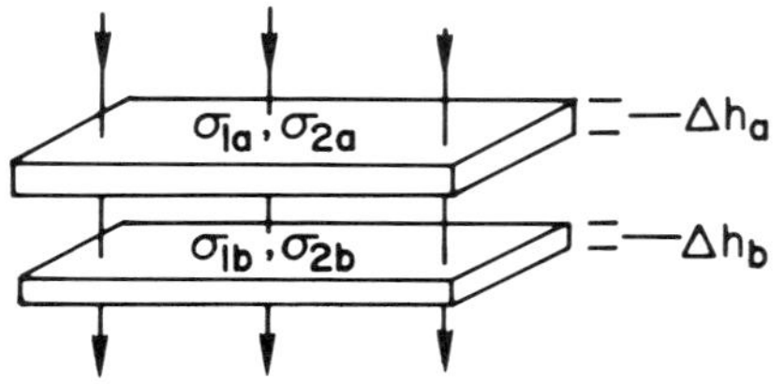

Figure 7. Illustration of two slabs of conducting ionosphere, connected by perfectly conducting field line.

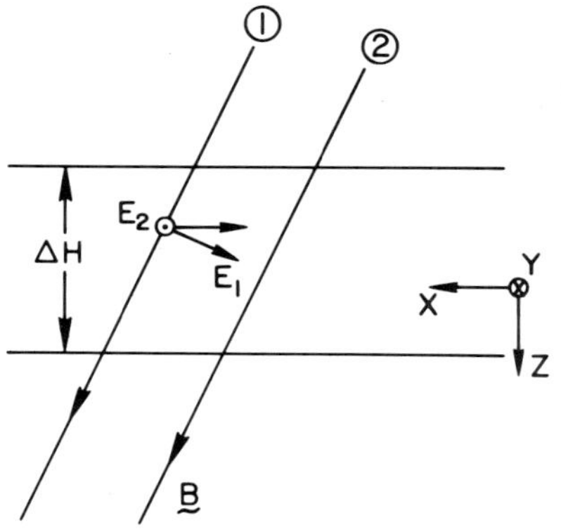

Figure 8. Illustration of magnetic-field dip-angle effect, for a slab of thickness ΔH.

rent is given by

$$\underset{\sim}{j} = [\int \sigma_1 dh \, \underset{\sim}{E} + \int \sigma_2 dh \, \hat{b} \times \underset{\sim}{E}] \quad (49)$$

However, to get a fully operational prescription for height-integrated electric current, we must consider the dip-angle effect, namely the effect of having the magnetic field not exactly vertical. Consider the case shown in Figure 8, where the magnetic field lines are in the xz plane, tilted downward by angle I from the x-direction. Consider first the Hall current density due to the field $\underset{\sim}{E}_1$. Its magnitude is simply $\sigma_H E_1$ and it is in the $-\hat{y}$ direction. Expressing horizontal current in terms of horizontal $\underset{\sim}{E}$ field gives

$$J_{H1y} = - \frac{\sigma_2}{\sin I} \left(\frac{\partial V}{\partial x}\right)_{y,z}$$

Note that we have assumed that the electric field $\underset{\sim}{E}$ could be written as $-\nabla V$ at E- and F-region altitudes. One can show that induction electric fields are negligible in the auroral ionosphere, for time scales greater than a few minutes. (Such an argument is detailed by A. D. Richmond in the electrodynamics section of his chapter.) The height-integrated Hall current in the y direction, per unit length in the x-direction, is simply

$$j_{H2y} = - \left(\frac{\sigma_2 \, \Delta H}{\sin I}\right) \left(\frac{\partial V}{\partial x}\right) \quad (50)$$

Consider now the Pedersen current that transports charge from field line 1 to field line 2. Its density is $\sigma_1 E_1$, or, expressed in terms of horizontal gradients in V

$$J_{p1x} = - \frac{\sigma_1}{\sin I} \left(\frac{\partial V}{\partial x}\right)_{y,z}$$

This Pedersen current flows perpendicular to $\underset{\sim}{B}$, along a field line length $\Delta H/\sin I$. The amount of this Pedersen current flowing per unit length in the y direction is given by

$$j_{p1x} = - \left(\frac{\sigma_1 \Delta H}{\sin^2 I}\right) \left(\frac{\partial V}{\partial x}\right)_{y,z} \tag{51}$$

Exactly analogous argumentation leads to the following formulas describing the electric current response to an electric field applied in the y-direction:

$$j_{H2x} = \left(\frac{\sigma_2 \Delta H}{\sin I}\right) \left(\frac{\partial V}{\partial y}\right)_{x,z} \tag{52}$$

$$j_{p2y} = -(\sigma_1 \Delta H) \left(\frac{\partial V}{\partial y}\right)_{x,z} \tag{53}$$

Equations (50)-(53) can be summarized and generalized to an arbitrary height-distribution of conductivity as follows:

$$\underset{\sim}{j} = \underset{\approx}{\Sigma} \cdot (-\nabla_h V) \tag{54a}$$

where $\underset{\approx}{\Sigma}$, the 2×2 tensor representing height-integrated conductivity, is given by

$$\Sigma_{xx} = \int \sigma_1 dh/\sin^2 I \tag{54b}$$

$$\Sigma_{yx} = -\Sigma_{xy} = \int \sigma_2 dh/\sin I \tag{54c}$$

$$\Sigma_{yy} = \int \sigma_1 dh \tag{54d}$$

where ∇_h is the horizontal gradient operator. In (54a), V is the potential in the rest frame of the solid earth. If the neutral-wind velocity is not zero in that frame, there is another term in (54a), involving a height-integral over local conductivities and wind velocities. Formulae (54b) and (54c) become singular at the equator, where $I = 0$. This singularity causes the equatorial electrojet, which must be treated using careful integrations along curved field lines rather than simple height-integrals. The physics of the equatorial electrojet is discussed in the chapter by A. D. Richmond. It is common practice to stop magnetosphere-ionosphere calculations short of the equator, to avoid the singularity.

D. A Fundamental Equation of Ionosphere-Magnetosphere Coupling

The equation of conservation of current in the ionosphere is

$$\nabla_h \cdot [\underset{\approx}{\Sigma} \cdot (-\nabla_h V)] = J_{\parallel i} \sin I \tag{55}$$

where $J_{\parallel i}$ is, as before, the density of current directed downward along the magnetic field lines at ionospheric height. Combining with (44) gives the fundamental equation governing Birkeland currents, namely

$$J_{\parallel i} \sin I = -\frac{B_i \sin I}{2B_e} \hat{z} \cdot \nabla_e p \times \nabla_e [\int ds/B] = \nabla_h \cdot [\underset{\approx}{\Sigma} \cdot (-\nabla_h V)] \tag{56}$$

where the ∇_e and ∇_h refer to gradients at the equatorial and ionospheric ends of the same field lines; ∇_e is a two-dimensional gradient in the equatorial plane, but ∇_h is a horizontal gradient operator on a spherical surface at ionospheric height. Formula (56) can also be written in the form where $\nabla_e p$ is replaced by $\nabla_e K$ or $\nabla_e \sum_s \eta_s \lambda_s$ (cf., equations (34) and (46)).

As we mentioned in the process of going from (43) to (44), the quantity

$$\frac{1}{B} \hat{b} \cdot \nabla p \times \nabla [\int ds/B]$$

is constant along a given field line. If we choose to evaluate it at a point just above the ionosphere, we can convert the ∇_e's in (56) to ∇_i's. The result is

$$-\left(\frac{\sin I}{2}\right) \hat{b} \cdot \nabla_i p \times \nabla_i [\int ds/B] = \nabla_h \cdot [\underset{\approx}{\Sigma} \cdot (-\nabla_h V)] \tag{57}$$

Equation (56) was derived by Vasyliunas (1970), although a similar relation is contained implicitly in the paper by Fejer (1964). Detailed discussions of the formulas for height-integrated conductivities and associated dip-angle factors can be found in the paper by Fejer (1953).

V. CALCULATIONS OF CONVECTION IN THE INNER MAGNETOSPHERE

Many calculations have been done using equation (56) or its equivalent. We begin our discussion of these by reviewing the standard logic diagram for such calculations. I will then go through a simplified analytic model calculation that illustrates the basic stability of the system and also illustrates how the inner magnetosphere responds to convection. Section VI will review some results of the more complicated (but more realistic) computer calculations.

A. Vasyliunas' Logical Loop

Figure 9 illustrates the basic logical structure of calculations that have been performed by a variety of people in the last dozen years. It is a slightly modified version of a figure drawn by Vasyliunas (1970). We discuss the loop by going around it once.

We start at the top of the diagram, with an initial hot-particle distribution in the magnetosphere. Given a magnetic field model, we could compute the density of gradient/curvature-drift current from equation (33); we compute the Birkeland-current distribution from (44) or one of the various equivalent forms. That completes the left side of the diagram.

At the bottom of the diagram, we solve equation (57) or an equivalent form to get the ionospheric potential distribution V. This requires a model for the distribution of height-integrated ionospheric conductivity $\underset{\approx}{\Sigma}(\underset{\sim}{x})$, as well as appropriate boundary conditions on V. Normally, we place conditions on V at the high-latitude and low-latitude boundaries of the calculation. The high-latitude boundary necessarily lies at or equatorward of the boundary of field lines that lie completely in the region where $\underset{\sim}{B}$ can be reliably estimated. Normally, we

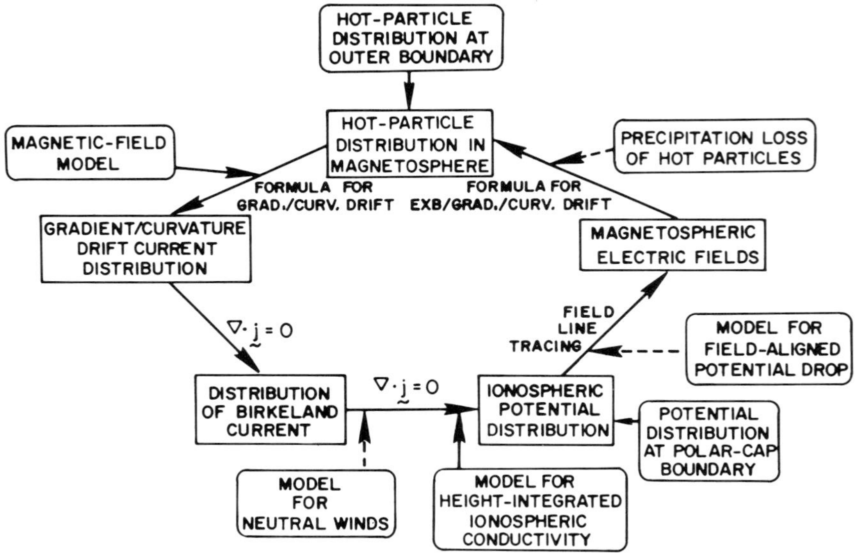

Figure 9. Logical loop for inner-magnetospheric convection calculations. It is a modified form of the loop proposed by Vasyliunas (1970). Boxes with rounded corners represent models that serve as input to the calculation. Rectangular boxes represent quantitites that are computed in the calculation. Arrows indicate flow of information. Dashed lines indicate connections that are neglected or trivialized in the present discussion, but could straightforwardly be included in the formalism.

specify the potential on that high-latitude boundary. The total potential drop across the curve is roughly the polar-cap potential drop (not exactly, however, since the high-latitude boundary of the model does not necessarily coincide with the polar-cap boundary). For the low-latitude boundary condition, we set the north-south current equal to zero at the equator. The requirement of zero current across the equator is rigorously correct if there is exact symmetry between hemispheres. If one were to try to do this problem for a case with a north-south asymmetry in conductivity, that could easily be done. If perfectly conducting field lines connect the northern and southern hemispheres, then the two hemispheres are effectively wired in parallel and Σ in equation (56) or (57) is interpreted as the average of the conductivity tensors at the northern and southern ends of the field line. We would still require $\Sigma \cdot \nabla V = 0$ at the equator, which then is interpreted as meaning zero net current into the equator. As we mentioned earlier, we frequently move the equatorward boundary to a small but non-zero latitude, to avoid the conductivity singularity at the equator. The equatorial electrojet is then excluded from these magnetosphere-ionosphere-coupling calculations.

In deriving equation (56) or (57), we assumed that the neutral-wind velocity was zero, in the rest frame of the solid earth. If we drop that assumption, we pick up a term of the form $\nabla_h \cdot \left[\int dh\, \sigma \cdot (v_n \times B)\right]$ in (56), where v_n = neutral-wind velocity, relative to the earth's surface below, and σ is an effective local conductivity tensor. Given a complete neutral-wind and conductivity model, this term could be computed and included in the calculation. However, due to uncertainties in v_n, this term is usually neglected, i.e., we pretend that $v_n = 0$.

We can map the ionospheric potential distribution computed from (56) out to the magnetospheric equatorial plane, for deduction of magnetospheric electric fields. In doing this mapping, however, we normally make two adjustments. First, equation (56) holds in the rest frame of the solid earth. The natural rest frame for the magnetosphere is one in which the sun-earth line is stationary, i.e., one which does not partake of the earth's 24-hour rotation. The electric field in this "magnetosphere rest frame" is related to the electric field in the local rest frame of the rotating earth by

$$E_m = E - (\omega \times r) \times B$$

For simplicity, we pretend that the magnetic field near the earth is dipolar with axis parallel to the rotation axis ($\hat{z}$-axis). Then we have

$$E_m = E + \omega r \sin\theta \left(B_\theta \hat{r} - B_r \hat{\theta}\right) = E + \frac{M\,\omega \sin\theta}{r^2}\left(-\sin\theta\, \hat{r} + 2\cos\theta\, \hat{\theta}\right)$$

where θ is colatitude and M is the equatorial magnetic field at the earth's surface times the cube of the radius of the earth. The corresponding transformation for the potential is

$$V_m = V - \frac{M\omega \sin^2\theta}{r} \approx V - (92{,}000 \text{ volts}) \sin^2\theta_i$$

where θ_i = ionospheric colatitude of the magnetic field line. The second correction deals with induction electric fields. If the magnetic field model varies in time, then the mapping also changes in time, and the corresponding induction electric field can also be included in the computed magnetospheric electric field. The $\underset{\sim}{E} \times \underset{\sim}{B}$ drift of plasma in the equatorial plane is the sum of the velocity due to time changes in the mapping and $(\nabla_e V_m) \times \underset{\sim}{B}_e/B_e^2$. The total drift velocity for a given species is the sum of the $\underset{\sim}{E} \times \underset{\sim}{B}$ drift and gradient/curvature drift. We then use this knowledge of computed total drift velocity to advance the plasma distribution in time. The plasma that was at $\underset{\sim}{x}$ moves to $\underset{\sim}{x} + \underset{\sim}{v}\Delta t$, undergoing adiabatic compression. Particle loss by precipitation, charge exchange, or whatever, should be included here, and has been included approximately in some calculations.

We have now completed one time step in the calculation, which now evolves in time. The various input models can be varied with time to simulate some specific kind of event, or we can hold the input steady, to allow the system to come to equilibrium.

B. Simple Analytic Model

There have been basically two approaches to theoretical implementation of the logical system shown in Figure 9. One approach is to make many simplifying assumptions and to do analytic calculations. The other is to be as realistic as possible and do the calculations numerically. In this presentation I will do a simple analytic calculation that illustrates most of the major features.

This analytic calculation will involve the following simplifications:

1. The modeling region in the ionosphere is treated in terms of simple polar coordinates, rather than spherical polar. The high-latitude boundary is at $r = a$, the low-latitude boundary at $r = \infty$.
2. Plasma-sheet plasma is visualized as consisting of a single ion species with energy invariant λ and density invariant $\eta(\underset{\sim}{x},t)$. The density invariant is constant, except for a single jump at

   $$r_1 = b + \delta r(\phi,t)$$

   where $b > a$, $|\delta r| << b$, and ϕ is an azimuthal angle (0 at noon, $\pi/2$ at dusk,...). The density invariant is given by

   $$\eta(\underset{\approx}{x}_i,t) = \begin{cases} \eta_1 \text{ (constant) for } r < b + \delta r(\phi,t) \\ \eta_2 \text{ (constant) for } r > b + \delta r(\phi,t) \end{cases} \qquad (58)$$

Electrons are taken to have $\lambda = 0$, so that they do not gradient/curvature drift.

3. The magnetic field is independent of ϕ and t. Specifically the flux-tube volume $\int ds/B$ is a function of r only.
4. Particles at the density jump gradient/curvature drift at angular rate $\dot{\phi}$, where $\dot{\phi}$ is independent of ϕ and t. Corotation is neglected, and $\underset{\sim}{E} \times \underset{\sim}{B}$ drift in the computed electric field is assumed to be much slower than gradient/curvature drift.
5. The height-integrated ionospheric conductivity is taken to be spatially uniform. The dip angle I is taken to be $\pi/2$.

Figure 10 shows the simplified system, both in the ionosphere and in the equatorial plane. Note the inversion involved in the mapping. (Large ionospheric radius r corresponds to small geocentric distance R, and vice versa.)

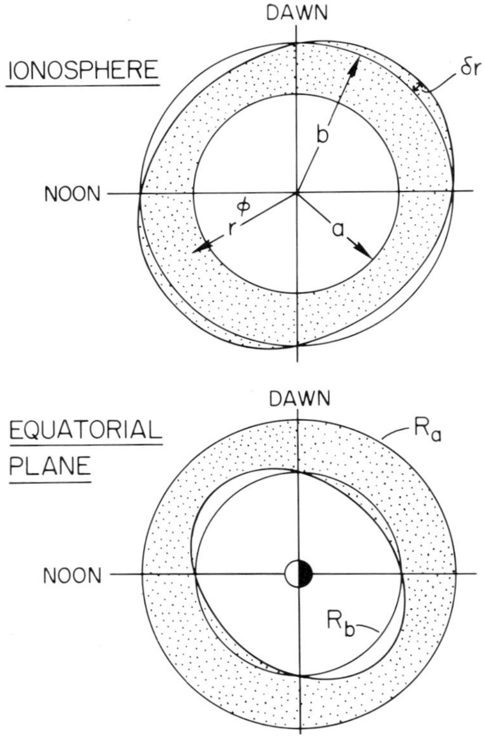

Figure 10. The top diagram shows the ionospheric coordinate system for the analytic convection calculation. The dotted region maps to the plasma sheet ($\eta = \eta_1$). The region equatorward of the plasma sheet has $\eta = \eta_2$. The lower diagram shows the same configuration mapped to the equatorial plane, on a very different scale. The ionospheric circles of radii a and b map to equatorial circles of radii R_a and R_b, respectively.

For a one-species, isotropic plasma (assumption 2) we can, according to (36), write the pressure as follows

$$p = \frac{2}{3} \eta\lambda (\int ds/B)^{-5/3} \tag{59}$$

Substituting into the basic ionosphere-magnetosphere coupling relation (57) gives

$$\nabla_h \cdot [\underset{\approx}{\Sigma} \cdot (-\nabla_h V)] = -\frac{1}{3} \lambda \left(\int ds/B\right)^{-5/3} \hat{b} \cdot \nabla_i \eta \times \nabla_i \int ds/B, \tag{60}$$

where we have set $I = \pi/2$ and assumed $\hat{b}$ = unit vector down (assumption 5). Differentiating (58) gives

$$\nabla_i \eta = (\eta_2 - \eta_1)\, \delta\big(r - b - \delta r(\phi)\big) \left(\hat{r} - \frac{1}{r}\frac{\partial \delta r}{\partial \phi}\hat{\phi}\right) \tag{61}$$

From assumption 3, we have

$$\left(\nabla_i \int ds/B\right)_b = \left(\frac{d \int ds/B}{dr}\right)_b \hat{r} \tag{62}$$

Substituting (61) and (62) and in (60) yields

$$\nabla_h \cdot [\underset{\approx}{\Sigma} \cdot (-\nabla_h V)] = \frac{\lambda(\eta_2 - \eta_1)}{3b} \left(\int ds/B\right)_b^{-5/3} \delta(r - b) \left(\frac{d \int ds/B}{dr}\right)_b \frac{\partial \delta r}{\partial \phi} \tag{63}$$

where we have considered δr small and kept only the term that is first-order in δr. Now let's work on the left side of (63). With the uniform-conductivity assumption (assumption 5) the Hall conductivity terms drop out, because

$$\nabla_h \cdot (\hat{z} \times \nabla_h V) = 0$$

and (63) becomes

$$\nabla_h^2 V = -G\, \delta(r - b) \frac{\partial \delta r}{\partial \phi} \tag{64a}$$

where the constant G is given by

$$G = -\frac{\lambda(\eta_1 - \eta_2)}{3b\Sigma_p} \left(\int ds/B\right)_b^{-5/3} \left(\frac{d \int ds/B}{dr}\right)_b \tag{64b}$$

and $\Sigma_p \equiv \Sigma_{xx} = \Sigma_{yy}$ for $I = \pi/2$. One solves (64) to find the potential distribution V generated by a certain plasma distribution (characterized here in terms of the δr function).

We close the system of equations by relating $d\delta r/dt$ to δr and V. The particles that lie on the inner edge $\underset{\sim}{E} \times \underset{\sim}{B}$ drift as well as gradient/curvature drift. The gradient/curvature drift is large but is purely azimuthal. The $\tilde{E} \times \tilde{B}$ drift is regarded as first-order small, but can be radial as well as azimuthal. The radial velocity of an inner-edge particle is thus given by

$$\frac{d\delta r}{dt} = \frac{1}{B_i} \left(\frac{1}{r}\frac{\partial V}{\partial \phi}\right)_b \tag{65}$$

Considering δr as a function of ϕ and t, rather than just t, gives

$$\left(\frac{\partial}{\partial t} + \dot{\phi}\frac{\partial}{\partial \phi}\right) \delta r = \frac{1}{bB_i} \left(\frac{\partial V}{\partial \phi}\right)_b \tag{66a}$$

where $\dot{\phi}$ is the azimuthal drift velocity of a boundary particle. With assumption 4 and (32), we have

$$\dot{\phi} = \frac{\lambda}{qB_eR_b} \left(\frac{d[\int ds/B]^{-2/3}}{dR}\right)_{R_b}$$

where R_b = geocentric distance in the equatorial plane corresponding to the ionospheric condition r = b. Since $B_e(R_b d\phi)dR = B_i(bd\phi)\,dr$ because of magnetic-flux conservation, we can write $\dot{\phi}$ in ionospheric terms:

$$\dot{\phi} = -\frac{\lambda}{qB_ib} \left(\frac{d[\int ds/B]^{-2/3}}{dr}\right)_b \tag{66b}$$

It will be regarded as a known constant.

The linear equations (64) and (66) can be solved for V and δr, as functions of both position and time. We start by solving (64) for $V(\tilde{x}_i,t)$, pretending $\delta r(\tilde{x}_i,t)$ is known for time t.

The general solution to $\nabla_h^2 V = 0$ in polar coordinates is of the form

$$V = \sum_m (a_m r^m + b_m r^{-m})\, e^{im\phi} \tag{67}$$

First consider the solution for arbitrary m and homogeneous boundary conditions (V = 0 at r = a and at r = ∞). Assume

$$\delta r(\phi,t) = \Delta r(t) e^{im\phi} \tag{68}$$

The solution then looks like

$$V = a_m\left[\left(\frac{r}{a}\right)^m - \left(\frac{a}{r}\right)^m\right] e^{im\phi} \quad \text{for } a < r < b$$

$$V = a_m\left[\left(\frac{b}{a}\right)^m - \left(\frac{a}{b}\right)^m\right] \left(\frac{b}{r}\right)^{|m|} e^{im\phi} \quad \text{for } r > b \tag{69}$$

This form satisfies $\nabla_h{}^2 V = 0$ for $a < r < b$ and $r > b$, has V continuous at $r = b$, and has $V = 0$ at $r = a$ and $r = \infty$. The constant a_m is determined by requiring that

$$\left(\frac{\partial V}{\partial r}\right)_{b+\varepsilon} - \left(\frac{\partial V}{\partial r}\right)_{b-\varepsilon} = -G \frac{\partial \delta r}{\partial \phi} = -G \; im \; \Delta r e^{im\phi}$$

The result is

$$a_m = \frac{Gib\Delta r}{2} \left(\frac{a}{b}\right)^{|m|}, \tag{70}$$

which essentially completes the solution to (64). To substitute into (66), we note that

$$\left(\frac{\partial V}{\partial \phi}\right)_b = i|m| \left(\frac{Gib\Delta r}{2}\right) \left(1 - \left(\frac{a}{b}\right)^{2|m|}\right) e^{im\phi}$$

so that (66a) becomes, with (68),

$$\left(\frac{\partial}{\partial t} + im\dot{\phi}\right) \Delta r = - \frac{G\Delta r|m|}{2B_i} \left[1 - \left(\frac{a}{b}\right)^{2|m|}\right] \tag{71}$$

The solution is simple, given an initial value of Δr:

$$\Delta r(t) = \Delta r(0) \exp \left\{\left[-im\dot{\phi} - \frac{G|m|}{2B_i} \left(1 - \left(\frac{a}{b}\right)^{2|m|}\right)\right]t\right\}. \tag{72}$$

The ripple decays with time if $G > 0$, grows with time if $G < 0$. Referring to the definition of G in (64b), we note that the sign of G is the sign of $(\eta_1 - \eta_2)$. (The derivative $d(\int ds/B)/dr$ is normally always negative. The further the ionospheric end of the field line is from the pole, the shorter the field line is, and the larger its equatorial B is.)

Our basic conclusion is the following: if the high-latitude ($r < b$) region has larger flux tube content than the lower-latitude ($r > b$) region, i.e., if $\eta_1 > \eta_2$, then the boundary at $r = b$ is stable. Ripples decay. If the low-latitude flux tubes are more heavily populated, i.e., $\eta_2 > \eta_1$, the boundary is unstable and ripples grow exponentially in time.

The physics of this instability can easily be seen qualitatively. In the left diagram of Figure 11, gradient/curvature drifting plasma exists outside an inner edge, which bulges earthward near local midnight. Gradient/curvature-drift currents flow westward along contours of constant $\int ds/B$, which are here represented by circular arcs. Net gradient/curvature drift current flows westward from the eastward edge of the bulge to the westward edge. Birkeland and ionospheric Pedersen currents must complete the circuit. Pedersen currents thus tend to flow eastward across the bulge, corresponding to an eastward field there. Given northward $\underset{\sim}{B}$ (out of the page), this eastward $\underset{\sim}{E}$ corresponds to flow away from the earth, i.e., such as to reduce the amplitude of the bulge. Thus the system is stable. In the right diagram of Figure 11, plasma exists earthward of an outer edge, which bulges outward. Gradient/curvature-drift currents again flow westward across the bulge, which generates an eastward $\underset{\sim}{E}$, and outward $\underset{\sim}{v}$. But this outward $\underset{\sim}{v}$ now represents an increase in bulge amplitude, i.e., the system is unstable.

This is an instance of the well-known interchange instability, which represents a quite general property of plasmas. A plasma in magnetostatic equilibrium tends to be unstable if high-content flux tubes have smaller volumes $\int ds/B$ than lower-content flux tubes. The plasma essentially avoids regions of large $|\underset{\sim}{B}|$. That is why the earliest mirror machines failed to confine plasmas. Magnetic-field strength tended to decline with distance from the axis of the machine. More

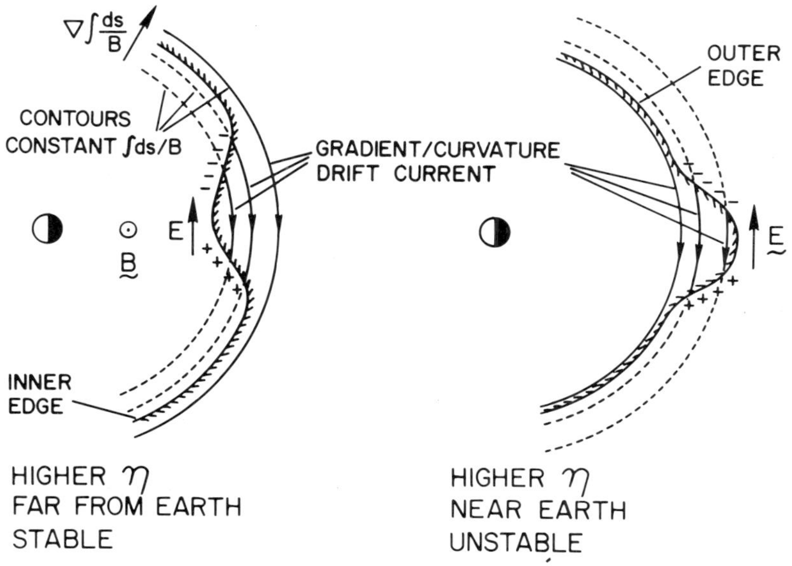

Figure 11. Illustration of the conditions for interchange instability in the magnetosphere. The view is of the equatorial plane. The left diagram illustrates the situation where plasma exists anti-earthward of an inner-edge, which is pushed earthward near local midnight. The right diagram pertains to the case where plasma exists earthward of an outer edge.

heavily populated flux tubes had smaller flux-tube volumes than less populated tubes, and the systems were interchange unstable.

The magnetosphere is basically stable against this interchange instability. The density invariant, or, more generally, $K = p(\int ds/B)^{5/3}$, tends clearly to increase with increasing geocentric distance. (An interesting possible exception to this tendency toward stability may occur when the plasmapause is very sharp. Although the energy density in the plasmasphere is small compared to the energy density in the more energetic particles of the trapped radiation belts, a sharp plasmapause may cause $p(\int ds/B)^{5/3}$ to decrease locally with increasing radial distance, causing a local instability. This situation has been analyzed quantitatively by Richmond (1973). (Richmond's paper gives a good general discussion of interchange instability in the magnetosphere, and references to the early work.)

Returning to (72), the decay rate of a ripple is given by

$$\frac{1}{\tau} = \frac{G|m|}{2B_i}\left[1 - \left(\frac{a}{b}\right)^{2|m|}\right]$$

$$= \frac{\lambda(\eta_1 - \eta_2)|m|}{6b\Sigma_p B_i}\left(\int ds/B\right)_b^{-5/3}\left[1 - \left(\frac{a}{b}\right)^{2|m|}\right]\left[-\frac{d\int ds/B}{dr}\right]_b$$

or, using (66)

$$\frac{1}{\tau} = |\dot{\phi}|\ |m|\left[\frac{(\eta_1 - \eta_2)q}{4\Sigma_p}\right]\left[1 - \left(\frac{a}{b}\right)^{2|m|}\right]. \tag{73}$$

The decay rate is equal to the drift rate $|\dot{\phi}|$ multiplied by $|m|$, and by a geometrical factor of order unity, and by a dimensionless factor $(\eta_1 - \eta_2)q/(4\Sigma_p)$. If we put η_1 equal to a typical plasma-sheet value ($\sim 2 \times 10^{21}$ weber^{-1}), we find that

$$\frac{e\eta_1}{2} \sim 160 \text{ mhos} \tag{74a}$$

which can be viewed, within this system of equations, as an effective conductivity of the inner plasma sheet. The effective Pedersen conductivity of the northern and southern ionospheres together, $2\Sigma_p$, is typically $\sim$ 2 to 20 mhos. If $\eta_2 \ll \eta_1$, we then have, for the dimensionless factor

$$\frac{(\eta_1 - \eta_2)q}{4\Sigma_p} \sim 8 \text{ to } 80 \tag{74b}$$

The fact that this ratio is large compared to unity suggests that the inner plasma sheet has greater effective current-carrying capacity than the ionosphere. The decay time of a ripple is of the order of the time required for an inner-edge particle to drift through one radian of phase in the ripple, divided by the dimensionless ratio (74b).

To understand how convection works in the inner magnetosphere, it is useful to solve (64) and (66) again, but this time for an inhomogeneous boundary condition. Namely, assume

$$V(a,\phi) = \frac{i}{2} V_{pc} \, e^{i\phi}, \tag{75}$$

the real part of which corresponds to standard convection with V_{pc} = polar-cap potential drop (V maximum at dawn ($\phi = 3\pi/2$), minimum at dusk ($\phi = \pi/2$)). We need only find one solution to this equation with inhomogeneous boundary conditions. To find the general solution, we simply add the general solution for homogeneous boundary conditions, which we have already found (equation (72), applied for arbitrary m). We choose to search for the time-independent solution to the problem with boundary condition (75).

For this example, we also set $\eta_1 = \eta$ and $\eta_2 = 0$. The boundary at $r = b + \delta r$ is then interpreted as the inner (i.e., low-latitude) edge of the plasma sheet ions.

For the time-independent case, (66a) reduces to simply

$$\frac{\partial \delta r}{\partial \phi} = \frac{1}{\dot{\phi} b B_i} \left(\frac{\partial V}{\partial \phi}\right)_b \tag{76a}$$

and (64a) becomes

$$\nabla_h^{\,2} V = - \frac{G\,\delta(r-b)}{\dot{\phi} b B_i} \left(\frac{\partial V}{\partial \phi}\right)_b = - \frac{iG\,\delta(r-b) V_b}{\dot{\phi} b B_i} \tag{76b}$$

if $V \propto e^{i\phi}$. The solution to $\nabla_h^{\,2} V = 0$ can be written

$$V = \frac{iV_{pc}}{4} \left[\left(\frac{r}{a}\right) + \left(\frac{a}{r}\right)\right] e^{i\phi} + C_1 \left[\frac{r}{a} - \frac{a}{r}\right] e^{i\phi} \quad \text{for } a < r < b \tag{77a}$$

For $r > b$, we have

$$V = e^{i\phi}\left(\frac{b}{r}\right) \left\{\frac{iV_{pc}}{4} \left[\frac{b}{a} + \frac{a}{b}\right] + C_1 \left[\frac{b}{a} - \frac{a}{b}\right]\right\} \tag{77b}$$

This satisfies the boundary conditions at $r = a$ and $r = \infty$, and has V continuous at $r = b$. Integrating (76b) across $r = b$ and using (77) yields

$$\left(\frac{\partial V}{\partial r}\right)_{b+\varepsilon} - \left(\frac{\partial V}{\partial r}\right)_{b-\varepsilon} = e^{i\phi}\{\left(-\frac{1}{b}\right)\left[\frac{iV_{pc}}{4}\left(\frac{b}{a}+\frac{a}{b}\right) + C_1\left(\frac{b}{a}-\frac{a}{b}\right)\right]$$

$$- \frac{iV_{pc}}{4b}\left[\frac{b}{a}-\frac{a}{b}\right] - \frac{C_1}{b}\left(\frac{b}{a}+\frac{a}{b}\right)\}$$

$$= -\frac{iG}{\dot{\phi} b B_i} e^{i\phi}\{\frac{iV_{pc}}{4}\left(\frac{b}{a}+\frac{a}{b}\right) + C_1\left(\frac{b}{a}-\frac{a}{b}\right)\}$$

Solving for C_1 gives

$$C_1 = -\frac{iV_{pc}}{4}\frac{(iG/2\dot{\phi}B_i)(1+(a^2/b^2)) - 1}{(iG/2\dot{\phi}B_i)(1-(a^2/b^2)) - 1} \tag{78}$$

We need only substitute this back in the earlier expressions (76a) and (77) to find $V(r,\phi)$ and $\delta r(\phi)$ in steady state. Results for V and δr are the following:

$$V(r,\phi) = \frac{1}{2}\{iV_{pc} e^{i\phi}\left[\frac{r}{a}\left(\frac{iGa^2}{2\dot{\phi}B_i b^2}\right) + \frac{a}{r}\left(1 - \frac{iG}{2\dot{\phi}B_i}\right)\right]\}\left[1 - \frac{iG\,(1-a^2/b^2)}{2\dot{\phi}B_i}\right]^{-1} \tag{79}$$

for $a < r < b$, and

$$V(r,\phi) = \left(\frac{iV_{pc}}{2}\right)\left(\frac{a}{r}\right) e^{i\phi}\left[1 - \frac{iG\,(1-a^2/b^2)}{2\dot{\phi}B_i}\right]^{-1} \quad \text{for } r > b \tag{80}$$

$$\delta r = \left(\frac{iV_{pc}}{2\dot{\phi} b B_i}\right)\left(\frac{a}{b}\right) e^{i\phi}\left[1 - \frac{iG\,(1-a^2/b^2)}{2\dot{\phi}B_i}\right]^{-1} \tag{81}$$

From (64b) and (66b), we note that

$$-\frac{G}{2\dot{\phi}B_i} = +\frac{q\eta}{4\Sigma_p}, \tag{82}$$

where the right side is the ratio that was discussed earlier (equation (74)) and found to be $\gtrsim 8$. The potential for $r > b$, i.e., on the low-latitude side of the inner edge, is smaller, by a factor of

$$\left[1 + \frac{iq\eta\,(1-a^2/b^2)}{4\Sigma_p}\right]^{-1}$$

than it would have been if there had been no inner edge. This is called the shielding effect. The inner edge of the plasma sheet ions tends to shield the region earthward of it from the convection electric field, because $q\eta/2$, which represents the effective conductance of the inner plasma sheet or ring current in this geometry, normally substantially exceeds the effective ionospheric conductivity $2\Sigma_p$.

The Birkeland current flowing down into the ionosphere, per unit length along the r = b line, is given by

$$j_{\parallel b}(\phi) = \Sigma_p\left[\left(\frac{\partial V}{\partial r}\right)_{b-\varepsilon} - \left(\frac{\partial V}{\partial r}\right)_{b+\varepsilon}\right]$$

Differentiating (79) and (80), and taking the limit $e\eta/\Sigma_p \to \infty$, we obtain

$$j_{\parallel b}(\phi) = -\frac{i\Sigma_p V_{pc} e^{i\phi}}{b\left(\frac{b}{a} - \frac{a}{b}\right)} \tag{83}$$

Assume for simplicity that the auroral-zone conductivity falls off rapidly for $r < a$, so that the current flowing into $r = a$ from the large r side goes immediately up along magnetic field lines. We then have

$$j_{\parallel a}(\phi) = -\Sigma_p\left(\frac{\partial V}{\partial r}\right)_{a+\varepsilon} - \frac{\Sigma_H}{a}\left(\frac{\partial V}{\partial \phi}\right)_a$$

which becomes, in the limit $e\eta/\Sigma_p \to \infty$,

$$j_{\parallel a}(\phi) = \frac{i\Sigma_p V_{pc} e^{i\phi}}{2a\left(\frac{b}{a} - \frac{a}{b}\right)}\left\{\left(\frac{a}{b} + \frac{b}{a}\right) - i\frac{\Sigma_H}{\Sigma_p}\left(\frac{b}{a} - \frac{a}{b}\right)\right\}. \tag{84}$$

Note that, aside from the Hall-conductivity term, which the geometrical factor makes relatively small, $j_{\parallel a}(\phi)$ is in phase with the applied potential drop, as specified in (75). Current flows down onto the positive-potential side of the boundary $r = a$, and up from the negative side. The Birkeland current on $r = a$, identified as the region-1 Birkeland current (chapter by T. W. Hill), is connected to the "generator" or "battery" of the circuit. The currents on $r = b$, identified as the region-2 Birkeland current contact only the shielded potential at $r = b$, and, in any case, are 90° out of phase with $V(b,\phi)$. (To see this, take the limit $e\eta/\Sigma_p \to \infty$ in (80).) Thus in this steady-state solution, the region-2 current has no net impact on the energetics. Figure 12 qualitatively shows the Birkeland-current distribution in the ionosphere, as implied by (83) and (84).

The physical situation, as viewed from the equatorial plane, is sketched qualitatively in Figure 13. The top diagram shows the inner edge from (81), but mapped to the equatorial plane. For our azimuthally symmetric magnetic-field model, the inner edge is closest to the earth (lowest ionospheric latitude) near midnight, furthest from the earth at noon. The gradient/curvature-drift current, which flows westward along contours of constant flux-tube volume, is greatest near midnight. Some of this drift current flows out of the dawnside inner edge, past midnight, and into the duskside inner edge. Birkeland current flows up from ionosphere at r = b to the magnetosphere on the dawn side, down from the magnetosphere to the ionosphere at r = b on the dusk side.

The lower diagram in Figure 13 is a sketch of the potential distribution implied by (79) and (80), expressed in the equatorial plane. The corotation potential is not included. The shielding greatly reduces the strength of convection in the near-earth region of the equatorial plane. It also rotates the near-earth electric field by about 90° relative to the overall sunward convection.

For other analytic convection calculations, see papers by Vasyliunas (1972), Jaggi and Wolf (1973), Wolf and Jaggi (1973), Southwood and Wolf (1978), Crooker and Siscoe (1981), and Siscoe (1982).

VI. COMPUTER MODEL OF INNER MAGNETOSPHERIC CONVECTION

A. Introductory Comments

In this section, I will try to give some indication of the present state-of-the-art in convection modeling, and for this purpose, we will discuss results from a set of computer models. These models use the basic formulas derived in Sections III and IV, and follow the computa-

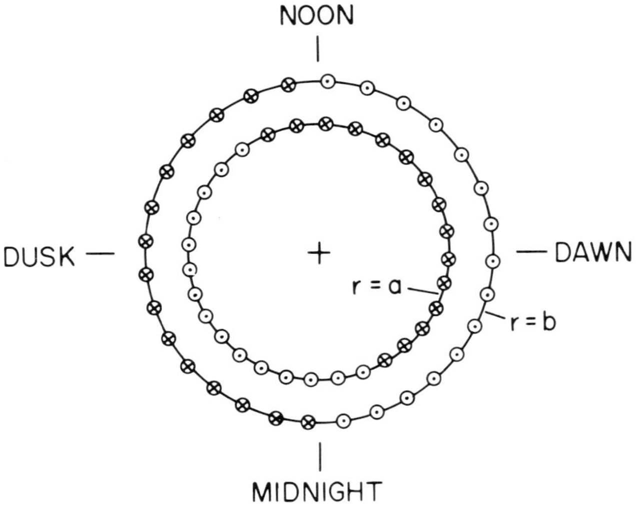

Figure 12. Birkeland current distribution from equations (83) and (84), for the case where $\Sigma_H/\Sigma_p = 2$. The symbols ⊙ and ⊗ represent, respectively, currents up from and down into the ionosphere.

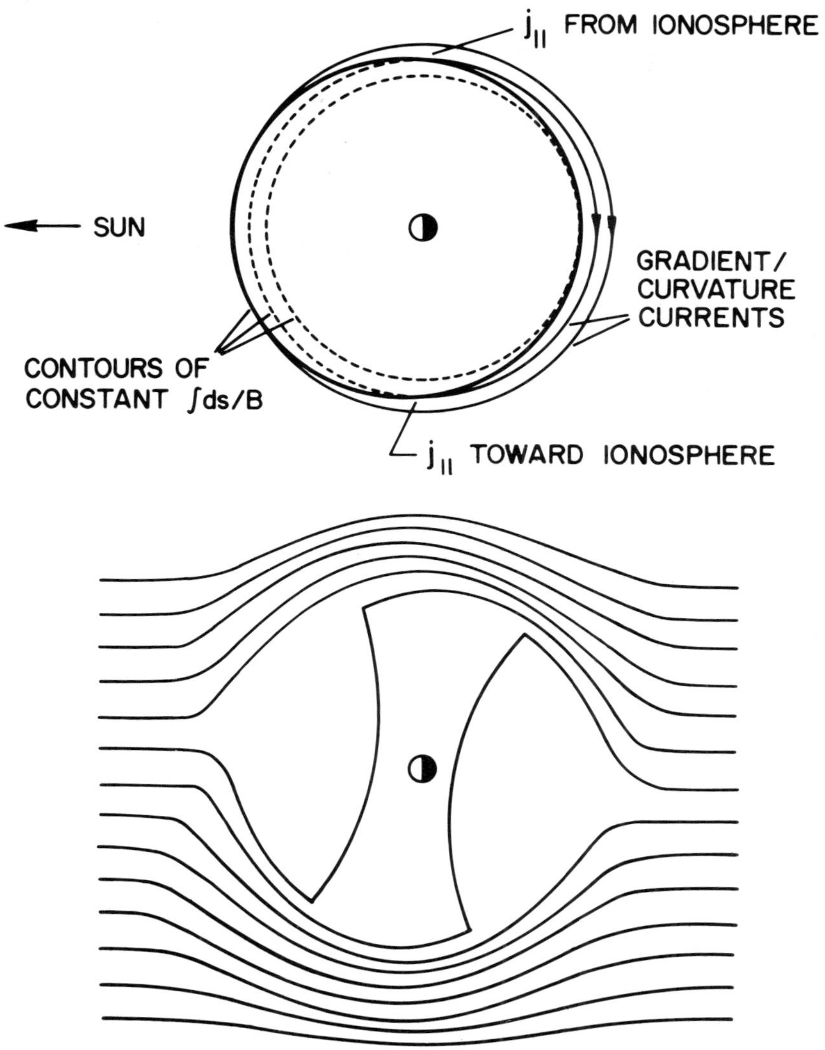

Figure 13. Sketches of steady-state convection in the inner magnetosphere (equatorial-plane view). The top diagram shows the relationship between the inner edge of the ion plasma sheet and contours of constant flux-tube volume. The lower diagram shows equipotentials, not including corotation.

tional scheme outlined at the beginning of Section V (Figure 9). However, many of the simplifying assumptions that were made for the analytic models of Section V can be eliminated in the computer models, specifically the following: flat ionosphere; monoenergetic plasma-sheet ions; zero-energy plasma-sheet electrons; nearly circular plasma-sheet inner edge; azimuthally symmetric, time-independent magnetic field; convection velocity and corotation velocity small compared to gradient/curvature drift; spatially uniform height-integrated ionospheric conductivity, with dip angle I equal to to 90°. The computer models described do, however, still neglect the effects on magnetospheric convection, of several significant physical processes: magnetic-field-aligned electric fields and injection of ionospheric ions into the magnetosphere (both discussed in the chapter by W. J. Burke); E- and F-region neutral winds and the equatorial electrojet (all discussed in the chapter by A. D. Richmond); and ion loss from the magnetosphere (discussed in Section VII of this chapter).

We can discuss here only a small fraction of the physical questions that the computer models have addressed, and a very small fraction of the comparisons that have been made with observations. The topics in this section were chosen with three aims in mind: 1. to show that the main qualitative conclusions of the analytic convection models of Section V are verified by the more realistic computer models; 2. to give the reader a feeling for how well convection-theory calculations agree with observations, given input information that is reasonably realistic; and 3. to give the reader some general feeling for how the inner magnetosphere behaves.

B. Input to the Computer Models

In our effort to make detailed quantitative comparisons with observations, we have chosen to model a few well-observed magnetospheric events. We estimate our required input data from observations of the events, and then check our model predictions against other observations of the same event. Of course, we try to choose events to model that are reasonably clear and "typical" examples of magnetospheric substorms or magnetic storms, because these are the magnetosphere's two clearest, large-scale modes of disturbance.

Inaccurate input information is, unfortunately, a major source of error (probably the major source of error) in these model calculations. Direct observations are never available to provide all of the detailed information that we must specify for the program. As indicated in Figure 9, the scheme requires four main inputs: 1. The number density and energy distributions of particles at our outer (or high-latitude) boundary; 2. Magnetic-field-model information, specifically flux-tube volumes and mappings from ionosphere to equatorial plane; 3. Model of height-integrated conductivity; and 4. Values of the potential V on our outer (or high-latitude) boundary. For the events discussed below, no information was available on plasma-sheet particles, so Maxwell-Boltzman distributions were assumed, with standard values for n_e, T_e, T_i; η values were assumed to be uniform over the parts of the far boundary where plasma flowed into the modeling region. Olson-Pfitzer (1974) quiet-time magnetic-field models were used, very crudely modified for the events studied; for the substorm of September 19, 1976, an extra current loop was added to represent the collapse of tail-like field lines to more dipolar form near local midnight; for the magnetic storm of July 29, 1977, the Olson-Pfitzer quiet-time model was compressed at the time of the sudden commencement, partially re-expanded later when the observed solar wind ram pressure dropped. The time-dependent auroral enhancement of height-integrated conductivities was estimated from electron fluxes measured by polar-orbiting satellite S3-2, in the case of the September 19, 1976 substorm, and from an empirical fitting formula, in the case of the July 29, 1977 storm. To provide the reader with an idea of what average distributions of height-integrated conductivities typically look like, we show, in Figure 14, Pedersen and Hall conductances for cases of low and high values of the auroral electrojet index (AE). (That index indicates the sum of the maximum strengths of the eastward and westward electrojets, as measured by a set of ground magnetometers. For definitions of this and other magnetic indices, see the review by Rostoker (1972).) The total potential drop across our high-latitude boundary was estimated from direct measurements from the S3-2 satellite, in the case of the September 19, 1976 substorm. For the July 29, 1977 storm, for which few direct polar-cap electric field measurements were available, the potential drop was estimated from the measured interplanetary magnetic field and an empirical formula discussed by Reiff et al. (1981). The formula is essentially a quantification of the observed fact that the polar-cap potential drop is largest when the interplanetary field is southward. More details concerning model inputs are described by Harel et al. (1981a).

C. Computer Models — Selected Results

Figure 15 shows computed equipotential diagrams in the equatorial plane for four times during the storm of July 29, 1977. The first diagram represents the situation before the sudden commencement. The computer model had been run for several hours magnetosphere time with essentially constant input, to come to approximate equilibrium. Note that the region near the earth is well shielded. The second diagram represents the situation just after the sudden commencement, in which the magnetosphere was drastically compressed. The magnetic field configuration was rearranged considerably, as was the inner edge of the plasma sheet. Consequently, the shielding was destroyed. Intense substorm activity began just after the compression, and that caused ionospheric conductivities to be high and shielding time scales to be long (cf., equation (73)). The third diagram in Figure 15 shows that in the computer model, there was still little shielding an hour after the sudden commencement, despite the fact that the inner edge of the plasma sheet had moved in to about 5 R_E on the nightside. Shielding did eventually reassert itself: the final diagram shows the well shielded configuration that existed after the initial substorm activity had died down. In summary, Figure 15 verifies the existence of strong shielding in steady-state situations. However, it points out that time variations can cause the shielding to breakdown and that, at least in extreme cases, the violations may persist for more than an hour.

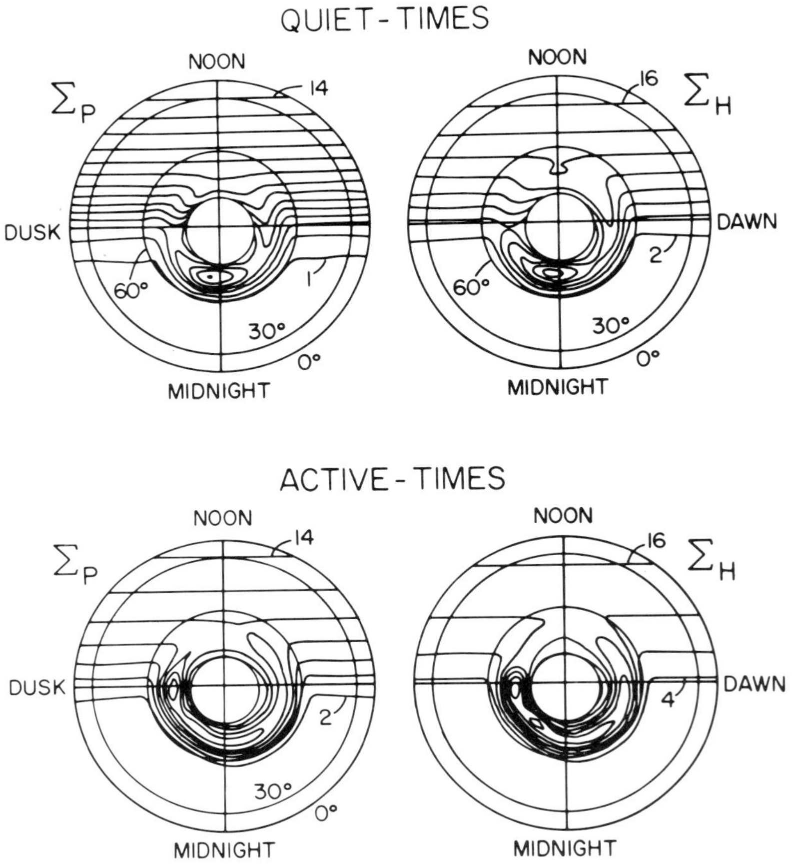

Figure 14. Typical models of Pedersen and Hall conductances (in mhos), used as input for simulations. The top two diagrams give Pedersen and Hall conductances for quiet times. The lower diagrams give similar information for active times (AE ≈ 1000 γ). The intervals between contours are 1 mho (quiet, Pedersen), 2 mhos (quiet, Hall, and active, Pedersen), and 4 mhos (active, Hall). Contours are not shown for the polar cap, which is outside our modeling region. The auroral enhancements implied by these models are based on average electron fluxes measured by the AE-C and AE-D spacecraft. (See Spiro et al., 1982.)

Figure 16 compares an average Birkeland-current pattern observed by magnetometer on the TRIAD spacecraft with a typical pattern from our computer model. Note that there is basic agreement between the computer model and the data, and that both show the same pattern as the analytic calculation (Figure 12). The computer-model current pattern extends to

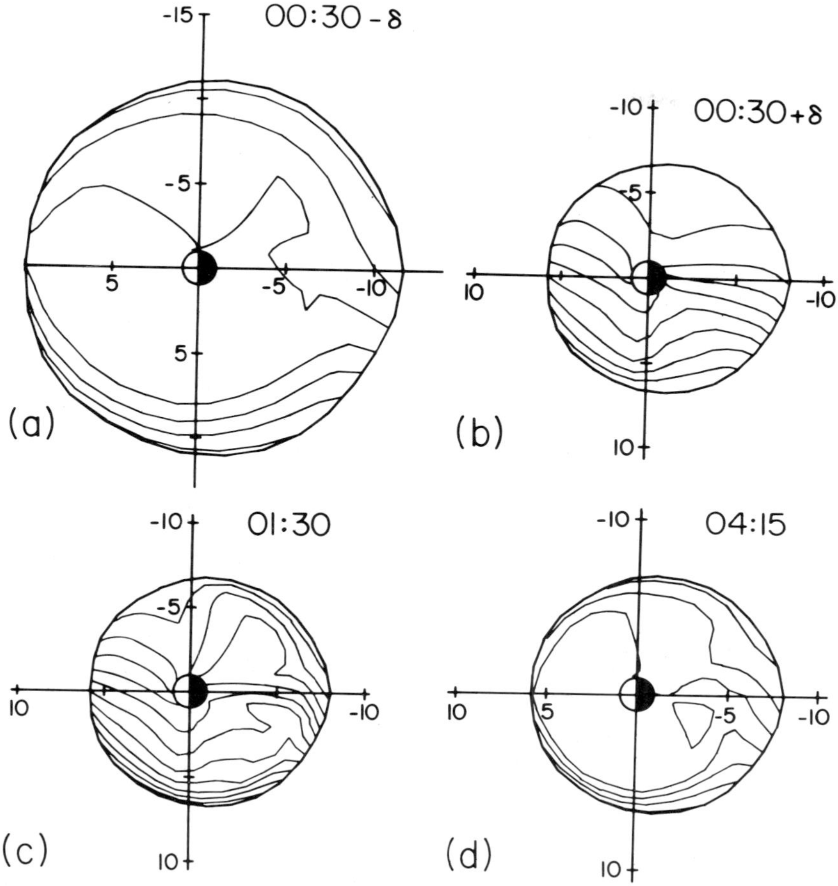

Figure 15. Equipotential plots, in the equatorial plane, for four times in the simulation of the July 29, 1977 magnetic storm. The corotation potential is not included. The potential difference between contours is 6 kV. The top two diagrams pertain to times just before and just after the sudden commencement. The sudden commencement destroyed the shielding, which took more than an hour to reassert itself. (From Wolf et al., 1982.)

lower latitude than the observational-average pattern, because the computer result pertains to a major storm, with a substantial magnetospheric compression and a strong ring current.

Figure 17 shows an ionospheric-current pattern computed for a time of high substorm activity during the event of September 19, 1976. This pattern shows a strong westward electrojet on the dawn side; however, it extends, at high latitudes, past midnight into the early-evening sector. The eastward electrojet flows around the dusk side and is generally less

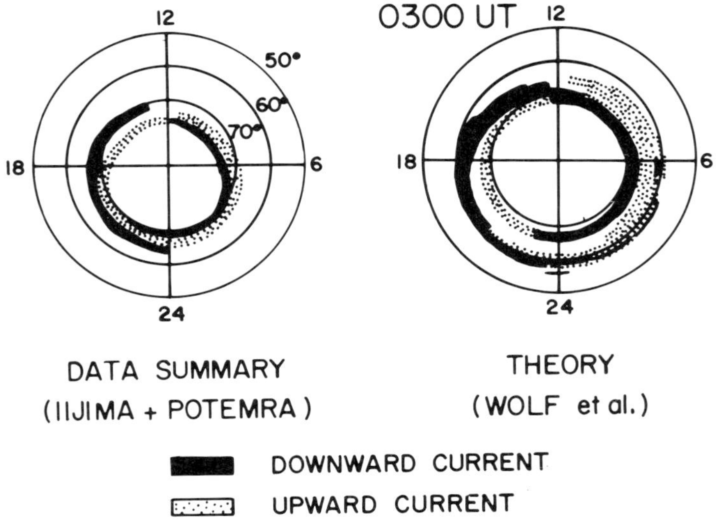

Figure 16. Comparison of an observational-average Birkeland current pattern with one computed for 0300 UT, between the two large substorms of the July 29, 1977 magnetic storm. The observational diagram was redrawn from the paper of Iijima and Potemra (1978) and pertains to AE > 100 γ.

intense. These eastward and westward electrojets are mainly Hall current. In addition, Pedersen currents flow equatorward on the dawn side, poleward on the dusk side. These Pedersen currents connect to the main region-1 and region-2 Birkeland currents (shown in Figure 16, though for a different event). The same basic substorm-time current pattern has now been derived in a number of ways, involving different observational inputs and different analysis techniques, and it seems to be fairly generally agreed upon.

Figure 18 shows the injection of plasma deep into the magnetosphere to form part of the trapped radiation, in our simulation of the magnetic storm of July 29, 1977. The picture shows the time development of the plasma-sheet inner edge for electrons that have about 50 keV energy at 5 R_E geocentric distance. Because of their strong eastward gradient drift, they formed a tongue structure, wrapping around past local noon. This formation of a long tongue turns out to depend delicately both on particle energy and on the strength of convection at the time the particles drift past local noon.

Figure 19 compares measured and theoretical values of the Dst index, which is a measure of total ring current strength. Dst is essentially a weighted average, over a set of low-latitude magnetometer stations that are well scattered in longitude, of the difference between

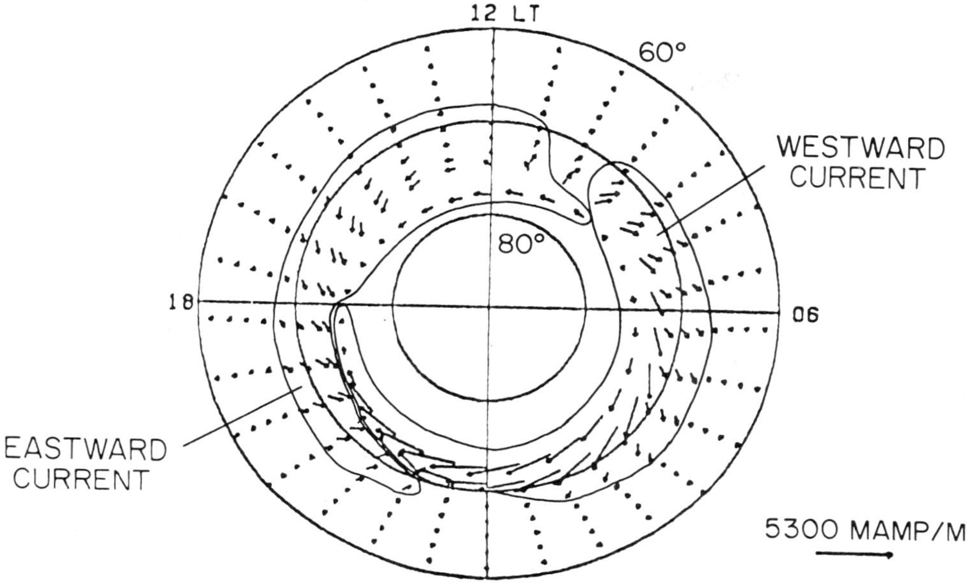

Figure 17. Ionospheric total current density, computed for a time near the peak of the substorm of September 19, 1976. (From Karty et al., 1982.)

the instantaneously measured northward magnetic field and the northward field expected on a quiet day. A westward-drifting ring current around the earth causes Dst to decrease, as it did in this storm. Boundary currents at the magnetopause (the so-called "Chapman-Ferraro currents) also contribute to Dst. The model Dst was computed by integrating the Biot-Savart law over all the magnetospheric currents within the modeling region. Currents flowing in the magnetopause were also included, and they were, in fact responsible for most of the sudden increase that occurred at 0030 UT, when the magnetosphere was compressed, and the abrupt decrease at 0430 UT, when the magnetosphere partially re-expanded. In the model, part of the depression in Dst came from newly injected full rings of current like the ones shown in Figure 18, but most came from plasma sheet particles that were injected deep into the magnetosphere but did not yet form complete rings. More complete rings would form if we modeled the recovery phase of the storm. The results shown in Figure 19 imply that enhanced convection and magnetospheric compression can inject sufficient plasma-sheet plasma deep into the magnetosphere to correspond to a classic storm-time ring current.

We should remark that measurements of ion composition in the plasma sheet and of the low-energy part of the ring current (e.g., Young, 1979) clearly indicate that simple convection from the plasma sheet is not the only process involved in ring-current injection. There clearly is sub-

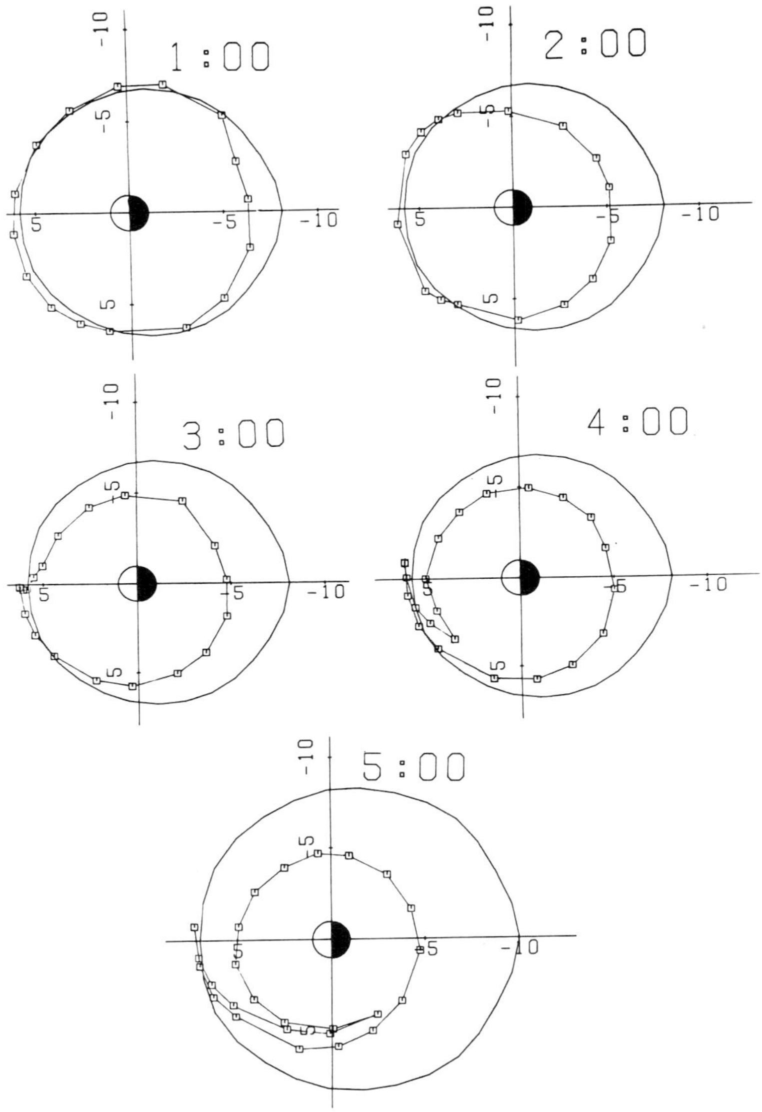

Figure 18. Injection of one component of the plasma sheet, to form part of the trapped ring current. The component shown consists of electrons with $\lambda = 3900$ eV $(R_E/\gamma)^{2/3}$, which corresponds to approximately 8 keV at L = 10, 32 keV at L = 6. (From Wolf et al., 1982.)

stantial direct exchange of ions with the ionosphere. Some ionospheric ions are accelerated up along the field lines (e.g., Sharp et al., 1982), eventually forming part of the ring-current and plasma sheet.

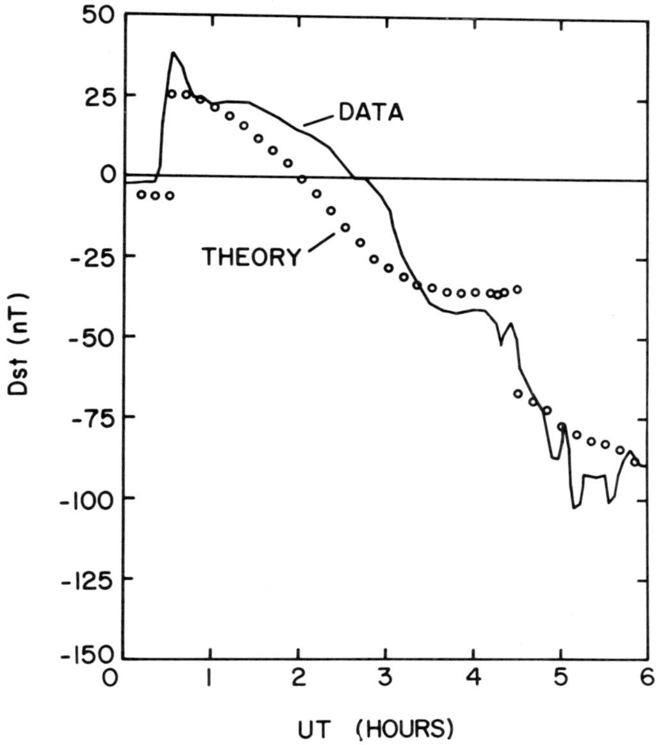

Figure 19. Comparison of observed and predicted Dst index. The zero level of the theoretical curve was chosen arbitrarily. The model sudden commencement occurred at 0030 UT, and a partial re-expansion occurred at 0430 UT. Observed Dst values were provided by M. Sugiura through the National Space Science Data Center. (From Wolf et al., 1982.)

To prevent the reader from mistakenly concluding that our computer model always agrees with observations, we present Figure 20, a comparison between observed and computed electric fields and transverse magnetic variations for one auroral-zone pass of satellite S3-2. The quantity ΔB gives essentially the line integral of $j_{\parallel}$ along the path of the spacecraft (approximately poleward in this case). Note that the model correctly predicts the sense and order of magnitude, and the fact that E and $j_{\parallel}$ are small at low latitudes. However, the detailed shape is wrong — partly due to subtle effects that are still under investigation.

Figures 21 and 22 illustrate the relationship between convection and the plasmapause. The two sides of Figure 21 show equipotential patterns, in the equatorial plane, for two times in the simulation of the September 19, 1976 substorm. The two times were chosen as approximate equilibrium states, because the total potential drop had been held

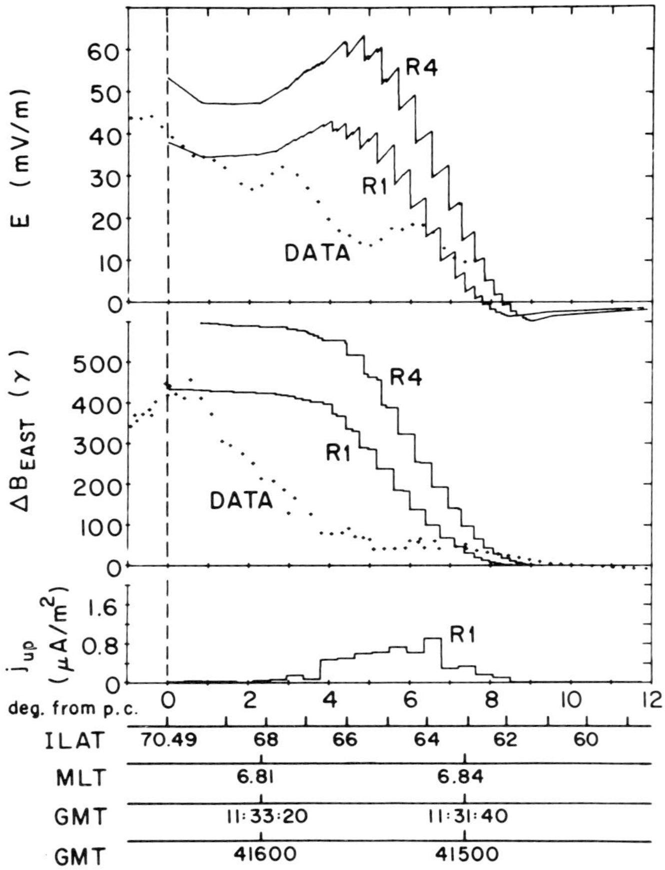

Figure 20. Comparison of theory and data for a dawnside auroral crossing of satellite S3-2. The upper panel compares the equatorward components of the observed and the model electric fields. "R1" and "R4" refer to two different computer runs. The dashed vertical line represents the poleward boundary of the calculation. The second panel compares observed and predicted eastward deflections of the spacecraft magnetometer. The bottom panel shows computed density of upward Birkeland current. Scales at the bottom show degrees invariant latitude and magnetic local time; Greenwich Mean Time in hours, minutes, and seconds; and Greenwich Mean Time in seconds. (From Harel et al., 1981b.)

steady for two hours prior to each of these times (about 20 kV for 0900 UT, about 80 kV for 1300 UT). The corotation potential is included in these displays, so that the equipotentials represent drift paths for low energy particles. If the electric-field pattern were to hold steady for several days, then the plasmapause would form at the last closed equipotential. Flux tubes earthward of this equipotential would fill up

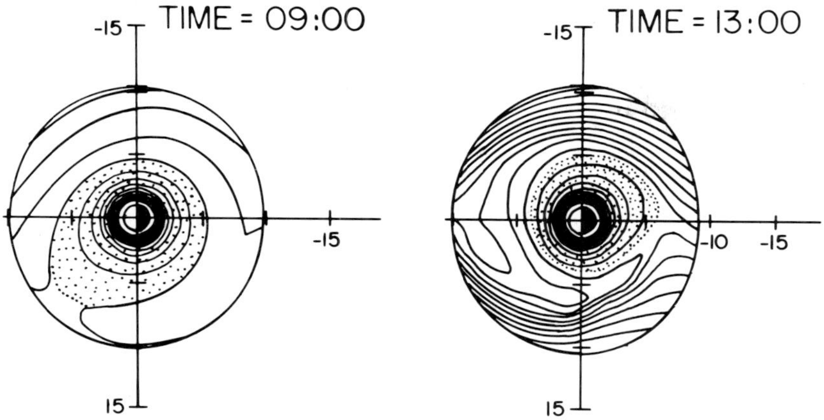

Figure 21. Equatorial equipotential diagrams for two times during the simulation of the September 19, 1976 substorm. The left diagram represents an approximate equilibrium for potential drop ≈ 18 kV, and the right diagram represents a state approaching equilibrium for potential drop ≈ 80 kV. The corotation potential is included in this display, so that the equipotentials represent $\underset{\sim}{E} \times \underset{\sim}{B}$ drift paths. Plotted equipotentials are 5 kV apart. The dotted regions are those lying on closed equipotentials.

with cold plasma evaporated up from the ionosphere, until they came into equilibrium with the ionosphere. Flux tubes on drift paths outside the last closed equipotential would presumably be emptied into the solar wind every few hours and thus would never build up high densities of cold ionospheric plasma. Note that the average radius of the last closed equipotential is smaller in the higher-convection case than in the lower-convection case, which is generally true. When convection increases, the inner edge of the plasma sheet, i.e., the shielding layer, moves earthward on the nightside. This can be seen in equation (81) (recall that $\dot{\Phi} < 0$ for positive shielding particles) or Figure 13, but is more striking in the computer-simulation results of Figure 15. Because the shielding layer usually represents a sharp gradient in the convection electric field, the last closed equipotential usually has about the same geocentric radius as the nightside shielding layer. Thus it is easy to explain the oberved fact that the plasmasphere radius correlates inversely with convection (Section II). Figure 22 shows the time evolution of the plasmapause through the model substorm. Pretending that convection had been steady for a long period before 0900 UT, we defined the plasmapause at 0900 to correspond approximately to the last closed equipotential at that time. We then followed the particles on that plasmapause boundary as they drifted during the substorm. Although the nightside plasmapause resembles the last closed equipotential by the end of the event, the dayside plasmapause does not. Much of the afternoon-sector part of the plasmasphere computed for 1300 UT in Figure 22

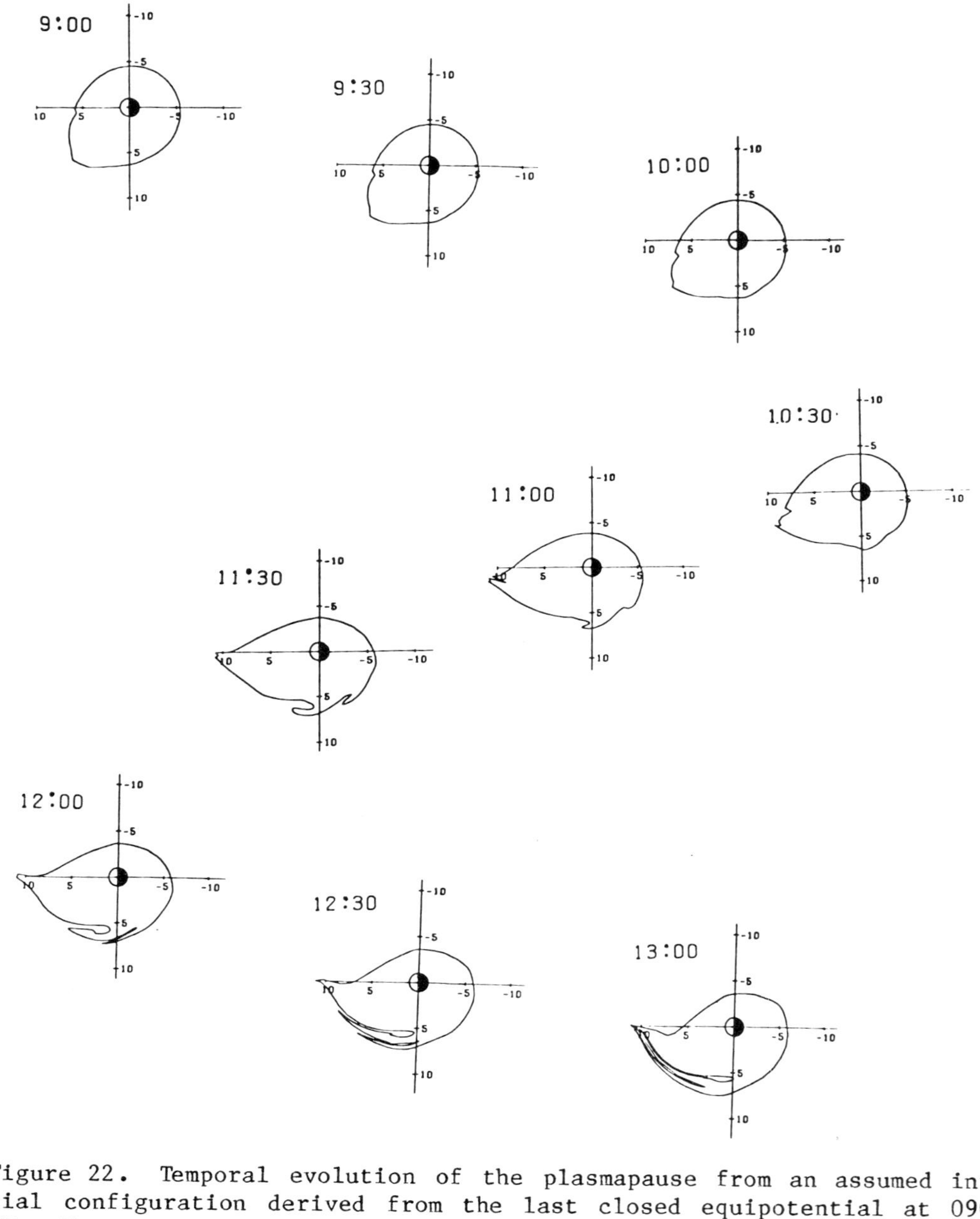

Figure 22. Temporal evolution of the plasmapause from an assumed initial configuration derived from the last closed equipotential at 0900 UT. The view is from above the magnetospheric equatorial plane with the sun to the left. The time development of the plasmapause is computed by following the $\underset{\sim}{E} \times \underset{\sim}{B}$ drift motion of plasma flux tubes from the initial boundary. (From Spiro et al., 1981.)

does not lie on closed drift paths, but is, instead, drifting to the magnetopause. A major substorm like the one modeled tears cold plasma off the outer plasmasphere and carries it away. For summaries of classic observations of the plasmasphere and the development of the theoretical ideas discussed here, see the review papers by Chappell (1972) and Carpenter and Park (1973). I should remark, however, that recent observations of plasmasphere composition and temperature structure now being made by the Dynamics Explorer spacecraft are raising new questions about the details of plasmasphere dynamics (Chappell, 1982).

Finally, Figure 23 shows the computed $\underset{\sim}{E} \times \underset{\sim}{B}$-drift velocity in the equatorial plane, near the peak of the September 19, 1976 substorm. This plot is presented to demonstrate that our computed results are consistent with our basic assumption of slow flow. The figure indicates that the peak flow velocity is about 40 km s^{-1}, whereas the sound speed and Alfvén speed in the inner plasma sheet (roughly equal since $\beta \sim 1$) are typically of the order of 300 to 1000 km s^{-1}. Thus the inertial term

$$\frac{\underset{\sim}{B} \times [\rho(\underset{\sim}{V} \cdot \nabla)\underset{\sim}{V}]}{B^2}$$

in the current equation (40) is smaller than the pressure-gradient term

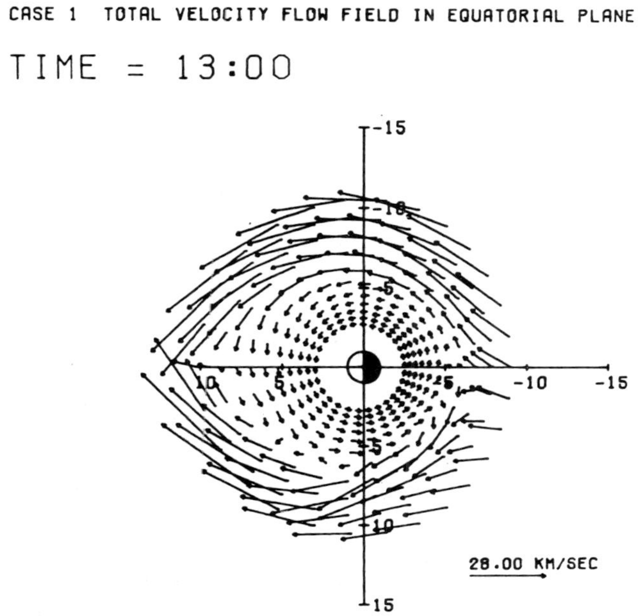

Figure 23. $\underset{\sim}{E} \times \underset{\sim}{B}$ drift velocity in the equatorial plane for a time near the peak of the September 19, 1976 substorm. The sun is to the left.

$$\frac{\underset{\sim}{B} \times \nabla p}{B^2}$$

by a factor ~ $(40/300)^2$.

Summary. In this review of computer-simulation results, I have tried to do the following: 1. Verify that the essential predictions with regard to shielding and Birkeland currents of the analytic convection model are supported by the more realistic computer models; 2. Illustrate some established features of the magnetosphere and ionosphere, including global patterns of ionospheric conductance and magnetosphere-driven ionospheric currents, ring current injection, and the relationship between the plasmapause and convection; 3. Give an indication of the level of agreement between the computer models and observations; and 4. Verify the self-consistency of our slow-flow assumption. Several physical effects still need to be included in the model, particularly magnetic-field-aligned electric fields, neutral winds, upstreaming ionospheric ions, and ion loss due to charge exchange and precipitation. The most serious need is for a magnetic-field model that is physically consistent with the convection model. These last two topics are discussed in the final section.

VII. PARTICLE LOSS AND MAGNETIC FIELD MODELS

In this section, we discuss two topics in magnetospheric theory that are closely related to the convection problem.

A. Loss of Particles from Magnetospheric Flux Tubes

Plasma-sheet and ring-current ions can leave the magnetosphere by precipitating into the dense layers of the atmosphere, or by charge exchange with a neutral atom at high altitudes (and thereby leaving the magnetosphere as energetic neutrals), or by drifting through the magnetopause. Electrons can be lost by precipitation or by drift through the boundary. Loss by drift through the magnetopause can be handled by the drift equations developed in Section III. Here we discuss estimation of the rates of precipitation and charge exchange.

Precipitation. The charged particles on a given flux tube that are subject to immediate loss are those with mirror points at altitudes of a few hundred kilometers or less. Using (10), we can write the criterion for precipitation as

$$\mu < \frac{E_K}{B_i} \tag{85}$$

where B_i is the magnetic field at the loss altitude (typically 0.5 gauss = 50,000 nT in the auroral zone) and E_K is the particle kinetic energy.

Particles that will precipitate before they mirror satisfy the following condition (expressed for an arbitrary point on the field line):

$$\frac{v_\perp^2}{v_\perp^2 + v_\parallel^2} \equiv \sin^2 \alpha < \frac{B}{B_i} \qquad (86)$$

where we have assumed, as before, that $\underset{\sim}{E} \cdot \underset{\sim}{B} = 0$, so that kinetic energy is constant along the field line. Since the magnetic field B out in the magnetosphere tends to be quite small, the particles due to be lost are confined to small pitch angles in the magnetosphere. In velocity space, these particles occupy a narrow conical region, called the "loss cone." If all particles conserved their first and second adiabatic invariants (μ and J, equations (10) and (12)) then the small number of particles in the loss cone would disappear into the atmosphere, and the rest would keep bouncing back and forth between hemispheres, conserving their μ and J. A large fraction of the plasma-sheet particles in a flux tube will be lost by precipitation only if their pitch angles are altered by violation of μ and/or J. Pitch-angle scattering obviously occurs at a substantial rate in the magnetosphere, for both electrons and ions in the plasma sheet. Both electrostatic and electromagnetic waves occur at frequencies high enough to cause violations of both μ and J. It is difficult to make quantitatively accurate predictions of loss rates. The loss rate due to pitch-angle scattering by a given type of waves is basically proportional to the square of the amplitude of the scattering waves, and this amplitude may be determined by nonlinear plasma processes that exchange energy among different wave modes, and by the rate at which wave energy propagates away from any given region, due to spatial inhomogeneities. Despite all this complexity, there is a useful and easily calculable rough upper limit on the precipitation rate. Namely, consider strong pitch-angle scattering. Assume that particles have their pitch angles scattered on each bounce by an amount large compared to the width of the loss cone. Assume further that this strong pitch-angle scattering makes the pitch angle distribution isotropic, so that the distribution function f is a function only of energy, and not of direction, at any point on the entire field line. This is then true specifically at the point of minimum B, through which all particles pass in their bounce motion. Liouville's theorem (see chapter by G. L. Siscoe) then implies that f = f(E) for the entire field line, independent of position along the field line. The distribution is isotropic everywhere on the field line, and the number density is also constant on the line. The rate at which particles of energy E, number density n are lost from a flux tube of unit magnetic flux is given by

$$\ell = \left(\frac{2}{B_i}\right) \left(\frac{n}{2}\right) \left[\frac{1}{2} \left(\frac{2E}{m}\right)^{1/2}\right] \qquad (87)$$

The factor ($2/B_i$) is the effective loss area for the tube of unit magnetic flux; the factor of 2 accounts for there being two hemispheres. The factor n/2 is the number density of downward-moving particles;

$(2E/m)^{1/2}$ is the particle velocity; and the factor 1/2 accounts for the fact that, for the downward-moving half of an isotropic particle distribution, the average downward velocity is 1/2 the total velocity. Since the number of particles in the unit flux tube is $\eta = n \int ds/B$, the inverse of the mean precipitation lifetime τ is given by

$$\frac{1}{\tau} = \frac{\ell}{\eta} = \left(\frac{E}{2m}\right)^{1/2} \left[B_i \int ds/B\right]^{-1} \tag{88}$$

One can get an idea of the magnitude of τ by working out the square-bracketed geometrical factor for a dipole magnetic field. The magnetic field due to the earth's dipole is given by

$$\underset{\sim}{B} = \frac{B_0 R_E{}^3 \left[\hat{z} - 3\hat{r}\ (\hat{r} \cdot \hat{z})\right]}{r^3} \tag{89}$$

where B_0 is the equatorial magnetic field at the earth's surface. A dipole field line is given by

$$r = L\ R_E \sin^2 \theta \tag{90}$$

where θ = colatitude. We consider field lines that extend far out in the magnetosphere ($L \gg 1$). The flux-tube-volume integral is given by

$$\int \frac{ds}{B} = 2 \int_{\theta_m}^{\pi/2} \frac{r d\theta}{B_\theta} = 2 \int_{\theta_m}^{\pi/2} \frac{\left(L\ R_E \sin^2 \theta\right)^4 d\theta}{B_0 R_E{}^3 \sin \theta}$$

where θ_m = ionospheric colatitude of the field line = $\sin^{-1} \sqrt{1/L}$. Since $\theta_m \ll 1$ for $L \gg 1$, we replace the lower limit of the integral by 0 and obtain

$$\int ds/B = 2\ L^4\ R_E\ B_0{}^{-1} \int_0^{\pi/2} \sin^7 \theta\ d\theta$$

$$\int ds/B\Big|_{\text{dipole}} = \left(\frac{32}{35} \cdot L\ R_E\right) \left(\frac{L^3}{B_0}\right) \tag{91}$$

Of course, the equatorial magnetic field is $B_0 L^{-3}$. The flux-tube volume is the inverse of the equatorial magnetic field times an effective length that is 32/35 times the geocentric distance of the equatorial crossing point. In the large-L approximation, we set $B_i = 2B_0$, and (88) becomes

$$\frac{1}{\tau} = \left(\frac{E}{2m}\right)^{1/2} \left[\frac{32}{35}\ L^4 R_E\right]^{-1} \tag{92}$$

Setting m = electron mass gives

$$\frac{1}{\tau}\bigg|_{electron} = \frac{1}{(3.5\ \text{hours})}\ \frac{E(\text{keV})^{1/2}}{(L/10)^4} \qquad (93)$$

For an ion with atomic weight A, we have

$$\frac{1}{\tau}\bigg|_{ions} = \frac{1}{(150\ \text{hours})}\ \frac{E(\text{keV})^{1/2}}{A^{1/2}(L/10)^4} \qquad (94)$$

Equations (93) and (94) (which were proposed by Kennel (1969)) or the more general formula (88) provide a handy approximate upper limit on the rate of precipitation.

Since a plasma sheet flux tube should typically convect through the magnetosphere in a few hours, and the inner edge of the plasma sheet is typically at $L \approx 10$, we see that a large fraction of plasma-sheet kilovolt electrons may be lost by precipitation. The same is not true of plasma sheet ions, unless they move into $L \lesssim 5$.

Charge Exchange. Kilovolt magnetospheric ions can also be effectively lost from the magnetospheric population by charge exchange with atoms of the neutral atmosphere at high altitudes. The reactions are of the form

$$X^+ + Y \rightarrow X + Y^+ \qquad (95)$$

where the X^+ on the left is an energetic ion from the plasma sheet or trapped radiation belts, the Y on the left is part of the neutral atmosphere (predominantly H) and has negligible kinetic energy. The electron is exchanged between the two atoms in an elastic collision. The X on the right is an energetic neutral that either escapes from the magnetosphere (having kinetic energy far above gravitational energy) or blows into the dense lower atmosphere. The Y^+ on the right has negligible kinetic energy, and thus does not contribute appreciably to the plasma pressure or gradient/curvature drift current. Figure 24 shows the product of the cross section for charge exchange and the ion velocity, for various ions and various energies. Table 1 shows lifetimes computed by Tinsley (1976) for a standard neutral-atmosphere model. Note that, for the most common ions, (H^+, O^+, He^+, He^{++}), charge-exchange lifetimes exceed typical convection times (a few hours) for $L \gtrsim 4$.

B. Magnetic Field Models

Computer models of the magnetic field in the earth's magnetosphere have developed through a line of research that has been quite separate from the convection models. The early magnetic-field models (e.g., Mead and Beard, 1964) included the earth's dipole and assumed zero current

		Charge Exchange Lifetimes, hours														
Height,	n_H,*	Energy of H^+, keV					Energy of He^{++}, keV					Energy of He^+, keV				
R_E	cm^{-3}	3	10	30	50	100	3	10	30	50	100	3	10	30	50	100
3	800	2.2	1.7	3.2	7.6	40	14	2.5	1.4	1.2	1.4	141	62	18	10.9	9.9
3.5	470	3.7	2.9	5.4	12.9	69	23	4.7	2.4	2.1	2.3	240	106	30	18.5	16.8
4	300	5.7	4.6	8.4	20	110	37	7.4	3.8	3.3	3.7	376	166	47	29	26
4.5	210	8.2	6.5	12.0	29	153	52	11	5.5	4.7	5.2	538	237	67	41	38
5	150	11.5	9.1	16.8	40	215	73	15	7.6	6.6	7.3	753	332	94	58	53

*Neutral hydrogen concentration for exospheric temperature of 950°K.

Table 1. Lifetimes (hours) against charge exchange for ions with 45° pitch angle at the magnetic equator as a function of energy and height (L value). (Taken from Tinsley, 1976.)

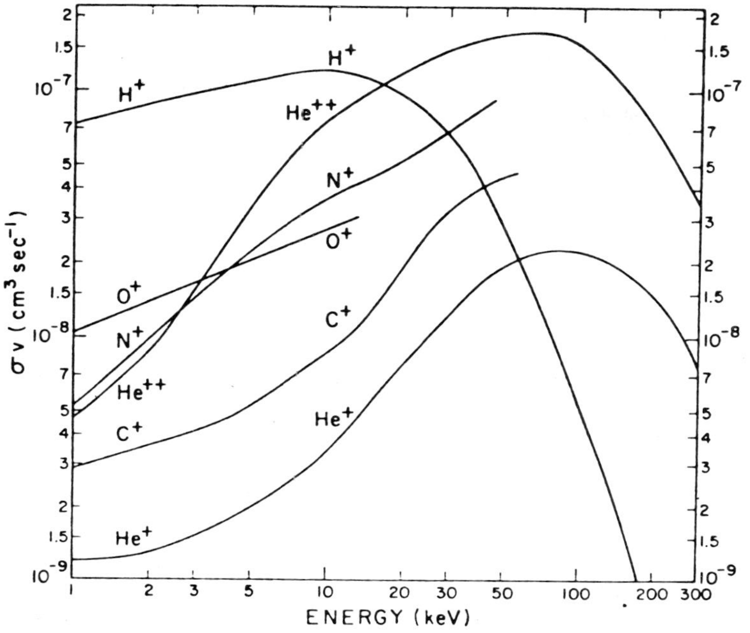

Figure 24. The product of cross section for charge exchange with H times velocity of the ion for H^+, He^{++}, N^+, O^+, C^+, and He^+ ions in the range of ion energy 1-300 keV. (From Tinsley, 1976.)

density in the magnetosphere. They also included a free magnetopause boundary, which is a tangential discontinuity separating field lines that connect to the earth from field lines that do not. (See chapter by V. M. Vasyliunas.) The magnetic field at the magnetopause satisfies

$$\underset{\sim}{B}_{mp} \cdot \hat{n} = 0 \tag{96}$$

where $\hat{n}$ = unit vector normal to the magnetopause. Magnetic pressure just inside the magnetopause is assumed to balance the total pressure just outside, the latter being estimated as $k\rho_{sw}(\underset{\sim}{V}_{sw} \cdot \hat{n})^2$, where ρ_{sw} and $\underset{\sim}{V}_{sw}$ are the mass density and velocity of unshocked solar wind, and k is a constant that is of the order of unity. We then have

$$\frac{B_{mp}^{\ 2}}{2\mu_0} = k\rho_{sw}(\underset{\sim}{V}_{sw} \cdot \hat{n})^2 \tag{97}$$

For given ρ_{sw} and $\underset{\sim}{V}_{sw}$, the solution to the equations

$$\nabla \cdot B = 0 \tag{98}$$

and $\nabla \times \underset{\sim}{B} = 0$, and conditions (97) and (98), yields the position and shape of the magnetopause, as well as $\underset{\sim}{B}$ everywhere inside. Actually, since the dipole has no length associated with it, these field models simply scale. The magnetic field just inside the magnetopause is a constant times the value of the dipole field at that point. At the subsolar point ($\hat{n} \parallel -\underset{\sim}{V}_{sw}$), the left side of (97) varies as r_{mp}^{-6}, where r_{mp} = geocentric distance of the subsolar magnetopause (often called the magnetopause standoff distance), because the dipole field varies as r^{-3}. Thus the relationship between solar-wind pressure and standoff distance is

$$r_{mp} \propto (\rho_{sw} V_{sw}^{2})^{-1/6} \tag{99}$$

As the magnetosphere was explored, the vacuum magnetic-field models were upgraded to include observed internal currents (namely tail currents and ring currents like those shown in Figure 3). The magnetic field was then required to satisfy

$$\nabla \times \underset{\sim}{B} = \mu_0 \underset{\sim}{J} \tag{100}$$

where $\underset{\sim}{J}$ was set equal to a reasonable distribution of ring and tail currents. Such models (e.g., Olson and Pfitzer, 1974) have been very successful in fitting quiet-time data. (See review by Walker, 1979.) They can be more accurately described as data-fits than as theories, since the three-dimensional current distribution $\underset{\sim}{J}(\underset{\sim}{x},t)$ is simply chosen to agree with averaged quiet-time magnetic-field data.

This semi-empirical approach has recently been extended by Olson and Pfitzer (1982) to describe a specific magnetic storm. The magnetopause standoff distance was varied according to (99), using observed values of $\rho_{sw} V_{sw}^{2}$. Total ring-current strength was adjusted in accord with the observed Dst index. Tail current strength was adjusted in accord with the Akasofu ε-index (e.g., Yeh et al., 1981), which has proved to be a good predictor of substorm activity.

In summary, observation-based quantitative magnetic-field modeling of magnetospheric storms and substorms has begun. It is not yet highly developed, but represents a promising research area.

Steps have recently been taken toward making the magnetic field models more theoretical, less empirical. For the slow-flow-region, the inertial term in the MHD momentum equation can be neglected, yielding

$$0 = -\nabla p + \underset{\sim}{J} \times \underset{\sim}{B} \tag{101}$$

Equation (101) implies that the particle pressure p is constant along a magnetic field line. One could easily envision a "new" scheme in which (98), (100), and (101) are solved for magnetospheric boundary conditions (including, of course, the earth's magnetic dipole). Instead of specifying the complete vector-current distribution for all $\underset{\sim}{x}$, as in the traditional magnetic-field models that solve only (98) and (100), one specifies only the pressure distribution in the equatorial plane as a boundary condition. The "new" scheme uses less input data because an additional physical equation (101) is solved. The quotation marks around "new" are required because magnetic field models that represent solutions to (98), (100), and (101) have been constructed for various laboratory magnetic-confinement devices and for other applications as well.

This "new" scheme has only been carried out very partially for a fully three-dimensional magnetosphere, although the problem is clearly within the range of present computer capacities. Voigt (1981) developed a set of quite realistic three-dimensional magnetic-field models that satisfy (98) and (100) accurately for magnetospheric boundary conditions, but these models solve (101) accurately only near the x-axis. Wu et al. (1981) used a three-dimensional MHD code, run long enough to come to equilibrium, to generate aproximate solutions to (98), (100), and (101). Other similar calculations are described in this volume (chapter by C. C. Wu). However, each computer run yields a magnetic-field model for one specific pressure distribution for the inner magnetosphere, and it is expensive to generate large sets of equilibrium magnetic field models in this way.

While the magnetic-field-modeling problem is computationally difficult in three dimensions, it turns out that (98), (100), and (101) can be formulated particularly elegantly and simply in two dimensions. We let $\underset{\sim}{A} = A(x,z)\,\hat{y}$, using the vector potential guarantees satisfaction of (98), and equations (100) and (101) become, respectively,

$$J_y = -\nabla^2 A/\mu_0 \tag{102}$$

and

$$\nabla p = J_y \nabla A \tag{103}$$

In this chosen two-dimensional representation, the vector $\underset{\sim}{B} = \nabla A \times \hat{y}$ is in the xz plane and is perpendicular to ∇A. Thus A is constant along a field line and can be used as a field-line label. Equation (101) or (103) imply p is also constant along a field line, so that we could write $p(x,z) = p(A(x,z))$. Then (102) and (103) combine to the relation

$$\nabla^2 A = -\mu_0 \frac{dp}{dA} \tag{104}$$

(We remark that a more general form of (104), applicable to the case where a gravitational term is included in the momentum equation (101), is derived in the chapter by A. Hundhausen.) Given dp/dA for all x and z, the equation for A is simply the Poisson equation. The essential difficulty lies in the evaluation of dp/dA as a function of x and z. Fuchs and Voigt (1979) solved the two-dimensional magnetosphere problem for the simple case where dp/dA is proportional to A. In general, the function p(A) must be specified as input to the calculation.

C. Combined Convection and Magnetic-Field Models

Basically, the "new" magnetic-field models require, as essential input, information on how much plasma is contained in each flux tube. Of course, that is information that the convection model could, in principle, provide. The convection model prescribes the history of each flux tube since it crossed the model-region boundary (where flux-tube content is prescribed by boundary condition) or since the beginning of the simulation (when flux-tube content is specified as an initial condition).

One might expect to be able to find a magnetic field model that is roughly consistent with a typical convection model, without enormous difficulty, by appropriately choosing values of parameters that are input to magnetic-field models (e.g., ring-current strength and location, inner edge of tail current, strength of tail current). For the region of the inner edge of the plasma sheet and ring current, such a parameterization approach looks like it will work, although there are annoying technical problems. Further out in the plasma sheet, however, it appears to be very difficult to mate standard magnetic-field models with the idea of convection. The central difficulty is illustrated by the results of Erickson and Wolf (1980), who tested a variety of standard magnetic-field models for consistency with the idea of sunward, adiabatic convection. Results are summarized in Figure 25, in terms of a pressure ratio p_a/p_e. The pressure $p_a(x_e)$ is given by adiabatic compression of a flux tube that crosses the equatorial plane at $x_e = -60$ (near the orbit of the moon):

$$p_a(x_e) = p_e(-60)\left[\frac{\int ds/B\ (x_e = -60)}{\int ds/B\ (x_e)}\right]^{-5/3}$$

The pressure p_e is computed from force balance in the magnetic field model. Using (100) and vector identities, we rewrite (101) in the form

$$\nabla p = -\nabla\left(B^2/2\mu_0\right) + (\underset{\sim}{B}\cdot\nabla)\underset{\sim}{B}/\mu_0$$

and integrate the z-component from the equatorial plane, where $B_x = 0$, up into the tail lobe, where $p = 0$. The result is

$$p_e(x_e) = \frac{B(x_e,z_e)^2}{2\mu_0} - \frac{1}{\mu_0}\int_0^{z_0} dz\ B_x(x_e,z)\ \frac{\partial}{\partial x}\ B_z(x_e,z)$$

(In the magnetotail, the situation is basically a one-dimensional force balance between equatorial particle pressure and tail-lobe magnetic pressure. Near the earth, the second term on the right becomes non-

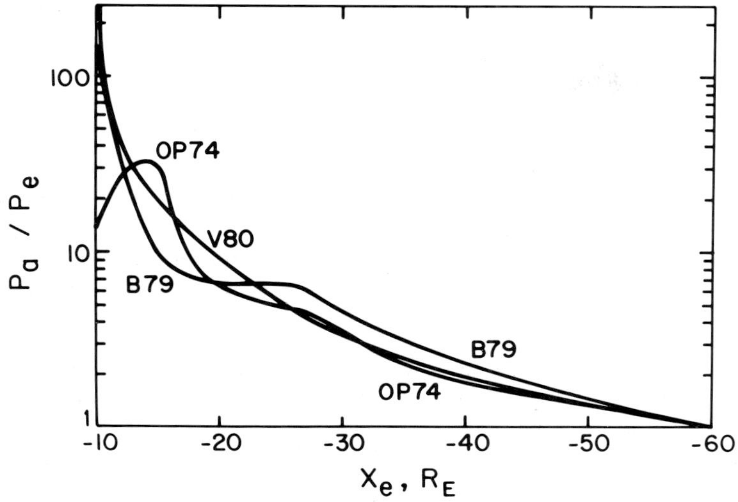

Figure 25. Pressure discrepancy between varius magnetic-field models and the adiabatic compression relation, assuming steady, sunward convection. The ratio $P_a(x_e)/P_e(x_e)$ is plotted against distance out into the magnetotail. The earth is at $x_e = 0$. the pressure $P_a(x_e)$ is the equatorial pressure at x_e, expected on the basis of adiabatic compression of plasma from $x_e = -60$. The pressure $P_e(\tilde{x}_e)$ is the pressure required to balance the lobe pressure in the magnetic field model; "OP74," "B79," and "V80" refer to the magnetic field models of Olson and Pfitzer (1974), Beard (1979), and Voigt (1981), respectively. P_a and P_e were defined to agree at $x_e = -60$.

negligible.) Figure 25 indicates that the magnetic-field models, which are mostly based on averaged observations, are drastically inconsistent with the idea of sunward, adiabatic, lossless convection. Schindler and Birn (1982), and also G.-H. Voigt and I, have constructed two-dimensional magnetic-field models for the magnetotail that are forced to be consistent with adiabatic, lossless convection; however, those models so far are grossly inconsistent with average magnetic-field observations (e.g., Behannon, 1970).

Straightforward mechanisms of particle loss do not seem to resolve the discrepancy. According to equation (93), a substantial fraction of plasma-sheet electrons may be precipitated before they come close to the earth, but the electrons carry only ~ 20% of the plasma-sheet pressure anyway — hardly enough to remove a discrepancy of more than an order of magnitude. According to equation (94) and Figure 24, ion lifetimes are much longer than convection times beyond about 6 R_E geocentric distance.

We conclude that there appears to be a fundamental physical difficulty with constructing self-consistent magnetic-field and convection models. The resolution of the difficulty is far from clear. It might lie in release of plasma by formation of a neutral line in the near-earth plasma sheet, with plasma ejection down the tail, as suggested by various people on completely different grounds, usually in connection with magnetospheric substorms (see, e.g., Hones, 1979, or Schindler, 1979).

In summary, I would say that the theories of convection and magnetic structure for the slow-flow region of the magnetosphere are reasonably well developed and have had some success in explaining and interpreting observations. More development is clearly needed, in magnetic-field modeling of magnetic storms and substorms, and in adding parallel electric fields, ion loss, ionospheric-ion injection, and neutral winds to the convection models. The biggest challenge at present lies in the self-consistent modeling of convection and magnetic fields.

Acknowledgments

I am grateful to G.-H. Voigt, R. W. Spiro, J. L. Karty, T. W. Hill, P. H. Reiff, and G. M. Erickson for helpful comments on the manuscript. Many people have contributed to the computer-simulation results presented, including those named above and also M. Harel and C.-K. Chen. This work was supported in part by the National Science Foundation under grants ATM-8206026 and ATM-8017316, by the Air Force Geophysics Laboratory under contract F19628-80-C-0009, and by the National Aeronautics and Space Administration under grant NGR44-006-137.

REFERENCES

Akasofu, S.-I.: 1979, "Physics of Magnetospheric Substorms," D. Reidel Publ. Co., Hingham, Mass.
Balsiger, H.: 1982, in B. Hultqvist and T. Hagfors (eds.), "High Latitude Space Plasma Physics," Plenum Publ. Corp., London.
Beard, D. B.: 1979, J. Geophys. Res. 84, pp. 2118-2122.
Behannon, K. W.: 1970, J. Geophys. Res. 75, pp. 743-753.
Carpenter, D. L., and Park, C. G.: 1973, Rev. Geophys. Space Phys. 11, pp. 133-154.
Chappell, C. R.: 1972, Rev. Geophys. Space Phys. 10, pp. 951-979.
Chappell, C. R.: 1982, in B. Hultqvist and T. Hagfors (eds.), "High Latitude Space Plasma Physics," Plenum Publ. Corp., London.
Crooker, N. U., and Siscoe, G. L.: 1981, J. Geophys. Res. 86, pp. 11201-11210.
Erickson, G. M., and Wolf, R. A.: 1980, Geophys. Res. Lett. 7, pp. 897-900.
Fejer, J. A.: 1953, J. Atmospheric Terrest. Phys. 4, pp. 184-203.
Fejer, J. A.: 1964, J. Geophys. Res. 69, pp. 123-137.
Fuchs, K., and Voigt, G.-H.: 1979, in W. P. Olson (ed.), "Quantitative Modeling of Magnetospheric Processes," Am. Geophysical Union, Washington, D. C., pp. 86-95.

Goldstein, H.: 1959, "Classical Mechanics," Addison Wesley Publ. Co., Reading, Mass.
Harel, M., Wolf, R. A., Reiff, P. H., Spiro, R. W., Burke, W. J., Rich, F. J., and Smiddy, M.: 1981a, J. Geophys. Res. 86, pp. 2217-2241.
Harel, M., Wolf, R. A., Spiro, R. W., Reiff, P. H., Chen, C.-K., Burke, W. J., Rich, F. J., and Smiddy, M.: 1981b, J. Geophys. Res. 86, pp. 2242-2260.
Hones, E. W., Jr.: 1979, Space Sci. Rev. 23, pp. 393-410.
Iijima, T., and Potemra, T. A.: 1978, J. Geophys. Res. 83, pp. 599-615.
Jaggi, R. K., and Wolf, R. A.: J. Geophys. Res. 78, pp. 2852-2866.
Karty, J. L., Chen, C.-K., Wolf, R. A., Harel, M., and Spiro, R. W.: 1982, J. Geophys. Res. 87, pp. 777-783.
Kennel, C. F.: 1969, Rev. Geophys. 7, pp. 379-419.
Landau, L. D., and Lifshitz, E. M.: 1960, "Mechanics," Addison-Wesley Publ. Co., Reading, Mass.
Mal'tsev, Yu. P.: 1974, Geomag. and Aeron. 14., pp. 128-129.
Mead, G. D., and Beard, D. B.: 1964, J. Geophys. Res. 69, pp. 1169-1179.
Northrop, T. G.: 1963, "The Adiabatic Motion of Charged Particles," Interscience Publ., New York.
Olson, W. P., and Pfitzer, K. A.: 1974, J. Geophys. Res. 79, pp. 3739-3748.
Olson, W. P., and Pfitzer, K. A.: 1982, J. Geophys. Res. 87, pp. 5943-5948.
Reiff, P. H., Spiro, R. W., and Hill, T. W.: 1981, J. Geophys. Res. 86, pp. 7639-7648.
Richmond, A. D.: 1973, Radio Science 8, pp. 1019-1027.
Roederer, J. G.: 1970, "Dynamics of Geomagnetically Trapped Radiation," Springer-Verlag, New York.
Rossi, B., and Olbert, S.: 1970, "Introduction to the Physics of Space," McGraw-Hill Book Co., New York.
Rostoker, G.: 1972, Rev. Geophys. Space Phys. 10, pp. 935-950.
Schindler, K.: 1979, Space Sci. Rev. 23, pp. 365-374.
Schindler, K., and Birn, J.: 1982, J. Geophys. Res. 87, pp. 2263-2275.
Schulz, M., and Lanzerotti, L. J.: 1974, "Particle Diffusion in the Radiation Belts," Springer-Verlag, New York.
Sharp, R. D., Ghielmetti, A. G., Johnson, R. G., and Shelley, E. G.: 1982, in R. G. Johnson (ed.), "Energetic Ion Composition in the Earth's Magnetosphere," Center for Academic Publications and Japan Scientific Society Press, Tokyo.
Siscoe, G. L.: 1982, J. Geophys. Res. 87, pp. 5124-5130.
Southwood, D. J., and Wolf, R. A.: 1978, J. Geophys. Res. 83, pp. 5227-5232.
Spiro, R. W., Harel, M., Wolf, R. A., and Reiff, P. H.: J. Geophys. Res. 86, pp. 2261-2272.
Spiro, R. W., Reiff, P. H., and Maher, L. J.: 1982, to be published in J. Geophys. Res.
Stiles, G. S., Hones, E. W., Jr., Bame, S. J., and Asbridge, J. R.: 1978, J. Geophys. Res. 83, pp. 3166-3172.
Tinsley, B. A.: 1976, J. Geophys. Res. 81, pp. 6193-6196.
Vasyliunas, V. M.: 1970, in B. M. McCormac (ed.), "Particles and Fields in the Magnetosphere," D. Reidel Publ. Co., Hingham, Mass., pp. 60-71.

Vasyliunas, V. M.: 1972, in B. M. McCormac (ed.), "Earth's Magnetospheric Processes," D. Reidel Publ. Co., Dordrecht, Holland, p. 29.
Voigt, G.-H.: 1981, Planet. Space Sci. 29, pp. 1-20.
Walker, R. J.: 1979, in W. P. Olson (ed.), "Quantitative Modeling of Magnetospheric Processes," Am. Geophysical Union, Washington, D.C., pp. 9-34.
Wolf, R. A., Harel, M., Spiro, R. W., Voigt, G.-H., Reiff, P. H., and Chen, C.-K.: 1982, J. Geophys. Res. 87, pp. 5949-5962.
Wolf, R. A., and Jaggi, R. K.: 1973, Comm. Astrophys. Space Phys. 5, pp. 99-107.
Wu, C. C., Walker, R. J., and Dawson, J. M.: 1981, Geophys. Res. Lett. 8, pp. 523-526.
Yeh, T., Kan, J. R., and Akasofu, S.-I.: 1981, Planet. Space Sci. 29, pp. 425-429.
Young, D. T.: 1979, in W. P. Olson (ed.), "Quantitative Modeling of Magnetospheric Processes," Am. Geophysical Union, Washington, D.C., pp. 340-363.

GLOBAL MHD MODEL OF THE EARTH'S MAGNETOSPHERE

C.C. Wu
Department of Physics
University of California
Los Angeles, California 90024

A global MHD model of the earth's magnetosphere is defined. An introduction to numerical methods for solving the MHD equations is given with emphasis on the shock-capturing technique. Finally, results concerning the shape of the magnetosphere and the plasma flows inside the magnetosphere are presented.

I. INTRODUCTION

In a global MHD model, MHD equations are used to describe the solar wind interaction with the magnetosphere. In recent years considerable efforts have been directed towards numerically solving these highly-nonlinear, time-dependent, three-dimensional equations. In this paper an introduction to the global MHD model will be presented, with about half of the paper devoted to the numerical methods used in the model and the other half devoted to results from my calculations. These results concern the shape of the magnetosphere and the plasma flows inside the magnetosphere.

The first numerical solutions of a global model were carried out by Spreiter and his coworkers (Spreiter and Alksne, 1969). However, these calculations modelled only the flow outside the magnetosphere. Recent global MHD models have been studied by many researchers including Leboeuf et al. (1979), Lyon et al. (1980), and Wu et al. (1981). The discussions in this lecture will closely follow my own work.

II. GLOBAL MHD MODEL

Our model of the earth's magnetosphere is based on an MHD description of the interaction of the solar wind and the geomagnetic field. The ideal MHD equations are:

$$\frac{\partial \rho}{\partial t} = -\vec{\nabla} \cdot (\rho \vec{v}), \tag{1}$$

R. L. Carovillano and J. M. Forbes (eds.), Solar-Terrestrial Physics, 369–391.

$$\frac{\partial(\rho\vec{v})}{\partial t} = -\vec{\nabla}\cdot[\rho\vec{v}\,\vec{v} + \overleftrightarrow{I}(p + \frac{B^2}{2}) - \vec{B}\,\vec{B}] \tag{2}$$

$$\frac{\partial\vec{B}}{\partial t} = \vec{\nabla}\times(\vec{v}\times\vec{B}), \tag{3}$$

$$\frac{\partial\varepsilon}{\partial t} = -\vec{\nabla}\cdot[(\frac{1}{2}\rho v^2 + \frac{p}{\gamma-1} + p)\vec{v} - (\vec{v}\times\vec{B})\times\vec{B}] . \tag{4}$$

Here the mass density, pressure, velocity and magentic field are denoted by ρ, p, $\vec{v}$ and $\vec{B}$ respectively, γ is the ratio of specific heats, and the energy density (ε) is given by $\varepsilon = \rho v^2/2 + B^2/2 + p/(\gamma-1)$. In Eqs. (1-4) we have chosen units such that factors of 4π and c do not appear. Here we have written the equations in the conservative form, the importance of which will be discussed later. In our calculations the geomagnetic field was approximated by a dipole field,

$$\vec{B}(\vec{r}) = \frac{3\hat{r}(\vec{\mu}\cdot\hat{r}) - \vec{\mu}}{r^3} \tag{5}$$

where $\vec{\mu}$ is the dipole moment and $\vec{r}$ is the position relative to the dipole center.

Since we are treating the model as an initial- and boundary- value problem, both initial conditions and boundary conditions must be specified. There are two kinds of boundary conditions: the ionosphere boundary is a physical one where the ionosphere-magnetosphere interaction is treated; the four boundaries marked by either "inflow" or "outflow" in Fig. 1 are numerical boundaries which are introduced to limit the computational domain. The physical boundaries are located at infinity. Initially the plasma is in a static equilibrium with the geomagnetic dipole field. At time = 0 the solar wind is introduced at the inflow boundary. Subsequent time evolution is obtained by integrating the MHD equations. The set-up of this model is close to that of wind tunnel experiments. Exact specifications of the initial and boundary conditions used in the calculations will be given in Section IV.

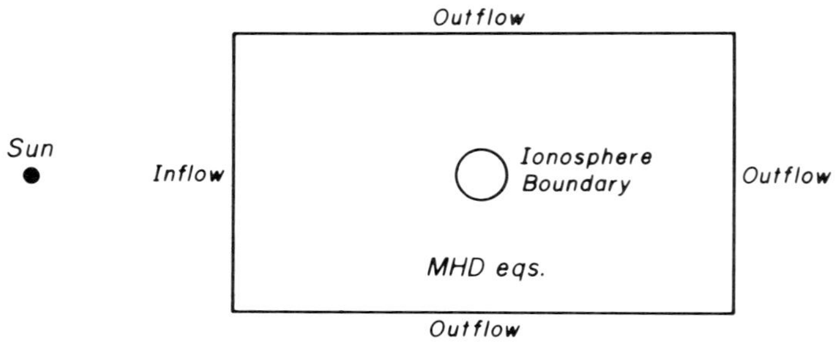

Figure 1. Global MHD model.

III. NUMERICAL METHODS

In our global MHD model bow shock and magnetopause boundaries are captured in the calculations, and explicit jump conditions are not required. Although the capturing technique has been extensively used in gas dynamics, for instance in studying reentry problems for space vehicles, the technique is relatively new to researchers in magnetospheric physics. For illustration we consider the one-dimensional gas flow problem. The governing equations are

$$\frac{\partial \rho}{\partial t} = -\frac{\partial}{\partial x}(\rho v_x) \tag{6}$$

$$\frac{\partial(\rho v_x)}{\partial t} = -\frac{\partial}{\partial x}(\rho v_x^2 + p) \tag{7}$$

$$\frac{\partial \varepsilon}{\partial t} = -\frac{\partial}{\partial x}\left[\left(\frac{1}{2}\rho v_x^2 + \frac{p}{\gamma-1} + p\right)v_x\right] \tag{8}$$

with $\varepsilon = \frac{1}{2}\rho v_x^2 + \frac{p}{\gamma-1}$. This system can be rewritten in the form

$$\frac{\partial u}{\partial t} + \frac{\partial}{\partial x}F = 0 \tag{9}$$

with u = column vector $(\rho, \rho v_x, \varepsilon)$ and F = column vector $(\rho v_x, \rho v_x^2 + p, (\varepsilon+p)v_x)$.

When discontinuities develop, the meaning of Eqs. (6-8) must be carefully considered.

The set of Eq. (9) represents the conservation laws of mass, momentum and energy. They are derived from the integral equation (in 3D notation)

$$\frac{\partial}{\partial t}\int_V u \, d\tau + \int_\sigma \vec{F} \cdot d\vec{A} = 0 \tag{10}$$

where V represents volume and σ is the boundary surface. These equations indicate that the change of mass, say, in the volume is equal to the flux crossing the surrounding boundary.

It is obvious that the solution of the differential Eq. (9) is also the solution of the integral Eq. (10); however the solution of the integral equations need not be the solution of the differental equations. Discontinuous solutions are admissable for integral equations but not allowed for differential equations. Therefore if one is interested in having discontinuous solutions, one must solve the integral equations.

However, integral equations allow solutions that are not physical. Let us derive the jump conditions for a steady normal shock. From equation (10) we have

$$F_1 = F_2 \tag{11}$$

where subscripts 1 and 2 represent conditions before and after the shock boundary. Particles move from region 1 to region 2 across the shock. From this equation we can relate quantities on one side of the shock to quantities on the other side by

$$\frac{\rho_2}{\rho_1} = \frac{\frac{\gamma+1}{\gamma-1}\frac{p_2}{p_1} + 1}{\frac{\gamma+1}{\gamma-1} + \frac{p_2}{p_1}} = \frac{v_1}{v_2} \tag{12}$$

These relations are called Rankine-Hugoniot equations. Furthermore, in terms of Mach number M, defined as the ratio of the fluid velocity to the fluid sound speed, $M = v/(\gamma p/\rho)^{1/2}$, we have the relation

$$\frac{p_2}{p_1} = 1 + \frac{2\gamma}{\gamma+1}(M_1^2 - 1) \tag{13}$$

and

$$\frac{\rho_2}{\rho_1} = \frac{(\gamma+1)M_1^2}{(\gamma-1)M_1^2 + 2} \tag{14}$$

According to the second law of thermodynamics, the entropy can only increase following a particle. For an ideal gas $s = C_v \log(p/\rho^\gamma)$. It can be shown that the entropy increases across the shock if and only if $p_2 > p_1$. A shock is always compressive. It then follows from Eqs. (12-14) that $M_1 > 1$, $\rho_2 > \rho_1$, $v_2 < v_1$, and $T_2 > T_1$, where temperature $T = p/\rho$. When $M_1 < 1$, the jump conditions given in Eqs. (12-14) refer to a nonphysical rarefaction shock. For $M_1 < 1$ the physical solutions should be smooth.

Now we will discuss numerical methods of solving gas flow problems. In one dimension, discretization of Eq. (10) gives, referring to Figure 2,

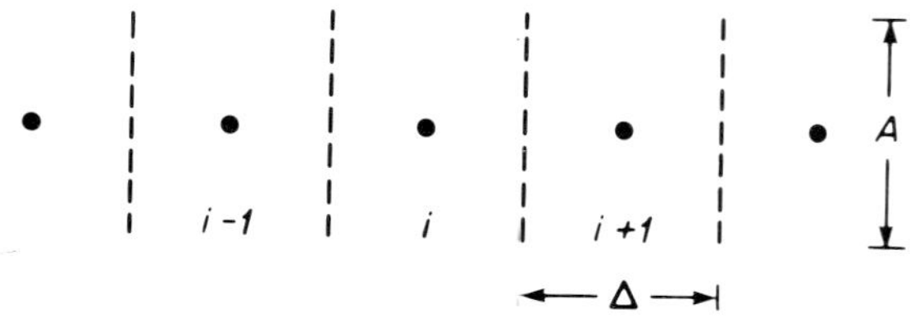

Figure 2. Discrete representation of x variable.

$$\frac{\partial}{\partial t} \; (u_i \cdot \Delta \cdot A) \; + \; F_{i+1/2} \cdot A \; - \; F_{i-1/2} \cdot A \; = \; 0 \qquad (15)$$

where A is the cross sectional area and Δ is the grid spacing. Dividing by ΔA in Eq. (15) gives

$$\frac{\partial}{\partial t} u_i \; + \; \frac{F_{i+1/2} - F_{i-1/2}}{\Delta} \; = \; 0 \qquad (16)$$

This is the finite difference form for the differential Eq. (9). Thus solving for differential Eq. (9) in the conservative form is in effect solving for integral Eq. (10).

To ensure an entropy-increase condition, dissipation is required. By using numerical dissipation, we will obtain Rankine-Hugoniot conditions, but not the details of the shock transition layer. The shock transition region will occupy 3 to 6 grid points depending on the numerical scheme (see review by Sod, 1978).

The Lax scheme (Lax, 1954), which is the simplest method of solving the conservation laws, replaces the time derivative in Eq. (16) by a forward difference in which averaging is introduced at the earlier point, i.e.,

$$\frac{\partial u_i}{\partial t} \; \rightarrow \; \frac{u_i^{t+\Delta t} - \frac{1}{2}\left(u_{i+1}^t + u_{i-1}^t\right)}{\Delta t} \qquad (17)$$

After a slight modification of Eq. (16), the difference equation in the Lax scheme becomes

$$\frac{u_i^{t+\Delta t} - \frac{1}{2}\left(u_{i+1}^t + u_{i-1}^t\right)}{\Delta t} \; + \; \frac{F_{i+1}^t - F_{i-1}^t}{2\Delta} \; = \; 0 \qquad (18)$$

The difference equation is equivalent to the equation

$$\frac{\partial u}{\partial t} + \frac{\partial F}{\partial x} \; = \; \frac{\Delta^2}{2\Delta t} \; \frac{\partial^2 u}{\partial^2 x} \; + \; \ldots . \qquad (19)$$

The dissipation term makes the scheme numerically stable and also provides dissipation for ensuring entropy condition. The integration time step size is limited by the Courant-Friedrichs-Lewy (CFL) condition that

$$\Delta t \; \leq \; \frac{\Delta x}{v} \qquad (20)$$

where v is the maximum characteristic speed. This means that the time step size should be small enough that a disturbance cannot travel more than one grid spacing in that time interval.

Let us emphasize the main points of our discussions in the following. If the solution of the gas flow problem is smooth then one can seek a solution from Eq. (9) or from another set of Eq. (9), such as

$$\frac{\partial \rho}{\partial t} + \frac{\partial}{\partial x}(\rho v_x) = 0 \tag{21}$$

$$\frac{\partial v_x}{\partial t} + v_x \frac{\partial}{\partial x} v_x + \frac{1}{\rho}\frac{\partial p}{\partial x} = 0 \tag{22}$$

$$\frac{\partial p}{\partial t} + \frac{\partial}{\partial x}(p v_x) + (\gamma-1) p \frac{\partial}{\partial x} v_x = 0 \tag{23}$$

and one shall obtain the same results. But if there are discontinuities in the solution, one should use the conservation laws of mass, momentum and energy. In addition one should require entropy conditions across the shock. Although numerical dissipation is used to obtain entropy conditions, sometimes a nonphysical shock can result in some numerical scheme (Osher and Solomon, 1982). It is advisable to check the jump conditions in the solution.

In the following I will show two examples of gas flow problems in which the Lax scheme was used. The first one is a typical shock tube problem. In this problem the initial conditions consist of two constant states separated by a diaphragm. At time $t = 0$ the diaphragm was abruptly removed. The initial data are $\rho = 1$, $p = 1$, $v_x = 0$ for $x \leq 0.5$ and $\rho = 0.125$, $p = 0.1$, $v_x = 0$ for $x > 0.5$. In this example $\gamma = 1.4$ and 1600 points were used in the calculations. The configuration at $t = 0.206$ is

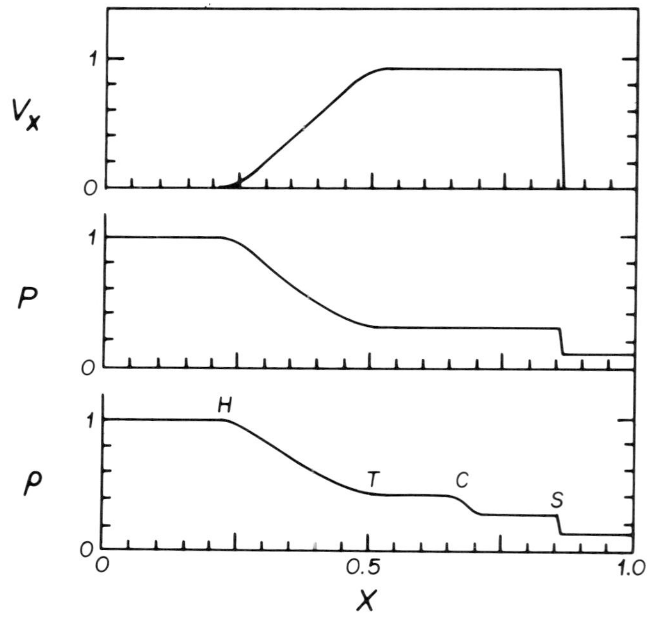

Figure 3. Shock tube problem. S, C, T and H denote the positions of shock, contact discontinuity, tail and head of a rarefaction wave, respectively.

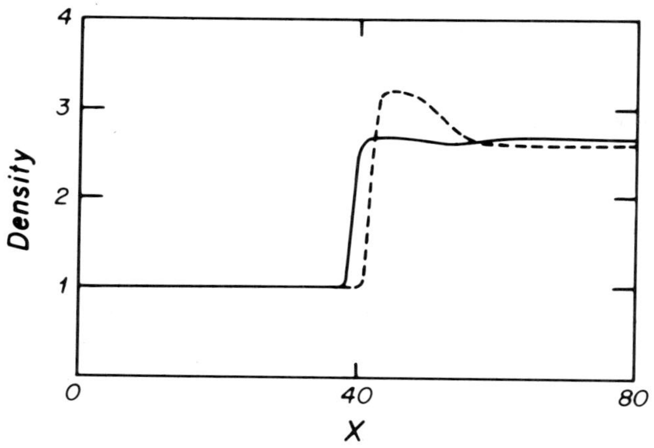

Figure 4. Comparisons between two formulations. The solid line gives the density distribution using conservative form, Eqs. (6-8), and the dashed line gives the density distribution using nonconservative form, Eqs. (21-23).

shown in Fig. 3. It contains five regions separated by a shock front, a contact surface, the tail and the head of a rarefaction wave. The results agree with the analytical solution. The transition of the shock occupied 4 to 6 grid points and the transition of the contact discontinuity occupied a wider region. Further discussions can be found in the review by Sod (1978), where a survey of finite difference schemes was given.

The second example is given to show the importance of using the conservative form of Eqs. (6-8). The initial data are $p = 0.5$, $\rho = 1$, $v = 4$ for $x \leq 40$, $p = 10.5$, $\rho = 8/3$, $v = 1.5$ for $x > 40$, and $\gamma = 2$. These two constant states satisfy the Rankine-Hugoniot jump conditions. The calculations were carried out with 400 points. Figure 4 shows the density distributions at $t = 9.6$ in two different calculations, one with Eqs. (6-8) and the other with Eqs. (21-23). The solution from the conservation laws shows consistency with the Rankine-Hugoniot relations. The solution from Eqs. (21-23), however, differs greatly from the Rankine-Hugoniot jump conditions. Since the pressure equation, (23), does not correspond to any conservation law, the jump condition is not defined. As shown in Figure 4, density reaches 3.2 after the discontinuity, while the Rankine-Hugoniot condition gives a limit of $\rho_2/\rho_1 \leq 3$ for $\gamma = 2$.

The MHD equations were written in the conservative form in Eqs. (1-4). The discontinuities in the MHD equations are much more complicated than the fluid dynamics. In addition to the entropy condition an additional condition, a so-called evolutionary condition, is required. It is argued that the evolutionary condition will be satisfied if the problem is treated as an initial value problem (Chu and Taussig, 1967). Thus, the numerical technique used in the fluid dynamics will be applied directly.

The global MHD modelling is a very large scale numrical work. The solutions have two discontinuities: bow shock and magnetopause, as well as two reconnection regions, one on the dayside and one on the nightside (following Dungey's reconnection model). To obtain accurate results, a mesh is required of about 100 by 50 by 50 covering the quadrant $25 \geq x \geq -200$, $y \geq 0$ and $z \geq 0$ in the solar magnetospheric coordinate system. Not only does the MHD modelling require such a large system, it also has to deal with several time scales.

In our magnetospheric model the fastest speed is the Alfvén speed near the earth, which is two orders of magnitude larger than the solar wind velocity. The shock formation develops in the time scale determined by the solar wind; but according to the CFL condition, Eq. (20), the numerical integration time step size is limited by the Alfvén speed. Therefore it is very important to alleviate this severe limitation. In realizing that the large Alfvén speed is localized near the earth, since the magnetic dipole field drops off as r^{-3}, one can avoid the limitation by devising a scheme which employs two calculation regions: (1) a small region around the earth with time step size limited by the Alfvén speed, and (2) a much larger surrounding region with the time step limited by the solar wind flow velocity. This procedure permits a two order-of-magnitude gain in computer speed over the previous case.

In addition, on can employ a nonuniform grid system in which grid points are concentrated in those areas (such as at the bow shock, the magnetopause and the ionosphere-magnetosphere boundary) where a more detailed spatial resolution is necessary.

A typical run with 94 by 53 by 53 grid points to produce a steady magnetospheric configuration (presented in the next section) took 2 to 3 hours on the Cray-1 computer.

IV. RESULTS

In the following some results from my recent calculations will be presented. The results concern the shape of the magnetopause and the flow pattern in the magnetosphere. The solar magnetospheric coordinate system is used.

We used Rusanov's scheme (Rusanov, 1962) in these calculations. Rusanov's method is a variant of Lax's scheme in which the artificial viscosity is maintained at a minimum value except where a large value is needed. Instead of the diffusion term in Eq. (19), the Rusanov scheme uses

$$\frac{\Delta^2}{\Delta t} \frac{\partial}{\partial x} \left(\alpha \frac{\partial u}{\partial x}\right) \qquad (24)$$

where α in the MHD model is proportional to $(v+c_s+c_A)/(v+c_s+c_A)_{max}$, with sound speed c_s and Alfven speed c_A. The diffusion term is therefore

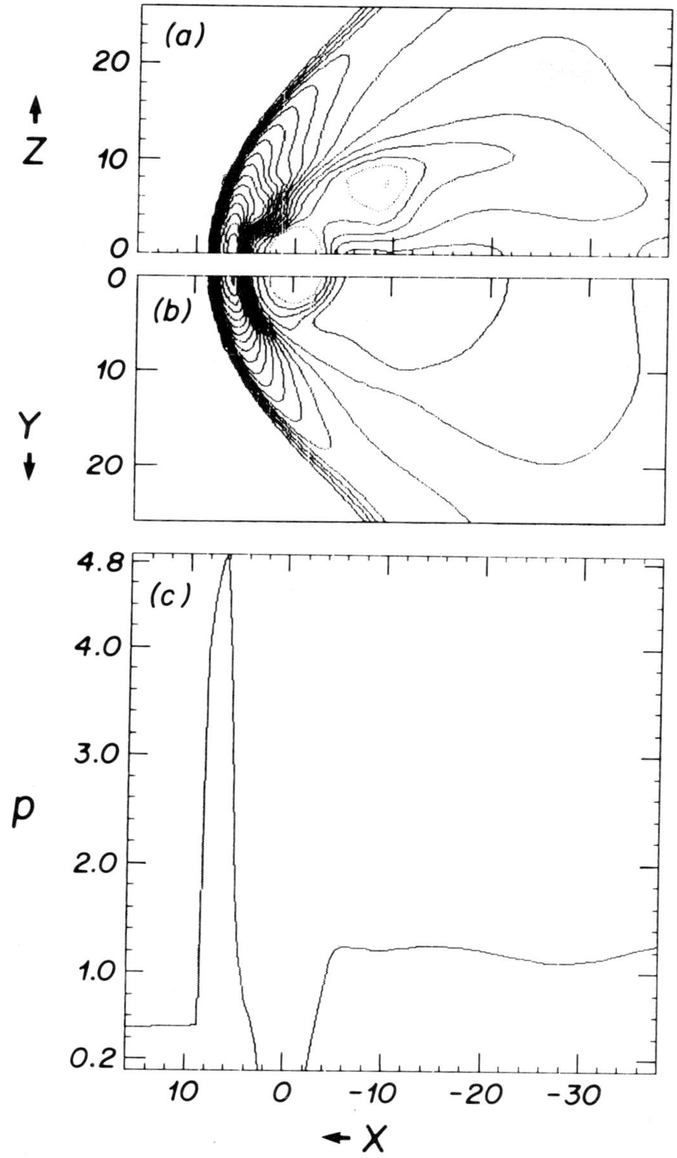

Figure 5. Pressure contours on both the noon-midnight and the equatorial planes in (a) and (b) respectively, and the pressure distribution along the earth-sun line in (c) for the case IMF = 0. The solid lines denote $p \geq p_{s.w.}$ and the dotted lines $p < p_{s.w.}$.

large only in the regions near the bow shock and magnetopause where the change of v is large and in the dipolar region where C_A changes rapidly.

For t = 0 the initial data are p = 0.2, ρ = 0.1, $\vec{v} = 0$, and $\vec{B} = \vec{B}_{dipole}$. At the inflow boundary the values of all variables are fixed to the solar wind parameters: p = 0.5, ρ = 1, $\vec{v} = -2.5\,\hat{x}$ and $\vec{B} = \vec{B}_{IMF}$. The boundary conditions in both the y and z directions are treated by linear extrapolation along the direction 45° from the x-axis. Linear extrapolation was used at the -x outflow boundary. The boundary conditions at the earth are specified by fixing the values of all variables at their initial values. This does not realistically model the ionospheric effects. (The problem of including the ionosphere is being investigated.)

The 3D calculations were performed with 94 x 53 x 53 grid ponts covering the quadrant $-38 \le x \le 16$, $0 \le y \le 34$, and $0 \le z \le 34$. All quantities were normalized with respect to the solar wind parameters: ρ = 1, p = 0.5 and γ = 2. Thus the solar wind sound speed equals 1. The solar wind Mach number was chosen to be 2.5. The geomagnetic field is given by Eq. (5) with $\vec{\mu} = -350\,\hat{z}$. By the pressure-balance principle, we will have the magnetopause at r = 5.8 along the sun-earth line, following the relation

$$\rho_{sw} v_{sw}^{2} = \frac{1}{2}(2\, B_{dipole})^2_{m.p.} = 2\,\frac{\mu^2}{r^6_{m.p.}} \tag{25}$$

If this position is identified at 10 R_E then 1 spatial unit is about 1.72 R_E, and 1 unit of magnetic field is about 17.3 gammas.

Figures 5(a) and 5(b) show pressure contours on both the noon-midnight and the equatorial planes. The pressure distribution along the earth-sun line is plotted in Figure 5(c). Similar plots for density distributions are given in Figure 6. These results are qualitatively similar to those given previously (Wu et al., 1981). The accuracy of our results can be evaluatd by checking against the Rankine-Hugoniot jump conditions. According to Eqs. (13-14), $p_2/p_1 = 8$ and $\rho_2/\rho_1 = 2.27$ for our case. This gives p = 4 and ρ = 2.27 just inside the shock along the sun-earth line. The calculated values, see Figures 5 and 6, agree well. The calculated stagnation pressure is 4.8 which gives k = 0.8 for the Newtonian formula $p_{st} = k\rho_{sw}v_{sw}^2$. The bow shock is located at 8.2 and the magnetopause is at 5.8. The ratio of the standoff distance to the magnetopause is then 0.41, which is larger than the observed average value of 0.33. This difference is due to the small Mach number used in our calculation.

Because of its improved accuracy, the new code enables us to study in detail the shape of the magnetopause, the flow patterns in the magnetosphere, the effects of IMF and other topics. Thus, instead of repeating our previous results (Wu et al., 1981) which showed that the MHD model reproduced many of the observed magnetospheric features including the bow

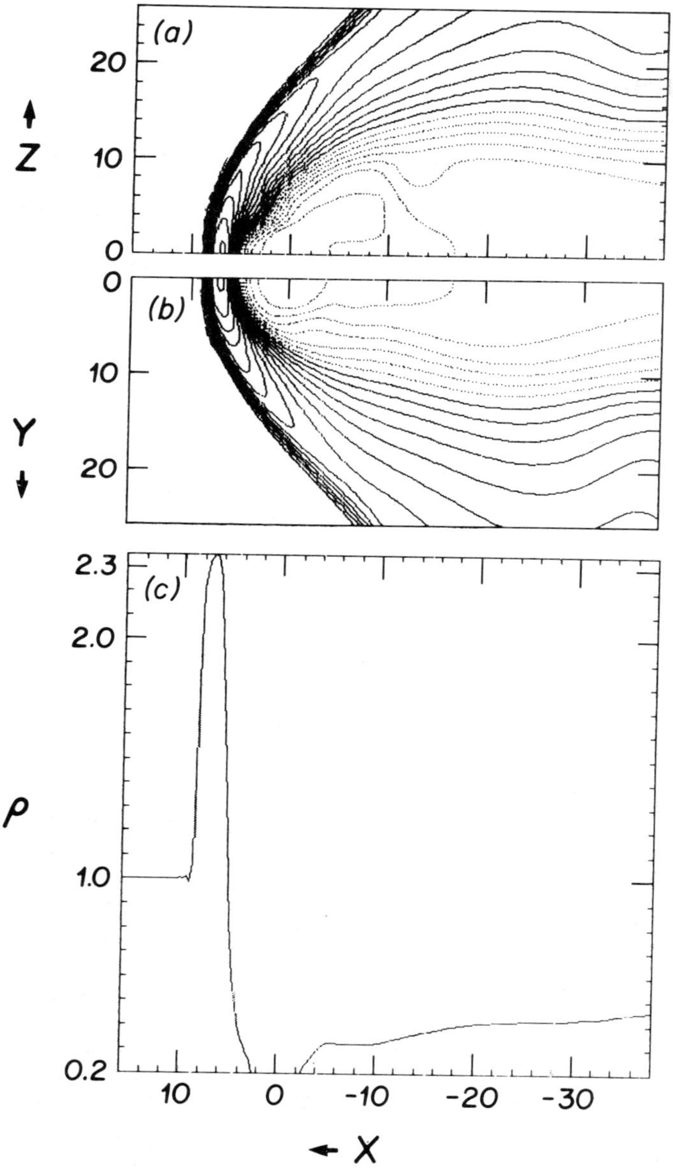

Figure 6. Density contours on both the noon-midnight and the equatorial planes in (a) and (b) respectively, and the density distribution along the earth-sun line in (c) for the case IMF = 0. The solid lines denote $\rho \geq \rho_{s.w.}$ and the dotted lines $\rho < \rho_{s.w.}$.

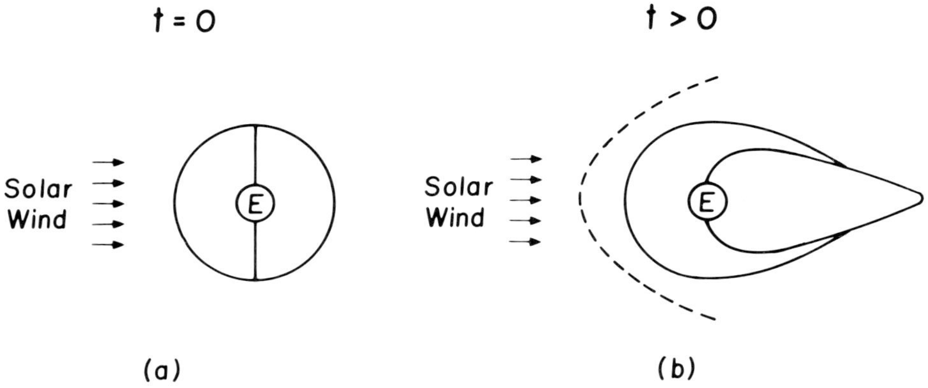

Figure 7. A sketch of the time development of the interaction of the solar wind with a line dipole. The dashed line in (b) represents the bow shock.

shock, magnetopause and plasma sheet, I will concentrate on two areas: the shape of the magnetopause and the flow patterns.

1. Shape of the Magnetosphere

I will start the discussion by considering 2D calculations. The time development of the interaction of the solar wind with a line dipole is schematically sketched in Figure 7. Figure 7(a) represents the initial state at t = 0, where plasma of uniform density and uniform pressure is in static equilibrium with the geomagnetic dipole field. The topology of the magnetic dipole field is represented by the 90° field lines which close at infinity. Figure 7(b) represents a state at a later time when the bow shock and magnetopause have been formed. The dayside magnetosphere was compressed by the solar wind and the field lines were stretched and draped around the nightside magnetosphere. The nightside magnetosphere was compressed by the dayside magnetosphere, and the field lines were stretched by the plasma motion within the magnetosphere. A tangential discontinuity was developed at the interface between the solar wind plasma and the dayside magnetosphere and will be referred to hereafter as the magnetopause boundary. Due to the stretching of the nightside magnetic field lines, a cross-tail current sheet was formed. Similarly, we found a current sheet was formed inside the dayside magnetosphere just above the cusp region due to the stretching of the dayside magnetic field lines. This current sheet will be called the cusp current sheet. Based on the pressure-balance principle, the plasma pressure has a local maximum along the tail current sheet and the cusp current sheet.

The field line pattern and the current structure of the resulting magnetosphere (Figure 7(b)) are sketched in Figure 8(a). The magnetopause surface current ($M_1M_2M_3$) flows eastward, the cusp current (C_1C_2) flows west-

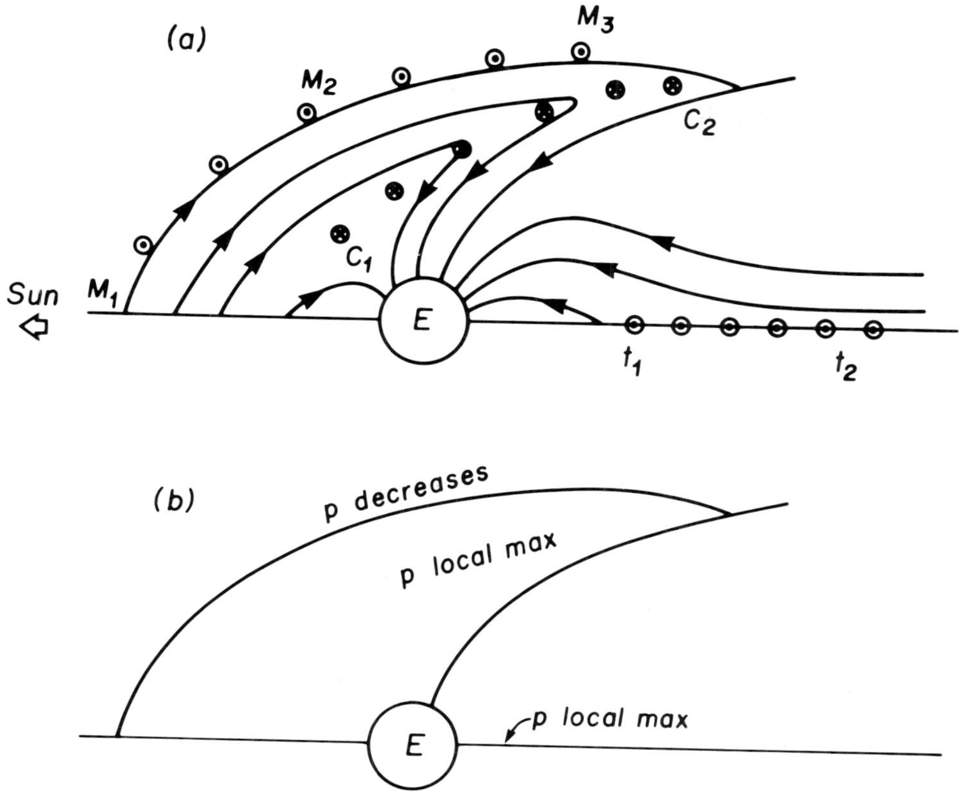

Figure 8(a). Field line pattern and current structure for the configuration Fig. 7(b). $M_1M_2M_3$ denotes the magnetopause surface, C_1C_2 the cusp current sheet and T_1T_2 the tail current sheet. C_1 denotes the location of the cusp in our MHD model.

8(b). Pressure distribution for the same configuration.

ward and the cross tail current (T_1T_2) flows eastward. Currents also flow in the bow shock eastward with southward IMF and westward with north IMF. The pressure distribution is sketched in Figure 8(b). The plasma pressure increases across the bow shock, decreases across the magnetopause and has local maxima at the tail current sheet and the cusp current sheet.

The distributions of the magnetopause surface current and the cusp current sheet are plotted as functions of the polar angle, θ, from the sun-earth line in Fig. 9. The magnetopause surface current is maximal at $\theta = 0°$ and drops off as θ increases but remains eastward. The cusp current starts near the cusp where a sheared magnetic field begins to form. It flows westward with a maximum near the cusp and drops off as θ increases further. The sum of these two current systems was also plotted in the figure. This total current system may be referred to as the Chapman-Ferraro current. Note the cusp in this MHD model refers to the region marked by

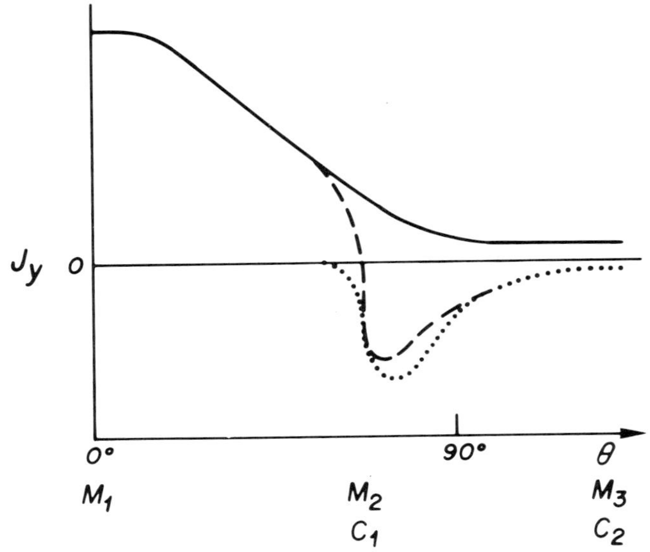

Figure 9. A sketch for the distributions of the magnetopause surface current (solid line) and the cusp current sheet (dotted line) as functions of θ. The total current distribution is given by the dashed curve.

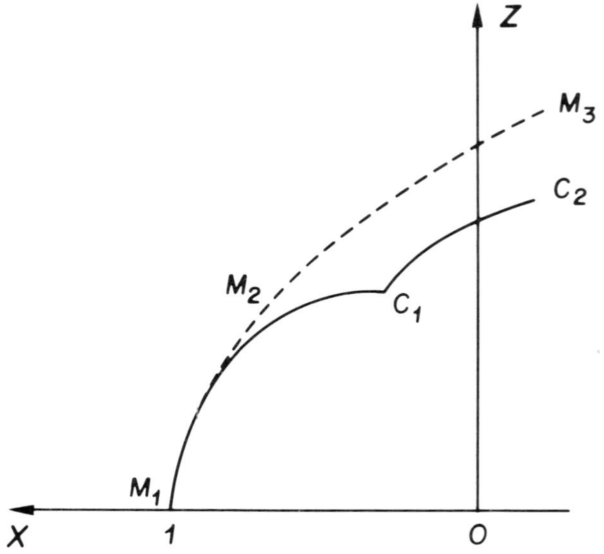

Figure 10. Shape of the magnetosphere in 2D. Curve $M_1M_2C_1C_2$ represents the shape based on the Chapman-Ferraro model. Curve $M_1M_2M_3$ represents the magnetopause surface and curve C_1C_2 shows the cusp current sheet from the MHD model.

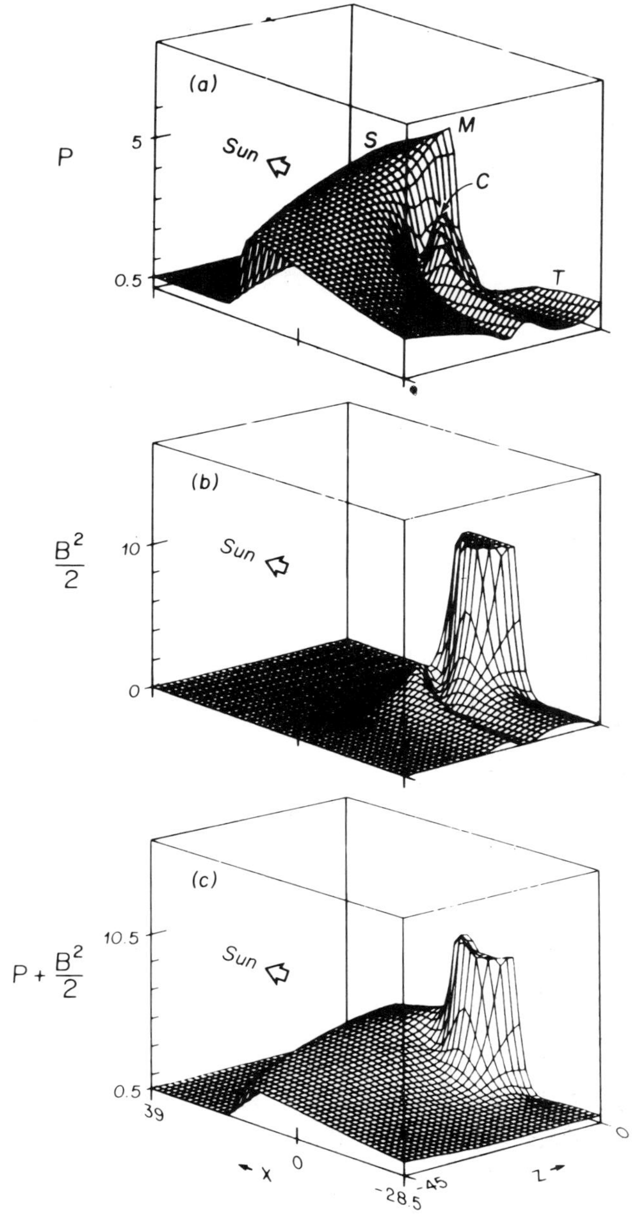

Figure 11. Perspective plots of p, $B^2/2$ and $p+B^2/2$ for the case of IMF=0 in 2D calculations. S, M, C and T denote bow shock, magnetopause, cusp current sheet and tail current sheet, respectively.

C_1 in Fig. 8(a) where field lines converge at the earth. However the cusp does not expose itself directly to the solar wind. The 90° field lines which might be identified as the cusp at t = 0 are compressed tailwards and extend to infinity in the final state.

The above picture of the magnetopause looks quite different from that of the Chapman-Ferraro model, but it can be reconciled in the following way. The Chapman-Ferraro model uses the pressure-balance principle across the boundary and assumes (a) a very simple pressure law on the outside of the boundary, (b) zero plasma pressure on the inside and (c) a zero magnetic field on the outside. The 2D Chapman-Ferrano model has been solved analytically. The shape is given in Figure 10 (also a sketch). Curve ($M_1M_2C_1C_2$) represents the shape based on the Chapman-Ferraro model, curve ($M_1M_2M_3$) represents magnetopause surface and curve (C_1C_2) shows the cusp current sheet from the MHD model. It is clear from this figure that the Chapman-Ferraro model represents the boundary of the magnetopause surface of our model from M_1 to M_2 and the cusp current sheet from C_1 to C_2. The magnetopause surface curent from M_2 to M_3 is neglected as it is small in comparison with the cusp current sheet. It is not surprising that the current distributions in both the MHD model and the Chapman-Ferraro model should agree. Both models use the pressure balance principle, and the magnetopause currents are used to shield the dipole fields.

The MHD results presented above were first suggested by our 2D calculations (Wu, 1982). The results shown in the following were carried out with 122 by 73 grid points covering the lower half of the xz plane. Figure 11(a) to 11(c) show the perspective plots of p, $B^2/2$ and $p+B^2/2$, respectively. The discontinuities are obvious in the p and $B^2/2$ plots. The $p+B^2/2$ plot shows the pressure balance principle. Figure 12 shows the field lines and Figure 13 shows the current distribution.

In 3D the dayside flux can be convected to the night side. At first it was thought that increasing the flux on the night side would push the dayside magnetosphere sunwards and that a magnetosphere shape like that of

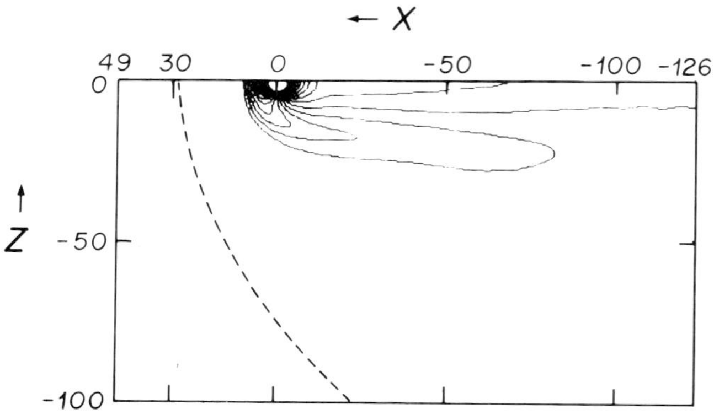

Figure 12. Magnetic field lines. The dashed line shows the location of the bow shock.

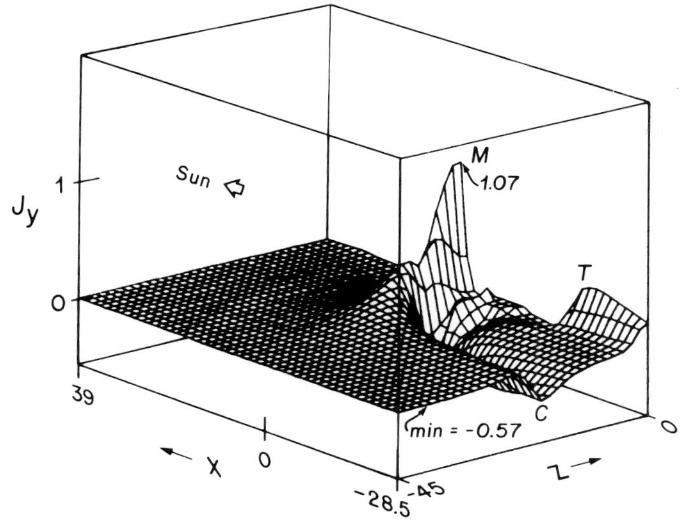

Figure 13. Perspective plot of currents for the case IMF=0. M, C and T denote magnetopause, cusp current sheet and tail current sheet respectively.

the Chapman-Ferrano model would result. This cannot happen if the 90° field line is fixed at the earth. The resulting configuration would have a cusp originating at 90° instead of ~ 75° of the Chapman-Ferraro model. But if the 90° field line is allowed to move or if reconnection processes are permitted, then a configuration of the Chapman-Ferraro model might result. However, moving the original 90° field lines to the 75° location requires a large change in the magnetic flux, and it is intuitively impossible.

Our 3D calculations show a structure similar to the 2D case on the noon-midnight meridian plane. Although more 3D calculations are required to check the dependence of the magnetopause structure on ionospheric boundary conditions and other parameters, the calculated structure seems reasonable from our arguments in the preceding paragraphs.

Here we show results when IMF is northward and has a magnitude of 0.5. The northward IMF will strengthen the cusp current and weaken both the magnetopause surface current and the cross-tail current. The southward IMF has just the opposite effect. Thus we choose the case with northward IMF to illustrate the results. Figures 14(a)-14(c) show the perspective plots of p, $B^2/2$ and $p+B^2/2$ on the noon-midnight plane, respectively. Features similar to those of the 2D case are clearly seen. The shape of the magnetosphere as obtained from our model is shown in Figure 15. Also shown in the figure are the shock positions. The predictions of the Chapman-Ferraro model are also included. Figures 16 and 17 show the field

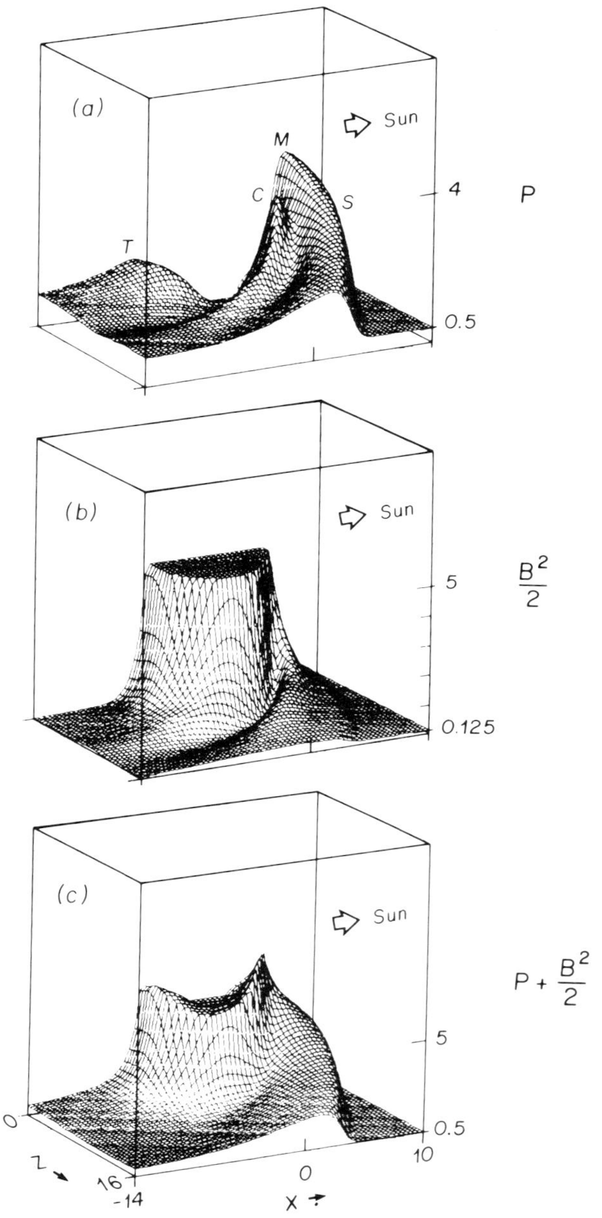

Figure 14. Perspective plots of p, $B^2/2$ and $p+B^2/2$ on the noon-midnight meridian plane for the case with northward IMF in the 3D calculations. S, M, C and T denote bow shock, magnetopause, cusp current sheet and tail current sheet, respectively.

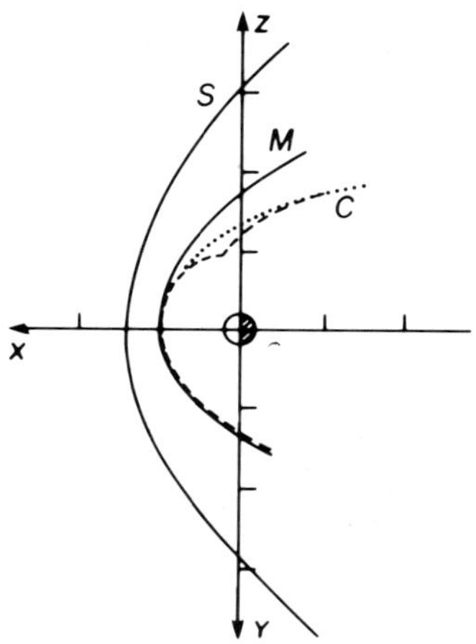

Figure 15. Shape of the magnetosphere on the noon-midnight and equatorial planes. The solid lines S and M represent the bow shock and magnetopause surface, respectively, as predicted from the MHD model for the case IMF=0. The dotted line C represents the cusp current sheet from the MHD model. The dashed line represents the prediction for the magnetopause in a Chapman-Ferraro model by Choe et al., 1973.

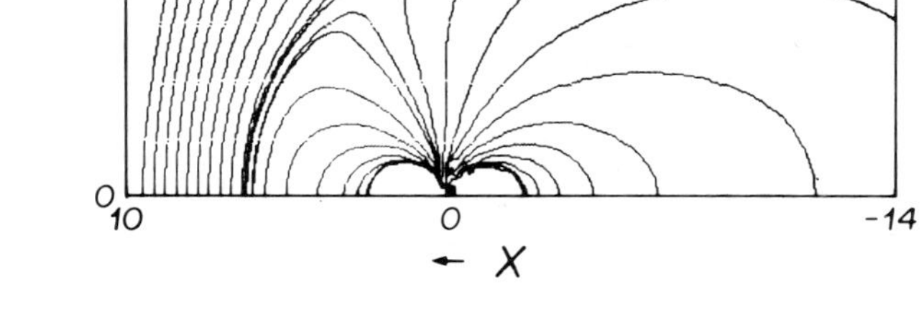

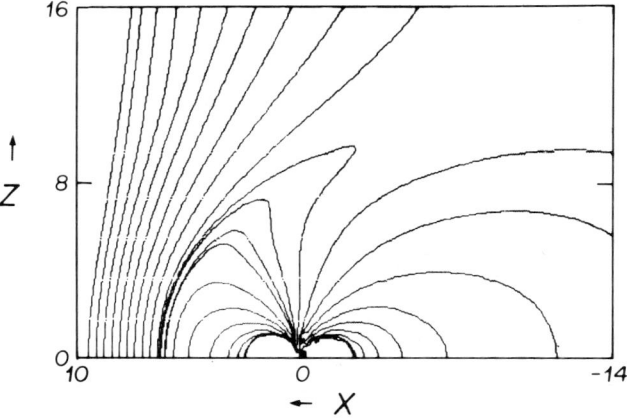

Figure 16. Magnetic field lines on the noon-midnight meridian plane from the MHD model with northward IMF.

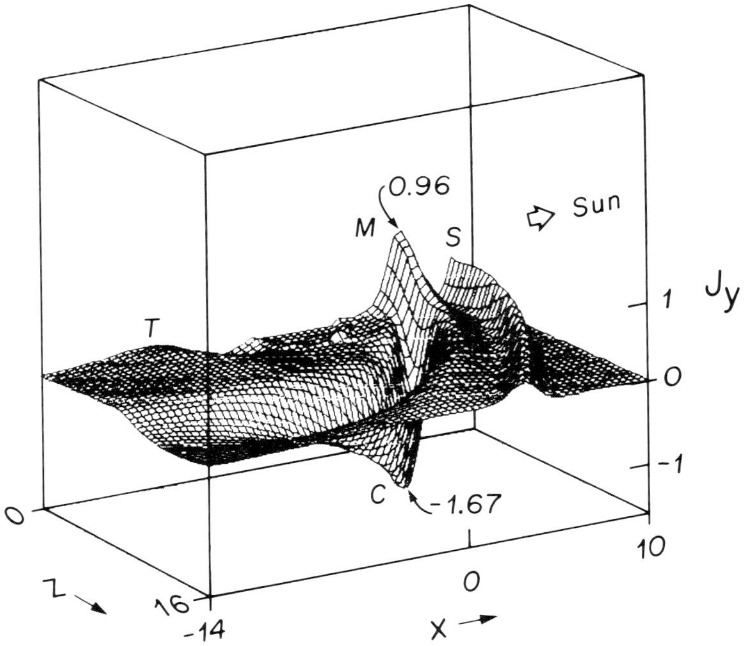

Figure 17. Perspective plot of current (J_y) on the noon-midnight meridian plane for the case with northward IMF in the 3D MHD model. S, M, C and T denote bow shock, magnetopause, cusp current sheet and tail current sheet, respectively.

lines on the noon-midnight meridian plane and the current distribution, respectively.

Although the MHD model and the Chapman-Ferraro models agree in the total magnetopause current distributions, the MHD model presents a very different picture from that of the Chapman-Ferraro model. First, the cusp in the MHD model does not expose to the solar wind plasma. Thus the solar wind plasma cannot directly enter the ionosphere through the cusp as expected from the Chapman-Ferraro model. Second, in the MHD model the magnetopause boundary is a smooth surface, and the solar wind plasma is expected to flow smoothly around it. In the Chapman-Ferraro model, it is still questionable whether a shock would be formed near the neutral point or if a smooth flow would result (Walters, 1966; Spreiter et al., 1968). Third, in the MHD model there exists a cusp current sheet. The current is strengthened when the IMF is northward, and one can expect the reconnection process to take place. In the Chapman-Ferraro model it is suggested that the northward IMF reconnects with the tail fields (Russell, 1972). Fourth, in the MHD model the dayside magnetosphere is closed when the IMF is zero or northward and is open when the IMF is southward.

The difference between the MHD model and the Chapman-Ferraro model is due to the fact that the Chapman-Ferraro model is a vacuum model while the MHD model is a fluid model. In a vacuum model the magnetic field can adjust itself by merging to reach a lower energy state. In the fluid model, the "frozen-in" condition is important, and magnetic field merging can occur only if conditions such as field line configuration, fluid flow pattern and presence of resistivity are met.

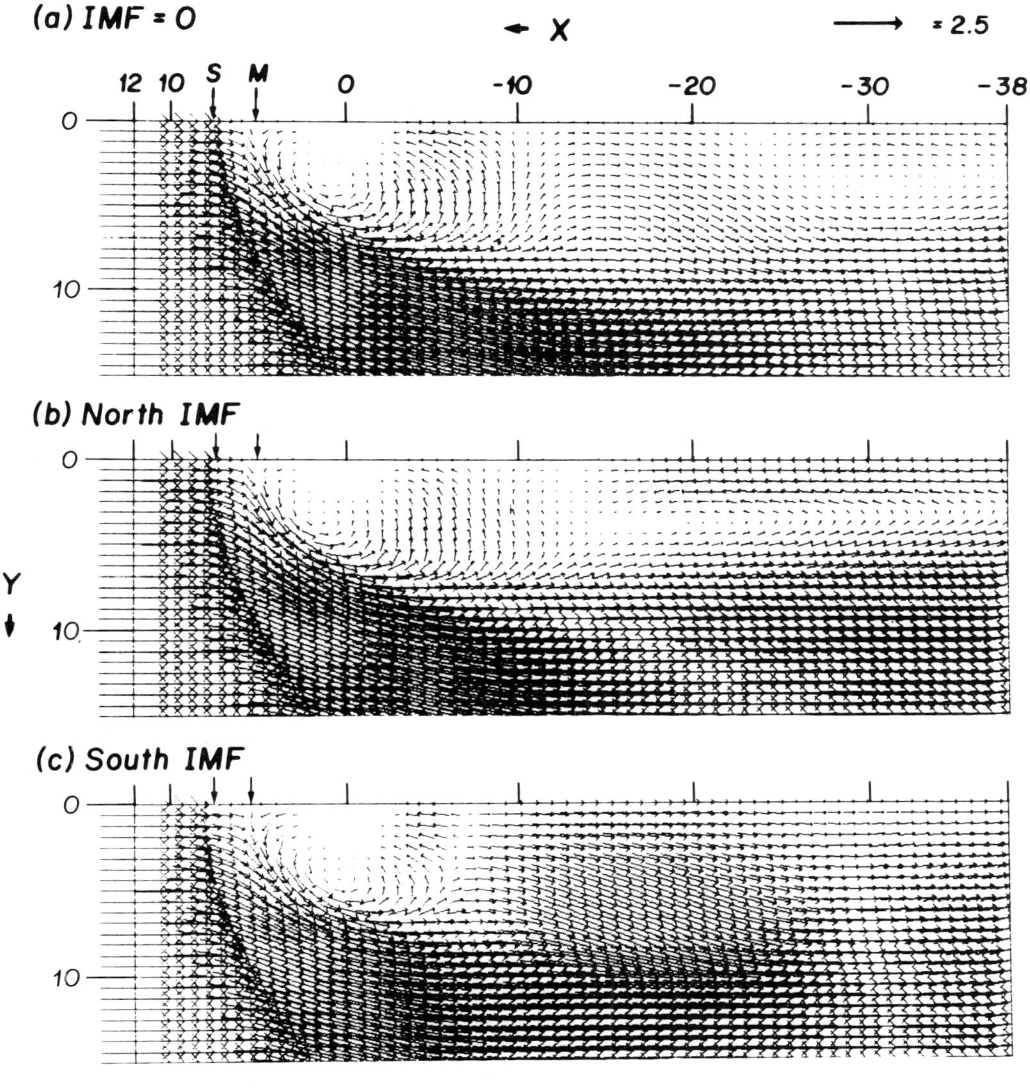

Figure 18. Plasma flow patterns in the equatorial plane for three cases: (a) IMF=0, (b) North IMF=0.5 and (c) South IMF=-0.5. S and M denote bow shock and magnetopause positions.

The configuration of high-latitude magnetic fields was recently reviewed by Fairfield (1977). The observations indicate that the transition between cusp and magnetosheath is very complex, being characterized by high levels of fluctuations. However the analysis by Mencke Hansen et al. (1976) suggests that closed magnetosphere models are in better agreement with the observations than are the open models. This may be considered to support the MHD model. Needless to say, more observations in the outer cusp region should be made. Knowledge of the nature of the cusp is of great importance in understanding the entry of charged particles into the magnetosphere. For the MHD model more calculations should be made to study the effects of east-west IMF and the stability of the configurations.

2. Plasma Flow Patterns in the Magnetosphere

Our numerical calculations suggest the following magnetospheric structure. In the inner magnetosphere (extending from 10 R_E on the dayside to -15 R_E on the night side) the magnetic pressure is larger than the material pressure, all field lines are closed, and the plasma is in convective motion perpendicular to the dipole axis. These features exist independent of IMF. In the tail region, however, the solar wind effect is important. With IMF = 0 or northward, a closed magnetosphere results, and a convection flow pattern exists. With IMF southward, the magnetosphere is open, and the flow is almost in the anti-sunward direction. Figure 18 shows the plasma flow patterns in the equatorial plane for the three cases. The convection flow patterns found in these calculations may explain the vortex flow patterns observed recently (Hones et al., 1981). Since in the MHD model the tail current sheet is formed by stretching the nightside magnetic field, the cross-tail current distribution has a maximum at x ~ -15 R, and reconnection is found to occur there for a southward IMF. Further discussion of the magnetospheric current system and the effects of IMF will be presented elsewhere.

V. CONCLUSION

In this paper an introduction to the global MHD model was given. Despite the efforts of many researchers in the last seeral years, it is fair to say that this mathematical problem has not been solved with complete satisfaction. However, because of the advances in numerical techniques and the availability of more powerful computers, one can expect rapid progress in the field of quantitative MHD modeling. It is hoped that these models will be invauable in our understanding of magnetospheric physics.

Acknowledgements

It is a pleasure to acknowledge valuable discussions with Drs. F.V. Coroniti, C.F. Kennel, L.C. Lee and C.T. Russell. I would also like to thank Daryl Ann Dutton Cody for preparing the manuscript and Marjorie Ishiwata for preparing the graphics. This work was supported by National Science Foundation Grant ATM 79-26492 and NASA Solar Terrestrial Theory Grant NAGW-78 and the MHD code was developed with support by the Department of Energy, DE-AM03-76SF00010 PA26, Task VIB.

REFERENCES

Choe, J.Y., Beard, D.B., and Sullivan, E.C.: 1973, Planet. Space Sci. 21, pp. 485-498.
Chu, C.K., and Taussig, R.T.: 1967, Phys. of Fluids 10, pp. 249-256.
Fairfield, D.H.: 1977, Rev. Geophys. Space Phys. 15, pp. 285-298.
Hones, E.W., Birn, J., Bame, S.J., Asbridge, J.R., Paschmann, G., Sckope, N., and Haerendel, G.: 1981, J. Geophys. Res. 86, pp. 814-820.
Lax, P.D.: 1954, Comm. Pure Appl. Math. 7, pp. 159-193.
Leboeuf, J.M., Tajima, T., Kennel, C.F., and Dawson, J.M.: 1978, Geophys. Res. Lett. 5, pp. 609-612.
Lyon, J.S., Brecht, H., Fedder, J.A., and Palmadesso, P.J.: 1980, Geophys. Res. Lett. 7, pp. 721-724.
Mencke Hansen, A., Bahnsen, A., and D'Angelo, N: 1976, J. Geophys. Res. 81, pp. 556-561.
Osher, S., and Solomon, F.: 1982, Math. Comp., to appear.
Russanov, V.: 1962, Nat. Research Council of Canada, Translatin No. 1027.
Russell, C.T.: 1972, Critical Problems of Magnetospheric Physics, ed. E.D. Dyer, National Academy of Sciences, Washington, D.C., pp. 1-16.
Sod, G.A.: 1978, J. Comp. Phys. 27, pp. 1-31.
Spreiter, J.R., and Alksne, A.Y.: 1968, Rev. Geophys. Space Phys. 7, pp. 11-50.
Spreiter, J.R., Alksne, A.Y., and Summers, A.L.: 1968, Physics of the Magnetosphere, ed. R.L. Carovillano, D. Reidel Publ. Co., Dordrecht, Holland.
Walters, G.K.: 1966, J. Geophys. Res. 71, pp. 1341-1344.
Wu, C.C., Walker, R.J., and Dawson, J.M.: 1981, Geophys. Res. Lett. 8, pp. 523-526.
Wu, C.C.: 1982, Yosemite '82, Origins of Plasmas and Electric Fields in the Magnetosphere, pp. 57.1-57.4.

ORIGINS AND CONSEQUENCES OF PARALLEL ELECTRIC FIELDS

William J. Burke
Air Force Geophysics Laboratory
Hanscom AFB, MA 01731

Michael Heinemann
Boston College
Chestnut Hill, MA 02164

I. INTRODUCTION

Over the past two decades there have been serious debates concerning the roles of parallel electric fields ($E_\parallel$) in the formation of aurorae. That such electric fields exist has been observationally established only in the last five years (Mozer et al., 1980). Early proponents of the pro-$E_\parallel$ case pointed to the nearly monoenergetic beams of keV electrons observed above auroral arcs. Opponents argued against $E_\parallel$ on both theoretical and observational grounds. Theoretically, it was argued that if significant field-aligned potential drops ever develop, then highly mobile ionospheric or magnetospheric electrons would move quickly to discharge them. Since the discharge time would be in the order of seconds and arcs persist on time scales of up to an hour, $E_\parallel$ must not be a major source of auroral energy. Plasma theorists were quick to point out that collective effects can significantly alter the mobility of electrons along magnetic field lines. Parallel potential drops arising from anomalous resistivity and double-layer phenomena are frequently observed in laboratory plasmas. However, questions of scaling differences between laboratory and space plasmas rendered such arguments conclusive mainly to their proponents.

O'Brien (1970) summed up what he considered three decisive observational arguments against the existence of $E_\parallel$: (1) both electrons and protons of similar energies are detected simultaneously in the auroral zone, (2) the spectra of auroral and plasma sheet electrons are quite similar, suggesting that the magnetosphere acts more like a leaky bucket, than an electron gun, and (3) the auroral monoenergetic beams are superimposed on low energy backgrounds. How could $E_\parallel$ accelerate some but not all electrons? Figure 1 gives a typical measurement of a monoenergetic beam of electrons superimposed on a low energy background.

R. L. Carovillano and J. M. Forbes (eds.), Solar-Terrestrial Physics, 393–424.

The currently accepted interpretation of measurements from the polar-orbiting satellites at altitudes at less than 3500 km concedes the first two arguments to O'Brien and disputes the third. In the auroral oval these satellites regularly detect distinct regions of diffuse auroral and inverted-V precipitation. The diffuse auroral region is characterized by relatively uniform, nearly isotropic fluxes of electrons and protons from the central plasma sheet. The thermal distributions measured in this region suggest that $E_{\parallel}$ acceleration may not be an important process for these precipitating particles. For polar orbiting satellites, inverted-V structures are characterized by fluxes of precipitating electrons that rise in average energy from several hundred eV to several keV then return to several hundred eV. Their representations on energy-time spectrograms resemble inverted-V's. In these structures the flux of precipitating protons is usually immeasurably small. Within inverted-V structures electron spectra are similar to that shown in Figure 1. Analysis of these electrons show that the "monoenergetic beam" has the characteristics of a few hundred eV population that has undergone a field-aligned acceleration of several keV. Particles with energies below the spectral peaks are now regarded as energy-degraded primaries and secondary electrons that are trapped between an electrostatic barrier above and a magnetic mirror below (Evans, 1974). For simplicity we use the terms primary and secondary, respectively, to describe the high- and low-energy populations shown in Figure 1.

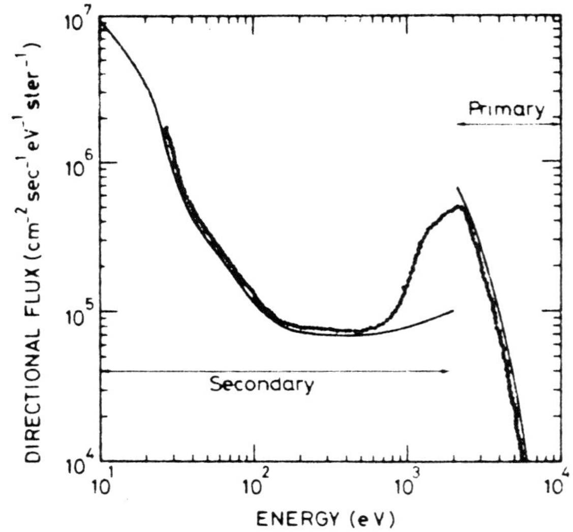

Figure 1. Auroral electron spectrum. The solid line represents a model calculation by Evans (1974) for 0° pitch angle precipitating electrons observed just above the atmosphere. The discontinuity separates electrons of atmospheric origin from those of magnetospheric origin. (Fridman and Lemaire, 1980)

At altitudes greater than 4000 km the S3-3 satellite regularly detected inverted-V structures extending for $\geqslant 1°$ in latitude co-extensive with regions of wave turbulence (Mozer, et al., 1980). Embedded within these structures are very narrow ($< 0.1°$) regions in which the electric field perpendicular to the magnetic field have very high intensities (up to 800 mV/m) and undergo rapid variations. Particle distributions measured within these regions of rapid E-field variations consist of downward accelerated auroral electrons and up-going, field-aligned beams of O^+ and H^+.

Undoubtedly observations from the Dynamic Explorer satellite will provide further clarifying insights concerning auroral electrical coupling processes. However, the fact that significant field-aligned potential drops occasionally occur above the auroral ionosphere is no longer a point of dispute. Thus, present day theoretical attention is concerned with mechanisms that help us understand how observed potential drops may be created and sustained.

The remainder of this chapter is divided into two major sections. The first section concerns the origins of $E_\parallel$ above the aurorae. In this section four mechanisms are identified with special emphasis placed on the so-called quasi-neutral and double layer theories. The second major section concerns the consequences of $E_\parallel$ on the moments of particle distribution functions. Such considerations are necessary to understand the structure of field-aligned currents measured in the vicinity of discrete arcs and the fluid characteristics of the auroral generation regions in the magnetosphere.

The chapter is written with a beginning graduate student rather than an expert audience in mind. Simple examples were chosen that illustrate physical mechanisms. In nowise does the chapter purport to be a review of electric field theory. References to the technical literature are kept at a minimum. Those given, however, may be used by interested students to reconstruct our present understanding of $E_\parallel$ in near-Earth space to any desired degree of complexity. The comprehensive study of parallel electric fields by Stern (1981), on which much of this chapter relies, is especially recommended to students because of the way various effects are added one by one.

II. ORIGINS OF $E_\parallel$

The problem to be addressed in this section can be illustrated in terms of a magnetic flux tube that intersects the Earth at auroral latitudes (Figure 2). In the equatorial region the plasma is of low density ($\leqslant 1$ cm^{-3}) and high temperature (~ 1 keV). At the ionospheric end of the flux tube the plasma is relatively dense and cold (< 1 eV). In studying the distribution of electrical potential (Φ) along the magnetic field (B) we describe distances along the field line by the symbol s. Arbitrarily we set $s = 0$ and $\Phi = 0$ at the magnetic equator. At the ionospheric end of the flux tube, we define a distance $s = L$ beyond which collisional effects are important.

The potential at $s = L$ is designated Φ_L. If there is no field-aligned potential drop $\Phi_L = 0$. To accelerate magnetospheric electrons to form an auroral arc $\Phi_L > 0$.

In describing the plasma in the flux tube of Figure 2 we make the following assumptions:

(1) Except in the case of anomalous resistivity, discussed below, the plasma is collisionless over the range $0 \leqslant s \leqslant L$.

(2) The total energy, $W = mv^2/2 + q\,\Phi$ and the magnetic moment $\mu = mv_\perp^2/2B$ of all plasma sheet particles are conserved. Because W and μ are conserved it is useful to introduce a parameter $\gamma(s) = B(s)/B(o)$. At $s = 0$, $\gamma = 1$ and at $s = L$ $\gamma = \gamma_L$.

(3) If W and μ for a given particle allow the particle to exist at two points along a field line then the particle has access to all intervening points. This condition of particle accessibility implies that Φ is a monotonic function of s in our considerations. Nature of course is not so limited.

(4) Distribution functions for all plasma species are gyrotropic, and depend on $v_\perp$ and $v_\parallel$, the components of velocity perpendicular and parallel to B, but not on the gyrophase angle.

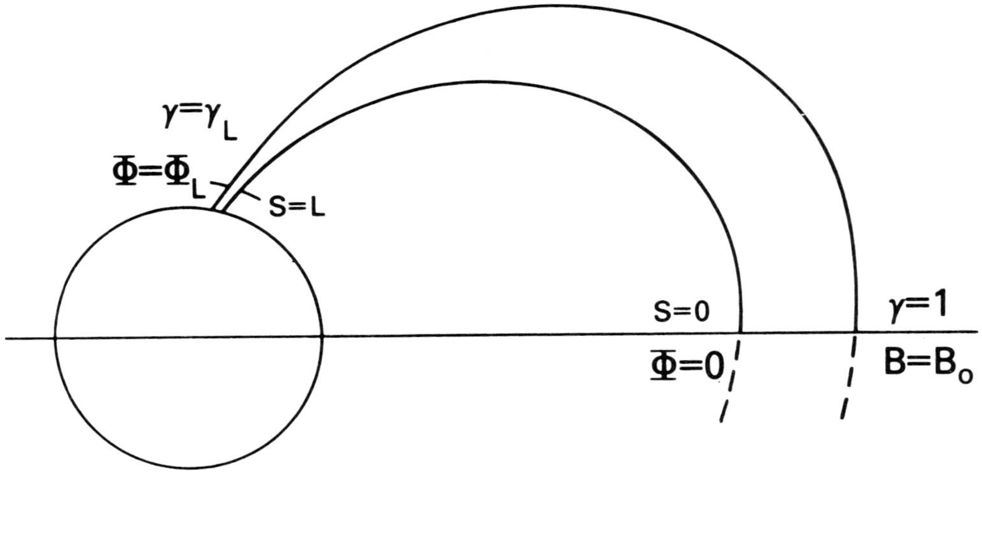

Figure 2. Magnetic field line configuration with $s = 0$ representing the magnetic equator and $s = s_L$ the position along the field line where the last reflected particles mirror. (Stern, 1981)

By definition the density of the σ species is:

$$n_\sigma(s) = \int f_\sigma(\vec{v}, s)\, d^3v$$
$$= \pi \int f_\sigma(v_\perp, v_\parallel, s)\, dv^2_\perp\, dv_\parallel \qquad (1)$$

We next introduce a one-dimensional distribution function.

$$G_\sigma(v_\parallel, s) = \pi \int f_\sigma(v_\perp, v_\parallel, s)\, dv^2_\perp \qquad (2)$$

so that,

$$n_\sigma(s) = \int G_\sigma(v_\parallel, s)\, dv_\parallel \qquad (3)$$

As we show below the potential distribution along s primarily results from boundary conditions assumed for $G_\sigma(v_\parallel)$ of the various plasma species. In the case of auroral arcs we must include a total of five species; three types of electrons and two types of positive ions. The electron populations are: (1) "hot" plasma sheet electrons that are accelerated into the arc, (2) cold ionospheric electrons, and (3) relatively hot, secondary electrons that are trapped between electrostatic and magnetic-mirror barriers. The ion populations are: (1) hot, plasma sheet protons that are largely reflected by the combined electrostatic and magnetic mirror forces, and (2) cold O^+ and/or H^+ ions that are accelerated out of the ionosphere.

Shawhan et al. (1978) distinguished four distinctive mechanisms that lead to the generation of field-aligned potential drops. These are: (1) anomalous resistivity, (2) the thermoelectric effect, (3) the magnetic mirror effect, and (4) double layers. The magnetic mirror effect is also frequently referred to as the quasi-neutral potential mechanism. The remainder of this section deals with these four mechanisms in turn.

II.1 Anomalous Resistivity

Concepts of "anomalous resistivity" are based on experience with the current-voltage characteristics of certain laboratory plasmas. If the current-carrying demands on these plasmas exceed certain critical levels, wave turbulence arises in the plasma which impedes the generation of higher current levels. To understand "anomalous" resistivity it is useful to review briefly the simplest concept of classical resistivity derived from one-dimensional currents carried by free electrons in a solid conductor. The current density j is:

$$\vec{j} = -ne\langle\vec{V}\rangle$$

where n and $\overline{V}$ are the electron gas density and mean drift velocity. Ohm's law relates current densities to electric field through the relationship

$$\vec{j} = \sigma \vec{E}$$

where σ, the conductivity of the medium, is the reciprocal of the resistivity η. Electrons accelerate until they collide with lattice sites. The average velocity is the directed velocity achieved by the electrons before they collide with the lattice.

$$\langle\vec{V}\rangle = -e\vec{E}/m\nu$$

where ν is the collision frequency.

Substituting $\langle\vec{V}\rangle$ in the expression for $\vec{j}$ gives:

$$\vec{j} = [ne^2/m\nu]\,\vec{E}$$

Thus,

$$\sigma = ne^2/m\nu = \varepsilon_o \omega^2_{pe}/\nu$$

or

$$\eta = \nu/\varepsilon_o \omega^2_{pe}$$

In an ionized plasma the classical collision frequency is determined from the various Coulomb, electron-neutral and ion-neutral collision cross sections.

On both large and small scales field-aligned currents enter and leave the high-latitude ionosphere. Since at topside altitudes Coulomb collision frequencies are very low it might be expected that a vanishingly small $E_{\parallel}$ is needed to satisfy ordinary $j_{\parallel}$ needs. Kindel and Kennell (1971) showed that as the ratio of $\langle\vec{V}\rangle/a_e$, where a_e is the thermal speed of current - carrying electrons exceeds a critical value the topside plasma becomes unstable to the onset of electrostatic ion-cyclotron turbulence.

Turbulent electrostatic waves bunch ions together so as to dramatically increase the rate at which electrons are scattered. We may define an effective or anomalous resistivity

$$\eta_{eff} = \nu_{eff} / \varepsilon_o \omega^2_{pe}$$

From numerical simulations, Biskamp and Chodura (1973) estimated that

$$\frac{\nu_{eff}}{\omega_{pe}} \simeq 0.1 \quad \varepsilon_0 E_{rms}^2/nkT_e$$

where E_{rms} is the root mean squared amplitude of the wave.

Thus, $j_\parallel \simeq [nkTe\omega_{pe} / 0.1 \, E_{rms}^2] \; E_\parallel$

Consider the case of a field-aligned current of 1 $\mu A/m^2$ carried by cold electrons moving upwards through the topside ionosphere. We assume a topside electron density of 100 cm^{-3} and a temperature of 3000°K. In regions of ion-cyclotron turbulence in the topside ionosphere measured wave amplitudes typically range from 20 to 100 mV/m^2. To support such currents requires $E_\parallel$ to be in the range of 0.1 to 0.5 mV/m. Thus, to generate significant field-aligned potential drops through anomalous resistivity requires that $E_\parallel$ extend great distances along magnetic field lines.

It should be pointed out that current-driven instabilities have a tendency to heat the current carrying electrons. This in turn tends to drive the $\langle V\rangle/a_e$ ratio below the critical level, thus quenching turbulence. The observation of ion-conic distributions throughout the auroral oval has suggested that current driven, ion-cyclotron turbulence is an important mechanism for heating ions perpendicular to the magnetic field by up to a factor of one hundred. In this picture the current carrying cold electrons would continually be replenished from the ionosphere and would never spend enough time in the turbulent region to be heated enough to quench the turbulence. In any case we expect anomalous resistivity to be an important mechanism in situations where cold, ionospheric electrons are the chief carriers of field-aligned, auroral currents. We note that anomalous resistivity may play a "seeding" role for strong parallel electric fields even when hot, magnetospheric electrons are the chief current carriers (Hudson and Mozer, 1978).

II.2 Thermoelectric Effect

The thermoelectric effect occurs in regions of contact between two plasmas with different temperatures. Hultquist (1971) suggested that this effect could produce significant field-aligned potential differences at the interface between the warm plasma sheet and the cold, topside ionospheric plasmas. The potential difference arises because there is a higher flux of electrons from the warm to the cold plasma. The current-carrying capacity of ions is considered to be negligible. As a result of the flux imbalance a net negative charge builds up in the cold plasma region. For Hultquist a steady state is achieved when the electrostatic potential exactly restores flux balance between the two plasmas, i.e., $j_\parallel = 0$. The electric field produced in this case is directed into the ionosphere. The mechanism is probably most effective in the diffuse auroral region where field-aligned currents are of low intensity or where a net upward flux of cold, ionospheric electrons is required for global current balance.

II.3 Quasi-neutral Potentials

Persson (1963, 1966) and Alfven and Falthammer (1963) first pointed out that parallel electric fields are not generated in a collisionless, magnetized plasma if and only if either the ions and electrons have the same pitch angle distributions or the magnetic field is uniform. Since it is unlikely that the pitch angle diffusion rates for magnetospheric ions and electrons are the same, and the Earth's magnetic field is not uniform, the existence of $E_{\parallel}$ in the magnetosphere is probably the rule rather than the exception. Consider a plasma near the magnetospheric equatorial plane whose ions and electrons have different pitch-angle anisotropies. One species mirrors deeper in the Earth's field than the other. Space-charge separations must develop in the vicinity of the mirror points. To maintain the plasma in a state of quasineutrality an electric field with a component along B is required. The situation is analogous to that of the topside ionosphere where ions and electrons have different scale-heights. To prevent electrons from running away from the ions an $E_{\parallel}$ is generated. However, where potentials on the order of five volts are required to maintain ionospheric charge neutrality, several kilovolts may be required for the magnetospheric plasma.

The degrees of sophistication with which magnetospheric, quasi-neutral potentials are analyzed can be quite different. Chiu and Schulz (1978) have used very realistic distributions of all five plasma species to solve numerically for the self-consistent potential distribution along magnetic field lines and have applied their results to well-defined auroral events. Here we summarize, in some detail, the simple case of two, hot magnetospheric populations that have slightly different pitch angle anisotropies. The example was devised by Stern (1981) to illustrate the basic physics of quasi-neutral potentials. We then compare the simple-case results with those of Chiu and Schulz using the more realistic particle representations.

We consider the potential distribution along magnetic field lines, self-consistently arising from two monoenergetic populations of ions and electrons. The total energy of ions (W_i) and electrons (W_e) is:

$$\begin{aligned} W_i &= M\,[v_{\parallel}^2(o) + v_{\perp}^2(o)]/2 = M\,V_i^2(o)/2 \\ &= M\,[v_{\parallel}^2(s) + v_{\perp}^2(s)]/2 + e\,\Phi(s) \end{aligned} \tag{4a}$$

$$\begin{aligned} W_e &= m\,[v_\parallel^2(o) + v_\perp^2(o)]/2 = m\,V_e^2(o)/2 \\ &= m\,[v_\parallel^2(s) + v_\perp^2(s)]/2 - e\,\Phi(s) \end{aligned} \tag{4b}$$

where M and m represent the ion and electron masses, respectively and e is an elemental unit of charge. In the equatorial plane the distribution functions for ions (f_i) and electrons (f_e) are:

$$f_i(v_\perp, v_\parallel, 0) = \begin{cases} f_{oi}\,\delta\,(v_\parallel^2(0) + v_\perp^2(0) - V_i^2(0)), & 0 \leqslant v_\parallel^2(0) \leqslant v_{iu}^2(0) \\ 0, & \text{otherwise.} \end{cases} \tag{5a}$$

$$f_e(v_\perp, v_\parallel, 0) = \begin{cases} f_{oe}\,\delta\,(\nu_\parallel^2(0) + \nu_\perp^2(0) - V_e^2(0)), & 0 \leqslant v_\parallel^2(0) \leqslant v_{eu}^2(0) \\ 0, & \text{otherwise.} \end{cases} \tag{5b}$$

Information concerning differential pitch-angle distributions is contained in the upper-bound parallel velocity v_{iu} and v_{eu} allowed to the ions and electrons, respectively. Upper bounds on the parallel velocities of ions and electrons are chosen so that the combined electrostatic and magnetic mirror forces cause all particles to mirror at or above s = L (Figure 2). Values of $v^2_{iu}(0)$ and $v^2_{eu}(0)$ may be determined from the conservation of W and μ.

$$W_i = Mv_{iu}^2(0)/2 + \mu\,B_o = \mu\,B_L + e\,\Phi_L$$

Solving for $v^2_{iu}(0)$ we get

$$v_{iu}^2(0) = 2\,[\,W_i\,(1 - 1/\gamma_L) + e\,\Phi_L/\gamma_L\,]/M \tag{6a}$$

where, as defined above $\gamma_L = B_L/B_o$. Note that since we have assumed $\Phi_L > 0$, if $e\,\Phi_L > W_i$ then

$$v_{iu}^2(0) = 2\,W_i/M$$

By a similar argument it can easily be shown that

$$v_{eu}^2(0) = 2\,[\,W_e\,(1 - 1/\gamma_L) - e\,\Phi_L/\gamma_L]/m \tag{6b}$$

At any point s along the field line $v_{\parallel}^2$ for both species is confined to some finite interval.

$$v_{im}^2 (s) \leq v_{\parallel}^2 \leq v_{iu}^2 (s)$$

and

$$v_{em}^2 (s) \leq v_{\parallel}^2 \leq v_{eu}^2 (s)$$

For monotonically increasing $\Phi(s)$ the minimum parallel ion velocity is zero. Depending on the strength of $E_{\parallel}$ this need not be true for electrons. The one-dimensional distribution (eq. (2)) for the σ species is

$$G_{\sigma}(v_{\parallel}, s) = \pi f_{o\sigma} \int \delta \left(v_{\parallel}^2 + v_{\perp}^2 + \frac{2q_{\sigma}}{m_{\sigma}} [\Phi(s)] - V_{\sigma}^2 (0) \right) dv_{\perp}^2$$

Such distributions have "box-car" shapes.

$$G_{\sigma}(v_{\parallel}, s) = \begin{cases} \pi f_{o\sigma} & v_{\sigma m}^2 (s) \leq v_{\parallel}^2 \leq v_{\sigma u}^2 (s) \\ 0 & \text{otherwise.} \end{cases} \qquad (7)$$

They may be integrated quite simply to determine the density of σ species.

$$n_{\sigma}(s) = 2 \pi f_{o\sigma} [v_{\sigma u}(2) - v_{\sigma m}(s)] \qquad (8)$$

Quasi neutrality at any point s demands that

$$n_i (s, \Phi(s)) = n_e (s, \Phi(s)) \qquad (9)$$

To determine the quasi-neutral potential explicitly it is useful to define two parameters

$$\alpha(s) \equiv [W_e + e\Phi(s)] / W_e \gamma(s) \qquad (10a)$$

and

$$\beta(s) \equiv [W_i - e\Phi(s)] / W_i \gamma(s) \qquad (10b)$$

Since $\gamma(0) = 1$ and $\Phi(0) = 0$ the values of both α and β at the equator are 1. Equation (10a) can be written:

$$\gamma(s)\, W_e\, \alpha(s) = 1/2\, m\, [v_{\parallel}^2(s) + v_{\perp}^2(s)] \tag{11}$$

For electrons mirroring at s = L

$$\gamma_L\, W_e\, \alpha_L = 1/2\, m\, v_{\perp L}^2 \tag{12}$$

$$= 1/2\, m\, v_{\perp}^2(s)\, \gamma_L/\gamma(s)$$

Combining equations (11) and (12) we can solve for the maximum "parallel kinetic energy" allowed to electrons at s

$$m\, v_{eu}^2(s)/2 = \gamma(s)\, W_e\, [\alpha(s) - \alpha_L] \tag{13a}$$

Similarly for ions

$$m\, v_{iu}^2(s)/2 = \gamma(s)\, W_i\, [\beta(s) - \beta_L] \tag{13b}$$

For simplicity we assume that the magnetic mirror force is sufficient to keep the minimum parallel velocity ($v_{\sigma m}(s)$) of both species equal to zero.

The quasi-neutrality condition can be expressed by squaring both sides of equation (9).

$$f_{oL}^2\, v_{iu}^2(s) = f_{oe}^2\, v_{eu}^2(s) \tag{14}$$

Substituting equations (13a, b) into (14) gives

$$f_{oi}^2\, W_i\, [\beta(s) - \beta_L]/M = f_{oe}^2\, W_e\, [\alpha(s) - \alpha_L]/m \tag{15}$$

Charge neutrality at the equator can be expressed as

$$f_{oi}^2\, W_i\, [1 - \beta_L]/M = f_{oe}^2\, W_e\, [1 - \alpha_L]/m \tag{16}$$

Dividing equation (15) by (16) gives a very simple expression for the quasi-neutrality condition.

$$[\ \alpha(s) - \alpha_L\ ] / [\ 1 - \alpha_L\ ] = [\ \beta(s) - \beta_L\ ] / [\ 1 - \beta_L\ ] \qquad (17)$$

By substituting the values of α and β defined in equations (10a) and (10b) into equation (17) it is readily shown that

$$\Phi(s) = \Phi_L\ [\ \gamma(s) - 1\ ] / [\ \gamma_L - 1\ ] \qquad (18)$$

This indicates that the potential increases linearly with B from zero at the equator to Φ_L. The parallel electric field is directly proportional to the magnetic mirror force $\partial B/\partial s$. It is for this reason quasi-neutral potentials are also referred to as magnetic-mirror potentials.

Although the expression for potential shows the linear dependence of $\Phi(s)$ on $B(s)$ it does not show explicitly how differential pitch angle distributions contribute to the potential. Alfven and Falthammer (1963) show this effect quite simply using monoenergetic distributions for ions and electrons that differentially fixes their magnetic moments.

$$f_\sigma(0) = [N_0/\pi B_0]\ [m_0/2]^{3/2}\ [W_\sigma - \mu_\sigma B_0]^{1/2}\ \delta(W - W_\sigma)\ \delta(\mu - \mu_0)$$

$$\sigma = e,\ i \qquad (19)$$

They show that for quasi-neutrality

$$E_\parallel = \partial B/\partial s\ \ [\ (w_{e\parallel} w_{i\perp} - w_{e\perp} w_{i\parallel}) / (w_{e\parallel} + w_{i\parallel})\ B_0]/e$$

where

$$w_{\sigma\parallel} = (\ m_\sigma\ \ v_{\sigma\parallel}^2\ )/2 \qquad (20a)$$

and

$$w_{\sigma\perp} = (\ m_\sigma\ \ v_{\sigma\perp}^2\ )/2 \qquad (20b)$$

at s = 0. From eq (2) it is seen that if the equatorial pitch angle distributions of ions and electrons is the same then $E_\parallel = 0$.

It may be objected that perhaps the Φ versus B relationship expressed in Equation (18) is deceivingly simple. The distribution functions used to represent plasma sheet ions and electrons were not realistic and both secondary electrons and ionospheric particles were ignored. Gravitational effects on massive ions are of particular concern. The escape energy of O^+ is 10.4 eV. Depending on the detailed structure of the electrostatic potential in the topside ionosphere these ions may or may not have access to the main acceleration region. Stern (1981) stresses the fact that equilibrium solutions can only be achieved when the ionosphere is included in the calculations.

Using realistic destribution functions for all five plasma species renders the quasi-neutral, potential problem analytically unsolvable. However, Chiu and Schulz (1978) have devised realistic representations of the various species and numerically solved for the quasi-neutral, potential distribution. Figure 3 is a plot of the directional, differential energy flux measured by Mizera et al. (1976) over an inverted-V structure. As indicated in the figure, measurements were taken at pitch angles of 0° and 180°. The solid and dashed lines give the functions used by Chiu and Schulz to represent downcoming and backscattered electrons, respectively. The two sets of measurements

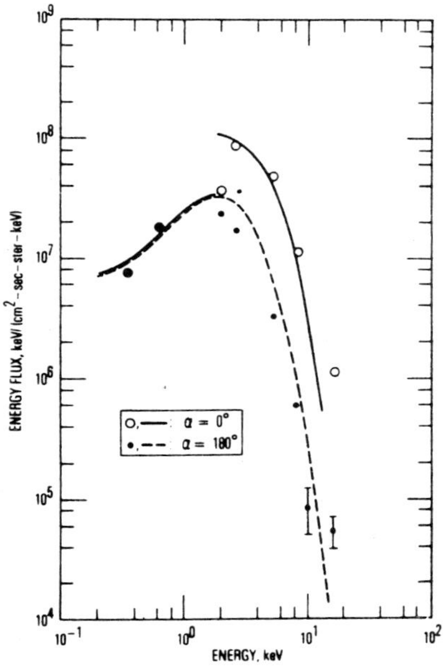

Figure 3. Electron energy fluxes at 0° and at 180° pitch angles observed in an 'inverted V' structure (Mizera et al., 1976) are shown as functions of electron energy. The difference between the down-going and up-going fluxes at 1.76 keV indicates the presence of a 'mono-energetic' beam of precipitating energetic electrons. (Chiu and Schulz, 1978)

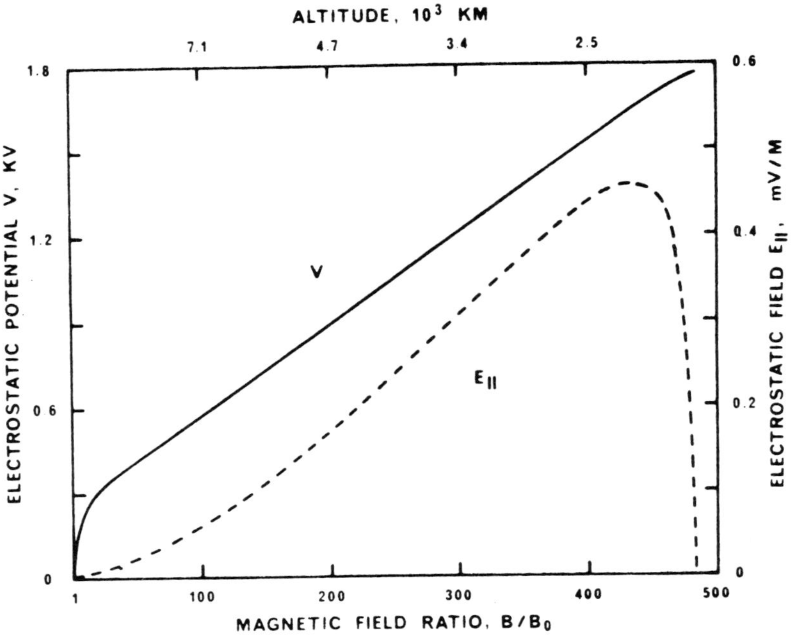

Figure 4. The self-consistent electrostatic potential and parallel electric field computed for the case corresponding to the observed electron flux of Figure 3 is shown as a function of magnetic field ratio and altitude. (Chiu and Shulz, 1978)

indicate that the parent, magnetospheric population had a density of 0.6 cm^{-3} with parallel and perpendicular thermal energies of 1.23 keV and 4.67 keV, respectively. This population was accelerated through a potential drop of 1.76 kV. The self-consistent, quasi-neutral values of Φ and $E_\parallel$ calculated by Chiu and Schulz are plotted in Figure 4 as functions of B/B_0 and altitude. Although the potential has a concave shape, it is linearly related to B over most of the computational region. Thus, the main features of Equation (18) are preserved with realistic particle distributions. In this model $E_\parallel$ has a broad maximum value of 0.45 mV/m at altitudes of above 2000 km.

II.4. Double Layers

The final acceleration mechanism to be considered is the double layer. Phenomena of this type are frequently observed in laboratory plasmas. Unlike quasi-neutral potentials, double layers derive from the breakdown of space-charge neutrality. Stern (1981) has suggested

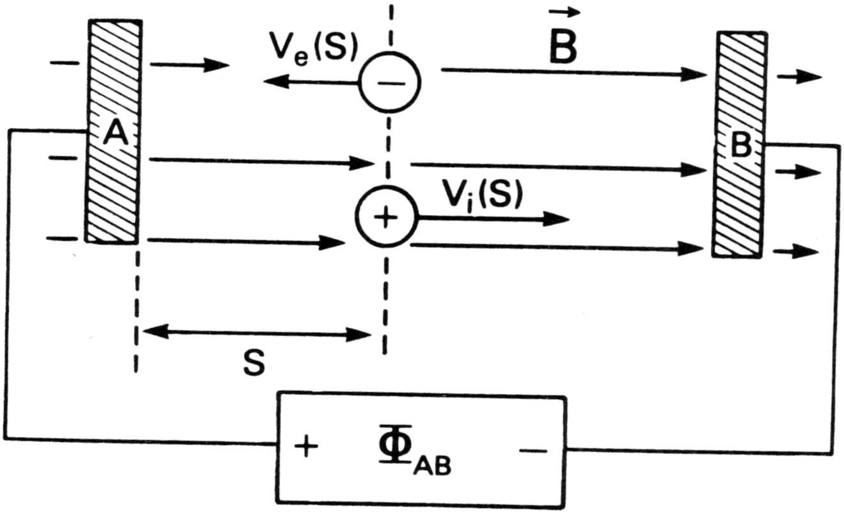

Figure 5. Collision free plasma in a homogeneous magnetic field subject to a fixed potential drop Φ_{AB}. (Stern, 1981)

the following definition:

> "A double layer is a discontinuity of the electric field resulting from a tendency of the plasma to maintain charge neutrality in the regions which adjoin it."

The discontinuity of the electric field results from locally unbalanced space charges.

Consider a simple circuit consisting of an ion emitting anode A and an electron emitting cathode B shown in Figure 5. A potential Φ_{AB} is applied across the plates. The problem is reduced to one-dimensional by the presence of a uniform magnetic field directed from A to B. For simplicity assume that the emitted ions and electrons are monoenergetic. Thus, at any point s the velocities of ions $v_i(s)$ and electrons $v_e(s)$ have single values.

Considerations of the continuity equations for ions:

$$v_i(s)\,[dn_i/ds] + n_i(s)\,[dv_i/ds] = 0 \tag{21}$$

and electrons

$$v_e(s)\,[dn_e/ds] + n_e(s)\,[dv_e/ds] = 0 \tag{22}$$

shows that there exist no equilibrium in which charge neutrality holds everywhere. Equations (21) and (22) indicate that dv_i/ds and dn_i/ds have opposite signs, as do dv_e/ds and dn_e/ds. Because of the applied potential dv_i/ds and dv_e/ds also have opposite signs. If $n_i(s) = n_e(s)$ at some point s, they can be equal at no other points between A and B. Rather than having space charge widely distributed between A and B the plasma will become polarized. In the final equilibrium the electric field vanishes except over a relatively small interval

$$s_D < s < s_D + d$$

in which the plasma is not neutral. The width of the double layer d, over which the entire potential drop occurs, is much shorter than the distance AB.

In modelling possible double layers above the auroral ionosphere, it is necessary to include five plasma species. Figure 6 is a sketch of the five plasma species and the electrostatic potential distribution in the vicinity of a simple double layer. Particles entering the double layer from the low potential (high altitude) side consist of hot, plasma sheet ions and electrons. A large fraction of the ions are reflected inside the double layer back to the plasma sheet. Because the double layer's thickness is relatively small, electrostatic forces greatly exceed magnetic mirror forces inside the double layer. Thus, any plasma sheet electrons entering the double layer are accelerated into the ionosphere.

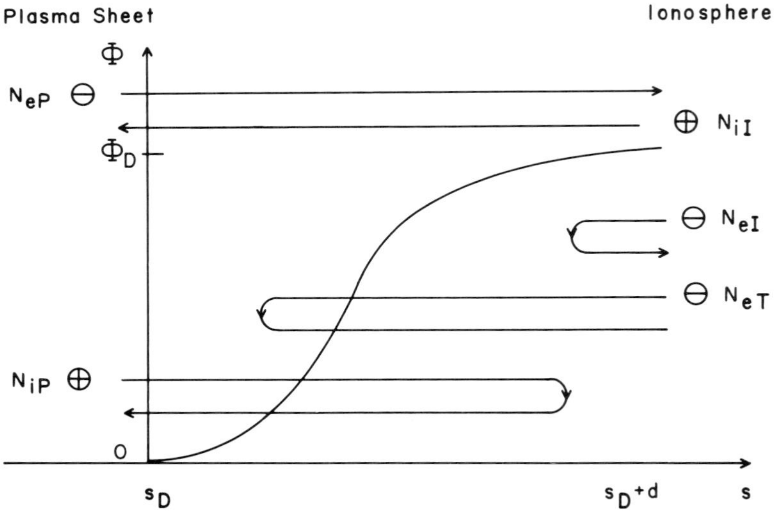

Figure 6. Schematic of potential distribution with simplified trajectories of five plasma species in the vicinity of a double layer.

On the high potential side of the double layer there are three species. Ionospheric ions that enter the layer are freely accelerated into the magnetosphere. Ionospheric and secondary electrons are reflected, at different altitudes by the double layer's potential barrier. The potential rises from $\Phi = 0$ at $s = s_D$ to $\Phi = \Phi_D$ at $s = s_D + d$. In the regions $s < s_D$ and $s > s_D + d$ the potential is assumed to be constant along magnetic field lines.

Kan and Lee (1980) introduced mathematically simple representations of the five plasma species to illustrate the physical criteria that must be satisfied for double layers to form above the auroral ionosphere. In this model plasma-sheet electrons are represented by one-dimensional, "water bag" distribution functions.

$$G_{eP}(v_{\parallel}) = \begin{cases} N_{eP}/2a_{eP} & V_e > a_{eP} \\ N_{eP}/(V_e + a_{eP}) & V_e < a_{eP} \end{cases}$$

if $v_{\parallel}$ is in the range of $v_{em} < v_{\parallel} < v_{eu}$ and $G_{eP}(v_{\parallel}) = 0$, otherwise. The symbols a_{eP} and V_e represent the thermal and field-aligned flow speeds of plasma sheet electrons just above the double layer, $s < s_D$. The minimum (v_{em}) and upper limit (v_{eu}) velocities of electrons at any point s along a field line are:

$$v_{em}(s) = \begin{cases} [(V_e - a_{eP})^2 + 2e\Phi(s)/m]^{1/2}, & V_e > a_{eP} \\ [2e\Phi/m] & V_e < a_{eP} \end{cases} \tag{23}$$

and

$$v_{eu}(s) = [(V_e - a_{eP})^2 + 2e\Phi(s)/m]^{1/2} \tag{24}$$

The density of plasma sheet electrons obtained by integrating $G_{eP}(v_{\parallel})$ is (equation 2):

$$n_{eP}(s) = \begin{cases} N_{eP}/2a_{eP} \ \{ [(V_e+a_{eP})^2 + 2e\Phi(s)/m]^{1/2} - [(V_e+a_{eP}) \\ \quad + 2e\Phi(s)/m]^{1/2}\} & V_e > a_{eP} \\ N_{eP}/(V_e+2a_{eP}) \ \{ [(V_e+a_{eP})^2 + 2e\Phi(s)/m]^{1/2} - [2e\Phi(s)/m]^{1/2}\} & \\ & V_e > a_{eP} \end{cases} \tag{25}$$

Equation (25) shows that n_{eP} depends on on s only through the potential, and that in the region $s < s_D$ where $\Phi(s) = 0$, $n_{eP} = N_{eP}$.

The distribution function representing plasma sheet ions is assumed to be Maxwellian. Thus, the density of these ions is

$$n_{iP}(s) = N_{iP} \exp - [e\Phi(s)/\Theta_{iP}] \tag{26}$$

where Θ_{iP} is the mean thermal energy of the plasma sheet ions and N_{iP} is their density in the $s < s_D$ region.

The degraded primary and secondary electrons are assumed to have a boxcar distribution and be trapped between the electrostatic barrier of the double layer and magnetic mirrors in the ionosphere. Since they are trapped their parallel velocity is restricted to the range $-[2e\Phi(s)/m]^{1/2} \leq v_{\parallel} \leq [2e\Phi(s)/m)]^{1/2}$.

The density of these electrons is represented by a function

$$n_{eT}(s) = N_{eT} \exp [\Phi(s)/\Phi_D]^{1/2} \tag{27}$$

which decreases from N_{eT} below the double layer to zero above it.

Ionospheric electrons are also assumed to have a Maxwellian distribution. Their density is

$$n_{eI}(s) = N_{eI} \exp [\Phi_D - \Phi(s)]/\Theta_{ei} \tag{28}$$

where Θ_{eI} represents the mean thermal energy of the cold, ionospheric electrons. Equation (28) shows that n_{eI} decreases from N_{eI} at $s = s_D + d$ to $N_{eI}e^{-e\Phi_D/\Theta_{eI}}$ at $s \leq s_D$. Since $e\Phi_D >> \Theta_{ei}$ most ionospheric electrons are electrostatically reflected close to the bottom of the double layer.

Upstreaming, ionospheric ions are also represented by water bag distribution functions. Their density at any point s is

$$n_{iI}(s) = \begin{cases} N_{iI}/2a_{iI}[(V_i+a_{iI})^2 + 2e(\Phi_D-\Phi)/M_i]^{1/2} - [2e\Phi_D - \Phi(s)/M_i]^{1/2}\} & V_i < a_{iI} \\ N_{iI}/(V_i+a_{iI})\ [(V_i+a_{iI})^2 + 2e(\Phi_D-\Phi)/M_i]^{1/2} - [2e(\Phi_D - \Phi(s)/M_i]^{1/2}\} & V_i < a_{iI} \end{cases} \tag{29}$$

where a_{iI}, V_i and M_i are the thermal speed, streaming speed and mass of of the ionospeheric ions. The mass of these ions generally is not the same as those of the plasma sheet ions.

Equations 25-29 show that the densities of the five species depend on s only through the potential. From the definition of a double-layer given above it is seen that to accelerate plasma sheet electrons the boundary conditions to be satisfied are:

(1) Charge neutrality in the regions above and below the double layer.

$$\sum_{\sigma=1}^{5} q_\sigma n_\sigma (\Phi = 0) = \sum_{\sigma=1}^{5} q_\sigma n_\sigma (\Phi = \Phi_D) = 0$$

(2) Negative space charge just inside the high altitude (low potential) boundary of the double layer

$$\sum_{\sigma=1}^{5} q_\sigma n_\sigma (\Phi = \varepsilon\Phi_D) \leqslant 0 , \qquad 0 < \varepsilon \ll 1$$

(3) Positive space charge just inside the low altitude (high potential) boundary of the double layer.

$$\sum_{\sigma=1}^{5} q_\sigma n_\sigma (\Phi = (1 - \varepsilon)\Phi_D) \geqslant 0$$

To satisfy these conditions with $e\Phi_D > 1/2\, m a_{eP}^2$ Kan and Lee (1980) have shown that

$$mV_e^2 \geqslant 2\Theta_{eP} + \Theta_{iP}, \qquad N_{eT} < N_c \qquad (30)$$

or

$$V_e > 0 \qquad N_{eT} \geqslant N_c \qquad (31)$$

where $\Theta_{eP} = 1/2\, m a_{eP}^2$ and N_c is a critical density.

$$N_c = N_{eP} [2e\Phi_D/m(V_e + a_{eP})^2]^{1/2} \qquad (32)$$

For low densities of trapped electrons the condition for the formation of double layers is recognized as a variant of the Bohm-Block condition (Bohm, 1949; Block, 1972). When the density of trapped electrons exceeds the critical level (Equation 32) the condition is relaxed significantly.

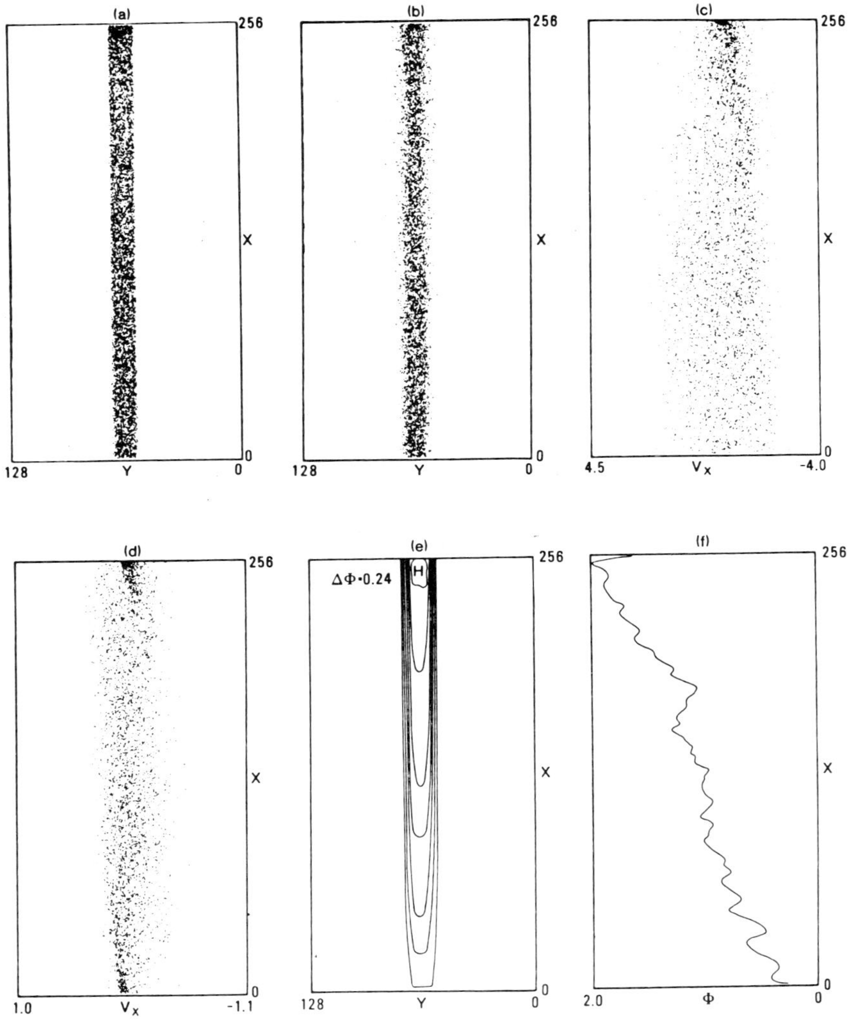

Figure 7. Results from simulation at time $\tau = 400\ \omega_{pe}^{-1}$ with an initial electron drift equal to zero; (a) and the electron positions with the ionospheric boundary at the bottom and the plasma sheet boundary at the top, (b) the ion positions, (c) the electron phase space density (d) the ion phase space density with all velocities in units of a_{ep}. (e) the electric potential contours, and (f) the potential as a function of altitude down the center of the current sheet. (Wagner et al., 1981)

Results of numerical simulations of double layers using the criteria of Kan and Lee are given in Figures 7 and 8 (Wagner et al., 1981). In these simulations auroral phenomena are treated as two-dimensional with X and Y representing distances along and latitudinally across magnetic field lines, respectively. It is assumed that: (1) the ionosphere has a finite Pedersen conductivity, (2) the five plasma species are present, (3) the magnetic field is dipolar with a 10% increase in strength over the 512 Debye-length region of simulation, and (4) a constant flux of plasma sheet particles is continually

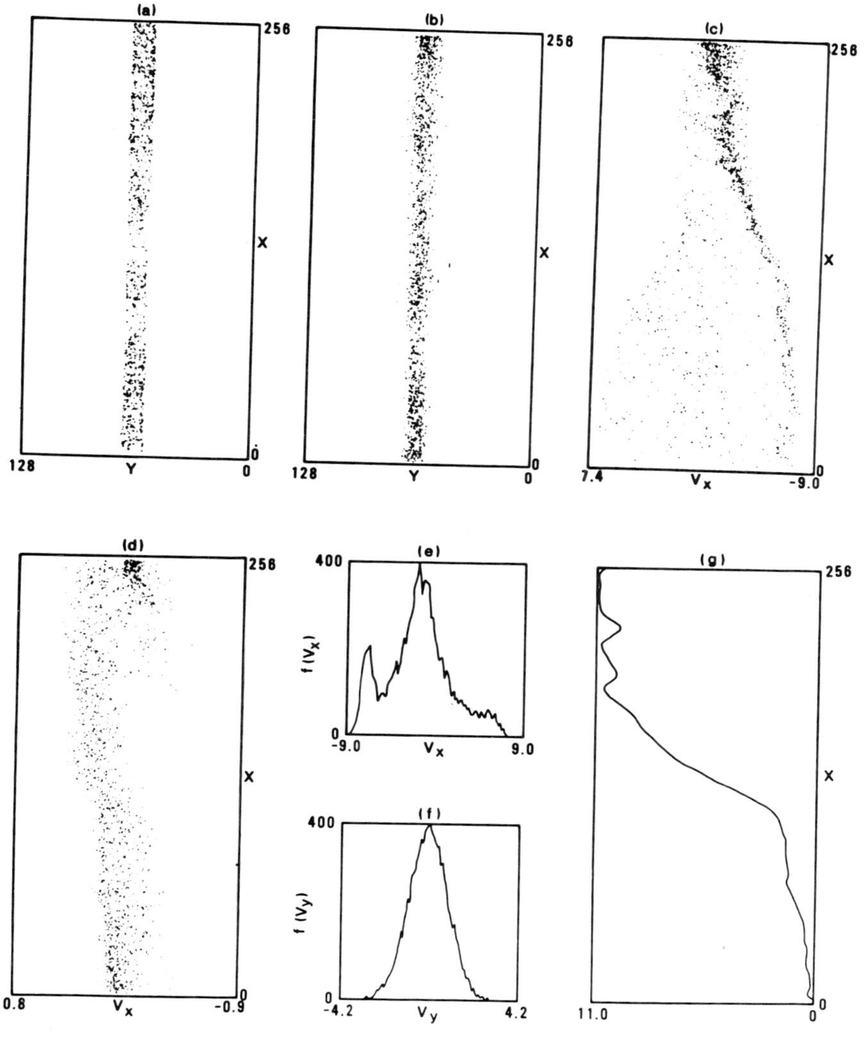

Figure 8. Results from a simulation at time $\tau = 400\ \omega_{pe}^{-1}$ with initial electron drift speed equal to 0.75 V_{th}. (a) the electron position, (b) the ion position, (c) the electron phase space, (d) ion phase space, (e) the electron velocity distribution in the x-direction, (f) the electron velocity distribution in the y-direction, and (g) the electric potential as a function of altitude. (Wagner et al., 1981)

introduced at the top boundary. The electrostatic potential is not imposed but it evolves self-consistently from the initial conditions.

Figures (7) and (8) give the distributions of electrons and ions in configuration and phase space as well as the potential at a time of 400 ω_{pe}^{-1} into the simulation. Initial conditions differ in that plasma sheet electrons have field-aligned drift speeds of zero (Figure 7) and 0.75 a_{eP} (Figure 8).

In the case of Figure (7f), with $V_e = 0$, the potential increases by 0.8 θ_{iP} in a way that is almost linear. The potential more closely resembles a quasi-neutral solution than that of a simple double layer sketched in Figure 6.

In the case of Figure (8g) the potential increases by 6 θ_{iP} over a relatively small distance. By examining the phase space distribution of electrons (8c) the accelerated electrons and the reflected secondaries are readily discerned. In Figure (8d) it is easily seen that most plasma sheet ions are reflected by the double layer. The relatively high number of ions in the upper left quadrant of Figure (8d) represent the accelerated beam ions.

III. CONSEQUENCES OF $E_\parallel$

In the previous section we examined several physical mechanisms by which parallel electric fields are supported above the high-latitude ionosphere. Here we assume the existence of field-aligned potential drops and consider their consequences for critical magnetospheric parmeters. The section is divided into two major subsections. The first concerns their effects on field-aligned current intensities at the top of the ionosphere. The second concerns the formulation of global magnetospheric models that include $E_\parallel$.

III.1 Field-Aligned Currents

Field-aligned currents into the ionosphere are mostly carried by upmoving, cold electrons. Currents out of the ionosphere are due mainly to precipitating electrons. On magnetic field lines with field-aligned potential drops there are essentially two free-streaming species, upgoing ions and downcoming electrons. The flux levels of electrons precipitating to form visible aurorae exceed 10^9 cm^{2}-sec^{1}. Observed flux levels of upgoing ion beams are in the 10^6-10^7 cm^2-sec^1 range (Mozer et al., 1980). Thus, while ion beams and conics are significant sources of heavy ions in the ring current they contribute insignificantly to the field-aligned currents.

It is useful to approximate the intensity of field-aligned currents from the down-coming flux of magnetospheric electrons.

$$j_{\parallel} = -e \int f_e(\vec{v}, L)\, v_{\parallel}\, d^3v. \tag{33}$$

Any magnetospheric electrons reaching altitudes below L are assumed to be lost. With assumed magnetospheric distribution functions and field-aligned potentials, values of $j_{\parallel}$ may be calculated. In applying the Liouville Theorem it is necessary to specify the elements of the parent population's phase space that have access to the ionosphere. For monotonic potentials particle trajectories are analyzed using methods similar to that of Section II.3. Whipple (1977) has worked out rules for specifying plasma boundaries in phase space. These rules, which are not repeated here, are very useful for inferring the distribution of field-aligned potentials from electron energy and pitch-angle spectra measured in the ionosphere. Fridman and Lemaire (1980) have integrated Equation (32) for the case of an isotropic, Maxwellian parent population with a density n and a temperature T. Lyons (1981) has presented their result in a formula that simply relates the field-aligned currents and potential drops.

$$j_{\parallel} = j_0\, (B_I/B_{\Phi})\, [1 - (1 - B_{\Phi}/B_I)\, \xi(\Phi)] \tag{34}$$

where $j_0 = -nq\,(kT/2\pi m)^{1/2}$ is the field-aligned current that would be produced by an isotropic electron distribution in the

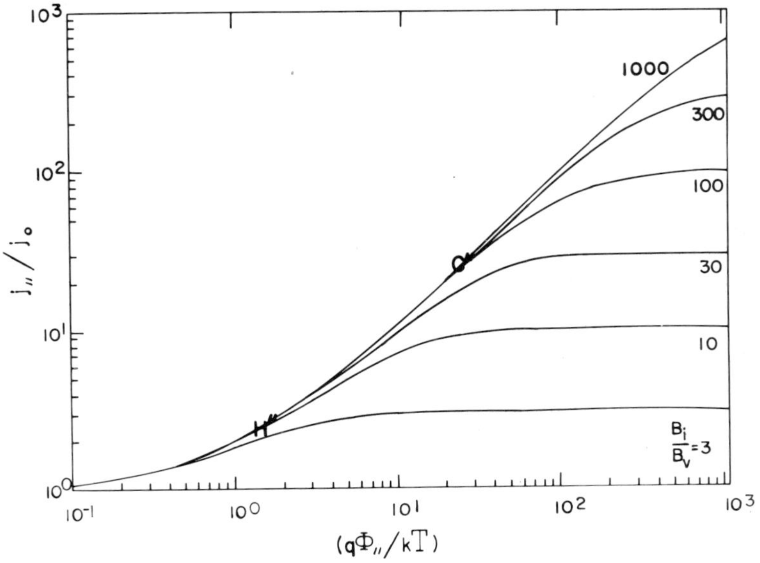

Figure 9. Current-voltage relationship based on Equation (34) with different assumed ratios of B_I/B_{Φ}.

absence of a field-aligned potential drop. B_I and B_Φ are the magnetic field strength at the ionosphere and at the top of the region of potential drop, respectively. The function containing the potential drop is

$$\xi(\Phi) = \exp\,(e\Phi/kT)\,[\,(B_I/B_\Phi) - 1\,] \tag{35}$$

Information about the distance over which the potential drop extends is contained only in the ratio B_I/B_Φ.

Figure 9 is a plot of $j_\parallel/j_0$ as a function of $e\Phi/kT$ for six different values of B_I/B_Φ. There are three aspects of these curves to which we direct attention:

(1) In the limit $B_I/B_\Phi = 1$, $j_\parallel = j_\phi$ no matter how large the potential drop. This corresponds to cases of narrow double layers situated just above the ionosphere. The flux of electrons into the ionosphere becomes highly field-aligned but the total current is not enhanced.

(2) In the limit $e\phi/kT$ goes to large values, $j_\parallel' = (B_I/B_\Phi)\,j_0$. The effect of a potential drop that extends for great distances along magnetic field lines is to allow more electron access to the ionosphere by opening the atmospheric loss cone of the parent population.

(3) For a significant range of $(e\Phi/kT)$ for any given values of (B_I/B_Φ) the relationship between $j_\parallel$ and Φ is linear. Lyons (1981) has found that for given auroral arcs $j_\parallel = K\,\Phi$ where K is an empirical constant. However, the constant changes from arc to arc. From Figure 9 we can understand variations in K in terms of different spatial extensions of Φ along the magnetic fields conjugate to the different auroral forms. Finally, we note that Burke, et al. (1982) have been able to explain magnetometer measurements of field-aligned currents of up to 145 $\mu A/m^2$ by S3-2 in terms of equation (34). Simultaneous measurements of down-coming electrons indicated that a two-Maxwellian population with densities of 0.92 and 1.10 cm^{-3} and temperatures of 200 eV and 4 keV, respectively, had been accelerated through a potential drop of 5 kV. The symbols C and H in Figure 9 represent the relative contributions of the 200 eV and the 4 keV populations to the total current.

III.2 Magnetospheric Models

In another chapter, Richard Wolf describes a magnetospheric convection code developed at Rice University. Many characteristics of electric fields and currents observed during selected magnetic storms and substorms have been successfully simulated with this code. Its prime areas of applicability are the central plasma sheet and conjugate regions of diffuse auroral precipitation. The magnetotail and the poleward portion of the auroral oval (Region 1) lie outside the domains of simulation. It is essentially an MHD model that does not include parallel electric fields. In this subsection we first show that the presence of $E_\parallel$ profoundly affects the main physical quantities calculated by the code. For this reason $E_\parallel$ cannot simply be tacked onto the present model. We then suggest procedures for self-consistently calculating magnetospheric particle pressures that included $E_\parallel$.

The Rice University code performs a chain of self-consistent calculations that describe the electrical coupling between the magnetosphere and the ionosphere. An initial particle pressure distribution is assumed. For the purpose of illustration the particle pressure (p) is assumed to be scalar. The magnetospheric force balance equation:

$$\vec{\nabla} p = \vec{j} \times \vec{B} \tag{36}$$

tells that the particle pressure is constant along $\vec{B}$ and that the current perpendicular to $\overline{B}$ is:

$$\vec{j}_\perp = (\vec{B} \times \vec{\nabla} p)/B^2 \tag{37}$$

From the continuity of current equation it follows that the field-aligned current at the top of the ionosphere is:

$$j_\parallel = - \int_0^L \vec{\nabla}_\perp \cdot \vec{j}_\perp (s) ds \tag{38}$$

The ionosphere is treated as a thin conducting layer with distributed, height integrated Pedersen and Hall conductivities Σ_p and Σ_H. These quantities are directly related to the energies and flux levels of locally precipitating particles. The global potential distribution is calculated using Ohm's law in the ionospheric, current-continuity equation. The electric fields derived in these calculations are mapped along equipotential, magnetic field lines to the magnetosphere. Particles are then allowed to $\overline{E} \times \overline{B}$ and gradient-curvature drift in the magnetospheric electric and magnetic fields. Particle pressures are recalculated and the cycle, beginning with Equation (36), is then repeated.

The presence of $E_\parallel$ affects model results in at least three junctures of the chain: (1) in mapping electric fields from the ionosphere to the magnetosphere, (2) in estimating the global distributions of Σ_p and Σ_H, and (3) in calculating magnetopsheric pressure distributions. In a qualitative way the third point may be seen by reflecting on the dependence of $j_\parallel$ on $E_\parallel$, shown in section II.1 and on p, implicit in equations (37) and (38).

An elementary method of computing steady state electric fields in the magnetosphere is to solve Poisson's equation, which relates the divergence of the electric field to the local charge density, subject to boundary conditions of given potentials in the conducting ionosphere. The difficulty in doing so is that the charge distribution depends on the potential and electric field and cannot be regarded as fixed as in classical electrostatics, but must be determined as part of the solution. This approach has been used by Swift (1975, 1976) and Chiu and Cornwall (1978) to study parallel electric fields in auroral arcs. In their treatments, the magnetospheric particles are subject to $\overline{E} \times \overline{B}$ and polarization drifts; the latter causes charge separation and consequantly influences the electric fields. In this section we outline a method of writing Poisson's equation for $\overline{E} \times \overline{B}$, gradient-curvature, and polarization drifting particles and indicate a variational principle from which it is derivable. The assumptions made are that: (1) the electric field is time-independent, (2) the magnetic field is fixed, (3) the plasma is collisionless, (4) the particles conserve their first adiabatic invariants except for the polarization drift, and (5) the particle distribution functions are gyrotropic. At first we neglect the polarization drift in our treatment and then take it into account as a correction.

A charged particle moving in an electrostatic field conserves its total energy

$$W = mv^2/2 + q\Phi$$

The velocity of the particle is composed of the velocity of the guiding center

$$\vec{U} = \vec{u}_\parallel + \frac{\vec{E} \times \vec{B}}{B^2} + \frac{\mu}{eB^2}[\vec{B} \times \vec{\nabla} B] + \frac{mu_\parallel^2}{eB^4}[\vec{B} \times (\vec{B} \cdot \vec{\nabla})\vec{B}] \tag{39}$$

plus the velocity of the particle about the guiding center, so that

$$v^2 = u^2 + U^2 + 2\vec{u} \cdot \vec{U}$$

The value of the last term averaged over a gyroperiod vanishes. The total energy can then be written:

$$W = mu_\parallel^2/2 + \mu B + q\Phi + W_\perp$$

where the conserved magnetic moment is:

$$\mu = m\, u_\perp^2/2B$$

and the energy associated with the perpendicular drift motions is

$$W_\perp = m\,U^2_\perp/2$$

$\vec{U}_\perp$ is given by the last three terms of Equation (39).

The square of the velocity parallel to the field is

$$u^2_\parallel = 2/m\,[\,W - W_\perp - \mu B - q\Phi\,].$$

The volume element in velocity space is

$$d^3v = \pi du_\parallel\, du_\perp^2$$

$$= \pi\,(B/2)\,(2/m)^{3/2}\,dW\,d\mu\,/\,\sqrt{W - W_\perp - \mu B - q\Phi}$$

For gyrotropic distribution functions, one can write the moments of the distribution function:

the number density is

$$n \equiv \int f(\vec{v})d^3v$$

$$= \pi\,(B/2)\,(2/m)^{3/2}\int f(W,\,\mu)dW\,d\mu\,/\,\sqrt{W - W_\perp - \mu B - q\Phi}$$

and the kinetic tensor is

$$K_{ij} \equiv m\int v_i v_j f(\vec{v})d^3v$$

$$= \pi\,(B/2)\,(2/m)^{3/2}\int v_i v_j f(W,\,\mu)dW\,d\mu\,/\,\sqrt{W - W_\perp - \mu B - q\Phi}$$

The component parallel to $\overline{B}$ is

$$K_\parallel = m\,\pi\,(B/2)\,(2/m)^{3/2}\int\sqrt{W - W_\perp - \mu B - q\Phi}\;f\,(W,\,\mu)dW\,d\mu$$

The integrations are over positive values of $u_\parallel^2$. Because the plasma is collisionless, $f(W,\mu)$ is constant along the trajectories of the guiding centers. The relation between the kinetic tensor and the pressure tensor is

$$K_{ij} = P_{ij} + nmV_iV_j$$

where $\vec{V}$ is the velocity of the species. The drift velocity is

$$\vec{V}_\perp = 1/n \int \vec{U}_\perp f(\vec{v}) d^3 v$$
$$= \vec{E} \times \vec{B}/B^2 + P_\perp /neB^3 [\vec{B} \times \vec{\nabla} B] + K_\parallel /neB^4 [\vec{B} \times (\vec{B} \cdot \vec{\nabla}) B]$$
$$- (1/ne)[\vec{\nabla} \times (P_\perp \vec{B}/B^2)]$$

and is composed of $\vec{E} \times \vec{B}$, gradient, curvature, and magnetization drifts, respectively.

The perpendicular pressure is

$$P_\perp = \frac{\pi B}{2} (2/m)^{3/2} \int (\mu B + W - mV_\perp^2) f(W,\mu)\, dW\, d\mu \,/ \sqrt{W - W_\perp - \mu B - q\Phi}$$

The expression for the number density can be used to write Poisson's equation only in a homogeneous plasma. In the presence of a diverging perpendicular electric field, the magnetic moment is not conserved (see Chiu and Cornwall, 1978). As a result, it is necessary to account for the fact that the guiding centers do not follow the trajectories implied by strict conservation. Taking account of this displacement of the guiding centers due to the polarization drift, we obtain, approximately, the number density

$$n = n_0 + \vec{\nabla} \cdot [mn/eB^2 (\vec{V} \times \vec{B})]$$

so that the charge density is

$$\eta = \eta_0 + \vec{\nabla} \cdot [\rho/B^2 \vec{V} \times \vec{B}]$$

where n_0 and η_0 are the number and charge densities computed without the polarization drift, ρ is the (ion) mass density.

Poisson's equation is then

$$\nabla^2 \Phi = - [\eta_0 + \vec{\nabla} \cdot \rho /B^2 (\vec{V} \times \vec{B})]/\varepsilon_0$$

The parallel kinetic tensor $K_\parallel$ depends on Φ and $\vec{\nabla}\Phi$; the following relations can be obtained by differentiating the argument and limits of integration of $K_\parallel$ and summing over species:

$$\partial K_\parallel / \partial \Phi = - \eta_0$$

$$\partial K_\parallel / \partial (\partial \Phi / \partial x_i) = - (\rho/B^2) (\vec{V} \times \vec{B})$$

These relations can be used to construct a variational principle which leads to Poisson's equation. A satisfactory variational principle is

$$\delta \int (\rho V_{\perp}^{2} + K_{\parallel} + \varepsilon_0 E^{2}/2)\, d^{3}x = 0$$

The Euler-Lagrange equation for this Lagrangian density is

$$[\partial/\partial x_i (\partial/\partial(\partial\Phi/\partial x_i)) - \partial/\partial\Phi](\rho V_{\perp}^{2} + K_{\parallel} + \varepsilon_0 (\vec{\nabla}\Phi)^{2}/2)$$

$$= \varepsilon_0 \nabla^{2}\Phi + \vec{\nabla} \cdot \rho/B^{2} (\vec{V} \times \vec{B}) + \eta_0 = 0$$

The potential is to be varied until

$$\int (\rho V_{\perp}^{2} + K_{\parallel} + \varepsilon_0 E^{2}/2)\, d^{3}x$$

is an extremum. In varying the potential, we have assumed that the mass density is held fixed as well as the pressure contributions to $\overline{V}_{\perp}$.

Strictly speaking, Poisson's equation and the variational principle as written here apply only in a homogeneous magnetic field because the contribution of the polarization drift is poorly determined. In a homogeneous field, that contribution to the Lagrangian density is $1/2\ \rho V_{\perp}^{2}$; in a dipole field it is better approximated by $1/2\ \xi\rho V_{\perp}^{2}$ where ξ is in the range 0.25 to 0.4 and depends on the magnetic field geometry and the nature of the electric field.

With this caveat, it is useful to look at the Lagrangian density in terms of polarizable media. Taking $\overline{V}_{\perp}$ as the $\overline{E} \times \overline{B}$ drift, the Lagrangian density can be written

$$\int \{ \rho V_{\perp}^{2}/2 + K_{\parallel} + \varepsilon_0 \vec{E} \cdot [\vec{E}_{\parallel} + (1 + (\rho/B^{2}\varepsilon_0))\vec{E}_{\perp}]/2 \}\, d^{3}x$$

It is closely related to the energy of the plasma. The first term, representing the kinetic energy, arises from the derivative of $K_{\parallel}$ with respect to $\overline{E}$ and is independent of the polarization drift. The second term represents (if parallel flows are ignored) the contributuion of the thermal energy of the plasma. And the last term has the form of the energy of a polarizable medium.

The significance of this type of approach is that it provides a unified view of the origin of electric fields in the magnetosphere. The potential in the magnetosphere is determined as a result of the balance of three types of energy in such a way that the sum of the separate contributions is minimized. Parallel electric fields can be expected to arise naturally in this formulation.

IV. FINAL COMMENTS

In this chapter we have discussed four mechanisms thought to be responsible for parallel electric fields above the auroral ionosphere. We have also considered some effects of these fields on field-aligned currents and magnetospheric pressure distributions. It would be unfair to close this chapter without a "caveat lector" and giving the impression that parallel electric fields are now well understood. The words of Hamlet "There are more things in heaven and earth..." apply equally well to space physicists as to philosophers.

What we have discussed above are steady state mechanisms that in some ways ignore the basic physics of how parallel electric fields evolve and are sustained. Goertz (1979) has stressed that $j_{\parallel}$ and $E_{\parallel}$ above auroral arcs are both directed out of the ionosphere. Thus, the region of $E_{\parallel}$ is an electrical load requiring a continuous input of energy. The source of energy appears in the above mechanisms only as a boundary condition.

Questions may also be raised concerning the stability and uniqueness of auroral double layers. Temerin et al. (1982) have reported observations of multiple double layers and solitons moving rapidly along magnetic field lines past the S3-3 satellite. These observed electrical structures more closely resemble double layers produced in computer simulations (Schunk, private communication, 1982) than the steady state potential sketched in Figure 6. Do the monoenergetic beams observed by satellites reflect merely the time integrated potential differences encountered by particles? Are there distinct types of aurorae, some characterized by near steady-state and others by flickering (Silevitch, 1981) field-aligned potentials? If so, what basic physics underlies the branching?

To end where we began: We have come a long way in the past two decades. The existence of parallel electric fields above auroral arcs is now accepted as an empirical given. From analogies with laboratory plasmas, mechanisms have been identified, which in as yet unspecified combinations, help us understand how these parallel electric fields might exist. There is enough evidence from satellite measurements and computer simulations to suggest that the most interesting phases of the parallel electric field debate may still lie ahead.

ACKNOWLEDGEMENTS

The authors wish to thank Mary Outwater of AFGL for typing the text and Delia Donatelli of Boston College for helpful comments on its organization. This work was supported in part by Air Force Contract Number F19628-82-K-0011 with Boston College.

REFERENCES

Alfven, H. and Falthammer, C.-G.: 1963, Cosmical Electrodynamics Oxford University Press Oxford 2nd edition, pp. 161-165.

Biskamp, D. and Chodura, R.: 1973, Phys. Fluids 16, 888.

Block, L.P.: 1972, Cosmic Electrodynamics 3, 349.

Bohm, D.: 1949, The Characteristics of Electric Discharges in Magnetic in Magnetic Fields, ed. by A. Guthrie and R.K. Wakerling, McGraw-Hill, New York.

Burke, W.J., Silevitch, M.B., and Hardy, D.A.: 1982, J. Geophys. Res. (submitted for publication).

Chiu, Y.T. and Schulz, M.: 1978, J. Geophys. Res. 83, 629.

Chiu, Y.T. and Cornwall, J.M.: 1980, J. Geophys. Res. 85, 543.

Evans, D.S.: 1974, J. Geophys. Res. 79, 2853.

Fridman, M. and Lemaire, J.: 1980, J. Geophys. Res. 85, 664.

Goertz, C.K.: 1979, Rev. Geophys. Sp. Phys. 17, 418.

Hudson, M.K. and Mozer, F.S.: 1978, Geophys. Res. Lett. 5, 131.

Hultquist, B.: 1971, Planet Sp. Sci. 19, 749.

Kan, J.R. and Lee, L.C.: 1980, J. Geophys. Res. 85, 788.

Kan, J.R. and Lee, L.C.: 1981, Physics of Auroral Arc Formation, ed. by S.-I. Akasofu and J.R. Kan, AGU, Geophysical Monograph 25, Washington, D.C., p. 206.

Kindel, J.M., and Kennel, C.F.: 1971, J. Geophys. Res. 76, 3055.

Lyons, L.R.: 1981, Physics of Auroral Arc Formation, ed. by S.-I. Akasofu and J.R. Kan, AGU, Geophysical Monograph 25, Washington, D.C., p. 252.

Mizera, P.F., Croby, D.R., Jr., and Fennel, J.F.: 1976, Geophys. Res. Lett. 3, 149.

Mozer, F.S., Cattell, C.A., Hudson, M.K., Lysak, R.L., Temerin, M. and Torbert, R.B.: 1980, Space Sci. Rev. 27, 155.

O'Brien, B.J.: 1970, Planet. Sp. Sci. 18, 1821.

Persson, H.: 1963, Phys. Fluids 6, 1756.

Persson, H.: 1966, Phys. Fluids 9, 1090.

Shawhan, S.D., Falthammer, C.-G. and Block, L.P.: 1978, J. Geophys. Res. 83, 1049.

Silevitch, M.B.: 1981, J. Geophys. Res. 86, 3573.

Stern, D.P.: 1981, J. Geophys. Res. 86, 5839.

Swift, D.W.: 1975, J. Geophys. Res. 80, 2096.

Swift, D.W.: 1976, J. Geophys. Res. 81, 3935.

Temerin, M., Cerny, K., Lotko, W. and Mozer, F.S.: 1982, Phys. Rev. Lett. (submitted for publication).

Wagner, J.S., Kan, J.R., Akasofu, S.-I., Tajima, T., Leboeuf, J.N. and Dawson, J.M.: 1981, Physics of Auroral Arc Formation, ed. by S.-I. Akasofu and J.R. Kan, AGU, Geophysical Monograph 25, Washington, D.C. p. 304.

Whipple, E.C., Jr.: 1977, J. Geophys. Res. 82, 1525.

ELECTROSTATIC ION ACOUSTIC WAVES

A. Hasegawa
Bell Laboratories
Murray Hill, New Jersey 07974

ABSTRACT

Linear and nonlinear properties of electrostatic ion acoustic waves are discussed. The subject matter includes the propagation and excitations of the wave, quasi linear and mode-coupling saturation of the instability, the formation of ion acoustic solitons and shocks and the generation of negative potential solitary waves and weak double layers.

I. INTRODUCTION

In this lecture, I attempt to illustrate certain aspects of plasma physics through a study of electrostatic ion acoustic wave. The collision-less ion acoustic wave is an old and a new subject in plasma physics. It is very rich in its contents involving various linear and nonlinear properties of plasmas.

The lecture is consisted of three Sections. Section II deals with linear properties of the ion acoustic wave including derivation of the dispersions relation with the effect of Landau damping and of an ambient magnetic field. The section also introduces the excitation processes of the ion acoustic wave due to an electron drift or to a stimulated Brillouin scattering. The nonlinear properties are introduced in Section III and IV. In Section III, incoherent nonlinear effects such as quasilinear and mode-coupling saturations of the instability are discussed. The coherent nonlinear effects such as the generations of ion acoustic solitons, shocks and weak double layers are presented in Section IV.

II. LINEAR PROPERTIES

In this section, we derive linear properties of the electrostatic ion acoustic wave, and its generation processes due to an electron drift and to a stimulated scattering of an electromagnetic wave.

A. Electrostatic, Collision-less Ion Acoustic Wave

Let us consider an electrostatic perturbation in a collision less plasma in a frequency range of ion inertia. First we consider a case without any ambient magnetic field.

We use the Vlasov equation to describe both electron and ion dynamics in the presence of an electrostatic potential field $\phi(\mathbf{x},t)$.

R. L. Carovillano and J. M. Forbes (eds.), Solar-Terrestrial Physics, 425–452.

$$\frac{\partial f^{(i)}}{\partial t} + \mathbf{v} \cdot \frac{\partial f^{(i)}}{\partial \mathbf{x}} - \frac{e}{m_i} \frac{\partial \phi}{\partial \mathbf{x}} \cdot \frac{\partial f^{(i)}}{\partial \mathbf{v}} = 0 \tag{2.1}$$

$$\frac{\partial f^{(e)}}{\partial t} + \mathbf{v} \cdot \frac{\partial f^{(e)}}{\partial \mathbf{x}} + \frac{e}{m_e} \frac{\partial \phi}{\partial \mathbf{x}} \cdot \frac{\partial f^{(e)}}{\partial \mathbf{v}} = 0 \tag{2.2}$$

Here the potential ϕ is related to the electron and ion distribution function $f^{(e)}(\mathbf{x},\mathbf{v},t)$ and $f^{(i)}(\mathbf{x},\mathbf{v},t)$ through the Poisson equation,

$$\nabla^2 \phi = -\frac{e n_o}{\epsilon_o} \left(\int_{-\infty}^{\infty} f^{(i)} d\mathbf{v} - \int_{-\infty}^{\infty} f^{(e)} d\mathbf{v} \right) , \tag{2.3}$$

where n_o(=const) is the average number density for each species, e(=1.6×10^{-19} Coulomb) and ϵ_o(=8.854×10^{-12} Farad/m) are the electron charge and space dielectric constant, m_e(=9.1×10^{-31} kg) and m_i are the electron and ion mass respectively. $\partial/\partial\mathbf{x}$ and $\partial/\partial\mathbf{v}$ are gradient operators in coordinate and velocity space.

Equations (1) to (3) are coupled nonlinear integral differential equations and are impossible to solve for a general case. To study a linear wave property, one usually linearizes the set of equations and applies the Fourier transformations. For the purpose of linearizations, let us expand dependent variables in the power of the electrostatic potential ϕ,

$$\phi = \phi_o + \epsilon\, \phi_1 + \epsilon^2 \phi_2 + \cdots \tag{2.4}$$

$$f^{(j)} = f_o^{(j)} + \epsilon\, f_1^{(j)} + \epsilon^2 f_2^{(j)} + \cdots . \tag{2.5}$$

$$j = i,e$$

Here $\phi_o(\mathbf{x})$ is the umperturbed potential which we assume to be zero and $f_o^{(j)}(\mathbf{x},\mathbf{v})$ is the umperturbed distribution functions which is normalized to unity. If ϕ_o=0, f_o satisfies, from Eqs. (2) and (3), $\partial f_o/\partial\mathbf{x}$=0. Hence f_o is a function of only the velocity $\mathbf{v}$. We usually assume a Maxwellian distribution for such a case

$$f_o^{(j)}(\mathbf{v}) = \frac{1}{\sqrt{2\pi}\, v_{Tj}} \exp\left(-\frac{v^2}{2v_{Tj}^2}\right)$$

$$j = i,e \ , \tag{2.6}$$

where v_T is the thermal velocity. v_T is related to the temperature T of the species through

$$m_j v_{Tj}^2 = T_j \tag{2.7}$$

which can be easily checked from

$$m \int v^2 f_o\, d\mathbf{v} = T$$

ϕ_1 and f_1 are related through,

$$\frac{\partial f_1^{(i)}}{\partial t} + \mathbf{v} \cdot \frac{\partial f_1^{(i)}}{\partial \mathbf{x}} - \frac{e}{m_i} \frac{\partial \phi_1}{\partial \mathbf{x}} \cdot \frac{\partial f_o^{(i)}}{\partial \mathbf{v}} = 0 , \tag{2.8}$$

$$\frac{\partial f_1^{(e)}}{\partial t} + \mathbf{v} \cdot \frac{\partial f_1^{(e)}}{\partial \mathbf{x}} + \frac{e}{m_e} \frac{\partial \phi_1}{\partial \mathbf{x}} \cdot \frac{\partial f_o^{(e)}}{\partial \mathbf{v}} = 0 \tag{2.9}$$

and

$$\nabla^2 \phi_1 = -\frac{e n_o}{\epsilon_o} \left(\int_{-\infty}^{\infty} f_1^{(i)} d\mathbf{v} - \int_{-\infty}^{\infty} f_1^{(e)} d\mathbf{v} \right) . \tag{2.10}$$

To study the linear response, it is convenient to use the Fourier-Laplace transformation. Let us consider a problem in which an initial value in $f_1(\mathbf{x},\mathbf{v},t=0)$ is given. The Fourier-Laplace transformation of f_1 is given by

$$f_{\mathbf{k},\omega}(\mathbf{k},\mathbf{v},\omega) = \int_{-\infty}^{\infty} d\mathbf{x} \int_{0}^{\infty} dt\, f_1(\mathbf{x},\mathbf{v},t) \exp[-i(\mathbf{k}\cdot\mathbf{v}-\omega t)] \tag{2.11}$$

where Im $\omega>0$ for the convergence at $t\to+\infty$. While f_1 is given by the inverse transformation,

$$f_1(\mathbf{x},\mathbf{v},t) = \frac{1}{(2\pi)^4} \int_{-\infty}^{\infty} d\mathbf{k} \int_{-\infty}^{\infty} d\omega\, f_{\mathbf{k},\omega}(\mathbf{k},\mathbf{v},\omega) \exp i(\mathbf{k}\cdot\mathbf{x}-\omega t) \tag{2.12}$$

with Im $\omega>0$. Here $f_{\mathbf{k},\omega}[=f(\mathbf{k},\mathbf{v},\omega)]$ is a function of the wave vector $\mathbf{k}$ and the frequency ω. The convenience of the use of the Fourier-Laplace transformation exists in that a differential equation becomes an algebraic equation. For example, Eq. (2.8) becomes,

$$- f_{\mathbf{k}}^{(i)}(t=0) - i\omega f_{\mathbf{k},\omega}^{(i)} + i\mathbf{k}\cdot\mathbf{v}\, f_{\mathbf{k},\omega}^{(i)} - \frac{e}{m_i} i\phi_{\mathbf{k},\omega}\, \mathbf{k}\cdot \frac{\partial f_o^{(i)}}{\partial \mathbf{v}} = 0 \tag{2.13}$$

or solving for $f_{\mathbf{k},\omega}^{(i)}$, we have

$$f_{\mathbf{k},\omega}^{(i)} = \frac{\dfrac{e}{m_i}\mathbf{k}\cdot\dfrac{\partial f_o^{(i)}}{\partial \mathbf{v}}\phi_{\mathbf{k},\omega}}{\mathbf{k}\cdot\mathbf{v}-\omega} + \frac{f_{\mathbf{k}}^{(i)}(t=0)}{i(\mathbf{k}\cdot\mathbf{v}-\omega)} . \tag{2.14}$$

Similarly

$$f_{\mathbf{k},\omega}^{(e)} = \frac{-\dfrac{e}{m_e}\mathbf{k}\cdot\dfrac{\partial f_o^{(e)}}{\partial \mathbf{v}}\phi_{\mathbf{k},\omega}}{\mathbf{k}\cdot\mathbf{v}-\omega} + \frac{f_{\mathbf{k}}^{(e)}(t=0)}{i(\mathbf{k}\cdot\mathbf{v}-\omega)} \tag{2.15}$$

and

$$k^2\phi_{\mathbf{k},\omega} = \frac{en_o}{\epsilon_o} \int_{-\infty}^{\infty} (f_{\mathbf{k},\omega}^{(i)} - f_{\mathbf{k},\omega}^{(e)})\, d\mathbf{v} . \tag{2.16}$$

If we substitute Eqs. (2.14) and (2.15), we have

$$\left[1 - \frac{\omega_{pi}^2}{k^2}\int_{-\infty}^{\infty} \frac{\mathbf{k}\cdot\partial f_o^{(i)}/\partial\mathbf{v}}{\mathbf{k}\cdot\mathbf{v}-\omega} d\mathbf{v} - \frac{\omega_{pe}^2}{k^2}\int_{-\infty}^{\infty} \frac{\mathbf{k}\cdot\partial f_o^{(e)}/\partial\mathbf{v}}{\mathbf{k}\cdot\mathbf{v}-\omega} d\mathbf{v}\right]\phi_{\mathbf{k},\omega}$$

$$\equiv \epsilon(\mathbf{k},\omega)\phi_{\mathbf{k},\omega} = \frac{en_o}{i\epsilon_o k^2}\int_{-\infty}^{\infty} \frac{f_{\mathbf{k}}^{(i)}(t=0)-f_{\mathbf{k}}^{(e)}(t=0)}{\mathbf{k}\cdot\mathbf{v}-\omega} d\mathbf{v} , \tag{2.17}$$

where $\omega_{pj}^2(=e^2n_o/\epsilon_o m_j)$ is the plasma frequency of the jth species. Solving for $\phi_{\mathbf{k},\omega}$, we have

$$\phi_{\mathbf{k},\omega} = \frac{1}{\epsilon(\mathbf{k},\omega)}\frac{en_o}{i\epsilon_o k^2}\int_{-\infty}^{\infty} \frac{f_{\mathbf{k}}^{(i)}(t=0) - f_{\mathbf{k}}^{(e)}(t=0)}{\mathbf{k}\cdot\mathbf{v}-\omega} d\mathbf{v} . \tag{2.18}$$

When $\phi_{\mathbf{k},\omega}$ is substituted for the inverse Fourier transformation expression, (2.12), the response in the real coordinates $\phi(x,t)$ is obtained. From the residue theorem, it is clear that the contribution from the pole(s) given by $\epsilon(\mathbf{k},\omega)=0$ decides the value of the integration (2.12). This means that $\epsilon(\mathbf{k},\omega)=0$ decides the property of the linear wave of the medium. $\epsilon(\mathbf{k},\omega)=0$ decides the relation between the frequency ω and the wave vector $\mathbf{k}$ and is called the dispersion relation. The dispersion relations for electrostatic waves in unmagnetized plasma is hence derived from Eq. (2.17) by writing $v = \mathbf{k}\cdot\mathbf{v}/k$ and integrating over the other two components of $\mathbf{v}$,

$$1-\frac{\omega_{pi}^2}{k^2}\int_{-\infty}^{\infty}\frac{\partial f_o^{(i)}/\partial v}{v-\omega/k}\,dv-\frac{\omega_{pe}^2}{k^2}\int_{-\infty}^{\infty}\frac{\partial f_o^{(e)}/\partial v}{v-\omega/k}\,dv=0\ . \tag{2.19}$$

To obtain the explicit relation $\omega=\omega(\mathbf{k})$, we must perform the integration over the velocity space. For this we must specify the distributions functions $f_o(\mathbf{v})$. Now the integral

$$\bar{g}(y)=\int_{-\infty}^{\infty}\frac{g(x)}{x-y}\,dx\ ,\ \mathrm{Im}\ y>0 \tag{2.20}$$

is called the Hilbert transformation of g(x). If Im y = +0, $\bar{g}$ is given by

$$\bar{g}(y)=i\pi g(y)+P\int_{-\infty}^{\infty}\frac{g(x)}{x-y}\,dx \tag{2.21}$$

where P is the principal value integral, the integral evaluated by skipping the pole. This expression indicates that the dispersion relation in general is a complex function of $\mathbf{k}$ and ω. ω has both real and imaginary parts for a given real value of $\mathbf{k}$. From Eqs. (2.19) and (2.21) it is also clear that if $\partial f_o/\partial v$ evaluated at $v=\omega/k$ has a large value, the imaginary part of the dispersion relation also becomes large. The imaginary part appears from the resonant interaction between particles with $v\simeq\omega/k$ and the wave with the phase velocity ω/k. Since $\partial f_o/\partial v$ is maximum near $v=v_T$ (= thermal speed), a wave whose phase velocity is near the thermal velocity has strong wave - particle interactions and is heavily damped (Landau damping). Since a sound wave has the phase velocity at near the thermal speed, a sound wave is heavily damped and does not exist in a collision less plasma. (Note that the ordinary sound wave propagates by means of interparticle collisions).

Most plasmas are under thermodynamically nonequilibrium and the electron and ion temperatures are different. It is because it takes the ion-electron collision time for electrons and ions to reach the same temperature, the time which is much longer compared with other characteristic time in a plasma. Let us suppose that the electron temperature T_e is much larger than the ion temperature T_i. Under this circumstance, the sound wave is expected to have a phase velocity somewhere in between the thermal velocities of electrons and ions,

$$v_{Ti}\ll\omega/k\ll v_{Te}\ . \tag{2.22}$$

If f_o is Maxwellian given by (2.6), it is easily seen that if ω/k satisfies (2.22), $\partial f_o^{(i)}/\partial v|_{v=\omega/k}$ is exponentially small. Then, integrating by parts,

$$\int_{-\infty}^{\infty}\frac{\partial f_o^{(i)}/\partial v}{v-\omega/k}\,dv\simeq-\int_{-\infty}^{\infty}\frac{f_o^{(i)}}{(v-\omega/k)^2}\,dv$$

$$\simeq\frac{k^2}{\omega^2}\int f_o^{(i)}\,dv=\frac{k^2}{\omega^2} \tag{2.23}$$

While for electrons

$$\int_{-\infty}^{\infty}\frac{\partial f_o^{(e)}/\partial v}{v-\omega/k}\,dv$$

$$=-\frac{1}{v_T^2}\int\frac{vf_o^{(e)}}{v-\omega/k}\,dv$$

$$\simeq-\frac{1}{v_T^2}\left[i\pi\frac{\omega}{|k|}\frac{1}{\sqrt{2\pi}v_{Te}}+1\right]$$

Thus, the dispersion relation becomes

$$1 - \frac{\omega_{pi}^2}{\omega^2} + \frac{\omega_{pe}^2}{k^2 v_{Te}^2}\left[1 + i\sqrt{\frac{\pi}{2}}\frac{\omega}{|k| v_{Te}}\right] = 0 \ . \tag{2.24}$$

At $\omega << \omega_{pi}$,

$$\omega \simeq kc_s \left[1 - i\sqrt{\frac{\pi}{2}}\sqrt{\frac{m_e}{m_i}}\right] , \tag{2.25}$$

where

$$c_s = \frac{\omega_{pi}}{\omega_{pe}} v_{Te} = \sqrt{\frac{Te}{m_i}} \tag{2.26}$$

is the ion sound speed.

Consequently we can see that if $T_e >> T_i$, there exists a weakly damped collision-less sound wave with the phase velocity given by $(T_e/m_i)^{1/2}$ and the Landau damping rate ($\mathrm{Im}\omega$) given approximately by $\omega(m_e/m_i)^{1/2}$. Such a collision-less sound wave is sometimes called the zeroth sound.

If $T_i \simeq T_e$, the ion acoustic wave is heavily damped by the Landau damping. For detail see Fried and Gould (1961). In an extremely long wavelength region, the ion-ion collision allows the wave to propagate free of the Landau damping like an ordinary sound wave if the ion-ion collision frequency ν_{ii} is larger than the Landau damping rate. The sound wave in such a regime is called the first sound. For an interesting transition from the first sound to the zeroth sound see Huang, et al. (1974). For a further note, if the ion acoustic wave becomes highly turbulent and the density wave of the wave intensity (phonons) propagates through wave-wave interactions (wave-wave collisions), the wave is called the second sound. See Kaner and Yakovenko (1970) for the reference. Although these names are familiar in phonons in liquid helium, we see that the zeroth, first and second sounds are all possible in plasmas.

B. Effect of An Ambient Magnetic Field

If there exists a stationary magnetic field in the plasma, the dispersion relation of the ion acoustic wave is modified because 1) the magnetic field confines the motion of electrons in one direction and 2) the ion cyclotron resonance appears. To see these effects, let us use the cold fluid equations for ions and the drift kinetic equation for electrons by assuming again $T_e >> T_i$. The fluid equations for ions read,

$$\frac{\partial \mathbf{v}^{(i)}}{\partial t} + (\mathbf{v}^{(i)} \cdot \nabla)\, \mathbf{v}^{(i)} = - \frac{e}{m_i} \nabla \phi + \mathbf{v}^{(i)} \times \omega_{ci} \tag{2.27}$$

$$\frac{\partial n^{(i)}}{\partial t} + \nabla \cdot (n^{(i)} \mathbf{v}^{(i)}) = 0 \ , \tag{2.28}$$

where $v^{(i)}(\mathbf{x},t)$ is the velocity field which varies in the direction parallel (z) as well as perpendicular ($\mathbf{x}$) to the magnetic field,

$$\omega_{ci} = \frac{e}{m_i} B_o \tilde{z} \tag{2.29}$$

is the vector ion cyclotron frequency directed parallel to the magnetic field.

The drift kinetic equation is an equation which is convenient to describe phase space dynamics of particles which are magnetized, i.e., the magnetic moment μ is constant. Here the parallel motion is assumed to be free but the perpendicular motion is given by the guiding center drifts. The equation hence has only one velocity space variable, the parallel velocity $v_{||}$, as compared with three components of velocity in the Vlasov equation.

For an electrostatic field in a uniform density and magnetic field, the drift kinetic equation reads,

$$\frac{\partial f^{(e)}}{\partial t} + v_{11} \frac{\partial f^{(e)}}{\partial z} + \nabla \cdot (\mathbf{v}_d f^{(e)}) + \frac{e}{m_e} \frac{\partial \phi}{\partial z} \frac{\partial f^{(e)}}{\partial v_{11}} = 0 \quad , \tag{2.29}$$

where the drift velocity $\mathbf{v}_d$ in a uniform plasma is given by

$$\mathbf{v}_d = \frac{-\nabla \phi \times \hat{z}}{B_o} \quad . \tag{2.30}$$

The electron density $n^{(e)}$ is then obtained from

$$n^{(e)}(\mathbf{x},t) = n_o \int_{-\infty}^{\infty} f^{(e)}(\mathbf{x},v_{11},t)dv_{11} \quad . \tag{2.31}$$

The Poisson equation remains the same as Eq. (2.3). To obtain the dispersion relation, we linearize Eqs. (2.27) to (2.29) and apply the Fourier transformation. Noting that the linear part of $(\mathbf{v}\cdot\nabla)\mathbf{v}$ and $\nabla\cdot(\mathbf{v}_d f)$ are zero, we have

$$-i\omega \mathbf{v}^{(i)}_{\mathbf{k},\omega} = -\frac{e}{m_i} i\, \mathbf{k}\, \phi_{\mathbf{k},\omega} + \mathbf{v}_{\mathbf{k},\omega} \times \omega_{ci} \tag{2.32}$$

$$n^{(i)}_{\mathbf{k},\omega} = n_o \frac{\mathbf{k}\cdot\mathbf{v}_{\mathbf{k},\omega}}{\omega} \, , \tag{2.33}$$

$$f_{\mathbf{k},\omega}{}^{(e)} = \frac{\frac{e}{m_e} k_{11} \frac{\partial f_o^{(e)}}{\partial v_{11}}}{\omega - k_{11} v_{11}} \phi_{\mathbf{k},\omega} \quad . \tag{2.34}$$

To obtain the ion density perturbation it is convenient to make a scalar and vector product of $\mathbf{k}$ to Eq. (2.32) and solve for $\mathbf{k}\cdot\mathbf{v}_{\mathbf{k},\omega}$. Substituting the result to Eq. (2.33), we have, for the ion density perturbation,

$$n^{(i)}_{\mathbf{k},\omega} = \frac{en_o}{m_i} \left[\frac{k_\perp^2}{\omega^2 - \omega_{ci}^2} + \frac{k_{11}^2}{\omega^2} \right] \phi_{\mathbf{k},\omega} \quad . \tag{2.35}$$

where $k_\perp(=k_x)$ and $k_{11}(=k_z)$ are the perpendicular and parallel wave numbers with respect to the direction of the magnetic field.

The electron density is obtained by integrating Eq. (2.34) over v_{11}. If we assume a Maxwellian distribution for $f_o(v_{11})$, we have

$$n^{(e)}_{\mathbf{k},\omega} = n_o \frac{e\phi_{\mathbf{k},\omega}}{T_e} \left[1 + i \sqrt{\frac{\pi}{2}} \frac{\omega}{|k_{11}| v_{Te}} \right] . \tag{2.36}$$

The dispersion relation is obtained by substituting Eqs. (2.35) and (2.36) into the Poisson equation,

$$1 + \frac{k_D^2}{k^2} \left(1 + i\sqrt{\frac{\pi}{2}} \frac{\omega}{|k_{11}| v_{Te}}\right) - \frac{\omega_{pi}^2}{k^2} \left(\frac{k_\perp^2}{\omega^2 - \omega_{ci}^2} + \frac{k_{11}^2}{\omega^2} \right) = 0 \quad , \tag{2.37}$$

where $k_D = \omega_{pe}/v_{Te}$ is the electron Debye wave number. For a plasma with reasonably high density such that $\omega_{pi} >> \omega_{ci}$, the dispersion relation near $\omega \simeq \omega_{ci}$ is obtained, by ignoring the Landau damping for simplicity,

$$\frac{k_{11}^2 c_s^2}{\omega^2} + \frac{k_\perp^2 c_s^2}{\omega^2 - \omega_{ci}^2} - 1 = 0 \ . \tag{2.38}$$

The wave given by this dispersion relation is sometimes called the electrostatic ion cyclotron wave. At a low frequency and long wavelength limit, $\omega << \omega_{ci}$, $k_\perp c_s/\omega_{ci} << 1$, this dispersion relation gives,

$$\omega = k_{11}\, c_s \ . \tag{2.39}$$

Hence the wave propagates only in the direction parallel to the magnetic field. This result is obtainable from the ideal MHD equation at a low β limit. Near the cyclotron resonance, the dispersion relation becomes, for $\omega \leq \omega_{ci}$

$$k_{11}^2 \simeq \frac{\omega^2}{\omega_{ci}^2 - \omega^2}\, k_\perp^2 \tag{2.40}$$

and for $\omega \gtrsim \omega_{ci}$

$$\omega^2 \simeq \omega_{ci}^2 + k_\perp^2 c_s^2/(1 - k_{11}^2 \rho_s^2). \tag{2.41}$$

where $\rho_s(=c_s/\omega_{ci})$ is the ion Larmor radius at the electron temperature.

At $\omega >> \omega_{ci}$ the dispersion relation becomes the same as the case without the magnetic field, Eq. (2.24).

C. Instability Due to An Electron Drift

In the presence of an electron drift, the ion acoustic wave is known to become unstable. The nature of instability is different depending whether the drift speed v_o is smaller or larger than the electron thermal speed v_{Te}.

Let us first consider the case $v_o < v_{Te}$. From Eqs. (2.17) and (2.21), we see that the imaginary part of the dispersion relation depends on $\partial f_o^{(e)}/\partial v$, evaluated at $\omega = \mathbf{k}\cdot\mathbf{v}(=kc_s)$ that is the derivative of the distribution function evaluated at electron velocity projected to the direction of the wavevector being equal to the ion sound speed. Hence the instability is expected to appear when $\partial f_o^{(e)}/\partial v|_{\mathbf{k}\cdot\mathbf{v}=\omega} > 0$. To see more detail, let us assume a drifted Maxwellian for $f_o^{(e)}$,

$$f_o^{(e)} = \frac{1}{\sqrt{2\pi}\, v_{Te}} \exp\left[-\frac{(v - v_o)^2}{2v_{Te}^2}\right] \ . \tag{2.42}$$

Then the dispersion relation, Eq. (2.24), is easily seen to be modified to

$$1 - \frac{\omega_{pi}^2}{\omega^2} + \frac{\omega_{pe}^2}{k^2 v_{Te}^2}\left[1 + i\sqrt{\frac{\pi}{2}}\,\frac{1}{v_{Te}}\,\frac{1}{|k|}(\omega - \mathbf{k}\cdot\mathbf{v})\right] = 0 \tag{2.43}$$

or by solving for ω,

$$\omega \simeq kc_s\,(1 + i\sqrt{\frac{\pi}{2}}\,\frac{1}{|k|}\,\frac{\mathbf{k}\cdot\mathbf{v}_o - kc_s}{v_{Te}}) \ . \tag{2.44}$$

Thus the instability appears for $v_o > c_s$. The wave particle interaction, sometimes called the inverse Landau damping, is responsible to the instability. We note that if $v_o >> c_s$, a wide range of spectrum with the angle of propagation given by $1 \geq \cos\theta > c_s/v_o$ is excited, where $\theta = \cos^{-1}(\mathbf{k}\cdot\mathbf{v}_o/kv_o)$.

If the drift speed exceeds the electron thermal speed, $v_o > v_{Te}$, the imaginary part of ω becomes very large and the Dirac relation, Eq. (2.21), is no longer applicable. In this case, the fluid instability called the two stream instability (which is caused by the negative energy wave carried by the electron beam) is responsible to excite the instability.

If $v_o >> v_{Te}$, substituting Eq. (2.42) into the dispersion relation, Eq. (2.19), we have

$$1 - \frac{\omega_{pi}^2}{\omega^2} - \frac{\omega_{pe}^2}{(\omega - kv_o)^2} = 0 \ . \tag{2.45}$$

The instability occurs near $k = \omega_{pe}/v_o$ whereas the dispersion relation becomes

$$\omega^3 \simeq -\frac{\omega_{pe}^3}{2}\frac{m_e}{m_i} \tag{2.46}$$

and the growth rate γ of the instability is given by

$$\gamma \simeq (\frac{m_e}{m_i})^{1/3}\omega_{pe} \ , \tag{2.47}$$

This instability was first derived by Pierce (1948), but is often called the Buneman instability (Buneman, 1958).

Note that the wave excited here is not the ion acoustic wave.

D. Instability Due to A Decay of An Electromagnetic Wave

The ion acoustic wave can also be excited by a decay of a high frequency wave. The high frequency wave can either be a transverse wave (an electromagnetic wave) or a longtitudinal wave (a Langmuir wave).

An electromagnetic wave can decay to a Langmuir wave and an ion acoustic wave (the decay instability) or to an ion acoustic wave and another electromagnetic wave (the stimulated Brillouin scattering). A Langmuir wave can decay to another Langmuir wave and an ion acoustic wave.

Instead of treating all these cases, let us consider an example of the stimulated Brillouin scattering (Gorburnov, 1969). Consider a large amplitude electromagnetic wave with the electric field $E_o \cos(k_o x - \omega_o t)$ propagating through a plasma. We express this electric field by a vector Fourier amplitude notation such as

$$\mathbf{E}_o = \frac{1}{2}[\mathbf{E}_{k_o} \exp i(k_o\, x - \omega_o t) + \text{c.c.}] \ . \tag{2.48}$$

Note that $\mathbf{E}_{k_o}$ is a complex Fourier amplitude of the electric field and has the dimension of the electric field. Thus E_{k_o} is dimensionally different from the Fourier transformed electric field such as the example shown in Eq. (2.11). The electric field induces oscillating motion to the plasma electrons in the direction of E_o given by

$$\mathbf{v}_{k_o} = \frac{e}{i\omega_o m_e}\mathbf{E}_{k_o}. \tag{2.49}$$

Also note that the magnetic field of the wave is given by

$$\mathbf{B}_{k_o} = \frac{1}{i\omega_o}(\mathbf{k}_o \times \mathbf{E}_{k_o}). \tag{2.50}$$

Here ω_o, $\mathbf{k}_o$ are the frequency and the wave vector of the incident and electromagnetic wave (often called a pump). For the wave to be able to propagate through a plasma $\omega_o \geq \omega_{pe}$ should be satisfied.

When this wave excites the ion acoustic wave, since the ion acoustic wave has a momentum $\hbar k \simeq \hbar k_D$ which is usually much larger than the momentum of the electromagnetic wave, $\hbar k_o \simeq \hbar\omega_{pe}/c$, the scattered electromagnetic wave is reflected backward, leaving a momentum of $2\hbar k_o$ to the ion acoustic wave and an energy of $\hbar(\omega_o - \omega_s)$ where ω_s is the frequency of the scattered electromagnetic wave. The radiation source of the scattered wave is a nonlinear current density J_s^n induced in the plasma by the beating between the density modulation due to the ion acoustic wave n_a and the velocity modulation v_o due to the incident wave,

$$J_s^n = -en_a v_o. \tag{2.51}$$

The nonlinear Lorentz force, $F_a^n = -e(\mathbf{v}_s \times \mathbf{B}_o + \mathbf{v}_o \times \mathbf{B}_s)$, due to the beating between the incident wave (subscript zero) and the scattered wave (subscript s) in turn induces an electron density modulation of the ion acoustic wave. The scattered wave has a wave number k_o and propagates in the direction opposite to the incident wave at the frequency $\omega_s = \omega_o - \omega$ where ω is the frequency of the induced ion acoustic wave. Hence the wave electric field is given by

$$\mathbf{E}_s = \frac{1}{2}[\mathbf{E}_{k_s} \exp i(k_o x + \omega_s t) + \text{c.c.}] \ . \tag{2.52}$$

Expressing $\mathbf{v}_s$ and $\mathbf{B}_s$ in terms of the Fourier amplitude of the scattered wave, $\mathbf{E}_{k_s}$ in the manner similar to Eqs. (2.49) and (2.50), we have

$$\begin{aligned}
F_a^n &= -\frac{e}{4}(\mathbf{v}_{k_s} \times \mathbf{B}_{k_o} + \mathbf{v}_{k_o} \times \mathbf{B}_{k_s})\exp i(kx-\omega t) + \text{c.c.} \ , \\
&= \frac{-e^2}{4i\omega_o\omega_s m_e}[\mathbf{E}_{k_s} \times (\mathbf{k}_o \times \mathbf{E}_{k_o}) \\
&\quad + \mathbf{E}_{k_o} \times (\mathbf{k}_s \times \mathbf{E}_{k_s})]\exp i(\mathbf{k}x-\omega t) + \text{c.c.} \\
&= \frac{-e^2}{2i\omega_o\omega_s m_e}(\mathbf{E}_{k_s} \cdot \mathbf{E}_{k_o})\mathbf{k}_o \exp i(kx-\omega t) + \text{c.c.} \\
&= \frac{1}{2}F_k^n \hat{x} \exp i(kx-\omega t) + \text{c.c.}, \ (k = 2k_o) \ .
\end{aligned} \tag{2.53}$$

Hence the nonlinear force is maximum if the scattered wave is polarized in the same direction as the incident wave and the force is directed in the direction of the electromagnetic wave vector. Note that $\mathbf{F}_a^n$ has two components of frequencies and wave numbers which originate from the products of the exponent of E_{k_o} and E_{k_s}, i.e., $\exp i(2k_o x - \omega t)$ and $\exp[-i(\omega_s + \omega_o)t]$. Naturally only the former is coupled to ion acoustic mode. The electron response due to the nonlinear force is obtained by adding $\mathbf{F}_a^n$ to the electron Vlasov equation. The resultant electron distribution function becomes

$$f_k^{(e)} = \frac{\frac{e}{m_e} k \frac{\partial f_o}{\partial v}}{\omega - kv}\phi_k + \frac{\frac{F_k^n}{m}\frac{\partial f_o}{\partial v}}{i(\omega - kv)} \ , \tag{2.54}$$

where $\mathbf{k}$ is taken in the direction of $\mathbf{F}_k^n$. The density perturbation becomes, assuming a Maxewillian distributions for f_o,

$$n_k^{(e)} = n_o\left[\frac{e\phi_k}{T_e}\left(1 + i\sqrt{\frac{\pi}{2}}\frac{\omega}{|k|v_{Te}}\right)\right] + n_o\frac{F_k^n}{iT_e k} \ . \tag{2.55}$$

The ion density is not affected by the electromagnetic waves because of their large mass. Assuming cold ions and using Poison's equation with Eq. (2.55), the induced potential perturbation satisfies

$$\epsilon_a(\omega,k)\ \phi_k = \frac{1}{ke}\frac{k_D^2}{k^2} F_k^n$$

$$= \frac{k_D^2}{k^2}\frac{e}{\omega_o\omega_s m_e} E_{k_s} E_{k_o} \ , \tag{2.56}$$

where

$$\epsilon_a(\omega,k) = 1 - \frac{\omega_{pi}^2}{\omega_2} + \frac{k_D^2}{k^2}\left(1 + i\sqrt{\frac{\pi}{2}}\frac{\omega}{|k|\ v_{Te}}\right) \ . \tag{2.57}$$

To obtain the stimulated process we should calculate the response of the scattered electric field due to the nonlinear current induced in the plasma, Eq. (2.51). To describe the propagation of an electromagnetic wave, it is sufficient to use cold electron dynamics since $c >> v_{Te}$ and assume that the ions are inmobile in such a high frequency field.

From the Maxwell equation, the scattered wave may be expressed

$$(k_o^2 - \frac{\omega_s^2}{c^2})\ E_{k_s} = i\omega_s\mu_o(J_s + J_s^n) \tag{2.58}$$

where J_s is the linear induced current given by

$$J_s = -en_o v_s = -\frac{e^2 n_o}{i\omega_s m_e} E_{k_s}, \tag{2.59}$$

while the nonlinear induced current is given from Eqs. (2.49) and (2.51)

$$J_s^n = -\ en_a v_o$$

$$= -\frac{1}{4}\ e\ n_k\ v_{k_o}^* \exp i(k_o x+\omega_s t) + c.c.$$

$$= -\frac{1}{4}\frac{e^2 n_o}{T_e}\ \phi_k\ \frac{e}{i\omega_o m_e}\ E_{k_o}^* \exp i(k_o x+\omega_s t) + c.c.$$

$$= -\frac{1}{4}\frac{\omega_{pe}^2\epsilon_o e}{i\ T_e\omega_o}\ \phi_k\ E_{k_o}^* \exp i(k_o x+\omega_s t) + c.c. \ . \tag{2.60}$$

where $E_{k_o}^*$ which is the complex conjugate of E_{k_o} is used such that J_s^n has a phase dependency of $\exp i[k_o x + (\omega_o-\omega)t] = \exp i[k_o x + \omega_s t]$.

Substituting Eqs. (2.59) and (2.60), we have the response of the scattered wave in the following form,

$$\epsilon_s(\omega_o-\omega,-k_o)E_{k_s} = -\frac{1}{2}\frac{\omega_s}{\omega_o}\frac{\omega_{pe}^2}{c^2}\frac{e\phi_k}{T_e} E_{k_o}^* \tag{2.61}$$

where

$$\epsilon_s(\omega,k) = k^2 - \frac{1}{c^2}(\omega^2-\omega_{pe}^2) \ . \tag{2.62}$$

Equations (2.56) and (2.61) show the coupling between the scattered electromagnetic wave, E_{k_s} and the induced ion acoustic wave ϕ_k through the incident wave E_{k_o}. The resonant decay occurs when

$$\epsilon_a(\omega,2k_o) \simeq 0 \tag{2.63}$$

$$\epsilon_s(\omega_o - \omega \cdot - k_o) \simeq 0 \ , \tag{2.64}$$

That is when

$$\omega \simeq kc_s = 2k_oc_s$$

$$k_o^2 - \frac{1}{c^2}[(\omega_o - 2k_oc_s)^2 - \omega_{pe}^2] \simeq 0$$

is satisfied (by the choice of the incident wave frequency). When the résonant condition is satisfied, the growth rate of the ion acoustic wave, γ, is obtained by writing

$$\omega = \omega_a + i\gamma$$

and expanding ϵ_a as well as ϵ_s around $\epsilon_a = \epsilon_s = 0$, we have from Eq. (2.56),

$$\left[\frac{2i\gamma}{\omega_a} + i\left(\frac{\pi m_e}{2m_i}\right)^{1/2}\right] \phi_k = \frac{e}{\omega_o\omega_s m_e} E_{k_s} E_{k_o} \tag{2.56'}$$

and from Eq. (2.61), using $\omega_o \simeq \omega_s \simeq \omega_{pe}$,

$$\frac{2i\gamma}{\omega_s} E_{k_s} = -\frac{1}{2}\frac{e\phi_k}{T_e} E_{k_o}^* \tag{2.61'}$$

If we eliminate ϕ_k and E_{k_s} from Eqs. (2.56′) and (2.61′), we have

$$\frac{4\gamma^2}{\omega_s\omega_a} - \frac{2\gamma}{\omega_s}\left(\frac{\pi m_e}{2m_i}\right)^{1/2} - \frac{1}{2}\frac{e^2}{\omega_o^2 m_e T_e}|E_{k_o}|^2 = 0 \tag{2.65}$$

The growth rate γ of the stimulated Brillouin scattering then becomes, ignoring the second term,

$$\gamma \simeq \frac{1}{2\sqrt{2}}(\omega_a\omega_s)^{1/2}\left[\frac{e^2|E_{k_o}|^2}{T_e m_e \omega_o^2}\right]^{1/2}$$

$$= \frac{1}{2\sqrt{2}}\left(\omega_a\omega_s\right)^{1/2}\frac{|v_{k_o}|}{v_{Te}} \ . \tag{2.66}$$

III. INCOHERENT NONLINEAR EFFECTS

In this Section, we discuss weak incoherent nonlinear effects associated with the excitation of the ion acoustic wave by drifting electrons. The "weak" here means that the nonlinear effect changes the linear property only by a small amount and that the excitations can still be expressed in terms of ensembles of linear waves. "Incoherent" means that the effects do not depend on the phase of the waves. When ion acoustic waves are excited by drifting electrons, the excited waves in turn scatter electrons and induce diffusions of electrons in velocity space. This reduces the value $\partial f_o/\partial v|_{\mathbf{k}\cdot\mathbf{v}=\omega}$ and reduces the growth rate (quasi linear effects). The excited waves also tend to cascade into different regime in $\omega - \mathbf{k}$ space by nonlinear wave-wave or nonlinear wave-particle interactions (nonlinear Landau damping). We treat their effects in this Section.

A. Quasi Linear Effects

When the ion acoustic wave is weakly driven such that $c_s < v_o \ll v_{Te}$, the excited waves can be considered as consisted of ensemble of ion acoustic waves with random phase. (It will be shown in Section IV that if $v_o \gtrsim 0.6 v_{Te}$, the excited ion acoustic waves change their properties significantly and lead to formation of double layers). The excited ensemble of waves in turn change the distribution function by wave particle scattering. In the presence of an ensemble of waves, the quasi-time-stationary distribution function $<f>$ for electrons satisfies the Vlasor equation,

$$\frac{\partial\langle f\rangle}{\partial t} + \mathbf{v}\cdot\frac{\partial\langle f\rangle}{\partial \mathbf{x}} + \frac{e}{m_e}\left\langle\frac{\partial\phi_1}{\partial \mathbf{x}}\cdot\frac{\partial f_1}{\partial \mathbf{v}}\right\rangle = 0 \quad , \tag{3.1}$$

where $\langle\partial\phi_1/\partial\mathbf{x}\cdot\partial f_1/\partial\mathbf{v}\rangle$ is the ensemble average of the product of the first order electric field and the derivative of the distribution function. If the ensemble is spatially uniform $\partial\langle f\rangle/\partial x$ is zero. Furthermore, if the Fourier amplitude expression is used such that

$$\phi_1(\mathbf{x},t) = \frac{1}{2}\left[\phi_{\mathbf{k}}\exp i(\mathbf{k}\cdot\mathbf{x}-\omega t) + \text{c.c.}\right] \ , \tag{3.2}$$

$$\left\langle\frac{\partial\phi_1}{\partial \mathbf{x}}\cdot\frac{\partial f_1}{\partial \mathbf{v}}\right\rangle = \frac{1}{4}\mathrm{Re}\sum_{\mathbf{k}}\left[-i\phi_{\mathbf{k}}^*\mathbf{k}\cdot\frac{\partial f_1}{\partial \mathbf{v}} + \text{c.c.}\right] \ . \tag{3.3}$$

From Eqs. (2.23) and (3.3), it is clear that only the resonant particles contribute to the ensemble average. Substituting the resonant particle response (and replacing f_o with $\langle f\rangle$) to Eqs. (3.3) and (3.1), we have

$$\frac{\partial\langle f\rangle}{\partial t} = \frac{\pi}{2}\left(\frac{e}{m_e}\right)^2\sum_{\mathbf{k}}\mathbf{k}\cdot\frac{\partial}{\partial\mathbf{v}}\left(\delta(\mathbf{k}\cdot\mathbf{v}-\omega)|\phi_{\mathbf{k}}|^2e^{2\gamma_{\mathbf{k}}t}\mathbf{k}\cdot\frac{\partial\langle f\rangle}{\partial\mathbf{v}}\right) \tag{3.4}$$

This is a diffusion equation for the electron distribution function in velocity space with the diffusion coefficient proportional to the square of the wave amplitude. This type of diffusion is called a quasi linear diffusion because the linear response is used to derive the equation. The equation indicates that the distribution function diffuses in velocity space and reduces the growth rate (which is proportional to $\partial\langle f\rangle/\partial v$). We also note that the diffusion coefficient is always positive even if the growth rate is zero meaning that if the waves are driven by other means the diffusion still results. If the wave is excited by the instability of drifting electrons, the growth rate $\gamma_{\mathbf{k}}$ also evolves in response to the change in the distribution function according to

$$\gamma_{\mathbf{k}} = \frac{\pi}{2}v_{Te}^2\omega_{\mathbf{k}}\int\delta(\mathbf{k}\cdot\mathbf{v}-\omega)\mathbf{k}\cdot\frac{\partial\langle f\rangle}{\partial\mathbf{v}}d\mathbf{v} \tag{3.5}$$

The coupled equations consisted of the growth rate expression $\gamma_{\mathbf{k}}$ as a function of $\langle f\rangle$, Eq. (3.5), and the change of $\langle f\rangle$, Eq. (3.4), due to the fluctuating field $\phi_{\mathbf{k}}$ which grows at a rate of $\gamma_{\mathbf{k}}$ give a consistent quasi-linear diffusion process provided that the excited spectrum stays linear, that is, $|\phi_{\mathbf{k}}|^2$ is given by

$$\frac{\partial|\phi_{\mathbf{k}}|^2}{\partial t} = 2\gamma_{\mathbf{k}}|\phi_{\mathbf{k}}|^2 \tag{3.6}$$

However, if nonlinear mode couplings induce the wave number spectrum of $\phi_{\mathbf{k}}$ to evolve continuously, the quasi linear equation becomes no longer closed, Equation (3.6) should be modified to take into account the effects of mode couplings.

B. Nonlinear Mode Couplings

As is shown in the example of the stimulated Brillouin scattering in Section II-D, waves can couple nonlinearly and cascade energy. In the example of the Brillouin scattering, the photon energy cascades by scattering off the ion acoustic wave.

Such a cascade process is called a nonlinear wave-wave interaction. For a nonlinear wave-wave interaction, both the frequency and wave number matching condition should be satisfied among interacting three (or more) waves, such that

$$\omega_1 = \omega_2 + \omega_3 \tag{3.7}$$

$$\mathbf{k}_1 = \mathbf{k}_2 + \mathbf{k}_3 \tag{3.8}$$

with $\omega_1(\mathbf{k}_1)$, $\omega_2(\mathbf{k}_2)$, $\omega_3(\mathbf{k}_3)$ all satisfying the dispersion relation. Noting that $\mathbf{k}$ is a three dimensional vector, the matching conditions (3.7) and (3.8) are satisfied only when the dispersion relation $\omega = \omega(\mathbf{k})$ is convex downward in ω—k plane. Such a dispersion relation is sometimes called a decay type. When the dispersion relation is a decya type, as is found in the example of the stimulated Brillouin scattering, a spectrum which is concentrated at some frequency range cascades down in ω plane. The ion acoustic wave becomes a decay type in the presence of a magnetic field, (see Hasegawa, 1976). However, in the absence of a magnetic field it is not of a decay type.

Even if a dispersion relation is not of a decay type, the decay of spectrum still occurs through the nonlinear wave particle interactions. Consider two waves with frequency ω_1 and ω_2 with $\omega_1(\mathbf{k}_1) > \omega_2(\mathbf{k}_2)$. A virtual wave with the frequency and the wave number given by $\omega_1 - \omega_2$ and $\mathbf{k}_1 - \mathbf{k}_2$ resonates with a particle with velocity $\mathbf{v}$ such that

$$\omega_1 - \omega_2 = (\mathbf{k}_1 - \mathbf{k}_2) \cdot \mathbf{v} \tag{3.9}$$

is satisfied. Then the wave at frequency ω_1 damps while the wave at frequency ω_2 grows and the wave energy cascades toward the lower frequency. Such a process is called the nonlinear Landau damping and is analogous to the stimulated Compton scattering of a photon.

The spectrum of the ion acoustic wave cascades due to the nonlinear Landau damping. Let us look at this process here. In a Vlasov-Poisson system, the only nonlinear term is $\partial\phi/\partial\mathbf{x} \cdot \partial f/\partial\mathbf{v}$ term in the Vlasov equation. In the presence of many waves with $\phi_\mathbf{k}$, $f_\mathbf{k}$, this product generates combinations of waves with wave numbers $\mathbf{k}' + \mathbf{k}''$. Thus the response in the $\mathbf{k}$ mode distribution function becomes

$$(\mathbf{k} \cdot \mathbf{v} - \omega_\mathbf{k}) f_\mathbf{k}^{(j)} = \frac{q_j}{m_j} \phi_\mathbf{k} \mathbf{k} \cdot \frac{\partial f_0^{(j)}}{\partial \mathbf{v}} + \frac{q_j}{m_j} \sum_{\mathbf{k}' \neq \mathbf{k}} \phi_{\mathbf{k}'} \mathbf{k}' \cdot \frac{f_{\mathbf{k}-\mathbf{k}'}}{\partial \mathbf{v}} , \tag{3.10}$$

$$k^2 \phi_\mathbf{k} = \sum_{j=i,e} \frac{q_j n_0}{\epsilon_0} \int_{-\infty}^{\infty} f_\mathbf{k}^{(j)} d\mathbf{v} . \tag{3.11}$$

By iteration of these coupled equations, we can obtain the following nonlinear wave equation for $\phi_\mathbf{k}$:

$$\epsilon_\mathbf{k}^{(1)} \phi_\mathbf{k} + \sum_{\mathbf{k}=\mathbf{k}'+\mathbf{k}''} \epsilon_\mathbf{k}^{(2)}(\mathbf{k}',\mathbf{k}'') \phi_{\mathbf{k}'} \phi_{\mathbf{k}''}$$
$$+ \sum_{\mathbf{k}=\mathbf{k}'+\mathbf{k}''+\mathbf{k}'''} \epsilon_\mathbf{k}^{(3)}(\mathbf{k}',\mathbf{k}'',\mathbf{k}''') \phi_{\mathbf{k}'} \phi_{\mathbf{k}''} \phi_{\mathbf{k}'''} + \cdots = 0 \tag{3.12}$$

where $\epsilon^{(1)}$ is the linear dielectric constant given by

$$\epsilon_{\mathbf{k}}^{(1)} = 1 + \sum_{j=i,e} \frac{\omega_{pj}^2}{k^2} \int d\mathbf{v} \frac{\mathbf{k} \cdot \dfrac{\partial f_0^{(j)}}{\partial \mathbf{v}}}{\omega_{\mathbf{k}} - \mathbf{k} \cdot \mathbf{v} + i0} \quad . \tag{3.13}$$

The 2nd and the 3rd order dielectric constants are, after symmetrizing,

$$\begin{aligned} \epsilon_{\mathbf{k}}^{(2)}(\mathbf{k}',\mathbf{k}'') = & -\frac{1}{2} \sum_{j=i,e} \frac{\omega_{pj}^2}{k^2} \frac{q_j}{m_j} \int d\mathbf{v} \frac{1}{\omega_{\mathbf{k}} - \mathbf{k} \cdot \mathbf{v} + i0} \\ & \cdot \left\{ \mathbf{k}' \cdot \frac{\partial}{\partial \mathbf{v}} \frac{1}{\omega_{\mathbf{k}''} - \mathbf{k}'' \cdot \mathbf{v} + i0} \mathbf{k}'' \cdot \frac{\partial}{\partial \mathbf{v}} \right. \\ & \left. + \mathbf{k}'' \cdot \frac{\partial}{\partial \mathbf{v}} \frac{1}{\omega_{\mathbf{k}'} - \mathbf{k}' \cdot \mathbf{v} + i0} \mathbf{k}' \cdot \frac{\partial}{\partial \mathbf{v}} \right\} f_0^{(j)} \quad , \end{aligned} \tag{3.14}$$

$$\begin{aligned} \epsilon_{\mathbf{k}}^{(3)}(\mathbf{k}',\mathbf{k}'',\mathbf{k}''') = & \frac{1}{3} \sum_{j=i,e} \frac{\omega_{pj}^2}{k^2} \left(\frac{q_j}{m_j} \right)^2 \int d\mathbf{v} \frac{1}{\omega_{\mathbf{k}} - \mathbf{k} \cdot \mathbf{v} + i0} \\ & \cdot \left\{ \mathbf{k}' \cdot \frac{\partial}{\partial \mathbf{v}} \frac{1}{\omega_{\mathbf{k}''} + \omega_{\mathbf{k}'''} - (\mathbf{k}'' + \mathbf{k}''') \cdot \mathbf{v} + i0} \right. \\ & \cdot \left[\mathbf{k}'' \cdot \frac{\partial}{\partial \mathbf{v}} \frac{1}{\omega_{\mathbf{k}'''} - \mathbf{k}''' \cdot \mathbf{v} + i0} \mathbf{k}''' \cdot \frac{\partial}{\partial \mathbf{v}} \right. \\ & \left. + \mathbf{k}''' \cdot \frac{\partial}{\partial \mathbf{v}} \frac{1}{\omega_{\mathbf{k}''} - \mathbf{k}'' \cdot \mathbf{v} + i0} \mathbf{k}'' \cdot \frac{\partial}{\partial \mathbf{v}} \right] \\ & \left. + \text{(2 other permutations of k's} \right\} f_0^{(j)} \quad . \end{aligned} \tag{3.15}$$

The nonlinear Landau damping is a process of energy transfer from one mode to the other through an interaction with a particle. Hence we consider two modes which satisfies the linear dispersion relation, $\omega = \omega(\mathbf{k})$ and $\omega' = \omega'(\mathbf{k}')$, such that

$$\epsilon_{\mathbf{k}}^{(1)}(\omega) = \epsilon_{\mathbf{k}'}^{(1)}(\omega') = 0 \quad . \tag{3.16}$$

We also assume that the beat mode with wave number $\mathbf{k}-\mathbf{k}'$ does not satisfy the linear dispersion relation, that is,

$$\epsilon_{\mathbf{k}-\mathbf{k}'}^{(1)}(\omega-\omega') \neq 0 \quad . \tag{3.17}$$

(Note if $\epsilon_{\mathbf{k}-\mathbf{k}'}^{(1)}(\omega-\omega') = 0$, it means that the beat mode is also a well defined wave, thus the nonlinear wave-wave interaction is possible. Since we consider a case such that the dispersion relation is not of a decay type, Eq. (3.17) is always satisfied).

First, we consider only to the order of $\epsilon^{(2)}$ and write down the equations for $\phi_{\mathbf{k}}$,

$$\epsilon_{\mathbf{k}}^{(1)}\phi_{\mathbf{k}} + \sum_{\mathbf{k}'}\epsilon_{\mathbf{k}}^{(2)}(\mathbf{k}',\mathbf{k}-\mathbf{k}')\phi_{\mathbf{k}'}\phi_{\mathbf{k}-\mathbf{k}'} = 0 \ . \tag{3.18}$$

Now, for $\phi_{\mathbf{k}-\mathbf{k}'}$ mode we have

$$\epsilon_{\mathbf{k}-\mathbf{k}'}^{(1)}\phi_{\mathbf{k}-\mathbf{k}'} + \sum_{\mathbf{k}''}\epsilon_{\mathbf{k}-\mathbf{k}'}^{(2)}(\mathbf{k}'',\mathbf{k}-\mathbf{k}'-\mathbf{k}'')\phi_{\mathbf{k}''}\phi_{\mathbf{k}-\mathbf{k}'-\mathbf{k}''} = 0 \ ,$$

where, if we put $\mathbf{k}'' = \mathbf{k}$,

$$\epsilon_{\mathbf{k}-\mathbf{k}'}^{(1)}\phi_{\mathbf{k}-\mathbf{k}'} + \epsilon_{\mathbf{k}-\mathbf{k}'}^{(2)}(\mathbf{k},-\mathbf{k}')\phi_{\mathbf{k}}\phi_{-\mathbf{k}'} = 0 \ . \tag{3.19}$$

If we substitute (4.144) into (4.142), we have

$$\epsilon_{\mathbf{k}}^{(1)}\phi_{\mathbf{k}} = \sum_{\mathbf{k}'}\frac{\epsilon_{\mathbf{k}}^{(2)}(\mathbf{k}',\mathbf{k}-\mathbf{k}')\epsilon_{\mathbf{k}-\mathbf{k}'}^{(2)}(\mathbf{k},-\mathbf{k}')}{\epsilon_{\mathbf{k}-\mathbf{k}'}^{(1)}}|\phi_{\mathbf{k}'}|^2\phi_{\mathbf{k}} \ . \tag{3.20}$$

If we permute $\mathbf{k}'$ with $\mathbf{k}-\mathbf{k}'$ as well as $\mathbf{k}$ and $-\mathbf{k}'$, we have four of these terms. Similarly, from $\epsilon^{(3)}$, there are three of the $\epsilon_{\mathbf{k}}^{(3)}(\mathbf{k}',\mathbf{k},-\mathbf{k}')$ terms. Thus combining all of these, we have

$$\epsilon_{\mathbf{k}}^{(1)}\phi_{\mathbf{k}} = \sum_{\mathbf{k}'}\left[\frac{4\epsilon_{\mathbf{k}}^{(2)}(\mathbf{k}',\mathbf{k}-\mathbf{k}')\epsilon_{\mathbf{k}-\mathbf{k}'}^{(2)}(\mathbf{k},-\mathbf{k}')}{\epsilon_{\mathbf{k}-\mathbf{k}'}^{(1)}} - 3\epsilon_{\mathbf{k}}^{(3)}(\mathbf{k}',\mathbf{k},-\mathbf{k}')\right]|\phi_{\mathbf{k}'}|^2\phi_{\mathbf{k}} \ . \tag{3.21}$$

Note that because the beat mode is non-resonant ($\epsilon_{\mathbf{k}-\mathbf{k}'}^{(1)} \neq 0$, we must retain $\epsilon_{\mathbf{k}}^{(3)}$ contribution here. $\epsilon_{\mathbf{k}}^{(1)}\phi_{\mathbf{k}}$ may be written as

$$\epsilon_{\mathbf{k}}^{(1)}\phi_{\mathbf{k}} = i\frac{\partial\epsilon}{\partial\omega}\left[-\gamma_{\mathbf{k}} + \frac{\partial}{\partial t}\right]\phi_{\mathbf{k}} \ . \tag{3.22}$$

Hence, from Eqs. (3.21) and (3.22),

$$\frac{\partial\epsilon}{\partial\omega}\Big|_{\mathbf{k}}\left(\frac{\partial}{\partial t} - \gamma_{\mathbf{k}}\right)\phi_{\mathbf{k}} = \mathrm{Im}\sum_{\mathbf{k}'}\left[\frac{4\epsilon_{\mathbf{k}}^{(2)}(\mathbf{k}',\mathbf{k}-\mathbf{k}')\epsilon_{\mathbf{k}-\mathbf{k}'}^{(2)}(\mathbf{k},-\mathbf{k}')}{\epsilon_{\mathbf{k}-\mathbf{k}'}^{(1)}} - 3\epsilon_{\mathbf{k}}^{(3)}(\mathbf{k}',\mathbf{k},-\mathbf{k}')\right]|\phi_{\mathbf{k}'}|^2\phi_{\mathbf{k}} \tag{3.23}$$

Equation (3.23) shows the rate of modal energy transfer from $\mathbf{k}'$ to $\mathbf{k}$ mode through the nonlinear interactions with particles.

In case of the ion acoustic wave, the nonlinear Landau damping due to ions is found to be important in which the resonant condition is given by

$$\omega-\omega' \simeq |\mathbf{k}|v_{Ti}$$

where $|\mathbf{k}-\mathbf{k}'| \equiv |\mathbf{k}''| \sim |\mathbf{k}|$. In this case, the mode coupling equation reduces to (Sagdev and Galeev, 1969).

$$\frac{\omega_{pi}^2}{\omega_{\mathbf{k}}^2}\frac{\partial}{\partial t}|\phi_{\mathbf{k}}|^2 - 2\gamma_{\mathbf{k}}|\phi_{\mathbf{k}}|^2$$

$$+\frac{16\pi^2 e^4 n_o \omega_k}{m_i^3 k^2}\sum_{\mathbf{k}'}\frac{(\mathbf{k}\cdot\mathbf{k}')^2}{\omega_{\mathbf{k}}^3\omega_{\mathbf{k}'}^3}$$

$$\frac{(\mathbf{k}\times\mathbf{k}')^2 v_{Ti}^2}{k''^2}\int d\mathbf{v}\,\delta(\omega_{\mathbf{k}}-\omega_{\mathbf{k}'}-\mathbf{k}''\cdot\mathbf{v})$$

$$\mathbf{k}''\cdot\frac{\partial f_o^{(i)}}{\partial \mathbf{v}}|\phi_{\mathbf{k}'}|^2|\phi_{\mathbf{k}}|^2 = 0 \ . \tag{3.24}$$

we note that the transfer rate is proportional to $\partial f_o/\partial v$ evaluated at the beat phase velocity $\mathbf{v} = (\omega_{\mathbf{k}}-\omega_{\mathbf{k}'})/(\mathbf{k}-\mathbf{k}')$ thus if $\partial f_o/\partial v<0$, the energy cascade to a low frequency.

IV. COHERENT NONLINEAR EFFECTS

In Section III, we discussed processes in which the evolution of the system is given by $|\phi_{\mathbf{k}}|^2$ times the characteristic time of the system. We also learned that the use of Fourier amplitudes is a powerful tool. In this section we consider processes in which the time scale is proportional to (or shorter than) ϕ times the characteristic time scale. In such a case, it is often convenient to study the wave dynamics in the real coordinate since the phase evolution becomes important.

A. Ion Acoustic Shock

Let us consider a large amplitude ion acoustic wave in a plasma with $T_e >> T_i$. We describe ions by cold fluid equations (2.27) and (2.28) while electrons by Boltzmann distribution,

$$n^{(e)} = n_o \exp(e\phi/T_e) \ , \tag{4.1}$$

where n_o is the plasma density at $\phi = 0$. We look for a simple wave solution, a solution which is stationary in a coordinate moving at a speed u. For this purpose we use the new coordinate

$$\xi = x - ut \tag{4.2}$$

and assume all the variables to depend only on ξ i.e.,

$$\frac{\partial}{\partial t} = -u\frac{d}{d\xi} \ , \ \frac{\partial}{\partial x} = \frac{d}{d\xi} \ . \tag{4.3}$$

Then, from the continuity equation of ions,

$$n^{(i)} = \frac{n_o u}{u - v^{(i)}} \ , \tag{4.4}$$

where we assume $n^{(i)} \to n_o$ for $v^{(i)} \to 0$. From the ion equation of motion (2.27), we have

$$\frac{d}{d\xi}\left(\frac{v^{(i)2}}{2} - uv^{(i)} + \frac{e}{m_i}\phi\right) = 0 \ .$$

If we assume $v^{(i)} \to 0$ as $\phi \to 0$, this gives

$$v^{(i)} = u \pm (u^2 - \frac{2e\phi}{m_i})^{1/2} \ . \tag{4.5}$$

Substituting Eq. (4.5) to (4.4), we have

$$n^{(i)} = \frac{u\,n_o}{(u^2 - 2e\phi/m_i)^{1/2}} \ . \tag{4.6}$$

The Poisson equation becomes then, using Eqs. (4.1) and (4.6),

$$\frac{d^2\phi}{d\xi^2} = -\frac{en_o}{\epsilon_o}\left[\frac{u}{(u^2-2e\phi/m_i)^{1/2}} - \exp(\frac{e\phi}{T_e})\right] \ . \tag{4.7}$$

If we write $d\phi/d\xi = p$, $d^2\phi/d\xi^2 = dp/d\xi = p\ dp/d\phi = dp^2/2d\phi$. Hence we can integrate Eq. (4.7) once to give

$$\frac{1}{2}\left(\frac{d\phi}{d\xi}\right)^2 = \frac{n_o}{\epsilon_o}\left[um_i(u^2-\frac{2e\phi}{m_i})^{1/2} + T_e\exp(\frac{e\phi}{T_e})\right] + C$$

$$\equiv - V(\phi,u) \tag{4.8}$$

Eq. (4.8) is equivalent to a motion of a particle with the kinetic energy $\frac{1}{2}\left(\frac{d\phi}{d\xi}\right)^2$ in a potential field given by $V(\phi,u)$ (if we replace ϕ by x and ξ by t). $V(\phi)$ is often called the Sagdeev potential. Depending on the structure of $V(\phi,u)$ Eq. (4.8) can have a periodic or non-periodic solution.

A specially interesting solution is a periodic solution with an infinite period. This can be found by a choice of the integrations constant C such that $d\phi/dx$ approaches zero as ϕ goes to zero. This gives

$$C = -\frac{n_o}{\epsilon_o}(m_iu^2+T_e) \ . \tag{4.9}$$

As ϕ is increased V decreases from zero and reaches to its minimum value, say, $-V_o$, and increases again as ϕ is further increased. V then becomes zero again at $\phi = \phi_m$ (see Fig. 1). Thus when ϕ is solved as a function of ξ by integrating Eq. (4.8) its value moves from zero and approaches to ϕ_m (when $V = 0$) then moves back again to zero, indicating a single humped solitary wave structure in ξ coordinates. This shows that the ion acoustic wave becomes a solitary wave as the amplitude is increased. As will be shown later that this is the consequence of the balance between the steepening of the wave due to the nonlinear phase velocity of the wave and the wave dispersion.

The speed of the solitary wave is a function of the amplitude of the solitary peak. The peak amplitude is obtained from $V(\phi,u) = 0$ at $\phi \neq 0$. Writing the peak amplitude ϕ_m, the relation between u and ϕ_m is obtained from $V(\phi_m,u) = 0$ as

$$u^2 = \frac{T_e}{2m_i}\frac{[\exp(e\phi_m/T_e)-1]^2}{\exp(e\phi_m/T_e)-1-e\phi_m/T_e} \ . \tag{4.10}$$

Equation (4.10) shows that as the maximum amplitude of the solitary wave approaches to zero, u^2 becomes T_e/m_i. That is, u approaches to the linear wave phase velocity.

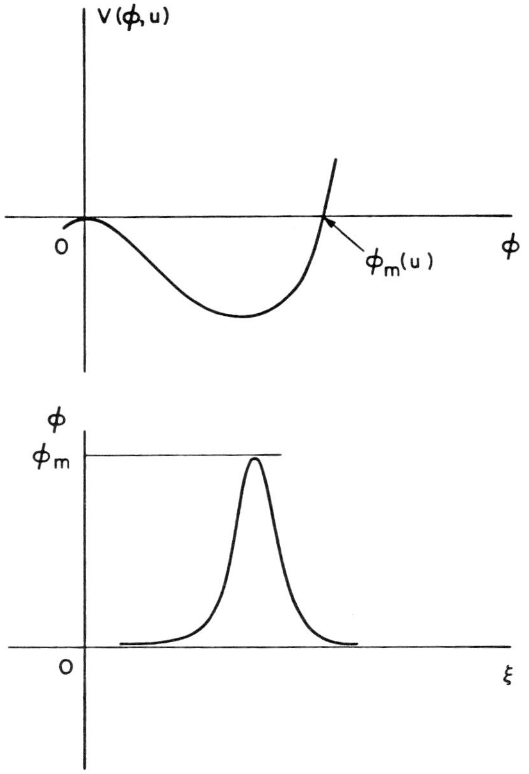

Fig. 1 The Sagdeev potential $V(\phi,u)$ and the corresponding solitary wave $\phi(x)$.

As ϕ_m is increased, u increases, indicating that the solitary wave propagates at a super-sonic speed. The Mach number of the wave is then defined as

$$M = u/c_s = u/(T_e/m_i)^{1/2} \ .$$

Since ϕ_m can not exceed $u^2m_i/(2e)$ as seen from Eq. (4.7) there exists a maximum value in u such that the solitary wave can exist. This value is obtained by using $\phi_m = u^2m_i/(2e)$ in Eq. (4.10) and solving for u (Sagdeev, 1966), as

$$M = u/c_s \simeq 1.6 \ . \tag{4.11}$$

If u exceeds this value, the ion velocity becomes multivalued indicating a total reflection of ion fluid by the wave.

If the thermal motion of ions are introduced, the reflections of ions start to appear even at u smaller than the critical value given by Eq. (4.11). The partial reflection of ions in their distribution function produces a potential structure which is not symmetric. Sagdeev (1969) argues that the reflection of ions by a train of solitons forms a collisionless shock. The ion acoustic shock is thus produced for a Mach number between unity and 1.6.

B. Ion Acoustic Solitons

It was shown that a large amplitude ion acoustic wave can form solitary waves which lead to collisionless acoustic shock by reflecting ions. Strictly speaking a solitary wave is not necessarily a soliton. A soliton is defined now as a solitary wave whose stability is proved by the integrability of the nonlinear equation which produces the solitary wave.

The ion acoustic wave, when its amplitude is small, in fact can be shown to become solitons. The proof is made here by showing that the wave can be described by the Korteweg-deVries equation which is known to be integrable through the inverse scattering method (Gardner, et al. 1967).

We follow the method presented by Washimi and Taniuti (1966). The method is often called the reductive perturbation. The method utilizes the fact that amplitude dispersion which causes the steepening of the wave is balanced by the linear dispersion of the wave to form a soliton. Hence we obtain a nonlinear equation as a perturbation on the linear dispersionless wave.

We use again cold-ion Boltzman-electron model described by

$$\frac{\partial v^{(i)}}{\partial t} + v^{(i)}\frac{\partial v^{(i)}}{\partial x} + \frac{e}{m_i}\frac{\partial \phi}{\partial x} = 0 , \quad (2.27)$$

$$\frac{\partial n^{(i)}}{\partial t} + \frac{\partial}{\partial x}(n^{(i)}v^{(i)}) = 0 \quad (2.28)$$

and

$$n^{(e)} = \exp(\frac{e\phi}{T_e}) . \quad (4.1)$$

$n^{(i)}$ and $n^{(e)}$ are connected through the Poisson equations

$$\frac{\partial^2 \phi}{\partial x^2} = -\frac{e}{\epsilon_o}(n^{(i)} - n^{(e)}) .$$

We expand $n^{(i)}$, $n^{(e)}$, ϕ, and $v^{(i)}$ in their powers,

$$\begin{aligned} n^{(i)} &= n_o + \epsilon n_1^{(i)} + \epsilon^2 n_2^{(i)} + \cdots \\ n^{(e)} &= n_o + \epsilon n_1^{(e)} + \epsilon^2 n_2^{(e)} + \cdots \\ \phi &= 0 + \epsilon\phi_1 + \epsilon^2\phi_2 + \cdots \\ v^{(i)} &= 0 + \epsilon v_1^{(i)} + \epsilon^2 v_2^{(i)} + \cdots \end{aligned} \quad (4.12)$$

where ϵ is proportional to the amplitude of the wave.

In addition we use coordinates suitable to describe phenomena which are slowly varying as compared with the linear wave such that the speed of variation is related to the nonlinearity of the system. Knowing that a dispersionless ion wave propagates at c_s, it is found to be convenient to use the coordinates ξ and τ defined

$$\epsilon^{1/2}(x - c_s t)/\lambda_D \equiv \xi \quad (4.13)$$

and

$$\epsilon^{3/2} t\omega_{pi} \equiv \tau . \quad (4.14)$$

where $\lambda_D = c_s/\omega_{pi}$ is the electron Debye wave length (without 2π).

Equation (4.12) shows that ξ is a moving coordinate at the ion sound speed c_s normalized to the Debye lenth *and* measured with a unit $\epsilon^{-1/2}$ longer than the original unit. Similarly τ is a slow time coordinate measured with a unit $\epsilon^{-3/2}$ slower than the real time scale. We further note that

$$\frac{\partial}{\partial t} = -\omega_{pi}\epsilon^{1/2}\frac{\partial}{\partial \xi} + \omega_{pi}\epsilon^{3/2}\frac{\partial}{\partial \tau} \ , \tag{4.15}$$

$$\frac{\partial}{\partial x} = \epsilon^{1/2}\frac{1}{\lambda_D}\frac{\partial}{\partial \xi} \ . \tag{4.16}$$

It is also convenient to normalize n, ϕ, and v such that

$$\begin{aligned} N &\equiv n^{(i)}/n_o \\ \Phi &= e\phi/T_e \\ n &= n^{(e)}/n_o \\ V &= v^{(i)}/c_s \ . \end{aligned} \tag{4.17}$$

If we now substitute Eqs. (4.12) to (4.17) to the original set of equations, we have, from the ion equation of motion,

$$\epsilon^{5/2}\frac{\partial V_1}{\partial \tau} + \epsilon^{5/2}V_1\frac{\partial V_1}{\partial \xi} - \epsilon^{3/2}\frac{\partial V_1}{\partial \xi} - \epsilon^{5/2}\frac{\partial V_2}{\partial \xi}$$

$$+ \epsilon^{3/2}\frac{\partial \Phi_1}{\partial \xi} + \epsilon^{5/2}\frac{\partial \Phi_2}{\partial \xi} = 0 \ . \tag{4.18}$$

Dividing this to $O(\epsilon^{3/2})$ and $O(\epsilon^{5/2})$ we have, from $O(\epsilon^{3/2})$

$$\frac{\partial}{\partial \xi}(V_1 - \Phi_1) = 0$$

$$V_1 = \Phi_1 \equiv \psi \tag{4.19}$$

and from $O(\epsilon^{5/2})$, using Eq. (4.19)

$$\frac{\partial \psi}{\partial \tau} + \psi\frac{\partial \psi}{\partial \xi} - \frac{\partial V_2}{\partial \xi} + \frac{\partial \Phi_2}{\partial \xi} = 0 \ . \tag{4.20}$$

Similarly from the continuity equation, at $O(\epsilon^{3/2})$,

$$\frac{\partial}{\partial \xi}(N_1 - V_1) = 0$$

or

$$N_1 = V_1 = \psi \tag{4.21}$$

and at $O(\epsilon^{5/2})$,

$$\frac{\partial \psi}{\partial \tau} + \frac{\partial}{\partial \xi}(\psi^2) - \frac{\partial N_2}{\partial \xi} + \frac{\partial V_2}{\partial \xi} = 0 \ . \tag{4.22}$$

From the Boltzmann distribution of electron density, at $O(\epsilon)$,

$$n_1 = \Phi_1 = \psi \tag{4.23}$$

and at $O(\epsilon^2)$,

$$n_2 = \Phi_2 + \psi^2/2 \ . \tag{4.24}$$

From the Poisson equation at $O(\epsilon)$

$$n_1 = N_1 = \psi$$

which is consistent with Eq. (4.23) and (4.21), and at $O(\epsilon^2)$,

$$\frac{\partial^2 \psi}{\partial \xi^2} = n_2 - N_2 \ . \tag{4.25}$$

Thus we left with Eqs. (4.20), (4.22), (4.24) and (4.25) which contain variables V_2, Φ_2, n_2, N_2 and ψ. These equations should be compatible among each other which requires ψ to satisfy a certain equation.

The compatibility condition leads to

$$\frac{\partial \psi}{\partial \tau} + \psi \frac{\partial \psi}{\partial \xi} + \frac{1}{2}\frac{\partial^3 \psi}{\partial \xi^3} = 0 \ . \tag{4.26}$$

This is the famous Korteweg-deVries equation. This equation is known to be integrable, that is its solution is obtainable from an arbitrary initial condition (Gardner et al. 1967). In particular if the initial condition is a localized function of ξ, $\psi = \psi_o(\xi,\tau=0)$, the asymptotic solution is given by a set of solitons and dispersive waves. The number of the solitons is given by the number of eigenvalues of a Schrödinger equation in which $\psi_o(\xi)$ is the potential. The amplitudes of the solitons are given by the eigenvalues. In particular, one soliton solution is obtained from Eq. (4.26)

$$\psi = 3\lambda \operatorname{sech}^2 \sqrt{\frac{\lambda}{2}}(\xi - \lambda\tau) \ . \tag{4.27}$$

It is easily recognized that the soliton amplitude of the Korteweg-deVries equation depends on the soliton speed. Recalling that ξ is the coordinate moving at the sound speed c_s, λ in Eq. (4.27) shows the incremental increase of the soliton speed over c_s. Therefore, the Mach number defined in Section IV-Ais related to λ through

$$M = 1 + \lambda \tag{4.28}$$

In fact the soliton solution Eq. (4.27) is obtainable from Eq. (4.10) by writing $e\phi_m/T_e = 3\lambda$, expanding $\exp(e\phi_m/T_e)$, and integrating the result with respect to ϕ.

We note here that although the Korteweg-deVries equation is solvable for its initial condition, it does not mean that the initial value problem in the original coordinates, x, is solvable. In this respect the solution is called a far field solution which is valid only after the true initial condition develops into left and right going waves.

C. Weak Double Layers

Sato and Okuda (1980) discovered in their computer simulation that the ion acoustic turbulence excited by a drifting electron with the drift speed v_o larger than 0.6 v_{Te} can spontaneously lead to the formation of weak double layers even if $v_o < v_{Te}$, that is even in the regime of ion acoustic instability (rather than Pierce-Buneman regime). The double layer is found to be formed stationary in the ion frame, that is, it has a property distinctively different from the solitons or shocks which, as were found in previous sections, move at supersonic speed. It was also recognized by Hasegawa and Sato (1982) that the double layer is formed by a negative potential solitary wave which is known to be absent in a fluid plasma as was demonstrated in previous sections.

In this section we attempt to describe the formation of the negative potential solitary wave and the weak double layer in a system of drifting electrons. The deviations is based on the recent work by Nishihara et. al., (1982).

Let us first study the energy exchange process between a streaming particle and a localized potential. The equation of motion for a charged particle with the charge q mass m and velocity v in a electrostatic potential $\phi(x,t)$ is given by

$$m\frac{dv}{dt} = -q\frac{\partial\phi}{\partial x} \quad . \tag{4.29}$$

Multiplying $v = dx/dt$, we have

$$\frac{d}{dt}(\frac{m}{2}v^2 + q\phi) = q\frac{\partial\phi}{\partial t} \quad . \tag{4.30}$$

If the potential is localized along the trajectory of the particle such that $\phi = 0$ at the beginning and the end of the path, Eq. (4.30) shows that the energy exchange occurs only when ϕ is an explicit function of time. That is, the energy exchange rate is proportional to the time rate change of ϕ during the interaction.

The change of the kinetic energy of the particle due to the interaction,

$$\Delta K = \frac{1}{2}mv(t=\infty)^2 - \frac{1}{2}mv(t=-\infty)^2$$

is obtained by integrating Eq. (4.30) and noting $\phi=0$ at $t=\pm\infty$,

$$\Delta K = \int q\frac{\partial\phi}{\partial t}dt \quad . \tag{4.31}$$

In a particular case in which ϕ is a function moving at the ion acoustic speed, c_s,

$$\phi(x,t) = \phi(x-c_s t) \equiv \phi(\xi) \tag{4.32}$$

Eq. (3) becomes

$$\Delta = \int q(-c_s)\frac{d\phi}{d\xi}\frac{d\xi}{v-c_s}$$

$$= -qc_s\int\frac{d\phi}{v(\phi)-c_s} \quad . \tag{4.33}$$

Equation (4.33) can be evaluated approximately for a particle with $v >> c_s$ such that the interaction is almost adiabatic, using $\frac{1}{2}mv^2 + q\phi = \epsilon_o \simeq$ const,

$$\Delta K \simeq mc_s \int_{v_{-\infty}}^{v_{\infty}} \frac{vdv}{v-c_s} \quad .$$

For a reflected particle $v_\infty \simeq -v_{-\infty} \equiv v_o$. Therefore

$$\Delta K \simeq -2mc_s v_{-\infty} \tag{4.34}$$

where $c_s < v_{-\infty} \leq (-2q\phi_m/m)^{1/2}$.

This expression indicates that the potential gains (loses) energy from (to) the particle if the particle travels initially in the direction of (opposite to) the motion of the potential. Therefore if the ion acoustic waves are excited by drifting electrons, the same electrons can keep pumping energy the negative potential by being reflected by the potential.

Knowing that the particle reflection is the essential ingredient for the growth of the negative potential, let us now derive the nonlinear wave equation including the effects of electron reflections. This task essentially amounts to a modification of the Korteweg-deVries equation which was obtained in Section IV-B by including the effect of the reflected electrons.

In the presence of a small amount of reflected electrons, the modification can be made by adding the electron density n_r corresponding to the reflected electrons to Eq. (4.1). $n_r^{(e)}$ is obtained by integrating the phase space density of the reflected electrons $f_r(v,\phi)$, where f_r is obtained by expanding the distribution function around the resonant velocity u

$$f_r(v,\phi) = \frac{\partial f_o}{\partial u}\Big|_{u=0} u \quad , \tag{4.35}$$

where $f_o = f_o(u)$ is the unperturbed distribution function and u represents the particle speed at $x = -\infty$ in the moving (ξ) coordinate, i.e.,

$$\frac{1}{2}m_e u^2 = \frac{1}{2}m_e v^2 - e\phi \quad , \tag{4.36}$$

where v is the local velocity in the ξ coordinate. We note that if the interaction between the particle and potential is purely adiabatic, $f = f(u)$ is an exact solution of the Vlasov equation. The range of v^2 which produces the reflected electrons is from 0 to $2e\phi/m_e - 2e\phi_m/m_e$, where $\phi_m(<0)$ is the minimum value of the potential.

Therefore,

$$n_r^{(e)} = \int_0^{\sqrt{2e(\phi-\phi_m)/m_e}} f_r dv$$

$$= \frac{\partial f_o}{\partial u}\Big|_{u=0} \int_0^{\sqrt{-2e(\phi-\phi_m)/m_e}} \left(v^2 - \frac{2e\phi}{m_e}\right)^{1/2} dv \quad . \tag{4.37}$$

If we use the same normalization as the previous section, and use f_o in the laboratory frame,

$$n_r = v_T \frac{\partial f_o}{\partial v|_{v=c_s}} \left[\sqrt{-\psi_m}\sqrt{-\psi_m+\psi} - \psi \ln \frac{\sqrt{-\psi_m}+\sqrt{-\psi_m+\psi}}{\sqrt{-\psi}} \right] . \tag{4.37'}$$

Adding n_r to n in Eq. (4.12), we obtain the Korteweg-deVries equation which includes the effect of reflected electrons,

$$\frac{\partial\psi}{\partial\tau} + \psi\frac{\partial\psi}{\partial\xi} + \frac{1}{2}\frac{\partial^3\psi}{\partial\xi^3} - \frac{\partial f}{\partial v|_{v=c_s}}\frac{\partial}{\partial\xi}$$

$$\left[\sqrt{-\psi_m}\sqrt{-\psi_m+\psi} - \psi \ln \frac{\sqrt{-\psi_m}+\sqrt{-\psi_m+\psi}}{\sqrt{-\psi}} \right] \tag{4.38}$$

This equation is no longer integrable, however, the numerical solution of this equation has revealed clearly that the negative potential grows and slows down simultaneously as shown in Fig. 2 while the pure Korteweg-deVries equations has given a damped diffusive solution as expected.

As the negative potential solitary wave grows, it reflects more electrons and builds up positive charge layers behind the negative potential. The result of the computer simulation by Nishihara et al. (1982) shows that, as the peak of the negative potential solitary wave grows, the solitary wave slows down and stops in the ion frame when it reaches to the maximum value as shown in Fig. 3. The potential profile at the peak value of the negative potential is shown in Fig. 4. The figure clearly shows that the existence of a negative potential spike in front of the double layer. This shows that the reflections of electrons by the negative potential spike leads to a formation of a weak double layer (Hasegawa and Sato, 1982).

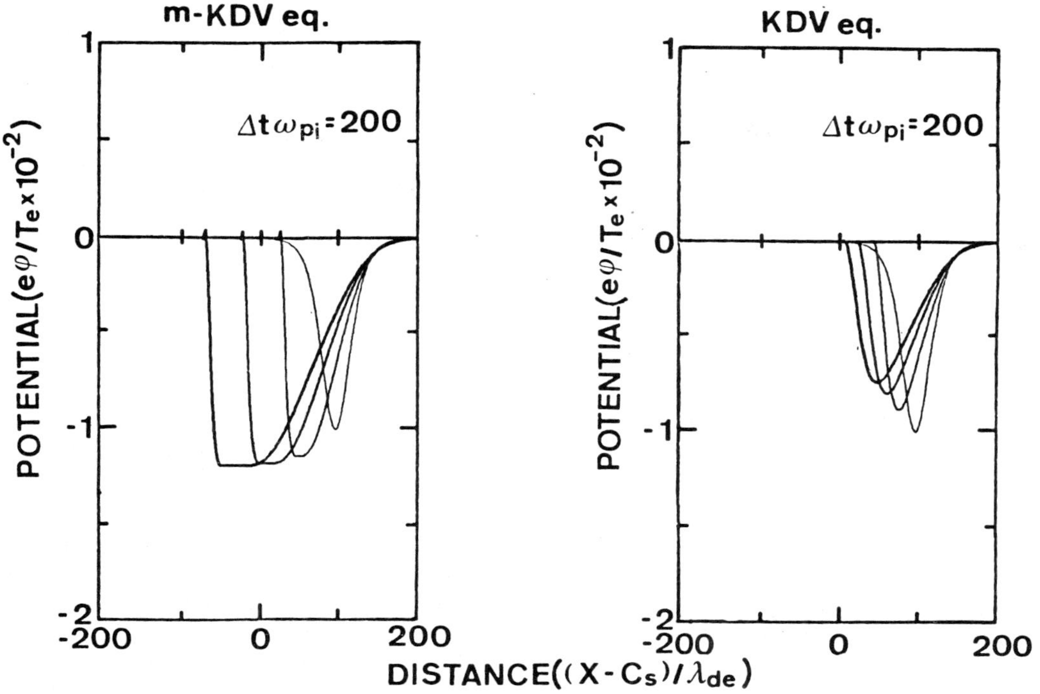

Fig. 2 Numerical solution of Eq. (4.38) (left) as compared with that of the KdV equation (right).

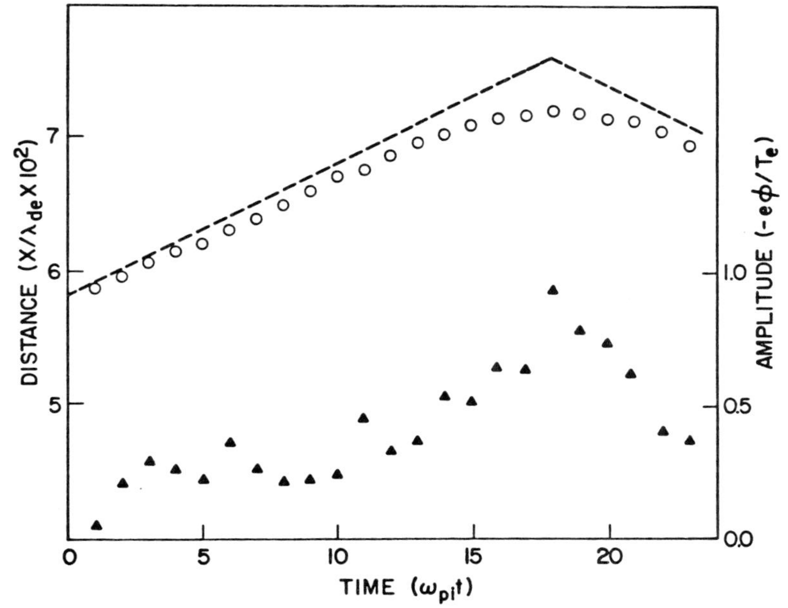

Fig. 3 Variation and trajectory of the peak amplitude of the negative potential solitary wave.

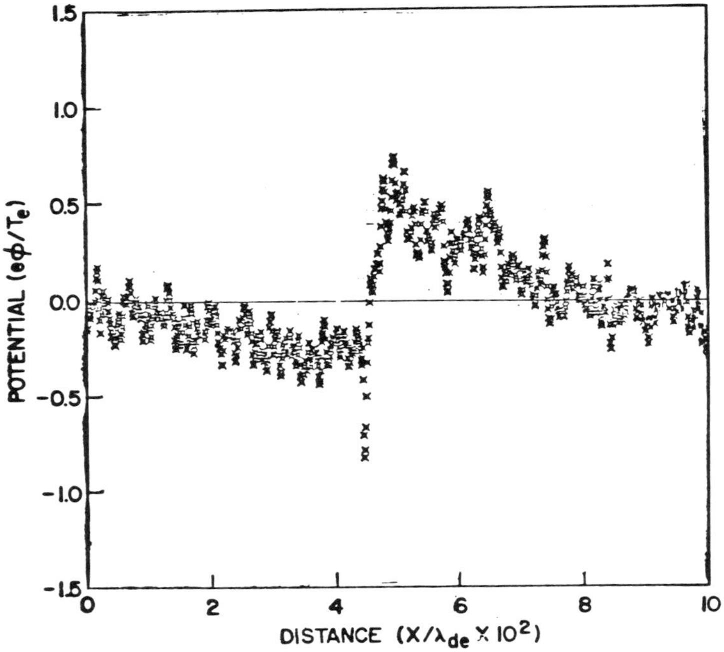

Fig. 4 The result of computer simulation of formation of a weak double layer.

V. CONCLUSION

The ion acoustic wave is, as many other waves in plasmas, a very rich subject and has many faces depending on the level of excitation. Proper understanding of the wave properties, which are still being discovered, is indispensible in studying collisionless plasma kinematics in various areas of space plasmas.

REFERENCES

Buneman, O. Instability, Turbulence and Conductivity in Current Carrying Plasma, Phys. Rev. Lett. *1*, 8 (1958).

Fried, B. D. and R. Gould, Longitudnial Ion Oscillation in a Hot Plasma, Phys. Fluids *4*, 139 (1969).

Gardner, C. S., J. M. Green, M. D. Kruskal, R. M. Miura, Method of Solving the Korteweg-deVries Equation, Phys. Rev. Letters, **19**, 1095 (1967).

Garbonov, L. M., Perturbations of a Median by a Field of a Strong Electromagnetic Wave, Sov. Phys. JETP *28*, 1220 (1969).

Hasegawa, A. Decay of Ion Acoustic Wave Magnetized Plasma, Physics Letters *57A*, 143 (1976).

Hasegawa, A. and T. Sato, Existence of a Negative Potential Solitary-Wave Structure and Formation of Double Layer, Phys. Fluids *25*, 632 (1982).

Huang, T. Y. L. Chen and A. Hasegawa, Collision Effects on Non-Equilibrium Ion Acoustic Wave, Phys. Fluids *7*, 1744 (1974)

Kaner, E. A. and V. M. Yakovenko, Weak Turbulence Spectrum and Second S und in Plasma , Sov. Phys. JETP *31*, 316 (1970).

Nishihara, K., H. Sakagami, T. Taniuti and A. Hasegawa, Formation of Weak Double Layers in Ion Acoustic Turbulence, to be published (1982).

Pierce, J. R., Possible Fluctuations in Electron Streams Due to Ions, J. Appl. Phys. *19*, 231 (1948).

Sagdeev, R. Z. and A. A. Galeev, Nonlinear Plasma Theory, Benjamin, Inc. New York, (1969) p. 99.

Sagdeev R. Z., **Reviews of Plasma Physics**, Ed. by M. A. Leontovich, Consultants Bureau, New York, 1966, p. 52.

Sato, T. and H. Okuda, Ion-Acoustic Double Layers, Phys. Rev. Lett. *44*, 740 (1980).

Washimi, H. and T. Taniuti, Propagation of Ion Acoustic Waves of Small Amplitude, Phys. Rev. Letters, **17**, 996 (1966).

HYDROMAGNETIC WAVES IN THE MAGNETOSPHERE

W.J. Hughes
Astronomy Department
Boston University
Boston, MA 02215

INTRODUCTION

Solar-Terrestrial Physics is a subject in which theory has usually followed observation. The first reported observation of a pulsation (Stewart, 1861) occurred about eighty years before Alfvén showed that a type of electromagnetic wave could propagate through a conducting fluid. Stewart's work did help to bring about the realisation that there must be a current carrying layer in the upper atmosphere. Other observations followed and it soon became clear that small amplitude oscillations of both the strength and direction of the Earth's magnetic field with periods of seconds to minutes, were common. They are now known as geomagnetic pulsations. They are the ground signature of hydromagnetic waves in the magnetosphere.

About ten years after Alfvén's work, Dungey (1954) (see also Dungey, 1963) first proposed that the long but regular periods of these oscillations might be the result of standing Alfvén waves being excited on geomagnetic field lines. A good theory took another twenty years to develop (Southwood, 1974: Chen and Hasegawa, 1974), but the idea was essentially correct. An isolated flux tube can act as a 1-dimensional resonant cavity because transverse Alfvén mode energy flows only along field lines and the ionosphere at the ends of the flux tube reflects Alfvén waves well. The problem of coupling between neighboring flux tubes is a tricky one, as we shall see later. However, coupling is often weak and the notion of waves standing on field lines is extremely useful in understanding long period hydromagnetic waves in the magnetosphere.

Figure 1 shows the configuration of a field line for the two lowest frequency normal modes. The ionosphere has been taken as a perfect conductor which means that field lines cannot move through it so their ends are fixed. This is a good first approximation but one we will relax later. Note that if we assume the background field is symmetric about the equator the fundamental mode has field line displacement (and hence field line velocity and electric field as $\underset{\sim}{E} = -\underset{\sim}{u} \times \underset{\sim}{B}$ in the hydromagnetic

R. L. Carovillano and J. M. Forbes (eds.), Solar-Terrestrial Physics, 453–477.

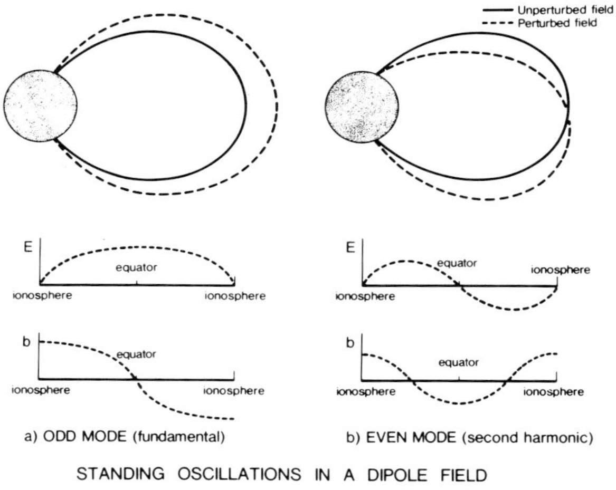

Figure 1. An idealised picture of the two lowest frequency resonances of a geomagnetic field line assuming a perfectly conducting ionosphere. Electric field, E, is proportional to field line displacement and magnetic perturbation, b, to field line tilt.

limit) symmetric about the equator while the magnetic perturbation, b, is antisymmetric. The reverse is true for the second harmonic. At the bottom of the figure the electric and magnetic perturbations are shown on a stretched out field line.

HM MODES IN COLD PLASMA

In an earlier chapter Siscoe has shown that three modes of hydromagnetic wave exist in a uniform magnetised plasma. If the plasma pressure is insignificant compared to the magnetic pressure, $B^2/2\mu_o$, the plasma is described as cold and the slow mode disappears. The fast and intermediate (or transverse) modes remain. Although the magnetosphere cannot be treated as a uniform medium, some properties of these modes are important. The fast mode has $\underset{\sim}{j}.\underset{\sim}{B} = 0$, so can carry no current parallel to the ambient field, $\underset{\sim}{B}$. However the magnetic perturbation vector, $\underset{\sim}{b}$, does in general have a component parallel to B so the fast mode can transmit pressure variations. The fast mode dispersion relation,

$$\omega^2 = k^2 V_A^2 \quad \text{where } V_A^2 = \frac{B^2}{2\mu_o \rho} \text{ is the Alfvén velocity,}$$

shows that the mode propagates isotropically. The group velocity is always parallel to the phase velocity. The transverse mode on the other hand carries a finite current along $\underset{\sim}{B}$, but the magnetic perturbations are non-compressional so $\underset{\sim}{B}.\underset{\sim}{b} = 0$. The transverse dispersion relation is

$$\omega^2 = k_{\shortparallel}^2 \, V_A^2$$

where $k_{\shortparallel}$ is the component of $\underset{\sim}{k}$ parallel to $\underset{\sim}{B}$, and it is straightforward to show that the group velocity and Poynting vector are always parallel to $\underset{\sim}{B}$ regardless of the direction of $\underset{\sim}{k}$. For this reason this mode is sometimes referred to as the guided mode, for energy is guided along the ambient field direction.

A typical magnetospheric Alfvén velocity is 1000 km/s while typical periods of geomagnetic pulsations are in the range 10-600s. This means typical wavelengths range from a few R_E to tens of R_E (Earth radii), scales comparable to the size of the magnetosphere itself. Thus uniform cold plasma theory is not a good description. The fact that inhomogeneity of the background medium must be considered makes pulsation theory both intriguing and non-trivial.

The basic equations are the hydromagnetic form of Ohms Law,

$$\underset{\sim}{E} = -\underset{\sim}{u} \times \underset{\sim}{B}$$

Maxwell's equations,

$$\nabla \times \underset{\sim}{E} = -\partial \underset{\sim}{B}/\partial t$$

$$\nabla \times \underset{\sim}{B} = \mu_o \underset{\sim}{j}$$

and the fluid momentum equation which simplifies in a cold plasma to

$$\rho\frac{\partial \underset{\sim}{u}}{\partial t} = \underset{\sim}{j} \times \underset{\sim}{B} = \frac{-1}{\mu_o} \underset{\sim}{B} \times (\mathrm{curl}\ \underset{\sim}{B}) = \frac{1}{\mu_o} (\underset{\sim}{B}.\nabla)\ \underset{\sim}{B} - \frac{1}{2\mu_o} \nabla B^2$$

If we further assume that the magnetic field can be represented as a background field $\underset{\sim}{B}$ plus a perturbation field $\underset{\sim}{b}$ and that $\underset{\sim}{B}$ is time stationary and curl-free we can write

$$\frac{\partial \underset{\sim}{b}}{\partial t} = -\nabla \times \underset{\sim}{E} = \nabla(\underset{\sim}{u} \times \underset{\sim}{B}) = (\underset{\sim}{B}.\nabla)\underset{\sim}{u} - \underset{\sim}{B}(\nabla.\underset{\sim}{u}) - (\underset{\sim}{u}.\nabla)\underset{\sim}{B} \qquad (1)$$

and

$$\mu_o \rho \frac{\partial \underset{\sim}{u}}{\partial t} = -\ \underset{\sim}{B} \times \mathrm{curl}\ \underset{\sim}{b} \qquad (2)$$

Given an ambient field, $\underset{\sim}{B}$, equations (1) and (2) are a closed set with $\underset{\sim}{u}$ and $\underset{\sim}{b}$ as unknowns. However they are not easy to solve. In a later section we will derive wave equations in a simple geometry; here we will just quote the results of Dungey (see Dungey, 1963, 1968) who derived wave equations for an axisymmetric $\underset{\sim}{B}$ with $B_\phi = 0$. (A dipole is of course a special case of this.) In cylindrical coordinates (r,ϕ, z) he obtained

$$\{\omega^2\mu_o\rho - \frac{1}{r}(\underset{\sim}{B}.\nabla)\, r^2(\underset{\sim}{B}.\nabla)\}\, (\frac{u_\phi}{r}) = \omega m(\frac{\underset{\sim}{B}.\underset{\sim}{b}}{r}) \tag{3}$$

$$\{\omega^2\mu_o\rho - rB^2(\underset{\sim}{B}.\nabla)\frac{1}{r^2B^2}(\underset{\sim}{B}.\nabla)\}\, (rE_\phi) = i\omega B^2(\underset{\sim}{B} \times \nabla)_\phi\, (\frac{\underset{\sim}{B}.\underset{\sim}{b}}{B^2}) \tag{4}$$

$$i\omega\underset{\sim}{B}.\underset{\sim}{b} = \frac{1}{r}(\underset{\sim}{B} \times \nabla)_\phi (rE_\phi) - imB^2\frac{u_\phi}{r} \tag{5}$$

where a variation of the form $\exp(im\phi - i\omega t)$ was assumed. Historically a lot of attention has been directed towards these equations, so it's worth discussing what they mean.

The LHS of (3) and (4) both have the form of a one-dimensional wave equation with the only spatial operator being $\underset{\sim}{B}.\nabla$, a derivative along the direction of $\underset{\sim}{B}$. The equations are coupled by the terms on the RHS which depend on $\underset{\sim}{B}.\underset{\sim}{b}$, the compressional part of the magnetic perturbation. Equation (5) shows how $\underset{\sim}{B}.\underset{\sim}{b}$, E_ϕ and u_ϕ are related and closes the set. It is tempting to think of the LHS as representing pseudo-transverse mode oscillations as the transverse mode dispersion relation depends only on $k_{\parallel}$, and the RHS representing coupling due to the fast mode which has a compressional component.

Dungey showed that these equations decouple in two limits. If the wave is axisymmetric ($m = 0$) the RHS of (3) vanishes. The LHS of (3) then describes a mode in which the electric field is purely radial and the magnetic and velocity perturbations are azimuthal. Magnetic shells (L-shells) decouple and each shell oscillates azimuthally independently of all others. This is the toroidal mode. In this limit equation (4) describes a mode in which the whole magnetospheric cavity pulsates coherently. The other limit occurs as $m \to \infty$. For the RHS of (3) to remain finite, $\underset{\sim}{B}.\underset{\sim}{b} \to 0$ so the RHS of (4) vanishes. Equation (4) then describes a mode in which $\underset{\sim}{E}$ is azimuthal and $\underset{\sim}{u}$ and $\underset{\sim}{b}$ are contained in a meridian plane. Now the oscillations of each meridian plane decouple. This is the poloidal mode.

In these limits (3) and (4) reduce to one dimensional eigen-equations and it is fairly trivial to solve them, at least numerically, along a field line. A boundary condition must be specified at the ionosphere; traditionally $\underset{\sim}{E} = 0$ was used. The result is a set of eigenfrequencies which are a function of L shell and the mass density of the plasma along the field line. These results do give a useful zero order fit to observed pulsation frequencies, but the limits of $m = 0$ and $m = \infty$ are obviously not usually applicable. Coupling occurs and is important. However before we develop this further we must first consider the ionosphere. For as we shall see, the boundary condition we apply there is crucial to the whole problem.

THE EFFECT OF THE IONOSPHERE

Before we can set about solving equations (1) and (2) we need a boundary condition to apply at the ionosphere. The simplest approach is to assume the ionosphere is a very good conductor so electric fields vanish there as was done in figure 1. This approach leads to difficulties which disappear if the ionosphere is taken to have a finite conductivity. In this section we consider how transverse Alfvén waves reflect off a conducting, horizontally layered ionosphere. Detailed derivations can be found elsewhere (Hughes and Southwood, 1976a; Southwood and Hughes, 1978); here the major points will be derived from simple arguments.

Consider the model shown in figure 2. The ambient field is uniform and directed vertically downwards, $\underset{\sim}{B} = B\hat{z}$. Four slab regions are differentiated: the magnetosphere in which the hydromagnetic solutions hold; a thin anisotropically conducting ionosphere with Pedersen and Hall conductivities, σ_P and σ_H; the atmosphere which is an insulator; and the ground which has an isotropic conductivity σ_g. Now any signal received on the ground will have a horizontal variation whose length scale is much shorter than the Alfvén wavelength of a pulsation, $V_A/\omega \gtrsim$ few R_E. This variation is critical to the problem.

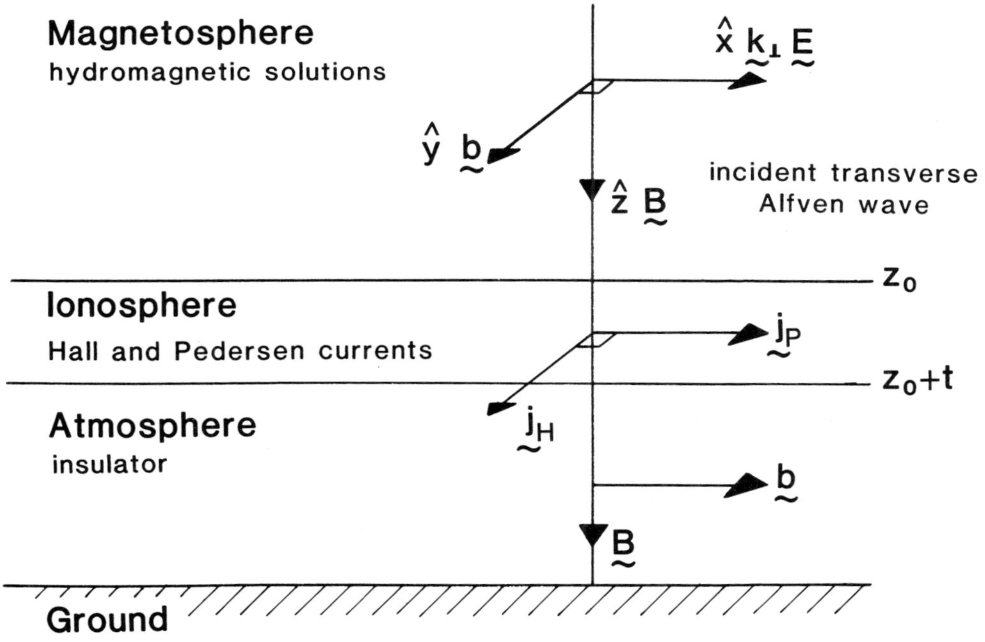

Figure 2. A schematic representation of a magnetospheric transverse Alfvén wave incident on a horizontally stratified model of the ionosphere, atmosphere and ground.

We represent it by assuming wave fields vary as $\exp(ik_\perp x - i\omega t)$. We assume that a transverse mode Alfvén wave propagates down the field lines. As transverse mode waves have $\underset{\sim}{k}_\perp \perp \underset{\sim}{b}$ and $\underset{\sim}{b} \perp \underset{\sim}{B}$, $\underset{\sim}{b} = b\hat{y}$ in the magnetosphere. In the atmosphere no currents can flow (displacement currents can be neglected) so in particular

$$j_z = \frac{1}{\mu_o} (\underset{\sim}{\nabla} \times \underset{\sim}{b})_z = \frac{i}{\mu_o} \underset{\sim}{k}_\perp \times \underset{\sim}{b}_{horizontal} = 0$$

Thus $\underset{\sim}{b}_{horizontal}$ must either be zero or parallel to $\underset{\sim}{k}_\perp$ which is in the x-direction. So without explicitly discussing the ionosphere we have shown that it must either act to screen the incident magnetic perturbation from the atmosphere and hence the ground, or cause the magnetic perturbation vector to rotate through 90°.

As the incident magnetic perturbation must be shielded by currents flowing in the ionosphere induced by the electric field of the wave which is in the x-direction, we have

$$\frac{\partial b_y}{\partial z} = \mu_o j_x = \mu_o \sigma_P E_x$$

Integrating this through the thin ionosphere and realising E_x varies very slowly with z we obtain

$$\Delta b_y = \mu_o E_x \int_{z_o}^{z_o+t} \sigma_P dz = \mu_o \Sigma_P E_x$$

where Σ_P is the ionospheric integrated Pedersen conductivity. Since $b_y = 0$ below the ionosphere, $\Delta b_y = b_{ym}$ where the subscript m denotes magnetosphere. The wave electric field, E_x, will also cause ionospheric Hall currents which will give rise to a magnetic perturbation in the x-direction

$$\Delta b_x = \mu_o \Sigma_H E_x$$

where Σ_H, the ionospheric integrated Hall conductivity, is defined similarly to Σ_P. Now, quite coincidently, in the terrestrial ionosphere $\Sigma_H/\Sigma_P \sim 1$, so $\Delta b_x \sim \Delta b_y$. The Hall current produces oppositely directed b_x perturbations above and below the ionosphere. Above the ionosphere, in the magnetosphere, b_x is associated with the fast mode in which $\underset{\sim}{k}$, $\underset{\sim}{b}$ and $\underset{\sim}{B}$ are coplanar. The fast mode dispersion relation is

$$\omega^2 = k^2 V_A^{\ 2} = (k_x^{\ 2} + k_z^{\ 2}) V_A^{\ 2}$$

But we have already seen that $k_x^{\ 2} >> \omega^2/V_A^{\ 2}$ so $k_x^{\ 2} \simeq -k_z^{\ 2}$. This means the fast mode produced by the Hall current must be evanescent; its amplitude will drop exponentially with altitude on a scale corresponding to $k_\perp$. Similarly, below the ionosphere where the mode is a pure

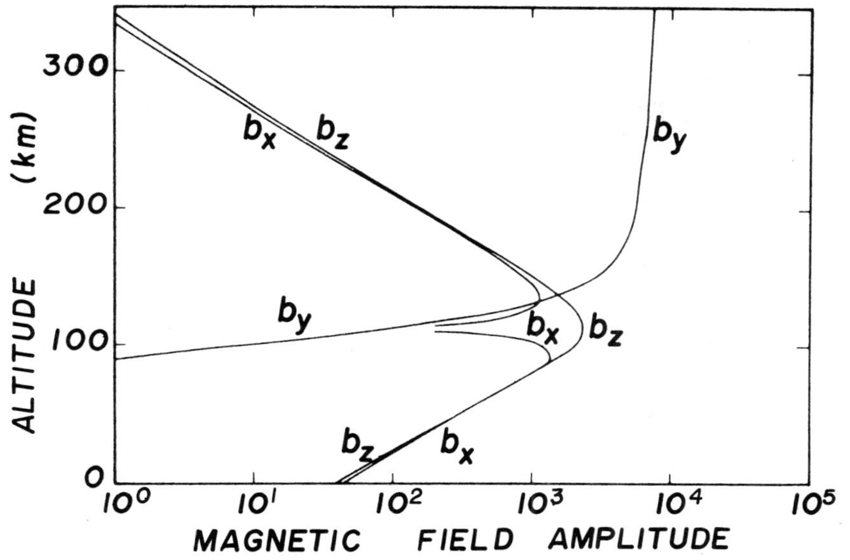

Figure 3. The variation with altitude of the wave magnetic field components that occur when a transverse Alfvén wave is incident on the ionosphere. These values were calculated numerically using a realistic model ionosphere. The change in direction of the dominant component occurs in the E region around 120 km. (After Hughes and Southwood, 1976a.)

electromagnetic wave, as $k_\perp >> \omega/c$, b_x will drop off exponentially away from the ionosphere. This behaviour is illustrated in figure 3 which shows the results of a full wave numerical integration of the problem using a realistic ionospheric conductivity model. A value of $1/k_x = 20$ km was used to emphasize the evanescent nature of b_x. The ionospheric conductivity reaches its maximum in the E region around 120 km which means that the largest changes in the magnetic field components occur there. b_y is totally shielded from the ground by these currents. Another important feature can now be understood. If $1/k_\perp \lesssim 120$ km, the E-region altitude, then the magnetic perturbation is smaller than the magnetospheric one, and it is very effectively shielded when $1/k_\perp << 120$ km.

So now we have as an ionospheric boundary condition

$$\underset{\sim}{b}_m \times \hat{z} = \mu_o \Sigma_P \underset{\sim}{E}_m \tag{6}$$

which derives from the fact that the ionospheric Pedersen currents totally shield the incident magnetic perturbation. It can be understood physically in another way by multiplying (6) by E. In magnitude only we get

$$\frac{1}{\mu_o} EB = \Sigma_P E^2 = J_P E$$

which can be interpreted as a balance between the downward Poynting flux in the wave and Joule heating in the ionosphere.

Before leaving the ionosphere we need to derive one more parameter, a reflection coefficient. Since the boundary condition (6) will not generally be satisfied by the incident wave a partial reflection will occur. In the magnetosphere both the incident and reflected wave are governed by the transverse mode dispersion relation

$$b = \frac{k_{\parallel}}{\omega} E = \pm \frac{1}{V_A} E$$

so

$$b_i + b_r = \frac{1}{V_A}(E_i - E_r)$$

where the subscripts i and r denote incident and reflected waves. At the ionosphere (6) must hold so

$$b_i + b_r = \mu_o \Sigma_P (E_i + E_r)$$

Combining these two expressions we get a reflection coefficient R

$$\frac{E_r}{E_i} = \frac{1 - \mu_o \Sigma_P V_A}{1 + \mu_o \Sigma_P V_A} = R \qquad (7)$$

We can immediately see that if $\Sigma_P = \infty$ then $E_r = -E_i$, reflection is perfect, the electric field in the ionosphere is zero and field line feet are fixed. In the other extreme, if $\Sigma_P = 0$ reflection is again perfect but now b is zero in the ionosphere. This means the ends of a field line are free to move but cannot tilt. Normally $\Sigma_P \gtrsim 1/\mu_o V_A$. If $\Sigma_P = 1/\mu_o V_A$ the incident wave is totally absorbed and no standing wave can be set up. As we shall see, the ionosphere is a significant sink of hydromagnetic wave energy and ionospheric conductivity plays an important part in determining the structure of a field line resonance.

FIELD LINE RESONANCE

We are now in a position to go back to the problem of the coupled equations derived earlier. The key to understanding the physics contained in them is to split the problem in two. First we shall investigate how signals vary across $\underset{\sim}{B}$ in a model where variation along $\underset{\sim}{B}$ is made simple, so simplifying the LHS of (3) and (4). Having done this we will discuss briefly solutions for the wave form along $\underset{\sim}{B}$ ignoring the coupling terms. These latter are best done numerically.

The essential feature of the resonance problem is that the field line Alfvenic eigenfrequency varies with position, particularly equatorial radial distance from Earth or L shell. We can retain this feature in a uniform field model provided the plasma mass density, ρ, varies across $\underset{\sim}{B}$. We take a model in which the magnetic field is uniform and in the z direction, $\underset{\sim}{B} = B\hat{z}$. The field lines are all the same length and end at $z = \pm\ell$ in an ionosphere with integrated Pedersen conductivity Σ_P. The plasma mass density ρ is a function of x, $\rho = \rho(x)$ and we will look for solutions of the hydromagnetic equations of the form $\exp(i\lambda y - i\omega t)$.

First we apply the ionospheric boundary condition (6) at $z = \pm\ell$ to derive the variation in z. We take E to vary as

$$E = E(x)\exp(i\lambda y - i\omega t)\,\{\exp(ikz+\kappa z) \pm \exp(-ikz-\kappa z)\}$$

where k and κ are both real. The plus and minus signs imply symmetry and antisymmetry about the equator so describe odd and even harmonics respectively (cf. figure 1). The magnetic perturbation is obtained using $i\omega b_y = -\partial E_x/\partial z$ etc. Applying the boundary condition and assuming $\mu_o\Sigma_P >> k\omega$ gives

$$k = n\pi/2\ell$$

$$\kappa = k/\ell\omega\mu_o\Sigma_P$$

where n is an odd or even integer depending on whether a plus or minus sign was chosen. The approximation $\mu_o\Sigma_P >> k\omega$ implies that ionospheric reflection is good and the approximately fixed-end condition holds (see last section). This is normally the case.

We can now write

$$\underset{\sim}{E} = (E_x(x),\ E_y(x),\ 0)\exp(i\lambda y + ik^*z - i\omega t) \qquad \text{where} \qquad k^* = k - i\kappa$$

and similar expressions for u and b. This has reduced the problem to 1 dimension; we are left only to investigate the variation in x. We start with the hydromagnetic momentum equation,

$$\rho\frac{\partial \underset{\sim}{u}}{\partial t} = \underset{\sim}{j} \times \underset{\sim}{B} = -\frac{1}{\mu_o}\underset{\sim}{B} \times \text{curl}\ \underset{\sim}{b}$$

Writing down x and y components gives

$$i\omega\mu_o\rho(x)\ u_x = -ik^*Bb_x + B\frac{db_z}{dx} \tag{8}$$

$$i\omega\mu_o\rho(x)\ u_y = +i\lambda Bb_z - ik^*Bb_y \tag{9}$$

$\underset{\sim}{u}$ and $\underset{\sim}{b}$ are also related by Faraday's law

$$i\omega\underset{\sim}{b} = \text{curl}\ \underset{\sim}{E} = -\text{curl}\ (\underset{\sim}{u} \times \underset{\sim}{B})$$

which gives

$$\omega b_x = -k^*Bu_x \tag{10}$$

$$\omega b_y = -k^*Bu_y \tag{11}$$

$$i\omega b_z = i\lambda Bu_y + B\frac{du_x}{dx} \tag{12}$$

Using (10) and (11) to eliminate b_x and b_y from (8) and (9) we obtain

$$(\omega^2\mu_o\rho(x) - k^{*2}B^2)\ u_x = -i\omega B\frac{db_z}{dx} \tag{13}$$

$$(\omega^2\mu_o\rho(x) - k^{*2}B^2)\ u_y = \omega\lambda Bb_z \tag{14}$$

Equations (12), (13) and (14) form a complete set which are analogous to the coupled equations for an axisymmetric field (3), (4) and (5). The main difference is that here the operator on the LHS is now an algebraic expression. Equations (13) and (14) can be rewritten

$$(K^2 - k^{*2})\ u_x = \frac{-i\omega}{B}\frac{db_z}{dx} \tag{15}$$

$$(K^2 - k^{*2})\ u_y = \frac{\omega\lambda}{B}\ b_z \tag{16}$$

where $K^2(x) = \omega^2\mu_o\rho(x)/B^2 = \omega^2/V_A{}^2(x)$

where $V_A(x)$ is the local Alfvén velocity which varies with x since $\rho = \rho(x)$. Now in general there will be a field line at some position $x = x_o$ such that $K^2(x_o) = \text{Real}\ (k^{*2})$. This is the field line at which field line resonance occurs and if k^* were purely real the transverse dispersion relation would be identically satisfied. The RHS of (15) and (16) are proportional to b_z, the compressional component of the wave magnetic field which is characteristic of the fast mode. We can see this by eliminating u_x and u_y between (12), (15) and (16) which gives

$$\left(K^2-k^{*2}-\lambda^2+\frac{d^2}{dx^2}\right) b_z = \frac{2K}{K^2-k^{*2}} \frac{dK}{dx} \frac{db_z}{dx} \tag{17}$$

The LHS of (17) is the fast mode dispersion relation; the RHS introduces the inhomogeneity. If K were not a function of x, the RHS of (17) would be zero and we would be left with the fast mode dispersion relation for a uniform plasma. Note again that if k* were purely real, a singularity would occur in the RHS at the resonant shell, $x = x_o$, where $K^2 = k^{*2}$. We have avoided the singularity by introducing the ionospheric boundary condition (6) which ensures k* has an imaginary part.

If the imaginary part of k*, $\kappa \ll k$, and $\rho(x)$ varies monotonically, we can obtain an approximate equation for u_x near the resonance point $x = x_o$

$$\frac{d^2u_x}{dx^2} + \frac{1}{x-x_o-i\varepsilon}\frac{du_x}{dx} - \lambda^2 u_x = 0 \tag{18}$$

where $\quad \varepsilon = 2\kappa k/\frac{dK^2}{dx} \simeq \frac{2\omega}{\mu_o \ell \Sigma_P \, d(\ln\rho)/dx}$

The scale length ε represents the scale of variation of the solution near resonance and depends on both the amount of dissipation (Σ_P) and the scale of the inhomogeneity ($d(\ln\rho)/dx$). Equation (18) has the form of a modified Bessel equation; to obtain a solution we require boundary conditions in x which must include a wave source. This is usually done by requiring $u_x(a) = 0$ and $u_x(b) = u_o\exp(i\omega t)$ where $a<x_o<b$. Solutions can then be obtained. Figure 4 illustrates one solution and also how it would appear on the ground. The solid lines are an analytic solution of (18) plotted versus $(x-x_o)$ corresponding to latitude. The dashed lines were obtained by numerically mapping this solution to the ground using a model dayside ionosphere and a uniformly conducting Earth. Points to be noted are:

(i) The solution in space has a peak in b_y which is associated with a sharp phase change. The peak and phase change have a scale $\sim \varepsilon$ (parameter values are indicated at the top of the figure). b_x barely has a peak and has little phase change.

(ii) The signal on the ground is rotated through 90° so that b_x on the ground corresponds best with $-b_y$ above the ionosphere (both are shown in the top panel, note the 180° phase difference) and b_y on the ground corresponds best with b_x above the ionosphere.

(iii) The correspondence is no better than it is because the signal has been severely spatially filtered. Features with scale lengths less than about 120 km cannot be seen on the ground. The most dramatic example is the reduction in amplitude of the peak in b_x on the ground compared to the peak in b_y in space.

(iv) The appearance of b_z on the ground is due to the localized nature of the source.

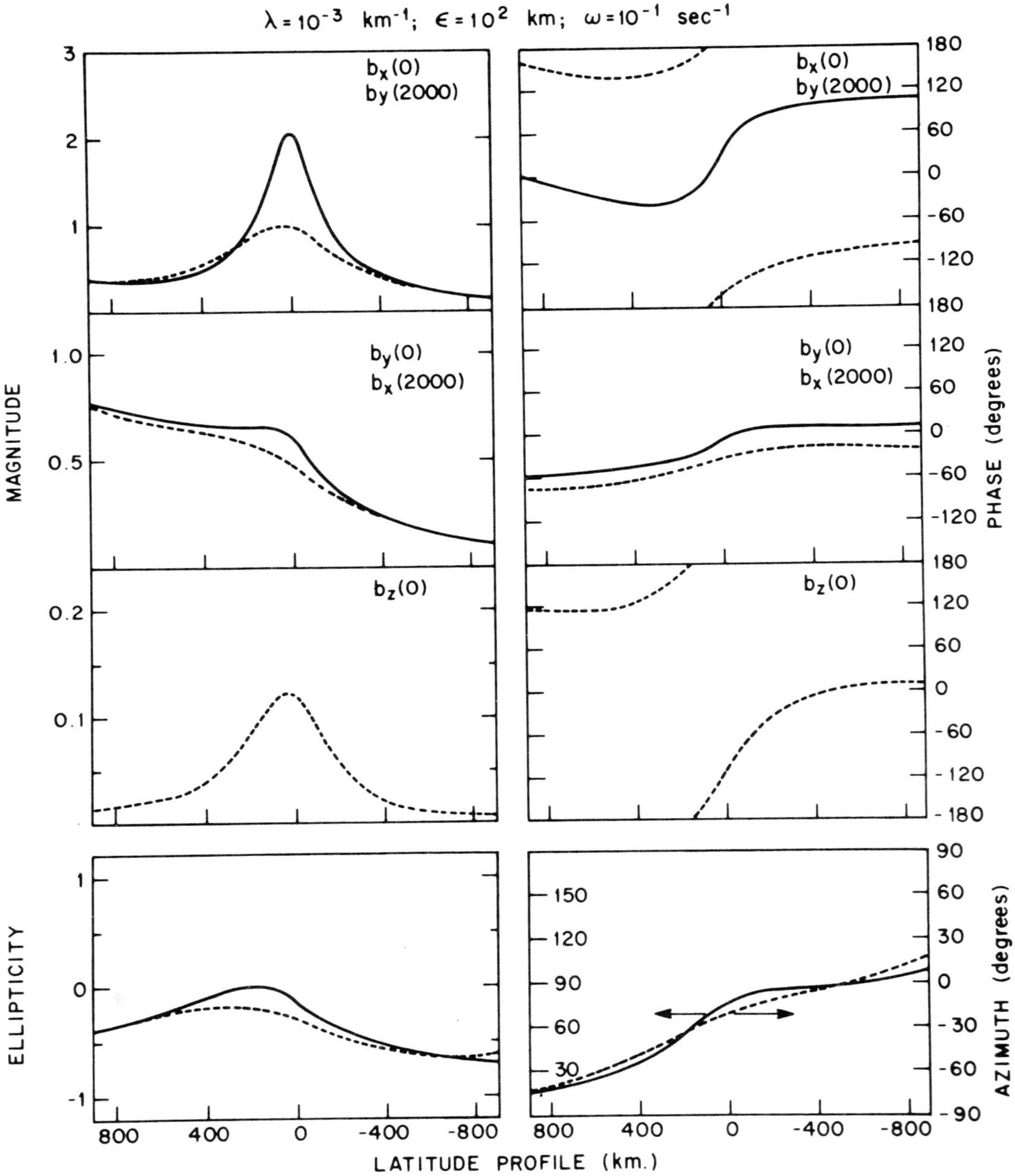

Figure 4. An L shell profile of the magnetic field perturbations associated with a field line resonance (solid lines) and a mapping of these fields through the ionosphere (dashed lines). The upper three pairs of panels show component amplitude and phase. The bottom pair shows polarization characteristics in the plane transverse to $\underset{\sim}{B}$. (After Hughes and Southwood, 1976b).

(v) The polarization reversal near the peak in the signal does not occur on the ground.

(vi) The polarization azimuth on the ground is 90° different from that in the magnetosphere.

SOLUTION ALONG $\underset{\sim}{B}$

Solutions for the wave form of a resonance along a realistic flux tube have to be made numerically, usually under the assumption of weak or non-existant coupling. This is because the variation of both field strength and mass density along the flux tube make analytic solutions impossible. Various solutions have been published, the earliest being those of Cummings et al., (1969). Newton et al. (1978) introduced boundary condition (6) and showed how loss of wave energy by Joule heating of the ionosphere alters the solution. Allan and Knox (1978) first examined the effects of asymmetric ionospheric boundary conditions while Singer et al. (1981) solved the equations in a more realistic geomagnetic model field. Schematically all the solutions have the same basic forms as those shown in figure 1, but the variation of direction, amplitude and phase of both $\underset{\sim}{E}$ and $\underset{\sim}{b}$ along the field line can be quite different.

RESONANCES IN WARM PLASMAS

So far we have only discussed cold plasma, that is situations where the plasma pressure is insignificant. However there are parts of the magnetosphere, especially the ring current and plasma sheet regions, where this is far from true. Inclusion of the plasma pressure terms introduces a new MHD mode, the slow mode, as Siscoe has shown. The slow mode has the slowest phase velocity of the three modes, hence its name. Its phase speed is usually comparable with the particle thermal speed. This normally results in its being efficiently Landau damped (see later section on wave damping) and so discussion of this mode would be academic except for some recent observations. One way of distinguishing between fast and slow modes is that in the fast mode plasma and magnetic pressure oscillate in phase so their effects reinforce each other while in the slow mode they oscillate in antiphase so their effects tend to cancel (see Siscoe's lectures). Recently compressional hydromagnetic waves with plasma and magnetic pressures oscillating in antiphase have been observed in the magnetosphere.

Including plasma pressure in the momentum equation changes the form of the RHS of the wave equations (3) and (4) so that they take the general form (Southwood, 1977)

$$\{\mu_o \rho \omega^2 - h_\alpha^{-1} (\underset{\sim}{B}.\nabla) h_\alpha^2 (\underset{\sim}{B}.\nabla)\} \left(\frac{u_\alpha}{h_\alpha}\right) = \frac{i\omega}{h_\alpha} \frac{\partial}{\partial \alpha} \left(\delta p_\perp + \frac{1}{\mu_o} \underset{\sim}{B}.\underset{\sim}{b}\right)$$

$$- \frac{h_\alpha \hat{n} \nabla_\alpha}{R} \left(\delta p_\perp - \delta p_{\shortparallel} + \frac{2}{\mu_o} \underset{\sim}{B}.\underset{\sim}{b}\right) \qquad (19)$$

where α is a coordinate perpendicular to $\underset{\sim}{B}$, h_α is a geometric factor, $\delta p_\perp$ and $\delta p_\parallel$ are the plasma pressure variations perpendicular and parallel to $\underset{\sim}{B}$, $\hat{n}$ the field line principal normal and R the field line radius of curvature. Note that the LHS of (19) is identical in form to the LHS of (4) and (5). The RHS is different in that the wave plasma pressure is added to the wave magnetic pressure variation, $\frac{1}{\mu_o} \underset{\sim}{B}.\underset{\sim}{b}$. This allows for a new possibility for if

$$\delta p_\perp \simeq - \frac{1}{\mu_o} \underset{\sim}{B}.\underset{\sim}{b}$$

the RHS will become small and an uncoupled resonance will result. It can be strictly shown that if this condition is satisfied, total energy flux perpendicular to B vanishes. This condition is just that expected in a slow mode wave, viz. plasma and magnetic pressures oscillating in antiphase, so such a mode might at first appear to be a slow mode though of course it is far removed from the slow mode of uniform plasma theory. What this does show is that modes with magnetic compression and yet energy guided along $\underset{\sim}{B}$ are possible in a warm plasma. Since modes which keep energy within a limited volume are more likely to reach a finite amplitude than modes which propagate isotropically, the long period compressional waves seen in the magnetosphere may well have this sort of structure.

FIELD-ALIGNED CURRENTS

Near the resonant field line a field line resonance has much in common with the transverse mode Alfvén wave, including the presence of field aligned currents. It is fairly straightforward to calculate these currents from observed field variations. Figure 5 shows a field line resonance observed in the ionospheric electric field measured by the STARE radar system together with a theoretical fit by a solution of (18). This is the north-south electric field component which must correspond to the east-west magnetic field component, b_y, above the ionosphere. The latitudinal variation of b_y can be represented by

$$b_y(x,y) = b(y) \frac{1-i\eta x}{(1+\eta^2 x^2)^{\frac{1}{2}}} \exp(-\lambda^2 x^2)$$

where x = 0 is the latitude of resonance.

In general the variation in longitude must include variations in both amplitude and phase so we can take

$$b(y) = b_o \exp(iky-\mu^2(y-y_o)^2)$$

where $y = y_o$ is the longitude of signal maximum. Next we obtain an expression for b_x using div $\underset{\sim}{b} = 0$ and realising that $\partial b_z/\partial z << \partial b_x/\partial x,\ \partial b_y/\partial y$,

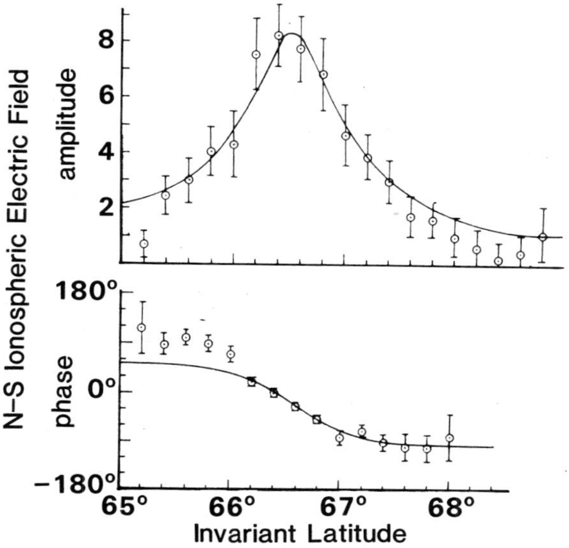

Figure 5. Latitudinal profiles of the amplitude (top panel) and phase (bottom panel) of the oscillating ionospheric electric field of a resonance region measured by the STARE radar. The solid line is a theoretical calculation. (After Walker, 1980).

$$b_x = - \int_{-\infty}^{x} \frac{\partial b_y}{\partial y} dx$$

Finally j_z is obtained from Ampere's law which gives

$$j_z = \frac{1}{\mu_o} \left(\frac{\partial b_y}{\partial x} - \frac{\partial b_x}{\partial y}\right)$$

Figure 6 shows the results of such a calculation. The current into and out of the ionosphere is shown at two times during a wave cycle. The current flows in approximately east-west aligned sheets which alternate in sign. Sheets appear to form at the equatorward edge (here taken as south) of the resonant region and to drift polewards first increasing then decreasing in amplitude. Figure 7 is a cartoon which shows how these currents close in the ionosphere. The field aligned currents are closed by Pedersen currents. The wave electric field also induces Hall currents which flow in closed circles and so are closed in the ionosphere. The magnetic field perturbation above the ionosphere is largely east-west between the current sheets. On the ground the magnetic field perturbations are the result of the Hall currents and are largely north-south.

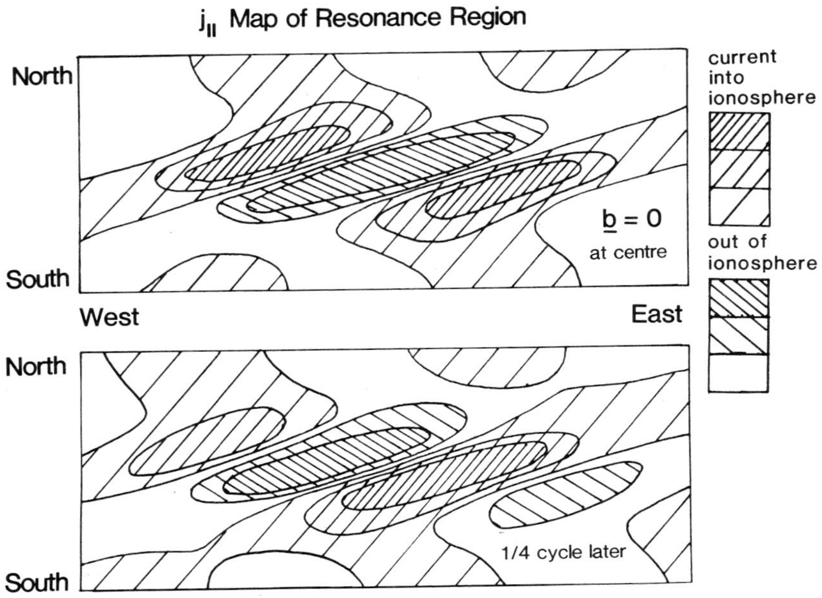

Figure 6. The field aligned currents flowing into and out of the ionosphere associated with a field line resonance region. The lower picture shows the situation a ¼ cycle later than the upper picture.

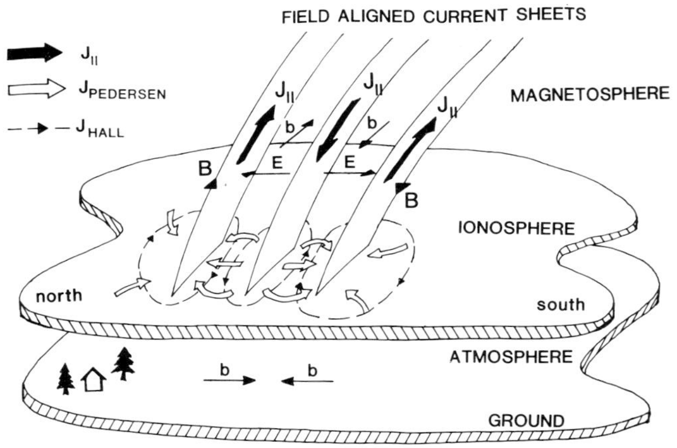

Figure 7. A schematic representation of how the field aligned currents in figure 6 are closed in the ionosphere. The field aligned and Pedersen currents form a solinoidal current which has no magnetic signature on the ground if the ionospheric conductivity is uniform. The ground signature is the result of the Hall currents which close in the ionosphere.

CONVECTION FLOW AND ALFVEN WAVES

Wolf showed in his lectures that plasma flow in the magnetosphere is largely governed by large scale electric fields induced by the solar wind flow past the magnetosphere. In order to maintain the movement of the base of the flux tubes through the dissipative ionosphere currents must be maintained in the ionosphere which are supplied by field aligned or Birkeland currents which link the magnetosphere and ionosphere. Wolf's formalism allowed him to set up a steady-state or slowly varying description of the global flow pattern on closed field lines. In neglecting the inertial term in the momentum equation he neglected all wave effects and his model is not valid for time variations comparable with the Alfvén wave transit time of his system, typically a time of a few minutes. If the convection flow is perturbed in some manner then eventually a new steady state will be realised. The process by which this new steady state is reached usually involves Alfvénic transients propagating around the system. In this section this will be illustrated using a particularly simple situation which is not meant to model any particular sort of event but rather to be illustrative of the processes involved.

It is important to remember that only the transverse mode Alfvén wave can carry a field aligned current. So any perturbation which necessitates changing the Birkeland current system requires the launching of a transverse Alfvénic impulse. Such a situation occurs when a sudden localised change occurs in the ionospheric conductivity in a region where a flow already exists. Figure 8a illustrates the field and current configuration shortly after the ionospheric conductivity is suddenly increased at one end of a field line. Before the change occurred there was a uniform electric field from left to right which drove a sheet Pedersen current, J_P, in the ionosphere given by $J_P = \Sigma_P E$ where Σ_P is the ionospheric integrated Pedersen conductivity. Flux tubes were uniformly flowing into the page at a speed $u = E/B$. When the conductivity of a limited region of the ionosphere is increased by $\Delta\Sigma_P$, either the electric field must reduce or current must flow into and out of the edges of the region in order for current to be continuous. The only path for new current is along field lines. In general a combination of the two occurs as is shown in figure 8a. The $J_{\parallel}$ sheet currents are closed in the magnetosphere by the sheet current shown as J_M which flows in an Alfvén speed V_A. Behind the wave front the electric field is decreased by ΔE, the electric field associated with the wave. Current conductivity at the ionosphere requires that

$$J_{\parallel} = \Delta J_P = \Delta\Sigma_P E - \Sigma_P \Delta E \tag{20}$$

ignoring second order terms. We can also relate J_M and ΔE using the momentum equation in the frame of the wave front

$$\rho V \frac{\partial u_{\perp}}{\partial z} = \underset{\sim}{j} \times \underset{\sim}{B}$$

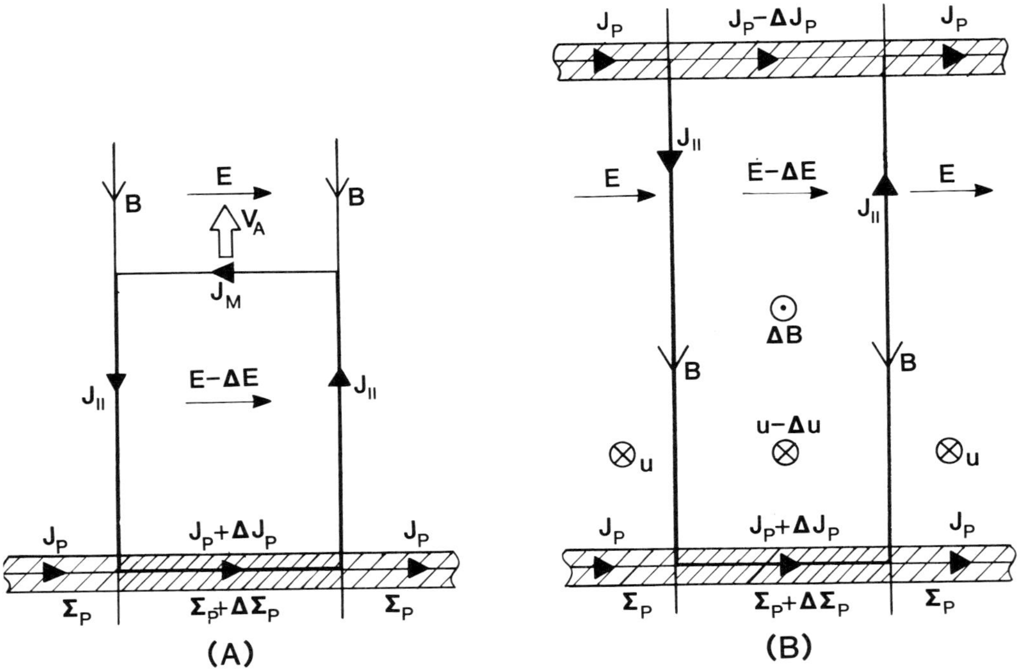

Figure 8. (a) The initial Alfvénic pulse produced by a sudden change in ionospheric conductivity, $\Delta\Sigma_P$, travels along the flux tube extending from the region where this change occurred. The field aligned currents flowing at the edges of this flux tube are closed by the current, J_M, flowing perpendicular to B in the wave front. (b) The final steady state arrived at after the Alfvenic pulse has reflected many times off the conjugate ionosphere. Current is diverted from the unchanged ionosphere and convection flow speed has been reduced.

Integrating this across the wave front gives

$$J_M = \int dz j_\perp = \rho V_A \frac{\Delta E}{B^2} = \Sigma_A \Delta E$$

where $\Sigma_A = 1/(\mu_o V_A)$ is the Alfvénic conductance of the field line.

What happens next depends on whether the field line is open or closed. We'll assume it is closed and that the ionosphere at the other end is identical to the initial unperturbed ionosphere. The wave will be reflected at the other ionosphere. The perturbation electric field behind the reflected wave will be

$$\Delta E^{(1)} = \Delta E^{(0)}(1-R) \qquad \text{where } R = \frac{\Sigma_P - \Sigma_A}{\Sigma_P + \Sigma_A}$$

is the ionospheric reflection coefficient (7) and we are assuming $\Sigma_A < \Sigma_P$ as is normally the case. Similarly

$$J_{\shortparallel}^{(1)} = \Sigma_A \Delta E^{(0)} \ (1+R)$$

The wave will be reflected again at the ionosphere where it originated and will continue to reflect back and forth between the conjugate ionospheres each time being reduced in amplitude. If we assume that the perturbation in Σ_P is not sufficient to significantly alter the reflection coefficient, after n reflections

$$\Delta E^{(n)} = \Delta E^{(0)} \ (1-R+R^2 \cdots + (-1)^n R^n) = \frac{\Delta E^{(0)} \ (1+(-1)^n R^{n+1})}{1+R}$$

$$J_{\shortparallel}^{(n)} = \Sigma_A \Delta E^{(0)} \ (1+R+R^2+ \cdots +R^n) = \frac{\Sigma_A \Delta E^{(0)} \ (1-R^n)}{1-R}$$

As $n \to \infty$ $\quad \Delta E^{(\infty)} = \Delta E^{(0)}/(1+R) = \frac{\Delta E^{(0)}}{2} \ (\frac{\Sigma_P+\Sigma_A}{\Sigma_P})$

$$J_{\shortparallel}^{(\infty)} = \Sigma_A \Delta E^{(0)}/(1-R) = \frac{\Delta E^{(0)}}{2} \ (\Sigma_P + \Sigma_A)$$

Thus $J_{\shortparallel}^{(\infty)} = \Sigma_P \Delta E^{(\infty)}$

and by eliminating $\Delta E^{(0)}$ using (20) we get

$$\Delta E^{(\infty)} = \Delta\Sigma_P E/2\Sigma_P$$

$$J_{\shortparallel}^{(\infty)} = \tfrac{1}{2} \ \Delta\Sigma_P E$$

The final steady state is illustrated in figure 8b. Half the excess current required is drawn from the opposite ionosphere. This current flows in sheets of field aligned current at both edges of the conductivity enhancement linking the ionosphere at either end of the field line. Between the current sheets E, and hence the convection velocity, is reduced because of the extra ionospheric drag caused by the enhanced conductivity. In addition the field lines are tilted in the direction of flow. This can be pictured as the field lines being held back by the ionosphere with the enhanced conductivity while the field aligned currents distribute the stress from one ionosphere to the other.

The time the system takes to come to this final state depends on the ratio of Σ_A to Σ_P and on the Alfvén travel time between ionospheres. If $\Sigma_P > \Sigma_A$, as will normally be the case, $R > 0$, the current in the initial pulse is less than the final current and at each reflection the current increases. We obtain a measure of the number of reflections the system takes to come to equilibrium, n_d, from the ratio of the initial current pulse to the final equilibrium current

$$n_d = 2\Sigma_A/(\Sigma_P + \Sigma_A)$$

An Alfvén mode hydromagnetic pulse will travel back and forth along the field line more than n_d times before new steady state is reached giving rise to a periodic disturbance in the magnetosphere with a fundamental equal to the lowest Alfvén mode eigen-frequency of the local field lines. The ratio of the damping time to the fundamental period of this signal is n_d. Thus the signal is damped least when Σ_A and Σ_P are most different. In the very unlikely case that $\Sigma_A = \Sigma_P$ the initial pulse carries just the current required in the final state, the reflection coefficient is zero, and no periodic signal results. The field line is analogous to a transmission line with a perfectly matched load (the ionosphere) on the end.

The transient hydromagnetic signal set up in this way would couple to other modes in the magnetosphere and as a result would be detectable throughout the magnetosphere. The classic observed magnetospheric transient signal is the pi2 type of pulsation which is associated with substorm onset.

SOURCES AND SINKS OF WAVE ENERGY

It would be inappropriate to end this discussion without saying something about the sources of free energy that can excite field line resonances. So far we have avoided the issue except for the impulsive excitation described in the last section and the implicit assumptions contained in the boundary conditions used earlier. This is not for want of ideas; many source mechanisms have been proposed and theories developed. However observational evidence in favour of one or another mechanism is not overwhelming. It's likely that several different mechanisms are important and excite different sorts of waves.

Another reason this subject has been left till last is that several of the proposed mechanisms cannot be understood using pure hydromagnetic theory. This is because they involve resonant interactions between a small subset of the plasma particle population and the wave. This sort of interaction requires a kinetic theory approach. As there is not space to fully develop the theory here, we will limit ourselves to a physical discussion.

The ultimate source of all magnetospheric hydromagnetic wave energy is of course the solar wind, though in some cases the connection is a tenuous one. The immediate sources concern us more and these can be split into two basic types, those external to the magnetosphere and those internal to it. The external sources are all closely linked to the solar wind flow around the magnetosphere. Excitation of a wind-over-water or Kelvin-Helmholtz instability on the magnetopause boundary has long been suggested as a possible source and there is increasing evidence that it is an important source of some longer period waves. Direct entry of waves produced at the bow shock or in the turbulent magnetosheath flow has also been proposed. Although detailed theories do not exist for this latter mechanism, the links between pulsation characteristics and solar wind and IMF parameters provide some observational support.

The mechanisms internal to the magnetosphere include various plasma instabilities. These are the result of non-equilibrium particle distribution functions which can exist because interparticle collisions are too rare to thermalise the plasma in the magnetosphere. Thermalisation is brought about by other processes; one of the most important is wave particle interactions by which wave fields and particle populations can exchange energy resulting in a redistribution of energy within the particle population and the growth or damping of waves. In the magnetosphere various pathological particle distributions are produced by the global convection flow driven by the solar wind interaction (see Wolf's lectures). These include temperature anisotropies, coexisting populations of differing energies, counter streaming particles and sharp spatial gradients. All these distributions have free energy available for wave growth.

When a wave has a phase velocity less than that of light there will always be some particles present whose velocity matches the wave phase velocity. These particles see the wave Doppler shifted to zero frequency which means that they see the wave as a time stationary spatial structure. These particles are resonant with the wave and will be secularly accelerated or decelerated by the wave fields. The resulting loss or gain of particle energy is matched by a gain or loss of wave energy. Significant exchanges occur when there is a sufficient population with nearly the right velocity or energy to be resonant. Particles with slightly lower energy will tend to be accelerated up to resonance, those with slightly higher energy decelerated. So whether there will be a nett flow of energy to or from the wave depends on the slope of the distribution function near the resonance point. In an equilibrium distribution, such as a uniform Maxwellian, the number of particles always decreases with energy so that wave damping always occurs. This is Landau damping. The more exotic distributions described earlier can have oppositely directed slopes so that wave growth occurs.

So far the discussion has been quite general and applies to most plasma wave excitation in the magnetosphere. Here we are primarily concerned with long period hydromagnetic waves and the corresponding long period bounce and drift motions of magnetospheric particles. Resonance occurs between these when

$$\omega - m\tilde{\omega}_d = N\omega_b \quad ; \text{N integer} \tag{21}$$

where ω, $\tilde{\omega}_d$ and ω_b are the wave frequency, bounce averaged particle drift frequency and particle bounce frequency (see Wolf's lectures) and m is the azimuthal angular wave number as before. We can see this graphically using figure 9 which is a schematic picture of particles traveling through a wave field. The field lines are represented as vertical lines with the northern and southern ionospheres at the top and bottom of the figure. The equator is the horizontal line in the center. The figure is drawn in the frame in which the wave east west phase velocity is zero; individual particles' drift and bounce motion in this frame are shown as heavy lines. Regions where the wave electric field is eastward (westward) are shaded (unshaded). The panel on the left depicts an electric field oscillation which is symmetric about the equator (fundamental mode) while the righthand panel depicts an anti-symmetric electric field disturbance (second harmonic).

Three different particle trajectories are shown. In the absence of a wave all particles bounce back and forth along field lines as well as drift east or west because of gradient and curvature drifts. The three trajectories shown are of particles with the same equatorial pitch angle (so they mirror at the same latitude) but all three have different equatorial crossing points. In addition the particle traveling along the dotted line has a different energy so that the ratio of its bounce period to drift period is different from the other two. All three orbits are special in that the particles drift an integral number of wavelengths in the time taken to make one full bounce.

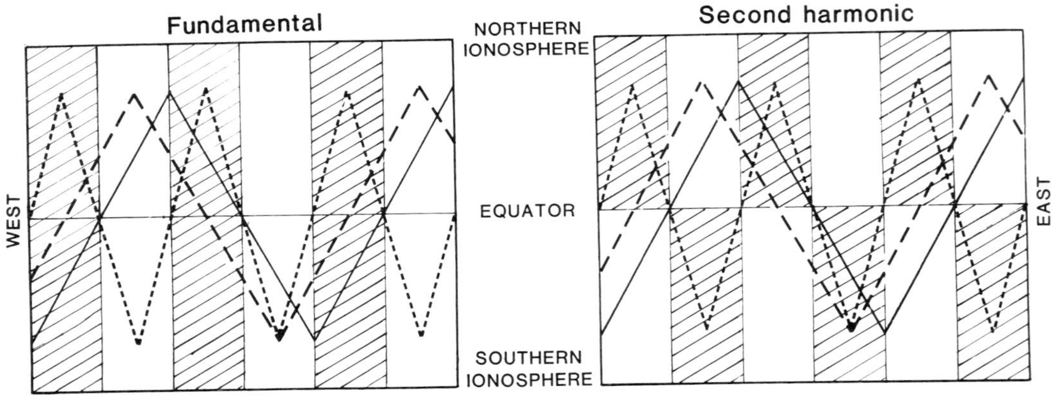

Figure 9. The drift paths of three ions through a fundamental (left) and second harmonic (right) field line resonance. Nett energy exchange occurs between the fundamental and the dashed trajectory and between the second harmonic and the dotted trajectory. (See text for details. Based on Southwood and Kivelson, 1982).

First follow the solid trajectory. In both panels this particle spends equal times in shaded and unshaded regions, so integrated over a full bounce it will be neither accelerated nor decelerated. The dashed trajectory is different. In the left hand panel it always crosses the equator in the center of a shaded region. As the electric field is stronger near the equator (cf. figure 1) this particle will suffer a nett eastward acceleration (assuming it to be an ion). However, the same trajectory in the right hand panel is like the solid trajectory in that it experiences equal amounts of eastward and westward field in one complete bounce. The dotted trajectory acts in the opposite manner. In the right hand panel it is always in shaded regions while in the left hand panel it spends equal times in shaded and unshaded regions, so it is accelerated by the even mode wave but unaffected by the fundamental mode.

To see how this picture relates to equation (21) we must convert from the frame of the wave to the earth's frame. A particle's east-west drift in the wave frame, Ω_d, is related to its drift in the earth's frame, $\tilde{\omega}_d$ by

$$\Omega_d = \tilde{\omega}_d - \omega/m \tag{22}$$

as ω/m is the east-west angular phase velocity of the wave. Now the solid and dashed line particles drift two wavelengths during each complete bounce, so we can equate the time taken to drift the two wavelengths, $4\pi/m\Omega_d$, to the bounce time, $2\pi/\omega_b$, which gives

$$m\Omega_d = 2\omega_b \tag{23}$$

Eliminating Ω_d between (22) and (23) we get

$$\omega - m\tilde{\omega}_d = -2\omega_b$$

which is equation (21) with N = -2. Similarly the dotted trajectory satisfies equation (21) with N = -1. In general odd harmonic waves resonate only with particles which drift an even number (or zero) wavelengths per bounce, corresponding to N being an even integer (or zero). Resonance with even harmonic waves (those with antisymmetric electric fields) occurs when N is an odd integer. This can be easily verified graphically. Although this technique shows when resonance occurs it is unable to predict wave growth or damping. This requires knowledge of the shape of the distribution function.

If N = 0 in equation (21), $m\tilde{\omega}_d = \omega$ and the particle does not move azimuthally in the frame of the wave. The particle trajectory becomes a vertical line in figure 9, so nett acceleration can only occur in the symmetric or fundamental mode. This is drift resonance, and it tends to concentrate particles into regions of higher and lower density which drift azimuthally at the wave's azimuthal phase velocity.

This graphical technique is not limited to resonant particle behaviour, but is equally applicable to non-resonant acceleration and deceleration. Even though over many bounces a particle may experience no nett acceleration, it may be accelerated over the last part of its bounce, or over the last few bounces. Although such effects are of no inherent theoretical interest, they can result in observations of particle flux modulation by spacecraft particle detectors which measure the instantaneous particle velocity or energy. This makes the interpretation of spacecraft flux modulations very difficult as most observations are probably not the result of resonant interactions.

As important as the source of wave energy is the ultimate sink of that energy. Again we have mentioned only one sink so far, Joule dissipation in the ionosphere. This is both because it can be easily understood and because it's probably the dominant sink. Another possible sink is Landau damping, which is a resonant interaction of the type just discussed in which energy is transferred from the wave to the particle population. A third possibility is that the wave structure evolves until $k_\perp$ becomes large, energy will then couple into the kinetic Alfvén mode, a mode in which finite ion Larmor radii effects are important and which has a finite $E_\parallel$. This mode is very efficiently damped if there are any cold electrons present. Each of these damping mechanisms results in a heating of a different plasma population: Joule heating heats the cold ionospheric plasma; Landau damping heats the resonant part of the ring current population; mode conversion heats the cold electrons along the entire flux tube. Mode conversion is probably the least important mechanism as a resonance must be confined to a very narrow flux tube of radius the order of a few thermal ion larmor radii for it to be important. We have already seen that the cross-field-line scale of a resonance is determined in part by the energy sinks in the system (cf. equation (18)) and it's unlikely that both Joule dissipation and Landau damping would be both so ineffective as to allow such a narrowly confined resonance to evolve.

CONCLUDING REMARKS

I hope I have shown that pulsations is an area where theory is relatively well developed. It is however, a field which is far from worked out. The importance of surface waves on both the plasmapause and magnetopause is not clear, and much work remains to be done in understanding the new spacecraft observations of particle flux modulations. Hydromagnetic wave electric fields can be quite large; direct electric field measurements on board GEOS show that they are often several times larger than the steady convection electric field. The effects of these fields on plasma flow in the magnetosphere has not been considered and it may have an important effect even for non-resonant particles. At high latitudes long period waves may be a significant source of ionospheric heating. Hydromagnetic waves must be considered as a part of the whole magnetospheric system, and although we have concentrated on how the system can produce and modify the waves, we should also consider the ways in which the waves might modify energy and plasma transport in the magnetosphere.

REFERENCES

Allan, W. and F.B. Knox, Interpretation of an ATS 6 Alfven wave using solutions with finite ionospheric conductivity, Geophys. Res. Lett., 5, 849, 1978.

Chen, L. and A. Hasegawa, A theory of long-period magnetic pulsations, 1, Steady state excitation of field line resonance, J. Geophys. Res., 79, 1024, 1974.

Dungey, J.W., Electrodynamics of the outer atmosphere, Penn. State Univ., Ionos. Res. Lab. Report, #69, 1954.

Dungey, J.W., The structure of the Exosphere or Adventures in Velocity Space, in Geophys. the Earth's Environment, C. Dewitt, Ed., p. 537, Gordon & Breach, New York, 1963.

Dungey, J.W., Hydromagnetic Waves, in Physics of Geomagnetic Phenomena, S. Matsushita and W.H. Campbell, Eds., p. 913, Academic Press, New York, 1968.

Hughes, W.J. and D.J. Southwood, The screening of micropulsation signals by the atmosphere and ionosphere, J. Geophys. Res., 81, 3234, 1976a.

Hughes, W.J. and D.J. Southwood, An illustration of modification of geomagnetic pulsation structure by the ionosphere, J. Geophys. Res., 81, 3241, 1976b.

Newton, R.S., D.J. Southwood and W.J. Hughes, Damping of geomagnetic pulsations by the ionosphere, Planet. Space Sci., 26, 201, 1978.

Singer, H.J., D.J. Southwood, R.J. Walker and M.G. Kivelson, Alfven wave resonances in a realistic magnetospheric magnetic field geometry, J. Geophys. Res., 86, 4589, 1981.

Southwood, D.J., Some features of field line resonances in the magnetosphere, Planet. Space Sci., 22, 483, 1974.

Southwood, D.J., Localised compressional hydromagnetic waves in the magnetospheric ring current, Planet. Space Sci., 25, 549, 1977.

Southwood, D.J., and W.J. Hughes, Source induced vertical components in geomagnetic pulsation signals, Planet. Space Sci., 26, 715, 1978.

Southwood, D.J., and M.G. Kivelson, Charged particle behaviour in low Frequency geomagnetic pulsations, 2: Graphical approach, J. Geophys, Res., 87, 1707, 1982.

Stewart, B., On the great magnetic disturbance which extended from August 2 to September 7, 1859 as recorded by photography at the Kew Observatory, Phil. Trans. Roy. Soc. Lond., 11, 407, 1861.

Walker, A.D.M., Modeling of pc5 pulsation structure in the magnetosphere, Planet. Space Sci., 28, 213, 1980.

COMPARATIVE MAGNETOSPHERES

Vytenis M. Vasyliunas
Max-Planck-Institut für Aeronomie
D-3411 Katlenburg-Lindau 3, Federal Republic of Germany

ABSTRACT

Magnetospheres are known to be associated with the planets Earth, Mercury, Jupiter, Saturn, Mars, and Venus (in the case of Venus and possibly Mars as well, one has a magnetosphere-like system rather than a true magnetosphere); they are also attributed to various astrophysical objects such as neutron stars, and magnetosphere-like systems have been suggested for comets. All these systems exhibit a considerable variety in sizes, plasma sources, and important physical effects. The plasma flow within a magnetosphere tends toward corotation with the central object, unless limited by one of several possible effects, related to finite ionospheric conductivity, flow of the external medium, or breakdown of radial stress balance, each of which becomes important beyond a particular characteristic distance.

INTRODUCTION

The magnetosphere of the Earth, although it has been the object of most attention, is by no means the only magnetosphere known or studied. Several other planets also possess magnetospheres which have been observed *in situ* by spacecraft instruments; in addition, magnetospheres or magnetosphere-like structures are attributed to a variety of astrophysical objects beyond the solar system. The study of comparative magnetospheres, which aims at a unified general description of magnetospheric phenomenology and physics applicable to many different objects, is important for a two-fold reason. First, it provides a test for the correctness and general applicability of our concepts and theories of magnetospheric physics, often developed in the first instance to fit specific phenomena of the terrestrial magnetosphere. Second, it provides a tool by which our detailed knowledge, based on extensive observations, of the Earth's magnetosphere may be used to gain insights into the properties of other, less accessible, magnetospheres where direct observations may be quite limited or even non-existent.

R. L. Carovillano and J. M. Forbes (eds.), Solar-Terrestrial Physics, 479–492.

Evidently, comparative magnetospheres is a very broad subject, and only a few aspects can be dealt with here. The first part of this chapter presents an overview of the various observed or postulated magnetospheres and magnetosphere-like systems and their principal properties. The second part discusses factors that determine the spatial extent of a magnetosphere and the character of plasma flow within it. (For a more extensive treatment of comparative magnetospheres see, e.g., Siscoe, 1979.)

SURVEY OF MAGNETOSPHERES

It is appropriate to consider first the definition of the concept "magnetosphere" in a general context. The essential elements of the definition, together with some important features commonly found in magnetospheres, are illustrated in Figure 1. We have a central object, a distinct well-defined body held to gether by its own gravity or internal cohesion (e.g. a planet, a star), immersed in a relatively tenuous external medium (e.g. the solar wind, the interstellar gas). The magnetosphere is then the region of space surrounding the central object within which a dominant influence on the local dynamical behavior is exerted by the magnetic field associated with the central object. Alternatively and equivalently, the magnetosphere may be defined as the region of space bounded by the magnetopause, which is defined in turn as the surface at which the direction of the local magnetic field changes from being governed principally by the magnetic field of the central object to being determined by the magnetic field of the external medium. This definition is universally used in the case of the Earth and other planetary magnetospheres, where an easy observational identification of the magnetopause is made possible by the continual variability of the interplanetary magnetic field contrasted with the relative constancy of the planetary magnetic dipoles.

A true magnetosphere, in the sense of the above definitions, will exist if the central object possesses a magnetic field of sufficient strength. (It is also taken for granted that the external medium is sufficiently ionized to behave as a plasma.) In nearly all cases of interest, the magnetic field of the central object is to a good approximation dipolar, and the condition of "sufficient strength" is met if the magnitude of the magnetic dipole moment exceeds a minimum value discussed further on. If the magnetic field of the central object is too weak to produce a true magnetosphere, the central object's interaction with the external medium may nevertheless in some cases give rise to structures similar to those found in magnetospheres; we shall refer to such cases as magnetosphere-like systems.

As the region of transition between the very different physical regimes of the central object and the external medium, a magnetosphere is by its very nature a spatially inhomogeneous system. Much of the complexity and richness of magnetospheric phenomena as well as the

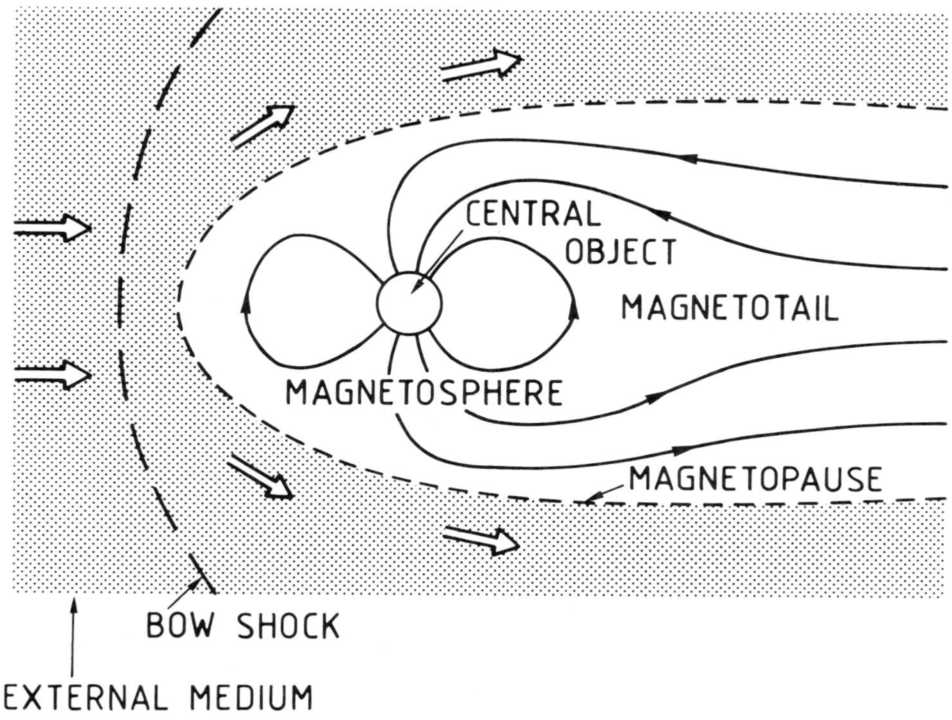

Figure 1. Schematic view of a magnetosphere.

difficulty of developing a comprehensive physical description of them may be traceable to the large ranges of values for many physical quantities and the overwhelmingly important role of gradients.

Before proceeding to a listing of known or suspected magnetospheres, we briefly mention some principal features by which they may be described or classified.

A structure characteristic of many magnetospheres and illustrated in Figure 1 is an extended magnetospheric tail (or magnetotail, for short). When the external medium has a pronounced flow (as in the case of the solar wind for all planetary magnetospheres), the magnetosphere on the upstream side of the central object, facing into the flow, has roughly a spherical shape with characteristic dimension R_M, while on the downstream side the magnetosphere is pulled out into an elongated tail-like shape aligned with the flow direction. The width of the magnetotail is typically comparable to or slightly larger than 2 R_M, but its length exceeds R_M by at least an order of magnitude and possibly more. In none of the magnetospheres studied to date has one been able yet to observe a definite termination of the magnetotail or to obtain a firm estimate of its length.

The flow of the external medium is strongly perturbed by the presence of the magnetosphere; in all observed cases, it is diverted wholly or partially around the volume occupied by the magnetosphere. When the unperturbed flow is supersonic, as in the case of the solar wind, this implies a detached bow shock wave upstream of the magnetospheric boundary.

The various magnetospheres differ considerably among themselves as to the dominant sources of plasma within them. Possibilities include (1) the external medium, (2) the central object, its atmosphere or ionosphere, and (3) bodies within the magnetosphere itself, such as moons (satellites) of the central object, rings, and neutral gas tori or clouds. In spite of the differences in plasma sources, many magnetospheres display a remarkable similarity in the spatial configuration of the plasma, namely a structure strongly concentrated toward the equator and described as a plasma sheet or disc.

What plasma motions can occur within the magnetosphere depends appreciably on the electrical conductivity in the outer layers of the central object that interact directly with the magnetosphere (the ionosphere/atmosphere if one exists, or the surface layer of the object if not). The coupling of the motions of these outer layers to the flow of plasma in the adjacent magnetosphere will be discussed further on. Qualitatively, one may characterize the conductivity of the outer layers as (1) poor, (2) good, or (3) effectively infinite, according to whether the two motions (1) can be relatively independent, (2) are strongly coupled, or (3) are constrained to be the same.

We now present in Table 1 a list of various magnetospheres and magnetosphere-like systems with their principal properties. The table gives the radius of the central object r_a, the order of magnitude of the ratio R_M/r_a (the size of the magnetosphere compared to the central object), the presence or absence of a bow shock, a well-defined magnetopause, and a magnetotail (indicated as yes, no, or - , no reliable information), the principal source or sources of magnetospheric plasma, a characterization of the conductivity of the outer layers associated with the central object, and identification of the phenomenon or physical effect that is of primary importance for the dynamics of the magnetosphere. The list does not claim to be a complete survey or a critical review.

The magnetosphere of the Earth is, on the upstream or dayside, some ten times larger in linear dimension than the planet itself; on the nightside it has a magnetotail that extends well beyond the orbit of the Moon. The principal source of magnetospheric plasma is the solar wind, but the ionosphere is also a plasma source of significant and at

Central object	External medium	r_a (km)	R_M/r_a	BS	MP	MT	Plasma sources	Conductivity of outer layers	Important physical effect
Earth	SW	6×10^3	10	Yes	Yes	Yes	EM, CO	Poor	SW flow
Mercury	SW	2×10^3	1.4	Yes	Yes	Yes	CO	Poor	SW flow
Jupiter	SW	7×10^4	10^2	Yes	Yes	Yes	WM	Poor	Rotation
Saturn	SW	6×10^4	20	Yes	Yes	Yes	WM	Poor	Rotation
Mars	SW	3×10^3	1	Yes	Yes	Yes	-	-	SW flow
Venus	SW	6×10^3	1	Yes	Yes	Yes	CO	Good	SW flow, internal outflow
Comet	SW	~1	$\sim10^4$	(Yes)	-	Yes	CO	Good	Outflow
Sun	Inter-stellar gas	7×10^5	$\sim10^4$	-	-	-	CO	∞	Outflow
Neutron star (isolated)	Inter-stellar gas	~10	~10	-	-	(No)	(CO)	∞	Radiation, relativistic particles
Neutron star (in binary system)	Mass flow from companion star	~10	$\sim10^2$	-	(Yes)	-	EM	∞	Gravity
Radio galaxy	Inter-galactic medium	$\sim10^{17}$	$\sim10^2$	-	-	(Yes)	(CO)	-	-

Abbreviations: SW solar wind, BS bow shock, MP magnetopause, MT magnetotail, EM external medium, CO central object, WM object within magnetosphere.

times possibly comparable importance. The dynamical processes within the magnetosphere are to a large extent governed by the flow of the solar wind, which in particular gives rise to a vast circulating flow of the plasma known as magnetospheric convection (see, e.g. Wolf, 1983 - this volume).

The magnetosphere of Mercury appears to be, in all aspects that have been studied on the basis of the (admittedly limited) available observations, to be essentially a scaled-down version of the terrestrial magnetosphere (Ogilvie et al., 1977). The main difference, aside from scale, is the relative size of the planet compared to the magnetosphere: the ratio R_M/r_a is here only about 1.4 instead of 10, so that the volume corresponding in the terrestrial case to geocentric distances less than about 7 R_E (and containing such phenomena as the radiation belts, the plasmasphere, and the inner ring current) here lies inside the planet. Mercury has no significant atmosphere or ionosphere, and hence the solar wind is presumably the sole major source of magnetospheric plasma.

The magnetosphere of Jupiter, by contrast, exceeds the linear size of the planet by a factor of 50-100; as has often been remarked, were the Jovian magnetosphere visible to us, it would appear on the sky as an object 2-3 times larger than the full Moon. The dominant source of magnetospheric plasma is here neither the solar wind nor the planet's ionosphere (although both contribute to some extent) but material from one of Jupiter's moons, Io; this is shown, besides other evidence, by the composition of the magnetospheric plasma which has a high proportion of heavy ions such as sulfur or oxygen. A dominant role in magnetospheric dynamics is played by Jupiter's rotation; the influence of the solar wind is largely confined to the outermost regions of the magnetosphere (and even there its precise extent is controversial; cf. Vasyliunas, 1983, and references therein).

The magnetosphere of Saturn is not as large as that of Jupiter, being only a factor of about 20 larger than the planet. It resembles the Jovian magnetosphere in the dominant role of planetary rotation and in having its major plasma source (as indicated again by compositional evidence) also in the interior of the magnetosphere rather than at either its inner boundary (the ionosphere) or its outer boundary (the solar wind). However, the precise identity of the source - whether it is the large moon Titan, one or more of the moons inward of Titan's orbit, the rings of Saturn, or any combination of these - has not been firmly established yet. The contributions of the ionosphere or of the neutral hydrogen torus associated with Titan as sources for the proton component of the plasma are also uncertain.

The above - Earth, Mercury, Jupiter, and Saturn - are the well-established true magnetospheres directly observed so far. Mars has apparently a magnetosphere which, however, on the dayside barely extends beyond the planet, with $R_M/r_a \sim 1$. The magnetic field of Mars therefore is at most marginally strong enough to support a magnetosphere, and it is still a controversial question whether Mars indeed has a true magnetosphere (albeit a small one) or whether instead its magnetic field is negligible and the observed magnetosphere-like structure arises by some other process, analogous to that occurring at Venus, for instance. Venus has associated with it a magnetosphere-like structure formed, it is now generally agreed, by the interaction between the solar wind and the planet's ionosphere and atmosphere (hence the boundary surface here is called the ionopause rather than the magnetopause). The magnetic field lines in the extended magnetotail structure result from amplification of the interplanetary magnetic field in the solar wind interaction with Venus; the magnetic field, if any, of Venus itself is too weak to play a detectable role.

Magnetosphere-like systems similar in many respects to what is found at Venus are thought to be formed in the interaction of comets with the solar wind. There are as yet no in situ observations of such systems; any inferences are based on indirect and theoretical arguments. With the cometary nucleus taken as the central objects, the estimated value of the ratio R_M/r_a is very large, in contrast to $R_M/r_a \sim 1$ at Venus; the large relative size is the result of the strong outflow of material from the comet. Channeling of this outflow by interaction with the solar wind, together with amplification of the interplanetary magnetic field, leads to a magnetotail structure that is identified with the observed ion tail of the comet. For a more detailed discussion see, e.g., the reviews in Part V of Wilkening (1982).

A magnetosphere-like system not usually discussed under that name is the heliosphere, the region extending from the sun (the central object) out to the (as yet undetected) boundary where the solar wind is effectively terminated by its interaction with the interstellar gas (the external medium). This system can be viewed in several respects as a greatly scaled-up version of that formed by a comet interacting with the solar wind: in both cases the dominant physical process is an outflow of matter from the central object. The magnetic field within the heliosphere, however, unlike the case either at a comet or at Venus, is derived from the magnetic field of the central object, although it has little or no dynamical importance beyond a distance of some tens of solar radii.

The remaining magnetospheres in Table 1 are associated with objects outside the solar system and are all more or less speculative. Pulsar magnetospheres (see Michel 1982 for a review) are thought to be dominated by the radiation-pressure and relativistic-particle effects associated with the rapid rotation and extremely strong magnetic fields of neutron stars, their central objects; the external medium, whatever it is, plays no appreciable role, according to most current views (for a somewhat different approach, see Michel and Dessler 1981). As a rough estimate of R_M one may take the distance to the light cylinder, where the speed of a particle corotating with the neutron star approaches the speed of light. Magnetospheres of X-ray sources in binary systems (see, e.g., review by Vasyliunas, 1979) also have neutron stars as central objects, but here the external medium, formed by a gravitationally accreted mass flow from the companion star, is all-important in determining the structure and properties of the magnetosphere. Finally, structures associated with some types of extended radio galaxies (see, e.g., Miley 1980) are occasionally interpreted in terms of an interaction of a galaxy with the intergalactic medium similar in some ways to the formation of a magnetotail; speculative and controversial though this idea is, it shows that concepts of magnetospheric physics may be applied up to galactic scales and beyond.

PLASMA FLOW: COROTATION AND ITS LIMITS

Determining the configuration of plasma flow within the magnetosphere is one of the central problems of magnetospheric physics. An extensive discussion of plasma flow in the terrestrial magnetosphere, including methods for constructing detailed models, is given by Wolf (1983 - this volume). Here we survey the basic principles that determine the plasma flow in the general context of comparative magnetospheres. A useful approach is to treat corotation of the plasma with the central object - which is the simplest mode of flow and one of dominant importance in many magnetospheres (though not in that of the Earth) - as a first approximation which holds if some specific conditions are met; the various departures of plasma flow from strict corotation may then be viewed as resulting from the breakdown of one or more of these conditions.

Corotation of magnetospheric plasma requires the following:

(1) The rotation of the central object must extend through its outer layers that interact directly with the magnetosphere. In the usual planetary case, this means that the neutral atmosphere within the height range of the ionosphere must corotate with the planet, i.e. planet-atmosphere coupling (maintained by a sufficiently effective vertical transport of momentum, through eddy viscosity or otherwise).

(2) The electrical conductivity of the outer layers, or more specifically of the ionosphere, must be sufficiently large to maintain effective atmosphere-ionosphere coupling. What this means may be best expressed in quantitative terms: the horizontal electric field $\underline{E}$ is related, on the one hand, to the height-integrated current density $\underline{I}$ within the ionosphere by the Ohm's law

$$\underline{I} = \underline{\underline{\Sigma}} \cdot (\underline{E} + \underline{V}_n \times \underline{B}/c) \qquad (1)$$

where $\underline{\underline{\Sigma}}$ is the height-integrated conductivity and $\underline{V}_n$ the neutral-atmosphere velocity, and, on the other hand, to the magnetospheric plasma flow velocity $\underline{V}$ just above the ionosphere by the MHD condition

$$\underline{E} = -\underline{V} \times \underline{B}/c \tag{2}$$

(Gaussian units are used, and c is the speed of light); the relative thinness of the ionosphere and the continuity of tangential components required by Faraday's law imply that the same $\underline{E}$ appears in both equations, whence by elimination

$$\underline{I} = \underline{\underline{\Sigma}} \cdot (\underline{V}_n - \underline{V}) \times \underline{B}/c \tag{3}$$

and corotation requires that Σ be sufficiently large to ensure $\underline{V} \cong \underline{V}_n$. What matters is clearly the value of Σ in relation to $\underline{I}$ and not in some absolute sense. Now the ionospheric currents are coupled, by virtue of the current continuity equation and via magnetic-field-aligned (Birkeland) currents, to the electrical current systems within the magnetosphere where the current density $\underline{j}$ is governed by the momentum equation (see, e.g. Siscoe, 1983 - this volume)

$$\rho d\underline{V}/dt + \nabla \cdot \underline{\underline{P}} = \underline{j} \times \underline{B}/c \tag{4}$$

Other things being equal, $\underline{I}$ scales with $\underline{j}$ which in turn scales with the mechanical stresses on the LH side of equation (4). Thus it is the ratio of the height-integrated conductivity in the ionosphere to the mechanical stresses in the magnetosphere that determines whether the ionosphere is sufficiently conducting to enforce corotation.

(3) The two conditions so far imply corotation of plasma just above the ionosphere, at the foot of a magnetic flux tube within the magnetosphere. To have corotational flow of plasma at all other points along the flux tube as well, it is necessary that equation (2) be valid everywhere; this is the MHD approximation whose presuppositions and implications are discussed in detail by Siscoe (1983 - this volume). This condition is sometimes referred to as ionosphere-magnetosphere coupling (in a narrow sense; the term is also used in a broader sense to describe the entire theory of plasma flow summarized here).

(4) If the preceding three conditions are satisfied, the components of plasma bulk velocity $\underline{V}$ perpendicular to $\underline{B}$ are equal to the perpendicular components of the corotational velocity $\underline{\Omega} \times \underline{r}$, but there may still be large differences in the flow parallel to $\underline{B}$. For the entire flow to be corotational, a necessary further condition is that of radial stress balance: the radial components of the $\underline{j} \times \underline{B}$ and the pressure gradient forces in equation (4) must be adequate to balance the centripetal acceleration of the plasma implied by corotation.

To the extent that these four conditions are satisfied in a particular region of the magnetosphere, the plasma flow there is azimuthal and has the angular velocity Ω of the rotation of the central object. As mentioned already, this corotating flow is a first approximation, and the actual plasma flow deviates from it by an amount that differs greatly among the various magnetospheres. Even a numerically small departure from corotation may be significant when it provides a radial component to the flow and hence a mechanism for transport of plasma from its source into other regions of the magnetosphere.

One type of non-corotational plasma flow in planetary magnetospheres arises from a breakdown of planet-atmosphere coupling: $\underline{V}_n$ in equations (1) and (3) differs from $\underline{\Omega} \times \underline{r}$ by the addition of the upper-atmosphere wind velocity field. The plasma in the magnetosphere then tends to follow the combined flow field of corotation plus winds. Such wind-induced plasma flows have been discussed particularly for the magnetosphere of the Earth as well as for that of Jupiter.

Much larger departures from corotation arise from effects related to the finite conductivity of the ionosphere. One effect, directly connected to the interaction of the magnetosphere with the external medium, is particularly important in the case of the Earth. Within the outermost boundary regions of the magnetosphere, mechanical stresses associated with plasma from the solar wind are sufficiently large in relation to the ionospheric conductivity so that the local plasma flow is determined primarily by the solar wind, with little or no effect from corotation or other atmospheric influences. The electric field associated with this flow by virtue of equation (2) continues from the boundary regions into the inner magnetosphere, modifying the plasma flow there and giving rise to the magnetospheric convection pattern which is discussed in Hill (1983 - this volume) and, with quantitative modelling, in Wolf (1983 - this volume). Here we give merely a rough estimate, adequate for comparative purposes, of the distance R_c at which corotation is effectively terminated by this process: assume the plasma flow speed near the magnetopause is some fraction ξ of the solar wind speed V_{sw} (empirically, $0.1 < \xi < 0.5$) and approximate the electric field associated with magnetospheric convection as uniform across the magnetosphere; then the convective flow speed varies as $1/B \sim r^3$ and the corotation speed as r, and setting the two equal at $r = R_c$ yields

$$(R_c/R_M)^2 \sim \Omega R_M/\xi V_{sw} \tag{5}$$

For the Earth, clearly, R_c/R_M is significantly less than 1 - corotation stops and solar-wind-driven magnetospheric convection becomes dominant well inside the magnetopause, a conclusion borne out by the much more accurate and detailed models reviewed in Wolf (1983 - this volume). For Jupiter, on the other hand, equation (5) predicts $R_c/R_M \gtrsim 1$: solar wind effects are too weak to dominate over corotation anywhere inward from the immediate boundary regions.

Nevertheless, corotation of plasma in the magnetosphere of Jupiter does not continue to the magnetopause but is self-limited by an effect that is also related to the finite ionospheric conductivity. As noted already, plasma is supplied to the Jovian magnetosphere principally from the Io torus, at a radial distance of 6 Jupiter radii; it is then transported outward, and its angular momentum per unit mass increases as Ωr^2 as long as corotation is maintained. This requires a torque, exerted by the $\underline{j} \times \underline{B}$ force associated with a system of radially directed electric currents whose closure through the ionosphere reduces, by virtue of equation (3), the plasma azimuthal flow speed to less than corotational values. A quantitative model by Hill (1979) shows that the angular speed decreases with increasing radial distance (partial corotation) and falls significantly below the planetary rotation rate at a characteristic distance R_H given by

$$R_H^{\ 4} = \pi(\Sigma/c^2)B_a^{\ 2}\ r_a^{\ 6}/S \tag{6}$$

where S is the net outward mass flux and $B_a r_a^3$ is the magnetic dipole moment of the planet. For Jupiter, observations suggest $R_H/r_a \sim 20$, compared with the magnetopause distance $R_M/r_a \sim 100$.

Other departures from corotation related to a finite ionospheric conductivity arise from stresses associated with azimuthal pressure gradients within the magnetosphere. These effects, included within the general modelling approach reviewed by Wolf (1983 - this volume), are significant in the case of the Earth and probably at Jupiter as well (see, e.g., Hill, Dessler, and Goertz, 1983). The radial flow required by the outward transport of plasma in the Jovian magnetosphere has been ascribed to the effect of such pressure gradients (either random or systematic), or alternatively to the effect of neutral atmospheric winds.

Departures from the MHD approximation along field lines are generally confined to small-scale structures, by the very nature of the approximation (cf. Siscoe, 1983 - this volume), and hence do not significantly affect the large-scale flow of plasma in the magnetosphere (except possibly in the rather exotic case of pulsar magnetospheres).

Non-corotational flows related to breakdown of radial stress balance can be expected when the magnetic stresses become smaller than the centrifugal stress $\rho\Omega^2 r^2$; in this case there is no longer a force large enough to provide the centripetal acceleration required for corotation, $\underline{j} \times \underline{B}$ (or equivalently the tension of the stretched-out magnetic field lines) being too small and the pressure gradient usually pointing in the wrong direction (the pressure decreasing outward). The critical distance R_o beyond which this occurs is often estimated roughly by setting the Alfvén speed equal to the corotational speed. A more precise estimate which takes into account the geometry of the magnetic field and the finite thickness of the plasma sheet is given by

$$R_o^{\ 4} \sim B_a^{\ 2} r_a^{\ 6}/2\pi\rho_s h_s \Omega^2 r_s^{\ 3} \tag{7}$$

where ρ_s is the plasma mass density and h_s the plasma sheet thickness at the plasma source location r_s, and it has been assumed that a unit magnetic flux tube has a constant plasma content independent of radial distance; for the magnetosphere of Jupiter from equation (7) one estimates $R_o/r_a \sim 24$. (See Vasyliunas, 1983, for a discussion of R_o as well as a general review of stress balance and its breakdown in the Jovian magnetosphere.) As plasma is transported outward beyond R_o, its flow is expected to change from corotation to a motion more nearly in a straight line, leading ultimately to a general outflow of plasma - a rotationally driven "planetary wind" - at least in directions such as the magnetotail where the flow is not impeded by the external pressure on the magnetopause.

Radial outflow, possibly channeled along a magnetotail, is the dominant motion of plasma in several magnetosphere-like systems where rotation of the central object is slow or negligible: Venus, comets, the heliosphere. The outflow in these cases in thermally rather than rotationally driven and, in terms of equations (4), arises from a dominant pressure gradient and resultant accelerations.

Finally, corotation must obviously stop, even if no other effect intervenes, before the light cylinder is reached

$$R_L = c/\Omega \tag{8}$$

where the corotational speed would equal the speed of light. It can be shown that this is also describable as the consequence of a breakdown in radial stress balance when relativistic flow speed are approached and it becomes necessary to take into account the mass density $B^2/8\pi c^2$ of the magnetic field as well as the relativistic increase of the inertia of the plasma.

MAGNETOSPHERIC DISTANCE SCALES

As described in the preceding section, each of the various mechanisms that lead to a flow of magnetospheric plasma distinctly different from corotation has associated with it a characteristic scale of distance from the central object, beyond which the non-corotational motion sets in or becomes appreciable. (An exception is neutral atmosphere winds, whose effects are usually distributed throughout the magnetosphere but are also in general small compared to corotation.) In a particular magnetosphere, the effective end of corotation will occur at whichever of the distances R_c, R_H, R_o, or R_L given in equations (5)-(8) is the smallest and will be due principally to the corresponding mechanism. If all these distances are larger than the size of the magnetosphere R_M, corotation is expected all the way out to the vicinity of the magnetopause.

The location of the magnetopause is determined primarily by pressure balance: the total pressure (plasma plus magnetic) on the inside should equal the pressure on the outside, which is generally nearly equal to the dynamic pressure $\rho_{sw} V_{sw}^2$ of the flowing external medium (the solar wind). An important related distance scale is the Chapman-Ferraro distance R_{CF}, obtained by setting $\rho_{sw} V_{sw}^2$ equal to the pressure of the planet's dipolar magnetic field:

$$R_{CF}^{6} = B_a^{2} r_a^{6} / 8\pi\rho_{sw} V_{sw}^{2} \tag{9}$$

This distance defines the size of the magnetosphere, $R_M \sim R_{CF}$, when the magnetic field is approximated (to within a factor of 2 or so) by the field of the planet's dipole even in the outer regions of the magnetosphere. If plasma pressures and/or inertial stresses within the magnetosphere are comparable to or larger than the magnetic pressure (this implies also that the magnetic field must deviate significantly from the dipolar values), they and the accompanying non-dipolar magnetic field pressures must be taken into account in determining R_M; in general, as illustrated in Figure 2, R_M becomes larger than R_{CF} (an effect

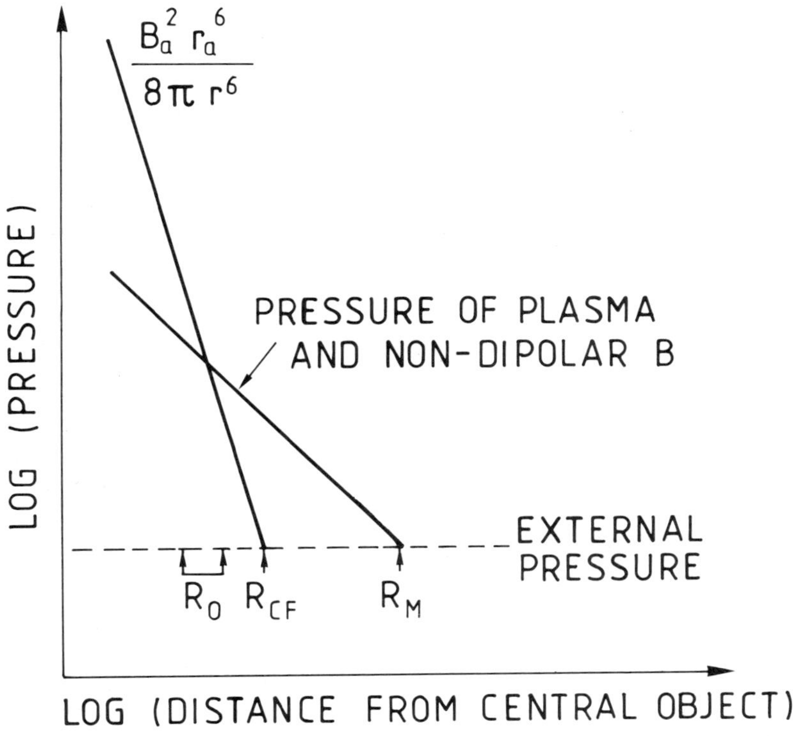

Figure 2. Variation of pressure with distance from the central object and its relation to the location of the magnetopause.

clearly observed at Jupiter). It should be noted that non-dipolar fields and significant plasma stresses are in any case present at distances larger than R_o (equation (7)).

Another distance scale is the gravitational radius

$$R_G = MG/V^2 \quad (10)$$

where MG is the gravitational mass of the central object and V a characteristic speed of plasma in the magnetosphere or the external medium. If $R_G > R_M$, gravitational effects are important throughout the magnetosphere; this occurs in the magnetospheres of accreting neutron stars associated with X-ray sources in binary systems. In all planetary magnetospheres, however, R_G is at most of the order of the planet's radius r_a; gravitational effects are mostly confined to the atmosphere and its immediate vicinity and play no appreciable role within the magnetosphere.

REFERENCES

Hill, T.W.: 1983, this volume.
Hill, T.W., Dessler, A.J., and Goertz, C.K.: 1983, in A.J. Dessler (ed)., "Physics of the Jovian Magnetosphere," Cambridge University Press, New York, pp. 353-394.
Michel, F.C.: 1982, Rev. Mod. Phys. 54, pp. 1-66.
Michel, F.C. and Dessler, A.J.: 1981, Astrophys. J. 251, pp. 654-664.
Miley, G: 1980, Ann. Rev. Astron. Astrophys. 18, pp. 165-218.
Ogilvie, K.W., Scudder, J.D., Vasyliunas, V.M., Hartle, R.E., and Siscoe, G.L.: 1977, J. Geophys. Res. 82, pp. 1807-1824.
Siscoe, G.L.: 1979, in C.F. Kennel, L.J. Lanzerotti, and E.N. Parker (ed.), "Solar System Plasma Physics," Vol. II, North Holland Publishing Co., Amsterdam, The Netherlands, pp. 299-402.
Vasyliunas, V.M.: 1979, Space Sci. Rev. 24, pp. 609-634.
Vasyliunas, V.M.: 1983, in A.J. Dessler (ed.), "Physics of the Jovian Magnetosphere," Cambridge University Press, New York, pp. 395-453.
Wilkening, L.L. (ed.): 1982, "Comets," University of Arizona Press, Tucson, Arizona, pp. 519-634.
Wolf, R.A.: 1983, this volume.

THE USE AND MISUSE OF STATISTICAL ANALYSES

Patricia H. Reiff
Center for Space Physics, Department of Space Physics and Astronomy, Rice University, Houston, Texas 77251

ABSTRACT

Statistical studies can be of significant benefit to guide or confirm theoretical development. For example, the open model of the magnetosphere has suggested numerous correlations between the Interplanetary Magnetic Field (IMF) and geophysical phenomena such as substorms, enhanced convection, polar cap size and the like. Conversely, trends and periodicities in experimental data spark progress in magnetospheric theory; for example, the discovery of upward ion beams and conics above the auroral zones has enhanced interest in the theory of ion cyclotron waves.

Since many of the most obvious periodicities (27 days, 11 years) and correlations have already been established, further progress demands more accurate statistical techniques and means of assessing more subtle effects. This paper presents several statistical techniques most commonly used in space physics, including Fourier analysis, linear correlation, auto- and cross-correlation, power spectral density and superposed epoch analysis, and presents tests to assess the significance of the results. When no test of significance is in common usage, a plausible test is suggested.

WHY USE STATISTICS?

"The purpose of computing is insight, not numbers" (Hamming, 1962).

Science is (or should be) a precise art. Precise, because data may be taken or theories formulated with a certain amount of accuracy; an art, because putting the information into the most useful form for investigation or for presentation requires a certain amount of creativity and insight.

Single case studies or single measurements from a single spacecraft are valuable when a scientific field is in an exploratory stage. That

R. L. Carovillano and J. M. Forbes (eds.), Solar-Terrestrial Physics, 493–522.

kind of knowledge is principally anecdotal, qualitative descriptions of phenomena. It sufficed that Columbus discovered that America and its inhabitants existed; however, before attempting to settle a new land, one would demand reasonable estimates of the numbers of the inhabitants and the friendliness of the tribes.

Statistical studies are of two fundamental types: exploratory and confirmatory. An exploratory study takes a data set and searches for patterns that can shed insight into as yet unknown processes; a confirmatory study uses a data set to confirm (or deny) an existing hypothesis. Both kinds of study have their usefulness, and their pitfalls.

Confirmatory Studies

Confirmatory studies are the easiest to perform, since the time scales and/or parameters to evaluate are suggested by the theory. Ever since Dungey (1961) put forth his idea of an "open magnetosphere," numerous studies have examined the correlation of B_S, the southward component of the Interplanetary Magnetic Field (IMF), or functions of B_S, with sundry magnetospheric parameters, and have gotten quite impressive, believable results. Significant correlations have now been observed between B_S and auroral occurrence and location, polar cap size, the Auroral Electrojet and Kp indices, magnetic flux in the magnetotail, and cross polar-cap potential drop (e.g., Arnoldy, 1971, Burch, 1974; Caan et al., 1975; Reiff et al., 1981). Figure 1 shows an example of a correlation between the cross-polar cap potential drop and a theoretical prediction of the merging rate (Vasylinas, 1982 — this volume), based on the IMF parameters.

Two common pitfalls should be kept in mind for this kind of study: time aliasing and intercorrelation of parameters. Time aliasing can

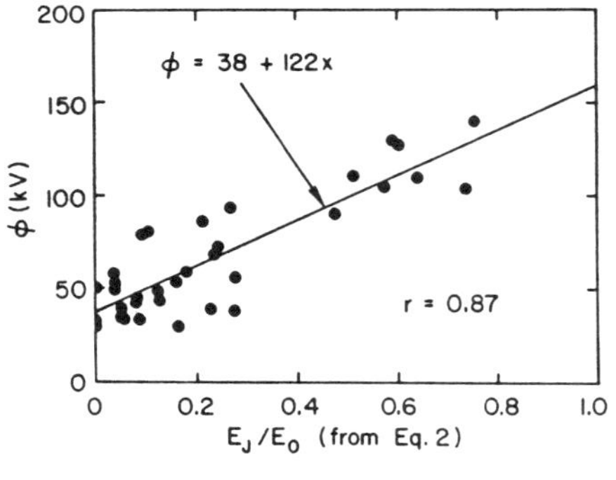

Figure 1. Correlation of the cross-polar cap potential drop Φ with a formula predicted from magnetic merging theory. The high correlation coefficient, r = 0.87, implies that 76% of the variability in Φ can be ascribed to variability in the IMF. (From Reiff et al., 1981).

occur because events in solar-terrestrial physics are rarely exactly periodic. The time between one sunspot maximum and the next may be 10.4 years one cycle and 13 years the next, yielding a long-term "11 year cycle" but with a certain amount of scatter. Thus, if one performs a standard test for periodicities (Fourier Analysis, autocorrelation, or Power Spectral Density (PSD)), one finds a peak at 11 years, but with significant signal at 9, 10, 12 and 13 year periods. Therefore, beware of scientists who propose a full-moon effect on geomagnetic activity and test their model by merely showing that the power spectral density at 29.5 days is larger than chance. One can calculate the complete PSD function of geomagnetic activity, and show that the residual power at 29.5 days is indeed statistically significant, but is merely a remnant of the huge peak at the 27-day solar rotation period (Rassbach et al., 1966).

Intercorrelation of parameters, "multicollinearity," confuses both confirmatory and exploratory statistical studies. The principle to recognize is that correlation is not causality: two occurrences may be highly correlated, but without direct causality — the causality involves an unknown third effect. One of the few 100% correlations in everyday experience is the headlight phenomenon: cars with their headlights on in parking lots are virtually always locked. This might lead to the hypothesis that people who are careless about their headlights are nevertheless careful about locking their cars, an uncomfortable hypothesis at best. The true causality of the matter lies in third parties: good neighbors who, if they see a car with its headlights on, will turn them off if the car is unlocked. (One can even quantify this phenomenon; people will generally turn off your headlights (if they can), but they generally won't roll up your car windows if it starts to rain.)

Multicollinearity becomes a problem in space data analysis because changes in varius solar wind parameters often occur together: passage of an interplanetary shock will simultaneously change the density, flow velocity, temperature, and field strength. Thus a response that is principally caused by changes in the solar wind velocity may also show a (negative) correlation with solar wind density. A high correlation coefficient merely indicates that it is unlikely that the two parameters are uncorrelated; it is no guarantee that you have tested the optimum functional form.

Exploratory Studies

Exploratory data analysis is, by its nature, an attempt to discover trends, dependencies and periodicities in a data set for which there is no good theory as a guide. The early studies relating sunspot number to auroral occurrence frequency and geomagnetic disturbance fall in this category. These showed highly significant correlations and periodicities (27 days, 11 years), and were extremely useful in guiding theoretical development (see Chapman and Bartels, 1940; Akasofu and Chapman, 1972). On the other hand, some of the exploratory sun-weather studies performed recently have indicated weak correlations. Since the magnitude of the solar-wind effect on the troposphere is expected to be small on energy grounds (Dessler, 1974), it is not obvious whether a weak correlation is a statistical fluke or a real indicator of a weak coupling mechanism.

Exploratory data analysis is subject to the same hazards as confirmatory data analysis, i.e., unknown third causes and time aliasing. In addition, there is a reasonable chance that a high correlation coefficient or large Fourier amplitude could arise entirely by chance. For example, the t-test, the bane of psychology students, can be used to put confidence limits on a result. Can we reject the null hypothesis (the hypothesis of no correlation) with 95% confidence? 99% confidence? ($t \equiv$ value of an inferred or fitted parameter divided by the uncertainty in that parameter; one attains 95% confidence with $t \simeq 2$ and 99% confidence with $t \simeq 3$). Nevertheless, a one-in-a-hundred correlation is likely to occur once in each hundred times you look; if ten researchers each examine ten correlations each year, then there should be "discovered" four spurious 95% correlations and one spurious 99% correlation each year. I'm reminded of a cartoon where the patient is reassured by the surgeon, "Don't worry, it's a 100 to 1 chance you'll turn out fine." "How many of these operations have you performed?", she asks. "Ninety-nine," he replies, "and each of them has been perfect."

COMMON-SENSE RULES

There are several statistical tests one can perform to judge significance of experimental results. These will be discussed in a later section. Here I will merely offer a few common-sense rules.

1. <u>Repeat the experiment</u>. A valid correlation is <u>repeatable</u>. If at all possible, take new data with the conditions as nearly the same as possible. This is <u>not</u> the same as adding more data points; it is taking new data and calculating separately a new correlation coefficient (or power spectrum). As we will see, a single data point far from the mean can have a huge influence. Often times a correlation disappears the month it is published, never to return (Vasyliunas and Dessler, 1981).

2. <u>Divide the data</u>. Sometimes it is not possible to take more data, because of the finite lifetimes of spacecraft, rockets, and balloons or because the AE index, for example, is unavailable. The next best thing is to divide the data into two or more sets and work with them separately. If it is a correlation study, <u>randomly</u> divide the data into two halves (no peeking!). If it's a study looking for periodicities, divide the data in two separate ways: first half versus second half, and odd-numbered points versus even-numbered points. The first division should show similarities in the high-frequency portion of the spectrum (but the lowest frequency information will be lost); the second test should show agreement in the low-frequency portion of the spectrum (but the high frequencies will be lost).

Sometimes there aren't enough data to divide, and there's no chance to repeat the experiment (at least, until the next launch). You have two basic options: "publish and pray" (that subsequent data analysis will bear you out) or "wait till the next launch." I've done both, and been lucky with "publish and pray." My correlation of polar cap potential drop with IMF parameters (Reiff et al., 1981) was completely confirmed by Doyle and Burke (1982). A couple of times when I've waited for more data, however, someone else has published the effect in the meantime. Publishing a spurious correlation, on the other hand, damages your reputation, and might confuse the theorists. It pays to be careful.

3. Does it make sense? Arthur Eddington said "Never believe an observational result until it is confirmed by theory" (quoted by Vasyliunas and Dessler, 1981). Naturally, an experimentalist shouldn't feel required to completely explain the data in order to publish them; however, it is helpful to ensure that the correlations and/or periodicities that arise from the statistical analysis make sense. Thus, when Chree (1912) found a 27-day periodicity in geomagnetic activity, it was natural to suspect the solar rotation (especially the appearance of sunspots). When Chapman and Bartels (1940) showed that sunspots alone were not enough, they invented "M regions" on the sun that had an influence on activity. We now know that the M regions are really coronal holes which emit high-speed streams (Sheeley and Harvey, 1981).

Other systems that show power at a 27 day cycle include the phases of the moon, the addition of chambers to a nautilus shell, tides (including tidally-produced atmospheric gravity waves), a woman's body temperature, the submission of mortgage or rent payments, etc. It wouldn't make sense to propose that these effects are caused by the solar rotation; however, there is strong evidence that all of these are caused by the moon's phase (even the menstrual cycles may have been under full-moon control, through the light-sensing action of the pineal gland). Beware: similar periods can stay nearly in phase for quite a while — it takes 14 months for a 28 day cycle to drift from being in phase to being out of phase with a 29 day cycle.

4. Make sure that the time markers for superposed epochs are independent of the studied data set. It is legitimate to use mid-latitude positive bay onsets as timing marks for studying substorm behavior in the magnetotail or to search for triggering mechanisms in the solar wind, as has been done profitably by Caan et al. (1975). It is also legitimate to use sector boundary crossings as time markers for tropospheric parameters (Roberts and Olson, 1973). It is hazardous to use features in the data stream of interest. One can mask a real effect, or create a spurious effect. For example, using sudden impulses to align equatorial magnetometer readings can give an incorrectly small "average geomagnetic storm," since some sudden impulses are followed by storms and some are not. Similarly, using pulses in a random data stream to superpose that data set upon itself will generally overemphasize those random pulses and de-emphasize the rest of the data.

5. Be careful about normalization. Superposed epochs with baselines removed by setting the value of the data equal to 0 (or normalized to 1) at the t = 0 time markers will have a zero standard deviation at t = 0. Any change is likely to appear significant. If one is superposing data ± 1 year from the time markers, for example, it is better to subtract the two-year mean from each subset of the data, rather than just the value at t = 0. The error bars will be more comparable, and statistical tests (such as the t test) will be valid.

6. Use appropriate time scales. It is a serious mistake to use a 3-hour index such as Kp to study auroral boundaries or average auroral fluxes, since both can change significantly in minutes. An hourly AE index is better, but better still is a one-minute AE index. The Kp index is probably adequate, however, for categorizing relatively slow processes such as particle density in the central plasma sheet (away from the boundaries) or radiation belt fluxes.

Similarly, it is pointless to use hourly-averaged interplanetary parameters to determine whether the magnetosphere is a "driven" or "storage" system, because the time lag is of the order of 40 minutes. Even one-minute ISEE-3 data are probably not adequate, since ISEE's distance from the bow shock is larger than a correlation distance (Russell et al., 1980). The only chance to resolve the controversy is to find a segment of data where the interplanetary parameters (IMF and solar wind density and velocity) are steady, and the IMF southward, on a one-minute time scale for at least a three hour span. If activity is continuous (perhaps after a time lag) then the magnetosphere is driven; if it is sporadic, then the magnetosphere stores and releases energy.

7. Be sure that you are comparing comparable things. Part of the resolution of the controversy between the "driven" versus "storage" schools of thought lies in the parameters that each group observes. When Akasofu (1980) states that the magnetosphere is "driven," he adds up several forms of energy gained by the magnetosphere: joule heating and ring current energy density as well as particle precipitation. The "storage" school of thought emphasizes that the energy contained in the magnetosphere changes state in a quasi-sporadic way, first appearing as enhanced tail field strength and then dumping energy in the form of particle precipitation and ring current injection. One may certainly argue whether all appropriate terms are included properly in each formalism, but so long as one group is discussing total energy and the other is discussing partitioning of energy, the controversy cannot be quantitatively settled (and may not, in fact, exist). (See also Clauer, 1982.)

8. Mean, median, mode. Most scientists recognize the difference between the mean (arithmetic average), the median (half of the distribution lies above, half below) and the mode (most common value). These differences can be significant, especially in a skewed system (the mean salary in a factory includes the effect of a few highly paid executives, whereas the median is more representative of the workers as a whole; see the clever discussion in How to Lie With Statistics (Huff, 1954)).

These differences creep into magnetospheric physics in subtle ways, for example in the oft-misused term "characteristic energy." The mean energy is defined to be the energy density divided by the number density, but some use the energy flux divided by the number flux, which can differ substantially from the mean if the distribution has a low-energy or high-energy tail (which it often does, because of atmospheric secondaries, for example). It can make a significant difference, for example, in the altitude of auroral energy deposition in the atmosphere. In fact, the more appropriate parameters are the bulk velocity, thermal velocities, and momentum flux tensor (Siscoe, this volume). For auroral electrons, the parallel potential drop and density and temperature of the primaries are most useful (Pulliam et al., 1981).

With these caveats in mind, let us look at several techniques used to establish correlations and periodicities in data streams and some examples of these common-sense principles at work.

USEFUL STATISTICAL TECHNIQUES

Statistics may be used, with due caution, to answer, with some confidence, the following questions:

1. Are two processes linearly correlated?
2. What time lag is best to optimize the correlation?
3. Will adding another term to a polynomial fit improve it significantly?
4. Are there any fundamental periodicities in the data?
5. What is the waveform at these periods?
6. Do external events elicit characteristic waveforms?
7. What are the best ways to smooth the data?

The Rice library has well over a thousand books on statistics to choose from. Eliminating the four shelves of "statistical mechanics" or "statistical physics" books (which, of course, are thermodynamics books in disguise), the four shelves of "statistical psychology" books (although some of these are quite helpful in defining terms or explaining the basics), we are left with two shelves each of computer-based statistics and engineering statistics, as well as another shelf of books on power spectra. The two most useful books are not there, however; someone has checked them out. One is Data Reduction and Error Analysis for the Physical Sciences by Philip R. Bevington (1969), which covers fitting data to arbitrary functions, distribution functions, smoothing, integrations, differentiation, and the like, with Fortran subroutines to use. The other book is The Measurement of Power Spectra, by R. B. Blackman and J. W. Tukey (1958). This book is not only useful, it is relatively inexpensive ($3.50 at this writing). A classic text on finite difference techniques is Hamming (1962). Most of the equations to follow can be found in one of these three books.

Linear Correlation Analysis

This is a statistical procedure to answer question #1: are two processes linearly correlated? With transformation of variables, linear correlation analysis can also be used to fit data to a power law or to an exponential, by using the logarithms of the dependent and independent variables, or of just the dependent variable, respectively. The linear correlation is the first (and often the only) statistical procedure most scientists perform, for several reasons: (1) it is easy to do (some hand calculators now perform it automatically), (2) it is easy to understand, and (3) most geophysical data cannot justify higher-order expansions.

Mean, variance, standard deviation. One should distinguish between true mean (μ) variance (σ^2) and standard deviation (σ) based on the parent population,

$$\begin{aligned} \mu &\equiv \lim_{N \to \infty} \left(\frac{1}{N} \sum x_i\right) \\ \sigma^2 &\equiv \lim_{N \to \infty} \left(\frac{1}{N} \sum (x_i - \mu)^2\right) \qquad (1) \\ &= \lim_{N \to \infty} \left(\frac{1}{N} \sum x_i^2\right) - \mu^2 \\ \sigma &\equiv \sqrt{\sigma^2} \end{aligned}$$

and their corresponding estimates $\bar{x}$, s^2, and s taken from a finite sample of N data points (all summations are assumed to run from $i = 1$ to N unless noted):

$$\begin{aligned} \bar{x} &\equiv \left(\sum x_i\right)/N \\ s^2 &\equiv \left(\sum (x_i - \bar{x})^2\right)/(N - 1) \qquad (2) \end{aligned}$$

The denominator of the variance s^2 is reduced from N to (N - 1) because one degree of freedom is "used" to estimate the mean. Similarly, all fitting functions yield only estimates of the true parent population. Nevertheless, we will agree that our sample is finite, so limited, but we will use our estimates as our best guess of the parent population, i.e., $\mu \simeq \bar{x}$; $\sigma^2 \simeq s^2$.

Principle of least squares. This fundamental method optimizes a fit to an arbitrary function $y = f(x)$ by minimizing the squares of the deviations of the "dependent" data variable y_i from their fitted or predicted values $y(x_i)$. The parameter to minimize is the reduced chi squared χ_ν^2, which is defined to be

$$\chi_\nu^2 \equiv \left(1/(\nu\sigma_y^2)\right) \sum \left(y_i - y(x_i)\right)^2 \qquad (3a)$$

where ν = number of degrees of freedom $\equiv N - n$, where n is the number of fitted parameters (= the order plus one if the fit is to a polynomial; = 3 for a fit to a gaussian distribution, etc.). Note that the squared deviations are used, so that positive and negative deviations do not cancel each other out. Large deviations are overemphasized, however, so that one wayward point can throw the entire fit off. Other fitting techniques use the absolute value of the deviations, either minimizing their sum, or minimizing the maximum absolute deviation. It is easy to show (Hamming, 1962) that, for finding a characteristic value, least-squares will give the mean, and the other two techniques will yield the median and the mid-range values, respectively.

Beware: some "canned" computer programs do not include σ^2 in the calculation of χ^2, so that χ^2 has units of y^2. Equation (3a) is valid if one does not know the uncertainties in the y-parameters or if the uncertainties are all the same. If we happen to know the uncertainty σ_i in each y-value independently, we use another form:

$$\chi_\nu^2 = \frac{1}{\nu} \sum \frac{(y_i - y(x_i))^2}{(\sigma_i)^2} \tag{3b}$$

For counting rate (Poisson) statistics, for example, $\sigma_i^2 = C_i + 1$, where C_i is the counting rate. For results of calculations the joint uncertainty must be used:

$$\sigma(y_i(z,t)) = \left|\frac{\partial y}{\partial z}\right| \sigma_z + \left|\frac{\partial y}{\partial t}\right| \sigma_t; \tag{4}$$

for measurement uncertainties, use the sum of the precision and the accuracy (a tape measure may be read to a <u>precision</u> of one-16th of an inch, but if it has been stretched, the <u>accuracy</u> may only be 20% of the measured value; the true uncertainty is the sum).

Uncertainties in the x-variable are not easily handled, and are generally assumed zero or at least constant. Unequal x-uncertainties can sometimes be made equal by a change of coordinate; e.g., energy band passes typically range from E_c/f to fE_c, where f is commonly 1.15 or so and E_c is the "center" energy. The x-uncertainties can be made equal by using the logarithm of the center energy as the dependent variable. The uncertainties in log x then become log f, a constant. Other times uncertainties in x can be added to uncertainties in y, if one has a good a priori guess to the functional fit:

$$\sigma_i(y) \approx \sigma(y_i) + |\partial f/\partial x|\, \sigma(x_i). \tag{5}$$

For a linear correlation, a fit to $y_i = a + bx_i$, in the case of uniform σ_i (no weighting), minimizing χ^2 is analytical (which is why it is available on hand calculators). Four sums need to be accumulated: Σx_i, Σy_i, Σx_i^2, and $\Sigma x_i y_i$. Then the solution is found by

$$a = (1/\Delta)\left(\sum x_i^2 \sum y_i - \sum x_i \sum x_i y_i\right)$$
$$b = (1/\Delta)\left(N \sum x_i y_i - \sum x_i \sum y_i\right) \quad (6)$$

where $\Delta = N\sum x_i^2 - (\sum x_i)^2$.

The uncertainties in the parameters can be found by

$$\sigma_a^2 = (\sigma^2/\Delta) \sum x_i^2$$
$$\sigma_b^2 = N\sigma^2/\Delta \quad (7)$$

where $\sigma^2 = (N - 2)^{-1} \sum (y_i - a - b\, x_i)^2$

$$= (N - 2)^{-1} \left(\sum y_i^2 + Na^2 + b^2\sum x_i^2 - 2a\sum y_i - 2b\sum x_i y_i + 2ab\sum x_i\right)$$

For data where the individual uncertainties σ_i are known, these results are (Bevington, 1969)

$$a = \frac{1}{\Delta}\left(\sum \frac{x_i^2}{\sigma_i^2} \sum \frac{y_i}{\sigma_i^2} - \sum \frac{x_i}{\sigma_i^2} \sum \frac{x_i y_i}{\sigma_i^2}\right)$$

$$b = \frac{1}{\Delta}\left(\sum \frac{1}{\sigma_i^2} \sum \frac{x_i y_i}{\sigma_i^2} - \sum \frac{x_i}{\sigma_i^2} \sum \frac{y_i}{\sigma_i^2}\right)$$

$$\Delta = \sum \frac{1}{\sigma_i^2} \sum \frac{x_i^2}{\sigma_i^2} - \left(\sum \frac{x_i}{\sigma_i^2}\right)^2 \quad (8)$$

$$\sigma_a^2 = (1/\Delta) \sum (x_i^2/\sigma_i^2)$$

$$\sigma_b^2 = (1/\Delta) \sum (1/\sigma_i^2)$$

The linear correlation coefficient, r, is defined to be the square root of the fraction of the variance that is "explainable" by the linear fit (Lewis-Beck, 1980). Therefore, $(1 - r^2)$ is the fraction of the variance that remains after the linear correlation. Thus, although $r = 0.5$ may sound like a "good fit," it means that 75% of the variance is still unexplained. The correlation coefficient is also the square root of bd, where b is the slope of the best linear fit of y upon x: $y = a + bx$, and d is the slope of the best linear fit of x upon y: $x = c + dy$. For no correlation, both b and d approach zero (therefore, so does r); for nearly perfect correlation, $b \approx 1/d$, and the correlation coefficient approaches unity (plus or minus; the sign of b (or d) is ascribed to r). In addition, $r \equiv \sigma_{xy}^2/(\sigma_x\sigma_y)$, the sample covariance divided by the product of the standard deviations. The sample covariance σ_{xy}^2 is defined to be $\left(1/(N - 2)\right) \Sigma (x_i - \bar{x}) (y_i - \bar{y})$. (Most texts assume that N is large and just use (1/N) as the averaging factor).

The value of the correlation coefficient r may be calculated by

$$r \equiv \frac{N \sum x_i y_i - \sum x_i \sum y_i}{[N \sum x_i^2 - (\sum x_i)^2]^{1/2} \, [N \sum y_i^2 - (\sum y_i)^2]^{1/2}} \tag{9}$$

Fits to polynomials or arbitrary functions, and multiple linear fits are similar, although they must be performed iteratively. Discussion and sample programs are given in Bevington (1969).

Tests of Significance

Chi-squared. For fitting to polynomials (or arbitrary functions), two tests of significance can be applied. The first involves χ^2. Generally, a value of χ_ν^2 less than 1 is a "good fit"; the data values, on average, are less than one standard deviation away from the fitted function. One can quantify the goodness of fit using the χ^2 distribution, the probability $P_\chi(\chi^2,\nu)$ that the value of χ_ν^2 would be exceeded by chance for ν degrees of freedom (Table A 1). Thus, 90% confidence in a fit is achieved with $\chi_\nu^2 \leqslant .824$ if $\nu = 100$; for $\nu = 10$, $\chi_\nu^2 \leqslant .487$ is required. WARNING: The value of χ^2 is inversely related to the variance of the data σ^2 if there is no weighting (unknown or equal uncertainties), or to the average experimental uncertainty $\left(\Sigma\ (1/\sigma_i^2)\right)^{-1}$ if the uncertainties are known. An easy way to make χ_ν^2 small is to overestimate the uncertainty in the data points!

T-test. A second test of significance, less dependent on ascribed data uncertainties, is the t-test (mentioned earlier). Each fitted parameter a is divided by its uncertainty σ_a and a t statistic for each parameter is formed. If $t \gtrsim 2$, one has a "good fit": zero is two standard deviations away from the estimated value, and a 95% confidence that the parameter is non-zero is attained. A value of $t \gtrsim 3$ yields 99% confidence that the parameter is "really" non zero. The values of t for 95 and 99% confidence levels depend weakly on ν; any statistical psychology text has tables of t versus ν in the back.

Correlation Coefficient. The linear correlation coefficient r can be used to assess significance in linear fits. Naturally, if $|r| = 1$, the probability that the two data sets are uncorrelated approaches zero. For $r = 0.5$, however, there is a $\leqslant .1\%$ chance of uncorrelation if N (the number of data points) $\geqslant 40$; a 1% chance for $N = 26$; 5% for $N = 16$, 10% for $N = 12$; and 20% for $N = 8$, and only a 50-50 chance of correlation for $N = 4$. Table A 2 (from Bevington, 1969) indicates the probability function for r. WARNING: r, like chi squared, is dependent on the variance of the data. A large single point, as we will show below, can inflate r (by increasing the variance).

Examples of Polynomial Fits. I wanted to show how even "completely" random data can always be fit with a straight line, which tends to horizontal. Therefore I generated 21 pairs of random numbers on my hand calculator. (Hand calculator "random" numbers are never completely random; they require a "seed" number to generate the next random number, which is then remembered as the next "seed." Continuous memory machines remember the seed even if the machine is turned off.)

The result of my first attempt is shown in Figure 2a, which is a linear fit to the random data points. The linear correlation coefficient is an impressive .448, which, from Table A 2, means that there is less than a 5% chance that the data are uncorrelated — a one-in-twenty chance that just happened to occur the first time I looked! On the other hand, $(1 - r^2) = .8$; 80% of the variance still remains. Multiplying the x and y values of the last data point by ten (Figure 2b) "improved" the correlation to .931; a less than 0.1% chance of uncorrelation! The reduced χ^2 for the two fits fell from a respectable 0.85 (65% confidence) to an incredibly good 0.14 (better than 99% confidence). So much for random data! The first fit failed the t-test for

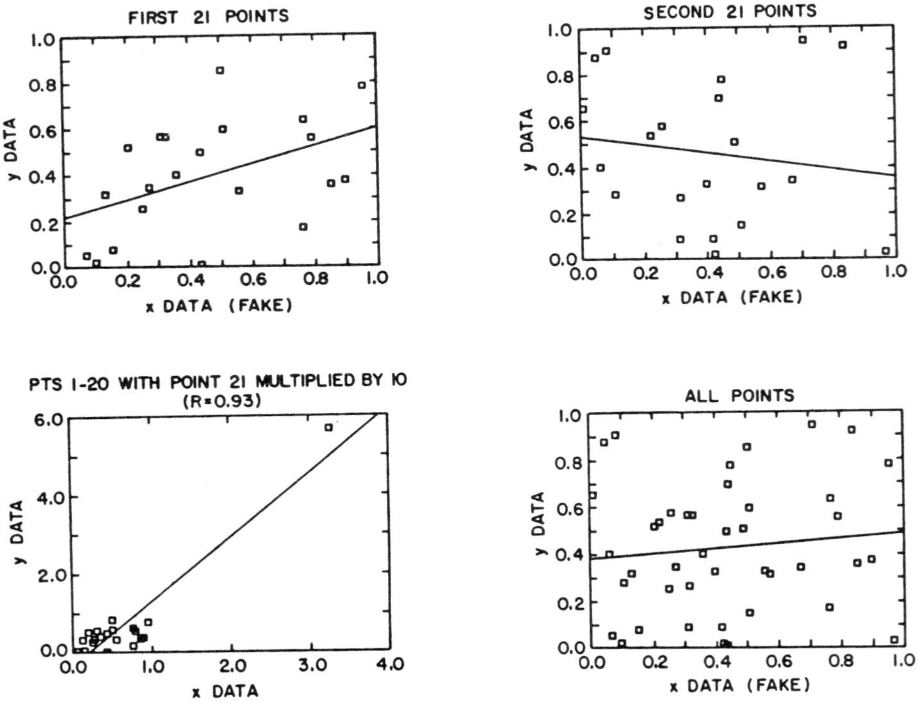

Figure 2. a (top left): Linear correlation of 21 random-number pairs. b (bottom left): The same data, with the last point multiplied by ten, "improves" the linear correlation from 0.448 to 0.93. c (top right): The next 21 random pairs. d (bottom right): All 42 random pairs.

the significance of the slope and intercept, which is comforting, but the second fit passed the t-test at the one percent level with flying colors. Our scientific curiosity may be piqued by the high correlations, so we will try common sense rule number 1: take new data. The next 21 pairs of points are shown in Figure 2c. The slope changes sign and the correlation coefficient goes to a comfortingly small -0.15 (less than 50% confidence). Adding the two data sets together (Figure 2d) yields a correlation coefficient of 0.10, which is just below the 50% level for 42 points (from Table A 2). In this case adding new data to the original data helped demolish the artificial correlation. However, merely adding the new data points to the high point example (Figure 2b) only decreased the correlation coefficient from 0.93 to 0.82. So BEWARE of points far from the median!

Question #2 will be deferred until the section on autocorrelation and cross-correlation.

Question #3, does adding terms improve the fit significantly, can be answered by means of the χ_ν^2 test. Adding new terms should (in principle) reduce the deviations from the fitted functions; however, it will also decrease ν (since the number of degrees of freedom is reduced). Thus χ_ν^2 may or may not decrease. In addition, the value of χ_ν^2 for a given confidence level decreases as ν decreases. Thus, fitting our random data with a second-order polynomial actually increases the reduced chi square, with values of .89, .89, .88, and .86 for second, third, fifth and eighth-order fits, respectively. No significant benefit was attained — if a second-order fit doesn't help, a third-order fit probably won't either. Higher order fits than the data deserve often lead to very humorous results (Figure 3, for the high data point case).

Techniques and programs for fitting data to arbitrary functions may be found in Bevington (1969).

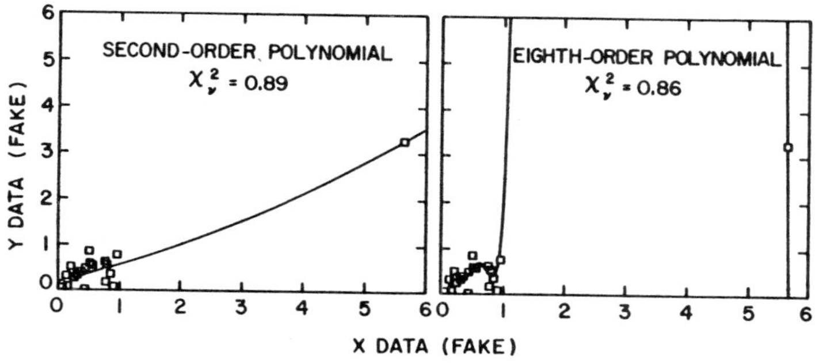

Figure 3. The results of fitting the data in (2b) with a second-order (left) or eighth-order fit.

SEARCHING FOR PERIODICITIES

Is it periodic?

Question #4 may be answered statistically by calculating any of three functions: Fourier transform, autocorrelation function, and power spectral density. Actually, the latter two are not independent but are Fourier transforms of one another. Superposed epoch analysis, which some researchers use to search for periodicities, is actually more appropriately used to answer questions 5 and 6. This will be discussed in a later section. Fourier series and power spectral density calculations are more precise for higher frequency information, whereas autocorrelation calculations are more precise for lower-frequency information.

Fourier Analysis

Fourier analysis is the most detailed spectral analysis that can be made of a data stream. It also permits analysis up to the highest frequency possible — half the sampling frequency. This highest frequency is called the Nyquist frequency. Phase information is retained because both sine and cosine terms are generated.

In principle, any function with a finite number of discontinuities can be Fourier analyzed, but in practice, data with many large discontinuities or randomly-spaced points far from the mean yield less reliable results. The analysis attempts to find the Fourier transform s(f) such that

$$y(t) = \int_{-\infty}^{\infty} s(f)e^{i\omega t}\,df \qquad (\omega = 2\pi f) \tag{10}$$

For N discrete data points spaced Δt apart, the fundamental frequency is $f_0 = 1/(N\Delta t)$ and the Nyquist frequency is $f_n = Nf_0/2 = 1/(2\Delta t)$. The Fourier series expansion is therefore ($m = N/2$; N must be an even number)

$$y_n = y(n\Delta t) = A_0/2 + A_m/2\cos(n\pi) + \sum_{j=1}^{m-1}\left(A_j\cos(2\pi jn/N) + B_j\sin(2\pi jn/N)\right) \tag{11}$$

The Fourier coefficients are found by (the sums run from $k = 0$ to $N-1$):

$$A_0 = (2/N)\sum_{k=0}^{N-1} y_k = 2\bar{y}; \qquad B_0 = 0$$

$$A_m = (2/N) \sum y_k \cos(\pi k); \qquad B_m = 0$$
$$A_j = (2/N) \sum y_k \cos(2\pi kj/N); \qquad j = 1, m-1 \qquad (12)$$
$$B_j = (2/N) \sum y_k \sin(2\pi kj/N); \qquad j = 1, m-1$$

One can show (Hamming, 1962) that all Fourier amplitudes may be calculated by only evaluating N cosine and sine terms cos $(2\pi j/N)$ and sin $(2\pi j/N)$, j = 1, m, and using trigonometric identities. This is the principle of the Fast Fourier Transform often seen in computer software packages.

Aliasing can cause difficulties in measurement of frequency spectra. Here aliasing (a slightly different definition than was used before) is caused by frequencies higher than the Nyquist frequency producing a signal in a lower-frequency channel, much as a stroboscope can halt motion at any integral fraction of the frequency of the motion. Figure 4 (adapted from Blackman and Tukey, 1958) shows how a high frequency signal can appear as a low-frequency signal because of aliasing (under-sampling). In some cases, we can rule out aliasing (we have no real fear of weekly fluctuations if we are looking at the number of pages published each month in the Journal of Geophysical Research; but if we find a signal in plasma measurements at a integral fraction of the gyrofrequency, we should worry about aliasing). "Pre-whitening" (smoothing the data before sampling) can help eliminate aliasing; however, real high-frequency information can be lost.

Significance. Assuming that aliasing is not a problem, the amplitudes A_j and B_j of the Fourier components may or may not be significant. Although I have not seen any discussion of the significance of Fourier components in any text, a reasonable test would be to compare the amplitudes A_j, B_j with the sample standard deviation σ. The zeroth term A_0 will, of course, be often larger than σ, since it is twice the mean value. The remaining terms, however, should generally be smaller than σ. It is easy to show that, for a pure sine wave $f(k) = A_j \cos(2\pi kj/N)$,

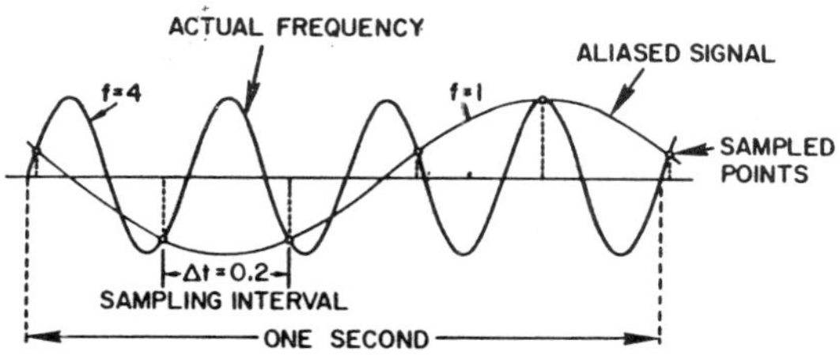

Figure 4. Aliasing (a high frequency signal appearing as a low-frequency signal) by undersampling of the data (adapted from Blackman and Tukey, 1968; used with permission).

the variance $\sigma^2 = A_j^2/2$. Thus one may invent a statistic $A_j^2/2\sigma^2$ which should be similar to a linear correlation coefficient. Ascribing statistical certainty is difficult, however, because a Fourier fit has no degrees of freedom! One may, however, perform a full Fourier transform, find the k largest coefficients, refit to only those k frequencies, and therefore have (N-k) degrees of freedom against which to test the k ratios $A_j^2/2\sigma^2$.

Predictability. Another excellent test of the goodness of a Fourier fit is to see how well it can predict future data. Since the phase information is present, one may simply take the next few data points and see how well the Fourier series predicts the data. If the waves are nearly coherent and reasonably significant, the prediction should be reasonable for at least a few periods of the highest-frequency significant term. Truly random data, however, will fall apart quickly. That is why computer jocks have never "made a killing" in the stock market, nor are sunspot numbers predicted more than a year ahead, despite the 11-year cycle. In fact, simply extending a Fourier reconstruction past the end time of the data will merely repeat the original

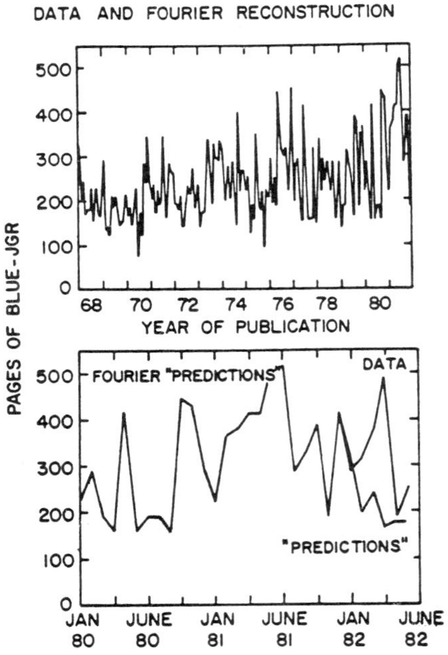

Figure 5. a (top): The number of pages per monthly issue in the Journal of Geophysical Research — Space Physics from 1968 through 1981. b (bottom): The last six data points are the Fourier predictions for 1982, with their corresponding actual values. As is obvious, the predictions are not particularly valid.

data from the beginning. A better way to attempt to "predict" the future would be to perform a Fourier analysis to ascertain the most significant frequencies, perform a least-squares fit to those frequencies alone, and continue them forward in time.

Example of a Fourier fit. Figure 5a shows the number of pages printed in the Journal of Geophysical Research – Space Physics monthly since 1968, along with a Fourier reconstruction of the data. The data before 1974 are multiplied by 0.7 to compensate for the change in page size. (The Fourier transform was computed using eqns. (12) and the data reconstituted using eqn. (11)). The fit is almost exact, with the data less than 0.01 pages from the reconstruction, on average. No one constructs χ^2 for a Fourier transform: with no degrees of freedom, it is not surprising that χ^2 is virtually zero. For continuous functions (as opposed to discrete data), it is reasonable to discuss the convergence of the Fourier fit to the function.

Figure 5b is a comparison of the Fourier "predictions" for 1982 and the corresponding data, showing how quickly it disintegrates. Figure 6a shows the cosine, sine, and ($\cos^2 + \sin^2$) amplitudes (6b) for each term (their abscissa is the frequency in units of f_0). At first glance, the peaks at 3, 6, 18, and 56 months appear to be significant (Figure 6b, top). For this example, however, the variance is 7158 page^2, so that $(A_j{}^2 + B_j{}^2)/2\sigma^2$ is only 7% for the 56-month signal and 5% or less for the remainder. So the signal, if any, is bound to be weak. Since the first three periods are multiples of one another, a complex eighteen-month pattern is suggested, with the higher frequencies creating a sub-waveform within the eighteen-month pattern. The five-year periodicity suggests the term of editorship as a significant influence. We will check both the 18-month and editorship trends by using superposed epoch analysis. The ($\cos^2 + \sin^2$) amplitudes $P_j = (A_j{}^2 + B_j{}^2)$ are called the Fourier Power Spectrum; we will see a quicker way to estimate these coefficients, the Power Spectral Densities, in our section on autocorrelation. Figure 6b (bottom) shows this latter estimate.

Superposed Epoch Analysis

The best way to discover the waveform (now that we know what frequencies to suspect) is the superposed epoch technique, first used in space data analysis by Chree (1912). If we have a dominant frequency (or several frequencies that are multiples of one another), we can calculate an average waveform by superimposing several subsets of the data, using the longest period. If the longest period is $m\Delta t$, then we divide the data into $k \equiv N/m$ blocks of m length each (discarding the remainder, if any). We then create the average of the first point in each subset, the second, etc. The average of the jth point in each subset is given by:

$$S(j) = \frac{1}{k}\sum_{i=1}^{k} x\big(j + (i-1)m\big) \qquad j = 1, 2, \ldots m \qquad (13)$$

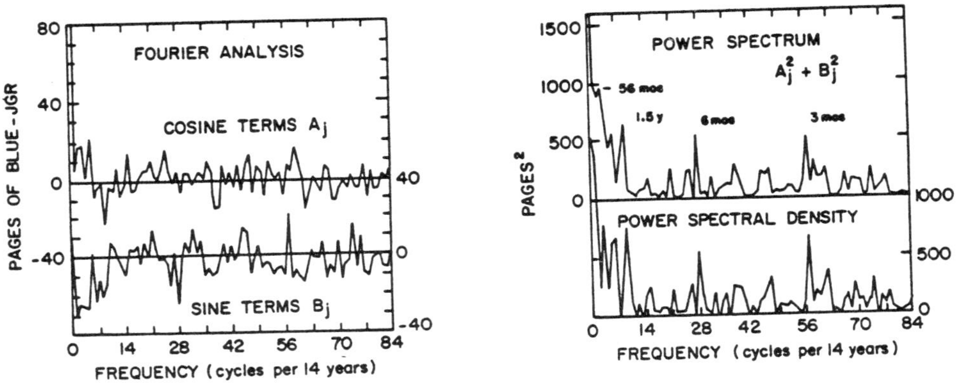

Figure 6. a (left): Cosine (A_j) and sine (B_j) terms for the Fourier transform of the JGR pages. b (right): The Fourier power spectrum = $A_j^2 + B_j^2$, and the Power Spectral Density, computed from the autocovariance.

Standard deviations σ_j for each of the S's may be calculated to assess whether the pattern is significant or not. The averages should span a range of $\gtrsim 2\ \sigma_{max}$. The values S(j) yield the average waveform (i.e., answer question 5). Figure 7 shows superposed epoch analyses for the JGR page data divided into 12-month and 18-month sections. The 18-month superposition shows a barely significant waveform, whereas the 12-month superposition is not significant, as suggested by the Fourier components.

<u>Triggering studies</u>. Question #6 asks whether external events (outside of the data stream) can trigger specific responses, and, if so,

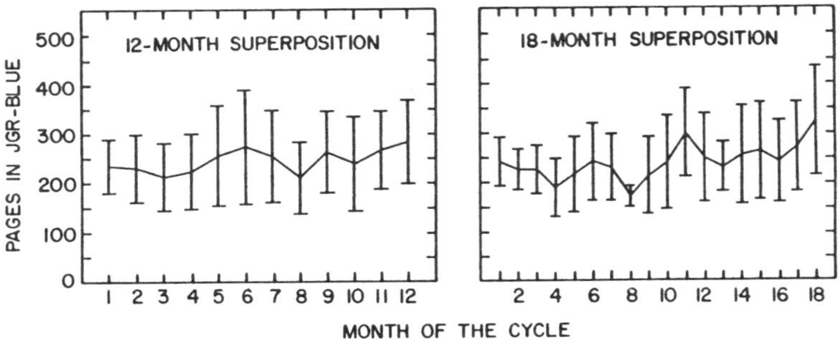

Figure 7. Superposed epoch analysis of twelve-month (left) and eighteen-month (right) subsets of the JGR page data.

what responses are elicited. For example, one may look for characteristic patterns in the solar wind velocity before and after a sector boundary crossing. We create a set of k timing marks, preferably using an independent data set. The length of time one chooses to observe before and after the timing markers is arbitrary (perhaps suggested by the data), but without loss of generality we can assume it runs from 0 to m steps after the markers. The superposed data can overlap; i.e., the separation between two markers may be less than m steps, if necessary. For example, several features in a 27-day solar rotation may each return in 27 days, and will obviously be less than 27 days from one another. The superposed averages S_j are, then

$$S_j = \frac{1}{k} \sum_{i=1}^{k} x(j + M(i)); \qquad j = 0, 1, \ldots m \qquad (14)$$

where M(i) is the set of marker times (substorm onsets, southward turnings of the IMF, sector crossings or the like). Again, standard deviations σ_j should be formed to test the significance of any changes.

Normalization. Linear trends can be a problem, if earlier subsets are much larger (or smaller) than later subsets. The simplest way to normalize is merely to subtract the value at the marker from each data point in that subset, i.e.,

$$S_j = \frac{1}{k} \sum_{i=1}^{k} \left[x(j + M(i)) - x(M(i))\right]; \qquad j = 0, 1, \ldots m \qquad (15)$$

I do not recommend that procedure, since it yields a zero standard deviation for the 0th point (and larger deviations farther away). It is safer to subtract the mean value in the subset from each member of the subset, i.e.,

$$S_j = \frac{1}{k} \sum_{i=1}^{k} x(j + M(i)) - \bar{x}(i); \qquad j = 0, 1, \ldots m \qquad (16)$$

where $\bar{x}(i) = \frac{1}{m + 1} \sum_{q=0}^{m} x(M(i) + q)$.

Examples of superposing the JGR data, with now the starting issues of three editors as the time markers, are shown in Figure 8a (normalized as in eqn. 15, by subtracting the zero-month value) and 8b (normalized as in eqn. 16, with 2-year means removed). Some significant signal is seen, with the number of pages nine months after editorship being significantly larger than the first few months after editorship. This "makes sense" (common sense rule #3), since nine months is approximately the gestation period for papers submitted to JGR to appear in print. The standard deviations are smaller with the means removed.

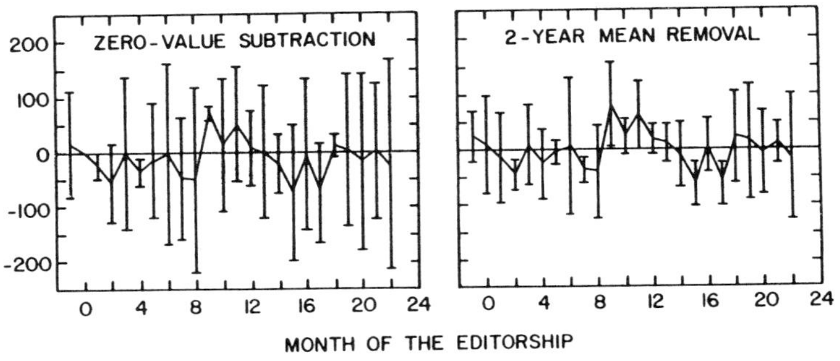

Figure 8. a (left): Superposed epoch analysis of JGR pages, using the first month of editorship as a timing mark (normalized by subtracting the value of the zeroth month's pages). (right): Same as (a), but normalized by removing the 2-year means.

AUTOCORRELATION AND CROSS-CORRELATION ANALYSES

Autocorrelation

Autocorrelation analysis may also be used to answer question #4: are there any fundamental periodicities in the data? The autocorrelation function C_j, and its Fourier transform, the power spectral density P_j, will show characteristic time scales: the first in the time domain, the second in the frequency domain.

Since there is no phase information (as in a Fourier transform), there is no danger of using these functions to "predict" the future. Similarly, the autocorrelation function, since it does not evaluate sines and cosines, is quick to do and always can be found (even with highly scattered data). Aliasing can also influence these functions: if the data repeat after 27 days, then the C_j's will also show signal at 54, 91 and 118 days, etc. Frequencies higher than the Nyquist frequency can masquerade as lower frequencies, as well.

The autocorrelation function (also called the autocovariance), C_j, is calculated by

$$C_j = \frac{1}{N-j} \sum_{q=1}^{N-j} Y_q \; Y_{q+j} \qquad j = 0,\ 1,\ 2 \ldots m, \text{ where } m < N \qquad (17)$$

where C_j is the autocovariance of time lag $\tau = j\Delta t$ and Y_q is either the time sequence of the data $Y_q = y_q$, $q = 1,N$ (spaced Δt apart) (Blackman and Tukey, 1958) or is the time sequence of the data with the mean removed $Y_q = y_q - \bar{y}$, (Peebles, 1980). I strongly recommend the latter procedure, since if one removes the mean, the zeroth term of the autocorrelation function is approximately the variance of the data. This will allow some estimate of the significance of the results. BEWARE of plots where the C_0 term is offscale. It generally means that none of the remaining terms are significant.

The autocorrelation function may be computed up to m = (N-1) terms; however, the larger time lags have fewer and fewer data points, so it is safer to use m = N/2. Our JGR page example is drawn in Figure 9. Here, as in the Fourier analysis, signal peaks are seen at 3 months and 6 months. In addition, there may be significant signal at 18 months, 22 months, 36 months, and 59 months. The time domain is obviously more useful for looking for signal at longer periods (since the lag time is linear in months); the frequency domain (Fourier or power spectral density) is most accurate for determination of shorter-period variations. Significance may be estimated by recalling that (if the mean is removed) $C_0 = \sigma^2$; thus if C_j/C_0 is $\gtrsim 0.5$, there is a significant signal at that time lag. Just as in the Fourier analysis, none of the peaks are significant.

One may have a more quantitative estimate of the significance of each term C_j by recalling the definition of C_j (with means removed):

$$C_j \equiv \frac{1}{N-j} \sum_{q=1}^{N-j} (y_q - \bar{y})(y_{j+q} - \bar{y}) \qquad j = 0, 1, \ldots m \tag{18}$$

which is formally the same definition as the covariance $\sigma_{xy}{}^2$, with y_{j+q} taking the part of x_q. Thus if we invent a statistic $p_j = C_j/C_0$, it is formally identical to the r (linear correlation coefficient) statistic, and should behave similarly. Thus we may use r-tables, such as A2, to quantify the significance of each term C_j. This is really quite powerful. In our example (Figure 9), $C_0 = 7158$. The 3-month lag value is 2372, making the p-statistic 0.33. One would assume that the number of degrees of freedom would be N - j = 165, making that statistic highly

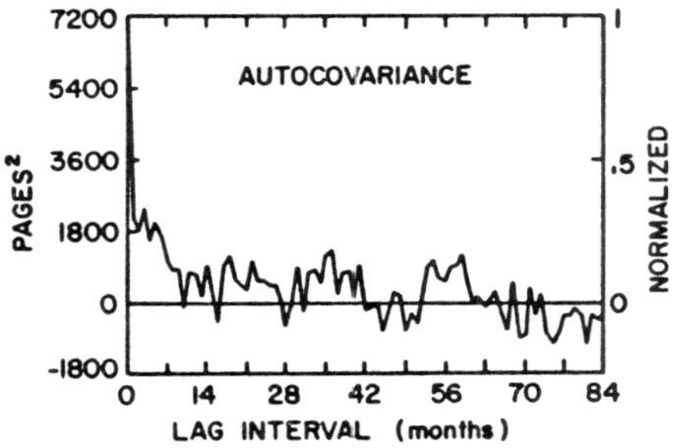

Figure 9. Autocorrelation function of the JGR page data. Note that low-frequency (large time lag) signals are relatively simple to determine. Since the mean has been subtracted, $C_0 = \sigma^2 = 7158$.

significant (> 99.9% confidence). On the other hand, random data have a tendency to persist, so that autocorrelations of truly random processes (such as these) often have a form similar to 1/j, with a peak at C_0, falling rapidly until $C_j \to 0$ as $j \to m$. This suggests using not (N - j) but j as the number of degrees of freedom (the number of data points of shift between the numerator and the denominator). Using that yields less than 50% confidence, which is more reasonable, considering our superposed epoch results (Figure 7).

Power Spectral Density

As indicated earlier, the power spectral density is just the cosine Fourier transform of the autocorrelation function. Since the autocorrelation function is symmetric about zero ($C(\tau) = C(-\tau)$), only cosine terms remain, and the C_j's are folded around zero. The power spectral density P_j is calculated by

$$P(jf_0) \equiv P_j = 2C_0/m + 2C_m/m \cos(2m\pi j/N) + \frac{4}{m}\sum_{q=1}^{m-1} C_q \cos(2q\pi j/N); \qquad j = 0, 1, \ldots m \tag{19}$$

where f_0 = the fundamental frequency = $1/(N\Delta t)$. One recovers $P_j \approx (A_j{}^2 + B_j{}^2)$ if $m = N/2$. Some authors multiply P_j by Δt; however, without it, the units of P_j are the same as C_j, i.e., (data units)^2. Thus, to test for significance, one may look at $P_j/2\sigma^2 = P_j/2C_0$ and see whether this value is $\gtrsim 0.5$. Figure 6b (bottom) shows the power spectral density for this page number example. The Power Spectral Density differs significantly from the sum of the squares of the Fourier components only in the lowest frequencies.

Cross-correlation analysis

Often two data streams are linearly correlated, but with one data set leading or lagging the other by an unknown time shift. Therefore to answer question #2, what time lag optimizes the correlation, one performs a cross-correlation analysis. Similar to an autocorrelation analysis, the data sets are multiplied together with a time lead or lag applied. Unlike the autocorrelation funtion, the cross-correlation function, also called the cross-covariance, $C_{xy}(\tau)$, is not generally symmetric around $\tau = 0$, so the Fourier transform of it, the cross-spectrum, will have both sine and cosine terms. The cross-covariance $C_{xy}(\tau)$ is calculated by

$$C_{xy}(\tau) = C_{xy}(j\Delta t) = \frac{1}{N-j}\sum_{q=1}^{N-j} X_q Y_{q+j} \tag{20a}$$

for positive j = (0, 1, 2...m) where m < N (i.e., X leads Y), and

$$= \frac{1}{N-|j|} \sum_{q=|j|+1}^{N} X_q \, Y_{q+j} \tag{20b}$$

for negative j's (-1, -2, -3...-m).

As before, X and Y may either be the data values (Blackman, 1965) or the data values with the individual means removed (Peebles, 1980). In my notation, a positive j (positive τ) indicates that X leads Y (Y lags X) by $\tau = j\Delta t$.

We saw that if the mean was removed, the zeroth autocorrelation function $C(0) \simeq \sigma^2$. Similarly, the zeroth cross-covariance $C_{xy}(0) \simeq \sigma_{xy}{}^2$ if the means are removed. Since you will recall that the linear correlation coefficient $r = \sigma_{xy}{}^2/\sigma_x\sigma_y = C_{xy}(0)/\sigma_x\sigma_y$, finding the time lag $j\Delta t$ with the highest cross-covariance $C_{xy}(j)$ will also maximize the linear correlation coefficient r. In fact, the correlation coefficient will be approximately $C_{xy}(\tau)/\sigma_x\sigma_y$. (The approximation occurs because σ_x and σ_y are calculated over the entire data set, not just the $N - |j|$ points.) One therefore may use the r-distribution to assess the significance of the C_{xy}'s. The number of degrees of freedom would be $N - |j| - 2$, for two means were used.

Example of Auto- and Cross-Correlation. Figure 10a (adapted from Slavin and Holzer, 1979), shows eroded magnetic flux $d\Phi_e/dt$ (transferred from the dayside to the nightside) (top), estimated from VB_s, and the returned magnetic flux $d\Phi_r/dt$ (transferred from the nightside to the dayside) (bottom), estimated from the AL index. It is apparent that both functions are quasiperiodic (with a period of roughly three hours) and are correlated with one another (with a time lag of about an hour). We may quantify these estimates by using auto- and cross-correlation analysis.

Figure 10b shows the autocorrelation statistic $p_j = C_j/C_0$ versus time for both the eroded (top) and returned (bottom) fluxes. Significant signal is seen at 3, 6, and 8.5 hours. One also sees clearly that the autocovariance, although it may be computed up to N-1 lags, is most stable only for N/2 lags. Figure 10c shows the Power Spectral Densities for the eroded and returned flux example. Unlike the JGR page example,, where the maximum power spectral density (excluding P_0) was roughly 1/10-1/7 C_0 (compare Figure 6b and 9), here the maximum PSD is on the order of C_0, showing a very strong periodicity at ~ 3 hours. Figure 10d shows the cross-correlation statistic $p_{xy}(j) = C_{xy}(j)/\sigma_x\sigma_y$ (solid) and r(j) (dashed). The difference occurs in the normalization: p_{xy} is normalized by $\sigma_x\sigma_y$, the sample standard deviation over the entire data set, whereas $r(j) = C_{xy}(j)/\sigma_x'\sigma_y'$, where $\sigma_x'(\sigma_y')$ is the sample standard deviation only over the subset of x(y) actually used in evaluating the crosscovariance. The agreement between $p_{xy}(j)$ and r(j), naturally, is less exact for larger time lags. The highest values occur at time lags of about -1.5 hours and -4.5 hours, i.e., the returned flux leading the

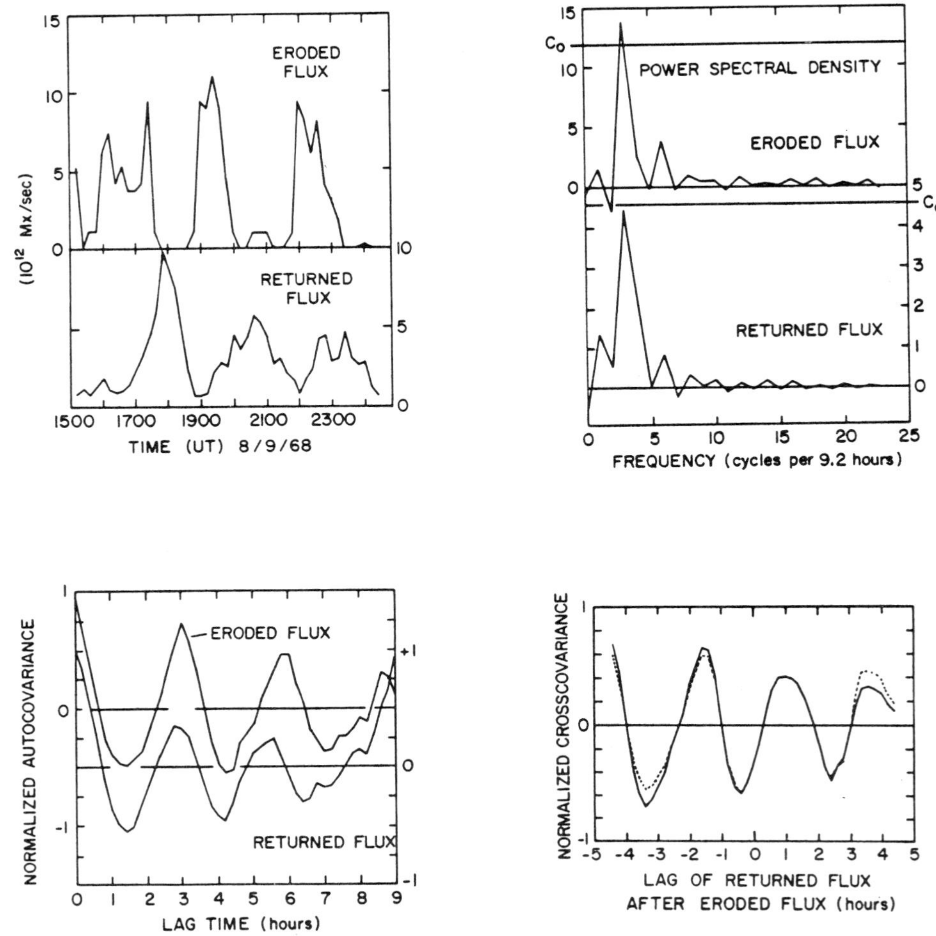

Figure 10. a (top left): Magnetic flux transferred from the dayside to the nightside $d\Phi_e/dt$, estimated from VB_s, and magnetic flux returned from the nightside to the dayside $d\Phi_r/dt$, estimated from AL (adapted from Slavin and Holzer, 1979). b (bottom left): Autocorrelation of $d\Phi_e/dt$ and $d\Phi_r/dt$. c (top right): Power Spectral Densities of $d\Phi_e/dt$ and $d\Phi_r/dt$. d (bottom right): Cross-correlation of $d\Phi_e/dt$ and $d\Phi_r/dt$. Solid: $p_{xy}(\tau)$; dashed, $r(\tau)$.

eroded flux. This is clearly contrary to our expectation of a positive time lag of about an hour, and, in fact, a significant signal is also seen at +1 hour. The reason that the results are unreasonable (common sense rule #3), is that both functions, for this short time period, are quite periodic, and it is impossible to determine a unique time lag for two exactly periodic functions. In this example, the largest cross-covariances occur when the sharp peak (the first) in the returned flux is multiplied by either of the two sharp peaks (the latter) in eroded flux, yielding the largest cross-covariance for an unreasonable time lag.

SMOOTHING THE DATA

The answer to question #7, "What is the best way to smooth the data", may be addressed by smoothing techniques. The definition of smoothing, according to Blackman and Tukey (1958) is the forming of moving linear combinations of unit total weight. "Smoothing and decimation" forms discrete moving linear combinations of k terms, but then eliminates all but every kth value. Averaging, then, at least formally is a smoothing technique.

One term that is often misused is "running average". A true running average does not reduce the number of data points. A five-point running average is defined to be

$$X'(j) = \frac{1}{5} \sum_{i=-2}^{2} X(j + i), \tag{21}$$

with suitable modifications at the end points. A simple or sequential five-point average, often incorrectly called a running average, on the other hand, does reduce the number of data points (by a factor of five):

$$X'(j) = \frac{1}{5} \sum_{i=1}^{5} X\big(5(j - 1) + i\big). \tag{22}$$

Thus a running average is a smoothing procedure; a simple average is a smoothing and decimation procedure.

The set of weights used in the above example is boxlike: {1/5, 1/5, 1/5, 1/5, 1/5}. When we then take the Fourier transform of the convolved data (data having undergone a smoothing function), the frequency response will have significant side lobes, since the Fourier transform of a box is (sin jx)/jx (Blackman and Tukey, 1958; Blackman, 1965; Peebles, 1980). In that sense the data won't really be any "smoother" than they were.

The solution to that problem is to use a set of weights that is peaked at the center, e.g., {1/16, 1/4, 3/8, 1/4, 1/16}, {1/4, 1/2, 1/4}, etc. Binomial expansions such as these are particularly simple and have predictable effects on the Fourier transform. The latter 3-point weight system is called a "Hanning window" after Julius von Hann. A smoothing function that is similar but better since it has side lobes 1/3 as high is {.23, .54, .23}, called a Hamming window after R. W. Hamming. If one performs a smoothing function before autocorrelation or Fourier analysis, it is called prewhitening. Prewhitening reduces high-frequency noise, so it can reduce aliasing; however, it also increases the effective Nyquist frequency. For a thorough discussion, see Blackman and Tukey (1958), Hamming (1962) or Blackman (1964).

CONCLUSIONS

Several statistical procedures useful for space data analysis (both exploratory and confirmatory) have been presented, including linear correlation, autocorrelation and cross-correlation, Fourier analysis and power spectral density, superposed epoch analysis, and smoothing. If possible, I have used authoritative tests to assess the significance of the results. In many of the procedures, however, no authoritative tests exist, and I have suggested ways for experimenters to determine (at least qualitatively) the significance of their results.

Data presented without some form of error analysis are often valueless to the readers, since they have no way to assess whether a particular bump on a spectrum or on a superposed epoch analysis is real or merely a statistical fluctuation. For example, although many linear correlations have been published, virtually none of these include the uncertainty in the intercept and slope σ_a and σ_b. Without these values, the significance of the fitted parameters cannot be assessed.

ACKNOWLEDGMENTS

I wish to thank D. L. Reasoner for inciting my interest in statistical analysis and to T. W. Hill for helpful comments on the manuscript. This work was supported by the Atmospheric Sciences Section of the National Academy of Sciences under grant ATM-8017316 and by the National Aeronautics and Space Administration under grant NGR-44-006-137.

REFERENCES

Akasofu, S.-I.: 1980, in S.-I. Akasofu (ed.), "Dynamics of the Magnetosphere," D. Reidel, Boston, pp. 447-460.

Akasofu, S.-I. and S. Chapman: 1972, "Solar-Terrestrial Physics," Oxford University Press, London.

Arnoldy, R. L.,: 1971, J. Geophys. Res. 76, pp. 5189-5201.

Bevington, P. R.: 1969, "Data Reduction and Error Analysis for the Physical Sciences," McGraw-Hill, New York.

Blackman, R. B.: 1965, "Linear Data-Smoothing and Prediction in Theory and Practice," Addison-Wesley, Reading, Mass.

Blackman, R. B. and J. W. Tukey: 1958, "The Measurement of Power Spectra," Dover, New York.

Burch, J. L.: 1974, Rev. Geophys. Space Phys. 12, pp. 363-378.

Caan, M. N., R. L. McPherron, and C. T. Russell: 1975, J. Geophys. Res. 80, pp. 191-194.

Chapman, S. and J. Bartels: 1940, "Geomagnetism," Oxford University Press, London.

Chree, C.: 1912, Phil. Trans. London (A) 212, pp. 75-116; 1913, 213, pp. 245-277.

Clauer, C. R.: 1982 (submitted), Planet. Space Sci.

Dessler, A. J.: 1974, in W. R. Bardeen and S. P. Maran (eds.), "Possible Relationships Between Solar Activity and Meteorological Phenomena," NASA SP-366, pp. 187-191.

Doyle, M. A. and W. J. Burke: 1982, in J. C. Foster and F. T. Berkey (eds.), "Origins of Plasmas and Electric Fields in the Magnetosphere," American Geophysical Union, Washington, pp. 34-1-34-3.

Dungey, J. W.: 1961, Phys. Rev. Lett. 6, p. 47.

Hamming, R. W.: 1962, "Numerical Methods for Scientists and Engineers", McGraw-Hill, New York.

Huff, D.: 1954, "How to Lie with Statistics," Norton & Co., New York.

Lewis-Beck, M. S.: 1980, "Applied Regression: An Introduction," Sage Publications, Beverly Hills, CA.

Peebles, P. Z.: 1980, "Probability, Random Variables, and Random Signal Principles," McGraw-Hill, New York.

Pulliam, D. M., H. R. Anderson, K. Stamnes, and M. H. Rees: 1981, J. Geophys. Res. 86, pp. 2397-2404.

Rassbach, M. E., A. J. Dessler, and A. G. W. Cameron: 1966, J. Geophys. Res. 71, pp. 4141-4146.

Reiff, P. H., R. W. Spiro and T. W. Hill: 1981, J. Geophys. Res. 86, pp. 7639-7648.

Roberts, W. O. and R. H. Olson: 1973, J. Atmos. Sci. 30, pp. 135-140.

Russell, C. T., G. L. Siscoe, and E. J. Smith: 1980, Geophys. Res. Lett. 7, pp. 381-184.

Sheeley, N. R., Jr. and J. W. Harvey: 1981, Solar Phys. 70, pp. 237-249.

Slavin, J. A. and R. E. Holzer: 1979, in "Quantitative Modeling of Magnetospheric Processes," ed. W. P. Olson, AGU, Washington, pp. 423-435.

Vasyliunas, V. M. and A. J. Dessler: 1981, J. Geophys. Res. 86, pp. 8435-8446.

ν \ P	0.99	0.98	0.95	0.90	0.80	0.70	0.60	0.50
1	0.00016	0.0063	0.00393	0.0158	0.0642	0.148	0.275	0.455
2	0.0100	0.0202	0.0515	0.105	0.223	0.357	0.511	0.693
3	0.0383	0.0617	0.117	0.195	0.335	0.475	0.623	0.789
4	0.0742	0.107	0.178	0.266	0.412	0.549	0.688	0.839
5	0.111	0.150	0.229	0.322	0.469	0.600	0.731	0.870
6	0.145	0.189	0.273	0.367	0.512	0.638	0.762	0.891
7	0.177	0.223	0.310	0.405	0.546	0.667	0.785	0.907
8	0.206	0.254	0.342	0.436	0.574	0.691	0.803	0.918
9	0.232	0.281	0.369	0.463	0.598	0.710	0.817	0.927
10	0.257	0.306	0.394	0.487	0.618	0.727	0.830	0.934
11	0.278	0.328	0.416	0.507	0.635	0.741	0.840	0.940
12	0.298	0.348	0.436	0.525	0.651	0.753	0.848	0.945
13	0.316	0.367	0.453	0.542	0.664	0.764	0.856	0.949
14	0.333	0.383	0.469	0.556	0.676	0.773	0.863	0.953
15	0.349	0.399	0.484	0.570	0.687	0.781	0.869	0.956
16	0.363	0.413	0.498	0.582	0.697	0.789	0.874	0.959
17	0.377	0.427	0.510	0.593	0.706	0.796	0.879	0.961
18	0.390	0.439	0.522	0.604	0.714	0.802	0.883	0.963
19	0.402	0.451	0.532	0.613	0.722	0.808	0.887	0.965
20	0.413	0.462	0.543	0.622	0.729	0.813	0.890	0.967
22	0.434	0.482	0.561	0.638	0.742	0.823	0.897	0.970
24	0.452	0.500	0.577	0.652	0.753	0.831	0.902	0.972
26	0.469	0.516	0.592	0.665	0.762	0.838	0.907	0.974
28	0.484	0.530	0.605	0.676	0.771	0.845	0.911	0.976
30	0.498	0.544	0.616	0.687	0.779	0.850	0.915	0.978
32	0.511	0.556	0.627	0.696	0.786	0.855	0.918	0.979
34	0.523	0.567	0.637	0.704	0.792	0.860	0.921	0.980
36	0.534	0.577	0.646	0.712	0.798	0.864	0.924	0.982
38	0.545	0.587	0.655	0.720	0.804	0.868	0.926	0.983
40	0.554	0.596	0.663	0.726	0.809	0.872	0.928	0.983
42	0.563	0.604	0.670	0.733	0.813	0.875	0.930	0.984
44	0.572	0.612	0.677	0.738	0.818	0.878	0.932	0.985
46	0.580	0.620	0.683	0.744	0.822	0.881	0.934	0.986
48	0.587	0.627	0.690	0.749	0.825	0.884	0.936	0.986
50	0.594	0.633	0.695	0.754	0.829	0.886	0.937	0.987
60	0.625	0.662	0.720	0.774	0.844	0.897	0.944	0.989
70	0.649	0.684	0.739	0.790	0.856	0.905	0.949	0.990
80	0.669	0.703	0.755	0.803	0.865	0.911	0.952	0.992
90	0.686	0.718	0.768	0.814	0.873	0.917	0.955	0.993
100	0.701	0.731	0.779	0.824	0.879	0.921	0.958	0.993
120	0.724	0.753	0.798	0.839	0.890	0.928	0.962	0.994
140	0.743	0.770	0.812	0.850	0.898	0.934	0.965	0.995
160	0.758	0.784	0.823	0.860	0.905	0.938	0.968	0.996
180	0.771	0.796	0.833	0.868	0.910	0.942	0.970	0.996
200	0.782	0.806	0.841	0.874	0.915	0.945	0.972	0.997

Table A1. χ^2 distribution. Values of the reduced chi-square $\chi_\nu{}^2 = \chi^2/\nu$ corresponding to the probability $P_\chi(\chi^2,\nu)$ of (continued to next page)

ν P	0.40	0.30	0.20	0.10	0.05	0.02	0.01	0.001
1	0.708	1.074	1.642	2.706	3.841	5.412	6.635	10.827
2	0.916	1.204	1.609	2.303	2.996	3.912	4.605	6.908
3	0.982	1.222	1.547	2.084	2.605	3.279	3.780	5.423
4	1.011	1.220	1.497	1.945	2.372	2.917	3.319	4.617
5	1.026	1.213	1.458	1.847	2.214	2.678	3.017	4.102
6	1.035	1.205	1.426	1.774	2.099	2.506	2.802	3.743
7	1.040	1.198	1.400	1.717	2.010	2.375	2.639	3.475
8	1.044	1.191	1.379	1.670	1.938	2.271	2.511	3.266
9	1.046	1.184	1.360	1.632	1.880	2.187	2.407	3.097
10	1.047	1.178	1.344	1.599	1.831	2.116	2.321	2.959
11	1.048	1.173	1.330	1.570	1.789	2.056	2.248	2.842
12	1.049	1.168	1.318	1.546	1.752	2.004	2.185	2.742
13	1.049	1.163	1.307	1.524	1.720	1.959	2.130	2.656
14	1.049	1.159	1.296	1.505	1.692	1.919	2.082	2.580
15	1.049	1.155	1.287	1.487	1.666	1.884	2.039	2.513
16	1.049	1.151	1.279	1.471	1.644	1.852	2.000	2.453
17	1.048	1.148	1.271	1.457	1.623	1.823	1.965	2.399
18	1.048	1.145	1.264	1.444	1.604	1.797	1.934	2.351
19	1.048	1.142	1.258	1.432	1.586	1.773	1.905	2.307
20	1.048	1.139	1.252	1.421	1.571	1.751	1.878	2.266
22	1.047	1.134	1.241	1.401	1.542	1.712	1.831	2.194
24	1.046	1.129	1.231	1.383	1.517	1.678	1.791	2.132
26	1.045	1.125	1.223	1.368	1.496	1.648	1.755	2.079
28	1.045	1.121	1.215	1.354	1.476	1.622	1.724	2.032
30	1.044	1.118	1.208	1.342	1.459	1.599	1.696	1.990
32	1.043	1.115	1.202	1.331	1.441	1.578	1.671	1.953
34	1.042	1.112	1.196	1.321	1.429	1.559	1.649	1.919
36	1.042	1.109	1.191	1.311	1.417	1.541	1.628	1.888
38	1.041	1.106	1.186	1.303	1.405	1.525	1.610	1.861
40	1.041	1.104	1.182	1.295	1.394	1.511	1.592	1.835
42	1.040	1.102	1.178	1.288	1.384	1.497	1.576	1.812
44	1.039	1.100	1.174	1.281	1.375	1.485	1.562	1.790
46	1.039	1.098	1.170	1.275	1.366	1.473	1.548	1.770
48	1.038	1.096	1.167	1.269	1.358	1.462	1.535	1.751
50	1.038	1.094	1.163	1.263	1.350	1.452	1.523	1.733
60	1.036	1.087	1.150	1.240	1.318	1.410	1.473	1.660
70	1.034	1.808	1.139	1.222	1.293	1.377	1.435	1.605
80	1.032	1.076	1.130	1.207	1.273	1.351	1.404	1.560
90	1.031	1.072	1.123	1.195	1.257	1.329	1.379	1.525
100	1.029	1.069	1.117	1.185	1.243	1.311	1.358	1.494
120	1.027	1.063	1.107	1.169	1.221	1.283	1.325	1.446
140	1.026	1.059	1.099	1.156	1.204	1.261	1.299	1.410
160	1.024	1.055	1.093	1.146	1.191	1.243	1.278	1.381
180	1.023	1.052	1.087	1.137	1.179	1.228	1.261	1.358
200	1.022	1.050	1.083	1.130	1.170	1.216	1.247	1.338

Table A1. χ^2 distribution (continued). exceeding χ^2 vs. the number of degrees of freedom ν. (From Bevington, 1969; used with permission).

N	P 0.50	0.20	0.10	0.050	0.020	0.010	0.005	0.002	0.001
3	0.707	0.951	0.988	0.997	1.000	1.000	1.000	1.000	1.000
4	0.500	0.800	0.900	0.950	0.980	0.990	0.995	0.998	0.999
5	0.404	0.687	0.805	0.878	0.934	0.959	0.974	0.986	0.991
6	0.347	0.608	0.729	0.811	0.882	0.917	0.942	0.963	0.974
7	0.309	0.551	0.669	0.754	0.833	0.875	0.906	0.935	0.951
8	0.281	0.507	0.621	0.707	0.789	0.834	0.870	0.905	0.925
9	0.260	0.472	0.582	0.666	0.750	0.798	0.836	0.875	0.898
10	0.242	0.443	0.549	0.632	0.715	0.765	0.805	0.847	0.872
11	0.228	0.419	0.521	0.602	0.685	0.735	0.776	0.820	0.847
12	0.216	0.398	0.497	0.576	0.658	0.708	0.750	0.795	0.823
13	0.206	0.380	0.476	0.553	0.634	0.684	0.726	0.772	0.801
14	0.197	0.365	0.458	0.532	0.612	0.661	0.703	0.750	0.780
15	0.189	0.351	0.441	0.514	0.592	0.641	0.683	0.730	0.760
16	0.182	0.338	0.426	0.497	0.574	0.623	0.664	0.711	0.742
17	0.176	0.327	0.412	0.482	0.558	0.606	0.647	0.694	0.725
18	0.170	0.317	0.400	0.468	0.543	0.590	0.631	0.678	0.708
19	0.165	0.308	0.389	0.456	0.529	0.575	0.616	0.662	0.693
20	0.160	0.299	0.378	0.444	0.516	0.561	0.602	0.648	0.679
22	0.152	0.284	0.360	0.423	0.492	0.537	0.576	0.622	0.652
24	0.145	0.271	0.344	0.404	0.472	0.515	0.554	0.599	0.629
26	0.138	0.260	0.330	0.388	0.453	0.496	0.534	0.578	0.607
28	0.133	0.250	0.317	0.374	0.437	0.479	0.515	0.559	0.588
30	0.128	0.241	0.306	0.361	0.423	0.463	0.499	0.541	0.570
32	0.124	0.233	0.296	0.349	0.409	0.449	0.484	0.526	0.554
34	0.120	0.225	0.287	0.339	0.397	0.436	0.470	0.511	0.539
36	0.116	0.219	0.279	0.329	0.386	0.424	0.458	0.498	0.525
38	0.113	0.213	0.271	0.320	0.376	0.413	0.446	0.486	0.513
40	0.110	0.207	0.264	0.312	0.367	0.403	0.435	0.474	0.501
42	0.107	0.202	0.257	0.304	0.358	0.393	0.425	0.463	0.490
44	0.104	0.197	0.251	0.297	0.350	0.384	0.416	0.453	0.479
46	0.102	0.192	0.246	0.291	0.342	0.376	0.407	0.444	0.469
48	0.100	0.188	0.240	0.285	0.335	0.368	0.399	0.435	0.460
50	0.098	0.184	0.235	0.279	0.328	0.361	0.391	0.427	0.451
60	0.089	0.168	0.214	0.254	0.300	0.330	0.358	0.391	0.414
70	0.082	0.155	0.198	0.235	0.278	0.306	0.332	0.363	0.385
80	0.077	0.145	0.185	0.220	0.260	0.286	0.311	0.340	0.361
90	0.072	0.136	0.174	0.207	0.245	0.270	0.293	0.322	0.341
100	0.068	0.129	0.165	0.197	0.232	0.256	0.279	0.305	0.324

Table A2. Linear correlation coefficient. The linear-correlation coefficient r vs. the number of observations N and the corresponding probability $P_c(r,N)$ of exceeding r in a random sample of observations taken from an uncorrelated parent population (ρ = 0). (From Bevington, 1969; used with permission).

THERMOSPHERIC DYNAMICS AND ELECTRODYNAMICS

Arthur D. Richmond
Space Environment Laboratory
National Oceanic and Atmospheric Administration
Boulder, Colorado 80303

1. INTRODUCTION

The thermosphere is that part of the earth's atmosphere at heights between about 90 km and 500 km. The lower boundary is defined to lie at the temperature minimum which occurs on the average around 90 km, called the mesopause. Figure 1 shows the mean temperature structure of the atmosphere according to the U.S. Standard Atmosphere, 1976. The temperature increases rapidly with height in the lower thermosphere, and becomes asymptotic to a value on the order of 1000 K at high altitudes. This asymptotic behavior is associated with very long atomic mean free

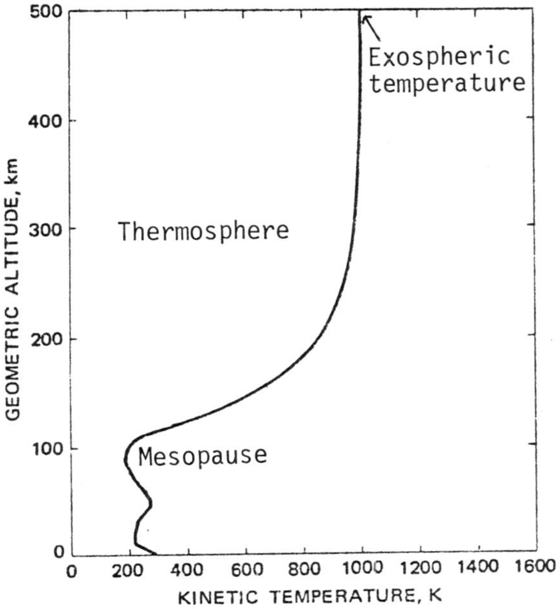

Figure 1. Atmospheric temperature structure.

R. L. Carovillano and J. M. Forbes (eds.), Solar-Terrestrial Physics, 523–607.

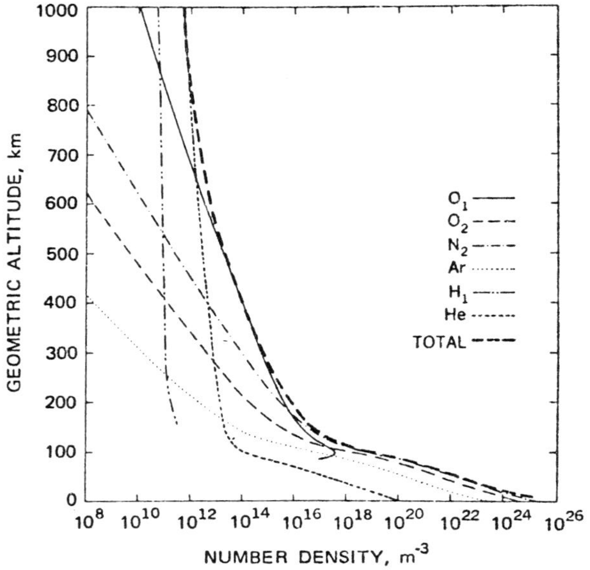

Figure 2. Atmospheric density and composition structure.

paths at high altitudes which allow very rapid vertical transfer of heat. Above about 500 km, the mean free paths become so long that atoms can go into ballistic orbits, and fluid treatments become inapplicable. This uppermost portion of the earth's neutral atmosphere, lying above the thermosphere, is called the exosphere, and will not be treated in these lectures. The asymptotic temperature of the thermosphere at high altitudes is referred to as the exospheric temperature.

Long mean free paths mean that heat conduction, viscosity, and diffusion become important processes in the thermosphere, unlike the lower atmosphere where they can usually be ignored. Diffusion, in fact, becomes so rapid that different constituents tend to separate in the gravitational field according to their mass. Figure 2 shows how the number densitites of different species vary with altitude in the atmosphere. Below the thermosphere the atmosphere is well mixed by turbulence, so that the decrease of density with height is similar for all important species (N_2, O_2, Ar, He). In the thermosphere some additional species become important, especially O and H, which are produced by dissociation of O_2 and of hydrogen-containing molecules. The rates at which densities decrease with altitude are different for each species, so that the composition varies with altitude, with O becoming the dominant species in the upper thermosphere and He and H dominating in the upper exosphere. The fact that all rates of density decrease with altitude are less steep in the thermosphere than in the lower atmosphere is related to the relatively high temperature, as will become apparent in Section 3.

In addition to the various neutral species shown in Fig. 2, the thermosphere also contains ionized atoms and molecules and free electrons, due to the ionizing action of solar ultraviolet radiation and of energetic particles impinging on the atmosphere from the magnetosphere. This ionized component is referred to as the ionosphere. The maximum density of electrons and of ions is on the order of 10^{12} m^{-3} at 300 km during the day, which is nearly three orders of magnitude less than the number density of neutrals at this height. In spite of their relatively low density, we shall see that the presence of ions has an important influence on the dynamics of the neutral gas. The ionosphere is electrically conducting, and carries measurable electric currents. These currents are particularly important at high latitudes, where they are coupled with currents flowing in the magnetosphere.

The high thermospheric temperature is produced by absorption of solar ultraviolet and X-ray radiation, and also by energy deposited from the magnetosphere. These energy sources are highly variable. The short-wavelength solar electromagnetic radiation changes substantially with the level of solar activity. Energy deposition from the magnetosphere changes by orders of magnitude between very quiet and higly active magnetospheric conditions. Consequently, the state of the thermosphere is also highly variable. The exospheric temperature, for example, can range from 600 K at night during low solar and magnetospheric activity, to 2000 K at day during high activity. The composition and the wind systems of the thermosphere are likewise highly variable. Figures 1 and 2 therefore represent only rather crudely the state of the thermosphere at any given time.

The practical interest of studying the thermosphere has been primarily for two reasons. First, the thermosphere has a strong influence on the ionosphere, which in turn influences the quality and even the feasibility of long-distance radio wave progagation. Second, near-earth satellites traverse the thermosphere and exosphere, and are affected by variations of thermospheric density. The drag on these satellites changes their orbits, and, eventually, causes them to reenter the denser atmosphere. In addition to these practical concerns, study of the thermosphere is of scientific interest because of the wide variety of physical phenomena which occur there.

This chapter focuses on the theory of dynamical and electrodynamical processes in the thermosphere. For a broader overview of the thermosphere, there exist a number of articles and books to which the reader is referred. A non-mathmematical description of thermospheric physics is presented by Roble (1977). There are several introductory books to the study of upper atmospheric phenomena, among which those by Rishbeth and Garriott (1969), Beer (1976), and Hargreaves (1979) will be found helpful. More advanced treatments of thermospheric theory can be found in books by Whitten and Poppoff (1971), Banks and Kockarts (1973a, b), Hines et al. (1974), and Kato (1980).

2. BASIC EQUATIONS

The basic equations needed to study thermospheric dynamics and electrodynamics can be derived from the theory of non-uniform gases, as developed, for example, by Chapman and Cowling (1970). We shall not go through the basic derivations here, but instead merely write down the rather general forms of the equations we shall need. Further development and discussion of the basic equations for thermospheric studies are contained in the books by Whitten and Poppoff (1971) and by Banks and Kockarts (1973b). A full list of notation appears at the end of this chapter. SI units are used throughout.

2.1. Mass continuity

For each species the time rate of change of mass density plus the divergence of mass flux must equal the net production rate associated with chemical production and loss processes. For the ith species

$$m_i \frac{\partial n_i}{\partial t} + \nabla\cdot(m_i n_i \vec{v}_i) = S_i \qquad (2.1)$$

where m_i is the mass of one particle, n_i is the number density, $\vec{v}_i$ the mean velocity, and S_i the net mass density production rate (or loss rate, if negative). The total mass density is

$$\rho = \sum_i m_i n_i \qquad (2.2)$$

and the mass-averaged velocity is defined as

$$\vec{v} = \frac{1}{\rho} \sum_i m_i n_i \vec{v} \qquad (2.3)$$

Then if (2.1) is summed over all species, we obtain

$$\frac{\partial \rho}{\partial t} + \nabla\cdot(\rho\vec{v}) = 0 \qquad (2.4)$$

since the net production rate of all species combined must cancel for mass to be conserved. Eq. (2.4) can also be expressed in the form

$$\frac{D\rho}{Dt} + \rho\, \nabla\cdot\vec{v} = 0 \qquad (2.5)$$

where the total time derivative D/Dt is defined as

$$\frac{D}{Dt} = \frac{\partial}{\partial t} + \vec{v}\cdot\nabla \qquad (2.6)$$

which represents the time derivative in the frame of reference of the moving fluid.

2.2. Momentum balance

Inertial reference frame. The momentum equation in a non-rotating frame of reference is

$$\rho \frac{D\vec{v}}{Dt} + \nabla \cdot \underline{\underline{P}} = \rho \vec{g}^{*} + \vec{J} \times \vec{B} + \rho_e \vec{E} \tag{2.7}$$

where $\underline{\underline{P}}$ is the generalized pressure tensor, $\vec{g}^{*}$ is the acceleration of gravity in the non-rotating frame, $\vec{B}$ and $\vec{E}$ are the magnetic field and electric field, and where $\vec{J}$ and ρ_e are the electric current and charge densities defined by

$$\vec{J} = \sum_i q_i n_i \vec{v}_i \tag{2.8}$$

$$\rho_e = \sum_i q_i n_i \tag{2.9}$$

with q_i being the electric charge of the ith species. For our purposes the last term in (2.7) involving the electric force on an electrostatic charge density is entirely negligible, essentially because quasi-static net charges are extremely small in a plasma due to near-perfect charge cancellation between positive and negative charges. We ignore this term hereafter. The pressure tensor $\underline{\underline{P}}$, which represents momentum flux by molecular motions, is dimensioned 3 x 3 and is symmetric about its diagonal (i.e., $P_{ij} = P_{ji}$). It contains the effects of both the usual scalar pressure and viscosity, as will be shown later. The term P_{ij} represents flux of the j component of momentum in the i direction. The divergence of $\underline{\underline{P}}$ is a vector quantity of the form

$$\nabla \cdot \underline{\underline{P}} = \hat{x}\left(\frac{\partial P_{xx}}{\partial x} + \frac{\partial P_{yx}}{\partial y} + \frac{\partial P_{zx}}{\partial z}\right) + \hat{y}\left(\frac{\partial P_{xy}}{\partial x} + \frac{\partial P_{yy}}{\partial y} + \frac{\partial P_{zy}}{\partial z}\right) + \hat{z}\left(\frac{\partial P_{xz}}{\partial x} + \frac{\partial P_{yz}}{\partial x} + \frac{\partial P_{zz}}{\partial z}\right) \tag{2.10}$$

Conversion to rotating reference frame. To a first approximation the atmosphere rotates with the earth. The high rotational velocity, on the order of the speed of sound, makes a dominant contribution to $\vec{v}$ and therefore can tend to obscure the smaller changes in $\vec{v}$ associated with the dynamical processes of interest to us. It is usually convenient to convert the equations to a frame of reference rotating with the earth, and to reinterpret $\vec{v}$ as velocity measured in this rotating frame of reference. For consistency, the time derivative D/Dt should also be taken with respect to this rotating frame. It is not immediately obvious that there should be any difference in this time derivative for different frames of reference, since the derivative is in any case defined to be with respect to the moving fluid. Indeed, for scalar quantities there is no difference in the total time derivative for rotating and non-rotating frames. However, we can show that rotation does affect D/Dt when operating on vector quantities like $\vec{v}$.

To see this, first imagine that the earth is immersed in a constant vector field $\vec{A}$, and that the earth is rotating at the angular rate $\vec{\Omega}$ (7.29×10^{-5} radians/second) where the direction of $\vec{\Omega}$ is from the center of the earth through the North Pole. To an observer sitting on the earth, the vector $\vec{A}$ will appear to be continually changing direction, unless it happens to be parallel to $\vec{\Omega}$. The component of $\vec{A}$ perpendicular to $\vec{\Omega}$ will appear to make a complete rotation in the time $2\pi/\Omega$, or one sidereal day. The time rate of change of $\vec{A}$ as seen by the rotating observer is

$$\left(\frac{d\vec{A}}{dt}\right)_{rot} = - \vec{\Omega} \times \vec{A} \tag{2.11}$$

If we now generalize the situation so that $\vec{A}$ may vary in space and time and so that the observer may move through the field, then the time rate of change of $\vec{A}$ he measures will not necessarily be zero in the non-rotating frame of reference. Nevertheless, it is easy to see that the time rate of change of $\vec{A}$ which he measures will be different depending on whether he is rotating or not rotating. In fact, the difference in $d\vec{A}/dt$ as measured by identically moving observers, one of whom is rotating, is given by (2.11), that is,

$$\left(\frac{d\vec{A}}{dt}\right)_{rot} = \left(\frac{d\vec{A}}{dt}\right)_{nonrot} - \vec{\Omega} \times \vec{A} \tag{2.12}$$

So far we have not specified what translational motion the observer is to have. For the total time derivative D/Dt the motion is defined to be at the mass mean velocity of the fluid, so that

$$\left(\frac{D\vec{A}}{Dt}\right)_{nonrot} = \left(\frac{D\vec{A}}{Dt}\right)_{rot} + \vec{\Omega} \times \vec{A} \tag{2.13}$$

If we let $\vec{A}$ be $\vec{r}$, the radius vector from the center of the earth, then (2.13) gives

$$\left(\frac{D\vec{r}}{Dt}\right)_{nonrot} = \left(\frac{D\vec{r}}{Dt}\right)_{rot} + \vec{\Omega} \times \vec{r} \tag{2.14}$$

The total time derivative of the position vector is just the velocity in the particular frame of reference, so (2.14) can also be written

$$\vec{v}_{nonrot} = \vec{v}_{rot} + \vec{\Omega} \times \vec{r} \tag{2.15}$$

If we let $\vec{A}$ be $\vec{v}_{nonrot}$, then (2.13) and (2.15) give

$$\left(\frac{D\vec{v}_{nonrot}}{Dt}\right)_{nonrot} = \left(\frac{D\vec{v}_{rot}}{Dt}\right)_{rot} + \vec{\Omega} \times \left(\frac{D\vec{r}}{Dt}\right)_{rot} + \vec{\Omega} \times (\vec{v}_{rot} + \vec{\Omega} \times \vec{r})$$

$$= \left(\frac{D\vec{v}_{rot}}{Dt}\right)_{rot} + 2\vec{\Omega} \times \vec{v}_{rot} + \vec{\Omega} \times (\vec{\Omega} \times \vec{r}) \tag{2.16}$$

The term $2\vec{\Omega} \times \vec{v}_{rot}$ represents Coriolis acceleration, which has the effect of tending to deflect the wind velocity. The term $\vec{\Omega} \times (\vec{\Omega} \times \vec{r})$ represents centripetal acceleration. Like gravity, it is a function only of position and can be expressed as the gradient of a potential. It is usual practice to combine it with $\vec{g}^*$ to produce $\vec{g}$, the effective gravitational acceleration in the rotating system:

$$\vec{g} \equiv \vec{g}^* - \vec{\Omega} \times (\vec{\Omega} \times \vec{r}) = - \nabla\Phi \tag{2.17}$$

where Φ is the geopotential.

To convert (2.7) to the earth's rotating frame, we note that the pressure tensor and the electromagnetic force remain unchanged. Using (2.16) and (2.17), and dropping the subscripts "rot", we obtain

$$\rho \frac{D\vec{v}}{Dt} + 2\rho\vec{\Omega} \times \vec{v} + \nabla\cdot\underline{\underline{P}} = \rho\vec{g} + \vec{J} \times \vec{B} \tag{2.18}$$

2.3. Energy balance

If we follow the motion of a unit mass of fluid and evaluate the energy crossing its surface as well as other energy added to it, we find the following expression for the time rate of change of energy density:

$$\rho \frac{D}{Dt} \left(u + \frac{v^2}{2}\right) + \nabla\cdot(\underline{\underline{P}}\cdot\vec{v} + \vec{q}) = \rho\vec{v}\cdot\vec{g} + \rho Q + \vec{J}\cdot\vec{E} \tag{2.19}$$

where u is internal energy per unit mass, $\vec{q}$ is the molecular flux of internal energy, Q is the net heating rate per unit mass due to chemical and radiative processes, and $\vec{E}$ is the electric field. The vector $\underline{\underline{P}}\cdot\vec{v}$, which in component form is

$$\begin{aligned}\underline{\underline{P}}\cdot\vec{v} = {} & \hat{x}\,(P_{xx} v_x + P_{xy} v_y + P_{xz} v_z) \\ & + \hat{y}\,(P_{yx} v_x + P_{yy} v_y + P_{yz} v_z) \\ & + \hat{z}\,(P_{zx} v_x + P_{zy} v_y + P_{zz} v_z)\end{aligned} \tag{2.20}$$

represents the flux of kinetic energy associated with the flux of momentum. The term $\rho\vec{v}\cdot\vec{g}$ represents work done by gravity, while $\vec{J}\cdot\vec{E}$ is the rate of energy transfer to the material medium by the electromagnetic field, both in the form of heating and acceleration of the gas. For all practical purposes, (2.19) can be applied in either a rotating or non-rotating frame of reference, provided that all quantities are defined consistently with the chosen reference frame.

If we express $\vec{g}$ as $-\nabla\Phi$ and note that

$$\vec{v}\cdot\vec{g} = -\vec{v}\cdot\nabla\Phi = -\frac{D\Phi}{Dt} \tag{2.21}$$

then we see that (2.19) can also be written

$$\rho\frac{D}{Dt}\left(u + \frac{v^2}{2} + \Phi\right) + \nabla\cdot(\underline{\underline{P}}\cdot\vec{v} + \vec{q}) = \rho Q + \vec{J}\cdot\vec{E} \tag{2.22}$$

In addition to having a relation for the total energy, as in (2.19), it is useful to have separate relations for the internal and the kinetic forms of energy. A relation for kinetic energy can be derived from the momentum equation (2.18) by taking its scalar product with $\vec{v}$:

$$\rho\frac{D}{Dt}\left(\frac{v^2}{2}\right) + \vec{v}\cdot\nabla\cdot\underline{\underline{P}} = \rho\vec{v}\cdot\vec{g} + \vec{v}\cdot\vec{J}\times\vec{B} \tag{2.23}$$

This states that the time rate of change of kinetic energy equals the rate of work done on the fluid by the various forces. Subtracting (2.23) from (2.19) we find

$$\rho\frac{Du}{Dt} = -\underline{\underline{P}}:\nabla\vec{v} - \nabla\cdot\vec{q} + \rho Q + \vec{J}\cdot(\vec{E} + \vec{v}\times\vec{B}) \tag{2.24}$$

where

$$\begin{aligned}\underline{\underline{P}}:\nabla\vec{v} \equiv\ & P_{xx}\frac{\partial v_x}{\partial x} + P_{xy}\frac{\partial v_y}{\partial x} + P_{xz}\frac{\partial v_z}{\partial x} \\ & + P_{yx}\frac{\partial v_x}{\partial y} + P_{yy}\frac{\partial v_y}{\partial y} + P_{yz}\frac{\partial v_z}{\partial y} \\ & + P_{zx}\frac{\partial v_x}{\partial z} + P_{zy}\frac{\partial v_y}{\partial z} + P_{zz}\frac{\partial v_z}{\partial z}\end{aligned} \tag{2.25}$$

The quantity $\vec{J}\cdot(\vec{E} + v\times\vec{B})$ represents <u>Joule heating</u> of the gas.

2.4. Pressure and internal energy

The balance equations for mass, momentum, and energy are quite general, but additional relations between the variables are also required to form a complete set of equations. We get these additional relations from the kinetic theory of gases. First of all, we can define a scalar pressure as

$$p = (P_{xx} + P_{yy} + P_{zz})/3 \tag{2.26}$$

For a perfect gas, it is related to density ρ and temperature T by

$$p = nkT = \rho RT \tag{2.27}$$

$$n \equiv \sum_i n_i \tag{2.28}$$

$$R \equiv kn/\rho \tag{2.29}$$

where k is the Boltzmann constant and R is the gas "constant," which in fact depends on the mean molecular mass. Henceforth in speaking of pressure the scalar pressure p will be meant; the remainder of the pressure tensor will represent viscous effects.

The internal energy density u is composed of the kinetic energy of the molecules plus energy associated with internal degrees of freedom of the molecules, in particular rotation, which we label u_I. We can write

$$\rho u = \frac{3}{2} nkT + \rho u_I = \frac{3}{2} p + \rho u_I \tag{2.30}$$

where (3/2) kT is the mean kinetic energy of a molecule. In a state of thermodynamic equilibrium u_I is a simple function of temperature; normally we have

$$\begin{aligned} \rho u_I &= 0 \text{ for atoms} \\ \rho u_I &= nkT \text{ for diatomic molecules} \end{aligned} \tag{2.31}$$

and we can write

$$u = c_v T \tag{2.32}$$

where the specific heat at constant volume is assumed constant, given by

$$\begin{aligned} c_v &= \frac{3}{2} R \text{ for atoms} \\ c_v &= \frac{5}{2} R \text{ for diatomic molecules} \end{aligned} \tag{2.33}$$

In general, if c_{vi} is the specific heat for the ith species, then the specific heat for the gas mixture is

$$c_v = \frac{1}{\rho} \sum_i n_i m_i c_{vi} \tag{2.34}$$

Hines (1977a,b) has discussed extensively the conditions under which (2.32) may or may not be valid. In particular, it loses validity if collisions are not sufficiently rapid to maintain an equilibrium distribution of energy between translational and internal degrees of freedom, as may occur in the thermosphere where collisions are much less frequent than in the lower atmosphere. However, we shall for simplicity assume that (2.32) is valid in our treatment of thermospheric dynamics.

For a mixture of perfect gases, the specific heat at constant pressure is related to that at constant volume by

$$c_p = c_v + R = \frac{1}{\rho} \sum_i n_i m_i c_{pi} \tag{2.35}$$

$$c_{pi} = c_{vi} + k/m_i \tag{2.36}$$

2.5. Diffusion

A sufficiently general expression for the diffusion velocity of the ith species, $\vec{v}_i - \vec{v}$, can be written as (Chapman and Cowling, 1970)

$$\vec{v}_i - \vec{v} = - \sum_j \frac{\Delta_{ij}}{p} \left[\nabla(n_j kT) - n_j m_j \vec{g} - n_j q_j (\vec{E} + \vec{v}_j \times \vec{B}) + n_j m_j \vec{F} \right] - D_{Ti} \frac{\nabla T}{T} \tag{2.37}$$

where the summation over j is for all species other than i; $\vec{F}$ is the sum of the pressure, gravitational, and electromagnetic forces per unit mass on the entire fluid; and Δ_{ij} and D_{Ti} are composition-dependent coefficients derivable from kinetic theory, when the interaction potentials between the various atoms, molecules, and charged particles are known. Although (2.37) is not entirely accurate when charged particles experience significant deflection by the magnetic field between collisions, as they generally do in the ionosphere, it is adequate for our purposes. For a more general treatment of charged particle diffusion in a magnetic field, see Shkarovsky (1961) or Chapman and Cowling (1970). It is usual practice in thermospheric theory to neglect terms in the summation for which $i \neq j$ and to use the simpler form

$$\vec{v}_i - \vec{v} = - D_i \left[\frac{\nabla n_i}{n_i} + (1 + \alpha_i) \frac{\nabla T}{T} - \frac{m_i}{kT} \vec{g} - \frac{q_i}{kT} (\vec{E} + \vec{v}_i \times \vec{B}) + \frac{m_i}{kT} \vec{F} \right] \tag{2.38}$$

where

$$D_i = \frac{\Delta_{ii} n_i}{n} \tag{2.39}$$

$$\alpha_i = D_{Ti}/D_i \tag{2.40}$$

The solid lines in Fig. 3 show the height variations of the molecular diffusion coefficients D_i for various species diffusing through N_2 (U.S. Standard Atmosphere, 1976).

The coefficients D_{Ti} and α_i represent thermal diffusion effects. Gas kinetic theory predicts, and experiments verify, that in the presence of a temperature gradient unlike gases will diffuse, even in the absence of density gradients and external forces. Generally, lighter molecules tend to diffuse toward warmer regions. In the thermosphere thermal

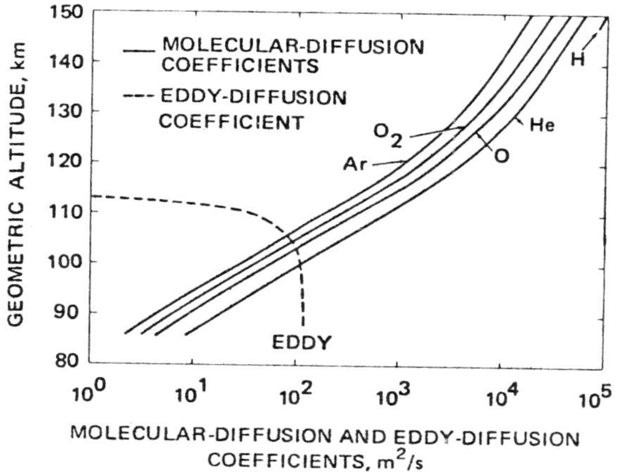

Figure 3. Diffusion coefficients.

diffusion turns out to be important primarily for the light species H and He, which are only minor constituents in the lower thermosphere where temperature gradients are significant. The U.S. Standard Atmosphere, 1976, uses values of $\alpha_i = -0.40$ for He and $\alpha_i = -.25$ for H.

If thermal diffusion is neglected, we see that the expression in brackets in (2.38) represents the negative of the net force per unit volume on the ith consitituent, divided by its partial pressure, in a frame of reference accelerating with the mean mass motion. This force can be thought of as being balanced by a frictional force caused by particle collisions. If the mean collision frequency for momentum loss by a diffusing particle is ν_i, then the force balance yields

$$\nu_i m_i n_i (\vec{v}_i - \vec{v}) = -\nabla (n_i kT) + n_i m_i \vec{g} + n_i q_i (\vec{E} + \vec{v}_i \times \vec{B}) - n_i m_i \vec{F} \quad (2.41)$$

so that

$$\nu_i = \frac{kT}{m_i D_i} \quad (2.42)$$

This effective collision frequency falls off rapidly with increasing altitude due to the decreasing air density, so that diffusion proceeds more rapidly as we go up in altitude. The collision frequency is often used for electrons and ions instead of the diffusion coefficient, as we shall see when discussing electrical conductivities.

The atmosphere is continually in motion on a wide range of scales, including small-scale turbulent motions. In studying atmospheric dynamics, it is usually convenient to treat atmospheric properties like n_i, $\vec{v}_i$, etc., as representing averages over some finite volume, perhaps on the order of 1 km³ in the thermosphere. Equations like (2.38) are strictly applicable only if n_i, $\vec{v}_i$, etc., represent local values, rather than volume-averaged values. In order to compensate for the inapplicability of molecular transport equations like (2.38) to volume-averaged quantities, it is convenient to introduce the concept of eddy transport by turbulence. This concept assumes that turbulent motions have the effect of mixing the atmosphere, tending to smooth out gradients of composition, momentum, and temperature in a manner somewhat, but not entirely, analogous to molecular transport processes. In the case of eddy diffusion, the mixing should tend to lead to uniformity of composition (which is not the case for molecular diffusion of species with different molecular weights in a gravitational field). The eddy diffusion velocity, which is to be added to the molecular diffusion velocity in (2.38), is written

$$\vec{v}_{Ei} = D_E \nabla \ln (m_i n_i/\rho) . \qquad (2.43)$$

The effective value of the eddy diffusion coefficient D_E depends on the volume of averaging for atmospheric properties, and hence the appropriate value to use depends on the scale size of phenomena under consideration. In addition, D_E appears to be quite variable in the atmosphere, depending on the level of turbulent activity. Observations of turbulence can give a rough idea of the appropriate magnitude of D_E. These observations generally show that turbulence dies out above about 110 km (the turbopause), presumably because molecular viscosity strongly damps it at higher levels. A reasonable estimate of the mean eddy diffusion coefficient as a function of height in the lower thermosphere appropriate for global-scale thermospheric structure is shown by the dashed line in Fig. 3.

2.6. Viscosity

The pressure tensor $\underline{\underline{P}}$ depends not only on density and temperature, but also on gradients of the velocity $\vec{v}$. Gas kinetic theory gives (e.g., Chapman and Cowling, 1970; Hines, 1977a)

$$P_{xx} = p - \mu \left(\frac{4}{3}\frac{\partial v_x}{\partial x} - \frac{2}{3}\frac{\partial v_y}{\partial y} - \frac{2}{3}\frac{\partial v_z}{\partial z} \right) \equiv p + P_{xx}^* \qquad (2.44)$$

$$P_{xy} = -\mu \left(\frac{\partial v_x}{\partial y} + \frac{\partial v_y}{\partial x} \right) \equiv P_{xy}^* \qquad (2.45)$$

where μ is the coefficient of viscosity and where $\underline{\underline{P}}^*$ is defined as the viscous momentum flux tensor. Other terms in the pressure tensor are determined by commuting the subscripts of (2.44) for diagonal terms and of (2.45) for off-diagonal terms.

The coefficients of viscosity for N_2, O_2, and O are similar. For O, Banks and Kockarts (1973b) give a formula equivalent to

$$\mu = 1.87 \times 10^{-5} \left(\frac{T}{273.15} \right)^{0.69} \mathrm{kg \cdot m^{-1} \cdot s^{-1}} \qquad (2.46)$$

We might note that this expression, if divided by ρ, is similar in magnitude to the molecular diffusion coefficients in Fig. 3.

Turbulence also adds to the effective viscosity, in the same manner that it adds to the effective diffusivity. In fact, momentum transport by turbulence is somewhat more effective than mass transport, since turbulent eddies have a slight tendency to return mass to its point of origin, whereas momentum is readily transferred between air parcels and their instantaneous environment. Crudely, we may add to the molecular viscosity coefficient an eddy viscosity coefficient estimated by

$$\mu_E = 2\rho D_E \tag{2.47}$$

2.7. Heat conduction

Assuming that the simplified form (2.38) for diffusive velocities is adequate we can write the heat flux as (after Chapman and Cowling, 1970)

$$\vec{q} = -\kappa\nabla T + \sum_i (n_i m_i c_{pi} + \alpha_i n_i kT)(\vec{v}_i - \vec{v}) \tag{2.48}$$

where κ is the heat conduction coefficient. Even in the absence of a temperature gradient heat can be transferred by diffusion, but normally the heat transport associated with κ is more important in the thermosphere.

The heat conduction coefficient is closely related to the viscosity coefficient and to the specific heat of a gas. For monatomic gases the following relation holds:

$$\kappa = 1.5\ c_p\mu \tag{2.49}$$

For N_2 and O_2 the multiplying factor of $c_p\mu$ is slightly temperature dependent, but (2.49) remains an adequate approximation in the thermosphere for these species as well.

Turbulence also transports energy. In considering the effective turbulent transport, one must take into consideration the fact that energy is added to downward moving air parcels by compression and removed from upgoing parcels by expansion, so that there exists a net downward energy transport even in the absence of temperature gradients. If we assume that eddy diffusion and eddy heat transport occur at comparable rates, then the eddy heat transport can be expressed as

$$\vec{q}_E = -\rho D_E\ [\nabla(c_p T) - \vec{g}] \tag{2.50}$$

which is to be added to the molecular transport of (2.48). The manner in which the form of the expression in brackets in (2.50) comes about will become clearer later, in the discussion of vertical heat transport by bulk motions.

The set of equations determining the number densities and velocities of the various species and the temperature of the thermosphere is now essentially complete. If we are given initial and boundary conditions, as well as the functions S_i, $\vec{E}$, and Q, we can in principle solve these equations. In practice additional simplifications must be made to make the problem tractable. Various simplifications are introduced at appropriate points in this chapter.

2.8. Electric conductivities

Our treatment of charged particle dynamics in this chapter is greatly simplified. First, we do not solve for charged particle number densities, but rather assume they are known. Second, we solve for charged particle diffusion velocities by applying (2.41) but neglecting the partial pressure gradient, gravitational, and bulk force ($\vec{F}$) terms. Then (2.41) can be written

$$\nu_i m_i n_i\ (\vec{v}_i - \vec{v}) - n_i q_i\ (\vec{v}_i - \vec{v}) \times \vec{B} = n_i q_i\ (\vec{E} + \vec{v} \times \vec{B}) \tag{2.51}$$

The component of (2.51) parallel to $\vec{B}$ yields

$$(\vec{v}_i - \vec{v})_{\parallel} = \frac{q_i}{m_i \nu_i}\ \vec{E}_{\parallel} \tag{2.52}$$

where the subscript "∥" refers to the component along $\vec{B}$. To solve (2.51) for the perpendicular component, first take its cross product with $\vec{B}$, yielding

$$n_i q_i B^2\ (\vec{v}_i - \vec{v})_{\perp} + \nu_i m_i n_i\ (\vec{v}_i - \vec{v}) \times \vec{B}$$
$$= n_i q_i\ (\vec{E} + \vec{v} \times \vec{B}) \times \vec{B} \tag{2.53}$$

The terms in $(\vec{v}_i - \vec{v}) \times \vec{B}$ can be eliminated from (2.52) and (2.53), yielding

$$(\vec{v}_i - \vec{v})_{\perp} = \frac{q_i \nu_i\ (\vec{E}_{\perp} + \vec{v} \times \vec{B})}{m_i\ (\nu_i^{\,2} + \Omega_i^{\,2})} + \frac{q_i \Omega_i\ (\vec{E} + v \times \vec{B}) \times \hat{b}}{m_i\ (\nu_i^{\,2} + \Omega_i^{\,2})} \tag{2.54}$$

$$\Omega_i \equiv \frac{q_i B}{m_i} \tag{2.55}$$

where $\hat{b}$ is a unit vector along $\vec{B}$ and Ω_i is the <u>angular gyrofrequency</u> of a charged particle in the magnetic field, defined here to be negative for negatively charged particles.

In order to examine the motion of the particles, let us place ourselves in the reference frame moving at the mass-mean velocity $\vec{v}$. The electric field measured in this frame is

$$\vec{E}' = \vec{E} + \vec{v} \times \vec{B} \tag{2.56}$$

and the charged particle velocity is

$$\vec{v}_i' = \vec{v}_i - \vec{v} \tag{2.57}$$

Then (2.52) and (2.54) give

$$\vec{v}_i' = \frac{q_i}{m_i \nu_i} \vec{E}_{\parallel}' + \frac{q_i \nu_i \vec{E}_{\perp}'}{m_i(\nu_i^2 + \Omega_i^2)} + \frac{q_i \Omega_i \vec{E}' \times \hat{b}}{m_i(\nu_i^2 + \Omega_i^2)} \tag{2.58}$$

If $\nu_i^2 >> \Omega_i^2$, this reduces to

$$\vec{v}_i' = \frac{q_i}{m_i \nu_i} \vec{E}' \tag{2.59}$$

Thus where collisions dominate, the charged particle moves parallel to the electric field (or antiparallel if q_i is negative), and is unaffected by the magnetic field. If $\nu_i^2 << \Omega_i^2$ (2.58) reduces to

$$\vec{v}_i = \frac{q_i}{m_i \nu_i} \vec{E}_{\parallel}' + \frac{\vec{E}' \times \hat{b}}{B} \tag{2.60}$$

The motion parallel to the magnetic field remains unaffected by $\vec{B}$, but the perpendicular motion is strongly affected. This perpendicular motion is simply the E × B drift, which is at 90° to the direction of the electric field, and is independent of the mass or charge of the particle. For ν_i^2 similar to Ω_i^2, the particle motion perpendicular to $\vec{B}$ lies between the direction given by (2.59) and (2.60), that is, it is neither parallel nor perpendicular to $\vec{E}'$. In the earth's atmosphere ν_i changes over orders of magnitude as a function of height, while Ω_i changes only little. The transition height where $\nu_i^2 = \Omega_i^2$ occurs around 75 km for free electrons and around 125 km for positive ions, as seen in Fig. 4 (Richmond, 1972). Well below the transition height particle motion is approximated by (2.59), while well above it the motion is approximated by (2.60).

When (2.52) and (2.54) are used in the definition (2.8) of current density, we find the following expression of Ohm's Law:

$$\vec{J} = \sigma_o \vec{E}_{\parallel} + \sigma_1 (\vec{E}_{\perp} + \vec{v} \times \vec{B}) + \sigma_2 \hat{b} \times (\vec{E} + \vec{v} \times \vec{B}) + \rho_e \vec{v} \tag{2.61}$$

$$\sigma_0 = \sum_i \frac{n_i q_i^2}{m_i \nu_i} \tag{2.62}$$

$$\sigma_1 = \sum_i \frac{n_i q_i^2 \nu_i}{m_i(\nu_i^2 + \Omega_i^2)} \tag{2.63}$$

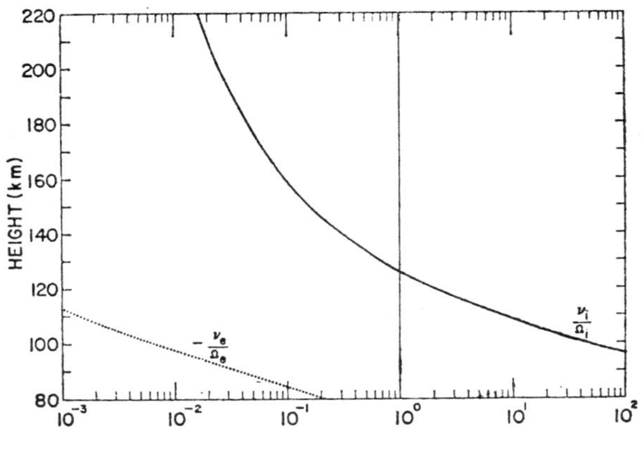

Figure 4. Height variations of collision frequency to gyrofrequency ratio for electrons (subscript e) and positive ions (subscript i).

$$\sigma_2 = - \sum_i \frac{n_i q_i^2 \Omega_i}{m_i(\nu_i^2 + \Omega_i^2)} \tag{2.64}$$

where σ_0, σ_1, and σ_2 are called the parallel, Pedersen, and Hall conductivities, respectively. The current represented by $\rho_e \vec{v}$ is entirely negligible because net charge densities are so small in the ionosphere. Note that for σ_0 and σ_1 all charged particles make a positive contribution, but for σ_2 electrons make a positive contribution, and positive ions make a negative contribution, with the electron contribution always having the larger magnitude, so that σ_2 is always positive. In the thermosphere electrons make by far the largest contribution to σ_0. Ions make the largest contribution to σ_1 above 100 km, but electrons dominate below about 100 km, where, however, σ_1 is relatively small. The electron contribution to σ_2 dominates below about 125 km, but ion motions become important for σ_2 above 125 km in the sense that the current carried by ions tends to cancel that by electrons.

Typical vertical profiles of the conductivities for the subsolar ionosphere are shown in Fig. 5 (Richmond, 1972). At night the conductivities are greatly reduced in proportion to the smaller E-region ionospheric densities. The conductivities play an important role both in the dynamics and electrodynamics of the thermosphere, as will become apparent later in this chapter.

A few notes of caution should be pointed out concerning our derivation of conductivities. It was mentioned that (2.37), and hence (2.41), is not entirely accurate when Ω_i^2 is comparable to or greater than ν_i^2. For electrons, the effective collision frequency for motions perpendicular to $\vec{B}$ must be increased about 40% from (2.42) for more accurate results when $\Omega_i^2 > \nu_i^2$; however this has little effect on the conductivities above 100 km. The effective collision frequency for ions is

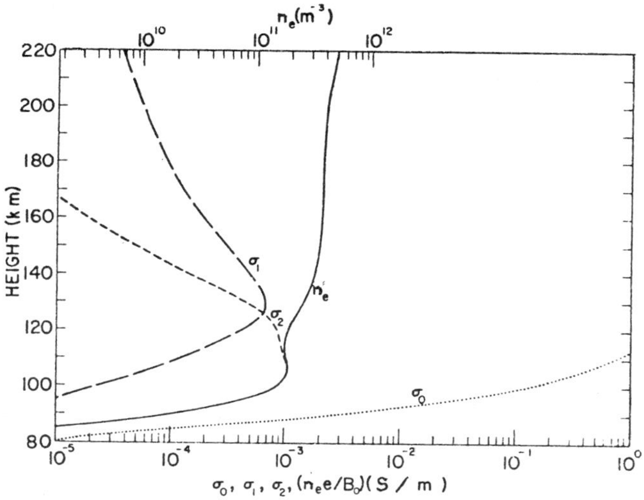

Figure 5. Typical low-latitude noontime profiles of electron density n_e and ionospheric conductivities.

only slightly affected by the magnetic field. Electron-ion collisions contribute to the electron collision frequency for motions parallel to $\vec{B}$ above about 200 km, but they barely influence motions perpendicular to $\vec{B}$ at these heights where electrons and ions have almost the same velocity $(\vec{E} \times \vec{B}/B^2)$ to $\vec{B}$. Thus electron-ion collisions can be neglected in calculating σ_1 and σ_2, but not σ_0.

The derivation of the conductivity expressions (2.62) - (2.64) implicitly assumed that the electron and ion drifts are relatively small in comparison with thermal velocities. This condition is sometimes violated in the ionosphere when the electric field becomes large. The theory for the electric current-electric field relation then becomes very complicated, such that the linear relation represented by Ohm's Law breaks down. In this chapter, however, (2.61) is assumed to be valid under all conditions.

3. LARGE-SCALE THERMOSPHERIC STRUCTURE AND DYNAMICS

The previous section covered most of the basic equations needed for studying the dynamical and electrodynamical behavior of the thermosphere. In this section we shall investigate how simplified forms of these equations can be used to examine large-scale thermospheric structure and dynamics. We begin by examining the large-scale vertical structure of the thermosphere, under the assumption that vertical gradients of the various parameters are much stronger than horizontal gradients, so that we can use one-dimensional forms of the equations. Our development here stresses basic theoretical concepts; for further development of the theory of thermospheric structure see Banks and Kockarts (1973a,b).

3.1. Hydrostatic equilibrium

Vertical accelerations within the atmosphere are generally quite small, particularly for global-scale motions. Hydrostatic equilibrium dominates, meaning that forces acting on the gas in the vertical direction essentially balance. Of these forces, gravity and the pressure gradient are by far the greatest. To a very good approximation, one can thus write the vertical component of (2.18) as

$$\frac{\partial p}{\partial z} = -\rho g \tag{3.1}$$

where z is the vertical (upward) coordinate, p is pressure, ρ is mass density, and g is the acceleration of gravity.

The perfect gas law (2.27) can be used to write (3.1) as

$$\frac{\partial p}{\partial z} = -\frac{p}{H} \tag{3.2}$$

$$H \equiv \frac{p}{\rho g} = \frac{RT}{g} = \frac{kT}{\bar{m}g} \tag{3.3}$$

$$\bar{m} \equiv \frac{\rho}{n} = \sum_i m_i n_i / \sum_i n_i \tag{3.4}$$

H is defined as the pressure scale height. If H is known as a function of altitude, (3.2) can be solved for p by integration to give

$$p(z) = p_o \exp\left(-\int_{z_o}^{z} \frac{dz'}{H}\right) \tag{3.5}$$

where p_o is the pressure at the reference height z_o. For an atmospheric region where H is constant in altitude (for example, if T is isothermal, if the composition is homogeneous so that $\bar{m}$ is constant, and if vertical variations of g can be ignored), (3.5) takes on the simple form

$$p = p_o \exp\left[-\frac{(z - z_o)}{H}\right] \tag{3.6}$$

showing that the pressure falls off exponentially with increasing altitude.

In the earth's thermosphere H varies by about an order of magnitude, from around 6 km at 90 km altitude to around 70 km at 500 km altitude for average conditions. The pressure and density vary much more, however, decreasing by factors of roughly 10^6 and 10^7, respectively, between 90 km and 500 km. Moreover, the exponential dependence of pressure and density on H makes them highly sensitive at upper altitudes to changes in H. At 500 km, for example, the density can change by almost three orders of magnitude between extreme conditions, for which the exospheric temperature varies from 550 to 2000 K.

3.2. Steady-state thermal structure

The thermal structure of the thermosphere can be examined with the aid of (2.24). Let us first simplify the form of this equation by defining a total heating rate Q^* which includes chemical, radiative, turbulent, Joule, and viscous heating:

$$\rho Q^* \equiv \rho Q + \vec{J} \cdot (\vec{E} + \vec{v} \times \vec{B}) - \underline{\underline{P}}^* : \nabla\vec{v} \tag{3.7}$$

so that (2.24) can be written

$$\rho \frac{Du}{Dt} + p\nabla\cdot\vec{v} + \nabla\cdot\vec{q} = \rho Q^* \tag{3.8}$$

By using the continuity equation (2.5) and the perfect gas law, we find that

$$p\nabla\cdot\vec{v} = -\frac{p}{\rho}\frac{D\rho}{Dt} = -\frac{p}{\rho}\frac{D}{Dt}\left(\frac{p}{RT}\right) = \rho\frac{D}{Dt}(RT) - \frac{Dp}{Dt} \tag{3.9}$$

so that (3.8) becomes

$$\rho\frac{D}{Dt}(u+RT) - \frac{Dp}{Dt} + \nabla\cdot\vec{q} = \rho Q^* \tag{3.10}$$

We are concerned here with steady-state conditions, for which the partial time derivative $\partial/\partial t$ vanishes and the total time derivative D/Dt becomes $\vec{v}\cdot\nabla$. Moreover, for large-scale structure horizontal gradients are small in comparison with vertical gradients, so that

$$\frac{D}{Dt} \approx v_z \frac{\partial}{\partial z} \tag{3.11}$$

In writing (3.11) we assume that vertical winds, even though small in comparison with horizontal winds, are nevertheless relatively large enough to be the dominant contributor to advective influences on the thermosphere. In most circumstances this is indeed a reasonable first approximation. The vertical component of heat flux q_z dominates over horizontal heat flux, so that (3.10) can be rewritten

$$\frac{\partial}{\partial z} q_z = \rho Q^* - v_z\left[-\frac{\partial p}{\partial z} + \rho\frac{\partial}{\partial z}(u+RT)\right] \tag{3.12}$$

For further simplification, let us neglect diffusion in calculating q_z from (2.48) and (2.50), and let us note further that from (3.1) $-\partial p/\partial z$ is just ρg and from (2.32) and (2.35) $u+RT$ is c_pT, so that (3.12) becomes

$$-\frac{\partial}{\partial z}\left[\kappa\frac{\partial T}{\partial z} + \rho D_E\left(g + \frac{\partial(c_pT)}{\partial z}\right)\right] = \rho Q^* - \rho v_z\left[g + \frac{\partial(c_pT)}{\partial z}\right] \tag{3.13}$$

At this point it is perhaps a bit clearer why turbulent heat conduction takes the form given in (2.50), that is, why it is proportional to $\vec{g} - \nabla(c_pT)$. The effective transport of heat by large-scale motions is also proportional to this quantity, as apparent in the advective term on the right-hand side of (3.13).

In the upper thermosphere, where ρ becomes very small and where D_E is negligible, (3.13) is approximately

$$- \frac{\partial}{\partial z} \left(\kappa \frac{\partial T}{\partial z} \right) \approx 0 \tag{3.14}$$

meaning that the conductive heat flux is nearly constant with height. There seem to be no significant heat sources or sinks at the top of the thermosphere, so the heat flux itself must nearly vanish:

$$- \kappa \frac{\partial T}{\partial z} \approx 0 \tag{3.15}$$

This leads to the condition that the upper thermosphere is nearly isothermal, as is apparent in Fig. 1.

In examining the global average vertical temperature structure we can drop the advective term on the right of (3.13), since global average vertical velocities are negligible. If we integrate (3.13) with respect to height from the altitude z to the top of the atmosphere and apply the upper boundary condition (3.15), we obtain

$$\kappa \frac{\partial T}{\partial z} + \rho D_E \left[g + \frac{\partial (c_p T)}{\partial z} \right] = \int_z^\infty \rho Q^* dz' \tag{3.16}$$

The net heating rate Q^* is positive at all but the lowest heights in the thermosphere, due to strong heat inputs and to the lack of significant radiative cooling at most heights. Where molecular heat conduction dominates over turbulent conduction in (3.16), the vertical temperature gradient must obviously be positive to provide the necessary downward conduction of heat. Where turbulent conduction dominates over molecular conduction (below about 110 km) the temperature gradient is still observed to remain positive down to about 90 km, but this gradient is determined not so much by its usefulness for downward heat conduction as it is by a balance of various heating and cooling processes.

In examining departures from global-average behavior the advective term in (3.13) becomes important. Because global dynamic motions have a tendency to maintain some degree of horizontal uniformity in the thermosphere, the steady-state conductive heat flux at any location does not differ greatly from its global-averaged value, and tends to be determined by the horizontally averaged heating rate $\overline{Q^*}$. As a rough approximation, the departures of Q^* from its horizontally averaged value tend to be balanced by the advective term:

$$\rho (Q^* - \overline{Q^*}) \approx \rho v_z \left[g + \frac{\partial (c_p T)}{\partial z} \right] \tag{3.17}$$

so that a crude approximation to the steady state vertical velocity is

$$v_z = (Q^* - \overline{Q^*}) / \left[g + \frac{\partial (c_p T)}{\partial z} \right] \tag{3.18}$$

Physically, (3.17) and (3.18) represent a balance between external heating (or cooling) and adiabatic cooling (or heating) of the gas produced by vertical motions which carry the air into regions of lower (or higher) pressure. Equation (3.18) says that regions of excess heating, such as subsolar latitudes or, during disturbed periods, the auroral zones, experience rising air motions, and that regions of deficient heating experience subsidence. Of course, there also exist horizontal winds in association with the vertical winds in order to maintain mass continuity. The horizontal winds generally blow from heated regions to cooled regions of the upper levels of the thermosphere, and in the reversed sense in the lower thermosphere. As we shall see, the effects of these vertical and horizontal winds are important not only for the thermal structure of the thermosphere, but also for its composition structure.

3.3. Thermospheric heating

There are several factors which can contribute to the net heating rate Q^*. Of these, the most important can be categorized under one of the following three classes of thermospheric heat sources.

Absorption of solar ultraviolet and X-ray (XUV) radiation. Although most of the sun's electromagnetic energy output can be fairly well represented by blackbody radiation at 5800 K peaking in the spectral region of visible light (400-700 nm wavelengths), the ultraviolet portion of the spectrum deviates from the blackbody curve and becomes dominated by narrow spectral emission lines in the extreme ultraviolet or EUV (10-100 nm wavelengths). Unlike the bulk of solar radiation which comes from the solar photosphere, extreme ultraviolet and X-ray (.01-10 nm) radiation have a substantial portion coming from the hotter chromosphere and corona, and are found to vary significantly with the level of solar activity.

In the thermosphere most of the solar radiation at wavelengths below 180 nm is absorbed by the main consituents, N_2, O_2, and O. Not all of the absorbed radiation shows up as heat in the thermosphere, however. Some of the energy is used to dissociate O_2 molecules into O atoms, and this chemical energy is released as heat only after the O has diffused downward to the mesopause region where it recombines to O_2. Some of the absorbed energy is reradiated to space, as excited atoms and molecules relax to lower energy states. On the average, about 50% of the absorbed solar energy goes into local heating of the gas (Torr et al., 1980).

Magnetospheric energy input. Heating of the thermosphere at auroral latitudes is a second important heat source, particularly during geomagnetic storms. The strong auroral electric currents produce Joule heating of the gas. Most of this heat goes into the height region where the Pedersen conductivity σ_1 is largest (115-130 km), but the Joule heating extends up into the upper thermosphere as well, where on a heat per unit mass basis it is every bit as important as at lower altitudes because of the exponential mass density decrease with altitude. In addition to this Joule heating, the auroral thermosphere receives magnetospheric energy by way of precipitating charged particles, primarily energetic electrons. The height at which these particles are absorbed depends on their energy, the more energetic particles penetrating deeper. The height range 100-130 km receives most of this energy, but at times the main energy absorption can occur at higher altitudes. It is also possible that precipitating energetic neutral hydrogen atoms can produce significant heating in the low-latitude upper thermosphere during magnetic storms (see Tinsley, 1979).

Heating by turbulence and atmospheric wave dissipation. Small-scale atmospheric motions are dissipated by viscosity and other effects in the thermosphere. These motions make the main contribution to the viscous heating term $-\underline{\underline{P}}^*:\nabla\vec{v}$ in (3.7) because the velocity shears associated with them are quite strong. The lower atmosphere is a huge reservoir of wave energy, some of which escapes into the thermosphere and is dissipated. Precise estimates of how much heating might be associated with dissipating waves are difficult, but there are indications that this heating may be comparable with other heat sources. It is worthwhile to note that dissipating waves not only produce heating but also contribute to the downward eddy heat transport (Walterscheid, 1981).

3.4. Mean meridional circulation of the thermosphere

The nonuniformity of thermospheric heating processes gives rise to vertical motions, as semiquantitatively expressed by (3.18), with associated horizontal motions set up to maintain mass continuity. Figure 6 shows the patterns of daily averaged motions in a latitude-height cross section of the thermosphere calculated for different heating distributions (adapted from Roble, 1977). For "quiet" magnetospheric activity conditions solar EUV heating prevails, giving rise to upward motions in low latitudes at equinox or upward motions in the summer hemisphere at solstice, with outward flow at upper levels and sinking motions in regions of heating deficiency. As the level of magnetospheric energy input increases toward "storm" conditions, upward motions at auroral latitudes begin to dominate the circulation pattern, with equatorward flows attained above 350 km, reversing the quiet flow pattern.

3.5. Steady-state composition

Let us now examine the steady-state distributions of number densities n_i for the various constituents. When partial derivatives with respect to time are dropped, the continuity equation (2.1) can be written in the form

$$m_i\nabla\cdot[n_i(\vec{v}_i - \vec{v})] = S_i - \nabla\cdot(m_in_i\vec{v})$$

$$= S_i - \nabla\cdot(\frac{m_in_i}{\rho}\rho\vec{v})$$

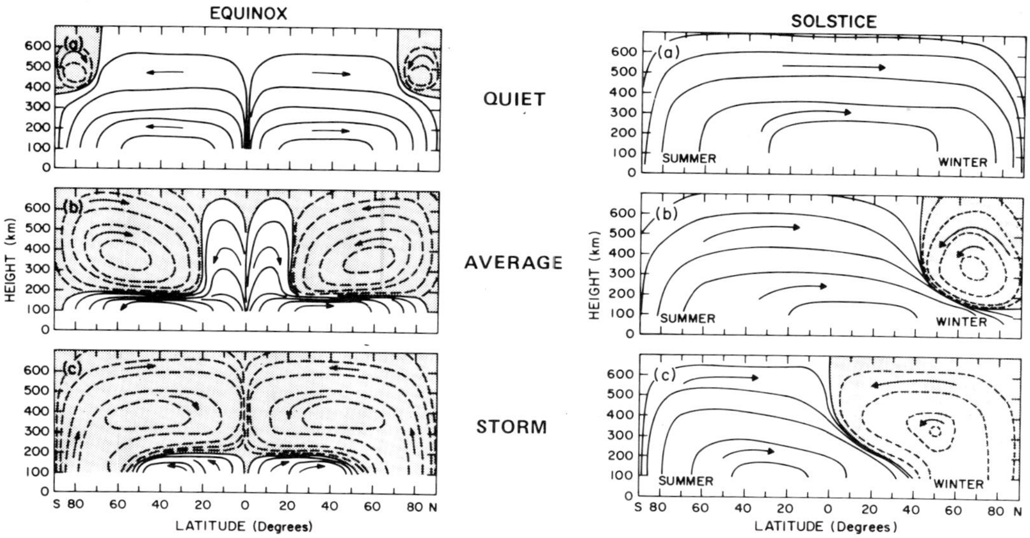

Figure 6. Mean meridional circulation patterns of the thermosphere for different heating conditions.

$$= S_i - \rho\vec{v}\cdot\nabla\left(\frac{m_i n_i}{\rho}\right) \tag{3.19}$$

where the last step makes use of the fact that $\nabla\cdot(\rho\vec{v})$ vanishes in a steady state. The diffusion velocity $\vec{v}_i-\vec{v}$ is given by (2.38). We consider here only uncharged particles, which experience no Lorentz force. For steady-state conditions the net force $\vec{F}$ on the medium can also be neglected. Because the thermosphere has a much smaller scale size in the vertical than in the horizontal, gradients of n_i and T will be nearly vertical, as will the diffusion velocity. With these simplifications (3.19) and (2.38) combine to yield

$$- \frac{\partial}{\partial z}\left\{ D_i m_i \left[\frac{\partial n_i}{\partial z} + (1+\alpha_i)\frac{n_i}{T}\frac{\partial T}{\partial z} + \frac{m_i n_i}{kT} g\right] + D_E \rho \frac{\partial}{\partial z}\left(\frac{m_i n_i}{\rho}\right)\right\} = S_i - \rho v_z \frac{\partial}{\partial z}\left(\frac{m_i n_i}{\rho}\right) \tag{3.20}$$

Analogous to our treatment of thermal structure, we can determine that in the upper thermosphere (3.20) simplifies to

$$- \frac{\partial}{\partial z}\left\{ D_i \left[\frac{\partial n_i}{\partial z} + (1+\alpha_i)\frac{n_i}{T}\frac{\partial T}{\partial z} + \frac{m_i n_i}{kT} g\right]\right\} = 0 \tag{3.21}$$

For the major species the net particle flux between the thermosphere and exosphere is generally negligible, so that the net vertical mass flux vanishes and (3.21) further simplifies to

$$\frac{\partial n_i}{\partial z} + (1 + \alpha_i) \frac{n_i}{T} \frac{\partial T}{\partial z} + \frac{m_i n_i}{kT} g = 0 \tag{3.22}$$

which expresses the state of diffusive equilibrium. This expression can be readily integrated to yield

$$n_i = n_{io} \left(\frac{T_o}{T}\right)^{1+\alpha_i} \exp\left(-\int_{z_o}^{z} \frac{dz'}{H_i}\right) \tag{3.23}$$

$$H_i \equiv \frac{kT}{m_i g} \tag{3.24}$$

where n_{io} and T_o are the number density and temperature at z_o. Equation (3.23) can be compared with the corresponding expression for total number density obtained from (3.5) and the perfect gas law:

$$n = n_o \frac{T_o}{T} \exp\left(-\int_{z_o}^{z} \frac{dz'}{H}\right) \tag{3.25}$$

Except for the thermal diffusion coefficient α_i, which is significant only for some minor constituents, (3.23) has nearly the same appearance as (3.25). The difference is that the relevant scale height H_i in (3.23) depends on the mass of the particular constituent. Heavy species like N_2 and O_2 therefore decrease in altitude much more rapidly than light species like He and H in the upper thermosphere, behavior which is apparent in Fig. 2.

Diffusive equilibrium gives us the rate of density decrease with height for the various species, but does not tell us the actual number densities. These can be obtained only by solving the more general equation (3.20) for all heights with appropriate boundary conditions. We shall not attempt such solutions here, but only discuss the importance of the two terms on the right, representing net sources and advection, and indicate what is the appropriate upper boundary condition.

Among the main thermospheric constituents, the source term S_i is most important for atomic oxygen. The main source of atomic oxygen is the atmospheric absorption of solar radiation at wavelengths shorter than 175 nm. A number of chemical reactions are involved but the net effect is that molecular oxygen is dissociated:

$$O_2 \xrightarrow[\text{radiation}]{\text{solar}} O + O \tag{3.26}$$

In order for O to recombine to form O_2, three-body collisions are necessary, and such collisions occur most frequently at low altitudes where the atmosphere is more dense. The recombination reaction can be written

$$O + O + M \rightarrow O_2 + M \tag{3.27}$$

where M is any type of molecule, such as N_2. The net source term S_i for

O is positive throughout most of the thermosphere where dissociation dominates, becoming negative only at the lowest levels where recombination dominates. It is clear that the net source term for molecular oxygen O_2 is just the negative of the source term for atomic oxygen.

The net flux of particles between the thermosphere and exosphere is negligible for most constituents, except for the light constituents He and H. These have such high thermal velocities, because of their light mass, that a small fraction of the atoms can escape the earth's gravitational attraction. For these atoms there is a net upward flux throughout the thermosphere, including its boundary with the exosphere. This flux, to the extent it can be quantified, can be imposed as an upper boundary condition for the solution of (3.20).

The advective term in (3.20) is important for understanding latitudinal variations of thermospheric composition. Let us for the sake of argument assume we are examining a region of excess heating so that v_z is positive (upward). For constituents whose mass density $m_i n_i$ decreases relatively more slowly than ρ with altitude $m_i n_i/\rho$ increases with altitude, so that the advective term in (3.20) is negative. This is generally the case for constituents with large scale heights, like He. Upward advection tends to act as a negative source term in this case, and therefore to deplete light constituents from this heated region of the thermosphere. Conversely, constituents whose mass density decreases with height more rapidly than ρ tend to be enhanced in thermospheric regions with enhanced heating. This behavior can be seen in empirical thermospheric models, based on satellite measurements near 300 km, as in Fig. 7 (Roble, 1977). The heated summer hemisphere tends to have an enhanced N_2 density, a relatively unaffected O density, and a depleted He density with respect to their densities at the equator.

3.6. Large-scale dynamics

So far we have considered steady-state conditions in the thermosphere. To examine diurnal and other types of variations we need to include time derivatives of quantities in the equations. Horizontal thermospheric winds, in particular, have large temporal variations, and so we shall examine them by including the inertial term in the equation of motion.

Let us rewrite (2.18) upon separating $\underline{\underline{P}}$ into its components of scalar pressure p and viscous momentum flux $\underline{\underline{P}}^*$, and by using Ohm's Law (2.61) to eliminate $\vec{J}$:

$$\rho \frac{D\vec{v}}{Dt} + 2\vec{\Omega} \times \vec{v} + \nabla p + \nabla \cdot \underline{\underline{P}}^* = \rho\vec{g} - \sigma_1 B^2 (\vec{v}_\perp - \vec{u}^e) - \sigma_2 B^2 \hat{b} \times (\vec{v} - \vec{u}^e) \quad (3.28)$$

$$\vec{u}^e \equiv \frac{\vec{E} \times \vec{B}}{B^2} \quad (3.29)$$

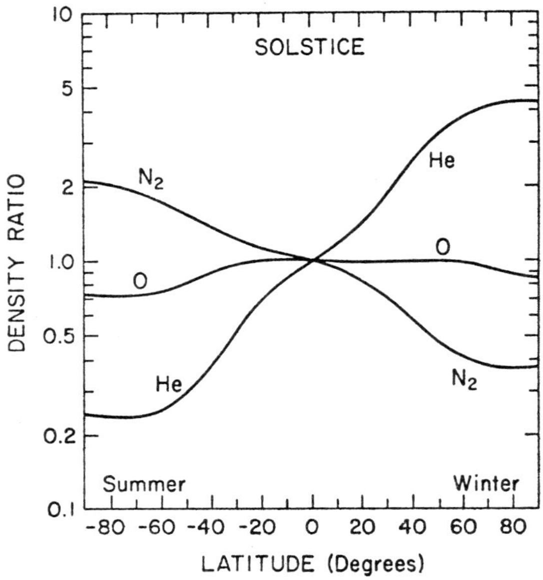

Figure 7. Relative density variations of N_2, O, and He at 300 km altitude.

where $\vec{v}_\perp$ is the component of $\vec{v}$ perpendicular to $\vec{B}$, and $\vec{u}^e$ is the electrodynamic E × B drift velocity discussed in connection with (2.60). At middle and low latitudes $\vec{u}^e$ tends to be on the order of 30 m/s, but it can exceed 1000 m/s at high latitudes. The wind velocity $\vec{v}$, on the other hand tends to be on the order 100 m/s. The relative importance of $\vec{u}^e$ and of $\vec{v}$ on the right hand side of (3.28) therefore changes strongly with latitude.

Let us also make the following simplifying assumptions:

1) The total time derivative D/Dt can be replaced by the partial derivative $\partial/\partial t$ because wind velocities are small enough to allow dropping the advective term $\vec{v}\cdot\nabla$. This is an advantage of working in a rotating frame of reference; for a non-rotating frame of reference the advective term would be of primary importance.

2) Vertical wind velocities are very small in comparison with horizontal velocities. This assumption is related to the fact that horizontal spatial scales are large in comparison with vertical spatial scales.

3) The only important term in the viscous momentum flux tensor is that involving vertical flux of horizontal momentum:

$$\nabla\cdot\underline{\underline{p}}^* = -\frac{\partial}{\partial z}\left(\mu\,\frac{\partial \vec{v}_h}{\partial z}\right) \tag{3.30}$$

where $\vec{v}_h$ is the horizontal component of wind velocity. This assumption comes from the dominance of vertical gradients over horizontal gradients and from the dominance of horizontal over vertical wind velocities.

4) The geomagnetic field is dipolar and aligned with the earth's rotational axis. Then

$$B_z = - B_p \cos\theta \tag{3.31}$$

$$B_\theta = - \frac{1}{2} B_p \sin\theta \tag{3.32}$$

where B_p is the polar field strength and θ is colatitude.

With these assumptions, the horizontal components of (3.28), multiplied by $1/\rho$, are:

$$\frac{\partial v_\theta}{\partial t} - \left(2\Omega - \frac{\sigma_2 B B_p}{\rho}\right)\cos\theta\, v_\phi + \frac{\sigma_1 B_p^{\,2}}{\rho} \cos^2\theta\, v_\theta - \frac{1}{\rho}\frac{\partial}{\partial z}\left(\mu \frac{\partial v_\theta}{\partial z}\right)$$

$$= - \frac{1}{\rho r}\frac{\partial p}{\partial \theta} + \frac{\sigma_1 B^2}{\rho} u_\theta^{\,e} + \frac{\sigma_2 B B_p}{\rho} \cos\theta\, u_\phi^{\,e} \tag{3.33}$$

$$\frac{\partial v_\phi}{\partial t} + \left(2\Omega - \frac{\sigma_2 B B_p}{\rho}\right)\cos\theta\, v_\theta + \frac{\sigma_1 B^2}{\rho} v_\phi - \frac{1}{\rho}\frac{\partial}{\partial z}\left(\mu \frac{\partial v_\phi}{\partial z}\right)$$

$$= - \frac{1}{\rho r \sin\theta}\frac{\partial p}{\partial \phi} + \frac{\sigma_1 B^2}{\rho} u_\phi^{\,e} - \frac{\sigma_2 B B_p}{\rho} \cos\theta\, u_\theta^{\,e} \tag{3.34}$$

where ϕ is east longitude. The structure of these equations is such that all terms explicitly involving wind velocity are on the left and everything else is on the right. The right-hand sides have often been treated as specified source functions for theoretical calculations of wind motions in the thermosphere.

Large scale wind motions in the lower atmosphere generally tend to be governed by a balance between the Coriolis terms and the pressure gradient terms in (3.33) and (3.34) (e.g., Holton, 1979). In the thermosphere the other terms in these equations can also be very important. Let us examine the nature of these other terms with the aid of idealized models.

3.7. Importance of ion drag

The forces proportional to σ_1 in (3.33) and (3.34) are often called <u>ion drag</u> forces. Microscopically they arise essentially from collisions between neutral molecules and ions drifting at different velocities. Above about 130 km ions tend to move at the velocity $\vec{u}^e$ perpendicular to $\vec{B}$, and to have the same component of velocity along $\vec{B}$ as the neutral wind. To see the effect of ion drag a little more clearly, let us ignore all other forces and assume a vertical geomagnetic field. Then (3.33) and (3.34) combine to give the vector equation

$$\frac{\partial \vec{v}_h}{\partial t} + \frac{(\vec{v}_h - \vec{u}^e)}{\tau_\sigma} = 0 \tag{3.35}$$

$$\tau_\sigma \equiv \frac{\rho}{\sigma_1 B^2} \tag{3.36}$$

where τ_σ is the time scale for ion drag to be effective. For simplicity, assume that τ_σ and $\vec{u}^e$ are constant, so that (3.35) has the solution

$$\vec{v}_h = \vec{v}_o\, e^{-t/\tau_\sigma} + \vec{u}^e\,(1 - e^{-t/\tau_\sigma}) \tag{3.37}$$

where $\vec{v}_o$ is the velocity at t = 0. As time progresses, the wind would approach the velocity $\vec{u}^e$, if it were valid to ignore all other forces. At middle latitudes, where $\vec{u}^e$ is usually smaller than $\vec{v}$, this effect generally tends to decelerate the wind, but the large ionization drifts at high latitudes can accelerate the wind to speeds of hundreds of meters per second. The time scale τ_σ is inversely proportional to the ionospheric plasma density, and can be as short as a fraction of an hour in the upper thermosphere around midday. (Some typical profiles of $1/\tau_\sigma$ are shown in the next section under gravity wave dissipation.) At 300 km altitude air motions tend to be governed primarily by a balance between ion drag and pressure gradient forces, with some influence of viscosity, so that winds tend to flow nearly antiparallel to pressure gradients rather than nearly perpendicular as in the lower atmosphere.

The influence of the terms proportional to σ_2 in (3.33) and (3.34) is somewhat more complicated than the influence of the σ_1 terms, but it turns out that the σ_2 terms are usually less important. On the left-hand sides of (3.33) and (3.34) we see that the σ_2 terms tend to reduce the Coriolis acceleration. Around 125 km at midday the σ_2 terms can in fact sometimes reverse the effective Coriolis parameter. However, this effect is limited in altitude and is not very important on a globally averaged basis. The terms involving σ_2 on the right-hand sides of (3.33) and (3.34) can be important at high latitudes where $\vec{u}^e$ is large, but their importance is again limited in altitude, between 110 km and 130 km. They represent a force on the wind in the direction of the electric field, associated with the component of ion motion in this direction.

3.8. Importance of viscosity

Just as we examined ion drag effects by balancing them against inertia, we can examine the importance of viscosity by assuming it to be the only force present. In this case, (3.33) and (3.34) give

$$\rho \frac{\partial \vec{v}_h}{\partial t} - \frac{\partial}{\partial z}\left(\mu \frac{\partial \vec{v}_h}{\partial z}\right) = 0 \tag{3.38}$$

In order to obtain an analytical solution to (3.38) let us assume that μ and H are constant and that ρ varies with z as

$$\rho = \rho_o \exp\left[-(z - z_o)/H\right] \tag{3.39}$$

as would be the case for an isothermal homogeneous atmosphere. One possible solution of (3.38) is

$$\vec{v}_h(z,t) = \vec{v}_o \frac{\tau_o}{t + \tau_o} \exp\left[-\tau_v(z)/(t + \tau_o)\right] \tag{3.40}$$

$$\tau_v(z) \equiv \rho(z)\, H^2/\mu \tag{3.41}$$

where τ_v can be considered a characteristic time scale for viscosity to be effective, a time scale which decreases exponentially with increasing altitude. This solution has the convenient properties that

$$\lim_{z \to \infty}\left(\mu \frac{\partial \vec{v}_h}{\partial z}\right) = 0 \tag{3.42}$$

$$\lim_{z \to -\infty} \vec{v}_h = 0 \tag{3.43}$$

Condition (3.42) means that there is no vertical momentum flux at the top of the atmosphere, a reasonable requirement.

In order for the solution (3.40) to be valid, the initial condition at t = 0 must be that the velocity profile is of the form

$$\vec{v}_h(z,0) = \vec{v}_o \exp\left[-\tau_v(z)/\tau_o\right] \tag{3.44}$$

where $\vec{v}_o$ and τ_o are arbitrary constants. For convenience, we can define ρ_o and z_o in (3.39) in terms of τ_o such that

$$\tau_v(z_o) = \tau_o \tag{3.45}$$

Figure 8 shows how the velocity profile evolves in time for the solution (3.40). Profiles are shown at times $t = 0$, $t = 3\tau_o$, and $t = 15\tau_o$. At

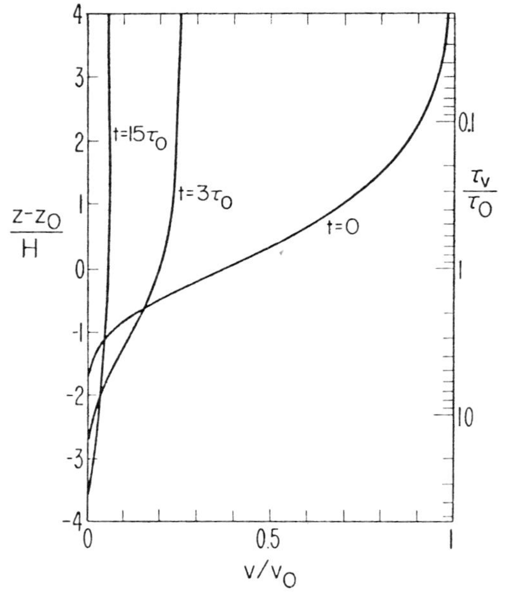

Figure 8. Velocity profiles at different times for viscous decay.

upper altitudes the velocity decays as $(t + \tau_o)^{-1}$. At altitudes below z_o the velocity first increases with time until $t = \tau_v - \tau_o$, after which it decays. The momentum is progressively transferred to lower and lower heights, where the velocity perturbation becomes progressively smaller because of the increasing mass density.

Table 1 shows viscous time scales along with other parameters obtained from the U.S. Standard Atmosphere, 1976, using (2.46) to parameterize μ. Although the isothermal model discussed above is not strictly valid for the thermosphere, the time scale $\rho(z)H^2(z)/\mu$ is still characteristic of viscous effects. We see that viscosity acts to smooth out velocity shears in less than an hour above 250 km, but takes about a day to be effective at 130 km. At the lowest heights eddy viscosity comes into play, so that the longest effective viscous time scales are a few days (roughly half of the parameter H^2/D_E, if we assume that (2.47) characterizes eddy viscosity).

Table 1 also lists a parameter $\rho H^2 c_p/\kappa$, which from (2.49) is just two-thirds of the viscous time scale. This parameter is a characteristic time scale for heat conduction to smooth out vertical temperature gradients, and can be derived in an entirely analogous manner to our derivation of τ_v by balancing temperature time variations against heat conduction effects. Thus the curves in Fig. 8 can also be used to represent the evolution of idealized temperature profiles if the abcissa and ordinate are appropriately redefined.

Table 1. Thermospheric parameters

z km	T K	H km	ρ kg/m^3	$\rho H^2/\mu$ hours	$\rho H^2 c_p/\kappa$ hours	H^2/D_E hours
90	187	5.64	3.42-6	2095	1397	73.5
100	195	6.01	5.60-7	379	253	89.3
110	240	7.72	9.71-8	94.1	62.7	499
120	360	12.1	2.22-8	40.0	26.6	
130	469	16.3	8.15-9	22.1	14.7	
140	560	20.0	3.83-9	13.9	9.28	
160	696	26.4	1.23-9	6.70	4.47	
180	790	31.7	5.19-10	3.73	2.48	
200	855	36.2	2.54-10	2.25	1.50	
250	941	44.9	6.07-11	.775	.517	
300	976	51.2	1.92-11	.310	.206	
350	990	55.8	7.01-12	.134	.089	
400	996	59.7	2.80-12	.060	.040	
450	998	63.6	1.18-12	.029	.019	
500	999	68.8	5.21-13	.015	.010	

3.9. Computer simulations of thermospheric dynamics

Computer models can be useful tools for theoretical investigations of the thermosphere. These models can incorporate quantitatively all of the major factors influencing thermospheric dynamics, factors which have widely varying time scales in different thermospheric regions. By comparing calculated winds, temperatures, and other quantities with observations, we can use computer models to help determine how well we understand the inputs, such as heating rates, forcing by auroral currents, and wave fluxes from the lower atmosphere. Simulations can also be run to help predict thermospheric behavior under altered input conditions, and can be used to provide much more thorough spatial coverage of thermospheric dynamics than observations alone can ever achieve.

Recently, results from three-dimensional time-dependent computer models of thermospheric dynamics have been presented (Fuller-Rowell and Rees, 1980, 1981; Dickinson et al., 1981; Roble et al., 1982). Figure 9 shows computed winds and temperature variations obtained for equinox conditions at solar cycle maximum by Dickinson et al. (1981). The simulation takes into account both solar radiation heating and Joule heating by high-latitude electric currents. At 300 km altitude the wind and temperature patterns are relatively simple. The temperature perturbation, measured from the global average temperature, maximizes in the afternoon at the equator, and has subsidiary maxima near the poles. The winds generally blow from the hotter to the cooler regions, and are stronger at night than at day because of the reduced nighttime ion drag. At 120 km altitude an important semidiurnal (12-hour period) component to the daily variations appears. This qualitatively different behavior

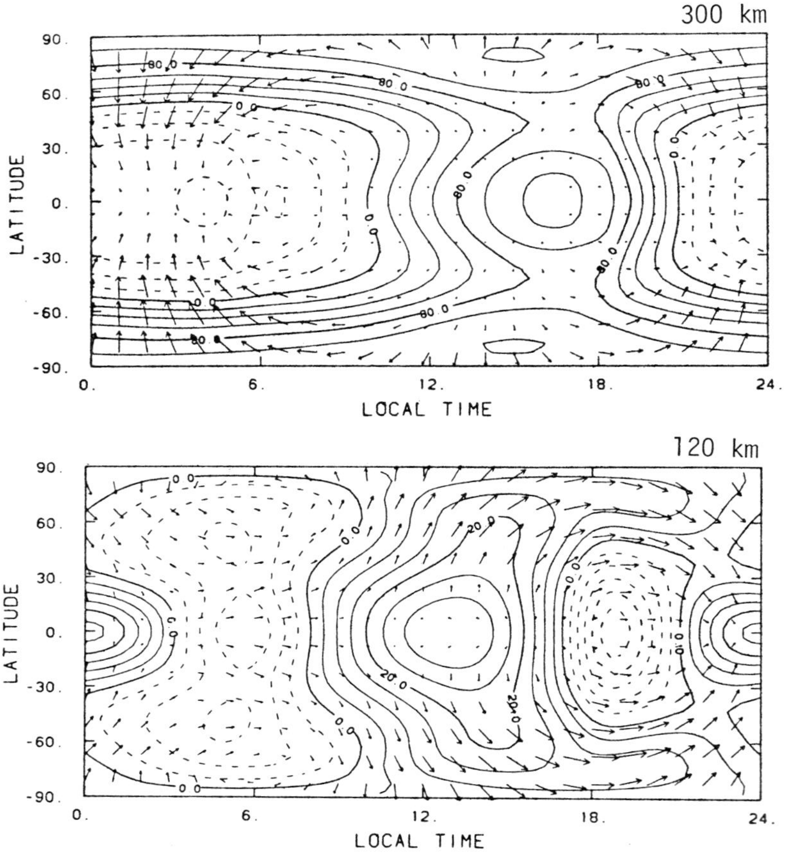

Figure 9. Calculated global distribution of perturbation temperature (in Kelvins) and winds along two constant pressure surfaces for equinox conditions during solar cycle maximum. Both solar heating and high-latitude heat sources are included. The maximum wind arrow is 130 m s^{-1} at 300 km and 90 m s^{-1} at 120 km.

between the two altitude regions is due to a number of factors, particularly the different shapes of local-time solar heating curves for the optically thin (300 km) and optically thick (120 km) cases, and also to the dominance of ion drag and viscosity at 300 km but not at 120 km. The semidiurnal oscillation at the lower height can be treated theoretically as one contribution to a class of global atmospheric waves known as atmospheric tides. Some more will be said about the theory of tides in the latter part of the next section.

4. ATMOSPHERIC WAVES

Atmospheric wave theory plays a central role in the theory of upper atmospheric dynamics. As we shall see, amplitudes of upward-propagating waves tend to grow exponentially in the atmosphere, so that many types of wave motions are apparent at high altitudes. Wave theory is most easily applicable to smaller-scale motions, for which horizontal gradients in background atmospheric conditions are presumably negligible, but even global-scale motions can be treated as waves, under the guise of atmospheric tidal theory. This section concentrates on the development of linear wave theory and its application to themospheric conditions. Further information about this subject can be found in Hines et al. (1974), Yeh and Liu (1974), Francis (1975), and Richmond (1978).

4.1. Entropy equation

Since to a first approximation wave motions are adiabatic, it becomes convenient to use specific entropy as a dependent variable instead of, say, internal energy or temperature. Specific entropy is conserved for thermodynamically reversible processes. Entropy considerations become complicated when diffusion is included, so for simplicity we shall impose the condition that diffusion is suppressed and that, therefore, composition is conserved following the fluid motion. The definition of specific entropy s (entropy per unit mass) from thermodynamics then requires that the change of entropy, multiplied by T, equals the change of internal energy plus the work done by pressure in expanding the fluid, or

$$\rho T \frac{Ds}{Dt} = \rho \frac{Du}{Dt} + p\nabla\cdot\vec{v} \qquad (4.1)$$

Then (3.8) gives

$$\rho \frac{Ds}{Dt} + \frac{\nabla\cdot\vec{q}}{T} = \rho \frac{Q^*}{T} \qquad (4.2)$$

or

$$\rho \frac{Ds}{Dt} + \nabla\cdot\left(\frac{\vec{q}}{T}\right) = -\frac{\vec{q}\cdot\nabla T}{T^2} + \rho \frac{Q^*}{T} \qquad (4.3)$$

Using (2.48) for $\vec{q}$ (neglecting diffusion and eddy heat conduction),

$$\rho \frac{Ds}{Dt} + \nabla\cdot\left(\frac{\vec{q}}{T}\right) = \kappa \frac{(\nabla T)^2}{T^2} + \frac{\rho Q^*}{T} \qquad (4.4)$$

from which it is clear that heat conduction leads to a net production of entropy. The quantity $\vec{q}/T$ is entropy flux.

We shall also need an equation relating entropy changes to pressure and density changes. We obtain this from (4.1) by making use of the continuity equation (2.5), the perfect gas law (2.27), the internal energy relation (2.32), and the specific heat relation (2.35). Then

$$\frac{Ds}{Dt} = \frac{1}{T}\frac{D}{Dt}(c_v T) + R\,\nabla\cdot\vec{v}$$

$$= \frac{1}{T}\frac{D}{Dt}\left(\frac{c_v}{R}\frac{p}{\rho}\right) - \frac{R}{\rho}\frac{D\rho}{Dt}$$

$$= \frac{c_v}{RT}\left(\frac{1}{\rho}\frac{Dp}{Dt} - \frac{p}{\rho^2}\frac{D\rho}{Dt}\right) - \frac{R}{\rho}\frac{D\rho}{Dt}$$

$$= \frac{c_v}{p}\frac{Dp}{Dt} - \frac{c_p}{\rho}\frac{D\rho}{Dt} \qquad (4.5)$$

4.2. Pressure change equation

Pressure is another convenient dependent variable in examining atmospheric waves. By a series of manipulations we can derive

$$\frac{Dp}{Dt} = \frac{D}{Dt}(\rho RT) = RT\frac{D\rho}{Dt} + \rho R\frac{DT}{Dt}$$

$$= -\rho RT\nabla\cdot\vec{v} + \frac{R}{c_v}\rho\frac{Du}{Dt}$$

$$= -p\nabla\cdot\vec{v} + \frac{R}{c_v}\left(\rho T\frac{Ds}{Dt} - p\nabla\cdot\vec{v}\right)$$

$$= -\frac{c_p}{c_v}p\nabla\cdot\vec{v} + \frac{R}{c_v}(\rho Q^* - \nabla\cdot\vec{q}) \qquad (4.6)$$

It is notationally useful to define the ratio of specific heats as

$$\gamma \equiv c_p/c_v \qquad (4.7)$$

so that

$$\frac{R}{c_v} = \gamma - 1 \qquad (4.8)$$

4.3. Linearized perturbation equations in a plane stratified atmosphere

We assume that we are dealing with small perturbations in an atmosphere whose basic state is static and varies only in the vertical direction. We can then linearize the momentum equation (2.18), the entropy equation (4.2), and the pressure equation (4.6). To do this, we first make replacements such as

$$p \rightarrow p + p' \qquad (4.9)$$

where a primed quantity represents a perturbation on the background quantity. We then collect together all terms which are effectively of first order in the perturbations. This yields

$$\rho \frac{\partial \vec{v}}{\partial t} + 2\rho\vec{\Omega} \times \vec{v} = - \nabla p' + \rho'\vec{g} - \nabla\cdot\underline{\underline{P}}^* + \vec{J} \times \vec{B} \tag{4.10}$$

$$\rho \frac{\partial s'}{\partial t} + \rho v_z \frac{ds}{dz} = - \frac{\nabla\cdot\vec{q}'}{T} + \frac{(\rho Q^*)'}{T} \tag{4.11}$$

$$\frac{\partial p'}{\partial t} - \rho g v_z + \gamma p \nabla\cdot\vec{v} = (\gamma-1) \left[- \nabla\cdot\vec{q}' + (\rho Q^*)'\right] \tag{4.12}$$

where we have not bothered to put primes on $\vec{v}$, $\underline{\underline{P}}^*$ or $\vec{J}$, as they are all assumed to have only perturbation components.

We actually have not yet defined the quantities s' and ds/dz, whose meanings in an atmosphere with height-varying composition are ambiguous. For convenience, let us define them by making use of (4.5), whose linearized form is

$$\frac{ds'}{dt} + v_z \frac{ds}{dz} = \frac{c_v}{p} \frac{\partial p'}{\partial t} - \frac{c_p}{\rho} \frac{\partial \rho'}{\partial t} + \frac{c_v}{p} v_z \frac{dp}{dz} - \frac{c_p}{\rho} v_z \frac{d\rho}{dz} \tag{4.13}$$

The simplest definitions of s' and ds/dz which satisfy (4.13) are

$$s' = \frac{c_v}{p} p' - \frac{c_p}{\rho} \rho' \tag{4.14}$$

$$\frac{ds}{dz} = \frac{c_v}{p} \frac{dp}{dz} - \frac{c_p}{\rho} \frac{d\rho}{dz} = \frac{1}{T} \left[g + \frac{c_p}{R} \frac{d}{dz}(RT)\right] \tag{4.15}$$

We can use (4.14) to eliminate ρ' in (4.10) so that the resulting equation together with (4.11) and (4.12) form a complete set to be solved for the variables $\vec{v}$, s', p', if $(\rho Q^*)'$ and $\vec{J}$ are known and if $\vec{q}'$ is also expressed as a function of s' and p'.

4.4. Wave energy equation

Intuitively, we often think of waves as containing free energy which is transported at the group velocity and which can be transformed into heat where the wave is dissipated. Mathematically, we can often define a wave-related quantity which has the units of energy density and whose time variations are balanced by a flux divergence term and by "production" and "dissipation" terms. We must be cautious, however, in identifying this mathematically defined "wave energy" with our intuitive notions of energy, especially in an atmosphere with non-uniform background temperature and wind structure. A loss of one unit of "wave energy" by dissipation does not necessarily produce exactly one unit of heat, for example. Fortunately, the differences between "wave energy" and real free energy are often small enough (less than a factor of two) that identifying the two with each other does not lead to serious quantitative errors. We shall make use of this fact in our development here in order to avoid excessive complication. Consistent with this concept, we shall use isothermal atmosphere approximations for mathematical quantities at points where these approximations provide significant notational simplifications.

To get a wave energy equation we take the scalar product of (4.10) with $\vec{v}$, multiply (4.11) by $gs'/(c_p ds/dz)$, multiply (4.12) by $p'/(\gamma p)$, add the resulting equations, and make some mathematical transformations, obtaining

$$\frac{\partial}{\partial t}\left(\rho\frac{v^2}{2}+\frac{p'^2}{2\gamma p}+\frac{\rho g}{2c_p ds/dz}s'^2\right)+\nabla\cdot(p'\vec{v})$$

$$=\left[\frac{gs'}{c_pT(ds/dz)}+\frac{\gamma-1}{\gamma}\frac{p'}{p}\right]\left[-\nabla\cdot\vec{q}'+(\rho Q^*)'\right]-\vec{v}\cdot\nabla\cdot\underline{\underline{P}}^*$$

$$+\vec{v}\cdot\vec{J}\times\vec{B} \qquad (4.16)$$

The quantity whose time derivative appears in (4.16) can be considered "wave energy density." The vector $p'\vec{v}$ represents an energy flux, while the right-hand side can be considered as net source terms.

The first quantity in brackets on the right-hand side of (4.16) is cumbersome to deal with. Fortunately, it simplifies greatly in a homogeneous isothermal atmosphere, for which

$$\frac{gs'}{c_pT(ds/dz)}+\frac{\gamma-1}{\gamma}\frac{p'}{p}=\frac{T'}{T} \qquad (4.17)$$

For simplicity, let us use this expression, even though it is not strictly valid in the height-varying thermosphere. Another simplifying assumption compatible with this is that

$$\vec{q}'=-\kappa\nabla T' \qquad (4.18)$$

which ignores perturbations in $\vec{q}$ caused by perturbations in κ interacting with the background temperature gradient.

The definition of wave energy flux is to some extent arbitrary. For example, making use of (4.17) and (4.18) and neglecting gradients of T, we can transform (4.16) to

$$\frac{\partial}{\partial t}\left(\frac{\rho v^2}{2} + \frac{p'^2}{2\gamma p} + \frac{\rho g}{2c_p(ds/dz)} s'^2\right) + \nabla\cdot\left(p'\vec{v} - \frac{\kappa T'\nabla T'}{T} + \underline{\underline{P}}^*\cdot\vec{v}\right)$$

$$= -\frac{\kappa}{T}(\nabla T')^2 + \underline{\underline{P}}^*:\nabla\vec{v} + \frac{T'}{T}(\rho Q^*)' + \vec{v}\cdot\vec{J}\times\vec{B} \qquad (4.19)$$

in which "wave energy flux" has a different implicit definition. Equation (4.19) has the conceptual advantage that molecular viscosity always leads to a local destruction of wave energy, since $\underline{\underline{P}}^*:\nabla\vec{v}$ is always negative, and heat conduction also destroys wave energy. The term on the right-hand side of (4.19) involving T'/T multiplied by the sum of electric, radiative, and chemical heating $(\rho Q^*)'$ also has a useful intuitive meaning. If heat is fed into a region of positive T', wave energy is generated. The atmosphere in this case acts as a sort of thermodynamic heat engine, converting a part of the thermal energy into wave energy, at the efficiency T'/T. The last term in (4.19) can represent either a wave energy source or loss, depending on the nature of the electric current and its relation to $\vec{v}$. If $\vec{J}\times\vec{B}$ represents only the ion drag force in the absence of an electric field, it opposes $\vec{v}$ and leads to wave energy loss. On the other hand, strong auroral-region currents driven by magnetospheric electric fields can tend to produce a velocity $\vec{v}$ in the same direction as $\vec{J}\times\vec{B}$, in which case wave energy is created.

4.5. Two simplified limiting cases

In the theoretical development of atmospheric waves with periods less than several hours, two characteristic parameters of the basic state atmosphere play a central role: the speed of sound and the buoyancy frequency. In order to gain some feel for the significance of these two parameters, let us examine two very simplified types of solutions to the set of equations (4.10)-(4.12). In both cases we shall neglect all source and dissipation effects, as well as Coriolis effects.

Sound waves. Let us first examine the case of small-scale, rapid perturbations for which gravitational effects and vertical gradients of the atmosphere can be neglected. Then (4.10) and (4.12) take the simplified forms

$$\rho\frac{\partial\vec{v}}{\partial t} + \nabla p' = 0 \qquad (4.20)$$

$$\frac{\partial p'}{\partial t} + \gamma p\nabla\cdot\vec{v} = 0 \qquad (4.21)$$

We can eliminate the velocity by taking the divergence of (4.20) and the time derivative of (4.21) and combining the results:

$$\frac{\partial^2 p'}{\partial t^2} - \frac{\gamma p}{\rho}\nabla^2 p' = 0 \qquad (4.22)$$

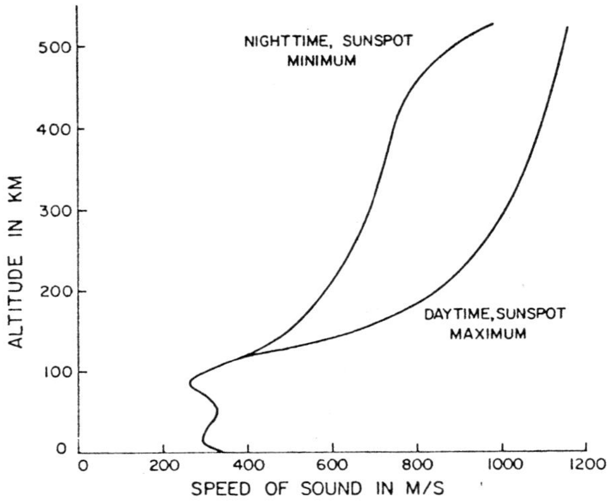

Figure 10. Atmospheric sound speed structure.

For plane waves, with gradients occurring only in the (arbitrary) x-direction, say, (4.22) becomes

$$\frac{\partial^2 p'}{\partial t^2} - c^2 \frac{\partial^2 p'}{\partial x^2} = 0 \tag{4.23}$$

$$c^2 \equiv \gamma p/\rho = \gamma g H \tag{4.24}$$

Solutions of (4.23) can be written generally in the form

$$p' = f_1(x + ct) + f_2(x - ct) \tag{4.25}$$

where f_1 and f_2 are any well-behaved mathematical functions. f_1 represents a waveform whose shape is unchanged in time as time progresses, but whose position moves in the -x direction with the speed c. f_2 represents another waveform moving in the +x direction at the same speed. This speed is defined as the sound speed. We see from (4.24) that it depends on the atmospheric scale height (as well as on γ), and that it therefore increases with height in the thermosphere. Figure 10 shows typical profiles of the sound speed with height (Yeh and Liu, 1974).

Buoyancy oscillations. Let us next examine the case of disturbances which produce a negligible pressure perturbation, and which represent a balance between acceleration and gravitational effects. For this case, (4.11) and the vertical component of (4.10) give

$$\frac{\partial s'}{\partial t} + \frac{ds}{dz} v_z = 0 \tag{4.26}$$

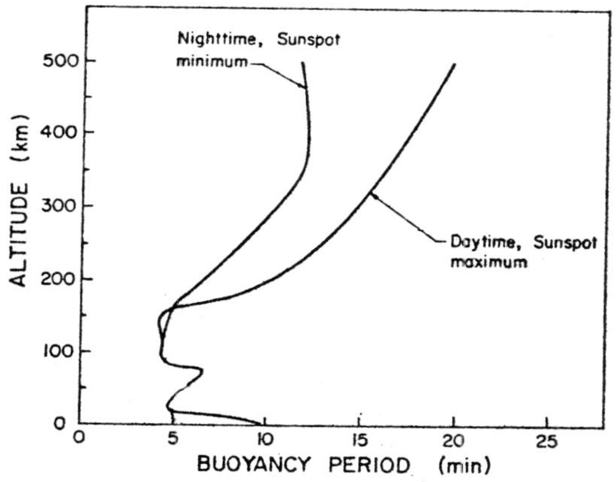

Figure 11. Atmospheric buoyancy period structure.

$$\frac{\partial v_z}{\partial t} + \frac{g}{c_p} s' = 0 \tag{4.27}$$

which can be combined to eliminate s', yielding

$$\frac{\partial^2 v_z}{\partial t^2} - N^2 v_z = 0 \tag{4.28}$$

$$N^2 \equiv \frac{g}{c_p}\frac{ds}{dz} = \frac{g}{c_p T}\left[g + \frac{c_p}{R}\frac{d(RT)}{dz}\right] \tag{4.29}$$

Solutions of (4.28) are of the form

$$v_z = A_1 e^{iNt} + A_2 e^{-iNt} \tag{4.30}$$

which represent oscillations at the angular frequency N, called the <u>buoyancy frequency</u> or <u>Brunt-Väisälä frequency</u>. This is a natural frequency at which air parcels tend to oscillate in vertical harmonic motion in the atmosphere. Figure 11 shows how the buoyancy period $2\pi/N$ varies with height in the atmosphere (Yeh and Liu, 1974).

4.6. Dispersion relation for waves in a dissipationless atmosphere, neglecting rotation

In the absence of sources, dissipation, and rotation (4.10)-(4.12) give

$$\rho \frac{\partial \vec{v}_h}{\partial t} + \nabla_h p' = 0 \tag{4.31}$$

$$\rho \frac{\partial v_z}{\partial t} + \frac{\partial p'}{\partial z} + \frac{p'}{\gamma H} - \frac{\rho g}{c_p} s' = 0 \tag{4.32}$$

$$\frac{\partial s'}{\partial t} + \frac{c_p N^2}{g} v_z = 0 \tag{4.33}$$

$$\frac{\partial p'}{\partial t} - \rho g v_z + \gamma p \left(\frac{\partial v_z}{\partial z} + \nabla \cdot \vec{v}_h \right) = 0 \tag{4.34}$$

where $\vec{v}_h$ is the horizonal component of velocity, and ∇_h is the horizontal component of the gradient operator. We assume plane waves varying in time and in horizontal distance $\vec{x}$ as

$$e^{i\omega t + i\vec{k}_h \cdot \vec{x}}$$

where ω is the angular frequency and $\vec{k}_h$ is the horizontal component of the wavenumber. Then (4.31)-(4.34) become

$$i\omega\rho\vec{v}_h + i\vec{k}_h\, p' = 0 \tag{4.35}$$

$$i\omega\rho v_z + \frac{\partial p'}{\partial z} + \frac{p'}{\gamma H} - \frac{\rho g}{c_p} s' = 0 \tag{4.36}$$

$$i\omega s' + \frac{c_p N^2}{g} v_z = 0 \tag{4.37}$$

$$i\omega p' - \rho g v_z + \gamma p \left(\frac{\partial v_z}{\partial z} + i\vec{k}_h \cdot \vec{v}_h \right) = 0 \tag{4.38}$$

We can use (4.35) and (4.37) to eliminate $\vec{k}_h \cdot \vec{v}_h$ in (4.38) and s' in (4.36), obtaining

$$\frac{\partial p'}{\partial z} + \frac{p'}{\gamma H} - \frac{i\rho(N^2 - \omega^2)}{gH\omega} v_z = 0 \tag{4.39}$$

$$\frac{\partial v_z}{\partial z} - \frac{v_z}{\gamma H} - \frac{i\omega}{\gamma p} \left(\frac{c^2 k_h^{\,2}}{\omega^2} \right) - 1 \quad p' = 0 \tag{4.40}$$

Analytic solutions to (4.39) and (4.40) are possible for the case where N, γ, and H are independent of altitude. The solutions are of the form

$$p' = P \exp \left(- \frac{z}{2H} + ik_z z + i\vec{k}_h \cdot \vec{x} + i\omega t \right) \tag{4.41}$$

$$v_z = V_z \exp\left(\frac{z}{2H} + ik_z z + i\vec{k}_h \cdot \vec{x} + i\omega t\right) \tag{4.42}$$

where the factors involving exp(z/2H) are needed to take account of the p-dependence of coefficients in (4.39) and (4.40). Inserting (4.41) and (4.42) into (4.39) and (4.40) yields

$$\left[ik_z + \left(\frac{1}{\gamma H} - \frac{1}{2H}\right)\right] P - \frac{i\gamma p_o\,(N^2-\omega^2)}{c^2\omega} V_z = 0 \tag{4.43}$$

$$-\frac{i\omega}{\gamma p_o}\left(\frac{c^2 k_h{}^2}{\omega^2} - 1\right) P + \left[ik_z - \left(\frac{1}{\gamma H} - \frac{1}{2H}\right)\right] V_z = 0 \tag{4.44}$$

where p_o is the atmospheric pressure at the reference level $z = z_o$. Non-trivial solutions are possible if the determinant of the coefficients in (4.43) and (4.44) vanishes, which gives

$$k_z{}^2 = \frac{N^2}{c^2}\left(1 - \frac{\omega^2}{N^2}\right)\left(\frac{c^2}{(\omega/k_h)^2} - 1\right) - \left(\frac{2-\gamma}{2\gamma H}\right)^2 \tag{4.45}$$

It is convenient to discuss the properties of wave solutions in terms of frequency ω (or period $2\pi/\omega$) and horizontal phase trace velocity ω/k_h. Vertically propagating waves, which are of most interest, must have real k_z, and therefore positive $k_z{}^2$. There are two regimes where this occurs, as can be readily seen by (4.45):

acoustic waves

$\omega > N$, $\omega/k_h > c$

gravity waves

$\omega < N$, $\omega/k_h < c$

These regimes are separated by a region of non-propagating waves where $k_z{}^2 < 0$.

4.7. Ducting

Because c and N vary with altitude, it is possible that waves with a given period and horizontal phase trace velocity can have real k_z at some heights and imaginary k_z at other heights. The height where $k_z{}^2$ goes to zero is a reflection level. It is possible for some types of waves to be trapped between two reflection levels, or between one atmospheric reflection level and the reflecting ground, and to be effectively "ducted" in the horizontal direction. For example, high-frequency acoustic waves can be ducted in cold atmospheric regions, where c is low, so long as their horizontal phase trace velocity is less than the speed of sound in the neighboring warm regions. Conversely, long-period gravity waves tend to propagate more readily in warm atmospheric regions, and can be reflected from cold layers. Gravity-wave ducting occurs mainly by reflections between the ground and the cold mesopause. A general analysis of wave ducting has been carried out by Francis (1973).

4.8. Vertical energy flux

From (4.35) onward, we have been treating p', v_z, etc., as complex quantities, from which real physical quantities are to be derived by taking the real part. With a star denoting the complex conjugate of a quantity, the real vertical wave energy flux, which from (4.19) is seen to be $p'v_z$ in the absence of heat flux and viscosity, is

$$\left(\frac{p'+p'^*}{2}\right)\left(\frac{v_z+v_z^*}{2}\right) = \frac{\mathrm{Re}(p'v_z)}{2} + \frac{\mathrm{Re}(p'v_z^*)}{2} \tag{4.46}$$

If we average over one wave period or one horizontal wavelength, the fact that both p' and v_z have a $e^{i\omega t + i\vec{k}_h\cdot\vec{x}}$ dependence means that the first quantity on the right-hand side of (4.46) will vanish but that the second term will in general not vanish (unless p' or v_z is zero or unless they are 90° out of phase). We shall therefore consider only the second term on the right of (4.46) in determining the vertical energy flux. From (4.41)-(4.43) we see that

$$v_z = \frac{c^2\omega}{\gamma(N^2-\omega^2)}\left[k_z - i\left(\frac{1}{\gamma H} - \frac{1}{2H}\right)\right]\frac{p'}{p} \tag{4.47}$$

so that

$$\frac{\mathrm{Re}(p'v_z^*)}{2} = \frac{c^2\omega\,\mathrm{Re}(k_z)}{\gamma(N^2-\omega^2)}\;\frac{p'p'^*}{2p} \tag{4.48}$$

We note first of all that vertical energy flux exists only if k_z is real, i.e., only if the wave has a vertical component of propagation. If k_z is positive, our definitions of phase factors imply downward phase propagation. That is, the argument $i\omega t + i\vec{k}_h\cdot\vec{x} + ik_z z$ of the exponent represents a constant phase if z decreases as t increases, for $\vec{x}$ held constant. When k_z is positive, (4.48) yields a negative (downward) flux when $\omega > N$ (acoustic waves), but a positive (upward) flux when $\omega < N$ (gravity waves). The direction of energy flux reverses when k_z changes sign. Therefore, vertical wave energy flux has the same sign as the vertical phase trace velocity for acoustic waves, but the opposite sign for gravity waves.

We note also that the wave energy flux is constant with altitude. This can be seen by multiplying (4.39) by v_z^*, multiplying the complex conjugate of (4.40) by p', adding the two results, and taking the real part to show that

$$\frac{\partial}{\partial z} \mathrm{Re}(p' v_z^*) = 0 \tag{4.49}$$

This result is valid even in a height-varying atmosphere, provided that source and dissipation terms are negligible.

4.9. Wave energy density

Let us also evaluate wave energy density in terms of $p'p'^*$, as we did for wave energy flux in (4.48). To do this we need to express $\vec{v}_h$, v_z, and s' in terms of p'. An expression for v_z already exists in (4.47). An expression for $\vec{v}_h$ is obtained from (4.35):

$$\vec{v}_h = -\frac{\vec{k}_h}{\rho\omega} p' = -\frac{\vec{k}_h c^2}{\gamma\omega} \frac{p'}{p} \tag{4.50}$$

An expression for s' is obtained by combining (4.37) and (4.47):

$$s' = \frac{c_p N^2 c^2}{\gamma g (N^2 - \omega^2)} \left[i k_z + \left(\frac{1}{\gamma H} - \frac{1}{2H} \right) \right] \frac{p'}{p} \tag{4.51}$$

From (4.47) and (4.50) we find the following expression for kinetic energy density:

$$KE = \frac{\rho v^2}{2} = \frac{\rho \mathrm{Re}(\vec{v}_h \cdot \vec{v}_h^* + v_z v_z^*)}{4}$$

$$= \frac{c^2}{4\gamma} \left[\frac{k_h^2}{\omega^2} + \frac{\omega^2}{(N^2-\omega^2)^2} \left| k_z - i\left(\frac{1}{\gamma H} - \frac{1}{2H} \right) \right|^2 \right] \frac{p'p'^*}{p} \tag{4.52}$$

With the aid of (4.51) we find the following expression for potential energy density:

$$PE = \frac{p'^2}{2\gamma p} + \frac{\rho g^2 s'^2}{2 c_p^2 N^2} = \mathrm{Re}\left(\frac{p'p'^*}{4\gamma p} + \frac{\rho g^2}{4 c_p^2 N^2} s's'^* \right)$$

$$= \frac{1}{4\gamma} \left[1 + \frac{N^2 c^2}{(N^2-\omega^2)^2} \left| i k_z + \left(\frac{1}{\gamma H} - \frac{1}{2H} \right) \right|^2 \right] \frac{p'p'^*}{p} \tag{4.53}$$

These two expressions simplify considerably when k_z is real. For this case, we can make use of the dispersion relation (4.45) to find

$$\left| k_z - i\left(\frac{1}{\gamma H} - \frac{1}{2H} \right) \right|^2 = \left| i k_z + \left(\frac{1}{\gamma H} - \frac{1}{2H} \right) \right|^2$$

$$= k_z^2 + \left(\frac{2-\gamma}{2\gamma H} \right)^2 = \frac{N^2}{c^2} \left(1 - \frac{\omega^2}{N^2} \right) \left(\frac{c^2}{(\omega/k_h)^2} - 1 \right) \tag{4.54}$$

which when used in (4.52) and (4.53) gives

$$KE = PE = \left(\frac{c^2}{(\omega/k_h)^2} - \frac{\omega^2}{N^2}\right)\left(1 - \frac{\omega^2}{N^2}\right)^{-1} \frac{p'p'^*}{4\gamma p} \tag{4.55}$$

This result shows that there is equipartition of wave energy between kinetic and potential forms for propagating waves. This condition is frequently, but not always, found in the theory of physical waves. For atmospheric waves, equipartition is found no longer to hold when we include Coriolis effects, as will be seen in the discussion of atmospheric tides.

4.10. Properties of long-period gravity waves

Gravity waves are generally more important than acoustic waves in the thermosphere. Since gravity waves also have less familiar properties than acoustic waves, let us concentrate on exploring gravity wave properties in somewhat more detail. For long-period gravity waves, $\omega^2 << N^2$, and the dispersion relation (4.45) simplifies to

$$k_z^2 = N^2 \frac{k_h^2}{\omega^2} - \frac{N^2}{c^2} - \left(\frac{2-\gamma}{2\gamma H}\right)^2 \tag{4.56}$$

which becomes upon use of (4.29) for N^2 (with $dH/dz = 0$)

$$k_z^2 = N^2 \frac{k_h^2}{\omega^2} - \frac{1}{4H^2} \tag{4.57}$$

We see that gravity waves can propagate vertically if their horizontal phase trace velocity ω/k_h is less than $2NH$, which for an isothermal atmosphere is slightly less than the speed of sound.

To get group velocities, we rewrite (4.57) as

$$\omega^2 = \frac{N^2 k_h^2}{k_z^2 + 1/(4H^2)} \tag{4.58}$$

Taking derivatives of (4.58) with respect to k_h and k_z (assumed real) yields

$$\frac{\partial\omega}{\partial k_h} = \frac{\omega}{k_h} = u_h \tag{4.59}$$

$$\frac{\partial\omega}{\partial k_z} = -\frac{\omega}{k_z}\frac{k_z^2}{k_z^2 + 1/(4H^2)} = -\frac{\omega}{k_z}\left(1 - \frac{u_h^2}{4N^2H^2}\right) \tag{4.60}$$

The horizontal group velocity is the same as the horizontal phase trace velocity, and both are denoted u_h. On the other hand, the vertical group velocity is in the opposite direction to, and is smaller in magnitude than, the vertical phase trace velocity ω/k_z. We can also determine that the vertical group velocity is small in comparison with the horizontal group velocity:

$$\left| \frac{\partial\omega/\partial k_z}{\partial\omega/\partial k_h} \right| = \frac{\omega}{N} \left(1 - \frac{u_h^{\,2}}{4N^2H^2}\right)^{\frac{1}{2}} \tag{4.61}$$

where we remember that $\omega << N$.

Although we shall not go through the mathematics here, it is fairly easy to show that the group velocity equals the wave energy flux divided by the wave energy density, for both the vertical and horizontal components. The energy flux of long-period gravity waves is therefore primarily horizontal, with the direction of energy flux differing from the horizontal by an angle no greater than $\tan^{-1}(\omega/N)$.

The air motions in long-period gravity waves are likewise mainly horizontal. From (4.52) and (4.54) we see that (with k_z real)

$$\frac{v_z\, v_z^*}{\vec{v}_h \cdot \vec{v}_h^*} = \frac{\omega^2}{N^2} \left(1 - \frac{u_h^{\,2}}{c^2}\right) \tag{4.62}$$

The ratio of vertical to horizontal wind velocity magnitudes is similar to the ratio of vertical to horizontal group velocity components unless u_h approaches its limiting value 2NH, in which case the vertical group velocity tends to vanish while v_z remains finite.

4.11. Gravity wave energy dissipation

Wave dissipative processes become very important in the thermosphere. Let us examine dissipation of gravity wave energy by evaluating the right-hand side of the wave energy equation (4.19) in the absence of heat sources and electric fields. We make use of the approximations that vertical gradients are much stronger than horizontal gradients and that the horizontal velocity is large compared to the vertical velocity. Then the right-hand side of (4.19) becomes

$$- \frac{\kappa}{T} \left(\frac{\partial T'}{\partial z}\right)^2 - \mu \left(\frac{\partial \vec{v}_h}{\partial z}\right)^2 - \sigma_1\, B^2\, \vec{v}_h \cdot \vec{v}_\perp \tag{4.63}$$

representing the negative of the wave energy dissipation rate per unit volume. A time scale for wave energy dissipation can be determined by comparing (4.63) with the wave energy density. To do this we assume for simplicity that dissipation is not strong enough to alter the properties of propagating waves. For long-period waves the wave energy density using (4.55) is

$$\text{KE} + \text{PE} = \frac{c^2}{u_h^{\ 2}} \; \frac{p'p'^*}{2\gamma p} \tag{4.64}$$

In order to compare (4.63) with (4.64) let us express the terms in (4.63) as functions of $p'p'^*$. Concerning T', we first make use of the fact that long-period gravity waves are hydrostatic. If we write (3.2) as

$$\frac{\partial}{\partial z}(\ell n\ p) = -\frac{g}{RT} \tag{4.65}$$

and take perturbations (assuming for simplicity that R and g are constant) we find

$$\frac{\partial}{\partial z}\left(\frac{p'}{p}\right) = \frac{g}{RT^2}\, T' = \frac{T'}{TH} \tag{4.66}$$

With the vertical variation of p' prescribed by (4.41), (4.66) yields

$$\frac{T'}{T} = \left(\frac{1}{2} + ik_zH\right)\frac{p'}{p} \tag{4.67}$$

Then for an isothermal background atmosphere

$$\frac{\partial T'}{\partial z} = \frac{T}{H}\left(\frac{1}{2} + ik_zH\right)^2 \frac{p'}{p} \tag{4.68}$$

and the first term in (4.63) becomes

$$\begin{aligned}
-\frac{\kappa}{T}\left(\frac{\partial T'}{\partial z}\right)^2 &= -\frac{\kappa}{2T}\,\text{Re}\left(\frac{\partial T'}{\partial z}\,\frac{\partial T'^*}{\partial z}\right) \\
&= -\frac{\kappa T}{2H^2}\left(\frac{1}{4} + k_z^{\ 2}H^2\right)^2 \frac{p'p'^*}{p^2} = -\frac{\kappa T}{2H^2}\left(\frac{N^2H^2}{u_h^{\ 2}}\right)^2 \frac{p'p'^*}{p^2} \\
&= -\frac{\kappa T N^2 \gamma}{c^2 p}\left(\frac{N^2H^2}{u_h^{\ 2}}\right)(\text{PE} + \text{KE})
\end{aligned} \tag{4.69}$$

This expression can be further simplified upon replacing N^2 by $g^2/(c_pT)$ from (4.29) (with RT constant), c^2 by γgH from (4.24), and p by ρgH from (3.3):

$$-\frac{\kappa}{T}\left(\frac{\partial T'}{\partial z}\right)^2 = -\nu_c\,(\text{PE} + \text{KE}) \tag{4.70}$$

$$\nu_c \equiv \frac{\kappa}{c_p \rho H^2}\,\frac{N^2H^2}{u_h^{\ 2}} \tag{4.71}$$

The quantity ν_c has the units of inverse time, and is the rate at which wave energy is dissipated by heat conduction. The viscous and ion drag terms in (4.63) can be similarly analyzed to yield

$$- \mu \left(\frac{\partial \vec{v}_h}{\partial z} \right)^2 - \sigma_1 B^2 v_h^2 = - (\nu_v + \nu_\sigma) (PE + KE) \qquad (4.72)$$

$$\nu_v = \frac{\mu}{\rho H^2} \frac{N^2 H^2}{u_h^2} \qquad (4.73)$$

$$\nu_\sigma = \frac{\sigma_1 B^2}{\rho} \qquad (4.74)$$

where we have assumed a vertical magnetic field so that $\vec{v}_\perp$ can be replaced by $\vec{v}_h$.

Wave energy dissipation by heat conduction and by viscosity proceeds at comparable rates. In fact, comparison of (4.71) and (4.73) upon use of (2.49) shows that

$$\nu_c = 1.5 \nu_v \qquad (4.75)$$

We can therefore consider ν_c and ν_v together and refer to their combined effect as "molecular dissipation." Inspection of (4.71) and (4.73) shows that the time scale for molecular dissipation is closely related to the time scales for viscous and heat conduction effects listed in Table 1. The dissipation rate increases rapidly with altitude because of the $1/\rho$ dependence. In addition, the molecular dissipation rate depends on the horizontal speed of the waves, u_h, with respect to the limiting wave speed, 2NH. Slower waves have shorter vertical wavelengths and are therefore dissipated more quickly. Figure 12 shows the altitude variation of the molecular dissipation coefficient $\nu_c + \nu_v$ for two wave velocities, 300 m/s and 600 m/s (Richmond, 1978). Because slow-moving waves are rapidly dissipated in the upper thermosphere, any waves observed above 200 km tend to have high horizontal speeds.

Wave dissipation by ion drag is proportional to the ion density, and therefore changes significantly with changing ionospheric conditions. Figure 12 shows typical dissipation rates for midlatitude daytime and nighttime conditions. Ion drag is usually somewhat less important than molecular dissipation. Note that the ion drag coefficient ν_σ in (4.74) is just the inverse of the time scale τ_σ in (3.36) derived for examining the importance of ion drag on large-scale dynamics.

4.12. Atmospheric tides

As we go to lower frequencies, the effects of the Coriolis force become important for gravity waves. As we go to longer horizontal wavelengths, we must take account of the fact that the earth's atmosphere has a finite horizontal extent, so that arbitrary wavelengths cannot be specified. Instead, a quantization of the horizontal wave structure occurs, essentially due to the fact that only integral numbers of wavelengths can exist in one circumference of the earth.

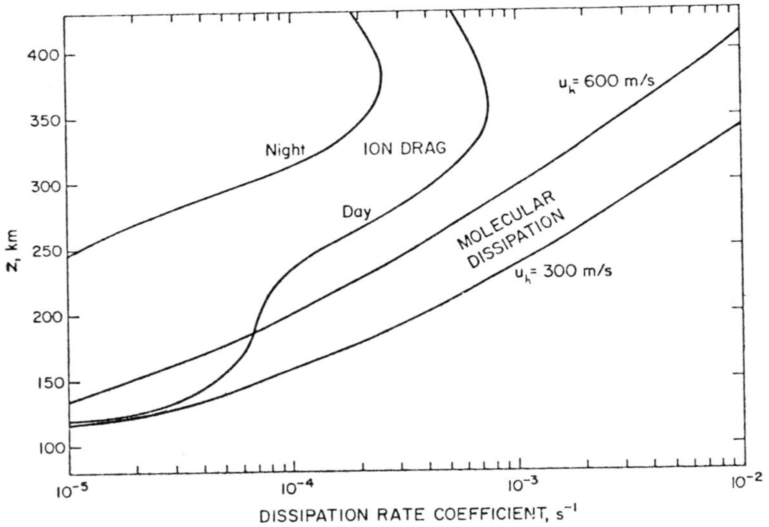

Figure 12. Gravity wave dissipation rate coefficients.

Atmospheric tides are global-scale gravity waves excited either by the gravitational attraction of the moon and sun, or, more importantly, by daily variations of atmospheric solar heating. The most important tides have periods of 24 or 12 solar hours or 12 "lunar hours" and they migrate westward around the earth with the apparent solar or lunar position. Tides are the dominant global-scale motions in the thermosphere, unlike the lower atmosphere where the dominant motions have significantly longer time scales. The strong daily variations of solar ultraviolet absorption in the thermosphere produce diurnal variations of the winds, as noted in the previous section. In addition, tides are generated at lower altitudes as solar radiation is absorbed by ozone and by water vapor. These tides can propagate upward into the thermosphere, whereupon their amplitudes tend to grow exponentially as the density decreases.

The general mathematical development of tidal theory is complicated by the facts that the earth is spherical and that the Coriolis force acting on a horizontal wind varies with latitude (see Chapman and Lindzen, 1970; Kato, 1980). For an idealized atmosphere which has no mean winds and whose temperature is horizontally uniform it is convenient to treat tides in terms of discrete modes, each of which has separable horizontal and vertical structure equations. Each mode varies in time and longitude as exp $(i\omega t + im\phi)$, where m is an integer. For the dominant tides, m = 1 for the diurnal (24 hour) period and m = 2 for the semidiurnal (12 hour) period, so that the westward phase velocity matches the apparent velocity of the sun or moon. For each period and value of m an infinite set of discrete latitudinal structures is mathematically allowed, although only a few of the lowest-order modes (those with the simplest

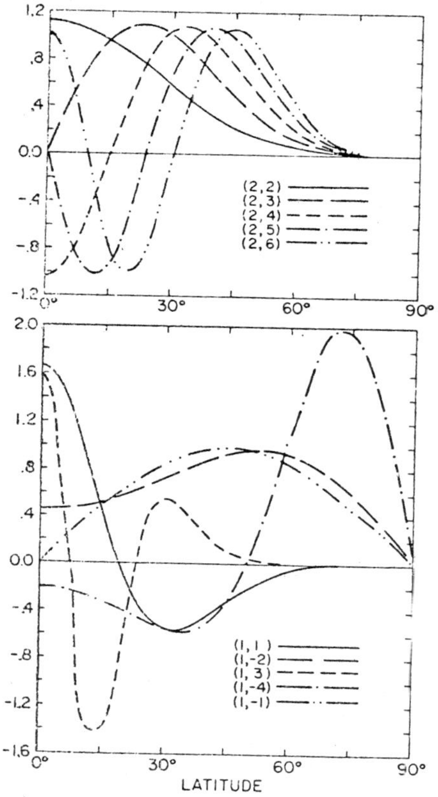

Figure 13. Semidiurnal (top) and diurnal (bottom) Hough functions.

structure) can be expected to dominate in the atmosphere. Figure 13 (Forbes and Garrett, 1979) shows the latitudinal structure of a number of solar tidal modes, called Hough functions. Each function is distinguished by a pair of integers [e.g., (1,-2)], of which the first gives m and the second labels the latitudinal structure of the mode. Negative and positive integers for the diurnal functions represent two classes of modes. These curves represent the relative latitudinal variations of temperature, pressure, density, and vertical velocity perturbations for individual modes. The actual amplitude and phase of any quantity are height-varying and depend on the solution of the vertical structure equation for that mode. In general several modes coexist and are usually assumed to superimpose linearly.

When realistic mean horizontal structure of the atmosphere is considered, distinct separable modes no longer exist. The modes effectively become coupled and calculations become considerably more difficult. Forbes and Garrett (1979) and Forbes (1982a,b) provide comprehensive coverage of sophisticated atmospheric tidal calculations.

Although tidal theory will not be explored here in detail, we can rather easily understand a few characteristics of tides by adding Coriolis effects to the theory of long-period gravity waves. Let us examine gravity waves in a plane, uniformly rotating atmosphere, rotating at the angular frequency $\vec{\Omega}^*$, which is meant to represent the vertical component of $\vec{\Omega}$ at some colatitude θ^*, i.e.,

$$\vec{\Omega}^* = \hat{z}\Omega\cos\theta^* \tag{4.76}$$

The wave energy equations (4.16) and (4.19) remain valid, as their derivations did not assume non-sphericity or non-rotation. Equation (4.35) gets an extra term to become

$$i\omega\rho\vec{v}_h + 2\Omega^*\rho\hat{z} \times \vec{v}_h + ik_h p' = 0 \tag{4.77}$$

Equation (4.40) must be replaced by

$$\frac{\partial v_z}{\partial z} - \frac{v_z}{\gamma H} - \frac{i\omega}{\gamma p}\left(\frac{c^2 k_h{}^2}{\omega^2 - 4\Omega^{*2}} - 1\right)p' = 0 \tag{4.78}$$

By comparing (4.40) and (4.78) we see that the dispersion relation can readily be obtained from (4.57) [the low-frequency limit of (4.45)] with the replacement of $k_h{}^2/\omega^2$ by $k_h{}^2/(\omega^2 - 4\Omega^{*2})$:

$$k_z{}^2 = \frac{N^2 k_h{}^2}{\omega^2 - 4\Omega^{*2}} - \frac{1}{4H^2} \tag{4.79}$$

The effect which rotation has on the dispersion relation can be very important. If $\omega^2 > 4\Omega^{*2}$, the introduction of rotation tends to make $k_z{}^2$ more positive, so that vertical wavelengths are shortened, or $k_z{}^2$ can be brought from negative to positive, so that vertical propagation is facilitated. If, however, rotation effects are so strong that $\omega^2 < 4\Omega^{*2}$, then we see that $k_z{}^2$ is invariably negative, so that vertical propagation is impossible. For diurnal tides $\omega^2 < 4\Omega^{*2}$ if the effective mean colatitude θ^* is less than about 60°, as is the case for modes like (1,-2) and (1,-4) which tend to be concentrated toward higher latitudes. Such modes invariably have negative values of $k_z{}^2$. Modes like (1,1) and (1,3) are concentrated at low latitudes where ω^2 is effectively greater than $4\Omega^{*2}$, and such modes propagate vertically with relatively large positive values of $k_z{}^2$.

Concerning energy relations for gravity waves in a rotating atmosphere, the vertical wave energy flux is still given by (4.48) and the potential energy density by (4.53). The kinetic energy density is no longer represented by (4.52), however. Instead of (4.50), the horizontal velocity is related to p' by the solution of (4.77), namely

$$\vec{v}_h = -\frac{\omega\vec{k}_h - 2i\Omega^*\hat{z} \times \vec{k}_h}{\omega^2 - 4\Omega^{*2}}\frac{c^2}{\gamma}\frac{p'}{p} \tag{4.80}$$

so that (with $v_z << v_h$)

$$KE = \frac{\rho \, Re(\vec{v}_h \cdot \vec{v}_h^*)}{4} = \frac{c^2}{4\gamma} k_h^2 \frac{(\omega^2 + 4\Omega^{*2})}{(\omega^2 - 4\Omega^{*2})^2} \frac{p' p'^*}{p} \tag{4.81}$$

If k_z is real,

$$KE = \frac{\omega^2(\omega^2 + 4\Omega^{*2})}{(\omega^2 - 4\Omega^{*2})^2} \; PE \tag{4.82}$$

and KE always exceeds PE. Rotation also alters the expression for group velocity of gravity waves, such that the vertical component of group velocity is less than that given by (4.60).

5. ELECTRODYNAMICS

5.1. Introduction

Up to this point we have treated the electric field $\vec{E}$ as an independently specified quantity as it enters into the equations of thermospheric dynamics. Actually, the electric field is interrelated with thermospheric winds, and at middle and low latitudes is largely determined by the wind distribution. This section focuses on the electric fields and currents set up by thermospheric winds. A review of recent literature is given by Richmond (1979a).

On the sunlit side of the earth there normally exist two large vortices of electric current in the ionosphere, as illustrated in Fig. 14. The coordinates are oriented with respect to the geomagnetic field. Current flows counterclockwise in the northern hemisphere and clockwise in the southern hemisphere, with 10^4A between contour lines. The total amount of current in each vortex is 170 kA. This figure comes from a numerical simulation, and does not include any effects associated with magnetospheric interactions at high latitudes. Even during geomagnetically quiet conditions magnetospheric effects tend to dominate the actual current pattern at latitudes above 65°, and so we should focus here only on the lower latitude portion of the figure. These currents are commonly referred to as "Sq currents", where S refers to solar daily variations (as opposed to weaker lunar daily variations which also exist) and q refers to geomagnetically quiet conditions. These currents were originally inferred in the last century from daily fluctuations observed in the ground-level magnetic field, fluctuations with amplitudes of only about 10^{-3} the magnitude of the main geomagnetic field.

Notice that there is a concentration of current contours near the magnetic equator. This concentration represents the equatorial electrojet, which occurs where the main geomagnetic field is nearly horizontal, and will be discussed more a little later. It should also be mentioned that the current pattern in Fig. 14 is for equinox conditions, assuming

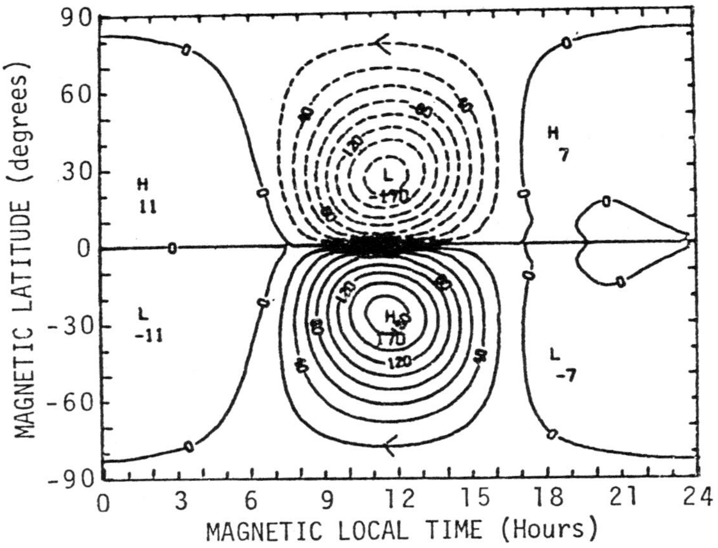

Figure 14. Height-integrated ionospheric currents in kA.

strict symmetry of the thermospheric winds and electrical conductivities about the magnetic equator. This symmetry breaks down when we examine solstice conditions, for example.

Electric fields are also associated with the ionospheric currents. Figure 15 (from Richmond et al., 1980) shows a pattern of the electrostatic potential for average quiet-day ionospheric conditions at equinox. This pattern was derived from actual measurements of plasma E × B drifts in the ionosphere at around 300 km altitude. These measurements represent conditions only below about 65° magnetic latitude, since no higher latitude measurements were included. As with the electric currents, the ionospheric electric fields at high latitudes tend to be dominated by magnetospheric interactions. The contour interval is 1000 V, and the total potential difference at low latitudes is about 6 kV.

5.2. Basic equations

To calculate ionospheric electric fields and currents we make use of Ohm's Law (2.61) and two of Maxwell's equations:

$$\nabla \times \vec{E} = -\frac{\partial \vec{B}}{\partial t} \tag{5.1}$$

$$\nabla \times \vec{B} = \mu_o \vec{J} + \mu_o \varepsilon_o \frac{\partial \vec{E}}{\partial t} \tag{5.2}$$

In the ionosphere we are dealing with current densities on the order of

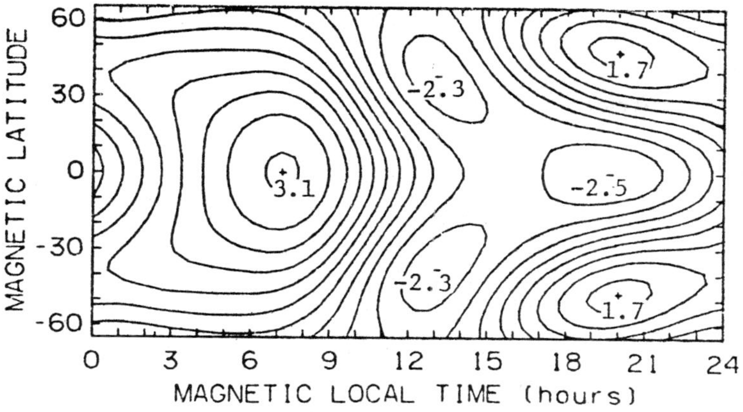

Figure 15. Ionospheric electrostatic potential in kV.

10^{-6} A/m² and electric fields on the order of 10^{-2} V/m or less. Comparing the order of magnitude of the two terms on the right of (5.2) for variations with a characteristic time scale τ, we find

$$\mu_o J = 4\pi \times 10^{-7} J \sim 10^{-6} \cdot 10^{-6} = 10^{-12} \tag{5.3}$$

$$\mu_o \varepsilon_o \frac{\partial E}{\partial t} \sim \frac{10^{-2}}{(3 \times 10^8)^2 \tau} \sim \frac{10^{-19}}{\tau} \tag{5.4}$$

We are normally dealing with time scales of seconds to hours ($\tau > 1$ s), so that (5.4) is in every case several orders of magnitude smaller than (5.3), and is henceforth ignored.

Strictly speaking, Maxwell's equations are valid only for a non-rotating frame of reference. It is more convenient, however, to work in a frame of reference rotating with the earth, not only because that is the frame in which most measurements are made (as, for example, those represented by Fig. 15), but also because the term $\partial \vec{B}/\partial t$ in (5.1) is quite large in a non-rotating frame due to zonal harmonics in the earth's main field. Fortunately, it turns out that (5.1) and (5.2) can be applied to a rotating frame of reference with more or less the same degree of validity as we had in neglecting the $\partial \vec{E}/\partial t$ term of (5.2). We can then neglect that part of $\partial \vec{B}/\partial t$ associated with the main geomagnetic field and consider only that part associated with the ionospheric currents.

Considerable simplification can be introduced if the currents change slowly enough that $\partial \vec{B}/\partial t$ can be neglected altogether in (5.1). For a change of ΔB on a time scale of τ over a spatial scale of L, the magnitude of the induced electric field is on the order of

$$E_{induced} \sim \frac{L}{\tau} \Delta B \tag{5.5}$$

Using typical values of $\Delta B = 30 \times 10^{-9}$ T and $L = 3 \times 10^6$ m this becomes

$$E_{induced} \sim \frac{10^{-1}}{\tau} \tag{5.6}$$

The normal electric fields at middle and low latitudes are on the order of 10^{-3} V/m in the rotating frame of reference, so we see from (5.6) that the induced field can be neglected if τ is long in comparison with 100 s. We shall confine our attention to such quasi-static conditions. Then $\vec{E}$ is curl-free, so that it can be represented by an electrostatic potential V:

$$\vec{E} = - \nabla V \tag{5.7}$$

In addition to (5.7), we shall use two more equations. The first of these is found by taking the divergence of (5.2):

$$\nabla \cdot \vec{J} = 0 \tag{5.8}$$

The second is just Ohm's Law (2.61), which we duplicate here for convenience:

$$\vec{J} = \sigma_0 \vec{E}_{\parallel} + \sigma_1 (\vec{E}_{\perp} + \vec{v} \times \vec{B}) + \sigma_2 \hat{b} \times (\vec{E} + \vec{v} \times \vec{B}) \tag{5.9}$$

It is normally assumed that $\vec{B}$ is well represented by the main geomagnetic field, so that if we are given the conductivities and $\vec{v}$, we can solve (5.7)-(5.9) for V, $\vec{E}$, and $\vec{J}$.

5.3. Horizontal and vertical current densities

Large-scale vertical current densities are generally much smaller than horizontal current densities within the dynamo region, defined as that altitude range where the transverse conductivities σ_1 and σ_2 are significant (see Fig. 5). At the base of the dynamo region the vertical current density J_z essentially vanishes. Above this level J_z can be calculated from (5.8):

$$J_z = -\int_0^z \nabla \cdot \vec{J}_h dz' \tag{5.10}$$

where $\vec{J}_h$ is the horizontal current density and where we have neglected the small variations of geocentric radial distance r with height. To estimate orders of magnitude, let L_h be a typical horizontal scale and L_z a typical vertical scale. Then

$$|J_z| \sim \frac{|J_h|}{L_h} L_z \tag{5.11}$$

The vertical scale is on the order of 30 km, while in midlatitude regions L_h is on the order of 3000 km, so that J_z/J_h is on the order of 10^{-2}. In the equatorial and auroral regions L_h can be smaller, more

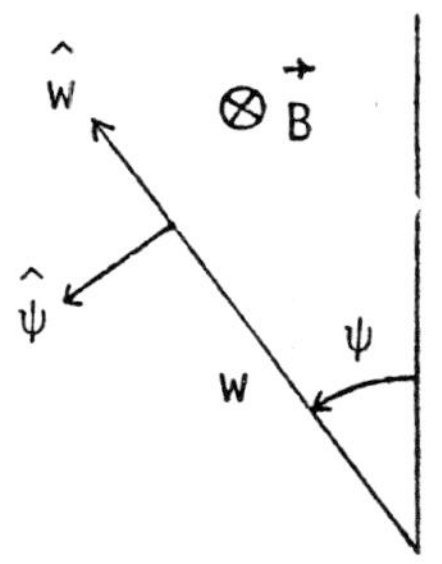

Figure 16. Polar coordinate definition.

like 300 km for the gross characteristics. For these regions J_z/J_h can be on the order of 10^{-1}. A typical midlatitude value of J_h is 10^{-6} A/m², so J_z is on the order of 10^{-8} A/m² at midlatitudes.

5.4. Dynamo action by winds

Thermospheric winds provide the energy source needed to drive the so-called ionospheric dynamo which maintains the system of electric currents and fields. To understand better how winds give rise to electric fields and currents in the presence of anisotropic conductivities, let us first examine two idealized cases. In both cases we assume a vertical magnetic field and uniform conductivities within a thin slab of infinite horizontal extent, with no current entering or leaving the slab from above or below. We use horizontal polar coordinates w,ψ as shown in Fig. 16, where w is the radial distance from the origin and ψ is the azimuthal angle.

Nondivergent rotational wind. Let the wind velocity be of the form

$$\vec{v} = v_\psi(w)\,\hat{\psi} \qquad (5.12)$$

where v_ψ is some function of w only and $\hat{\psi}$ is a unit vector in the ψ-direction. For simplicity of discussion let us assume that v_ψ is positive. This wind circulates counterclockwise around the origin of our coordinate system. If $\hat{b}$ is a unit vector directed downward, the wind generates a "dynamo" component of current given by

$$\vec{J}_d = -\sigma_1 v_\psi B\hat{w} + \sigma_2 v_\psi B\hat{\psi} \qquad (5.13)$$

The radial component of this current is convergent, and since we allow no current to leave the slab from above or below, the convergence cannot be offset by diverging vertical currents. Space charges very quickly are set up to the point where an additional electric-field-driven current is established whose divergence exactly cancels the convergence of the wind-driven current. In our geometry the electric field must be strictly radial in order to be electrostatic:

$$\vec{E} = E_w(w)\hat{w} \tag{5.14}$$

The current driven by this field is

$$\vec{J}_e = \sigma_1 E_w \hat{w} - \sigma_2 E_w \hat{\psi} \tag{5.15}$$

The sum of radial components of $\vec{J}_d$ plus $\vec{J}_e$ must vanish to have a divergence-free current, so that

$$E_w = v_\psi B \tag{5.16}$$

or in vectoral form

$$\vec{E} = -\vec{v} \times \vec{B} \tag{5.17}$$

We readily see that with an electric field of the form (5.17) <u>no</u> electric current will flow, either in the radial or in the azimuthal direction. The electrostatic field $\vec{E}$ exactly cancels the "dynamo" electric field $\vec{v} \times \vec{B}$.

This example is illustrative of the more general condition that a wind system will generate an electric field but no current if it everywhere satisfies

$$\nabla \times (\vec{v} \times \vec{B}) = 0 \tag{5.18}$$

or, equivalently,

$$\vec{v} \times \vec{B} = \nabla V \tag{5.19}$$

which gives the condition

$$\vec{v} = -\frac{\nabla V \times \vec{B}}{B^2} + v_\parallel \hat{b} \tag{5.20}$$

where $v_\parallel$ is an arbitrary velocity parallel to $\vec{B}$. We see that V represents the electrostatic potential generated by this wind system, and that the ionization E × B drift velocity is identical to the component of neutral velocity perpendicular to $\vec{B}$. Note that a wind system satisfying (5.18) has very special properties. From (5.19) we see that ∇V must be everywhere perpendicular to $\vec{B}$, and that V must therefore be constant along any given field line. (This requirement has nothing to do with high electrical conductivity along the field line.) The wind velocity $\vec{v}$ must therefore vary in a particular manner along $\vec{B}$; in the case of uniform $\vec{B}$, the perpendicular velocity would have to be constant along a field line. The special case we examined indeed satisfied (5.18), but thermospheric winds in general do not satisfy this condition.

<u>Irrotational divergent wind</u>. As the second example let us examine the dynamo consequences of a wind given by

$$\vec{v} = v_w(w)\hat{w} \tag{5.21}$$

which blows everywhere away from the origin of our polar coordinate system. The dynamo component of current is

$$\vec{J}_d = \sigma_1 v_w B\hat{\phi} + \sigma_2 v_w B\hat{w} \tag{5.22}$$

Again a radial polarization electric field will be set up which will drive a current of the form (5.15). Balancing the radial components of current in this case yields

$$E_w = -\frac{\sigma_2}{\sigma_1} v_w B \tag{5.23}$$

The total current does not vanish, but is found from (5.15), (5.22), and (5.23) to be

$$\vec{J} = \vec{J}_d + \vec{J}_e = (\sigma_1 + \frac{\sigma_2^{\,2}}{\sigma_1}) v_w B\hat{\phi} \equiv \sigma_3 v_w B\hat{\phi} \tag{5.24}$$

The total current in this example circulates around the origin in continuous circuits. The Hall current driven by the electrostatic field supplements the Pedersen current driven by the dynamo electric field. The effective conductivity σ_3, called the Cowling conductivity, is larger than the Pedersen conductivity σ_1. In the ionosphere, the height-averaged value of σ_2 is similar to the height-averaged value of σ_1, so the effective conductivity σ_3 for this type of wind system would be about twice the size of σ_1.

5.5. Approximation of infinitely conducting magnetic field lines

The parallel conductivity σ_0 is more than two orders of magnitude greater than the transverse conductivities σ_1 and σ_2 at all heights above 100 km. Modelling work has shown that the parallel and perpendicular components of current density are generally comparable in the ionosphere. Consequently, the parallel electric field must generally be at least two orders of magnitude smaller than the perpendicular electric field, and electric potential differences along a field line are quite small. This fact leads to a very useful approximation in calculating large-scale electric fields and currents: the assumption that magnetic field lines are equipotentials. Assuming that the parallel electric field is zero must be accompanied by an assumption of infinite parallel conductivity so that parallel currents can flow. The parallel electric field is then eliminated as a dependent variable in our equations. In order to determine $J_{\parallel}$, we cannot use Ohm's Law, but must instead integrate the divergence of the perpendicular current $\vec{J}_{\perp}$ along the field line, where $\vec{J}_{\perp}$ can still be obtained from Ohm's Law. That is,

$$\nabla\cdot\vec{J} = \nabla\cdot\left(J_{\parallel}\frac{\vec{B}}{B} + \vec{J}_{\perp}\right) = 0 \tag{5.25}$$

$$\vec{B}\cdot\nabla\left(\frac{J_{\parallel}}{B}\right) = -\nabla\cdot\vec{J}_{\perp} \tag{5.26}$$

$$J_{\parallel} = -B\int_{\ell_0}^{\ell} \frac{\nabla\cdot\vec{J}_{\perp}}{B}\, d\ell' \tag{5.27}$$

where $d\ell'$ is an element of distance along the magnetic field, and ℓ_0 is some point on the field line where $J_{\parallel}$ vanishes, e.g., a point in the atmosphere below the conducting ionosphere.

The approximation of infinitely conducting field lines is generally valid not only within the ionosphere, but also through the plasmasphere between magnetically conjugate points in the ionospheres of the northern and southern hemispheres. At latitudes below about 60°, the plasma along geomagnetic field lines is dense enough and cold enough that it can be treated as a collisional plasma. Assuming only electron-ion collisions to be important, with singly charged ions, the formula for σ_0 is (e.g., Shkarofsky, 1961)

$$\sigma_0 = \frac{0.591}{m_e^{1/2}} \frac{(kT_e)^{3/2}}{\ln\Lambda} \left(\frac{4\pi\varepsilon_0}{e}\right)^2 \tag{5.28}$$

$$\Lambda = \frac{3}{2e^3} \frac{(4\pi\varepsilon_0 kT)^{3/2}}{(\pi n_e)^{1/2}} \tag{5.29}$$

where m_e is the electron mass, e its charge, T_e the electron temperature, and n_e the electron density. Using nominal values of $T_e = 3000$ K and $n_e = 10^9$ m^{-3}, σ_0 in the plasmasphere is 140 S/m. Models of large-scale interhemispheric field-aligned currents yield current densities on the order of 10^{-8} A/m^2 at middle latitudes. The parallel electric field needed to drive such currents is on the order of 10^{-10} V/m, which even for a field line length of 10^7 m would mean a potential difference of only 10^{-3} V between the tops of the ionospheres at conjugate points. In reality, larger parallel electric fields than this exist in the upper ionosphere associated with the gravitational, pressure, and frictional forces which we have neglected in our derivation of conductivities. However, even these larger fields lead to potential differences of less than 1 V between conjugate points. Thus for large-scale electric fields conjugate points in the northern and southern hemispheres are essentially at the same potential.

5.6. Dipole coordinates

Because the electrical conductivity is so intimately related to the geomagnetic field direction, it is useful to do calculations in a coordinate system linked to the field. A general coordinate system for a realistic magnetic field which contains quadrupole and higher-order

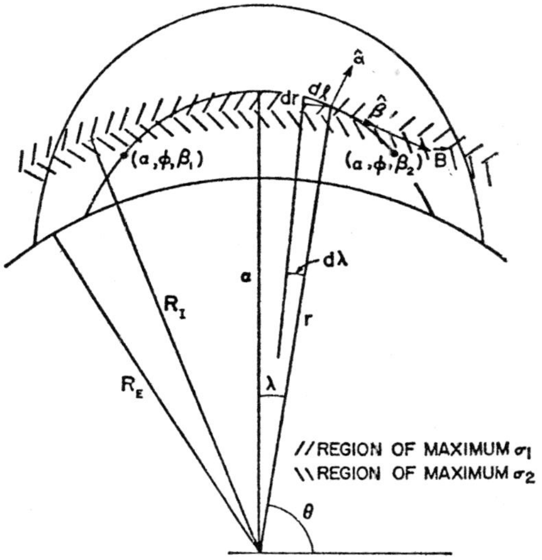

Figure 17. Dipole coordinate definition.

components is rather complicated, but for theoretical modelling it often suffices to treat the geomagnetic field as a pure dipole. Then we can use orthogonal dipole coordinates α, ϕ, β defined in terms of spherical coordinates r, θ, ϕ as

$$\alpha = r/\sin^2\theta \tag{5.30}$$

$$\phi = \phi \tag{5.31}$$

$$\beta = \cos\theta/r^2 \tag{5.32}$$

where $\theta = 0$ is defined to be the north geomagnetic pole and ϕ is geomagnetic east longitude. The associated metric coefficients are

$$h_\alpha = \sin^3\theta/(1 + 3\cos^2\theta)^{\frac{1}{2}} \tag{5.33}$$

$$h_\phi = r\sin\theta \tag{5.34}$$

$$h_\beta = r^3/(1 + 3\cos^2\theta)^{\frac{1}{2}} \tag{5.35}$$

The geometry of the dipole coordinates is shown in Fig. 17, in which the direction of increasing ϕ is directed out of the page. The unit vector $\hat{\beta}$ is directed along the magnetic field, and is thus equivalent to $\hat{b}$. In these coordinates (5.7)-(5.9) become (assuming $\partial V/\partial\beta = 0$)

$$\vec{E} = -\nabla V = -\frac{\hat{\alpha}}{h_\alpha}\frac{\partial V}{\partial\alpha} - \frac{\hat{\phi}}{h_\phi}\frac{\partial V}{\partial\phi} \tag{5.36}$$

$$\nabla\cdot\vec{J} = \frac{1}{h_\alpha h_\phi h_\beta}\left[\frac{\partial}{\partial\alpha}(h_\phi h_\beta J_\alpha) + \frac{\partial}{\partial\phi}(h_\beta h_\alpha J_\phi) + \frac{\partial}{\partial\beta}(h_\alpha h_\phi J_\beta)\right] = 0 \tag{5.37}$$

$$\vec{J} = \hat{\alpha}(\sigma_1 E_\alpha - \sigma_2 E_\phi) + \hat{\phi}(\sigma_2 E_\alpha + \sigma_1 E_\phi) + \hat{\alpha}(\sigma_1 v_\phi + \sigma_2 v_\alpha)B + \hat{\phi}(\sigma_2 v_\phi - \sigma_1 v_\alpha)B + \hat{\beta}J_\beta \tag{5.38}$$

5.7. Solution for V

If the wind and conductivity distributions are specified, (5.36)-(5.38) can be solved for the electric field and current density. One method of solution proceeds as follows.

First multiply (5.37) by $h_\alpha h_\phi h_\beta$ and integrate along a magnetic field line with respect to β between points β_1 and β_2 below the ionosphere (see Fig. 17). At these end points $J_\beta = 0$ so that terms involving J_β disappear:

$$\frac{\partial}{\partial\alpha}\int_{\beta_1}^{\beta_2} h_\phi h_\beta J_\alpha d\beta + \frac{\partial}{\partial\phi}\int_{\beta_1}^{\beta_2} h_\beta h_\alpha J_\phi d\beta = 0 \tag{5.39}$$

Next use (5.36) and (5.38) in (5.39) to obtain

$$\frac{\partial}{\partial\alpha}\left[S_\alpha \frac{\partial V}{\partial\alpha} - \Sigma_2 \frac{\partial V}{\partial\phi}\right] + \frac{\partial}{\partial\phi}\left[S_\phi \frac{\partial V}{\partial\phi} + \Sigma_2 \frac{\partial V}{\partial\alpha}\right] = \frac{\partial}{\partial\alpha}\left[\int_{\beta_1}^{\beta_2} h_\phi B(\sigma_1 v_\phi + \sigma_2 v_\alpha)d\ell\right] + \frac{\partial}{\partial\phi}\left[\int_{\beta_1}^{\beta_2} h_\alpha B(\sigma_2 v_\phi - \sigma_1 v_\alpha)d\ell\right] \tag{5.40}$$

$$S_\alpha \equiv \int_{\beta_1}^{\beta_2} \frac{h_\phi}{h_\alpha}\sigma_1\, d\ell \tag{5.41}$$

$$S_\phi \equiv \int_{\beta_1}^{\beta_2} \frac{h_\alpha}{h_\phi} \sigma_1 \, d\ell \tag{5.42}$$

$$\Sigma_2 \equiv \int_{\beta_1}^{\beta_2} \sigma_2 d\ell \tag{5.43}$$

where the element of length along the field line, $d\ell$, equals $h_\beta d\beta$.

Equation (5.40) is a second order elliptic partial differential equation for V. The inner boundary condition, at, say, $\alpha = R_E + 80$ km (the base of the conducting region), is

$$\int_{\beta_1}^{\beta_2} h_\phi J_\alpha d\ell = 0 \tag{5.44}$$

or

$$S_\alpha \frac{\partial V}{\partial \alpha} - \Sigma_2 \frac{\partial V}{\partial \phi} = \int_{\beta_1}^{\beta_2} h_\phi B(\sigma_1 v_\phi + \sigma_2 v_\alpha) d\ell \tag{5.45}$$

The outer boundary condition is more difficult to define realistically. As we go to values of α greater than about $4R_E$ (field lines which intersect the ionosphere higher than 60° latitude) the validity of our treatment begins to break down. First, the outer magnetosphere becomes increasingly distorted from the assumed dipolar field. Second, the treatment of field-aligned currents using Ohm's Law breaks down because the magnetospheric plasma is no longer collision-dominated and, in addition, parallel diffusion velocities can be comparable with thermal velocities. Significant field-aligned electric fields can result. Third, additional dynamo current sources appear on the outer part of the field lines associated with the thermal and kinetic energies of the magnetospheric plasma. In principle the divergence of these additional dynamo currents should be included on the right-hand side of (5.40). The simplest way to get around all these difficulties is just to ignore them, and to carry (5.40) all the way out to $\alpha = \infty$, corresponding to the field line at the geomagnetic pole. In this case the outer boundary condition is

$$\lim_{\alpha \to \infty} V = \text{constant} \tag{5.46}$$

where the constant may arbitrarily be chosen as zero or any other number. This boundary condition was used to calculate the electric field from which the height-integrated current in Fig. 14 was obtained.

5.8. Equatorial electrojet

The mathematical problem of calculating electric fields and currents from given distributions of winds and conductivities simplifies considerably when the number of dimensions is reduced. To a reasonably good approximation the equatorial electrojet can be treated as having negligible east-west gradients in comparison to north-south and vertical gradients, and it therefore provides a simplified instructive example of how electric fields and currents are set up in the ionosphere. For the equatorial electrojet the eastward electric field (the westward gradient of V) turns out to play an important role, and so we shall assume that E_ϕ is non-negligible, but that its east-west gradient can be neglected, along with east-west gradients of the wind and conductivities. The eastward electric field at the equator is in fact determined by global dynamo processes; since we are not evaluating the global processes here we must simply take E_ϕ as a given quantity.

It should be noted that our assumption of negligible east-west gradients of the electric field makes the quantity $h_\phi E_\phi$ (= $-\partial V/\partial\phi$) constant not only in longitude, but also in altitude and latitude. This fact results from the curl-free nature of the electric field, so that

$$\frac{\partial}{\partial\alpha}(h_\phi E_\phi) = \frac{\partial}{\partial\phi}(h_\alpha E_\alpha) = 0 \tag{5.47}$$

$$\frac{\partial}{\partial\beta}(h_\phi E_\phi) = \frac{\partial}{\partial\phi}(h_\beta E_\beta) = 0 \tag{5.48}$$

We therefore have a single value of $h_\phi E_\phi$ as a free parameter in our problem, a value which must be independent of α and β.

To solve for the electrodynamic structure of the electrojet we drop the second terms on the left and right of (5.40). This equation can then be immediately integrated with respect to α, and when we apply the boundary condition (5.45) at $\alpha = R_E + 80$ km we see that (5.45) is itself the general solution of (5.40) for all values of α above the base of the ionosphere. This is just the condition that the net current flowing through a surface bounded by magnetic field lines on top and by the base of the conducting region on the bottom must vanish, since we are allowing no divergence of the eastward current.

In order to simplify the equations further let us take advantage of the fact that the metric coefficients change only little over the equatorial region and can approximately be replaced by constants:

$$h_\alpha \approx 1 \tag{5.49}$$

$$h_\phi \approx R_I \tag{5.50}$$

$$h_\beta \approx R_I{}^3 \tag{5.51}$$

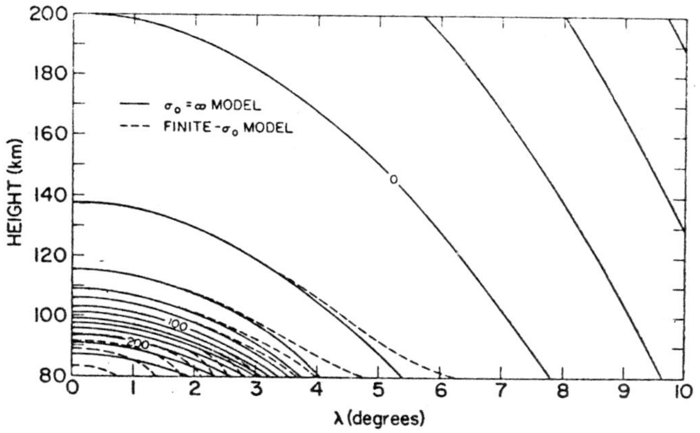

Figure 18. Electrostatic potential distribution in the equatorial electrojet, in volts.

where R_I is the mean radius of the conducting shell as seen in Fig. 17. We can then treat the eastward electric field E_ϕ as a constant, since both h_ϕ and $h_\phi E_\phi$ are constants. For simplicity, let us also drop the wind component v_α in comparison with v_ϕ, since the mainly horizontal winds have only a small component in the nearly vertical α-direction close to the magnetic equator. We can then solve for $E_\alpha (\approx -\partial V/\partial \alpha)$ as a function of α from (5.45):

$$E_\alpha = \frac{\Sigma_2}{\Sigma_1} E_\phi - \frac{1}{\Sigma_1}\int_{\beta_1}^{\beta_2} \sigma_1 v_\phi B d\ell \qquad (5.52)$$

$$\Sigma_1 \equiv \int_{\beta_1}^{\beta_2} \sigma_1 d\ell \qquad (5.53)$$

The polarization electric field E_α is thus equal to Σ_2/Σ_1 times E_ϕ, less the mean value of $v_\phi B$ along the field line, weighted by σ_1. In the lower E-region of the ionosphere σ_2 is much larger than σ_1, and the line integral Σ_2 is also much larger than Σ_1 for field lines peaking below 115 km, by a factor of up to about 30. Even for relatively weak eastward electric fields E_ϕ strong nearly vertical polarization fields E_α can develop. Figure 18 (Richmond, 1972) shows a model calculation of how V varies with height in the equatorial region for daytime conditions with an E_ϕ of 0.3 mV/m present but without winds. The height and latitude (λ) scales are different by about 6 to 1, and contours are spaced at 20 V. The strong nearly vertical gradient of V in the 90-110 km height range is apparent. The dashed lines show a solution for V in which the finite parallel conductivity is rigorously taken into account; it is clear that our assumption of infinite parallel conductivity is quite adequate above 90 or 100 km.

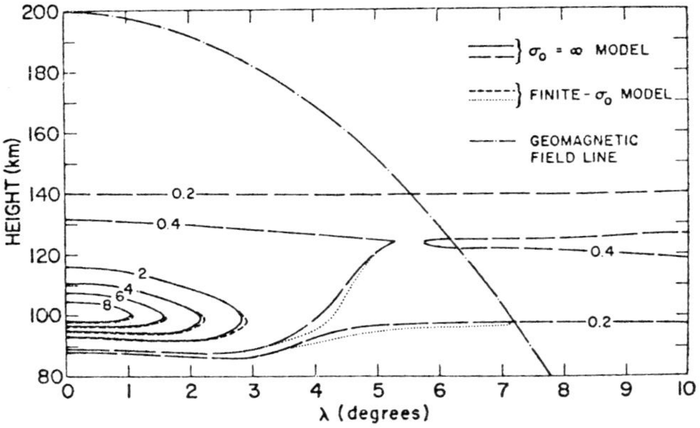

Figure 19. Eastward electric current density in the equatorial electrojet, in $\mu A/m^2$.

The eastward electric current in the electrojet can be very strong. From (5.38) and (5.52) we find

$$J_\phi = \left(\sigma_1 + \sigma_2 \frac{\Sigma_2}{\Sigma_1}\right) E_\phi + \sigma_2 \left(v_\phi B - \frac{1}{\Sigma_1}\int_{\beta_1}^{\beta_2} \sigma_1 v_\phi B d\ell\right) \qquad (5.54)$$

The effective conductivity for driving an eastward current by an eastward electic field, $\sigma_1 + \sigma_2 (\Sigma_2/\Sigma_1)$, has a strong similarity with the Cowling conductivity σ_3 in (5.24), but its interpretation is somewhat different in that Σ_2/Σ_1 is not the local ratio of Hall to Pedersen conductivities, but rather the ratio of their field-line integrals. In the present case the large ratio Σ_2/Σ_1 in the lower E region leads to strong eastward currents. Figure 19 (Richmond, 1972) shows a model calculation of the eastward current density for the electric field of Fig. 18, in units of 10^{-6} A/m^2. No winds were included in this calculation. Notice in (5.54) that eastward winds can also contribute to the eastward current density, but only if $v_\phi B$ varies along the field line. If $v_\phi B$ does not vary, we have a case satisfying (5.18), which results in a modification of the electric field but not of the current.

The equatorial electrojet has been a widely studied aspect of the global ionospheric dynamo system. A recent review of work in this field has been presented by Forbes (1981).

6. THERMOSPHERIC DISTURBANCES

The physical principles discussed in the previous sections can be employed to understand what happens in the thermosphere during periods when significant amounts of energy are deposited from the magnetosphere into auroral latitudes. The observed behavior of the thermosphere at these times can be quite complex, and numerical modelling studies can be helpful in trying to sort out and determine the relative importance of the various processes at play. This section concentrates on describing and interpreting results from some instructive modelling studies designed to represent thermospheric dynamics during geomagnetically disturbed times.

6.1. Two-dimensional time-dependent model description

Many of the disturbance effects described in this section are based on numerical simulations with a two-dimensional time-dependent model originally described in a paper by Richmond and Matsushita (1975). This computer model uses the basic equations of thermospheric dynamics to step forward in time and calculate variations of the winds, temperature, pressure, and electric field over a grid of points spread between the north pole and equator and between 80 km and 450 km altitude. The more important assumptions made in the model are as follows:

1. The atmosphere remains in hydrostatic equilibrium. This is a valid assumption for time scales of several minutes or more.
2. All quantities are independent of longitude.
3. All changes in the atmosphere occur symmetrically about the equator.
4. The initial state of the atmosphere is at rest and is independent of latitude.
5. The horizontal wind velocity, the gas energy density, and the pressure are constant with respect to time at an altitude of 80 km.
6. The composition of the atmosphere varies with altitude but remains a fixed function of pressure in time.
7. The atmosphere is sufficiently shallow with respect to its horizontal extent that quantities essentially on the order of H/R_E (pressure scale height/radius of the earth) can be neglected in comparison with quantities of order 1.
8. The geomagnetic field is dipolar, aligned with the earth's rotational axis.
9. The electric field is electrostatic and has no component in the east-west direction.
10. There exists a steady latitude-independent heat source of just the right vertical distribution to maintain the initial temperature structure in the absence of auroral energy input.

For modeling a disturbance, the following quantities are specified as functions of latitude and time: the vertical profile of ionospheric electron density, the vertical profile of particle precipitation heating, and the height-integrated equatorward electric current density. The winds, pressure changes, temperature changes, electric field and electric current density are then all calculated from the model over a grid spacing of 1° or less in latitude and 0.3 pressure scale height in altitude.

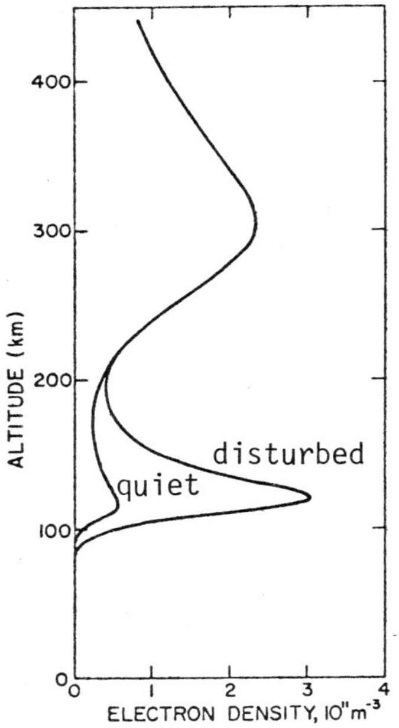

Figure 20. Model electron density profiles.

6.2. Substorm simulation

Richmond and Matsushita (1975) used this model to simulate the thermospheric response to a magnetic substorm. They started with a latitude- and time-independent background electron density (labelled "quiet" in Fig. 20) upon which was added an auroral enhancement between 66° and 74° latitude, growing and decaying over a period of two hours. The "disturbed" profile in Fig. 20 shows the electron density at 70°, 30 minutes after the start of the substorm, representing the latitude and time of maximum disturbance. A height-integrated equatorward current density K_θ was imposed over the same latitude range and time interval as the electron density enhancement. This current is supposedly fed by downward field-aligned currents between 70° and 74° and by upward field-aligned currents between 66° and 70° in such a manner that the entire current system is divergence-free. Figure 21 shows the latitude distributions of the height-integrated conductivities Σ_1 and Σ_2, the computed equatorward electric field E_θ, the height-integrated equatorward and eastward current densities K_θ and K_ϕ, and the Joule power density $K_\theta E_\theta$ at the time of maximum substorm intensity, 30 minutes after the start. These distributions are highly idealized representations of actual substorm conditions on the night side of the earth, but they suffice for generating the basic atmospheric response to be expected.

One advantage which numerical simulations have over observations of disturbances is their complete temporal and spatial coverage of the

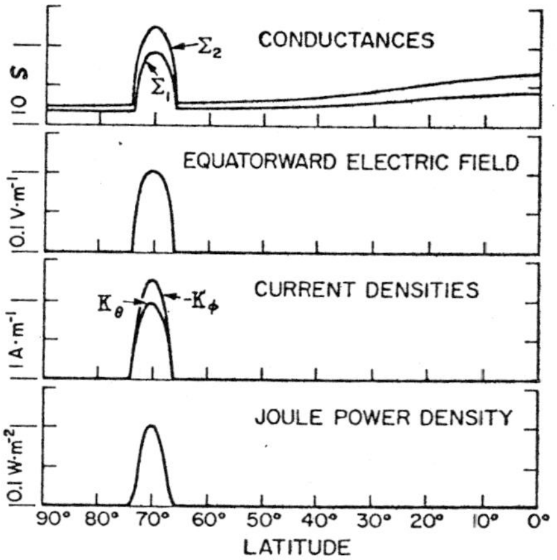

Figure 21. Latitudinal profiles of various parameters at t = 30 minutes.

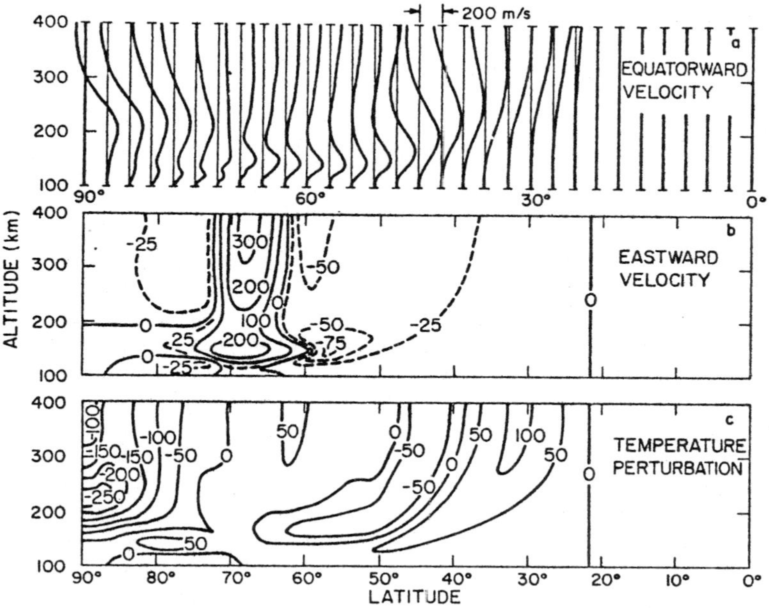

Figure 22. "Snapshot" of the disturbance at t = 2 hours. (a) Equatorward velocity profiles. (b) Eastward velocity (m/s). (c) Temperature perturbation (K).

event. We can thus examine successive "snapshots" of the disturbance as time progresses. Figure 22 shows a "snapshot" at a time two hours after the start of the simulation, which happens to coincide with the end of the auroral disturbance. A wave-like disturbance is seen to propagate away from the auroral zone, and is especially apparent in the profiles

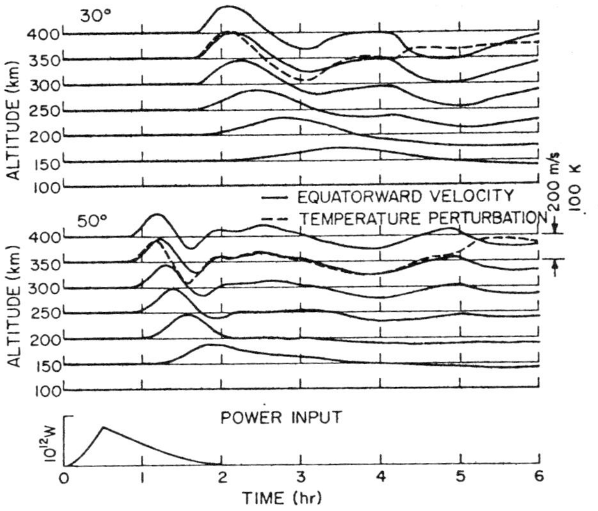

Figure 23. Time variations of equatorward wind at 30° and 50° latitude. The temperature perturbation is also shown at 350 km altitude. The hemispherical power input is shown at the bottom.

of equatorward velocity and in the temperature perturbation. This disturbance is caused by gravity waves generated by the auroral currents around 120 km altitude and propagating outward and upward into the upper thermosphere. Lines of constant phase, such as some of the 0 lines of the temperature perturbation, tend to fan out from the source, moving equatorward and downward as time progresses. Thus even though the gravity wave energy is mostly going upward, the phase surfaces are going downward, a result of the opposite directions of vertical group and phase velocities for gravity waves.

The eastward velocity in Fig. 22 shows primarily the effects of ion drag acceleration. The equatorward electric field set up in the auroral zone causes ions to drift rapidly toward the east. By collisional interaction with the neutrals, the drifting ions set the air into motion. The amount of momentum imparted to the air depends on the electron density; the two maxima in the eastward wind around 150 km and 350 km are related to the maxima in the disturbed electron density profile seen in Fig. 20.

Figure 23 shows time variations of the equatorward wind at various altitudes for latitudes of 30° and 50°, and also shows the hemispherical integral of the total power input, which represents primarily Joule heating by the auroral currents. Gravity wave features are clearly evident in the winds. The front of the disturbance propagates at a nearly constant velocity of 750 m/s, as determined from the arrival times at different latitudes. At a given latitude, the disturbance appears to move downward in time, as noted earlier. The disturbance has

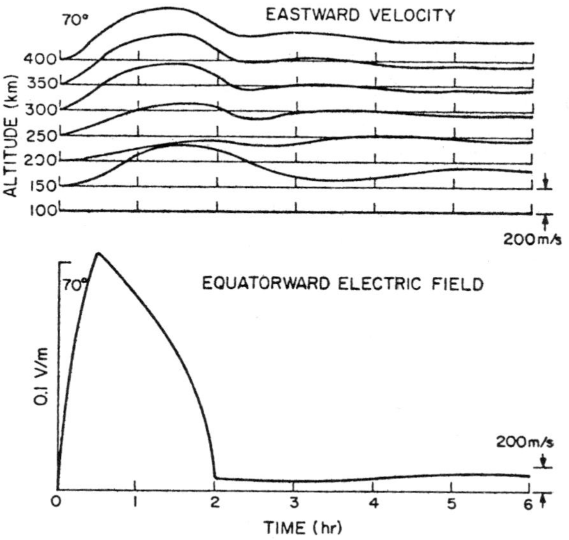

Figure 24. Time variations of the eastward wind and of the equatorward electric field at 70° latitude.

the general appearance of a strongly damped oscillation. The damping can be associated with the fact that slow-moving gravity waves are more strongly damped by molecular dissipation than are fast-moving waves, as noted in Section 4.11. In contrast, calculations of aurorally generated waves in a non-dissipative atmosphere show long trains of oscillations, because the undamped slow-moving waves continue to arrive at distant observation points long after the initial disturbance has passed. In Fig. 23 the apparent period of the oscillation increases both with increasing distance from the source and with decreasing altitude. This feature is related to the limiting ascent angle for gravity wave propagation. As noted in Section 4.10, the longer the wave period, the more nearly horizontal a gravity wave is constrained to move. We therefore expect shorter-period waves to reach the upper thermosphere nearer to the source than do longer-period waves.

Figure 23 also shows the temperature perturbation at 350 km. The perturbation appears to be nearly in phase with the velocity during the initial passage of the disturbance, but suddenly goes out of phase after a few hours. The out-of-phase behavior indicates a poleward propagating gravity wave disturbance. If we calculate the expected travel time of the disturbance front all the way to the equator and back to the particular latitude of observation, using the speed of 750 m/s, we find that it agrees with this interpretation. The disturbance has reflected off the equatorial boundary, or, in more physical terms, the mirror image of the disturbance generated in the opposite hemisphere has crossed the equator and is headed toward the pole. During this late stage of the disturbance, only very long period gravity waves remain, and these are

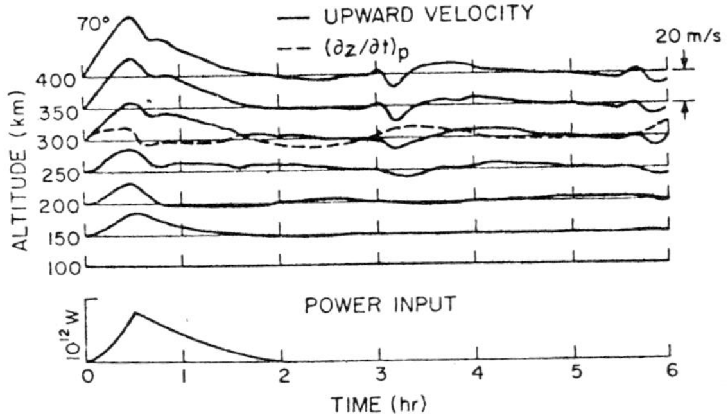

Figure 25. Time variations of the upward wind at 70° latitude. The vertical velocity of a constant-pressure surface is also shown at 300 km altitude. The hemispherical power input is shown at the bottom.

observationally difficult to separate from regularly occurring variations of the thermosphere.

Figure 24 shows variations of the eastward wind velocity at the center of the auroral zone (70° latitude), as well as variations of the calculated equatorward electric field. The scale for the electric field is chosen such that the resultant E × B drift velocity has the same scale as the wind velocity. The wind velocity does not approach the velocity of the strong ion drifts during the substorm, mainly because the characteristic time scale for ion drag to be effective, $\rho/(\sigma_1 B^2)$, is greater than two hours at all heights. Moreover, the wind depends not only on the local ion drag acceleration, but also on momentum transport between neighboring latitudes.

Notice that the electric field does not vanish at the end of the two-hour disturbance, even though the equatorward current has been turned off. The eastward wind in the dynamo region generates an electric field of such a magnitude that the ion drift velocity equals the mean neutral velocity (with a weighting function in altitude proportional to σ_1). The ion drifts during the substorm set the air into motion, and the air motion tends to act in the sense of maintaining the original ion drifts. This phenomenon has been called the "flywheel effect." For this simulation the coupling between the ion motions and the neutral "flywheel" is relatively weak, however.

The vertical wind velocity at 70° roughly follows the time history of the power input, as seen in Fig. 25. The manner of estimating the vertical velocity from the <u>local</u> heating rate discussed in Section 3.2 explains the basic features of the vertical motion. On the other hand, atmospheric expansion due to heating at <u>lower</u> levels plays only a minor role in driving vertical winds. To see the effect of atmospheric expansion, the vertical motion of a constant pressure surface $(\partial z/\partial t)_p$ is plotted in Fig. 25 by a dashed line at 300 km. If no horizontal motion were allowed there could be no mass movement through a given pressure surface, since in hydrostatic equilibrium the pressure is determined by the weight of the overlying air mass, which would have to remain constant in the absence of horizontal mass flux. The vertical motion of the pressure surface at 300 km is relatively small, indicating that vertical and horizontal circulation does occur and is tending to redistribute the heat horizontally.

6.3. Electrodynamic effects of stormtime heating

Blanc and Richmond (1980) used the model to examine global electrodynamic effects caused by thermospheric circulation changes during a magnetic storm. They simulated the effects of Joule heating by arbitrarily specifying a heat source in the auroral zone with a height distribution proportional to σ_1. However, they did not have any net equatorward current flowing, and hence no field-aligned currents and no strong auroral electric fields. The purpose of avoiding auroral electric fields was to avoid unrealistically large ion drag effects which would result from the longitudinally symmetic model with a steady ion drift. The heat source does, however, produce a strong meridional circulation extending down to middle latitudes. For the simulation presented here, the heat input is centered at 69° latitude and is held steady for 12 hours after a brief switch-on period. The magnitude of the heating is comparable to what might occur in a major magnetic storm.

Figure 26 shows the temporal variations of the wind velocity at 45° latitude, along with the eastward E × B drift velocity u_ϕ^e (which is negative and therefore westward) and the hemispheric integral of the heat input (8.24×10^{11}W). The auroral heating causes an upwelling and an outflow of air at altitudes above about 120 km. The equatorward wind v_θ at 45° is positive at the upper altitudes after the arrival of the disturbance front, and some wave features are noticeable, especially at 198 km. At 108 km the equatorward wind is very small, and becomes negative (poleward) after about four hours. This poleward wind returns mass towards the auroral zone to replace the mass lost through upper-level outflow. Mass continuity requires only a very small wind in the dense lower thermosphere to offset the mass flux by the large winds in the rarefied upper thermosphere.

The Coriolis acceleration associated with the equatorward wind above 120 km causes a westward wind to develop (negative v_ϕ). More accurately, we should explain the westward wind in terms of angular momentum transport out of the auroral region. As an elemental annulus of air encircling the earth's axis moves equatorward, its radius expands and its absolute rotational velocity decreases. Seen with respect to the rotating earth, the reduction of rotational velocity of the annulus corresponds to a westward wind. The advantage of examining the development of the westward wind in this manner is that it readily explains a "saturation" effect which is apparent in Fig. 26. That is, the westward wind will no longer increase after the maximum transfer of angular momentum is achieved, which occurs when air parcels originating in the

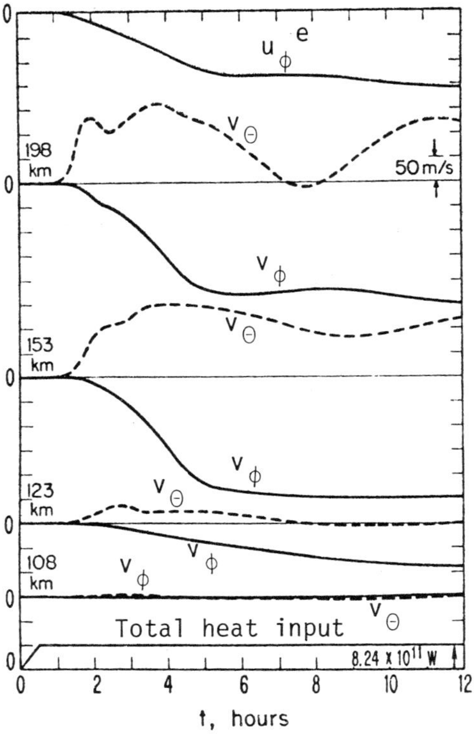

Figure 26. Time variations of the E × B drift velocity and neutral wind velocities at 45° latitude. The hemispherical heat input to the auroral zone is shown at the bottom.

auroral zone finally reach 45° latitude. Of course, ion drag and viscous forces also act on the wind, so that the effects observed in v_ϕ are not due solely to angular momentum transport. The slow buildup of a westward wind at 123 km, where v_θ is small, appears to be caused primarily by viscous coupling with the strong westward winds at higher levels.

The E × B drift velocity u_ϕ^e in Fig. 26 develops in a manner intermediate between the developments of the westward winds at 123 km and at 153 km. The poleward electric field generated by the winds can be calculated from (5.45) with $\partial V/\partial\phi$ set to zero. At middle latitudes where h_ϕ, h_α, and B are nearly constant along the line of integration through the dynamo region and where the conductivities and winds can be considered to have insignificant latitudinal variations but significant height variations along the integration path, (5.45) yields the following expression for u_ϕ^e:

$$u_\phi^e = \int(\sigma_1 v_\phi - \sigma_2 v_\theta \sin I)dz/\int\sigma_1 dz \tag{6.1}$$

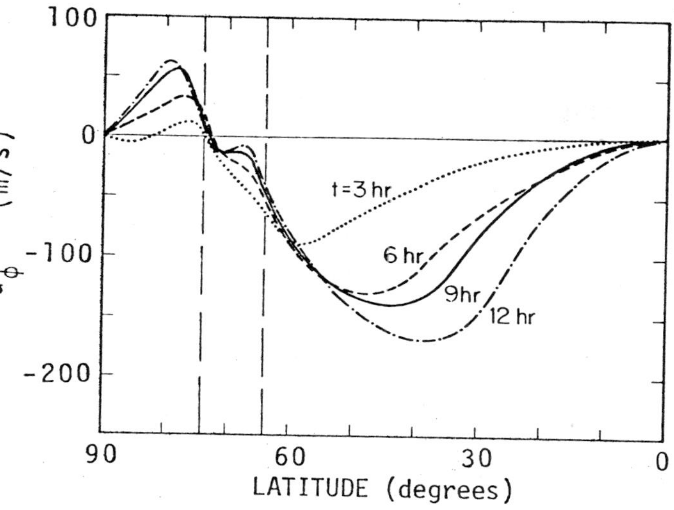

Figure 27. Latitude variations of the eastward E × B drift velocity at different times.

where the height integrals are over the thickness of the dynamo region, and where I is the angle of inclination of the magnetic field below the horizontal. Because σ_2 is small at altitudes where v_θ is significant, (6.1) shows that the westward electrodynamic drift velocity is essentially the vertical average of the westward wind velocity, weighted by σ_1. This simple result is partly a consequence of the model's neglect of longitudinal variations, but Blanc and Richmond (1980) showed that it gives a correct rough estimate of middle-latitude east-west E × B drifts even when realistic longitudinal gradients of the conductivities are included.

Figure 27 illustrates the latitudinal variations of u_ϕ^e as time progresses. The vertical dashed lines show the region of auroral heat input. Because the north-south wind v_θ is constrained to vanish at both the pole and equator, the transfer of east-west angular momentum also goes to zero at these latitudes, along with the electrodynamic drift velocity. At midlatitudes the saturation effect mentioned earlier in conjunction with v_ϕ is also apparent in u_ϕ^e: the westward drift builds up towards a maximum value which increases with increasing distance from the auroral zone. The time needed to reach saturation depends on the strength of heating, and also increases with increasing distance from the auroral zone. At 12 hr the westward E × B drift has saturated down to about 45° latitude.

6.4 Energy transfer to lower latitudes

In Section 3 it was discussed how heat can be transferred away from regions of net heating by meridional circulation. Upwelling in the region of heating causes decompressional cooling and horizontal outflow at higher levels; this horizontal outflow induces sinking motions at other latitudes which gives rise to local compressional heating as well as a return horizontal flow toward the heat source at lower altitudes. In Section 4 the dissipation of gravity wave energy was discussed. Gravity waves which propagate out from the disturbed auroral region and are dissipated in the upper thermosphere will also contribute to the net transfer of heat from the auroral region to lower latitudes. It is of interest to compare the relative importance of these two energy transfer mechanisms during magnetic storms: meridional circulation vs. gravity waves.

Richmond (1979b) started from the hypothesis that meridional circulation is relatively more important in transferring heat to latitudes below 45° than are gravity waves. To test this hypothesis he set up a simulation designed to be favorable for the generation of gravity waves. If gravity waves are less important than circulation for energy transfer even under these favorable conditions then the original hypothesis is supported.

Simulations like that discussed in Section 6.2 show that only low-frequency gravity waves, with periods longer than about one hour, propagate effectively from the auroral region to latitudes below 45°. It is therefore not necessary to include rapid fluctuations in specifying the temporal variation of auroral energy input for the purpose of testing the hypothesis. The amount of gravity wave energy generated tends to increase more rapidly with increasing magnitude of auroral currents than does the strength of the circulation, so a large-amplitude source is also necessary to test the hypothesis. Richmond (1979b) therefore specified an auroral current source with strong, low-frequency fluctuations.

Figure 28 shows the time history of the simulation. The equatorward height-integrated current density (shown in the middle) was specified to be negative for the first three hours and positive for the next three hours in order to cause a similar reversal in the sign of the equatorward electric field (shown at the bottom) and avoid unrealistically large buildup of ion-drag winds. The Joule power input to the entire hemisphere is shown between these two curves. The computed thermospheric temperature response at 400 km is shown for three different latitudes at the top. The tendency for the dominant frequencies to decrease with increasing distance from the source is clear. A strong spike which appears to propagate equatorward and reach the equator at t = 6 hours is probably due to non-linear wave steepening effects. Notice that at the end of the 12-hour simulation the temperature remains above the initial state at all three latitudes, a consequence of the heat transfer under consideration.

The complete temporal and spatial information available from the model about the disturbancé enable a detailed analysis of the heat transfer mechanisms. Figure 29 shows distributions of three quantities of

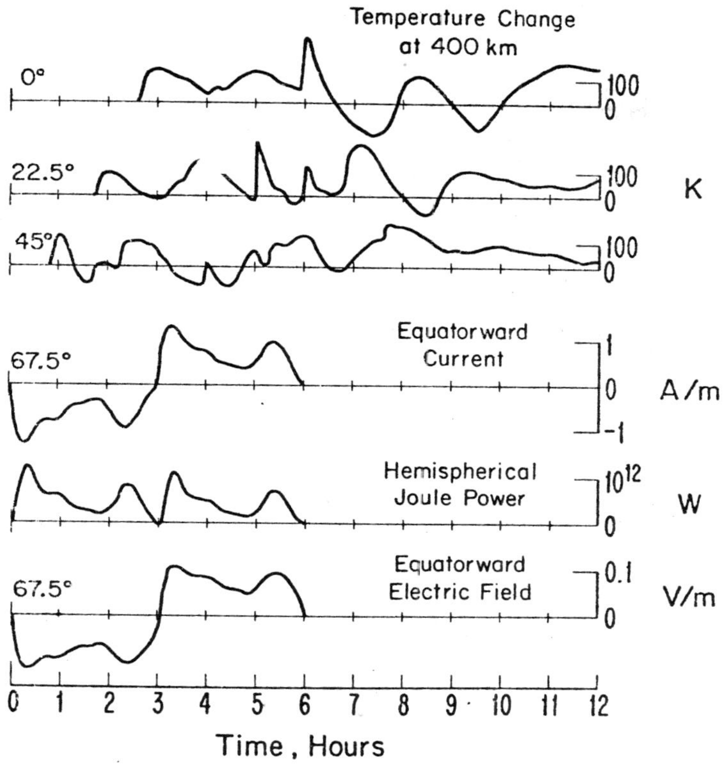

Figure 28. Time variations of various parameters in a thermospheric disturbance simulation.

interest, integrated in time over the entire 12-hour simulation. The vertical scale is given in terms of the number of pressure scale heights above the lower boundary at 80 km, since the model grid is linear with respect to this quantity. Because levels of constant pressure move up and down, the geometric height scale on the right is only approximate. The region of auroral heat input is shown by the two vertical lines on either side of 67.5° latitude in the middle and bottom frames. The crosses within this region represent the calculated centers of Joule heating (center frame) and wave energy production (bottom frame).

The top frame in Fig. 29 shows the mass flux associated with the meridional circulation, integrated not only in time but also in longitude around the earth. The central clockwise cell clearly shows the pattern of upflow in the region of heating, equatorward flow above about 121 km, subsidence at middle and low latitudes, and return flow below 121 km. A smaller counterclockwise cell toward the pole is the high-latitude counterpart of this. The remaining mass flux cells do not represent the quasi-steady circulation but rather residuals of the large transient motions which have occurred. The middle frame shows the heating associated with

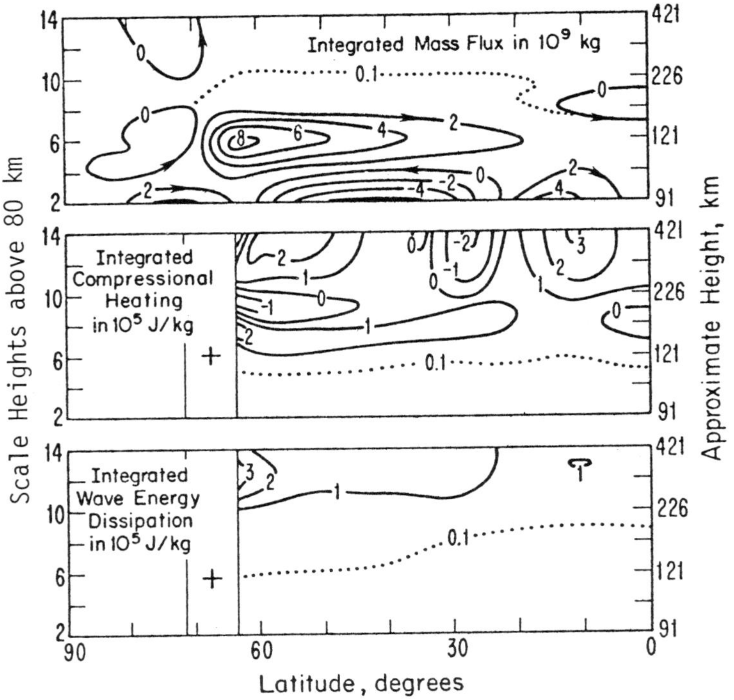

Figure 29. Latitude-height distributions of various quantities integrated over the 12-hour simulation.

the net sinking motions, with residual transient phenomena apparent especially at the higher altitudes. This heating is given as heat per unit mass. If advection and heat conduction were not important, 10^5 J/kg would result in approximately a 100 K temperature increase. The amount of heating per unit volume is relatively less at the upper altitudes than at the lower altitudes because of the exponentially decreasing density. The bottom frame shows the amount of heating associated with wave energy dissipation.

For analyzing the relative importance of compressional and wave heating it is useful to integrate some relevant energy parameters with respect to latitude. Figure 30 shows vertical profiles of some latitude-, longitude-, and time-integrated quantities. In contrast to Fig. 29, where energy rates are expressed as energy per unit mass, Fig. 30 expresses rates as energy per scale height, plotted linearly against the number of scale heights above 80 km so that visual integration in altitude is possible. On the left are profiles of Joule heat input to the entire auroral region and of the total wave energy production. On the right are integrals between 0° and 45° latitude of the time- and latitude-integrated compressional heating rate and wave energy dissipation rate. The horizontal scales on the left and right differ by a factor of five.

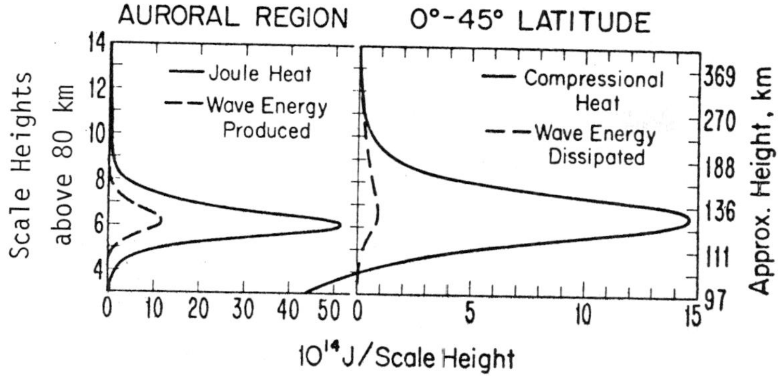

Figure 30. Altitude distributions of energy parameters integrated in time, longitude, and latitude.

We see that both the Joule heat input and the wave energy generation are concentrated within a few scale heights around 120 km, and that the heating at 0°-45° latitudes occurs somewhat higher. Wave energy dissipation, in particular, extends well into the upper thermosphere, where it is comparable in magnitude with compressional heating. In the lower thermosphere, where most of the energy resides, compressional heating strongly dominates over wave heating.

In Section 3 it was discussed how thermal conduction redistributes heat vertically on a characteristic time scale given by $\rho H^2 c_p/\kappa$. For the current simulation the mean time between the low-latitude heat input and the end of the simulation is roughly six hours. From Table 1 we see that conduction will effectively redistribute heat at all altitudes above about 150 km in this time, so that in examining the upper thermospheric temperature increase we should take account of all heat deposited above 150 km (about 7 scale heights above 80 km). The compressional heating is a few times larger than the wave energy dissipation above this level. Consequently, the original hypothesis of Richmond (1979b) is supported, in that gravity wave transport of heat to low latitudes is less important than meridional circulation for explaining upper thermospheric temperature increases, even in a simulation designed to favor gravity waves. On time scales longer than six hours the relative importance of compressional heating becomes even greater, as we can see by lowering the altitude above which total heating is evaluated in Fig. 30.

6.5 Stormtime composition changes

Section 3.5 discussed how meridional circulation of the thermosphere affects the distributions of different constituents, and Fig. 7 shows how seasonal changes in composition results in association with the summer-to-winter circulation in the upper thermosphere. Auroral heating in magnetic storms also causes important composition changes. Figure 31 (Roble, 1977) is a schematic representation of relative dens-

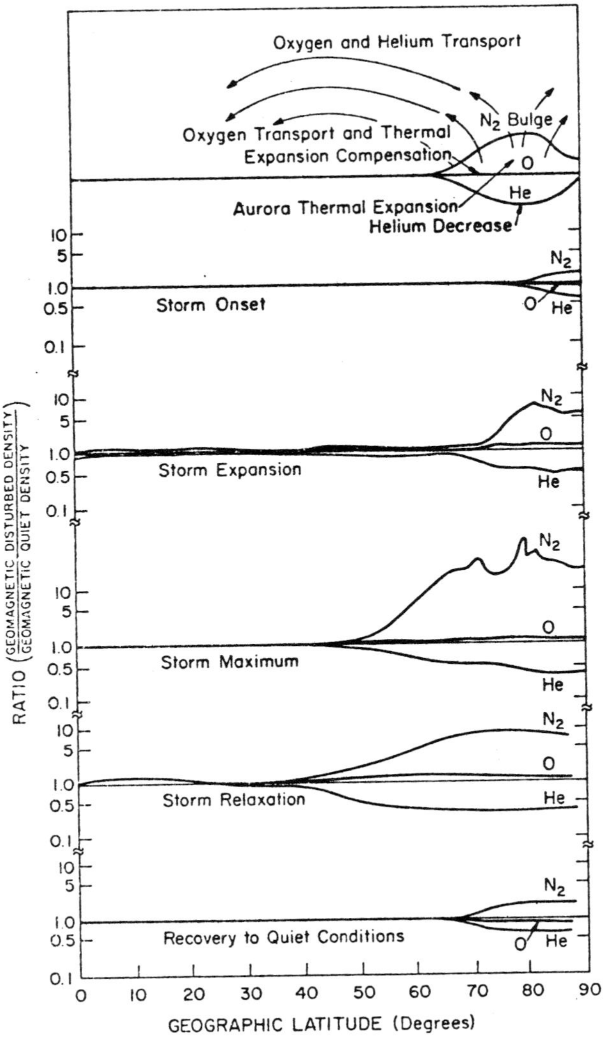

Figure 31. Schematic of the thermospheric compositional response to geomagnetic storms as a function of latitude at various times during the storm. The ratio is storm-time departure relative to geomagnetically quiet conditions.

ity changes in certain species in the upper thermosphere for several stages of a storm, based on satellite measurements and theoretical studies. The high latitude heating enriches the densities of heavy constituents like N_2 and depletes light constituents like He. These changes expand to middle latitudes by horizontal advection and then slowly relax as diffusion tends to restore diffusive equilibrium. Such composition changes play a major role in altering ionospheric chemistry processes during storms, leading to widespread ionospheric depletions.

6.6 Conclusions

The simulations described in this section show how a wide variety of wave, circulation, and electrodynamic processes come into play simultaneously during thermospheric disturbances. These processes have been examined theoretically only under rather idealized conditions so far, and considerable improvements in their analysis are possible.

Notation

$\vec{B}$ - geomagnetic field

B_z, B_θ - vertical and southward components of $\vec{B}$

B_p - polar magnitude of $\vec{B}$

ΔB - magnetic perturbation by ionospheric currents

$\hat{b}$ - $\vec{B}/B$

c - speed of sound

c_v, c_p - specific heats at constant volume, pressure

c_{vi}, c_{pi} - specific heats of ith constituent

D_E - eddy diffusion coefficient

D_i - molecular diffusion coefficient for ith constituent

D_{Ti} - thermal diffusion parameter in (2.37)

$\vec{E}$ - electric field

$\vec{E}'$ - $(\vec{E} + \vec{v} \times \vec{B})$

$\vec{E}_\parallel$, $\vec{E}_\perp$ - components of $\vec{E}$ parallel and perpendicular to $\vec{B}$

e - magnitude of electron charge

$\vec{F}$ - body force per unit mass

$\vec{g}$ - acceleration of gravity in rotating frame

$\vec{g}^*$ - acceleration of gravity in nonrotating frame

H - pressure scale height

H_i - $kT/m_i g$

h_α, h_β, h_ϕ - metric coefficients for dipole coordinates

I - inclination angle of geomagnetic field below horizontal

$\vec{J}$ - electric current density

J_z, $\vec{J}_h$ - vertical and horizontal components of $\vec{J}$

$\vec{J}_d$, $\vec{J}_e$ - current components driven by winds, electric field

$\vec{J}_\parallel$, $\vec{J}_\perp$ - current components parallel and perpendicular to $\vec{B}$

J_α, J_β - α, β components of $\vec{J}$

K_θ, K_ϕ - equatorward and eastward height-integrated current density

KE - kinetic energy density

k - Boltzmann constant

k_z, $\vec{k}_h$ - vertical, horizontal components of wave number vector

L, L_h, L_z - characteristic scale heights

ℓ - distance along $\vec{B}$

m_e - electron mass

m_i - mass of ith constituent

$\bar{m}$ - mean molecular mass

N - buoyancy frequency

n - particle number density

n_e - electron number density

n_i - number density of ith species

P - defined by (4.41)

$\underline{\underline{P}}$ - pressure tensor

$\underline{\underline{P}}^*$ - viscous momentum flux tensor

P_{xy}, etc. - elements of $\underline{\underline{P}}$

PE - potential energy density

p - pressure

p' - pressure perturbation

p'^* - complex conjugate of p'

Q - heating rate per unit mass

Q^* - heating rate per unit mass including Joule and viscous heat

$\overline{Q^*}$ - horizontally averaged value of Q^*

$\vec{q}$ - heat flux

$\vec{q}'$ - perturbation of $\vec{q}$

q_z - vertical component of $\vec{q}$

q_i - electric charge of ith species

R - gas "constant" ($kT/\bar{m}$)

R_E - radius of earth

R_I - radius of dynamo region

$\vec{r}$ - radius vector

S_i - net mass production rate of ith species

S_α, S_ϕ - defined by (5.41), (5.42)

s - specific entropy

s' - perturbation in s

T - temperature

T' - temperature perturbation

T_e - electron temperature

t - time

t_0 - constant

u - internal energy per unit mass

u_I - internal energy associated with internal degrees of freedom of molecules

u_h - horizontal gravity wave phase and group velocity

$\vec{u}^e$ - $\vec{E} \times \vec{B}/B^2$

u_θ^e, u_ϕ^e - θ, ϕ components of $\vec{u}^e$

V - electrostatic potential

V_z - defined by (4.42)

$\vec{v}$ - center of mass velocity

$\vec{v}_\perp$, $v_\parallel$ - components of $\vec{v}$ perpendicular, parallel to $\vec{B}$

$\vec{v}_h$, v_z - horizontal, vertical components of $\vec{v}$

v_α, v_θ, v_ϕ - α, θ, ϕ components of $\vec{v}$

v_w, v_ψ - w, ψ components of $\vec{v}$

$\vec{v}_h^*$, v_z^* - complex conjugates of $\vec{v}_h$, v_z

$\vec{v}_i$ - velocity of ith constituent

$\vec{v}_i{}'$ - $\vec{v}_i - \vec{v}$

$\vec{v}_{Ei}$ - eddy diffusion velocity of ith constituent

$\vec{v}_0$ - $\vec{v}$ at time t_0

w - radius in polar coordinates

$\vec{x}$, x - distance

$\hat{x}$ - unit vector in x direction

$\hat{y}$ - unit vector in y direction

z - altitude

$\hat{z}$ - upward unit vector

α - defined by (5.30)

$\hat{\alpha}$ - unit vector in α direction

α_i - thermal diffusion coefficient

β - defined by (5.32)

$\hat{\beta}$ - unit vector in β direction

β_1, β_2 - integration limits in Fig. 17

γ - c_p/c_v

Δ_{ij} - diffusion parameter in (2.37)

ε_0 - dielectric constant of free space

θ - colatitude

θ^* - colatitude used to define Ω^* in (4.76)

κ - thermal conduction coefficient

Λ - defined by (5.29)

λ - latitude

μ - viscosity coefficient

μ_E - eddy viscosity coefficient

μ_0 - permeability of free space

ν_i - collision frequency for ith constituent

ν_c, ν_v, ν_σ - wave dissipation rates by conduction, viscosity, ion drag

ρ - mass density

ρ' - perturbation in ρ

ρ_e - electric charge density

Σ_1, Σ_2 - field-line integrals of σ_1, σ_2

σ_0, σ_1, σ_2 - parallel, Pedersen, Hall conductivities

τ - time scale

τ_v, τ_σ - time scales for viscosity, ion drag to be effective

Φ - geopotential

ϕ - east longitude

$\hat{\phi}$ - unit vector in ϕ direction

ψ - azimuthal angle in polar coordinates

$\vec{\Omega}$ - angular rotation rate of earth

$\vec{\Omega}^*$ - defined by (4.76)

Ω_i - gyrofrequency of ith constituent

ω - angular frequency

References

Banks, P. M. and Kockarts, G.: 1973a, Aeronomy, part A, Academic Press, New York.
Banks, P. M. and Kockarts, G.: 1973b, Aeronomy, part B, Academic Press, New York.
Beer, T.: 1976, The Aerospace Environment, Wykeham Publ., London.
Blanc, M. and Richmond, A. D.: 1980, J. Geophys. Res. 85, p. 1669.
Chapman, S. and Cowling, T. G.: 1970, The Mathematical Theory of Non-Uniform Gases, Cambridge Univ. Press, Cambridge.
Chapman, S. and Lindzen, R. S.: 1970, Atmospheric Tides, D. Reidel, Dordrecht.
Dickinson, R. E., Ridley, E. C., and Roble, R. G.: 1981, J. Geophys. Res. 86, p. 1499.
Forbes, J. M.: 1981, Rev. Geophys. Space Phys. 19, p. 469.
Forbes, J. M.: 1982a, J. Geophys. Res. 87, p. 5222.
Forbes, J. M.: 1982b, J. Geophys. Res. 87, p. 5241.
Forbes, J. M. and Garrett, H. B.: 1979, Rev. Geophys. Space Phys. 17, p. 1951.
Francis, S. H.: 1973, J. Geophys. Res. 78, p. 2278.
Francis, S. H.: 1975, J. Atmospheric Terrest. Phys. 37, p. 1011.
Fuller-Rowell, T. J. and Rees, D.: 1980, J. Atmospheric Sci. 37, p. 2545.
Fuller-Rowell, T. J. and Rees, D.: 1981, J. Atmospheric Terrest. Phys. 43, p. 701.
Hargreaves, J. K.: 1979, The Upper Atmosphere and Solar-Terrestrial Relations: An Introduction to the Aerospace Environment, Van Nostrand Reinhold Co., New York.
Hines, C. O.: 1977a, Planetary Space Sci. 25, p. 1045.
Hines, C. O.: 1977b, Planetary Space Sci. 25, p. 1061.
Hines, C. O. and colleagues: 1974, The Upper Atmosphere in Motion, American Geophysical Union, Washington, D.C.
Holton, J. R.: 1979, An Introduction to Dynamic Meteorology, Academic Press, New York.
Kato, S.: 1980, Dynamics of the Upper Atmosphere, D. Reidel, Dordrecht.
Richmond, A. D.: 1972, 'Numerical Model of the Equatorial Electrojet', Air Force Cambridge Research Laboratories (Report AFCRL-72-0668).
Richmond, A. D.: 1978, J. Geophys. Res. 83, p. 4131.
Richmond, A. D.: 1979a, J. Geomag. Geoelectr. 31, p. 287.
Richmond, A. D.: 1979b, J. Geophys. Res. 84, p. 5259.
Richmond, A. D. and Matsushita, S.: 1975, J. Geophys. Res. 80, p. 2839.
Richmond, A. D., Blanc, M., Emery, B. A., Wand, R. H., Fejer, B. G., Woodman, R. F., Ganguly, S., Amayenc, P., Behnke, R. A., Calderon, C., and Evans, J. V.: 1980, J. Geophys. Res. 85, p. 4658.

Rishbeth, H. and Garriott, O. K.: 1969, Introduction to Ionospheric Physics, Academic Press, New York.

Roble, R. G.: 1977, in Geophysics Study Committee (ed.), The Upper Atmosphere and Magnetosphere, National Academy of Sciences, Washington, D.C., Ch. 3, p. 57.

Roble, R. G., Dickinson, R. E., and Ridley, E. C.: 1982, J. Geophys. Res. 87, p. 1599.

Shkarofsky, I. P.: 1961, Can. J. Phys. 39, p. 1619.

Tinsley, B. A.: 1979, J. Geophys. Res. 84, p. 1855.

Torr, M. R., Torr, D. G., and Richards, P. G.: 1980, Geophys. Res. Letters, 7, p. 373.

U.S. Standard Atmosphere, 1976, National Oceanic and Atmospheric Administration, Washington, D.C.

Walterscheid, R. L.: 1981, Geophys. Res. Letters 8, p. 1235.

Whitten, R. C. and Poppoff, I. G.: 1971, Fundamentals of Aeronomy, John Wiley & Sons, Inc., New York.

Yeh, K. C. and Liu, C. H.: 1974, Rev. Geophys. Space Phys. 12, p. 193.

THE TERRESTRIAL IONOSPHERE

R.W. Schunk
Center for Atmospheric and Space Sciences
Utah State University
Logan, Utah 84322

The theory relating to the basic physics governing the behavior of the terrestrial ionosphere is reviewed. The review covers the coupling of the ionosphere to both the neutral atmosphere and magnetosphere, the creation and transport of ionization in the ionosphere, and the ionospheric thermal structure. The review also covers the variation of the ionosphere with altitude, latitude, longitude, universal time, season, solar cycle, and geomagnetic activity. In addition, some unique ionospheric features are discussed, such as the polar ionization hole, the main electron density trough, the ion temperature hot spots, the high-latitude ionization tongue, the equatorial fountain, Appleton's peaks, and the polar wind.

1. INTRODUCTION

Ionospheric research during the last two decades has shown that the earth's ionosphere exhibits a significant variation with altitude, latitude, longitude, universal time, solar cycle, season, and geomagnetic activity. This variation is a consequence of the competition between the various forces acting within and on the ionosphere. Of particular importance are the forces that arise as a result of atmosphere-ionosphere-magnetosphere coupling; through currents, energetic particles, electric fields, and atmospheric drag. Because of the importance of this coupling, it is convenient to first briefly describe some of the more efficient coupling mechanisms and the ionospheric response to them. This is done in Section 2. In Section 3 the basic theory pertaining to the creation and transport of ionization in the ionosphere is described. Sections 4, 5, and 6 are devoted to a detailed discussion of the middle and low latitude ionosphere, the high-latitude ionosphere, and the polar wind, respectively.

2. IONOSPHERIC ENVIRONMENT

The earth's magnetosphere is shown schematically in Figure 1. The

R. L. Carovillano and J. M. Forbes (eds.), Solar-Terrestrial Physics, 609–676.

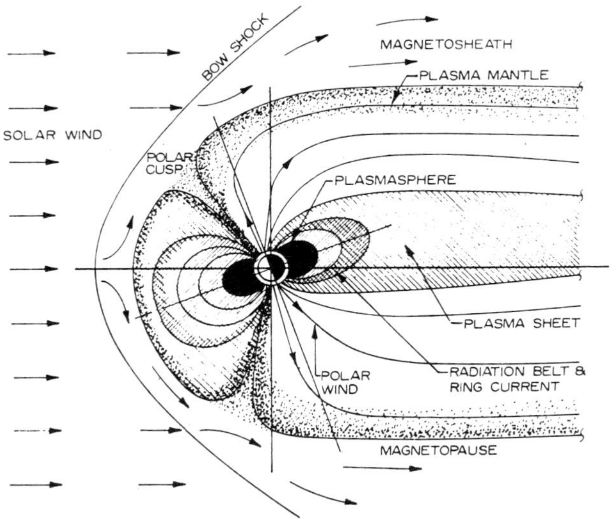

Figure 1. Schematic illustration of the earth's magnetosphere (from Bahnsen, 1978). Reprinted by permission of Pergamon Press.

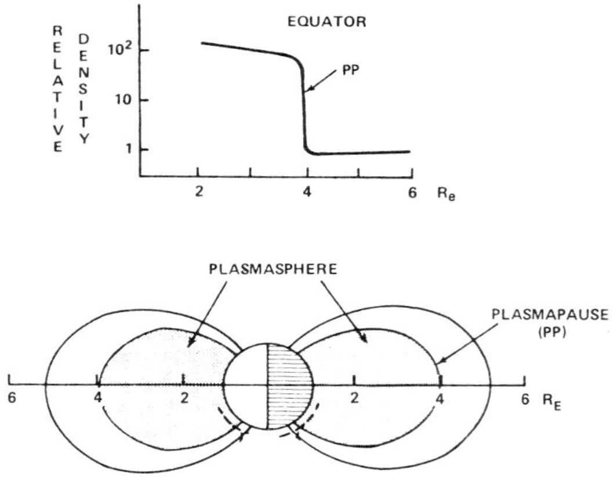

Figure 2. Schematic illustration of the plasmasphere and its bounding surface, which is called the plasmapause. The upper figure shows the electron density variation as a function of radial distance in the equatorial plane.

magnetosphere is the region contained within the magnetopause, which is the boundary layer that separates the confined geomagnetic field from the magnetized solar wind plasma in the magnetosheath. This large region is populated by thermal plasma and energetic charged particles of both solar wind and ionospheric origin. On the dayside, the solar wind plasma in the magnetosheath has direct access to the ionosphere in the vicinity of the polar cusp (or cleft). At ionospheric heights, the cleft occupies a narrow latitudinal band that is centered near noon but is extended in longitude. Within this band, energetic particle precipitation affects the ionospheric plasma through heating and ionization processes.

The plasma sheet and ring current are composed of magnetosheath plasma that has been injected into the magnetosphere and energetic O^+ and He^+ ions of ionospheric origin. The plasma sheet particles have direct access to the high-latitude ionosphere in the nocturnal auroral zone, a region which will be discussed later. Within the auroral zone, energetic particle precipitation from the plasma sheet leads to both the creation of ionization and plasma heating. The ring current is in part the earthward extension of the plasma sheet. It is composed of trapped energetic protons and electrons which grad-B drift in opposite directions, causing a ring of current to flow around the earth. During magnetic storms, ring current particles can transfer a substantial amount of energy to the ambient electrons, and this energy is conducted down along geomagnetic field lines to the ionosphere, causing elevated electron temperatures.

The plasmasphere, which is shown schematically in Figure 2, is a torus-shaped volume that surrounds the Earth and contains a relatively cool, high density plasma of ionospheric origin. The plasma in this region corotates with the Earth, but it can also flow along geomagnetic field lines from one hemisphere to the other. In the equatorial plane, the plasmasphere has a radial extent of about 4 R_E, and its boundary, called the plasmapause, is marked by a large decrease in electron density as you leave the plasmasphere. The plasmapause is essentially the boundary between plasma that corotates with the Earth and plasma that does not.

2.1. High-latitude ionosphere

At high latitudes, the plasma does not corotate with the Earth, but instead moves under the action of electric fields of magnetospheric origin. In addition, thermal plasma is capable of escaping from the topside ionosphere along 'open' geomagnetic field lines, a process termed the polar wind. This high-latitude region between about 120-3000 km and the same altitude range in the plasmasphere is the region of primary interest for this review. The physical mechanisms operating in this low altitude regime will be discussed in more detail in the paragraphs that follow.

Owing to the interaction of the shocked solar wind with the geomag-

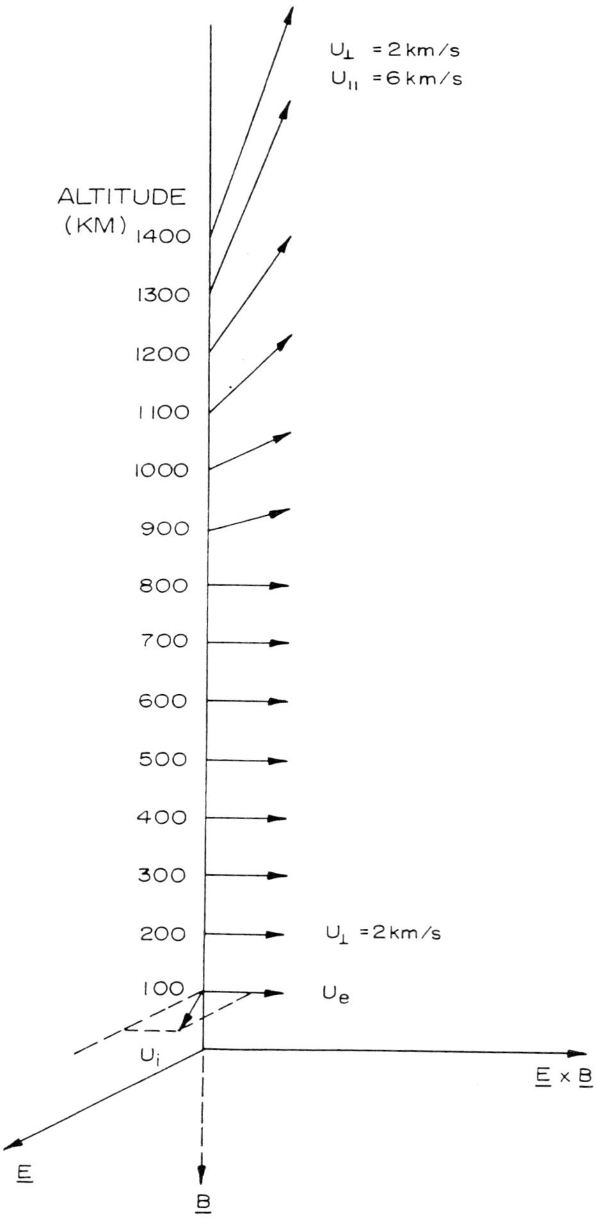

Figure 3. Ion and electron drift velocities as a function of altitude in the high latitude ionosphere.

netic field, an electric potential difference is generated across the tail of the magnetosphere, with the resulting electric field pointing from dawn to dusk. This cross-tail potential difference is typically 76 kilovolts, but can vary from 20 to 120 kilovolts depending on the level of geomagnetic activity. Except for isolated regions and brief periods, the geomagnetic field lines are equipotentials due to the high electrical conductivity along field lines. Consequently, this cross-tail potential difference is mapped into the high latitude ionosphere as an electric field that is directed perpendicular to the geomagnetic field. At ionospheric heights, this perpendicular (or convection) electric field is typically 25-50 mV m^{-1} in the polar cap, but can be much greater than 100 mV m^{-1} in restricted latitudinal bands. Only the high latitude ionosphere is influenced by the magnetospheric electric field, since most of the time the ring current provides an effective shield for the plasma within the plasmasphere.

The effect that the perpendicular electric field has on the ionosphere depends on altitude, as shown in Figure 3. At all ionospheric altitudes, the electron-neutral collision frequency is much less than the electron cyclotron frequency, and hence, the combined effect of the perpendicular field, $\underline{E}_{\perp}$, and the geomagnetic field, $\underline{B}$, is to induce an electron drift in the $\underline{E} \times \underline{B}$ direction. For the ions, on the other hand, the different ion-neutral collision frequencies are greater than corresponding cyclotron frequencies at low altitudes (E region), with the result that the ions drift in the direction of the perpendicular electric field. As altitude increases, the ion velocity vectors rotate toward the $\underline{E} \times \underline{B}$ direction owing to the decreasing ion-neutral collision frequencies. At F region altitudes (> 160 km), both ions and electrons drift in the $\underline{E} \times \underline{B}$ direction. At still greater heights (> 800 km), the plasma begins to flow out of the topside ionosphere with a speed that increases with altitude (the polar wind). As a consequence, the plasma flow tends to become field-aligned and supersonic at high altitudes.

A consequence of the electron and ion motion shown in Figure 3 is that above about 160 km the ionosphere drifts horizontally under the action of the magnetospheric electric field. The electric field induced drift pattern is a two-cell pattern with anti-sunward flow over the polar cap and sunward flow at lower latitudes. However, the ionospheric plasma at high latitudes also has a tendency to corotate with the earth. When the corotation velocity is added to this two-cell pattern, the resulting drift pattern for the ionospheric plasma is similar to that shown in Figure 4. Also shown in Figure 4 are the locations of the high latitude ionization hole, the main trough, and the quiet-time auroral oval. The ionization hole is a low density, low temperature region that appears during quiet geomagnetic activity. The auroral oval is a region where intense particle precipitation occurs, which leads to the creation of ionization and to electron, ion, and neutral gas heating. The main trough is a region situated just equatorward of the nightside auroral oval that is characterized by very low electron densities. Typically, peak electron densities in the main trough are a factor of 100 lower than those found on either side of the trough.

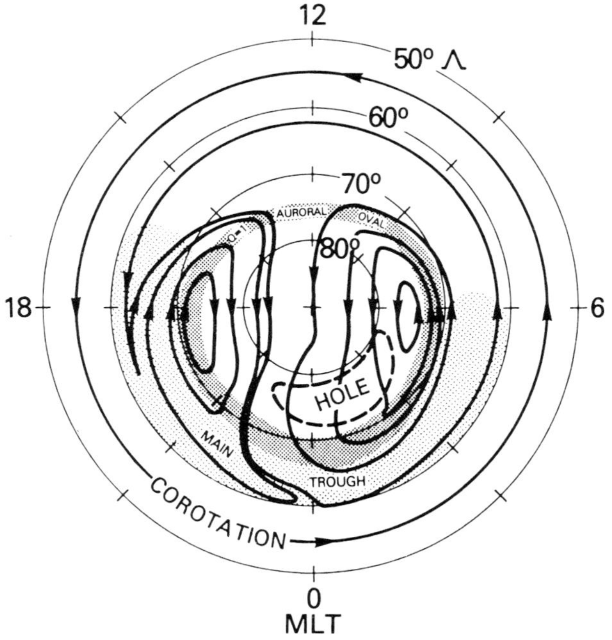

Figure 4. Schematic illustration of the earth's polar region showing the plasma convection trajectories at 300 km(solid lines) in a magnetic local time (MLT), invariant latitude reference frame. Also shown are the locations of the high-latitude ionization hole, the main plasma trough, and the quiet time auroral oval (from Brinton et al, 1978).

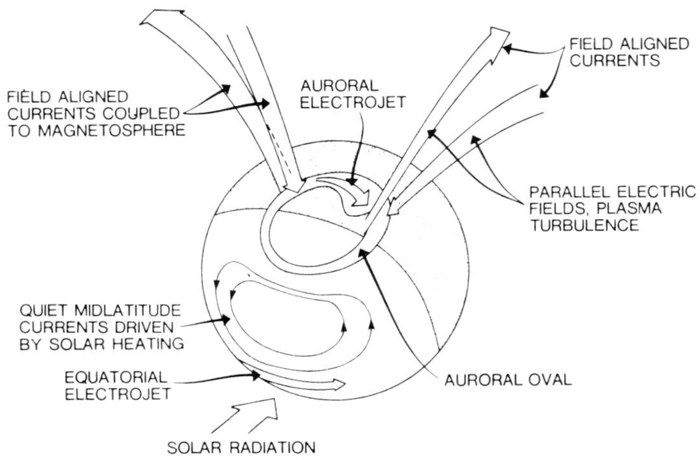

Figure 5. Schematic diagram showing the current systems in the terrestrial ionosphere. From "Space Plasma Physics: The Study of Solar-System Plasmas", National Academy of Sciences, 1978.

Another consequence of the electron and ion motion shown in Figure 3 is that the magnetospheric electric field drives a horizontal current through the E-region (≃ 110 km) owing to the differing electron and ion drifts. This horizontal current is coupled to the magnetosphere through field-aligned (Birkeland) currents, as shown schematically in Figure 5. The field-aligned current patterns for both quiet and disturbed geomagnetic conditions is shown in Figure 6 for the northern hemisphere. The field-aligned currents are concentrated in two principal areas which encircle the geomagnetic pole. The poleward current regions exhibit current flow into the ionosphere in the morning sector and away from the ionosphere in the evening sector, while the equatorward current regions contain current flows in the opposite directions at a given local time. The basic field-aligned current flow pattern is the same during geomagnetically quiet and active periods. The magnitudes of the currents in the poleward and equatorward regions are not well-known, but it appears that the net current is inward on the morningside and outward on the eveningside in the northern hemisphere.

Particle precipitation and Joule dissipation in the auroral oval produces heat for the neutral gas, and this heat input can significantly affect the thermospheric wind pattern. Without auroral heating, the thermospheric wind would blow across the polar cap from the dayside to the nightside due to solar heating. However, auroral heating causes the thermosphere in the vicinity of the oval to expand, as shown schematically in Figure 7. On the dayside, this expansion can retard or reverse the thermospheric wind depending on the auroral heating rate. On the nightside, auroral heating and momentum forcing by the convecting ionosphere act to produce the so-called midnight surge. This feature, which occurs during geomagnetic storms, corresponds to very strong equatorward winds near midnight, with the wind speed reaching 600 m/s on occasion.

2.2. Mid-latitude ionosphere

At mid-latitudes, where the geomagnetic field lines are inclined to the vertical direction, the effect of a meridional (north-south) wind is to force the F region ionization up or down field lines, depending on the direction of the wind. An equatorward wind drives the ionization to higher altitudes where chemical loss rates are lower, while the reverse is true for a poleward wind. The equatorward wind at night therefore acts to maintain F layer ionization, while the poleward wind during the day acts to depress the F layer.

Another transport process that affects the mid-latitude ionosphere is the interhemispheric flow of plasma along geomagnetic field lines. Figure 8 shows the direction of the interhemispheric flow for solstice conditions, as deduced from data obtained at Millstone Hill. In the summer hemisphere, the flow is upward and out of the topside ionosphere throughout the day and night. In the winter hemisphere, the flow is upward during the day and downward at night. This interhemispheric flow will be either a sink or a source of F region ionization, depending on whether the flow is out of or into the ionosphere.

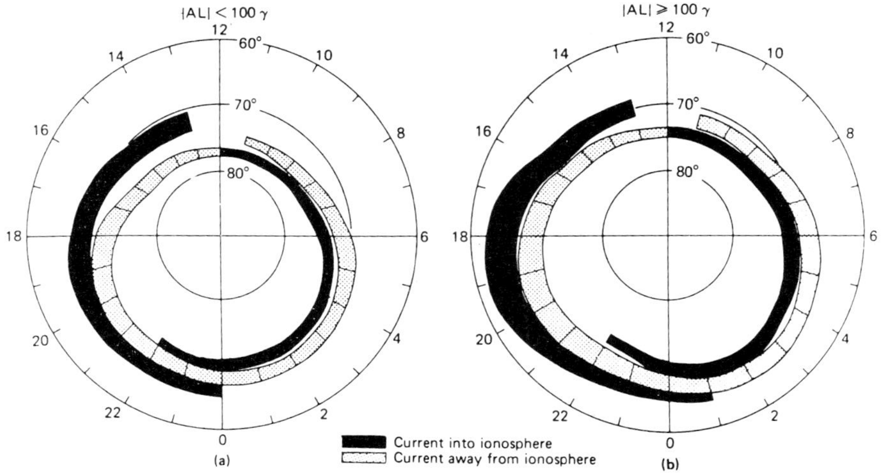

Figure 6. The distribution and flow directions of large-scale field-aligned currents determined from (a) data obtained from 439 passes of the Triad satellite during weakly disturbed geomagnetic conditions and (b) data obtained from 366 Triad passes during active periods (from Iijima and Potemra, 1978).

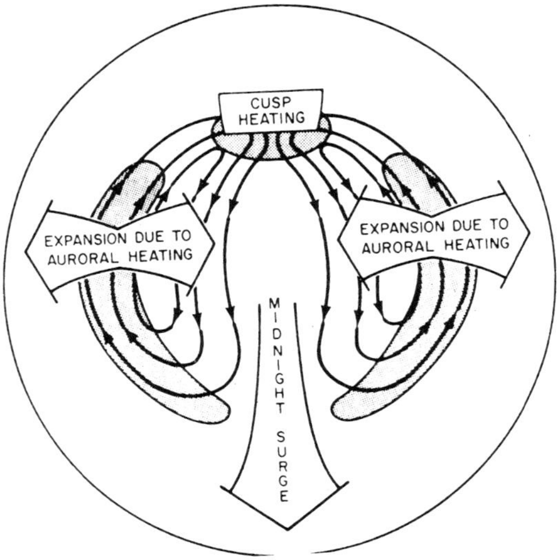

Figure 7. Schematic illustration showing the heating region in the earth's auroral zone and antisolar ion convection, which drives a thermospheric wind toward the equator near local midnight (from Babcock and Evans, 1979).

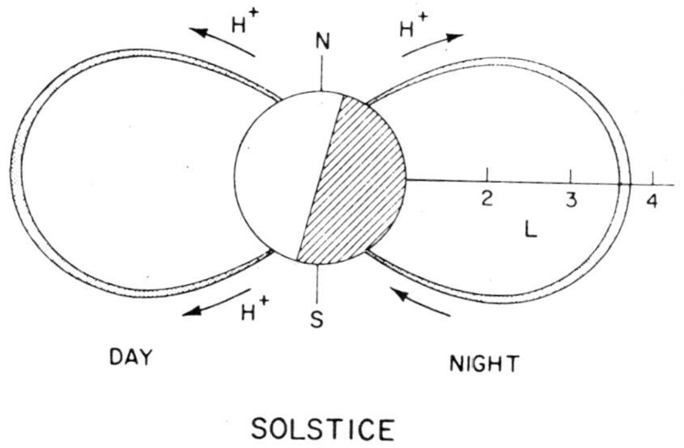

Figure 8. Schematic illustration showing the direction of the plasma flow along geomagnetic field lines at mid-latitudes for solstice conditions (from Evans and Holt, 1978). Reprinted by permission of Pergamon Press.

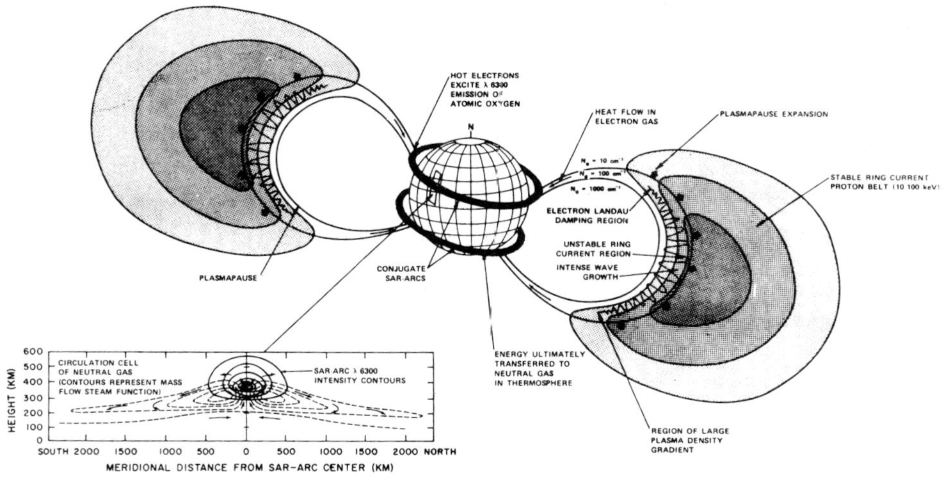

Figure 9. Schematic illustration showing the interaction of cold plasmaspheric plasma with hot ring current plasma. As a result of this interaction, heat is conducted through the electron gas along geomagnetic field lines to the ionosphere, leading to SAR-arc emission (from Rees and Roble, 1975).

During geomagnetic storms, the cold, high density plasma in the plasmasphere comes into contact with the hot, low density plasma in the ring current, as shown schematically in Figure 9. As a result of Coulomb collisions and/or wave-particle interactions (via ion cyclotron or hydromagnetic waves), energy is transferred from the ring current particles to the thermal electrons along the interaction region. This energy is then conducted down to the lower ionosphere along geomagnetic field lines, producing elevated electron temperatures. At altitudes between 300-400 km, the hot electrons have sufficient energy to collisionally excite atomic oxygen to a higher electronic state, from $O(^3P)$ to $O(^1D)$, which requires an energy of 1.97 eV. The excited atoms subsequently emit 6300 A photons, which is the characteristic red line of atomic oxygen. The emission generally occurs over a narrow latitudinal band equatorward of the auroral oval. The stability of the emission led to the term stable auroral red arc (SAR-arc) being used to describe the phenomenon.

2.3. Low latitude ionosphere

At low latitudes, the geomagnetic field lines are nearly horizontal and this introduces some unique transport effects. First, thermospheric winds blowing across the equator from the summer to the winter hemisphere can effectively induce an interhemispheric plasma flow along geomagnetic field lines, as shown schematically in Figure 10. As the ionospheric plasma rises on the summer side of the equator, it expands and cools, while on the winter side it is compressed and heated as it descends.

Another interesting transport effect at low latitudes is the so-called equatorial fountain. In the daytime equatorial F region, eastward electric fields associated with neutral wind induced ionosheric currents drive a plasma convection that is upward, as shown in Figure 11. The plasma lifted in this way then diffuses down the magnetic field

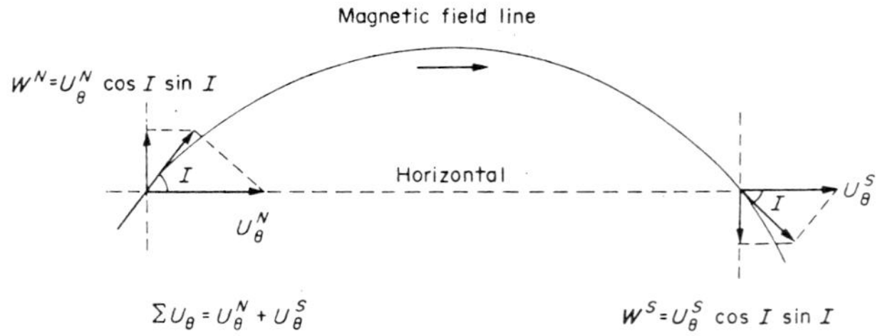

Figure 10. Schematic illustration showing the vertical ion drift (W) produced by a horizontal wind component in the magnetic meridian (U_θ). The magnetic field dip angle is denoted by I (from Bittencourt and Sahai, 1978). Reprinted by permission of Pergamon Press.

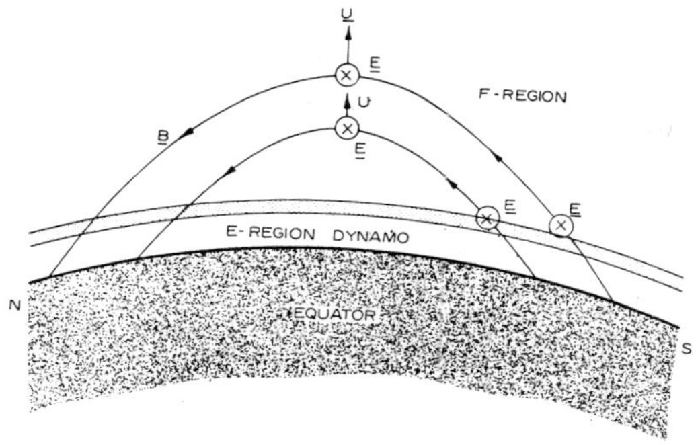

Figure 11. Near the equator, atmospheric winds in the E-region generate electric fields that are transmitted along the curved geomagnetic field lines to the F-region. During the daytime, these dynamo electric fields are eastward, which produces an upward motion of the equatorial F-layer.

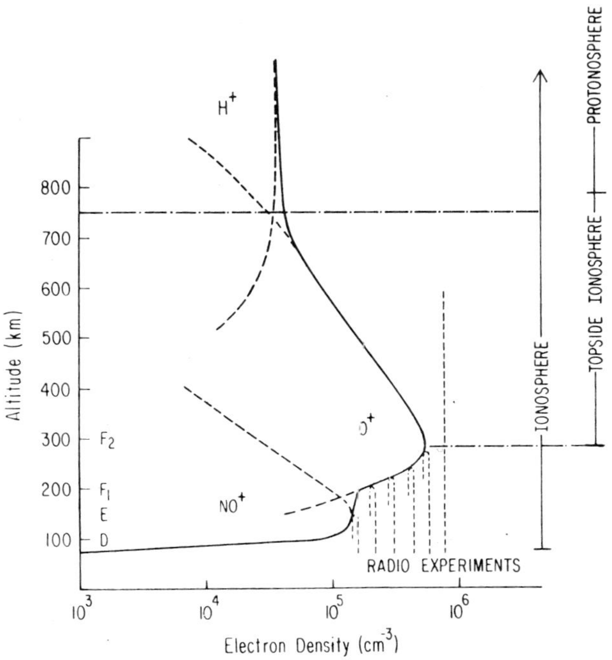

Figure 12. Schematic illustration of the ionosphere showing the various layers; D, E, F_1 and F_2 (from Banks et al, 1976). Reprinted by permission of Annual Reviews Inc.

lines and away from the equator due to the action of gravity. This combination of electromagnetic drift and diffusion produces a 'fountain-like' pattern of plasma motion. The result of this plasma transport is that ionization peaks are formed in the subtropics on each side of the magnetic equator (often called Appleton's peaks).

2.4. Ionospheric layers

Prior to the advent of rockets and satellites, the exploration of the ionosphere was based on the reflection of high frequency radio waves from this medium. Early radio experiments (1925-1930) indicated that the ionosphere is stratified into layers (D, E, F_1 and F_2), as shown schematically in Figure 12. Although the radio methods could not identify the ion composition, subsequent in situ measurements by rockets and satellites were not only useful for identifying the ion composition, but also for obtaining information on temperatures and flow velocities as well.

Table 1 shows typical parameters for the daytime ionosphere at mid-latitudes. The E-region ($\simeq$ 110 km) is dominated by molecular species, with the major ions being NO^+, O_2^+ and N_2^+ and the most abundant neutrals being N_2 and O_2. The total ion density is of the order of 10^5 cm^{-3}, while the neutral density is about 5×10^{11} cm^{-3}. The plasma is weakly-ionized, i.e., collisions between charged particles are not important. The ions, electrons, and neutrals in the E-region are basically in thermal equilibrium at a temperature of about 350° K. In the F-region (200-300 km) the atomic species dominate, with O^+ and O being the major ion and neutral species, respectively. At the F-region peak, the ion density is roughly a factor of 10 greater than that in the E-region, while the neutral density is more than two orders of magnitude lower. The plasma in the F-region is partially-ionized, i.e., collisions between different charged particles and between charged particles and neutrals must be considered. At F-region altitudes the ions, electrons, and neutrals are not in thermal equilibrium, and the temperatures are considerably higher than in the E-region. The topside ionosphere is generally defined to be the region above the F2-peak, while the protonosphere is the region where the lighter atomic ions (H^+, He^+) dominate. Although the neutrals outnumber the ions by a factor of 10 at 1000 km, the plasma is effectively fully-ionized owing to the long-range nature of Coulomb collisions. Thermal nonequilibrium exists in the protonosphere as it does in the F-region, but the species temperatures are higher. The D-region differs from the E-region in that negative molecular ions occur, as do three-body collisions. Because of a space limitation, the D-region will not be discussed further.

3. THE THEORY

From the above dicussion, it is apparent that the terrestrial ionosphere exhibits a complex behavior owing to the various forces acting within and on it. To understand this behavior it is necessary to take

Table 1. Typical Daytime Ionospheric Parameters at Mid-Latitudes

Parameter	E-Region	F-Region	Protonosphere
Ion Species	NO^+, O_2^+, N_2^+	O^+	H^+, He^+
Neutral Species	N_2, O_2	O	H, He
$n_e (cm^{-3})$	10^5	10^6	10^4
$n_n (cm^{-3})$	$5x10^{11}$	10^9	10^5
$T_e (^oK)$	350	1600	3000
$T_i (^oK)$	350	1200	2000
$T_n (^oK)$	350	990	1000
Plasma condition	Weakly-ionized	Partially-ionized	Fully-ionized

account of the following processes: photoionization, collisional impact ionization, chemical reactions, neutral composition changes, atmospheric winds, electrodynamic drifts, magnetic field-aligned diffusion, thermal coupling between species, thermal conduction, and local heat sources and sinks. In this section, the relevant physical processes will be described mathematically.

3.1. Transport equations

The quantitative study of the different flow situations that are found in the terrestrial ionosphere is normally begun through the use of conservation equations which describe the spatial and temporal evolution of the concentration, drift velocity, and temperature of the different species in the ionosphere. These conservation equations are obtained by taking velocity moments of Boltzmann's equation, and the three lowest order equations are the continuity, momentum, and energy equations, respectively,

$$\partial n_s/\partial t + \nabla \cdot (n_s \underline{u}_s) = P_s' - L_s' n_s \qquad (1)$$

$$n_s m_s (D_s \underline{u}_s/Dt) + \nabla p_s + \nabla \cdot \underline{\tau}_s - n_s m_s \underline{G}$$
$$- n_s e_s [\underline{E} + (1/c)\, \underline{u}_s \times \underline{B}] = \delta \underline{M}_s/\delta t \qquad (2)$$

$$(D_s/Dt)(3p_s/2) + (5/2)p_s (\nabla \cdot \underline{u}_s) + \nabla \cdot \underline{q}_s + \underline{\tau}_s : \nabla \underline{u}_s$$
$$= \delta E_s/\delta t + Q_s - L_s \qquad (3)$$

where $D_s/Dt = \partial/\partial t + \underline{u}_s \cdot \nabla$ is the convective derivative of species s, $p_s = n_s k T_s$ is the partial pressure, n_s is the number density, m_s is the mass, e_s is the charge, T_s is the temperature, $\underline{u}_s$ is the drift velocity, $\underline{q}_s$ is the heat flow vector, $\underline{\tau}_s$ is the stress tensor, P_s' is the ionization production rate, L_s' is the ionization loss frequency, Q_s is the heating rate, L_s is the cooling rate; $\underline{G}$ is the acceleration due to gravity, $\underline{E}$ is the electric field, $\underline{B}$ is the magnetic field, $\partial/\partial t$ is the time derivative, ∇ is the coordinate-space gradient, c is the speed of light, and k is Boltzmann's constant. The double-dot operator in equation (3) corresponds to the scalar product of the two tensors. The quantities $\delta \underline{M}_s/\delta t$ and $\delta E_s/\delta t$ represent the rate of momentum and energy exchange, respectively, in collisions between species s and the other species in the plasma.

3.2. Collision terms

In order to calculate collision terms for the transport equations (1) - (3), it is generally necessary to adopt an approximate expression for the species velocity distribution functions. Only for the simplest case of displaced Maxwellians have general collision terms been evaluated which apply to arbitrary interparticle force laws, large temperature differences, and large relative drifts between the interacting species. In this approximation, stress and heat flow are neglected and the behavior of the plasma is expressed in terms of just the species density, drift velocity, and the temperature. Since the drift velocity has three components, there are a total of five parameters describing each species in the plasma.

With this so-called 5-moment approximation, the appropriate collision terms take the form

$$\delta \underline{M}_s/\delta t = \sum_t n_s m_s \nu_{st} (\underline{u}_t - \underline{u}_s)\, \Phi_{st} \tag{4}$$

$$\delta E_s/\delta t = n_s m_s \Sigma_t \nu_{st}\, [3k(T_t - T_s)\Psi_{st} + m_t(\underline{u}_s - \underline{u}_t)^2 \Phi_{st}] / (m_s + m_t) \tag{5}$$

where Φ_{st} and Ψ_{st} are velocity dependent correction factors and ν_{st} is the momentum transfer collision frequency for gases s and t. The momentum transfer collision frequency is defined in a later subsection, and relevant expressions for Φ_{st} and Ψ_{st} are given by Schunk (1977). Here we merely note that $\Phi_{st} = \Psi_{st} = 1$ for Maxwell molecule interactions (interaction potential that varies inversely as the fourth power of the particle separation) and arbitrary relative drift speeds as well as for arbitrary interaction potentials and low relative drift speeds.

The level of approximation that includes both stress and heat flow is called the 13-moment approximation. Unfortunately, at this level of approximation, collision terms for arbitrary interaction potentials have been derived only for low-speed relative flows, where the species drift velocity differences are small compared to the species thermal speeds. In this limit, collision terms have been derived for both small and large temperature differences between the interacting species. The former are called Burgers' 'linear' collision terms and the latter

Table 2. Momentum Transfer Collision Frequencies for Electron-Neutral Interactions

Species	ν_{en}, s^{-1}
N_2	$2.33 \times 10^{-11} n(N_2)[1 - 1.21 \times 10^{-4} T_e] T_e$
O_2	$1.82 \times 10^{-10} n(O_2)[1 + 3.6 \times 10^{-2} T_e^{1/2}] T_e^{1/2}$
O	$8.9 \times 10^{-11} n(O)[1 + 5.7 \times 10^{-4} T_e] T_e^{1/2}$
He	$4.6 \times 10^{-10} n(He) T_e^{1/2}$
H	$4.5 \times 10^{-9} n(H)[1 - 1.35 \times 10^{-4} T_e] T_e^{1/2}$

From Schunk and Nagy (1978)

Burgers' 'semilinear' collision terms. For both cases the momentum and energy collision terms are given by

$$\delta \underline{M}_s/\delta t = - \sum_t n_s m_s \nu_{st}(\underline{u}_s - \underline{u}_t) + \sum_t \nu_{st} z_{st} \mu_{st} [\underline{q}_s - (\rho_s/\rho_t) \underline{q}_t]/kT_{st} \tag{6}$$

$$\delta E_s/\delta t = - \sum_t n_s m_s \nu_{st} \, 3k \, (T_s - T_t)/(m_s + m_t) \tag{7}$$

where $\rho_s = n_s m_s$ is the mass density, $\mu_{st} = m_s m_t/(m_s + m_t)$ is the reduced mass, and $T_{st} \equiv (m_t T_s + m_s T_t)/(m_s + m_t)$ is the reduced temperature. The quantity z_{st} is a pure number that is different for different combinations of species s and t; representative values are given by Schunk (1977). The heat flow terms that appear in the momentum collision term (6) account for thermal diffusion and the effect of heat flow on ordinary diffusion.

3.3. Collision frequencies

In the terrestrial ionosphere, the relevant collision processes include Coulomb interactions, nonresonant ion-neutral interactions, resonant charge exhange, and electron-neutral interactions. The appropriate momentum transfer collision frequencies for two-body elastic electron-neutral interactions are given in Table 2.

For Coulomb interactions, the momentum transfer collision frequency takes the form

$$\nu_{st} = (16\sqrt{\pi}/3)(\mu_{st}/2kT_{st})^{3/2}(e_s e_t/\mu_{st})^2 n_t m_t \ln\Lambda/(m_s + m_t) \tag{8}$$

where $\ln\Lambda$ is the Coulomb logarithm. For the ionosphere, $\ln\Lambda \sim 15$ and the Coulomb collision frequency can be approximated numerically by

$$\nu_{st} = 1.27\ z_s^2 z_t^2 A_{st}^{1/2}\ n_t/(A_s T_{st}^{3/2}) \tag{9}$$

where A_s is the particle mass in atomic mass units, A_{st} is the reduced mass in atomic mass units, z_s and z_t are the particle charge numbers, n_t is in cm^{-3}, and T_{st} is in ^oK. For ion-ion interactions, this expression reduces further to

$$\nu_{st} = B_{st} n_t / T_{st}^{3/2} \tag{10}$$

where B_{st} is a numerical coefficient; values are given in Table 3 for the ion species found in the ionosphere. Equation (9) also reduces further for electron-electron and electron-ion interactions,

$$\nu_{ei} = 54.5\ n_i z_i^2 / T_e^{3/2} \tag{11}$$

$$\nu_{ee} = 54.5\ n_e/[\sqrt{2}\ T_e^{3/2}] \tag{12}$$

where subscript e denotes electrons and subscript i is for ions.

Ion-neutral interactions can be either resonant or non-resonant. Resonant charge exchange occurs when an ion collides with its parent neutral or it can occur accidentally as in the case of the reaction $H^+ + O \rightleftarrows H + O^+$. The relevant resonant ion-neutral momentum transfer collision frequencies are given in Table 4. Nonresonant ion-neutral interactions occur between unlike ions and neutrals, and they correspond to a long-range polarization attraction coupled with a short-range repulsion. In this case, the ion-neutral momentum transfer collision frequency takes the form

$$\nu_{in} = 2.21\pi(\gamma_n e^2/\mu_{in})^{1/2} n_n m_n/(m_i + m_n) \tag{13}$$

where subscript n corresponds to neutrals and γ_n is the neutral atom polarizability. For a given ion-neutral pair, equation (13) takes a particularly simple form,

$$\nu_{in} = C_{in} n_n \tag{14}$$

where C_{in} is a numerical coefficient; values are given in Table 5 for the different ion-neutral combinations found in the ionosphere.

It should be noted that the momentum transfer collision frequencies are not symmetric with respect to a change of indices, but satisfy the relation

Table 3. The Collision Frequency Coefficients B_{st} for Ion-Ion Interactions

s \ t =	H^+	He^+	N^+	O^+	N_2^+	NO^+	O_2^+
H^+	0.90	1.14	1.23	1.23	1.25	1.25	1.25
He^+	0.28	0.45	0.56	0.57	0.59	0.60	0.60
N^+	0.088	0.16	0.24	0.25	0.28	0.28	0.28
O^+	0.077	0.14	0.22	0.22	0.25	0.26	0.26
N_2^+	0.045	0.085	0.14	0.15	0.17	0.17	0.18
NO^+	0.042	0.080	0.13	0.14	0.16	0.16	0.17
O_2^+	0.039	0.075	0.12	0.13	0.15	0.16	0.16

Table 4. Momentum Transfer Collision Frequencies for Resonant Ion-Neutral Interactions

Species	T_r, °K	ν_{in}, s^{-1}
H^+, H	50	$2.65 \times 10^{-10} n(H) T_r^{1/2} (1 - 0.083 \log_{10} T_r)^2$
He^+, He	50	$8.73 \times 10^{-11} n(He) T_r^{1/2} (1 - 0.093 \log_{10} T_r)^2$
N^+, N	275	$3.83 \times 10^{-11} n(N) T_r^{1/2} (1 - 0.063 \log_{10} T_r)^2$
O^+, O	235	$3.67 \times 10^{-11} n(O) T_r^{1/2} (1 - 0.064 \log_{10} T_r)^2$
N_2^+, N_2	170	$5.14 \times 10^{-11} n(N_2) T_r^{1/2} (1 - 0.069 \log_{10} T_r)^2$
O_2^+, O_2	800	$2.59 \times 10^{-11} n(O_2) T_r^{1/2} (1 - 0.073 \log_{10} T_r)^2$
H^+, O	300	$6.61 \times 10^{-11} n(O) T_i^{1/2} (1 - 0.047 \log_{10} T_i)^2$

$T_r = (T_i + T_n)/2$. From Banks and Kockarts (1973).

Table 5. The Collision Frequency Coefficients $C_{in} \times 10^{10}$ for Non-Resonant Ion-Neutral Interactions

i \ n =	H	He	N	O	N_2	O_2
H^+	r	10.6	26.1	r	33.6	32.0
He^+	4.71	r	11.9	10.1	16.0	15.3
N^+	1.45	1.49	r	4.42	7.47	7.25
O^+	r	1.32	4.62	r	6.82	6.64
N_2^+	0.74	0.79	2.95	2.58	r	4.49
NO^+	0.69	0.74	2.79	2.44	4.34	4.27
O_2^+	0.65	0.70	2.64	2.31	4.13	r

r means the collisional interaction is resonant

$$n_s m_s \nu_{st} = n_t m_t \nu_{ts} \tag{15}$$

3.4. Stress and heat flow expressions

The continuity, momentum, and energy equations (1) - (3) can be applied separately to each species in the plasma, but only after heat flow and stress tensor expressions are obtained do they constitute a closed set (cf. Schunk, 1977). For most ionospheric applications, simplified stress tensor and heat flow expressions can be adopted because the plasma is collision-dominated. For a fully-ionized electron-ion plasma, these simplified expressions are given by

$$\underline{\tau}_i + (5\Omega_i/6\nu_{ii}) \, [\underline{b} \times \underline{\tau}_i - \underline{\tau}_i \times \underline{b}] = -\eta_i[\nabla\underline{u}_i + (\nabla\underline{u}_i)^T - \tfrac{2}{3}(\nabla\cdot\underline{u}_i)\underline{I}] \tag{16}$$

$$\underline{\tau}_e - (5\omega_e/6\nu_e')[\underline{b} \times \underline{\tau}_e - \underline{\tau}_e \times \underline{b}] = -\eta_e[\nabla\underline{u}_e + (\nabla\underline{u}_e)^T - \tfrac{2}{3}(\nabla\cdot\underline{u}_e)\underline{I}] \tag{17}$$

$$\underline{q}_i - (5\Omega_i/4\nu_{ii})\underline{q}_i \times \underline{b} = -\lambda_i \nabla T_i \tag{18}$$

$$\underline{q}_e + (5\omega_e/4\nu_e)\underline{q}_e \times \underline{b} = -\lambda_e \nabla T_e - \beta_e \underline{J} \tag{19}$$

where the viscosity coefficients (η_i, η_e), thermal conductivities (λ_e, λ_i), and thermoelectric coefficient (β_e) are given by

$$\eta_e = 5p_e/(6\nu_e') \tag{20}$$

$$\eta_i = 5p_i/(6\nu_{ii}) \tag{21}$$

$$\lambda_e = 25kp_e/(8m_e\nu_e) \tag{22}$$

$$\lambda_i = 25kp_i/(8m_i\nu_{ii}) \tag{23}$$

$$\beta_e = (15\nu_{ei}/8\nu_e)(kT_e/e) \tag{24}$$

and where

$$\underline{J} = n_e \, e(\underline{u}_i - \underline{u}_e) \tag{25}$$

$$\nu_e = \nu_{ee} + 13\nu_{ei}/8 \quad (26)$$

$$\nu_e' = \nu_{ee} + \nu_{ei} \quad (27)$$

In equations (16)-(27), $\omega_e = eB/m_e c$, $\Omega_i = e_i B/m_i c$, $\underline{b}$ is a unit vector along $\underline{B}$, $\underline{I}$ is the unit dyadic, and $(\nabla\underline{u})^T$ represents the transpose of the second-order tensor $(\nabla\underline{u})$.

Upon eliminating the collision frequencies in equations (20)-(23) using the expressions given in the previous subsection, it is easy to show that for a fully-ionized plasma

$$\eta_i/\eta_e \sim (m_i/m_e)^{1/2}(T_i/T_e)^{5/2} \quad (28)$$

$$\lambda_i/\lambda_e \sim (m_e/m_i)^{1/2}(T_i/T_e)^{5/2} \quad (29)$$

The ion viscosity coefficient is therefore larger than the corresponding electron coefficient by approximately $(m_i/m_e)^{1/2}$. As it turns out, electron viscosity is not important in the ionosphere and ion viscosity is only important at certain altitudes in the polar wind. However, both electron and ion thermal conduction are important, with the electron thermal conductivity larger than the ion conductivity by approximately $(m_i/m_e)^{1/2}$. In the ionosphere, thermal conduction occurs primarily along geomagnetic field lines, so the $\underline{q} \times \underline{b}$ terms in equations (18) and (19) can generally be neglected. Finally, we note that electron thermal conduction is important at altitudes as low as 200 km, and therefore, the fully-ionized coefficient (22) must be corrected for electron-neutral collisions; this is accomplished by replacing (26) with the expression

$$\nu_e = \nu_{ee} + 13\nu_{ei}/8 + 5\sum_n \nu_{en} z_{en}'/4 \quad (30)$$

where z_{en}' is given by Schunk (1977).

3.5. Photoionization

Solar extreme ultraviolet (EUV) radiation photo-ionizes the neutral constituents of the upper atmosphere, producing free electrons and ions. The photoionization process occurs predominantly at the lower levels of the ionosphere, where the neutrals are abundant. Typically, the peak in the ionization rate occurs at about 150 km owing mainlyoto the absorption of radiation with wavelengths less than 796A (the ionization threshold of N_2). Photons with wavelengths in the range of 796-1027 A penetrate down into the E region. For the E and F regions of the ionosphere, the most important photoionization processes are

$$N_2 + h\nu \rightarrow N_2^+ + e \tag{31a}$$

$$\rightarrow N^+ + N + e \tag{31b}$$

$$O_2 + h\nu \rightarrow O_2^+ + e \tag{32}$$

$$O + h\nu \rightarrow O^+ + e \tag{33}$$

$$N + h\nu \rightarrow N^+ + e \tag{34}$$

where (31b) is produced with an efficiency of about 21%.

In order to illustrate the basic physics behind the photoionization process, it is convenient to make the following simplifying assumptions: (a) the radiation is monochromatic, (b) the atmosphere consists of a single absorbing gas, (c) the neutral scale height H_n is constant, and (d) the atmosphere is plane and horizontally stratified. With these assumptions, the ionizing radiation penetrates the atmosphere in the manner shown schematically in Figure 13. Let $\sigma^{(a)}$ be the absorption cross section, ζ be the ionizing efficiency (i.e., the number of photoelectrons produced per photon absorbed), I(z) be the photon flux, and n(z) be the neutral density, where z is altitude. For a simple gas illuminated by monochromatic radiation, the probability per unit time a

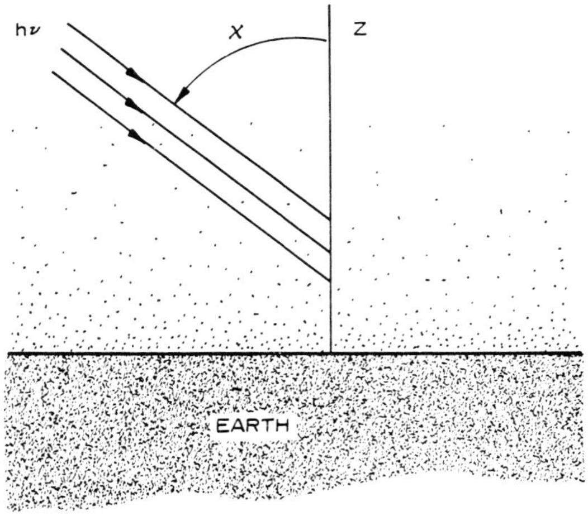

Figure 13. Schematic illustration showing monochromatic radiation penetrating a plane and horizontally stratified atmosphere. The solar zenith angle is measured from the vertical Z - axis.

given molecule absorbs a photon is $I\sigma^{(a)}$, and therefore, the probability per unit time of producing an ion-electron pair is $\zeta I\sigma^{(a)}$. Hence, the rate of production per unit volume is

$$P' = I \zeta n\sigma^{(a)} \tag{35}$$

As the radiation flux penetrates the atmosphere, it is attenuated owing to absorption. After traveling an elemental distance ds, the radiation flux will decrease by

$$dI = -In\sigma^{(a)}ds \tag{36}$$

Since the variation of altitude along the path of the radiation is given by $ds = -dz \sec \chi$, equation (36) can be expressed in the form

$$d/dz(\ln I) = n\sigma^{(a)}\sec \chi \tag{37}$$

The radiation that reaches the altitude z is obtained by integrating from z to infinity,

$$\ln(I_\infty/I) = \int_z^\infty dz\; n\sigma^{(a)}\sec \chi \tag{38}$$

where I_∞ is the unattenuated flux at the top of the atmosphere. Now sec χ does not vary with z owing to the plane earth assumption and $n = n_o \exp[-(z-z_o)/H]$, where the neutral scale height H is constant and z_o is a reference attitude. Therefore, equation (38) becomes

$$\ln(I_\infty/I) = Hn\sigma^{(a)}\sec \chi \equiv \tau \tag{39}$$

where τ is the optical depth. Hence, the photon flux varies as

$$I = I_\infty e^{-\tau} \tag{40}$$

Substituting equation (40) into equation (35) yields the following production rate:

$$P' = I_\infty \zeta n\sigma^{(a)}e^{-\tau} \tag{41}$$

Introducing the ionization cross section, $\sigma^{(i)} = \zeta\sigma^{(a)}$, and using equation (39) and the expression for n, the production rate can be expressed in the form

$$P' = P_o \exp\{1 - (z - z_o)/H - \exp[(z_o - z)/H]\sec \chi\} \tag{42}$$

where the reference level z_o has been chosen to be the level of unit optical depth ($\tau=1$) for overhead sun ($\chi = 0°$). At this level, the incident radiation is attenuated by a factor of e. The quantity P_o is the peak rate of production for overhead sun

$$P_o = \sigma^{(i)} I_\infty / [\sigma^{(a)} e\, H] \tag{43}$$

Equation (42) corresponds to the classical Chapman production function. Figure 14 shows the normalized Chapman production function versus $(z-z_o)/H$ for several values of the solar zenith angle χ.

For practical computations of ionization production rates, the simplifications that were needed to derive the Chapman production function are too restrictive. In particular, the absorption and ionization cross sections are a strong function of wavelength, as shown in Figure 15. Also, the radiation incident upon the top of the atmosphere is neither monochromatic nor a smooth function of wavelength, as shown in Figure 16. Furthermore, the absolute flux of solar radiation changes with solar activity; Figure 16 corresponds to solar minimum conditions. In addition, the neutral atmosphere is not plane and horizontally stratified; there is more than one neutral species; and the neutral scale height is not constant.

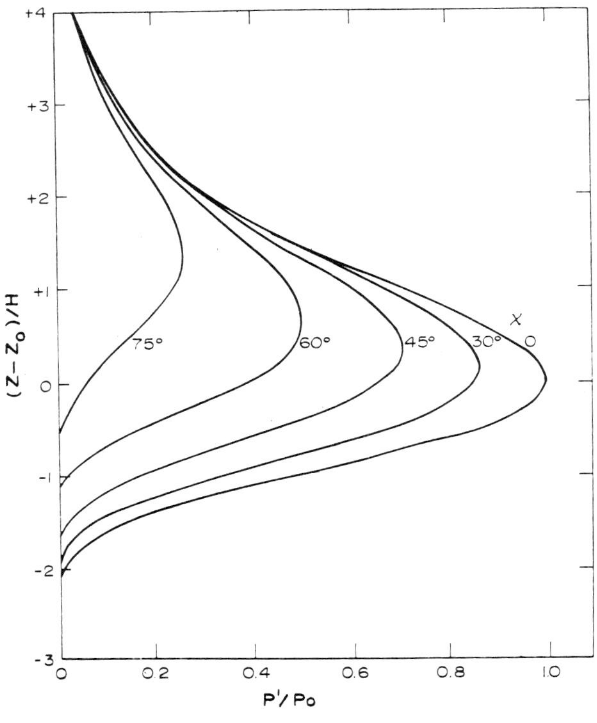

Figure 14. Normalized Chapman production function versus $(Z - Z_o)/H$ for several values of the solar zenith angle.

For practical applications, the photoelectron production rate, P_e' is given by

$$P_e'(E,z) = \sum_\ell \sum_n n_n(z) \int_0^\infty d\lambda I_\infty(\lambda)\sigma_n^{(i)}(\lambda)p_n(\lambda,E_\ell)\exp[-\tau(\lambda,z)] \quad (44)$$

where

$$\tau(\lambda,z) = \Sigma_n \sigma_n^{(a)}(\lambda)n_n H_n \,ch(R_n,\chi) \quad (45)$$

$$H_n = kT_n/m_n g \quad (46)$$

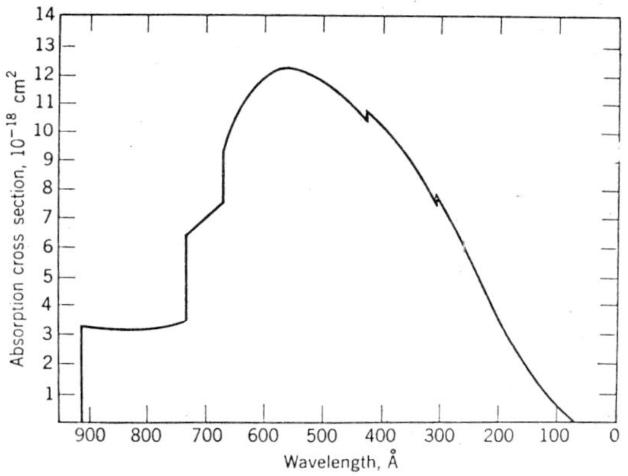

Figure 15. The absorption cross section for atomic oxygen (from Whitten and Poppoff, 1971). Reprinted by permission of John Wiley & Sons.

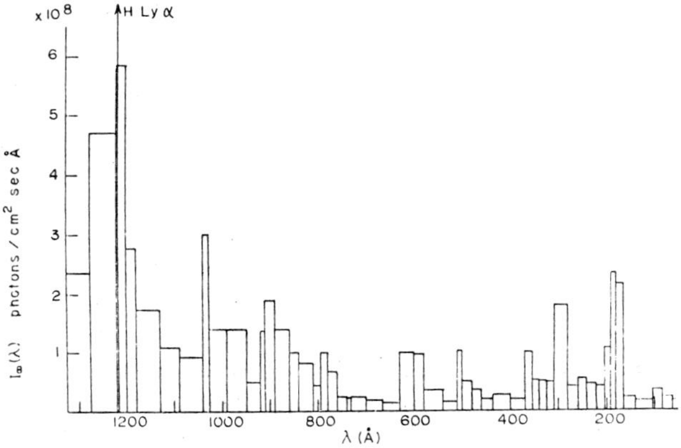

Figure 16. The solar radiation spectrum $I_\infty(\lambda)$ in the extreme ultraviolet region (from Takayanagi and Itikawa, 1970).

$$R_n = (R + z)/H_n \tag{47}$$

In equations (44) to (47), $E= E_\lambda - E_\ell$, E_λ is the energy corresponding to wavelength λ, E_ℓ is the ionization of energy of a given excited ion state ℓ, R is the planetary radius, χ is the solar zenith angle, $ch(R_n,\chi)$ is the Chapman grazing incidence function, and the $p_n(\lambda,E_\ell)$ are the branching ratios of the various excited ion states. For $\chi < 80°$, the Chapman function can be replaced by sec χ in every term in the summation in equation (45).

3.6. Chemical reactions

There are numerous chemical reactions that occur in the terrestrial ionosphere and they act to both create and destroy ions. The most important chemical reactions for the E and F regions are

$$O^+ + N_2 \rightarrow NO^+ + O, \; k_1 \tag{48}$$

$$O^+ + O_2 \rightarrow O_2^+ + O, \; k_2 \tag{49}$$

$$O^+ + H \rightleftarrows H^+ + O, \; \begin{array}{l} k_f = 2.5\text{x}10^{-11} T_n^{1/2} \\ k_r = 2.2\text{x}10^{-11} T_i^{1/2} \end{array} \tag{50}$$

$$N_2^+ + O_2 \rightarrow O_2^+ + N_2, \; 5\text{x}10^{-11}(300/T) \tag{51}$$

$$N_2^+ + O \rightarrow O^+ + N_2, \; 1\text{x}10^{-11}(300/T)^{0.23} \text{ for } T \leq 1500^\circ K \tag{52}$$

$$N_2^+ + O \rightarrow NO^+ + N, \; 1.4\text{x}10^{-10}(300/T)^{0.44} \text{ for } T \leq 1500^\circ K \tag{53}$$

$$N^+ + O_2 \rightarrow NO^+ + O, \; 2.6\text{x}10^{-10} \tag{54}$$

$$N^+ + O_2 \rightarrow O_2^+ + N, \; 3.1\text{x}10^{-10} \tag{55}$$

$$He^+ + N_2 \rightarrow N^+ + He + N, \; 9.6\text{x}10^{-10} \tag{56}$$

$$He^+ + N_2 \rightarrow N_2^+ + He, \; 6.4\text{x}10^{-10} \tag{57}$$

$$He^+ + O_2 \rightarrow O^+ + He + O, \; 1.1\text{x}10^{-9} \tag{58}$$

$$N_2^+ + e \rightarrow N + N, \; 1.8\times10^{-7}(300/T_e)^{0.39} \tag{59}$$

$$O_2^+ + e \rightarrow O + O, \; 1.6\times10^{-7}(300/T_e)^{0.55} \tag{60}$$

$$NO^+ + e \rightarrow N + O, \; 4.2\times10^{-7}(300/T_e)^{0.85} \tag{61}$$

where

$$k_1 = 1.533 \times 10^{-12} - 5.92 \times 10^{-13}(T/300) + 8.60 \times 10^{-14}(T/300)^2; \quad 300 \leq T \leq 1700^\circ K \tag{62a}$$

$$k_1 = 2.73 \times 10^{-12} - 1.155 \times 10^{-12}(T/300) + 1.483 \times 10^{-13}(T/300)^2; \quad 1700 < T < 6000^\circ K \tag{62b}$$

$$k_2 = 2.82 \times 10^{-11} - 7.74 \times 10^{-12}(T/300) + 1.073 \times 10^{-12}(T/300)^2 - 5.17 \times 10^{-14}(T/300)^3 + 9.65 \times 10^{-16}(T/300)^4; \quad 300 \leq T \leq 6000^\circ K \tag{63}$$

and where the reaction rate constants are in units of $cm^3 \; s^{-1}$ and T is the effective temperature (cf Schunk and Nagy, 1980). In equation (50), k_f is for the forward reaction, while k_r is for the reverse reaction. The various rate constants were determined in laboratory experiments.

In the F-region, the dominant ion O^+ is produced by photoionization of O (equation 33) and lost in reactions with N_2 and O_2 (equations 48 and 49). In this case, the calculation of the loss frequency, which enters into the O^+ continuity equation (1), is straightforward

$$L'(O^+) = k_1 n(N_2) + k_2 n(O_2) \tag{64}$$

In the topside ionosphere, the accidentally-resonant charge exchange reaction (50) is the main source of H^+. The main loss of the molecular ions is due to dissociative recombination (equations 59-61). At F-region altitudes, where the electron density is large, the recombination rates are very rapid.

3.7. Heating rates

The main source of energy for the terrestrial ionosphere is extreme ultraviolet radiation from the sun. The energy carried by the ionizing photons exceeds the energy required for ionization, with the excess energy going into electron and ion kinetic and ion excitation energy.

However, because the photoions are much more massive than the photoelectrons they acquire very little recoil energy during photoionization, and most of the excess energy goes to the photoelectrons. The initial photoelectron energy depends not only on the energy of the ionizing photon and the identity of the neutral species, but also on the ionization state of the photoion. Typically, photoionization produces photoelectrons with initial energies of some tens of electron volts.

Only a relatively modest amount of the initial photoelectron energy is deposited directly in the ambient electron gas. Most of the excess kinetic energy is lost in both elastic and inelastic collisions with neutral particles and in Coulomb collisions with the ambient ions. If the photoelectrons lose their energy at an altitude near where they are produced, the heating is said to be "local", while if the photoelectrons lose their energy over a distance greater than a neutral scale heights the heating is termed "nonlocal". Nonlocal heating effects occur mainly at high altitudes, where ambient densities are low, and at high photoelectron energies. Photoelectrons with sufficient energy can even escape from the ionosphere, travel along geomagnetic field lines and deposit their energy in the conjugate ionosphere. Typically, photoelectron energy deposition above about 300 km constitutes nonlocal heating.

Shortly after creation a given photoelectron undergoes a number of inelastic and elastic collisions. Briefly, for photoelectron energies greater than about 50 eV, ionization and optically allowed excitation of the neutral constituents are the dominant energy loss processes. At energies of about 20 eV, the energy loss via excitation of metastable levels of the major constituents is comparable to the energy loss through allowed transitions, becoming of increasing importance as the energy decreases. At photoelectron energies below about 5eV, energy loss through excitation of the vibrational levels of N_2 becomes important. Finally, below about 2 eV energy loss to the ambient thermal electrons through elastic collisions is the dominant photoelectron energy loss process, although loss due to excitation of the rotational levels of N_2 is not entirely negligible.

To calculate the total amount of energy received by the thermal electron gas from the photoelectrons, it is necessary to calculate the equilibrium photoelectron flux $\phi(E,z)$, where E is the photoelectron energy. The electron heating rate, $Q_e(z)$, has been calculated from the photoelectron flux with the aid of the following relationship:

$$Q_e(z) = \int_{E\ell}^{\infty} \phi(E,z) \left(\frac{dE}{dx}\right)_e dE \tag{65}$$

where

$$(dE/dx)_e = \text{Photoelectron - thermal electron loss rate} \tag{66}$$

In the initial studies of photoelectron fluxes and heating rates, the photoelectron energy deposition was assumed to be local (valid below about 300 km). In addition, the photoelectron energy loss was assumed to be continuous, thus permitting the derivation of stopping cross sections for the neutral constituents and ambient electron gas,

$$L(E,n) = (1/n_n)\ dE(n)/dx \tag{67}$$

$$L(E,e) = (1/n_e)\ dE(e)/dx \tag{68}$$

The rate at which an electron of energy E loses energy in traversing a background gas of density n_n is given by

$$-dE(n)/dx = n_n \sum_j \varepsilon_j \sigma_j(E) \tag{69}$$

where the summation in equation (69) extends over all possible transitions having electron impact excitation cross sections $\sigma_j(E)$ and excitation energies ε_j.

The photoelectron-thermal electron energy loss rate takes account of Coulomb collisions, Cerenkov wave generation, and quantum effects for energetic electrons. A very close approximation for this loss rate is given by the following simple expirical expression:

$$L(E,e) = \frac{3.37 \times 10^{-12}}{E^{0.94}\ n_e^{0.03}} \left(\frac{E - E_e}{E - 0.53E_e} \right)^{2.36} \tag{70}$$

where $E_e = 8.618 \times 10^{-5} T_e$, and T_e is the electron temperature.

With the local and continuous energy loss approximation, the equilibrium photoelectron flux is expressed in terms of stopping cross sections (67) and (68) and the initial photoelectron energy spectrum through the relation

$$\phi(E,z) = \frac{\int_E^\infty P_e'(E',z)dE'}{\sum_n n_n L(E,n) + n_e L(E,e)} \tag{71}$$

The photoelectron energy loss processes are mostly inelastic in nature (except for electron-electron collisions), therefore it is more appropriate to calculate the flux assuming discrete energy loss processes. The formulation of the equilibrium photoelectron flux, based on discrete energy loss processes, can be written as

$$\phi(E,z) = \frac{P_e'(E,z) + \sum_n\sum_j n_n \sigma_{nj}(E+\delta E_{nj})\phi(E+\delta E_{nj},z)}{\sum_n\sum_j n_n \sigma_{nj}(E)} \tag{72}$$

where σ_{nj} = cross section for species n for inelastic collision j which results in an energy loss of δE_{nj}; and where v(E) is the velocity of a photoelectron with energy E.

Photoelectron flux calculations using either equation (71) or (72) are based on the assumption that photoelectrons lose their energy near the same altitude where they are created. This "local" energy loss approximation holds below about 300 km, but at higher altitudes photoelectron transport becomes significant. Starting in the late 1960's, a variety of methods which account for this nonlocal effect have been developed. Because of a space limitation the various techniques cannot be outlined here. All the published photoelectron transport calculations can be divided into the following two basic groups: (i) statistical (Monte Carlo) methods and (ii) techniques based on the solution of some form of transport equations derivable from the Boltzmann equation (cf. Schunk and Nagy, 1978).

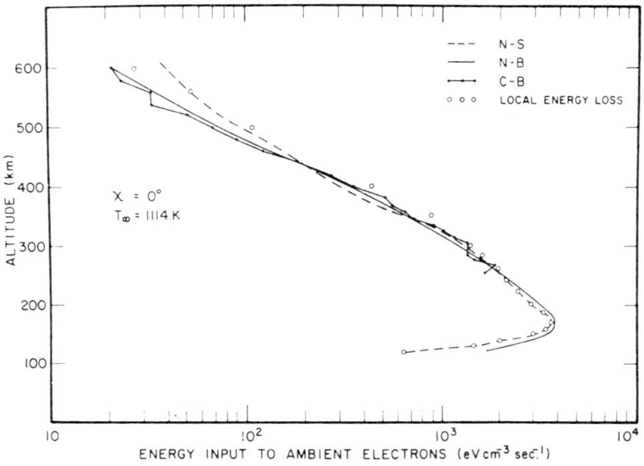

Figure 17. Thermal electron heating rates as a function of altitude for overhead sun calculated by four different numerical techniques (from Cicerone et al, 1973).

Figure 17 shows the ambient electron heating rates calculated by three different photoelectron transport methods and by using the local energy loss approximation. The calculations were for overhead sun at the location of the Arecibo Observatory. Not only are the three photoelectron transport methods in good agreement, but the local energy loss approximation is good at all altitudes for this particular set of conditions. As far as the heating rate profile is concerned, the dominant feature is the sharp peak at about 170 km.

In the daytime polar cusp and nocturnal auroral oval, energetic electron precipitation from the magnetosphere acts as a heat source for the ambient electron gas through Coulomb collisions. However, as with photoelectrons only a small fraction of the incident auroral electron energy ultimately appears as electron gas thermal energy. As the incident (or primary) electrons penetrate the Earth's atmosphere, they lose energy by ionizing and exciting the ambient neutral constituents. The ionizing collisions yield secondary electrons with energies in the tens of electron volts. These secondary electrons, in turn, lose energy in inelastic collisions with the neutral atmosphere and ionosphere, producing electronically, vibrationally, and rotationally excited molecules, atoms, and ions. Once the energy of the primary, secondary or tertiary electrons drops below about 5 eV, energy loss to the ambient electron gas becomes important. This latter mechanism acts to heat the ambient electrons and produces enhanced electron temperatures in the daytime cusp and nocturnal auroral oval.

To date, there have been a number of theoretical studies of the way energetic electrons interact with the Earth's upper atmosphere and ionosphere. However, since this interaction is similar to the way photoelectrons interact with the atmosphere and ionosphere, we will not repeat the details.

The primary heat source for the ion gas in the ionosphere is the ambient electron gas. Additional ion heat sources exist, such as electric field heating, heating by exothermic chemical reactions, and frictional heating by means of neutral winds. However, these heat sources are generally characteristic of certain regions of the ionosphere, and are seldom the primary heat source for the ions. The heat gained by the ions from the electrons is given by equation (7), with the ion-electron collision frequency given by equation (8). If the different ion species in the ionospheric plasma have separate temperatures, the energy exchange between the different ion gases is also given by equations (7) and (8).

If an ion species in the ionospheric plasma is flowing relative to the other ion species or to the neutral gas, such as in the polar wind and in the high-latitude F-region, frictional heating occurs through collisions as energy of directed motion is converted into random thermal energy. Likewise, if a thermospheric wind blows through a stationary ion gas, the ion gas can again be frictionally heated owing to ion-neutral collisions. This type of frictional heating is not described by the 'linear' collision term (7). When friction heating is important, this linear term should be replaced by the 5-moment collision term (5), which takes account of the frictional heating that arises as a result of a relative flow between interacting gases. The appropriate collision frequencies are given in Table 3 for ion-ion interactions, Table 4 for resonant ion-neutral interactions, and Table 5 for non-resonant ion-neutral interactions.

3.8. Cooling rates

There are a number of processes that are effective in cooling the electron gas. In the lower F-region, where the molecular species are abundant, rotational excitation of N_2 and O_2 and excitation of the fine structure levels of atomic oxygen are the most important cooling processes. However, at electron temperatures greater than $1500^\circ K$, vibrational excitation of N_2 and O_2 and electronic excitation of O and O_2 have to be considered. At high altitudes, Coulomb collisions with the ambient ions are an important energy loss mechanism for thermal electrons, but the effect on the electron temperature is negligible.

The calculation of the thermal electron cooling rates requires a knowledge of the excitation cross sections. These cross sections are either calculated or measured, as a function of electron energy. Average thermal electron cooling rates suitable for use in the energy equation (3) are then obtained by taking the appropriate integral over Maxwellian electron and neutral veolcity distributions. In some cases, the cooling rates are fitted with convenient analytic expressions. Some of the more important electron gas cooling rates are as following:

$$L(e,N_2)_{rot} = 2.9\text{x}10^{-14} n_e n(N_2)(T_e - T_n)/T_e^{1/2} \tag{73}$$

$$L(e,O_2)_{rot} = 6.9\text{x}10^{-14} n_e n(O_2)(T_e - T_n)/T_e^{1/2} \tag{74}$$

$$L(e,N_2)_{vib} = 2.99\text{x}10^{-12} n_e n(N_2) \exp\left[f(T_e - 2000)/2000T_e\right] \cdot \left[\exp\left(-g\frac{T_e - T_n}{T_e T_n}\right) - 1\right] \tag{75}$$

$$L(e,O_2)_{vib} = 5.2\text{x}10^{-13} n_e n(O_2) \exp\left[h(T_e - 700)/700T_e\right] \cdot \left[\exp\left(-2770\frac{T_e - T_n}{T_e T_n}\right) - 1\right] \tag{76}$$

$$L(e,O)_{fine} = \varepsilon\, n_e\, n(O)\, (T_e - T_n) \tag{77}$$

where the units are eV $cm^{-3}\,s^{-1}$. In equations (73) - (77) the quantities f, g, and h depend only on the electron temperature, while ε is a more complicated function (see Schunk and Nagy, 1978). The electron cooling rates for electron-ion and for elastic electron-neutral interactions can be obtained from equation (7) using the momentum transfer collision frequencies given in equation (11) and Table 2.

The relative importance of the various electron cooling rates is dependent upon the atmospheric and ionospheric conditions. Figure 18 shows a comparison of the cooling rates for typical daytime conditions at mid-latitudes. For the daytime ionosphere, the dominant electron cooling results from excitation of the fine structure levels of atomic oxygen below 220 km and from Coulomb interactions with the ambient ions above 220 km.

The neutral atmosphere provides the main cooling of the ion gas in the terrestrial ionosphere. The energy exchange between the ion and neutral gases is given by equation (7), and the relevant collision frequencies are given in Tables 4 and 5.

4. MIDDLE AND LOW LATITUDE IONOSPHERE

In this section, the basic physics governing the behavior of the middle and low latitude ionosphere is described, and a discussion is

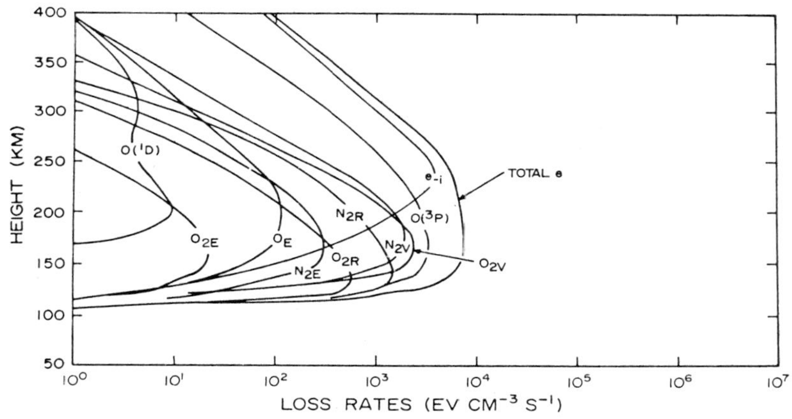

Figure 18. Electron energy loss rate profiles in the daytime ionosphere. The curves labeled N_{2V} and O_{2V} represent losses due to the excitation of the vibrational states of N_2 and O_2, respectively; N_{2R} and O_{2R} represent losses in the excitation of rotational levels of N_2 and O_2; and N_{2E}, O_{2E}, O_E, He, and H represent the losses due to elastic collisions between electrons and the neutral species. The curve labeled O(^{1}D) represents the losses in the excitation of O to the ^{1}D state, and that labeled O(^{3}P) the losses due to the excitation of fine structure levels in O. The curve labeled e-i represents the losses in Coulomb collisions between electrons and ions (from Perkins and Roble, 1978).

given on how the ionosphere varies with latitude, local time, altitude, season and solar activity.

4.1. Chapman layer

In the E and F1 regions, diffusion and wind-induced plasma transport are not important, and hence, the continuity equation (1) for the electron density reduces to

$$P'_e = L'_e n_e \tag{78}$$

for steady state conditions. Assuming that NO^+ is the dominant ion, $L'_e = \alpha n(NO^+) = \alpha n_e$, where α is the recombination rate constant given in equation (61). Therefore, equation (78) becomes

$$P'_e = \alpha n_e^2 \tag{79}$$

If the simple Chapman production function (42) is used in equation (79), the solution for the electron density is given by

$$n_e(z,\chi) = \left(\frac{P_o}{\alpha}\right)^{1/2} \exp\left\{1/2\left[1 - \frac{z-z_o}{H} - \exp\left(\frac{z_o-z}{H}\right)\sec\chi\right]\right\} \tag{80}$$

This is the expression for the idealized Chapman layer.

Near the peak of the layer ($z \simeq z_o$), the exponentials can be expanded in a Taylor series, and the expression for n_e reduces to

$$n_e(z,0^o) \simeq [P_o/\alpha]^{1/2} [1 - (z-z_o)^2/4H^2] \tag{81}$$

for overhead sun ($\chi = 0^o$). Hence, the electron density profile is parabolic near the peak of the layer.

4.2. Ambipolar diffusion

In the F-region, both wind-induced plasma drifts and field-aligned plasma diffusion are important in addition to chemical reactions. Since at F-region altitudes the ion and electron collision frequencies are much smaller than the corresponding cyclotron frequencies, the plasma is constrained to move along geomagnetic field lines. This field-aligned motion is influenced by gravity as well as vertical density and temperature gradients. Owing to the small electron mass, the effect of gravity is to cause a charge separation, with the lighter electrons tending to settle on top of the heavier ions. However, a polarization electrostatic field develops which acts to prevent a large charge separation. Once this electrostatic field has developed, the ions and electrons move together as a single gas under the influence of gravity and the density and temperature gradients. Such a motion is called ambipolar diffusion.

To illustrate the effects of ambipolar diffusion, it is convenient to make the following assumptions: (a) The magnetic field is vertical, and hence, wind-induced plasma drifts are not important; (b) Steady state conditions prevail ($\partial/\partial t = 0$); (c) The flow speed is subsonic; (d) The heat flow terms in the collision term (6) are negligible; (e) change neutrality exists ($n_e = n_i$); (f) The field-aligned current is zero ($\underline{u}_e = \underline{u}_i$); and (g) Stress tensor effects are negligible. The effect of the assumption of subsonic flow can be seen by comparing the nonlinear inertial term with the pressure-gradient term in the momentum equation (2). Assuming that $\nabla \sim 1/L$, where L is a characteristic scale length, the ratio of these two terms is

$$nm(\underline{u} \cdot \underline{\nabla})\underline{u}/\underline{\nabla} p \sim nmu^2/p = u^2/(kT/m) = M^2 \qquad (82)$$

where M is the Mach number of the flow. Therefore, if the plasma flow is subsonic, the nonlinear inertial term can be neglected.

With the above simplifying assumptions, the ion momentum equation parallel to $\underline{B}$ reduces to

$$\partial p_i/\partial z + n_i m_i g - e\, n_i E = -n_i m_i \nu_{in} u_i \qquad (83)$$

where g is gravity, E is the polarization electrostatic field, and z is the vertical coordinate. The electrostatic field can be obtained from the electron equation of motion, which is similar to (83). However, owing to the small electron mass several terms are negligible, and the electron momentum equation effectively reduces to

$$eE = -\,(1/n_e)\, \partial p_e/\partial z \qquad (84)$$

Substituting equation (84) into (83), setting $p_e = n_e kT_e$ and $p_i = n_i kT_i$, and applying charge neutrality, leads to the ambipolar diffusion equation

$$u_i = -D_a[(1/n_i)\, \partial n_i/\partial z + (1/T_p)\, \partial T_p/\partial z + 1/H_p] \qquad (85)$$

where the ambipolar diffusion coefficient (D_a), plasma scale height (H_p), and plasma temperature (T_p) are given by

$$D_a = k(T_e + T_i)/(m_i \nu_{in}) \qquad (86)$$

$$H_p = 2kT_p/(m_i g) \qquad (87)$$

$$T_p = (T_e + T_i)/2 \tag{88}$$

When the diffusion equation (85) is substituted into the ion continuity equation (1), the result is

$$\partial n_i/\partial t - (\partial/\partial z)\{D_a\ [\partial n_i/\partial z + (n_i/T_p)\partial T_p/\partial z + n_i/H_p]\} = P_i' - L_i' n_i \tag{89}$$

This equation is useful for estimating some characteristic time constants in the F-region. The time constant for the decay of the F-region is obtained by equating the first and last terms in equation (89). Setting $\partial/\partial t = 1/\tau_c$ (τ_c is the chemical time constant), we obtain

$$\tau_c = 1/L_i' \tag{90}$$

The diffusion time constant, τ_D, can be obtained by equating the first two terms in equation (89). Setting $\partial/\partial t = 1/\tau_D$ and $\partial/\partial z = 1/H_p$, where H_p is the characteristic scale length, we obtain

$$\tau_D = H_p^2/D_a \tag{91}$$

In the lower F-region $\tau_c << \tau_D$, and therefore, chemistry dominates. However, D_a increases exponentially with altitude owing to its dependence on the neutral density, and hence, at high altitudes diffusion dominates. The simple picture is that the height of the F-region peak, h_mF_2, is located at the altitude where $\tau_c \sim \tau_D$, and that diffusion dominates above and chemistry below h_mF_2. This is in sharp contrast to the situation in the E-region, where the peak in the electron density is determined solely by chemical processes (the Chapman layer).

4.3. Diffusive equilibrium

As noted above, diffusion rapidly dominates at altitudes above the F-region peak owing to the rapid increase of D_a with altitude. Therefore, above the peak the ion density can be obtained from equation (85),

$$(1/n_i)\ \partial n_i/\partial z = -1/H_p - (1/T_p)\partial T_p/\partial z - u_i/D_a \tag{92}$$

Since D_a is large, the last term on the right-hand side of equation (92) is negligible, and the resulting equation is the diffusive equilibrium equation.

For an isothermal ionosphere, equation (92) reduces to

$$(1/n_i)\partial n_i/\partial z = -1/H_p \tag{93}$$

If the variation of gravity with altitude is ignored, equation (93) can be integrated easily,

$$n_i = (n_i)_r \exp[-(z-z_r)/H_p] \tag{94}$$

where the subscript r corresponds to some reference altitude. Therefore, in the diffusive equilibrium region the ion (or electron) density should decrease exponentially with altitude, as shown schematically in Figure 12.

4.4. Wind-induced plasma drift

At mid-latitudes the geomagnetic field is inclined to the vertical, and this affects the ion motion in two ways. First, the effectiveness of diffusion is reduced because the density and temperature gradients are in the vertical direction, while the ions can move only along the inclined geomagnetic field lines like beads on a string. The inclination of $\underline{B}$ is described by the magnetic field dip angle I, which is defined relative to the horizontal direction ($I = 0^o$ at the magnetic equator and 90° at the poles). If u_i is the ambipolar diffusion velocity given by equation (85) for a vertical magnetic field, then $u_i \sin I$ is the diffusion velocity along $\underline{B}$ for an inclined field and $u_i \sin^2 I$ is its vertical component, which enters into the ion continuity equation. This effect can be taken into account simply by replacing D_a with $D_a \sin^2 I$ in the ambipolar diffusion equation.

Another consequence of an inclined $\underline{B}$ is that meridional (or north-south) neutral winds can induce a vertical plasma drift. If V is the southward component of the neutral wind, the induced plasma drift along $\underline{B}$ is V cos I and the associated vertical plasma drift is V cos I sin I. Therefore, for an inclined magnetic field the ambipolar diffusion equation (85) becomes

$$u_i = V \sin I \cos I - D_a \sin^2 I[(1/n_i)\partial n_i/\partial z + (1/T_p)\partial T_p/\partial z + 1/H_p] \tag{95}$$

During the day the wind blows from the equator to the poles, which induces a downward ionization drift, while the reverse occurs at night. The effect of such drifts on the F-layer is shown in Figure 19. The daytime northward wind acts to drive the F-layer downwards where chemical loss rates are greater, which leads to a decrease in the peak electron density (N_mF_2). The nighttime equatorward wind, on the other hand, raises the F-layer, and hence, acts to increase N_mF_2.

4.5. Decay of the F-layer

The decay properties of the ionosphere are shown in Figure 20.

Starting with a typical daytime ionosphere at t=0, the photoionization rates are set to zero and the decay of the E and F regions is followed for several hours. The E-region, which is populated by the molecular ions NO^+, O_2^+ and N_2^+, decays very rapidly owing to the fast dissociative recombination rates. The O^+ density distribution in the F-region decays exponentially with time in a shape-preserving fashion with a time constant that is approximately equal to the inverse of the O^+ loss frequency at the height of the F_2 peak. For the case shown in Figure 20, the initial O^+ peak density decreases by a factor of 10 in about 4 hours.

The ionospheric decay shown in Figure 20 occurs in the absence of ionization sources. However, it is not completely representative of nighttime conditions because ionization sources other than direct photoionization exist in the nocturnal mid-latitude ionosphere. The nocturnal F-region is maintained by a downward flow of ionization from the high altitude plasmasphere, while the nocturnal E-region is maintained by production due to resonantly scattered H Lyman α and H Lyman β solar radiation.

4.6. Thermal structure

At low altitudes, the electron energy balance is governed by local

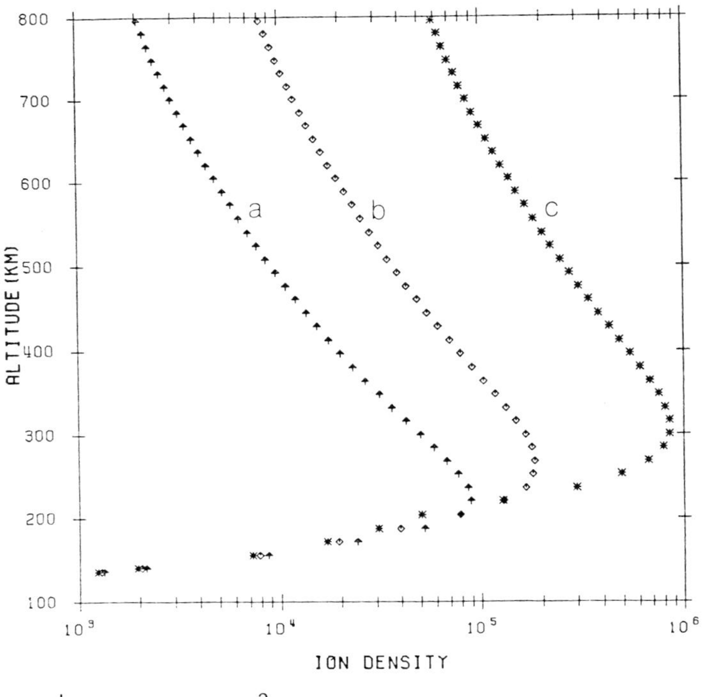

Figure 19. O^+ density (cm^{-3}) profiles for a northward wind (curve a), no wind (curve b), and a southward wind (curve c).

heating and cooling processes, and the energy equation (3) for electrons reduces to

$$\delta E_e / \delta t + Q_e = L_e \tag{96}$$

The altitude range over which this equation applies depends upon the ionospheric and atmospheric conditions. For example, for daytime solar minimum conditions, equation (96) may apply only below about 150 km, while for daytime solar maximum conditions it might apply to altitudes as high as 300 km. At higher altitudes, electron thermal conduction and

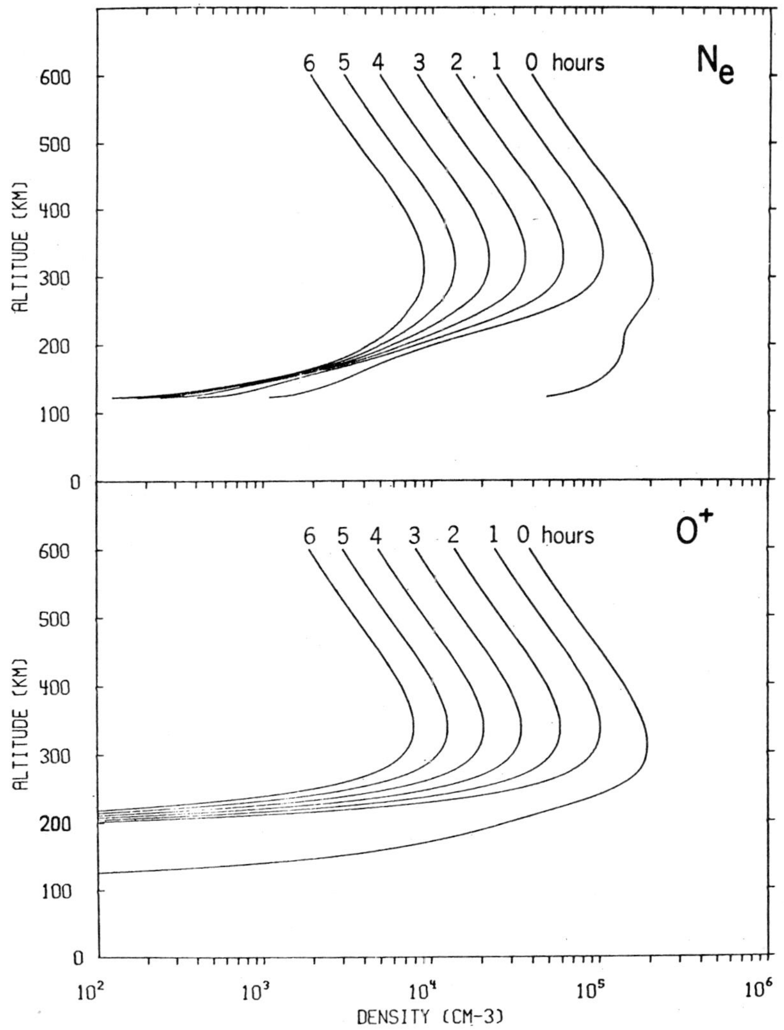

Figure 20. Electron and O^+ density profiles at selected times during an ionospheric decay (from Schunk et al, 1976).

temporal variations become important and the electron equation is given by

$$(3n_e k/2)\partial T_e/\partial t = \sin^2 I(\partial/\partial z)(\lambda_e \partial T_e/\partial z) + \delta E_e/\delta t + Q_e - L_e \qquad (97)$$

where the factor $\sin^2 I$ appears because the electron heat flow is along the inclined magnetic field, while the temperature gradient is in the vertical direction.

At altitudes above the F-region peak, thermal conduction dominates the electron energy balance and equation (97) reduces to

$$\partial/\partial z(\lambda_e \ \partial T_e/\partial z) = 0 \qquad (98)$$

Furthermore, above the F-region peak the plasma is effectively full-ionized so that λ_e is given by

$$\lambda_e = 7.7 \times 10^5 \ T_e^{5/2} \quad \mathrm{eVs}^{-1} \ \mathrm{cm}^{-1} \ \mathrm{K}^{-1} \qquad (99)$$

Equations (98) and (99) can be easily solved for the electron temperature variation as a functon of altitude,

$$T_e = [T_{eb}^{7/2} - (7/2)(q_{et}/7.7\mathrm{x}10^5)(z-z_b)]^{2/7} \qquad (100)$$

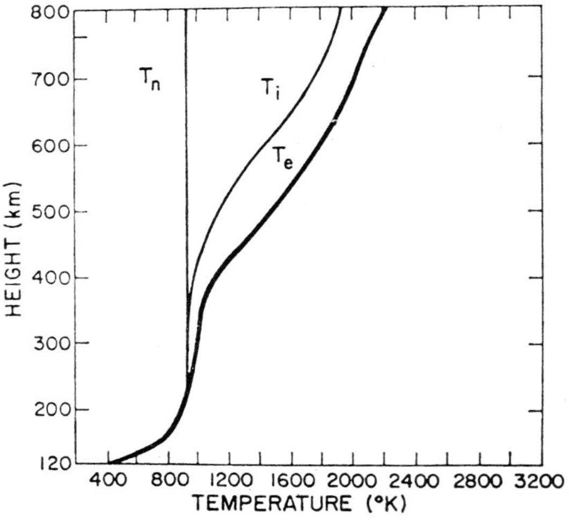

Figure 21. Calculated electron, ion, and neutral gas temperature profiles at 0222 LT over Millstone Hill on March 23-24, 1970 (from Roble 1975). Reprinted by permission of Pergamon Press.

where T_{eb} is the electron temperature at the bottom boundary of the thermal conduction region and q_{et} is the electron heat flow through the top boundary. Equation (100) shows that if there is a downward heat flow through the top boundary ($q_{et} < 0$), then T_e increases with altitude. If $q_{et} = 0$, $T_e = T_{eb}$ at all altitudes, i.e., the electron temperature is isothermal.

The equation describing the ion temperature at mid-latitudes is similar to the electron equation (97). However, owing to the smaller ion thermal conductivity, ion heat conduction is important only above about 600 km. Below this altitude, the ion temperature can be obtained by equating local heating and cooling rates (similar to equation 96).

Figure 21 shows typical mid-latitude electron, ion, and neutral temperatures as a function of altitude for nighttime conditions. During the night, the main heat source for the electrons is energy flowing down from the magnetosphere, which is capable of maintaining T_e greater than T_i and T_n at altitudes above 200 km. For the ions, the main heating is due to collisional coupling to the hot electrons, which is sufficient to elevate T_i above T_n at heights greater than 400 km. Below this altitude, ion-neutral collisional coupling is strong and $T_i = T_n$.

4.7. Diurnal variation of the ionosphere

In Figures 22-24, the calculated diurnal variations of electron density, electron temperature, and ion temperature over Millstone Hill on 23-24 March 1970 are compared with the diurnal variations measured at Millstone Hill. The physical processes that control the diurnal variation of the electron density change with altitude and local time. After

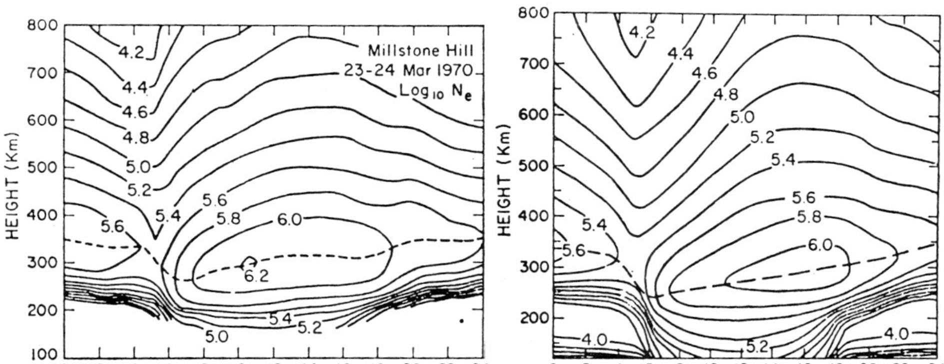

Figure 22. Contours of electron density (n_e, cm^{-3}) over Millstone Hill on March 23-24, 1970. Panel (a) shows the measured densities, while panel (b) shows the calculated densities. The dashed curve shows the F_2 - peak (from Roble, 1975). Reprinted by permission of Pergamon Press.

sunrise, the ionization increases rapidly from its nighttime minimum. The ionization in the F-region below about 300 km is under strong solar control, peaking at noon when the solar zenith angle is smallest and then decreasing symmetrically as the sun decreases. Ionization above 300 km, however, is influenced by other effects, such as neutral winds, electron and ion temperatures, electric fields, plasma flow between the magnetosphere and ionosphere, and neutral composition. Therefore, the electron density coutours in this region do not show a strong solar zenith angle dependence, and the maximum ionization occurs late in the afternoon near the time when the neutral exospheric temperature peaks.

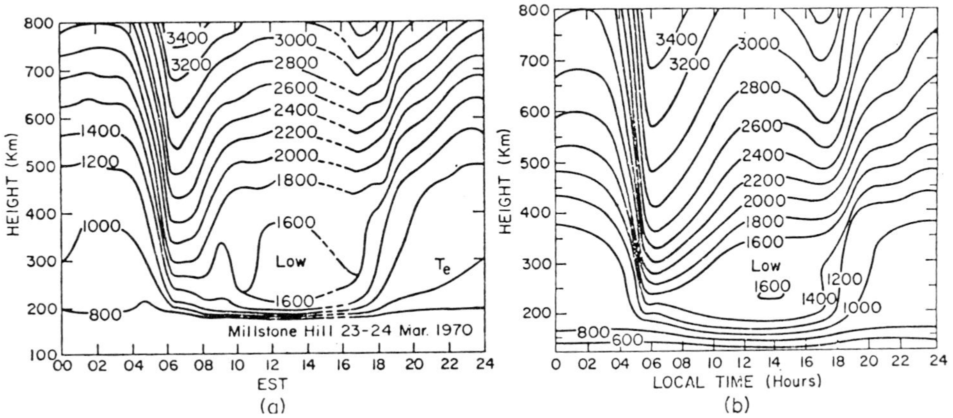

Figure 23. Contours of electron temperature (°K) over Millstone Hill on March 23-24, 1970. (a) Measured electron temperatures; (b) Calculated electron temperatures (from Roble, 1975). Reprinted by permission of Pergamon Press.

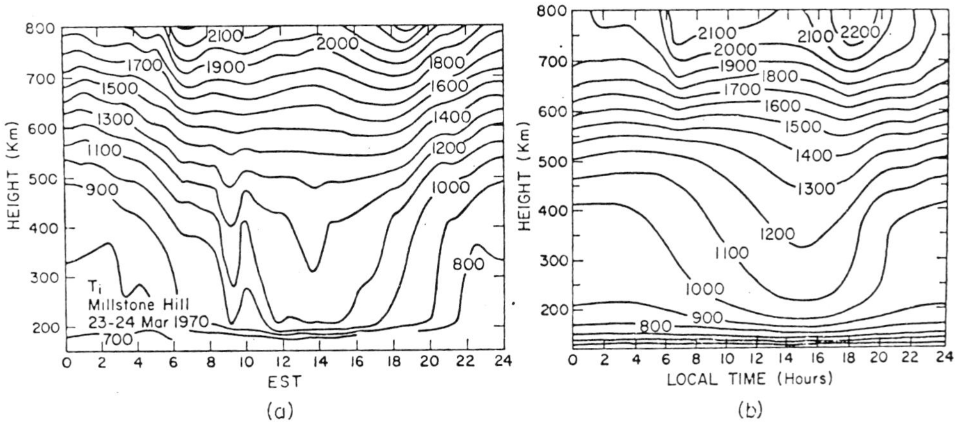

Figure 24. Contours of ion temperature (°K) over Millstone Hill on March 23-24, 1970. (a) Measured ion temperatures; (b) Calculated ion temperatures (from Roble, 1975). Reprinted by permission of Pergamon Press.

At night the electron density at the F-region peak is controlled by several processes, including a downward-directed ionization flux, neutral winds, vertical drifts due to perpendicular electric fields, and ambipolar diffusion in the lower F-region (below 300 km). However, the height of the F-region peak is controlled primarily by the neutral winds forcing ionization up and down the inclined geomagnetic field lines.

The electrons are heated by photoelectrons that are created in the photoionization process and by a flow of heat from the magnetosphere. In addition to these heat sources, the electron temperature is controlled by the electron density through elastic and inelastic collisions between the electrons and the ions and neutrals. When the electron density is high, cooling to the neutrals is greater than when the electron density is low. Consequently, for a given heating rate the electron temperature is inversely related to the electron density. It should also be noted that at night photoelectron heating is absent and the electron temperature is maintained by energy flow from the magnetosphere, which produces the positive gradient in the nocturnal electron temperature above 200 km.

With regard to the ion temperature, it basically follows the neutral temperature below about 400 km. Above this height, the ion temperature increases with altitude owing primarily to the increased thermal coupling to the hot electrons; there is also a small ion heat flow from the magnetosphere.

4.8. Solar cycle variation of the ionosphere

Figure 25 shows the solar cycle variation of the electron temperature and density for the daytime, mid-latitude ionosphere at equinox. At solar maximum, the solar EUV fluxes and the atomic oxygen densities are greater than at solar minimum, which in turn result in greater electron densities and lower electron temperatures. The greater electron densities at solar maximum are simply a result of the increased production rate, while the lower electron temperatures are a consequence of the enhanced thermal coupling to the cold ions which results from the increased electron densities. With regard to the shape of the T_e profile, at solar maximum there is a pronounced peak at about 250 km, while for solar minimum conditions the electron temperature increases monotonically with altitude. The T_e peak at solar maximum is again a consequence of the high electron densities, which cause electron-ion energy loss to dominate thermal conduction immediately above 250 km. This, in turn, causes the decrease in T_e between 250 and 380 km. Above 380 km, thermal conduction dominates and T_e increases with altitude in response to magnetospheric heat flow.

4.9. Seasonal variation of the ionosphere

Figure 26 shows the seasonal variation of the daytime, mid-latitude ionosphere. The most important feature to note is that N_mF_2 in winter is greater than N_mF_2 in summer despite the fact that the sun is more

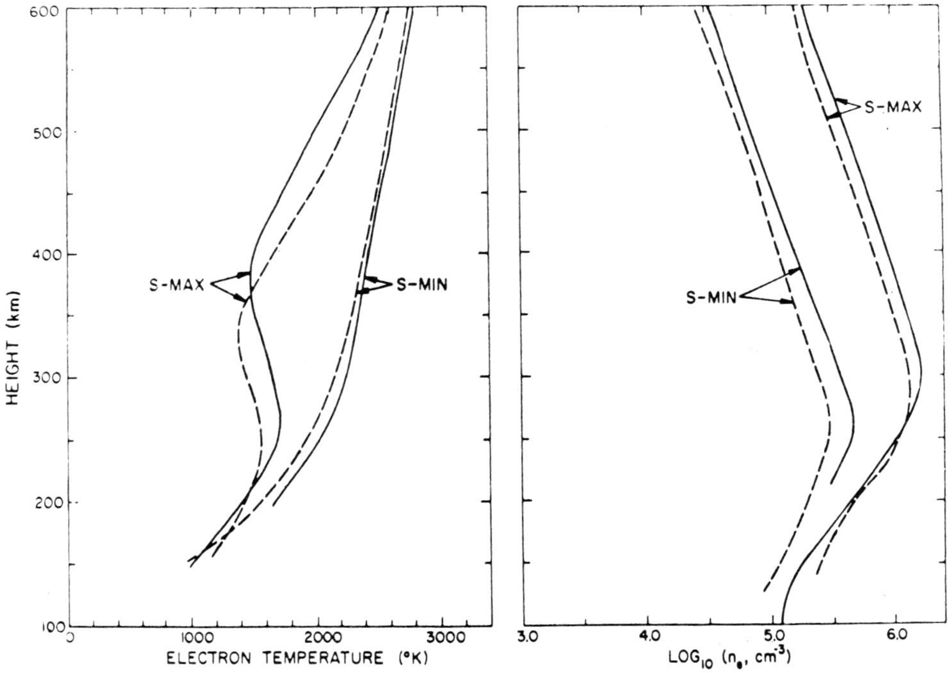

Figure 25. Vertical profiles of electron temperature and electron density for the daytime mid-latitude ionosphere at equinox for both solar minimum and solar maximum conditions. The solid curves are profiles measured at Millstone Hill, while the dashed curves are calculated (from Roble, 1976).

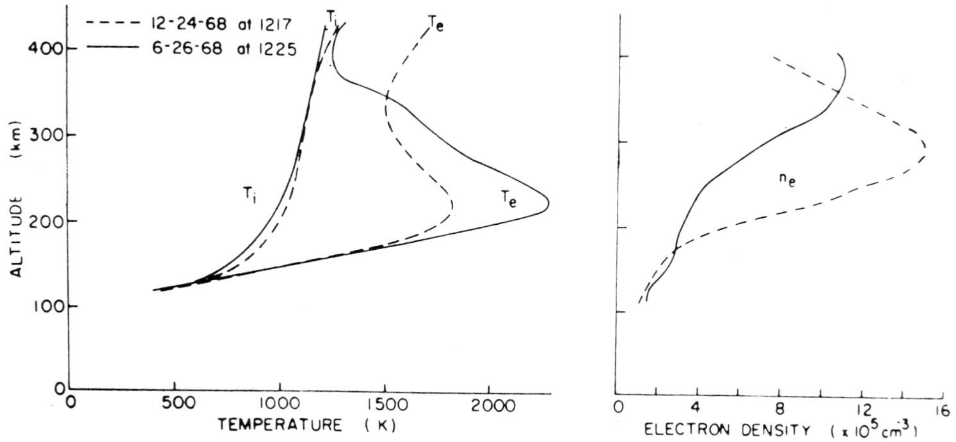

Figure 26. Summer (solid curves) and winter (dashed curves) profiles of T_e, T_i, and n_e for the daytime, mid-latitude ionosphere (from Swartz and Nisbet, 1973).

nearly overhead in summer than in winter. The phenomenon is called the 'winter anomaly' and results primarily from atmosphere composition changes. In winter the O/N_2 ratio is much greater than in summer, which acts to produce enhanced O^+ densities. This enchancement is more than enough to offset the less favorable solar zenith angle, and the net result is that N_mF_2 is greater in winter than in summer at mid-latitudes.

4.10. Ionospheric behavior at low latitudes

As noted earlier, in the daytime equatorial F-region eastward electric fields associated with neutral wind induced ionospheric currents drive a plasma convection that is upward. The plasma lifted in this way then diffuses down the magnetic field lines and away from the equator because of the action of gravity. The combination of electromagnetic drift and diffusion produces a 'fountainlike' pattern of plasma motion, as shown in Figure 27. The result of this plasma transport is that ionization peaks are formed in the subtropics on each side of the magnetic equator (called Appleton's peaks).

5. HIGH LATITUDE IONOSPHERE

The high-latitude ionosphere exhibits a more complex behavior than the mid-latitude ionosphere owing to a variety of physical processes

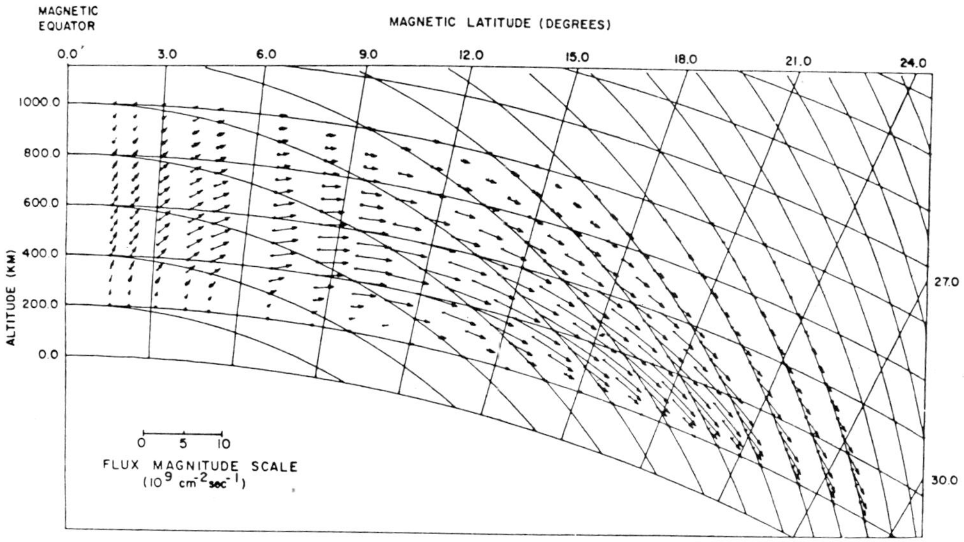

Figure 27. Plasma drift pattern at low latitudes due to the combined action of an electromagnetic drift across magnetic field lines and plasma diffusion along field lines. The magnetic field lines are shown every 200 km above the equator (from Hanson and Moffett, 1966).

that are unique to this region. Foremost among these are the electric fields associated with magnetospheric plasma convection, the presence of intense particle precipitation within the auroral oval, the polar wind escape of thermal plasma, and other features generally resulting from the auroral disturbance of the neutral atmosphere. In the subsections that follow, some of these processes will be discussed and their effect on the high-latitude ionosphere will be elucidated.

5.1. Plasma convection

As noted earlier, electric fields of magnetospheric origin are mapped down along geomagnetic field lines to the high latitude ionosphere. These electric fields are directed perpendicular to the geomagnetic field and cause the high latitude ionosphere to move horizontally across the polar region. In order to illustrate the basic physics involved, it is convenient to make the following simplifying assumptions: (a) The plasma is weakly-ionized; (b) the neutral atmosphere is stationary; (c) Steady state conditions prevail; (d) The flow is subsonic; (e) The ionosphere is isothermal; and (f) Stress effects are negligible.

With the above simplifying assumptions, the momentum equation (2) for the horizontal motion of a given ion species (subscript i) reduces to

$$kT_i \nabla n_i - e_i n_i [\underline{E} + (1/c)\underline{u}_i x \underline{B}] = -n_i m_i \nu_{in} \underline{u}_i \tag{101}$$

where $\underline{E}$ is the electric field of magnetospheric origin (the convection electric field). This equation can also be expressed in the following form:

$$\underline{u}_i = -(D_i/n_i)\nabla n_i + \mu_i \underline{E} + (\Omega_i/\nu_{in})\ (\underline{u}_i x \underline{b}) \tag{102}$$

where the ion diffusion coefficient, D_i, mobility coefficient, μ_i, and cyclotron frequency, Ω_i, are given by

$$D_i = kT_i/(m_i \nu_{in}) \tag{103}$$

$$\mu_i = e_i/(m_i \nu_{in}) \tag{104}$$

$$\Omega_i = e_i B/(m_i c) \tag{105}$$

Equation (102) can be readily solved by first expressing this equation in terms of the individual velocity components. The solution to this equation is given by

$$\underline{u}_i = -(D_{i\perp}/n_i)\nabla n_i + \mu_{i\perp} \underline{E} + (\underline{u}_E + \underline{u}_{iD})/(1+\nu_{in}^2/\Omega_i^2) \tag{106}$$

where the electromagnetic drift, $\underline{u}_E$, and diamagnetic drift, $\underline{u}_{iD}$, are given by

$$\underline{u}_E = c(\underline{E}x\underline{B})/B^2 \quad (107)$$

$$\underline{u}_{iD} = -c(\nabla p_i x\underline{B})/(e_i n_i B^2) \quad (108)$$

and where

$$D_{i\perp} = D_i/(1+\Omega_i^2/\nu_{in}^2) \quad (109)$$

$$\mu_{i\perp} = \mu_i/(1+\Omega_i^2/\nu_{in}^2) \quad (110)$$

An equation similar to (106) holds for the electrons.

At altitudes above about 160 km, $\nu_{in}/\Omega_i << 1$. Also, a comparison of the terms in equation (106) for altitudes above 160 km indicates that

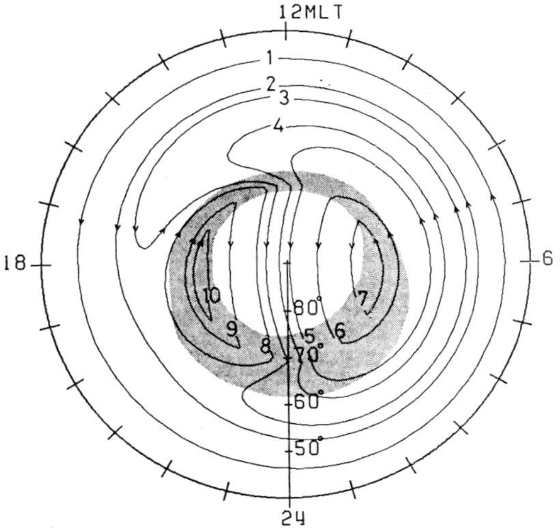

LABEL	1	2	3	4	5	6	7	8	9	10
PERIOD (DAY)	1.06	1.35	1.32	0.79	0.58	0.25	0.15	0.27	0.11	0.06

Figure 28. Plasma drift trajectories in the magnetic quasi-inertial frame for a cross-tail magnetospheric electric potential of 90 kV. The shaded region corresponds to an auroral oval for K_p=5. Magnetic local time (MLT) is indicated by tick marks at hourly intervals and magnetic latitude is indicated on the vertical scale. The trajectories have been numbered in order to indicate circulation times, which are tabulated in the lower part of the figure (from Sojka et al, 1981b).

the electrodynamic drift is by far the largest term for both ions and electrons. Therefore, at high latitudes the F-region plasma moves under the influence of the magnetospheric electric field with the well-known $\underline{E} \times \underline{B}$ drift velocity (107).

5.2. Magnetospheric convection pattern

Looking down on the polar region from a point fixed in space, the plasma convection pattern that results from the magnetospheric electric field is a two-cell pattern with antisunward flow over the polar cap and return flow at lower latitudes. The two cells may be symmetric or asymmetric with enhanced flow in either the dawn or dusk sectors of the polar region. In addition to this magnetospheric-driven flow, the high-latitude ionosphere also has a tendency to corotate with the earth. When this is taken into account, the resultant plasma convection pattern takes a form similar to that shown in Figure 28. In this figure, ten representative plasma drift trajectories are shown for an asymmetric magnetospheric electric field pattern with enhanced plasma flow in the dusk sector of the polar region. The total cross-tail potential is 90 kV. Also shown in Figure 28 is a representative auroral oval. When field tubes of plasma enter this region they are subjected to an ion production source owing to energetic electron precipitation.

Field tubes of plasma following trajectories 1 and 2 appear to corotate; however, only trajectory 1 corresponds to corotation. Field tubes following trajectory 2 take 1.35 days to complete a full circulation owing to speed variations along its path. Field tubes of plasma following trajectories 3 and 4 execute motions which result in a reversal of corotation in the afternoon sector. Field tubes following trajectories 5, 6 and 7 form a dawn cell rotating in a corotational sense, while field tubes following trajectories 8, 9 and 10 rotate in a counter corotation sense. In the centers of these two cells the plasma circulates extremely rapidly; 0.15 days for trajectory 7 and 0.06 days for trajectory 10. An evening sector stagnation region is present in an extended region from 1800 to 2200 MLT and from 58° to 62° latitude.

A cross-tail potential of 90 kV is fairly large, and therefore, large plasma convection velocities can be expected in certain regions of the high-latitude ionosphere. This is shown in Figure 29, where contours of the horizontal plasma convection speed are plotted in the magnetic quasi-inertial frame. Each contour is labeled with its appropriate speed in m/s; the region with speeds below 100 m/s is indicated by the shading. A region of high speed, reaching almost 2 km/s, is located in the dusk sector and it corresponds to the enhanced magnetospheric electric field on the dusk side of the polar ionosphere. Over the polar cap the horizontal speed lies in the 200 to 600 m/s range. In contrast, an extended low-speed region is present in the afternoon and evening sectors. The location of this extended low-speed region has a direct bearing on the location of the main or mid-latitude plasma density trough.

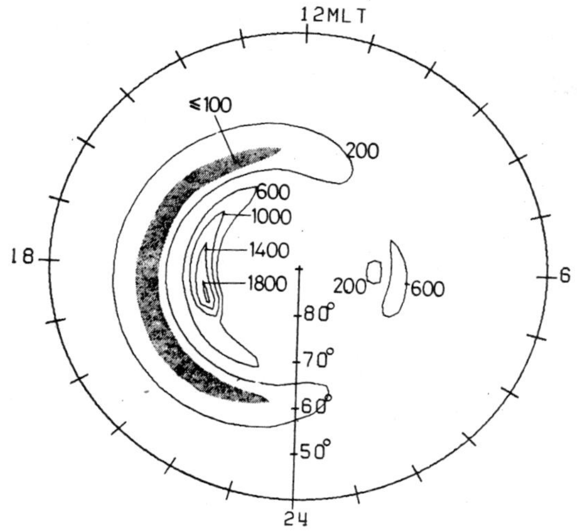

Figure 29. Contours of horizontal plasma drift speeds in the magnetic quasi-inertial frame. The contours are labeled in units of m s^{-1} and the shaded region corresponds to speeds below 100 m s^{-1} (from Sojka et al, 1981b).

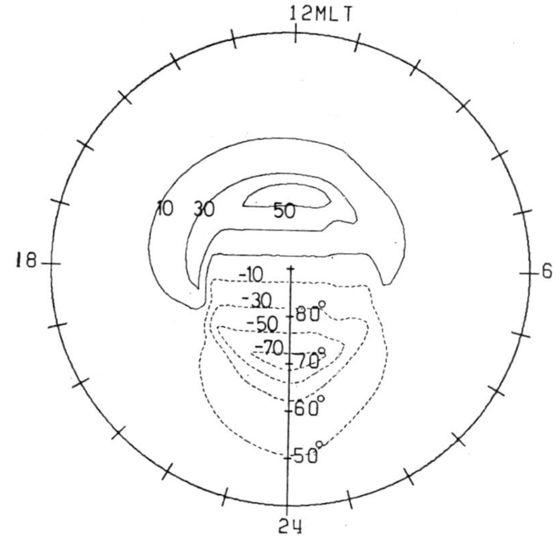

Figure 30. Contours of the vertical component of the electrodynamic plasma drift velocity displayed in the magnetic quasi-inertial frame. The solid contours correspond to upward drift and the dashed contours correspond to downward drift. The contours are labeled in units of m s^{-1} (from Sojka et al, 1981b).

Since the geomagnetic field lines at high latitudes are not completely vertical, an ExB plasma motion will have a vertical component. Figure 30 shows contours of the vertical component of the plasma convection velocity in the magnetic quasi-inertial frame. Each contour is labeled with the appropriate velocity in m/s; the dashed contours represent downward velocities, while the solid contours represent upward velocities. Upward electrodynamic drifts occur on the dayside where the plasma is convecting toward the magnetic pole, while downward electrodynamic drifts occur in the nightside where the plasma is convecting away from the magnetic pole. For the case considered, the vertical plasma drift ranges from +50 m/s to -70 m/s. Vertical drifts near the extremes of this range have a pronounced effect on both the F-region peak electron density, N_mF_2, and the altitude of the peak, h_mF_2.

The convection pattern shown in Figures 28-30 is the one seen in the magnetic reference frame. Because of the displacement between the geographic and geomagnetic poles, this pattern rotates about the geographic pole while continually pointing toward the sun. Therefore, the high latitude ionosphere moves toward and then away form the sun during the course of a day. This motion introduces a universal time dependence in the photoionization rate, and hence, electron density.

5.3. Electron density morphology

Comprehensive, time-dependent models of the convecting high-latitude ionosphere have been developed recently in an effort to determine the extent to which the various chemical and transport processes affect the ion composition and electron density at F-region altitudes. The numerical models take into account field-aligned diffusion, thermospheric winds, ExB drifts, polar wind escape, energy dependent chemical reactions, magnetic storm induced neutral composition changes, and ion production due to solar EUV radiation and energetic particle precipitation. Model studies have been conducted for both weak and strong plasma convection and for both summer and winter conditions. One of the important results that emerged for these studies was that high-latitude ionospheric features, such as the 'main trough', the 'ionization hole', the 'tongue of ionization', and the 'aurorally produced ionization peaks', are a natural consequence of the competition between the various chemical and transport processes known to be operating in the high-latitude ionosphere.

The response of the high-latitude ionosphere to weak plasma convection for winter solstice is shown in Figure 31, where the O^+ density is shown as a gray-scaled contour plot in a magnetic local time (MLT), magnetic latitude polar diagram for an altitude of 300 km. The gray scale range was chosen to emphasize low density regions. The contour plots clearly show the gross features of a mid-latitude (or main) ionization trough on the nightside at low latitudes, a region of enhanced ionization in the vicinity of the auroral oval which is situated just poleward of the main trough, and a high-latitude ionization hole in the polar cap around local dawn. However, the detailed characteristics of these features differ for the four times shown, indicating that the high-latitude ionosphere displays a marked universal time (UT) variation. This UT variation results from the rotation of the geomagnetic pole, and hence plasma convection pattern, about the geographic pole.

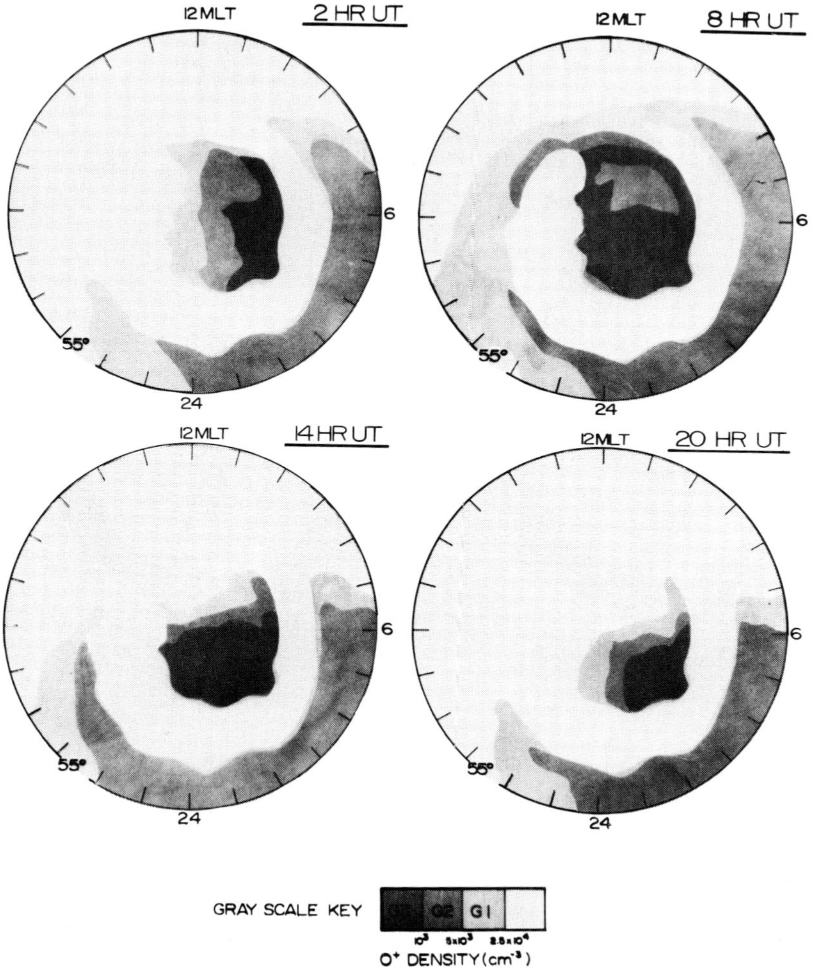

Figure 31. O^+ density contours at 300 km in a magnetic reference frame for four equally spaced universal times. MLT is shown as tick marks at 1 hour intervals and magnetic latitude varies linearly from the 55° circle to the magnetic pole at the center of each plot. The contour range was chosed to highlight the low density regions (from Sojka et al, 1981a).

Figure 32 shows O^+ density contours at 300 km for winter solstice and for strong convection. As was found for weak convection, the ionosphere displays a marked UT variation. Also, the auroral oval and main trough are still evident, but the trough is not as deep for strong convection. A feature that is clearly absent for strong convection is the 'ionization hole' in the polar cap. This ionization hole results from slow antisunward convection across the dark polar cap in combination with ordinary ionic recombination. However, for strong convection the ionization hole does not form owing to the much shorter transit times across the polar cap.

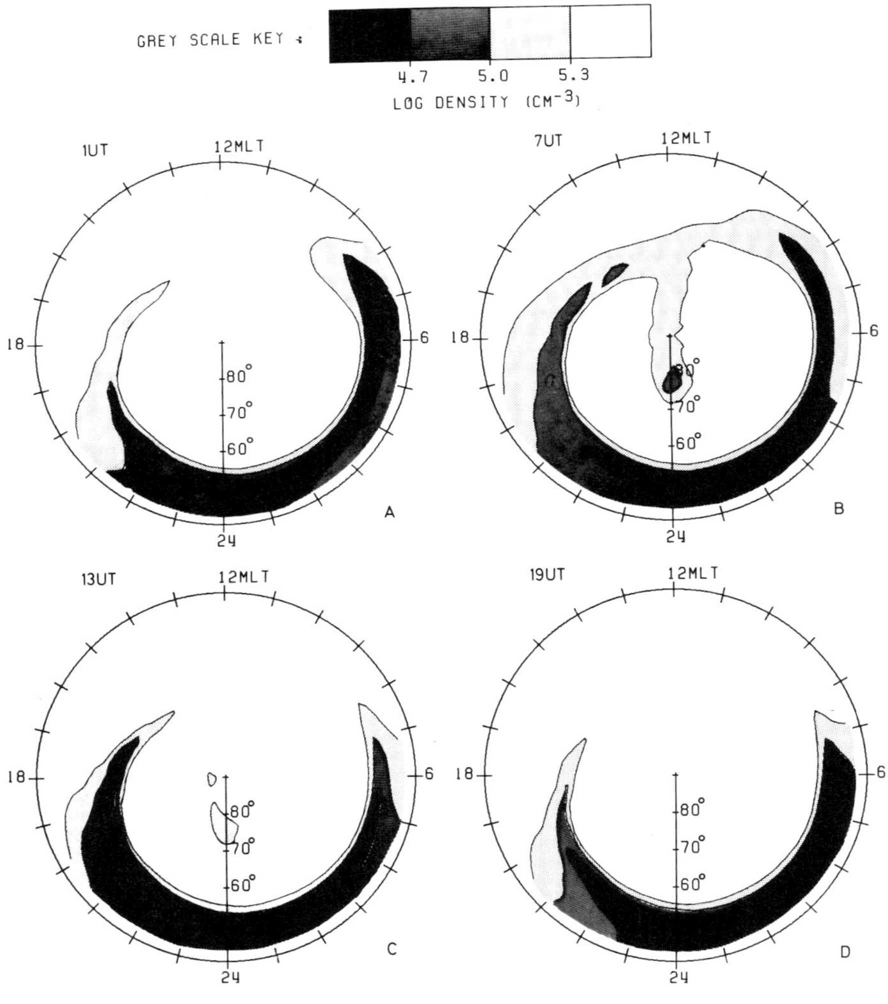

Figure 32. O^+ density contours at 300 km in a magnetic reference frame for four universal times. The contour plots are for winter solstice and strong plasma convection (from Sojka et al, 1981b).

Figure 33 shows O^+ density contours at 300 km for summer solstice and strong convection. For summer solstice, most of the polar cap is sunlit, and consequently, the main trough is not as deep and its MLT extent is not as great. However, the high-latitude ionosphere exhibits a significant UT variation, as was found in the previous cases.

5.4. Effect of electric fields on ion temperature

Because the electron densities are smaller at high latitudes than at middle and low latitudes, ion-electron energy coupling is not as im-

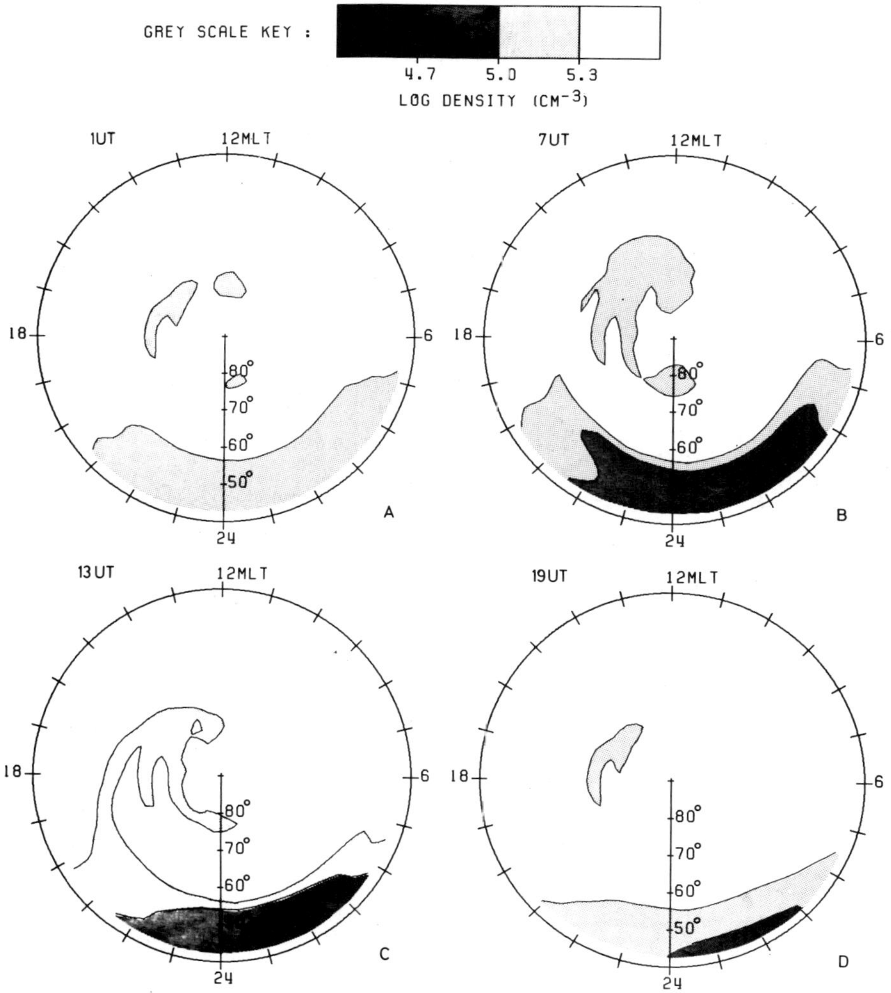

Figure 33. O^+ density contours at 300 km in a magnetic reference frame for four universal times. The contour plots are for summer solstice and strong plasma convection (from Sojka et al, 1982).

portant at high latitude as it is at the lower latitudes. To a good approximation, the ion temperature in the high-latitude F-region can be obtained simply by equating the ion-neutral collisional coupling terms in equation (5),

$$T_i = T_n + (m_n/3k)\ (\underline{u}_i - \underline{u}_n)^2 \tag{111}$$

where we have assumed that there is only one neutral species. Since the field-aligned velocity component is small, the ion-neutral relative velocity term in equation (111) can be calculated from the momentum equation (2) by assuming that $\underline{E} \times \underline{B}$ flow dominates. With this assumption, T_i can be expressed directly in terms of the magnetospheric convection electric field. Above 160 km, the expression for T_i reduces to

$$T_i = T_n + (m_n/3k)\ (E'c/B)^2 \tag{112}$$

where $\underline{E}'$ is the effective electric field

$$\underline{E}' = \underline{E} + (1/c)\ \underline{u}_n \times \underline{B} \tag{113}$$

Equation (112) indicates that high ion temperatures should be located in regions where the convection electric field is enhanced, such as in the dusk sector for the convection pattern shown in Figures 28-30. Using an ion energy equation that is more general than equation (112), Schunk and Sojka (1982) calculated the ion temperatures that are associated with the convection pattern shown in Figures 28-30. Figure 34 shows contours of the ion temperature in the magnetic quasi-inertial frame at 0 UT for altitudes of 360 and 800 km. At other universal times the results are similar, and consequently, are not presented. The intense heating in the strong convection cell in the dusk sector of the polar ionosphere is clearly seen at 360 km, there T_i reaches 3500°K near the center of the cell. At the center of the cell and at lower altitudes (~160 km) the ion temperature reaches 4400°K. This heating, which is a consequence of ion-neutral frictional interactions, is less evident at high altitudes (800 km), with the result that T_i decreases with altitude in the F-region in the strong convection cell. Also, at high altitudes, thermal conduction and horizontal transport effects become important, and the hot spot becomes less evident. A feature acting to mask the hot spot at high altitudes is the ring of enhanced ion temperatures that is associated with the auroral oval.

Figure 35 shows ion temperature profiles as a function of altitude at selected locations both inside and outside of the hot spot. Outside of the hot spot the ion temperature profiles display the characteristic behavior found at mid-latitudes. That is, at low altitudes the ion temperature is strongly coupled to the neutral temperature, while at higher altitudes the ion temperature first increases with altitude owing to ion-electron thermal coupling and then becomes constant with altitude owing to the dominance of thermal conduction. In the hot spot, on the other hand, a markedly different behavior is obtained in that the ion temperature either decreases with altitude or is approximately constant with altitude in the F-region, depending on the strength of the convection electric field and the location in the hot spot. At the locations where the ion temperature decreases with altitude, the ion heat flow is upward from the lower ionosphere to high altitudes, which is opposite to that normally found at middle and high latitudes.

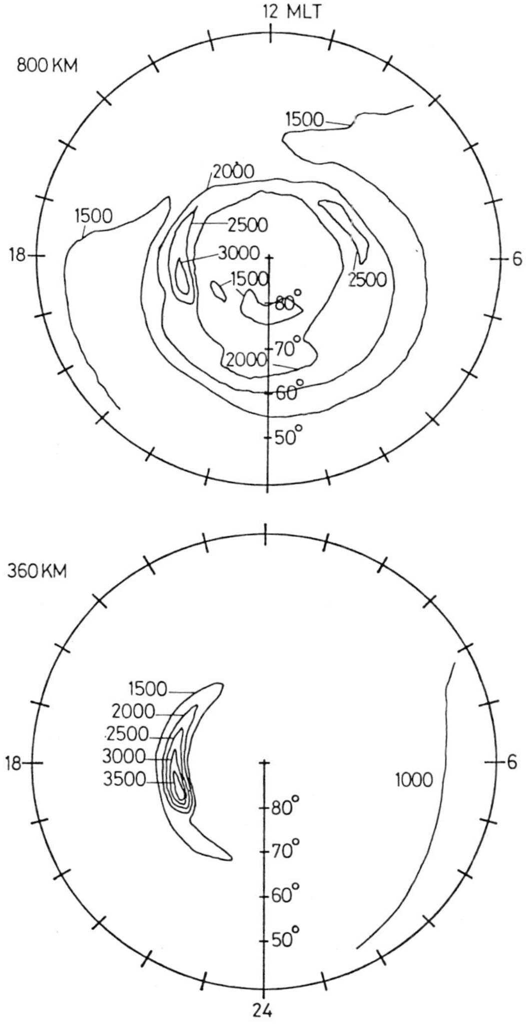

Figure 34. Contours of the ion temperature (^{o}K) in the magnetic quasi-inertial frame for altitudes of 360 km (bottom panel) and 800 km (top panel). Magnetic local time (MLT) is indicated by tick marks at hourly intervals and magnetic latitude is indicated on the vertical scale (from Schunk and Sojka, 1982).

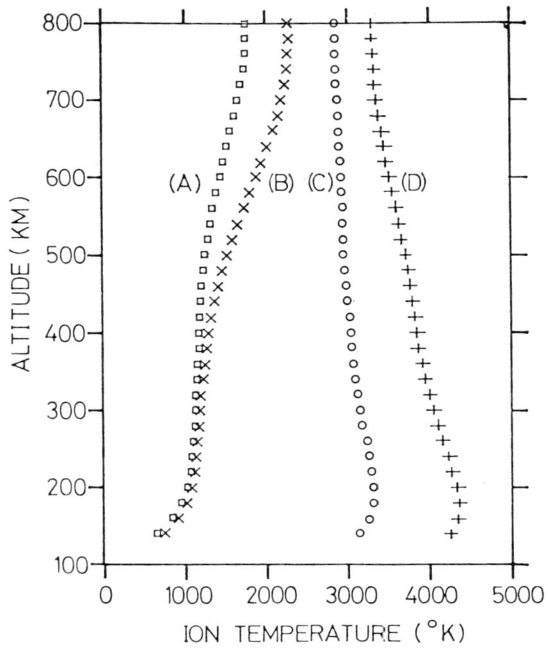

Figure 35. Ion temperatures as a function of altitude at selected locations. The locations are: (A) dayside, equatorward of auroral oval; (B) dayside, auroral oval; (C) edge of hot spot; and (D) center of hot spot (from Schunk and Sojka, 1982).

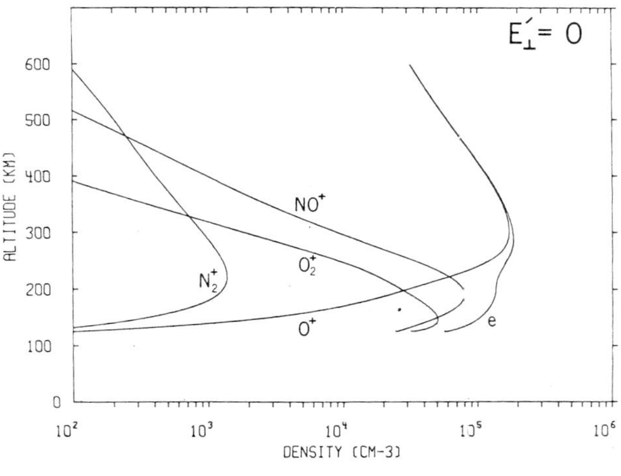

Figure 36. Ion and electron density profiles for the daytime high-latitude ionosphere and $E' = 0$ mV/m (from Schunk et al, 1975).

5.5. Effect of electric fields on ion composition

For large electric fields, equation (112) indicates that $T_i \sim (E')^2$. However, for high ion temperatures the rate coefficient for the reaction $O^+ + N_2 \rightarrow NO^+ + N$ varies as $k_1 \sim T_i^2$ (see equation 62b), so that $k_1 \sim (E')^4$. Therefore, in regions where the electric field is enhanced there should be an increased $O^+ \rightarrow NO^+$ conversion. This effect is shown in Figures 36, 37, and 38, where ion and electron density profiles are shown for convection electric field of 0, 50 and 100 mV/m, respectively. The profiles were calculated for daytime steady state conditions. For no electric field, the high-latitude ionosphere is similar to the mid-latitude ionosphere in that the molecular ions are dominant in the E-region, while O^+ is the major ion in the F-region. The transition from molecular to atomic ion dominance occurs at about 225 km. However, as the electric field strength increases, so does the $O^+ \rightarrow NO^+$ conversion rate. For 100 mV/m the conversion rate is sufficent to raise the atomic/molecular ion transition altitude to 330 km, making NO^+ an important F-region ion. Also, it should be noted that for 50 mV/m the electron density profile is nearly constant in the altitude range 160-360 km. An n_e altitude variation of this nature is unique to the high latitude ionosphere.

6. POLAR WIND

As noted in Section 2, the geomagnetic field lines at high latitudes extend deep into the tail of the magnetosphere. As a consequence, light ions, such as H^+ and He^+, are capable of escaping from the topside ionosphere, and this outflow has been called the polar wind. In this section, the basic physics behind the polar wind will be discussed as will its effect on the density and thermal structure of the high-latitude ionosphere.

6.1. Subsonic H^+ outflow

Before describing the polar wind mathematically, it is useful to note some important facts. First, the polar wind flow is field-aligned so that only the motion along $\underline{B}$ needs to be considered. Second, at the altitudes where the polar wind exists ($\gtrsim$800 km), production and loss processes are not important. Third, when H^+ and He^+ are flowing out of the topside ionosphere, they remain as minor ions to altitudes as high as 3000 km. That is, the minor ion (subscript x) and major ion (subscript i) densities are such that

$$n_x \ll n_i \tag{114}$$

Therefore, the major ion executes ambipolar flow with $n_i \simeq n_e$ and $\underline{u}_i \simeq \underline{u}_e$. Finally, it should be noted that in the altitude range 800-3000 km, the minor ion effectively collides only with the major ion owing to the long range nature of Coulomb collisions.

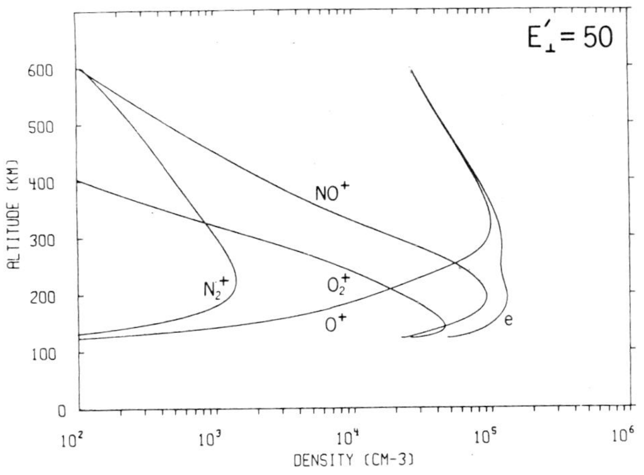

Figure 37. Ion and electron density profiles for the daytime high-latitude ionosphere and E′= 50 mV/m (from Schunk et al, 1975).

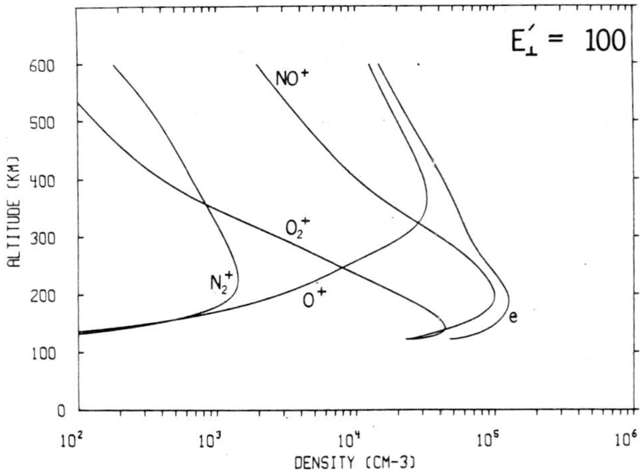

Figure 38. Ion and electron density profiles for the daytime high-latitude ionosphere and E′ = 100 mV/m (from Schunk et al, 1975).

Assuming steady state, subsonic outflow and taking account of the above facts, the minor ion continuity and momentum equations reduce to

$$\partial/\partial z(n_x u_x) = 0 \tag{115}$$

$$\partial p_x/\partial z + n_x m_x g - e_x n_x E = n_x m_x \nu_{xi}(u_i - u_x) \tag{116}$$

where E is the polarizaton electrostatic field set up by the major ions and electrons (equation 84). Integrating equation (115), it is apparent that the minor ion flux is constant with altitude,

$$n_x u_x = F_x \tag{117}$$

where the constant flux F_x is determined by the chemical processes occuring at low altitudes. Setting $p_x = n_x k T_x$, the momentum equation (116) can be written as

$$u_x = u_i - D_x[(1/n_x)\,\partial n_x/\partial z + (1/T_x)\partial T_x/\partial z - e_x E/(kT_x) + 1/H_x] \tag{118}$$

where the minor ion diffusion coefficient and scale height are given by

$$D_x = kT_x/(m_x \nu_{xi}) \tag{119}$$

$$H_x = kT_x/(m_x g) \tag{120}$$

Equation (118) indicates that the major ion affects the minor ion in three ways. First, if the major ion is flowing ($u_i \neq 0$), it tends to carry the minor ion along. However, the minor ion also tends to diffuse, but this diffusion is inhibited by the major ion through collisions. In addition, the polarization electrostatic field set up by the major ions and electrons exerts a force on the minor ions. Using equation (84) to eliminate the electrostatic field, the minor ion momentum equation can be written as

$$u_x = u_i - D_x[(1/n_x)\partial n_x/\partial z + 1/H_x + (1/T_x)\partial/\partial z(T_e+T_x) + (T_e/T_x n_e)\partial n_e/\partial z] \tag{121}$$

This is the classical diffusion equation for a minor ion in the terrestrial ionosphere.

To show the characteristic minor ion solutions, it is convenient to assume an isothermal ionosphere. With this assumption, the major ion (or electron) density is governed by equation (93), and equation (121) can be written as

$$F_x = -D_x[\partial n_x/\partial z + n_x(1/H_x - T_e/T_xH_p)] \tag{122}$$

where we used the fact that $u_i \simeq 0$ in the 800-3000 km altitude range. Taking the derivative of equation (122) yields the following second-order differential equation for n_x:

$$\partial^2 n_x/\partial z^2 + [1/H_p + (1/H_x - T_e/T_xH_p)]\partial n_x/\partial z$$
$$+(1/H_x - T_e/T_xH_p)(1/H_p)n_x = 0 \tag{123}$$

where we used the fact that $(1/D_x)\partial D_x/\partial z = 1/H_p$. The two linearly independent solutions to equation (123) are given by

$$n_x = (n_x)_r \exp[(T_e/T_xH_p - 1/H_x)(z-z_r)] \tag{124}$$

$$n_x = (n_x)_r \exp[-(z-z_r)/H_p] \tag{125}$$

The general solution for n_x is a linear combination of equations (124) and (125) with the appropriate integration constants.

It is instructive to consider separately the two solutions (124) and (125). Equation (124) indicates that if $T_e \sim T_x$ and if the minor ion is light, the density of the minor ion will increase exponentially with altitude above the reference level. This solution corresponds to diffusive equilibrium. The solution is valid until $n_x \sim n_i$, and then species x is no longer a minor ion.

Solution (125) indicates that the minor ion density decreases exponentially with altitude with the same scale height as the major ion. This solution corresponds to the maximum outflow that is possible. Since n_xu_x = constant, the flow velocity displays the opposite behavior to the density, i.e., it increases exponentially with altitude. For this solution, the minor ion always remains minor; however, at some altitude the flow becomes supersonic, and hence, the assumption of subsonic flow becomes invalid.

6.2. Supersonic H^+ outflow

For supersonic H^+ outflow, the nonlinear inertial term in the momentum equation (2) cannot be neglected and the situation becomes more complex. To illustrate this case, it is convenient to make the following simplifying assumptions: (a) The flow is ambipolar ($n_i = n_e$, $u_i = u_e$); (b) The ionosphere is isothermal; (c) Steady state conditions prevail; (d) The neutrals are stationary; and (e) There is only one ion species. With these assumptions, the sum of the ion and electron momentum equations is given by

$$n_i m_i u_i \partial u_i/\partial z + k(T_e+T_i)\partial n_i/\partial z + n_i m_i g = -n_i m_i \nu_{in} u_i \quad (126)$$

However, from the continuity equation,

$$\partial n_i/\partial z = -(n_i/u_i)\partial u_i/\partial z \quad (127)$$

The substitution of equation (127) into (126) yields,

$$(u_i^2 - v_{th}^2)(1/u_i)\partial u_i/\partial z + g = -\nu_{in} u_i \quad (128)$$

where v_{th} is the thermal speed of the ion-electron gas,

$$v_{th}^2 = k(T_e+T_i)/m_i \quad (129)$$

Introducing the Mach number, $M = u_i/v_{th}$, equation (128) can be cast in the following form:

$$\partial M/\partial z = -[g/v_{th}^2 + \nu_{in} M/v_{th}]M/(M^2-1) \quad (130)$$

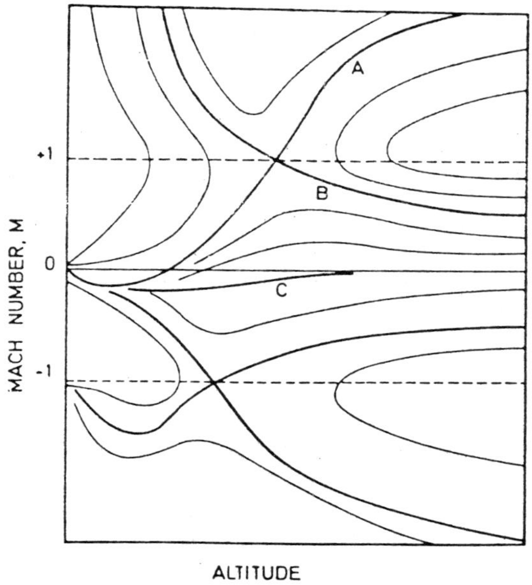

Figure 39. Ion Mach number diagram showing the transition from subsonic to supersonic flow for both outward and inward solutions (from Banks and Kockarts, 1973). Reprinted by permission of Academic Press.

This equation corresponds to a first-order ordinary differential equation for the Mach number. Note that the equation contains singularities at M = ±1; at the point of transition from subsonic to supersonic flow. For such a case, the different solutions to the Mach number equation are shown schematically in Figure 39. For M > 0, corresponding to polar

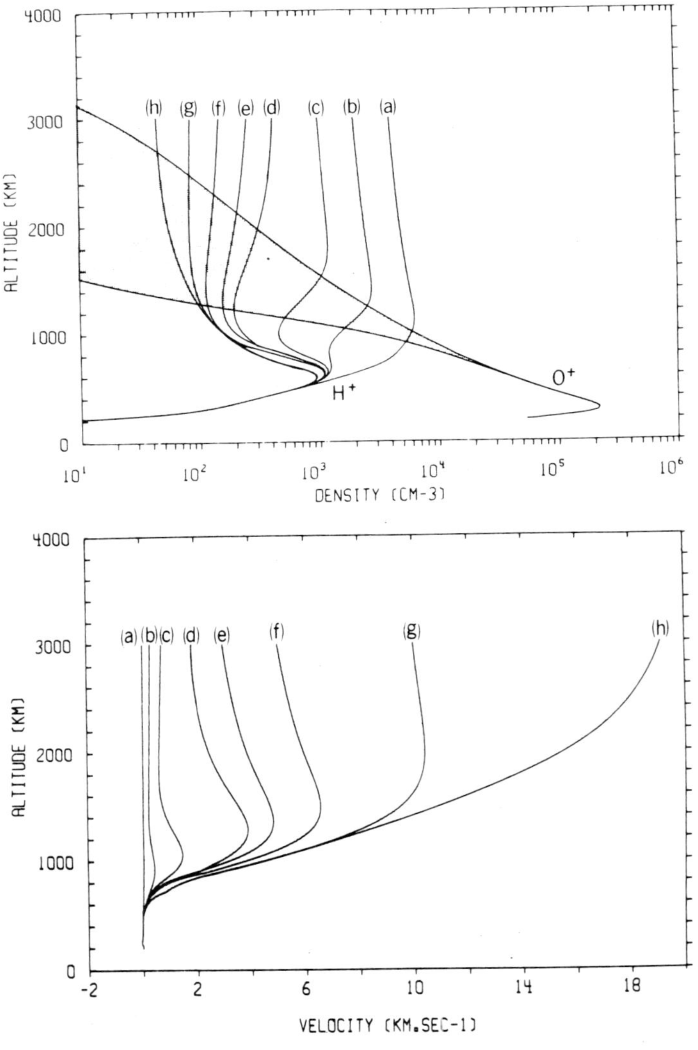

Figure 40. Theoretical H^+ density and field-aligned drift velocity profiles for the Earth's daytime high-latitude ionosphere. The different curves correspond to different H^+ escape velocities at 3000 km: (a) 0.06, (b) 0.34, (c) 0.75, (d) 2.0, (e) 3.0, (f) 5.0, (g) 10.0, (h) 20.0 km s^{-1}. The shaded region shows the range of O^+ densities (from Raitt et al, 1975). Reprinted by permission of Pergamon Press.

wind outflow, there are both subsonic ($M < 1$) and supersonic ($M > 1$) regons. All of the solutions that remain subsonic at all altitudes are valid, physical solutions. However, for supersonic outflow only the critical solution A is a physical solution. For this solution, the H^+ flow is subsonic at low altitudes, passes through the singular point $M = 1$, and then is supersonic at high altitudes.

6.3. Hydrodynamic solutions

Polar wind studies using the transport equations (1) to (3) are called hydrodynamic studies. This method has received the most attention over the last decade. Typical hydrodynamic results are shown in Figures 40 and 41 for the case when convection electric field effects are neglected. Figure 40 shows the effect on the H^+ and O^+ densities and H^+ field-aligned drift velocity of different H^+ escape velocities at 3000 km. Curve a represents near diffusive equilibrium, with H^+ becoming the dominant ion at 900 km (the O^+ density in this case follows the lower curve of the shaded region). As the upper boundary velocity is increased, the H^+ density is progressively reduced with a peak in the H^+ density profile appearing near 600-700 km altitude. Curve h, which represents a flow velocity of 20 km s^{-1} at 3000 km, corresponds to a supersonic flow of H^+ with an escape flux of 8.5×10^7 cm^{-2} s^{-1}.

The H^+ and O^+ temperatures are both affected by the H^+ flow, as shown in Figure 41. For O^+, the behavior is fairly simple in that as H^+ flows out of the topside ionosphere with an increasing velocity, the O^+ temperature at high altitudes decreases. This behavior results because as H^+ becomes a minor ion, the O^+-neutral thermal coupling becomes stronger. For H^+, on the other hand, the variation of temperature with escape velocity is more complicated. As the H^+ escape velocity is increased, the H^+ temperature at high altitudes first decreases, then increases, and then decreases again. This behavior is related to the relative contributions made to the H^+ thermal balance by convection, advection, thermal conduction, frictional heating, and collisional cooling, and we refer the reader to the paper by Raitt et al (1975) for a more detailed discussion of the H^+ energy balance.

A separate peculiarity of H^+ flow is its flux limiting character. As the outward flow velocity increases, the H^+ flux rapidly rises to a saturation limit. Figure 42 illustrates this behavior in terms of the H^+ density at 3000 km. At sufficiently high densities, the flux is inward (negative) but as the H^+ density is lowered, the outward flux is quickly established and soon saturates in magnitude. Thus, while arbitrarily large plasma inflow can occur in response to high densities within the protonosphere, the outflow is limited by various atmospheric constraints. These include the plasma temperatures, the neutral atomic hydrogen concentration, the O^+ density and the atmospheric composition. The limiting fluxes for several atomic hydrogen densities are included in Figure 42, showing that the H^+ flux changes proportionately to the density of H.

6.4. Collisionless solutions

The polar wind results that have been discussed in the previous subsections are valid at the altitudes where the H^+ and He^+ gases are collision-dominated. As a rough guide, the ion gases are effectively collision-dominated when

$$u_i/H_i\nu_i \ll 1 \tag{131}$$

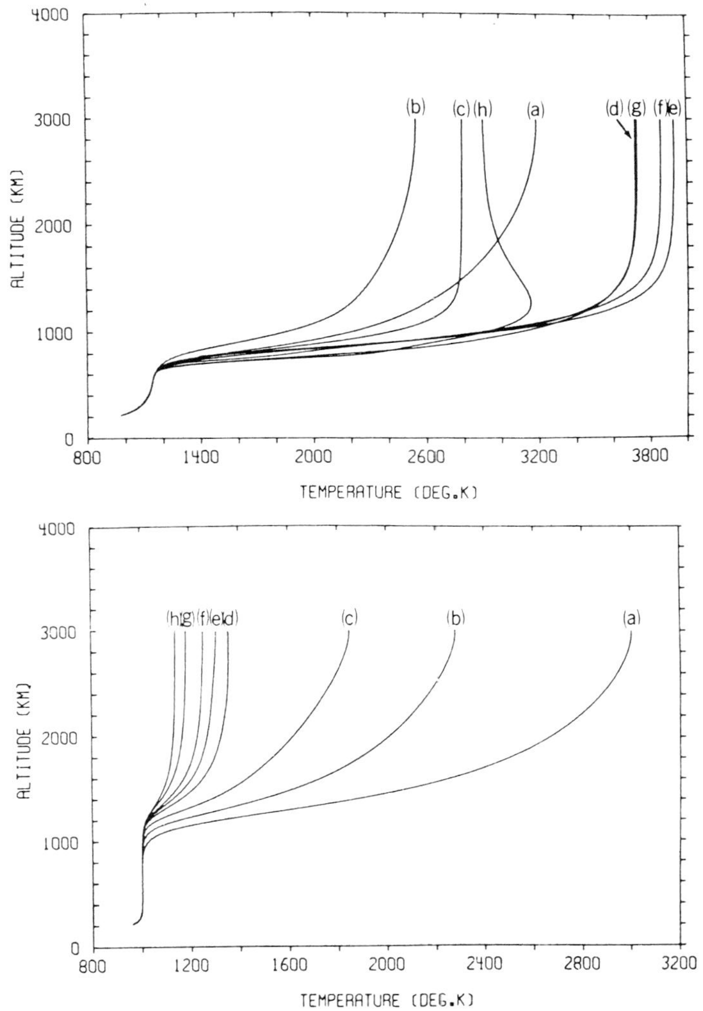

Figure 41. Theoretical H^+ (top) and O^+ (bottom) temperature profiles for the Earth's daytime high-latitude ionosphere. These profiles correspond to the density and drift velocity profiles shown in Figure 40 (from Raitt et al, 1975). Reprinted by permission of Pergamon Press.

where u_i is the ion field-aligned drift velocity, H_i is the ion density scale height, and ν_i is the appropriate ion collision frequency. For H^+, this condition generally begins to break down at 1000 km and is clearly violated at 2000 km. When the plasma is not collision-dominated, the H^+ pressure distribution becomes anisotropic and the H^+ heat flow vector is not simply related to the gradient in the H^+ temperature.

The effect of an anisotropic H^+ pressure distribution on the momentum balance enters through the stress term in equation (2). With allowance for this effect, the momentum equation for field-aligned flow becomes

$$n_i m_i u_i \partial u_i/\partial z + \partial/\partial z(n_i k T_i^{\parallel}) - n_i e_i E + n_i m_i g + n_i k(T_i^{\parallel} - T_i^{\perp})(1/A)\partial A/\partial z = -n_i m_i \nu_{in} u_i \tag{132}$$

where $T_i^{\parallel}$ and $T_i^{\perp}$ are the ion temperatures parallel and perpendicular to $\underline{B}$, respectively. The quantity $(1/A)\partial A/\partial z$ accounts for the divergence of the geomagnetic field with altitude. Equation (132) is the generalization of the classical hydrodynamic equation (126).

The collisionless characteristics of the polar wind can be described by kinetic, hydromagnetic and generalized transport models. For supersonic flow, these models produce density and drift velocity pro-

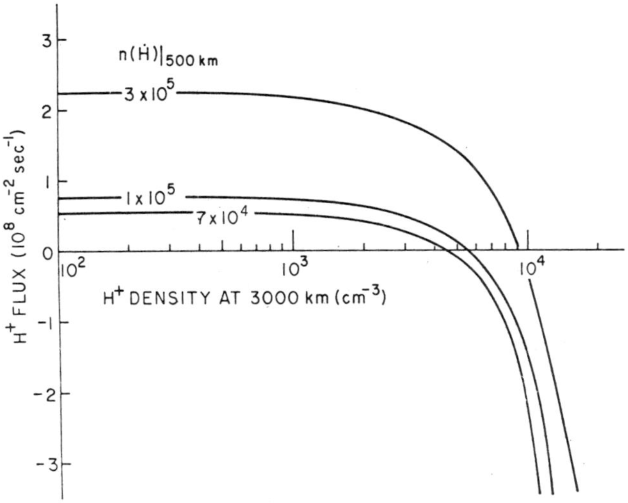

Figure 42. H^+ flux for different H^+ boundary densities and neutral hydrogen densities (from Banks, 1972).

files that are similar to those obtained from the hydrodynamic equations. However, the ion temperature distributions are different, with the collisionless models yielding large temperature anisotropies at high altitudes. Typical results are shown in Figure 43, where the H^+ and O^+ temperatures parallel and perpendicular to the geomagnetic field are plotted as a function of altitude for collisionless, supersonic H^+ outflow. The ion temperature distributions were calculated with both kinetic and hydromagnetic models and the results are similar. The parallel ion temperatures are essentially constant with altitude at high altitudes, while the perpendicular ion temperatures decrease monotonically with altitude. The net result is a parallel-to-perpendicular temperature anisotropy that grows with altitude, reaching nearly a factor of 50 for H^+ ions at a distance of 10 earth radii.

6.5. Self-similar solutions

To date, most of the polar wind solutions were obtained for steady state conditons. Recently, however, Singh and Schunk (1982) studied the collisionless expansion of an electrically neutral plasma into a vacuum

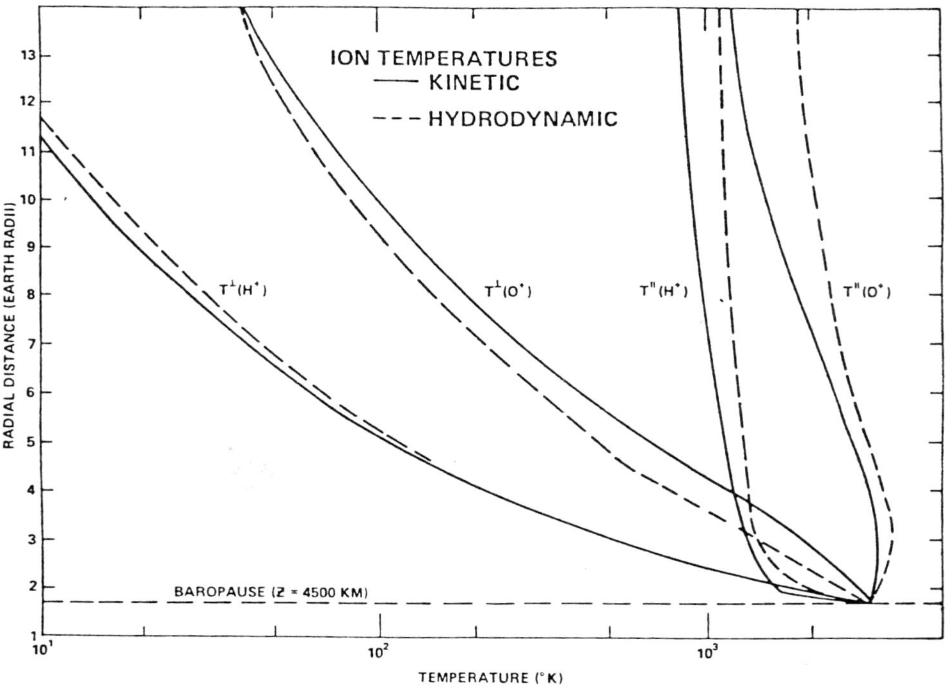

Figure 43. O^+ and H^+ temperatures parallel and perpendicular to the geomagnetic field obtained from kinetic (solid curves) and hydrodynamic (dashed curves) models of the collisionless, supersonic polar wind (from Holzer et al, 1971).

and their results may be relevant to the initial expansion of the polar wind. Figure 44 shows a schematic of the initial plasma density configuration that has been frequently used in studies involving expanding plasmas. At $t=0$, the half-space $x<0$ is filled with a semi-infinite, electrically neutral collisionless plasma. For $t>0$, the plasma is allowed to expand into the vacuum and the subsequent temporal evolution is followed.

The simplest treatment of the plasma expansion is obtained by assuming that the electrons always stay in an equilibrium with the developed electric fields; thus, they obey the Boltzmann distribution,

$$n_e(x) = z_i n_o \exp[e\phi(x)/k_B T_e] \tag{133}$$

where n_e denotes the electron density, n_o is the ion density in the unperturbed plasma, T_e is the electron temperature, k_B is the Boltzmann constant, z_i is the ion charge number, and ϕ is the electrostatic potential.

Assuming that the ions are cold, the ion continuity and momentum equations reduce to

$$\partial n_i/\partial t + \partial/\partial x(n_i u_i) = 0 \tag{134}$$

$$\partial u_i/\partial t + u_i \partial u_i/\partial x = -(z_i e/m_i)\partial\phi/\partial x \tag{135}$$

where u_i is the ion flow velocity, m_i is the ion mass and n_i is its density.

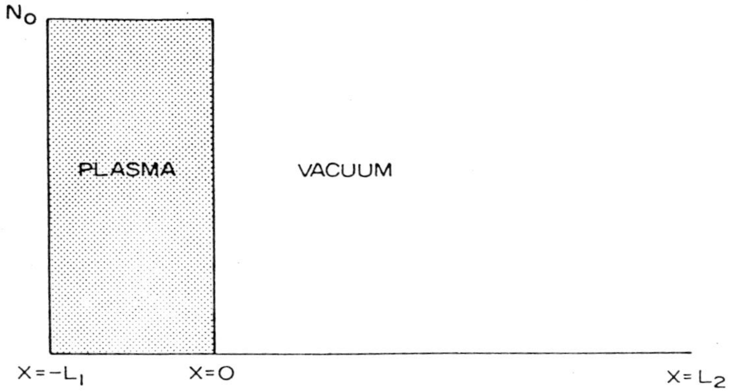

Figure 44. The numerical simulation region. At $t=0$ the region $-L_1 \le x \le 0$ is filled with plasma, while the region $0 \le x \le L_2$ is empty (from Singh and Schunk, 1982).

If quasi-neutrality is assumed, the set of equations (134) and (135) allow self-similar solutions, which depend only on the ratio (x/t) of the independent variables x and t;

$$n_e = z_i n_i = z_i n_o \exp[-(\xi+1)] \tag{136a}$$

$$u_i = C_s\ (\xi+1) \tag{136b}$$

$$\phi = -(k_B T_e/e)(\xi+1) \tag{136c}$$

for $(\xi+1)\geq 0$, where $\xi = (x/C_s t)$ is the self-similar variable and $C_s = (k_B T_e/m_i)^{1/2}$. For $\xi+1<0$, the plasma remains unperturbed.

The most fascinating aspect of the above solution is the possibility of ion acceleration, because $u_i \sim (x/t)$. This acceleration is caused by the polarization electric field,

$$E = -d\phi/dx = (k_B T_e/e)H(\xi+1)/C_s t \tag{137}$$

where $H(\xi+1)$ is the step function. Note that at a given time E is a constant for $x>-C_s t$.

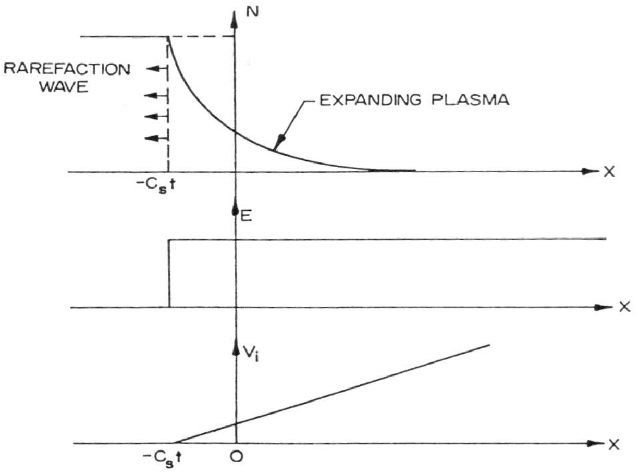

Figure 45. Self-similar (S-S) solution for the expansion of a plasma into a vacuum. (Top) S-S density profile. For t>0 a rarefaction wave propagates into the region x<0 as the plasma expands into the vacuum. (Middle) S-S electric field, which is uniform in the region $-C_s t \leq x < \infty$. (Bottom) S-S ion drift velocity profile. Note the linear increase in V_i with x. At a given x, V_i decreases as t^{-1} (from Singh and Schunk, 1982).

Figure 45 shows the characteristics of the self-similar solution for the collisionless expansion of a single ion plasma into a vacuum. As the expansion proceeds, a rarefaction wave propagates into the plasma at the ion-acoustic speed. The density profile in the expansion region is concave at all times, and the associated polarization electrostatic field does not vary with position, but its magnitude decreases inversely with time (see equation 137). Because of this electric field ion acceleration occurs, and some very energetic ions exist far from the expanding plasma front. This implies that the polar wind should be a source of energetic ions for the magnetoshpere.

Acknowledgement. This research was supported by NASA Grant NAGW-77 and NSF Grant ATM-8015497 to Utah State University.

REFERENCES

Babcock, R.R., and Evans, J.V.: 1979, J. Geophys. Res., 84, pp.5349.
Bahnsen, A.: 1978, J. Atmos. Terr. Phys., 40, pp.235.
Banks, P.M.: 1972, "Critical Problems of Magnetospheric Physics", ed. Dyer, E.R., pp.157-178, Washington, D.C., Inter-Union Comm. Solar Terr. Phys., National Academy of Sciences.
Banks, P.M., and Kockarts, G.: 1973, "Aeronomy", Academic, New York.
Banks, P.M., Schunk, R.W., and Raitt, W.J.: 1976, Annl. Rev. Earth & Planet. Sci., 4, pp.381-440.
Bittencourt, J.A., and Sahai, Y.: 1978, J. Atmos. Terr. Phys., 40, pp.669.
Brinton, H.C., Grebowsky, J.M., and Brace, L.H.: 1978, J. Geophys. Res., 83, pp.4767.
Cicerone, R.J. Swartz, W.E., Stolarski, R.S., Nagy, A.F., and Nisbet, J.S.: 1973, J. Geophys. Res., 78, pp.6709-6728.
Evans, J.V., and Holt, J.M.: 1978, Planet. Space Sci., 26, pp.727.
Hanson, W.B. and Moffett, R.J.: 1966, J. Geophys. Res., 71, pp.5559.
Holzer, T.E., Fedder, J.A., and Banks, P.M.: 1971, J. Geophys. Res., 76, pp.2453-2468.
Iijima, T., and Potemra, T.A.: 1978, J. Geophys. Res., 83, pp.599.
Perkins, F.W., and Roble R.G.: 1978, J. Geophys. Res., 83, pp.1611-1624.
Raitt, W.J., Schunk, R.W., and Banks, P.M.: 1975, Planet. Space Sci., 23, pp.1103.
Rees, M.H., and Roble, R.G.: 1975, Rev. Geophys. Space Phys., 13, pp.201-242.
Roble, R.G.: 1975, Planet Space Sci., 23, pp.1017-1030.
Roble R. G.: 1976, J. Geophys. Res., 81, pp.265-269.
Schunk, R.W.: 1977, Rev. Geophys. Space Phys., 15, pp.429-445.
Schunk, R.W., and Nagy, A.F.: 1978, Rev. Geophys. Space Phys., 16, pp.355-399.
Schunk, R.W., and Nagy, A.F.: 1980, Rev. Geophys. Space Phys., 18, pp.813-852.
Schunk, R.W., Raitt, W.J., and Banks, P.M.: 1975, J. Geophys. Res. 80, pp.3121.
Schunk, R.W., Banks, P.M., and Raitt, W.J.: 1976, J. Geophys. Res., 81, pp.3271.

Schunk, R.W., and Sojka, J.J.: 1982, Geophys. Res. Lett., in press.
Singh, N., and Schunk, R.W.: 1982, J. Geophys. Res., in press.
Sojka, J.J., Raitt, W.J., and Schunk, R.W.: 1981a, J. Geophys. Res., 86, pp.2206-2216.
Sojka, J.J., Raitt, W.J., and Schunk, R.W.: 1981b, J. Geophys. Res., 86, pp.6908-6916.
Sojka, J.J., Schunk, R.W., and Raitt, W.J.: 1982, J. Geophys. Res., 87, pp.187-198.
Swartz, W.E. and Nisbet, J.S.: 1973, J. Geophys. Res., 78, pp.5640.
Takayanagi, K., and Itikawa Y.: 1970, Space Sci. Rev., 11, pp.380-450.
Whitten, R.C. and Poppoff, I.G.: 1971, "Fundamentals of Aeronomy", Wiley, New York.

PHOTOCHEMICAL PROCESSES IN THE MESOSPHERE AND LOWER THERMOSPHERE

Jeremy R. Winick
Optical Physics Division
Air Force Geophysics Laboratory
Hanscom Air Force Base, Ma. 01731

Abstract. In the mesosphere and lower thermosphere (between 50-120 km), photochemical processes produce important minor constituents that can effect the thermal, radiative and electrical properties of the atmosphere. Since the subject of atmospheric photochemistry is very extensive, only an introduction to selected topics is given. We first introduce the basic concepts; the hydrostatic law, transport processes, the absorption of solar radiation, and chemical kinetic processes. Next the density profiles of the minor species, O, O_3, H, OH, and HO_2 in the mesosphere are illustrated by solving the one-dimensional continuity equation. The profiles of the longer-lived species NO, H_2O, and O in the lower thermosphere are also discussed. The final portion outlines the ion chemistry of the lower E-region and the D-region. The photochemical control of the ion composition is emphasized.

1. PHYSICAL AND CHEMICAL CONCEPTS

In this chapter we will show how photochemical processes help to determine the profiles of the important minor species in the mesosphere such as O, O_3, H_2O, and the hydrogen radical species H, HO_2, and OH. In the lower thermosphere the profiles of NO and O, the latter of which by 120 km becomes the second most common constituent, are determined by a competition between photochemical and transport processes. The temperature structure, and thus the basic state of the atmosphere, is determined by a number of processes: the lower boundary conditions, the absorption of solar ultraviolet (UV) radiation and the transport of infrared radiation, and the mean circulation. The radiative properties are in a large part controlled by the minor species CO_2, O_3 and NO. The absorption of solar UV leading to the dissociation of O_2 in the lower thermosphere is a major contribution to the large increase of temperature with altitude in this region. The complete solution for the temperature, density, and composition requires a complicated radiative and dynamical model that is coupled to the photochemistry. The theoretical solution to the full problem is too complex for even the largest computers to

R. L. Carovillano and J. M. Forbes (eds.), Solar-Terrestrial Physics, 677–732.

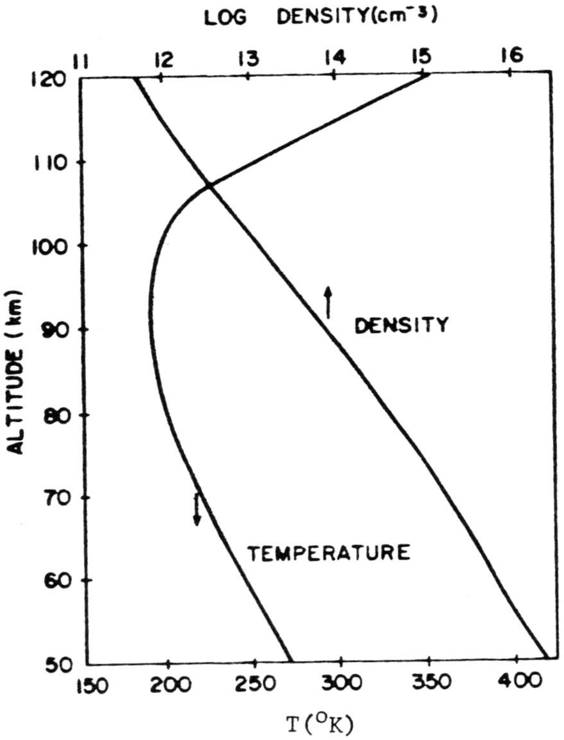

Figure 1. Temperature and number density from U.S. Standard Atmosphere 1976.

solve with inclusion of detailed photochemistry.

In this chapter we will explore the role of photochemistry in determining the minor neutral constituents and the ion composition. We will use the U.S. Standard Atmosphere 1976 as our basic state for the temperature, total number density, and mean molecular mass.

Figure 1 shows the temperature and number density profile in the 50-120 km region from the model atmosphere. The temperature declines from the stratopause at 50 km where UV absorption by ozone is the heat source. The broad minimum around 85-90 km is the mesopause region, the coldest portion of the earth's atmosphere. This temperature minimum is largely due to the fact that CO_2 IR radiation can escape to space (Forbes 1983, this volume). The rapid increase in temperature above 100 km is due to UV absorption by O_2. The standard atmosphere is probably closest to a mean low or middle latitude model. In the middle and high latitudes there are noticeable seasonal variations. For example, the summer polar mesopause at about 85 km can be very cold, about 150 K. Shorter term variations can also occur with the passage of gravity waves in the mesosphere which can lead to variations in density and pressure about the model values. We nevertheless will adopt this background atmosphere in order to determine photodissociation and reaction rates. For modelling of specific conditions that vary far from the model, one should use a more appropriate temperature and density profile.

The Hydrostatic Law

Table 1 is a compilation of pertinent parameters of the model atmosphere shown in Figure 1. The atmosphere is assumed to be in hydrostatic equilibrium, that is, the vertical pressure gradient balances the gravitational force at each altitude,

$$dp/dz = -\rho g \tag{1.1}$$

where p is the pressure, z the altitude, ρ the mass density, and g the acceleration of gravity. If we then substitute the expression for the density from the ideal gas law

$$p = \rho kT/m \tag{1.2}$$

where k is 1.38×10^{-16} erg K^{-1} (Boltzmann constant), m is the mean

Table 1. Model Atmosphere

Altitude z(km)	Density n(cm^{-3})	Temp T(°K)	Scale Ht. H(km)	Collision freq. (s^{-1})	Mean free path L (cm)	Mean mol. wt. m(g/mole)
50	2.14(16)*	270.7	8.05	5.62(6)	7.91(-3)	28.96
55	1.18(16)	260.8	7.77	3.05(6)	1.43(-2)	28.96
60	6.44(15)	247.0	7.37	1.62(6)	2.62(-2)	28.96
65	3.39(15)	233.3	6.97	8.29(5)	4.98(-2)	28.96
70	1.72(15)	219.6	6.57	4.08(5)	9.81(-2)	28.96
75	8.30(14)	208.4	6.24	1.92(5)	2.04(-1)	28.96
80	3.84(14)	198.6	5.96	8.66(4)	4.40(-1)	28.96
85	1.71(14)	188.9	5.68	3.76(4)	9.89(-1)	28.96
90	7.12(13)	186.9	5.64	1.56(4)	2.37(0)	28.91
95	2.92(13)	188.4	5.73	6.44(3)	5.79(0)	28.73
100	1.19(13)	195.1	6.01	2.68(3)	1.42(1)	28.40
105	5.02(12)	208.8	6.56	1.18(3)	3.36(1)	27.88
110	2.14(12)	254.9	7.72	5.48(2)	7.88(1)	27.27
115	9.69(11)	300.0	9.88	2.80(2)	1.75(2)	26.68
120	5.11(11)	360.0	12.09	1.63(2)	3.31(2)	26.20

From U.S. Standard Atmosphere 1976
* 2.14(16) means 2.14×10^{16}

molecular mass, and T is the absolute temperature, (1.1) becomes

$$dp/p = - dz/H. \quad (1.3)$$

Here

$$H = kT/mg \quad (1.4)$$

is the pressure scale height of the bulk atmosphere. Using the ideal gas law (1.2) with the number density $n = \rho/m$ equation (1.3) can be put into other useful forms;

$$dp/p = dn/n + dT/T, \quad (1.5)$$

$$p = p_o \exp(-\int dz/H) \quad (1.6)$$

$$n = n_o(T_o/T)\exp(-\int dz/H). \quad (1.7)$$

Thus the hydrostatic atmosphere has an exponential decrease of pressure with a characteristic scale length H. This parameter is listed in Table 1 for the multi-component atmosphere. The scale height is most strongly dependent upon the temperature and less so on the average molecular mass

$$m = \sum f_i m_i, \quad (1.8)$$

where m_i is the mass of species i and f_i is its fractional composition (volume mixing ratio). The average mass which is 28.96 amu below 80 km (78% N_2, 21% O_2, 1% Ar) decreases noticeably above 95 km as O_2 (mass 32) is dissociated into O. The decrease in mass above about 110 km is also due to diffusive separation. Below 95 km the average mass stays nearly constant since photochemical and diffusive separation effects are overwhelmed by turbulent mixing. The diffusive separation becomes increasingly important higher in the thermosphere.

The pressure scale height is relatively constant except for the steep increase starting near 105 km. Thus for small increments of altitude, an isothermal approximation is valid. In this case (since m and g are even more slowly varying than T) equations (1.6) and (1.7) become

$$p = p_o \exp[-(z-z_o)/H], \quad (1.9)$$

$$n = n_o \exp[-(z-z_o)/H], \quad (1.10)$$

and the pressure and density scale heights are the same. For a slowly varying temperature profile a better approximation can be obtained by using a linear expression for the scale height

$$H = H_o + \beta z \quad (1.11)$$

$$\beta = \partial H/\partial z, \quad (1.12)$$

$$p/p_o = (H/H_o)^{-1/\beta}, \tag{1.13}$$

$$n = n_o(H/H_o)^{-(1+1/\beta)}. \tag{1.14}$$

The reduced height can be defined in terms of the local scale height

$$\zeta = \int dz/H; \tag{1.15}$$

then

$$p=p_o\exp(-\zeta). \tag{1.16}$$

In this case the reduced height is the characteristic exponential scale length of the atmosphere. Equations (1.15) and (1.16) can be used to derive the column density of the atmosphere above height z_o.

$$\begin{aligned}\int n\,dz &= \int nH d\zeta = \int (nkT/mg)d\zeta \\ &= (n_oT_ok/mg)\int(p/p_o)d\zeta \\ &= (n_oT_ok/mg)\int\exp(-\zeta)d\zeta \\ &= n_oH_o\end{aligned} \tag{1.17}$$

Examination of Table 1 shows that the density decreases by almost five orders of magnitude between 50 km and 120 km. Correspondingly the collision frequency decreases by a similar amount and the mean free path increases by about the same factor. Thus we can expect processes that compete with collisions such as radiative de-excitation will be a strong function of altitude.

Transport Processes

The fact that the molecular mass is constant for altitudes below 80 km indicates that the atmosphere is homogeneous and except for minor constituents has a constant composition. The atmosphere is well mixed by turbulent or eddy diffusion processes that are much stronger than the molecular diffusion processes that tend to gravitationally separate species of different mass. In diffusive equilibrium each species will be distributed with its own scale height

$$H_i = kT/m_ig. \tag{1.18}$$

This becomes the case above about 110 km (see Figure 2) if one compares molecular diffusion coefficients with the eddy diffusion coefficient. For the ith minor species the molecular diffusion coefficient is determined from the expression

$$D_i^{-1} = (\sum f_j/D_{ij})^{-1} \tag{1.19}$$

where D_{ij} is the binary diffusion coefficient for species i through a gas of species j. Since (Banks and Kockarts, 1973)

$$D_{ij} = C(1+m_i/m_j)^{1/2} (8kT/\pi m_i)^{1/2}/n, \tag{1.20}$$

where C is a constant which is inversely proportional to the momentum transfer cross section, the lighter species diffuse faster and all diffusion increases approximately exponentially with height. Figure 2 shows molecular diffusion coefficients as a function of height for the U.S. Standard Atmosphere 1976. A wide range for the eddy diffusion coefficient K_z has been inserted. There is only one coefficient for all species. The eddy diffusion coefficient is a convenient method of representing transport for one-dimensional models where it must represent mixing that occurs by processes such as winds as well as turbulent eddies. Above 110 km turbulence is dissipated and molecular diffusion becomes the dominant vertical transport mechanism for the neutral atmosphere. In the region where it dominates, K_z must be determined empirically by solving the continuity equation

$$\partial n_i/\partial t = P_i - L_i - \nabla \cdot \Phi_i \tag{1.21}$$

for a tracer species with concentration n_i, where the vertical eddy flux is

$$\Phi_i = -K_z[M]\,\partial f_i/\partial z.$$

K_z is adjusted in the vertical eddy flux term until the modelled profile matches the measured profile. For this to be done unambiguously the chemical production and loss terms P_i and L_i must be well known as well as $n_i(z)$.

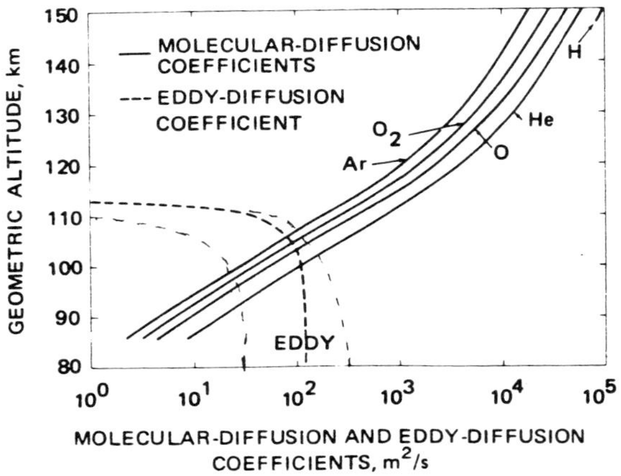

Figure 2. Molecular diffusion coefficient and eddy diffusion coefficient (K_z) from U.S. Standard Atmosphere 1976. Spread of K_z is author's estimate of the range of values expected. Note that units of K_z are $m^2\ s^{-1}$.

The impact of photochemistry on the atmospheric composition is incorporated in the chemical production P_i and loss L_i terms in equation (1.21). The last term on the right hand side is the divergence of the density flux of species i. The flux Φ_i can be written as

$$\Phi_i = n_i v_i, \tag{1.22}$$

and v_i is the velocity of species i where in general v_i is determined from molecular diffusion and the wind field on all scales down to turbulence. The full three-dimensional solution for v_i which is coupled to the chemistry through photochemical processes (such as dissociation and other heating processes and cooling via infrared radiation) is intractable. Even the most sophisticated general circulation models parameterize heating and cooling terms and use simplified chemistry. Since we are primarily interested in photochemical processes we will simplify transport terms as much as possible. It must be kept in mind that when transport effects become important in determining a species profile, our approximations can lead to large uncertainty in model predictions.

The first approximation we will make is to only consider the vertical coordinate. The vertical scale of the atmosphere, the scale height, varies from about 5 to 12 km in the mesosphere and lower thermosphere. Horizontal scale lengths are many orders of magnitude larger and variations much more gradual. The influence of photochemistry is much more noticeable in the vertical due to rapid variation in photodissociation and reaction rates with altitude. We will thus be concerned with obtaining the vertical profile of species i

$$\partial n_i(z)/\partial t = P_i(z) - L_i(z) - \partial \Phi_{i,z}/\partial z \tag{1.23}$$

where $\Phi_{i,z}=n_i$. The vertical component of the velocity $v_{i,z}$ can be broken into two parts: The molecular diffusion portion which is a solution a solution to (Banks and Kockarts, 1973)

$$v_i^D = -D_i[(1/n_i)dn_i/dz + 1/H_i + (1 + \alpha)(1/T)dT/dz] \tag{1.24}$$

where α is the thermal diffusion coefficient, a generally small term, and H_i is the scale height of species i, and the eddy diffusion portion is

$$v_i^K = -K[(1/n_i)dn_i/dz + (1/T)dT/dz + 1/H]. \tag{1.25}$$

Below 100 km it is usually sufficient to solve the continuity equation with only v_i^K. We will emphasize the region below 100 km in this chapter, but first it is important to consider the time scales of the terms in equation (1.21), in order to determine the relative importance of the the chemical and transport terms. The eddy diffusion velocity is

$$v_i^K \sim K(z)/H, \tag{1.26}$$

and the characteristic time for eddy diffusion is the characteristic distance divided by the velocity

$$\tau_K \sim H^2/K. \tag{1.27}$$

Using H from Table 1 and K from Figure 2 we can see that the mixing lifetime is the order of a few times 10^5 seconds in the 100 km region. For chemical lifetimes less than this the photochemical terms will determine the species density. For species with chemical lifetimes of the order of 10^5 seconds or longer, transport effects are important and may even dominate. Since we are primarily interested in the photochemistry, we will next examine the photochemical production and loss terms, before we return to solving the continuity equation for the minor constituents in the upper mesosphere.

Photochemical Production and Loss

Since equation (1.23) is a first order differential equation in time, the production and loss terms are chemical kinetic rate expressions. These terms strongly couple the reactive species, and are of three basic basic varieties: unimolecular, bimolecular, or termolecular. Table 2a displays the wide variety of gas phase chemical reactions that are encountered in the 50-120 km region. Table 2b gives rough estimates of the limiting values of the rate constants. Individual rate constants can be many orders of magnitude slower than the limiting "gas kinetic" rate even when the reaction is exothermic, if there is a large energy barrier between reactants and the products that leads to a large activation energy. A good example is reaction 36 in Table 3 which is exothermic by about 1.4 eV, but is very slow at temperatures near 300 K.

Since the chemical production and loss terms are the driving terms of the continuity equation, a knowledge of chemical kinetics is essential for the photochemical modeller. However, a review of chemical kinetics is beyond the scope of this chapter. We will just introduce a few of the ideas that are needed to describe the types of reactions that occur in the atmosphere. The reader is encouraged to consult the literature for further details (for example Johnston, 1966).

Chemical kinetic rate constants are determined primarily by laboratory experiments. Many of these rates are not known to great precision because the experiments are difficult. Many important reactions involve one or more unstable radical species. These radicals are difficult to generate in high concentrations and the determination of the concentration of the trace amounts often requires state-of-the-art techniques. Often many competing reactions occur and their effect must be considered. New measurements of the rates using new and improved techniques can lead to revised values of the rate coefficients that can greatly effect photochemical models. Revised rate constants such as the increase of rate constant for reaction 26 in Table 3 by more than a factor of ten (Howard and Evenson, 1977) decreased the catalytic ozone destroying capacity of NO in the lower stratosphere. Thus the knowledge of the values of the rate constants used in a photochemical models and their source is necessary if models are to be compared. NASA (NASA 1979, 1981) and other

Table 2a. Types of Chemical Reactions

Reaction	Rate	Type
Neutrals		
$NO + h\nu \longrightarrow NO^+ + e$	$J[NO] = [NO] \int I_\lambda \Phi_\lambda \sigma_\lambda \, d\lambda$	Photoionization
$O_2 + h\nu \longrightarrow O + O$	$J[O_2] = [O_2] \int I_\lambda \Phi_\lambda \sigma_\lambda \, d\lambda$	Photodissociation
$NO(X\ ^2\Pi) + h\nu \longrightarrow NO(A\ ^2\Sigma)$	$J[NO] = [NO] \int I_\lambda \Phi_\lambda \sigma_\lambda \, d\lambda$	Photoexcitation
$O_2(^1\Delta_g) \longrightarrow O_2(^3\Sigma_g^-) + h\nu$	$A[O_2(^1\Delta_g)]$	Fluorescence or luminescence
$O(^1D) + N_2 \longrightarrow O(^3P) + N_2$	$k[O(^1D)][N_2]$	Quenching of $O(^1D)$
		Vibrational excitation of N_2
$O_3 + H \longrightarrow OH + O_2$	$k[O_3][H]$	Bimolecular reaction
$O + O_2 + M \longrightarrow O_3 + M$	$k[O][O_2][M]$	Termolecular reaction
$N + O \longrightarrow NO^* \longrightarrow NO + h\nu$	$k[N][O]$	Radiative combination
$N_2 + CO_2 \longrightarrow N_2 + CO_2(001)$	$k[N_2][CO_2]$	Energy transfer
Ions		
$N_2^+ + O_2 \longrightarrow N_2 + O_2^+$	$k[N_2^+][O_2]$	Charge transfer
$N_2^+ + O \longrightarrow NO^+ + N$	$k[N_2^+][O]$	Ion-atom interchange
$NO^+ + e \longrightarrow N + O$	$\alpha\ [NO^+][e]$	Dissociative recombination
$O^+ + e \longrightarrow O + h\nu$	$k[O^+][e]$	Radiative recombination
$e + O_2 + O_2 \longrightarrow O_2^- + O_2$	$k[O_2][O_2][e]$	Three-body attachment
$e + O_3 \longrightarrow O^- + O_2$	$k[O_3][e]$	Dissociative attachment
$e + O \longrightarrow O^- + h\nu$	$k[O][e]$	Radiative attachment
$O_2^- + M \longrightarrow e + O_2 + M$	$k[O_2^-][M]$	Collisional detachment
$O^- + O \longrightarrow O_2 + e$	$k[O^-][O]$	Associative detachment
$O^- + h\nu \longrightarrow O + e$	$J[O^-] = [O^-] \int I_\lambda \Phi_\lambda \sigma_\lambda \, d\lambda$	Photodetachment
$H^+(H_2O)_n + NO_3^- \longrightarrow HNO_3 + nH_2O$	$\alpha\ [H^+(H_2O)_n][NO_3^-]$	Ion-ion recombination

Table 2b. Estimates of Reaction Rate Coefficients

Bimolecular neutrals	$2 \times 10^{-10}\ cm^3 molec^{-1} s^{-1}$
Termolecular neutral	$1 \times 10^{-30}\ cm^6 molec^{-2} s^{-1}$
Radiative combination	$1 \times 10^{-16}\ cm^3 molec^{-1} s^{-1}$
Ion -neutral charge transfer	$2 \times 10^{-9}\ cm^3 molec^{-1} s^{-1}$
Termolecular ions	$1 \times 10^{-27}\ cm^6 molec^{-2} s^{-1}$
Dissociative recombination	1×10^{-7} to $5 \times 10^{-6}\ cm^3 molec^{-1}$ for large cluster
Ion-ion recombination	6×10^{-8} to $1 \times 10^{-7}\ cm^3 molec^{-1} s^{-1}$
Associative attachment	$1 \times 10^{-10}\ cm^3 molec^{-1} s^{-1}$

agencies publish reports with updates and critical reviews (Baulch et al, 1979) of the chemical kinetic data.

Elementary gas phase chemical reactions are of three basic types. Unimolecular processes are decompositions in which the rate is independent of other species concentrations. The most important unimolecular rates are photoprocesses such as excitation, dissociation and ionization. For the dissociation reaction of molecule AB

$$AB + h\nu \longrightarrow A + B \qquad J_d \tag{1.28}$$

the rate law is

$$-d[AB]/dt = d[A]/dt = d[B]/dt = J_d[AB]. \tag{1.29}$$

In (1.29) J_d is the dissociation coefficient and [AB] is the concentration of molecule AB. Throughout this chapter we will use units of centimeters and seconds for rate constants, which is consistent with concentrations in number of particles per cm^3. Many other units are used for concentrations and the corresponding rate constants, especially by chemists. We have chosen units that seem to be most commonly used by photochemical modellers. Photoprocesses are the driving force of atmospheric chemistry. We will consider the role of the solar flux and the determination of the dissociation rates in the next section.

Bimolecular reactions involve binary collisions. A typical reaction is

$$AB + C \longrightarrow AC + B. \qquad k_b \tag{1.30}$$

The rate expressions for this reaction are

$$-d[AB]/dt = -d[C]/dt = d[AC]/dt = d[B]/dt = k_b[AB][C]. \tag{1.31}$$

For small molecules the limiting rate for binary collisions is $k \approx 2\times10^{-10}$ cm^{-3} s^{-1} when AB and C are neutrals and about 2×10^{-9} cm^{-3} s^{-1} for the case where AB or C is an ion (see Table 2b). This limiting rate occurs when every collision leads to product formation. The bimolecular rate constant is generally of the form

$$k = B \exp(-E_a/RT) = C \exp(-A/T) \tag{1.32}$$

where E_a is an activation energy. For many reactions which proceed appreciably only at high temperatures (large E_a), E_a is determined at elevated temperatures and the same form is assumed to hold at the lower temperatures of the atmosphere. This can lead to large errors in the rate constant if equation (1.32) is not valid when T is low. However, in most cases (1.32) is accurate and is probably the best estimate for k until k is measured at the required temperature.

Termolecular reactions involve a third body, and are often recombin-

TABLE 3. Chemical Reactions and Rate Coefficients†

1.	$O_2 + h\nu \longrightarrow O + O$	J_1, λ <2420 Å
2.	$O + O_2 + M \longrightarrow O_3 + M$	k_2=1.0(-34)exp(510/T)§
3a.	$O_3 + h\nu \longrightarrow O_2 + O(^1D)$	J_{3D}, λ <3100 Å
3b.	$O_3 + h\nu \longrightarrow O_2 + O(^3P)$	J_{3P}, λ <11400 Å
4.	$O + O_3 \longrightarrow O_2 + O_2$	k_4=1.9(-11)exp(-2300/T)
5.	$O + O + M \longrightarrow O_2 + M$	k_5=3.8(-30)exp(-170/T)(1/T)
6.	$O(^1D) + M \longrightarrow O(^3P) + M$	k_6=1.95(-11)exp(170/T)$[N_2]/[M]$ + 2.9(-11)exp(67/T)$[O_2]/[M]$
7.	$O + OH \longrightarrow O_2 + H$	k_7=4.0(-11)
8.	$O_3 + OH \longrightarrow HO_2 + O_2$	k_8=1.6(-12)exp(-940/T)
9.	$H + O_3 \longrightarrow OH + O_2$	k_9=1.4(-10)exp(-470/T)
10.	$H + O_2 + M \longrightarrow HO_2 + M$	k_{10}=2.1(-32)exp(290/T)
11	$O + HO_2 \longrightarrow O_2 + OH$	k_{11}=4.0(-11)
12.	$H_2O + O(^1D) \longrightarrow OH + OH$	k_{12}=2.3(-10)
13.	$H_2 + O(^1D) \longrightarrow OH + H$	k_{13}=9.9(-10)
14a.	$H + HO_2 \longrightarrow H_2 + O_2$	k_{14a}=.29·k_{14}
14b.	$H + HO_2 \longrightarrow H_2O + O$	k_{14b}=.02·k_{14}, k_{14}=4.7(-11)
14c.	$H + HO_2 \longrightarrow OH + OH$	k_{14c}=.69·k_{14}
15.	$OH + HO_2 \longrightarrow H_2O + O_2$	k_{15}=4.0(-11)
16.	$HO_2 + HO_2 \longrightarrow H_2O_2 + O_2$	k_{16}=2.5(-12)exp(-600/T)
17.	$O(^1D) + O_2 \longrightarrow O_2 + O$	k_{17}=2.9(-11)exp(67/T)
18.	$O(^1D) + CH_4 \longrightarrow CH_3 + OH$	k_{18}=1.4(-10)
19.	$O(^1D) + CH_4 \longrightarrow CH_2O + H_2$	k_{19}=1.4(-10)
20.	$NO_2 + OH + M \longrightarrow HNO_3 + M$	k_{20}=2.6(-30)$(T/300)^{-2.9}$
21.	$N(^2D) + O \longrightarrow N + O$	k_{21}=1.4(-10)
22.	$O(^1D) + O_3 \longrightarrow O_2 + O_2$	k_{22}=1.2(-10)
23.	$O(^1D) + O_3 \longrightarrow O_2 + 2O$	k_{23}=1.2(-10)
24.	$HO_2 + O_3 \longrightarrow OH + 2O_2$	k_{24}=1.4(-14)exp(-580/T)
25.	$OH + H_2O_2 \longrightarrow HO_2 + H_2O$	k_{25}=1.0(-11)exp(-750/T)
26.	$NO + HO_2 \longrightarrow NO_2 + OH$	k_{26}=3.4(-12)exp(250/T)
27.	$H_2O + h\nu \longrightarrow H + OH$	k_{27}=4.5(-12)exp(-275/T)
28.	$O_2(b^1\Sigma_g^+) + N_2 \longrightarrow O_2 + N_2$	k_{28}=2.0(-15)
29.	$O_2(b^1\Sigma_g^+) + O_2 \longrightarrow O_2 + O_2$	k_{29}=4.5(-16)
30a.	$N_2O + O(^1D) \longrightarrow NO + NO$	k_{30a}=6.6(-11)
30b.	$N_2O + O(^1D) \longrightarrow N_2 + O_2$	k_{30b}=5.1(-11)
31.	$NO + O_3 \longrightarrow NO_2 + O_2$	k_{31}=2.3(-12)exp(-1450/T)
32.	$NO_2 + h\nu \longrightarrow NO + O$	J_{32}, λ <3980 Å
33.	$NO_2 + O \longrightarrow NO + O_2$	k_{33}=9.3(-12)
34.	$NO + h\nu \longrightarrow N + O$	J_{34}, λ <1908 Å
35.	$N + NO \longrightarrow N_2 + O$	k_{35}=3.4(-11)
36.	$N(^4S) + O_2 \longrightarrow NO + O$	k_{36}=4.4(-12)exp(-3220/T)
37.	$N(^2D) + O_2 \longrightarrow NO + O$	k_{37}=6.0(-12)
38.	$N_2^+ + O \longrightarrow NO^+ + N$	k_{38}=1.4(-10)
39.	$N_2 + O^+ \longrightarrow NO^+ + N$	k_{39}=1.0(-12)
40.	$N + O_2^+ \longrightarrow NO^+ + O$	k_{40}=1.2(-10)
41a.	$N^+ + O_2 \longrightarrow NO^+ + O$	k_{41a}=3.0(-10)
41b.	$N^+ + O_2 \longrightarrow N + O_2^+$	k_{41b}=3.0(-10)

Table 3 (cont.)

42.	$N_2^+ + e \longrightarrow N + N$	α_{42}=1.0(-7)
43.	$NO^+ + e \longrightarrow N(^2D\ 76\%,\ ^4S\ 24\%) + O$	α_{43}=4.5(-7)(T/300)$^{-.7}$
44.	$N_2^+ + O_2 \longrightarrow N_2 + O_2^+$	k_{44}=4.3(-11)
45.	$O_2^+ + NO \longrightarrow NO^+ + O_2$	k_{45}=4.4(-10)
46.	$O_2^+ + O_2 + M \longrightarrow O_4^+ + M$	k_{46}=2.6(-30)(T/300)$^{-3.2}$
47.	$O_2^+ + H_2O + M \longrightarrow O_2^+(H_2O) + M$	k_{47}=2.5(-28)
48.	$O_4^+ + O \longrightarrow O_2^+ + O_3$	k_{48}=3.0(-10)
49.	$O_4^+ + H_2O \longrightarrow O_2^+(H_2O) + O_2$	k_{49}=1.5(-9)
50.	$O_2^+(H_2O) + H_2O \longrightarrow H^+(H_2O)(OH) + O_2$	k_{50}=1.0(-9)
51.	$O_2^+(H_2O)(OH) + H_2O \longrightarrow H^+(H_2O)_2 + OH$	k_{51}=1.4(-9)
52.	$O_2^+ + e \longrightarrow O + O$	α_{52}=3.5(-7)
53.	$NO^+ + H_2O + M \longrightarrow NO^+(H_2O) + M$	k_{53}=1.6(-28)
54.	$NO^+(H_2O)_3 + H_2O \longrightarrow H^+(H_2O)_3 + HNO_2$	k_{54}=7.0(-11)
55.	$NO^+ + CO_2 + M \longrightarrow NO^+(CO_2) + M$	k_{55}=2.5(-29)
56.	$NO^+ + N_2 + M \longrightarrow NO^+(N_2) + M$	k_{56}=2.0(-31)
57a.	$e + O_2 + O_2 \longrightarrow O_2^- + O_2$	k_{57a}=1.4(-29)(300/T)exp(-600/T)
57b.	$e + O_2 + N_2 \longrightarrow O_2^- + N_2$	k_{57b}=1.0(-31)
58.	$e + O_3 \longrightarrow O^- + O_2$	k_{58}=9.0(-12)(T/300)$^{1.5}$
59.	$O_2^- + O \longrightarrow O_3 + e$	k_{59}=1.5(-10)
60.	$O_2^- + O_2(^1\Delta_g) \longrightarrow 2O_2 + e$	k_{60}=2.0(-10)
61.	$O_2^- + h\nu \longrightarrow O_2 + e$	J_{61}
62.	$CO_3^- + NO \longrightarrow NO_2^- + CO_2$	k_{62}=1.1(-11)
63.	$CO_3^- + O \longrightarrow O_2^- + CO_2$	k_{63}=1.1(-10)
64.	$CO_3^- + h\nu \longrightarrow CO_2 + O^-$	J_{64}
65.	$NO_2^- + O_3 \longrightarrow NO_3^- + O_2$	k_{65}=1.2(-10)

† Rate coefficients have the following units: bimolecular, $cm^3 molec^{-1} s^{-1}$: termolecular, $cm^6 molec^{-2} s^{-1}$.

§ Read 1(-34) as 1×10^{-34}.

---

ation reactions. The reaction

$$A + B + C \longrightarrow AB + C \qquad k_t \qquad (1.33)$$

gives rise to the following rate expressions;

$$-d[A]/dt = -d[B]/dt = d[AB]/dt = k_t[A][B][C]. \qquad (1.34)$$

Since AB is a molecule bound with respect to A and B the binding energy must be removed by the third body C (often denoted by M for any third body; mostly the major atmospheric constituents N_2 or O_2) or AB will fall apart again into A and B. The rate of many three-body reactions increases as the temperature decreases. At low temperature there is more time for the third body to stabilize the collision. There isn't any standard form for the three-body rate constant but often it is expressed as

$$k_t = k_o(T/300)^{-c}. \quad (1.35)$$

Forms like equation (1.32) with negative E_a are also used, but a negative activation energy has no physical significance. Neutral three-body reaction rates are the order of 1×10^{-30} cm^{-6} s^{-1} or less.

An important parameter for photochemical modelling is the lifetime or characteristic time of a chemical species against destruction by chemical reaction. The lifetime of species A against reaction is the concentration of A divided by the destruction rate,

$$\tau_A = [A]/\text{destruction reaction rate}. \quad (1.36)$$

For reaction (1.33),

$$\tau_A = 1/(k_t[B][C]).$$

An important three-body reaction is the formation of ozone.

$$O + O_2 + M \longrightarrow O_3 + M. \qquad k_2 \quad (1.37)$$

Since O_2 constitutes 21% of the atmosphere below about 90 km the rate of (1.37) is proportional to the square of the pressure ($\sim[M]^2$). The lifetime of atomic oxygen against formation of O_3 is thus

$$\tau_O = (k_2[O_2][M])^{-1} \quad (1.38)$$

which rapidly decreases with increasing altitude. Earlier we saw that the transport lifetime was about 2×10^5 seconds. The transport and chemical lifetimes of O are equal when

$$2\times10^5 = [(1\times10^{-34})\exp(510/200)(0.21\ M^2)]^{-1}$$

where we have estimated the temperature to be 200 K. This leads to a value of M of 1.2×10^{15} cm^{-3} which occurs (Table 1) at about 75 km. Three body reactions start to become important around 75 km, higher when the rate constant is faster than the relatively small k_2, but possibly lower if the major constituent O_2 is replaced by a minor constituent. We will see later that the 80 km region is a transition region in the nature of both the neutral and ion chemistry. The importance of three-body reactions below this point is a major reason for the increasing complexity of the photochemistry in the mesosphere.

Table 3 is a compilation of chemical reactions that will be used in this chapter. It is by no means a complete set, but it is sufficient for highlighting the important photochemical processes in the 50-120 km region. More comprehensive tabulations are available (Baulch et al, 1979, NASA 1981). Most of the neutral rate coefficients in Table 3 are taken from NASA 1979. For the photoprocesses no values are listed since the photodissociation rates depend upon altitude and solar zenith angle which determine the opacity of the overlying atmosphere. In the next section we will consider the solar flux, its absorption by atmospheric constituents and the calculation of photodissociation rates.

Absorption of Solar Flux in the Mesosphere and Lower Thermosphere

Since N_2 and O_2 are the major constituents of the atmosphere, the possible interaction of these molecules with solar radiation is potentially most important. Throughout the region of interest, N_2 is the major constituent. Figure 3 shows potential energy curves of the most important electronic states of N_2. The lowest excited state, the $A\ ^3\Sigma_u^+$ at 6.22 eV (corresponding to about 12420/6.22=1997 Å)is the lowest state of a large triplet manifold which is uncoupled from the singlet ground state because a singlet-triplet transition requires a change of electron spin which is not a dipole-allowed process. The ^3A state is thus metastable with radiative lifetime of 2 s. The triplet states are most easily accessible by photoelectron and auroral electron processes that lead to well recognized emissions (Rees, this volume). The lowest excited singlet state, the $a\ ^1\Pi_g$ at 8.59 eV (about 1446 Å) is the upper state for the Lyman-Birge-Hopfield (LBH) band system. The LBH system (g-g) is forbidden for electric dipole radiation and is thus weak, the peak cross section being about 1×10^{-21} cm^2. Vibrational levels of the a

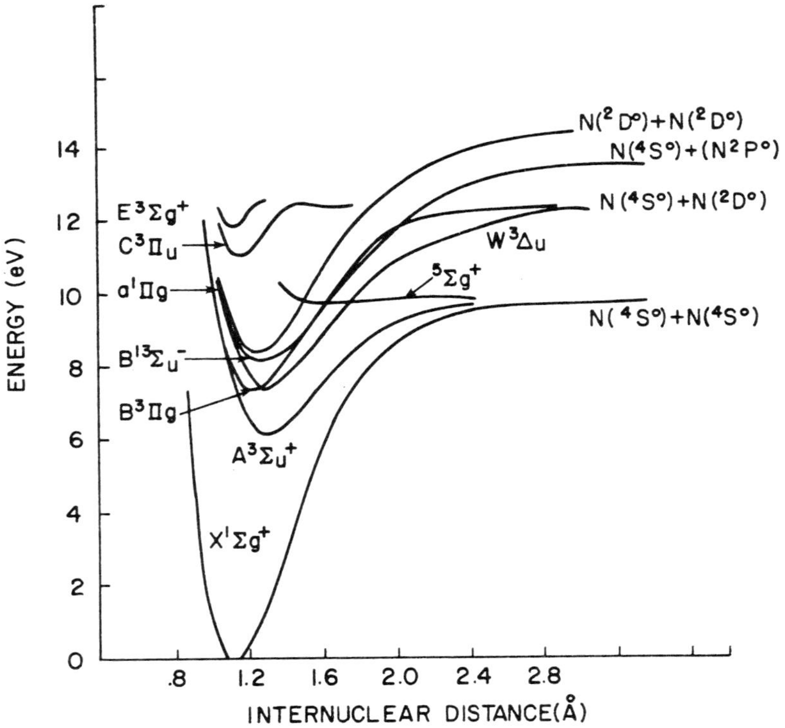

Figure 3. Potential energy curves of some important N_2 electronic states.

$^1\Pi_g$ at and above v'=6 (wavelength λ <1269 Å) can predissociate via the intersecting repulsive $^5\Sigma_g^+$ state, but photodissociation of N_2 by this path is quite small. There are no strong transitions in N_2 for wavelengths greater than 1000 Å. Below 1000 Å N_2 has stronger Rydberg transitions that can predissociate, but in this region O_2 (λ <1027 Å) and O (λ <910 Å) have large ionization cross sections. Below 796 Å however, N_2 can be ionized and the cross section is quite large, although the solar flux is decreasing except for a few prominent lines. Most of th N_2 photoionization occurs at altitudes above 120 km. As we will see later, N_2 dissociates to a much smaller degree than O_2 and the N_2 process is dominated by electron collisions and by ionization followed by chemical reaction.

Atomic nitrogen is a very minor constituent below 120 km due to the relative inefficiency of the N_2 dissociation. There are two low-lying excited states of N that are very important in atmospheric chemistry. The ^2P state is 3.58 eV above the ground $N(^4S)$ state, and the ^2D state is 2.38 eV above the ground state. These states are not dipole connected to the ground state (since they are doublet-quartet transitions). The ^2D has a radiative lifetime of about 26 hours (the transitions radiates the 5200 Å line) and thus is strongly quenched in the lower thermosphere. The ^2P state can radiatively cascade to the ^2D with a lifetime of 13 s, or radiate to the ground state with a lifetime of 185 s. It is believed that atomic oxygen rapidly quenches ($k=1\times10^{-11} cm^3 s^{-1}$) the $^2P \rightarrow ^2D$ transition. Thus $N(^2P)$ in the lower thermosphere will either be quenched or radiate to $N(^2D)$ or react in a similar manner to $N(^2D)$. $N(^2D)$ state is most significant because it reacts with O_2 to form NO at a rate orders of magnitude faster than that of $N(^4S)$.

Molecular oxygen potential energy curves are shown in Figure 4. As opposed to N_2, there are two very important low-lying metastable states, the a $^1\Delta_g$ at 0.977 eV and the b $^1\Sigma_g^+$ at 1.63 eV. These states have long radiative lifetimes and have very small absorption cross sections. The B $^3\Sigma_u^-$ state at 6.19 eV is the lowest state that is dipoled allowed from the X $^3\Sigma_g^-$ ground state. The transition B-X gives rise to the Schumann-Runge bands. The absorption cross section in these bands is highly structured with peak-to-valley variations of two orders of magnitude and the peak absorption varying from a few times 10^{-19} cm^2 at 1750 Å (the (17,0) band) to a few times 10^{-23} cm^2 at 2050 Å (the (0,0) band). This variation with vibrational band is due to the very small Franck-Condon factor for the (0,0) band caused by the large offset in the equilibrium nuclear distance between the two states. For higher vibrational states of the B state the transition can be more vertical and thus the Franck-Condon factors are larger. The O_2 absorption cross section is plotted in Figure 5. Figure 5c schematically shows the absorption for O_2 in the Schumann-Runge band (SRB) region of 1750-2050 Å. At higher resolution each vibrational band can be seen to be composed of many rotational lines that are not resolved in Figure 5c (see Kockarts, 1971). Above v'=4 the rotational lines are broadened, indicating predissociation. The highly structured cross section of O_2 in the SRB makes calculation of the penetrating solar flux

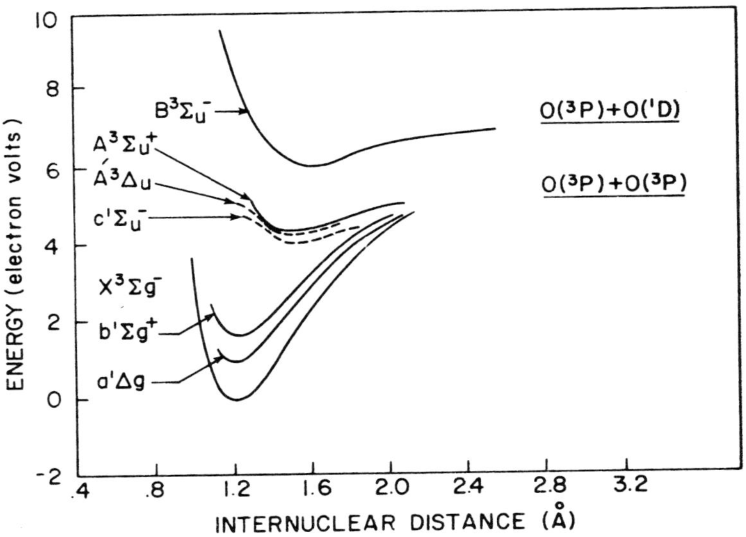

Figure 4. Potential energy curves of some important O_2 electronic states.

and O_2 dissociation difficult. For accurate work a very fine wavelength grid is needed. Moreover, the cross section structure is temperature dependent. Line-by-line calculations are computationally time consuming, especially for use in diurnal photochemical models where they must be repeated many times. Parameterizations have been developed that are much easier to use and that should be accurate enough for most uses (Kockarts 1976, Frederick and Hudson 1980b).

For wavelengths below 1750 Å the O_2 absorption cross section is shown in Figures 5a and 5b. The intense absorption in the Schumann-Runge continuum (SRC) between 1350 and 1750 Å leads to ground state $O(^3P)$ and $O(^1D)$ atoms (see Figure 7 for atomic oxygen levels). The absorption of these wavelengths occurs above 90 km and is largely responsible for the large proportion of O_2 that is dissociated into O in the lower thermosphere. Between 1300 and 1027 Å the O_2 cross section is highly structured with Rydberg transitions overlying the continuum. Of particular importance is the occurrence of a deep minimum in the absorption just at the wavelength of the Lyman-α solar line at 1215.6 Å. This fact is shown in Figure 5a and in more detail in Figure 6. This coincidence is of great aeronomic significance since it allows penetration of high energy radiation deep into the mesosphere. For accurate work the temperature dependence of the O_2 cross section (Carver et al, 1977) and the shape of the Lyman-α line must be treated accurately since the strength and shape of the line changes with the eleven year solar cycle.

Below 1027 Å O_2 can be ionized. The absorption is characterized by several bands superimposed on the continuum down to about 700 Å. above about 115 km, O is more abundant than O_2, and at higher altitudes

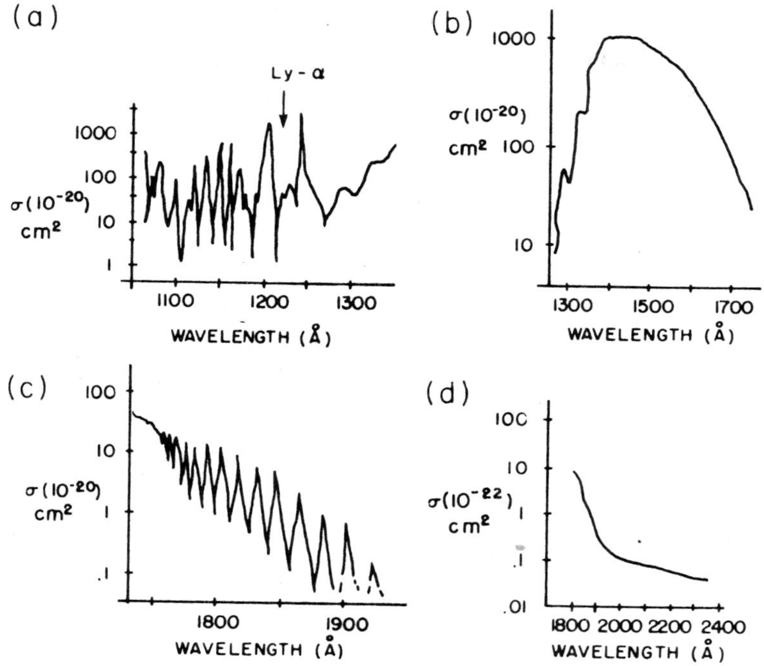

Figure 5. O_2 absorption cross section: (a) 1050-1350 Å including location of the Lyman-α line, (b) 1300-1750 Å showing Schumann-Runge continuum, (c) a low resolution schematic of the Schumann-Runge bands, (d) Herzberg continuum.

it becomes the major species. Atomic oxygen absorption is mostly important above 120 km. A simplified energy level diagram of O is presented in Figure 7. At wavelengths above the ionization threshold at 910 Å the absorption is significant only in a few allowed lines. The 1304 Å transition to the ^3S state is optically thick in the thermophere. The lowest two excited states $O(^1D)$ at 1.97 eV and $O(^1S)$ at 4.2 eV are metastable states that are not significantly produced by absorption of solar radiation by O. They are produced in notable amounts by dissociation of O_2 and O_3 and by chemical reactions. These states are the sources of the oxygen green line ($O(^1S) \rightarrow O(^1D)$ + 5577 Å) and the red line ($O(^1D) \rightarrow O(^3P)$ + 6300 Å) that are prominent airglow emissions. These metastable oxygen states are important reservoirs of energy and important sources of reactive radical species in the in the stratosphere since they react with H_2O whereas ground state atomic oxygen does not.

Ozone is one of the most important and well known minor constituents. Its maximum mixing ratio is only about 1×10^{-5} at around 30 km. However, it is a major absorber of ultraviolet and infrared radiation, and a source of $O(^1D)$ atoms. The ozone absorption spectrum is shown in

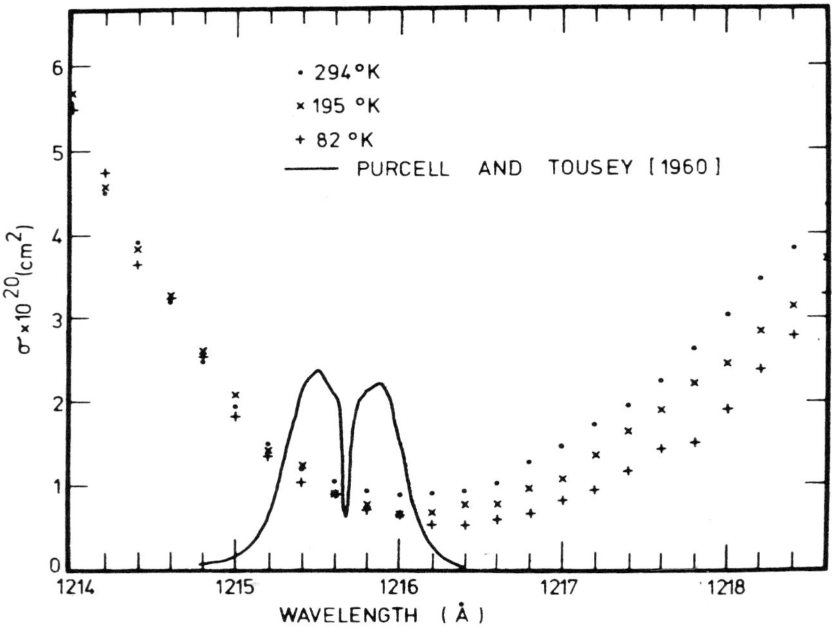

Figure 6. Detail of O_2 absorption cross section around the Lyman-α line (after Carver et al J. Geophys. Res. 82,p1955 copyright American Geophysical Union 1977). Shape of the Lyman-α line is also shown (after Purcell and Tousey, J. Geophys. Res. 65, p370, copyright American Geophysical Union 1960).

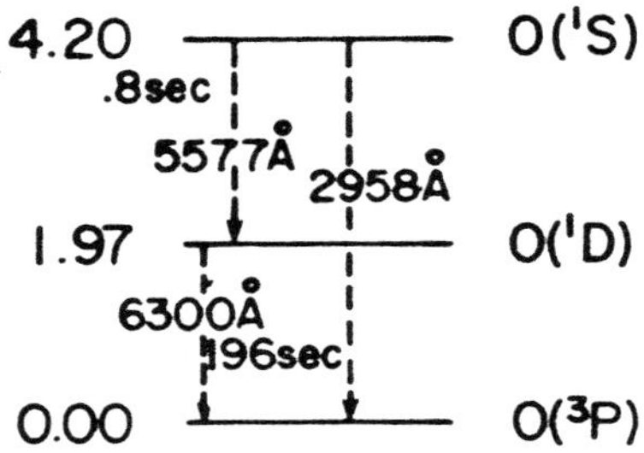

Figure 7. Low-lying metastable states of atomic oxygen. The ground state $O(^3P)$ is actually a triplet of closely spaced levels $^3P_0$, $^3P_1$, and $^3P_2$. The lowest level is indicated in the figure. Radiative lifetimes and wavelengths are indicated for various transitions.

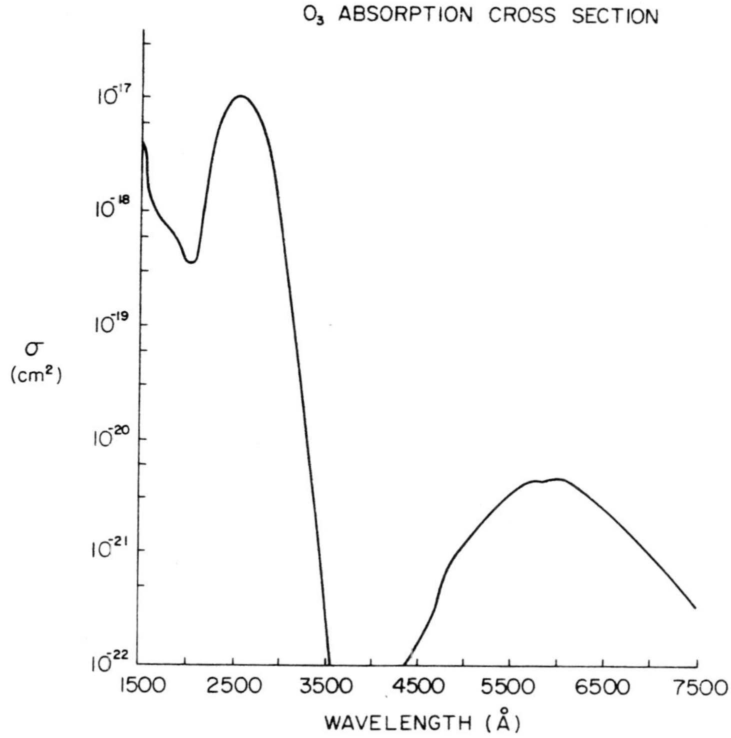

Figure 8. Ozone absorption cross section.

Figure 8. The dissociation energy of ozone into O_2 and O

$$O_3 \longrightarrow O_2 + O \tag{1.39}$$

is only 1.05 eV when the O_2 and O are in their ground states. This gives a possible photodissociation threshold at wavelengths over 1 μm. However, the absorption is very weak for wavelengths greater than about 3000 Å, although the weak secondary maximum at about 6000 Å can absorb out appreciable sunlight at twilight. The major absorption is in the Hartley bands, between 2000 Å and 3000 Å where the major dissociation channel is

$$O_3 + h\nu \longrightarrow O_2(^1\Delta_g) + O(^1D). \tag{1.40}$$

Ozone absorption in this band maximizes at $\sigma = 1\times10^{-17}$ cm^2 when $\lambda = 2550$ Å. This large absorption cross section occurs at a relatively long wavelength where there is a lot more solar photons (see Figures 8 and 10) than in the region below 1750 Å where O_2 strongly absorbs. The ozone dissociation process is a large source of energy into the atmosphere. A 2500 Å photon has about 5 eV of energy. In the process (1.40) 1.05 eV goes into O_3 dissociation, 0.98 eV goes into $O_2(^1\Delta_g)$ and 1.97 eV goes into $O(^1D)$, leaving about 1 eV going into $O_2(^1\Delta_g)$ vibration or into kinetic energy of the O_2 or O. At lower altitudes most of the energy of the metastable $O_2(^1\Delta_g)$ and the $O(^1D)$ is quenched and the energy goes into heat. The $O(^1D)$ releases some of its energy by chemically reacting with H_2O and forming hydrogen radical species that are important in mesospheric neutral chemistry.

Water vapor is a trace constituent of the atmosphere above the tropopause. It can be photodissociated by radiation with wavelengths less than 2400 Å, but the absorption above 1850 Å is very weak. The cross section is large at the wavelength of Lyman-α and below 1750 Å, and thus the dissociation rate will rapidly increase in the upper mesosphere. Below the upper mesosphere the O_2 SRB opacity determines the H_2O photodissociation rate (Frederick and Hudson 1980a).

Nitric oxide is another important minor species. The dissociation energy is 6.5 eV which corresponds to a threshold at 1908 Å. A simplified energy level diagram of NO is shown in Figure 9. The NO absorption spectrum is dominated by highly structured band systems (γ, δ, β, etc). The lowest energy dissociation occurs via predissociation in the γ, β, and γ, bands below about 1900 Å. This process corresponds to

$$NO + h\nu \longrightarrow N(^4S) + O, \tag{1.41}$$

which becomes appreciable in the upper mesosphere where radiation penetrates the O_2 SRB. Calculation of the overlap of the two highly structured cross sections is a tedious if not difficult problem (Frederick and Hudson, 1979). The ionization potential of NO is 9.25 eV which corresponds to a wavelength of 1340 Å. This is lower than any major or

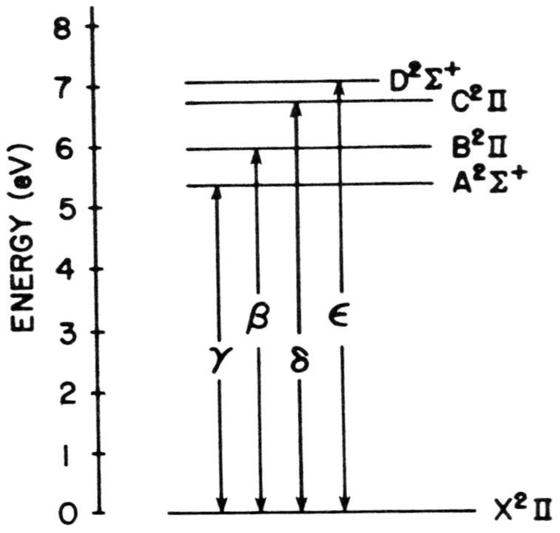

Figure 9. Lowest electronic levels of NO and corresponding allowed transitions.

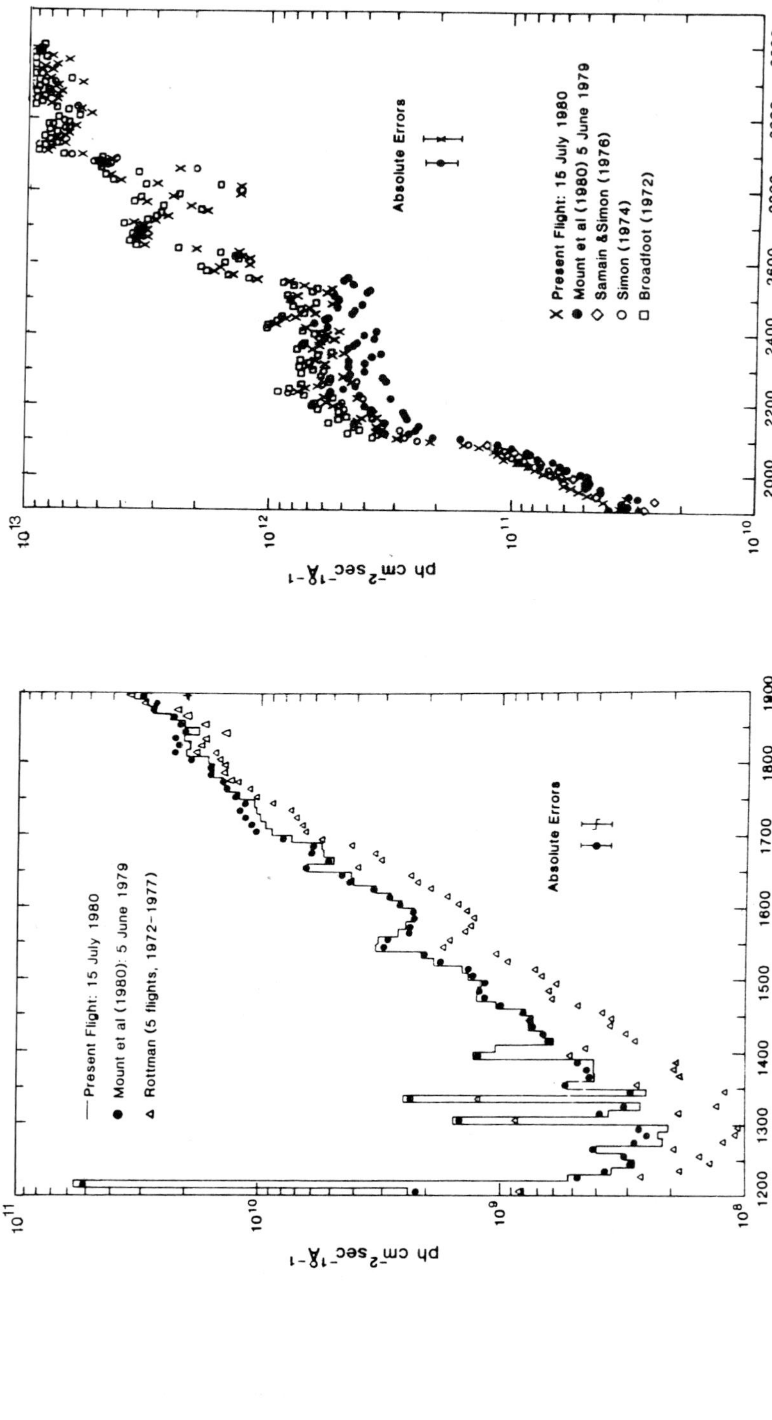

Figure 10. Solar flux. (a) 1200-1900 Å , (b) 1900-3200 Å (from G. H. Mount and G.J. Rottman, J. J. Geophys. Res. 86, p9195, copyright American Geophysical Union 1981)

comparable minor atmospheric constituent. The ionization cross section at the Lyman-α line is large enough so that NO can be ionized well into the mesosphere by solar radiation.

The solar flux below 3000 Å drives the photochemistry in the 50-120 km region. The solar flux in the 1200-3200 Å region is plotted in Figures 10a and 10b (Mount and Rottman, 1981). The flux roughly follows the emission of a blackbody at absolute temperatures between 4500 K and 5500 K with extra contributions from individual emission lines becoming increasingly important below 1500 Å. The continuum flux falls off by almost 5 orders of magnitude from 3000 Å to 1200 Å. The flux at the peak of the ozone absorption is about 3 orders of magnitude greater than that at the peak of O_2 SRC. The Lyman-α line contains about $2.5\text{-}5.0\times10^{11}$ photons $cm^{-2}\ s^{-1}$ varying with the solar cycle. This is more flux than is contained in the 1400-1500 Å region. Below 1200 Å the spectrum is dominated by emission lines. The most important for the lower thermosphere are Lyman-β at 1025.7 Å and C(III) at 977 Å.

The solar flux plotted in Figures 10a and 10b indicates a large variation in the flux below 1600 Å, most of which is believed to be due to a real difference between solar maximum and solar minimum. The scatter at longer wavelengths is indicative of the difficulty of making absolute irradiance measurements in the ultraviolet. It should be noted that these measurements have been made on brief rocket flights and are just snapshots. Studies of the effect of the 27 day solar rotation period are being done with satellites. Satellites, however, have difficulty with instrumental drifts in calibration and the instruments are not recovered for a post-experiment recalibration. The solar flux is thus probably known at best to about 20% and its short term variation is uncertain. This fact should be kept in mind when one worries about other uncertainties in the photochemistry.

Calculation of Photodissociation Rates

The photodissociation or photoionization rate coefficient J is calculated from the expression

$$J(z) = \int I_\lambda(z)\ \Phi_\lambda(z)\ \sigma_\lambda(z) d\lambda\ , \tag{1.42}$$

where $I_\lambda(z)$ is the irradiance at altitude z in the interval λ to λ + dλ, $\Phi_\lambda(z)$ is the quantum efficiency for the process (ionization,dissociation, or excitation to a specific state), $\sigma_\lambda(z)$ is the absorption cross section at λ . The absorption cross section and quantum efficiency can be altitude dependent if they are temperature or pressure dependent. The irradiance can be calculated from

$$I_\lambda(z) = I^o{}_\lambda \exp(-\tau_\lambda(z)), \tag{1.43}$$

where $I^o{}_\lambda$ is the solar flux incident at the top of the atmosphere, and $\tau_\lambda(z)$ is the optical depth at altitude z, compared to Ch(x,Θ).

The expression (1.42) for J(z) is an integral over wavelength. In the previous section we saw that the solar flux, although roughly following a blackbody curve, is structured and only known over a finite bandwidth. The absorption cross sections are even more structured and can not be fit to some simple function of wavelength. Thus the expression for J is approximated by a sum over wavelength intervals i

$$J = \sum \Phi_i \; \sigma_i I_i^{\,o} \exp(-\tau_i). \qquad (1.47)$$

As previously noted, the grid must be very fine in the structured regions such as 1750-2050 Å since average values will give incorrect penetration of radiation. Calculation of dissociation coefficients can be a major computational expense for diurnal photochemical models.

2. THE ONE-DIMENSIONAL CONTINUITY EQUATION IN THE UPPER MESOSPHERE

Below 100 km eddy diffusion dominates molecular diffusion so that equation (1.23) can be written for minor species i as

$$dn_i/dt = P_i - L_i + d/dz(K_z[M]\,df_i/dz) \qquad (2.1)$$

In the mesosphere we will consider species derived from the dissociation of O_2 and H_2O as well as NO. The major chemical reactions that involve these species are listed in Table 3. We will then have a continuity equation for all minor species that derive from H_2O, O_2 and NO: that is O, O_3, H, OH, HO_2, and NO_2 as well as H_2O and NO. We will not consider other species that are of marginal importance in the mesosphere such as HNO_3 and H_2O_2 but which become very important in the stratosphere. These continuity equations compose a set of coupled partial differential equations that are second order in the space coordinate. The equations are usually put into a finite difference form. For the space coordinate this is straightforward.

$$dn/dt=(\Delta z)^{-2}[K_{m+1/2}n_{m+1}-(K_{m+1/2} + K_{m-1/2})n_m$$
$$+K_{m-1/2}n_{m-1}) + P - L \qquad (2.2)$$

where we have dropped the subscript i for the ith species to simplify the notation. In equation (2.2) n_m is the value of the density at the mth level, Δz is the vertical grid spacing and $K_{m+1/2}$ is the geometric mean of K_m and K_{m+1}. Since the equation is second order in the spatial variable, two boundary conditions are required for solution. In aeronomic calculations upper and lower boundary values of either the density or the flux are usually specified, at a boundaries at least a scale height away from the region of interest. Error in the boundary value will usually not propagate much beyond a scale height unless the species being modelled is sensitive to transport, in which case the one-dimensional model is probably suspect anyway.

$$\tau_\lambda(z) = \sum \int n_j(s)\sigma_j(s)ds. \qquad (1.44)$$

Here the sum is over all absorbing species and s is the distance along the slant path toward the sun. As we have seen the major absorbing species for $\lambda > 1300$ Å is O_2 with O_3 important in the lower mesosphere. Below 1000 Å, O and N_2 are also important absorbers. At altitudes below 120 km molecular absorption is strong and a single scattering treatment of the radiation is adequate except in some atomic resonance lines such as O(I) line at 1304 Å.

For solar zenith angles Θ less than about 75 degrees, the atmosphere can be approximated by a plane parallel slab neglecting curvature. In this case equation (1.44) can be simplified with the use of (1.17) to

$$\tau_\lambda(z) = \sum n_j H^o_j \ \sigma_{j\ \lambda} \sec\Theta \qquad (1.45)$$

if we have assumed that σ is independent of altitude. This is a useful formula when O_2 is the dominant absorber. For $\Theta > 75^o$, Chapman's grazing incidence function must be used in place of $\sec\Theta$ to take into account the earth's curvature. If we set $x=(R+z)/H$ where R is the earth's radius, an approximation to the Chapman function is

$$Ch(x,\Theta) = (0.5\pi\ x \sin\Theta)^{0.5}\exp(0.5xy)\{1+\mathrm{erf}[xy)^{0.5}]\} \qquad (1.46)$$

where $y=\cos^2\Theta$ and erf is the error function. Figure 11 shows $\sec\Theta$

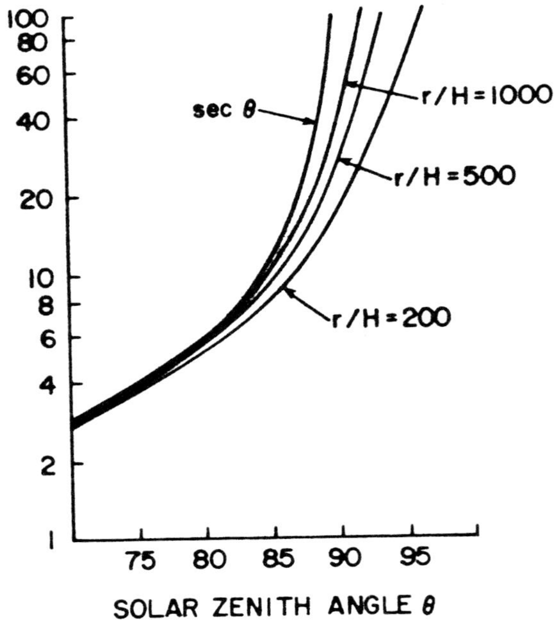

Figure 11. Approximation of the Chapman grazing incidence function $Ch(r/H,\Theta)$ compared with sec Θ.

The set of continuity equations (2.1) can be written

$$dn_i/dt = P_i - \bar{L}_i n_i - T_i \qquad (2.3)$$

where T_i is the transport term and $\bar{L}_i$ is the loss frequency which is just the loss rate divided by the density (L_i/n_i). The chemical production and loss terms couple these equations strongly and nonlinearly. The transport lifetimes are the order of 10^5 seconds, while the chemical lifetimes can be as short as 10^{-5} seconds. The large range of time constants of strongly coupled species means that these equations are numerically "stiff". This means that in order to obtain solutions at late times where long-lived species have converged, one must accurately compute at all times the evolution of already decayed states which may contribute little to the long-lived solution. For all explicit and many implicit methods this means time steps of the order of the shortest time constant must be used to insure stability. Explicit integration means that only values calculated from the last time step are used to obtain the next step. If we neglect the transport term for simplicity, the difference between implicit and explicit can be easily shown. Equation (2.3) then becomes

$$[n_i(j+1)-n_i(j)]/\Delta t = \zeta\,[P_i(j) - n_i(j)L_i(j)] + (1-\zeta)[P_i(j+1) - n_i(j+1)L_i(j+1)] \qquad (2.4)$$

where $n_i(j+1)$ is the density at the (j+1)th time step and $\zeta=1$ yields the totally explicit form and $\zeta=0$ yields the totally implicit form. The explicit form can easily be stepped forward in time, calculating P_i and L_i from the already obtained $n_i(j)$. This method is bound to become unstable when Δt is larger than the smallest time constant. The totally implicit form is much more stable, but the n(j+1) that are being sought are also present on the right-hand side of (2.4) and thus couple all of the equations. The resulting set of equations could be solved readily if it were linear. However, the production and loss terms are often bimolecular or termolecular rate expressions that involve the n_i quadratically. The equations are then not easily solved. One way to linearize the equations is use of P(j) and $n(j+1)\bar{L}(j)$ for the chemical production and loss terms in (2.3). Then the equation is semi-implicit, and its solution is straightforward, but error can result even though the solution will remain stable for larger time steps than for the explicit case. This error can arise since the resulting continuity equation is not guaranteed to conserve mass in a closed system the way (2.4) is. Another way to handle the nonlinear terms is by using an iterative method: use n(j) to calculate P(j) and L(j) and the first solution at the next time step n(j+1) which is used to recalculate P(j+1) and L(j+1) and then solve for n(j+1) again. Often the use of $(1/2)[n(j)+n(j+1)]$ or $[n(j)n(j+1)]^{1/2}$ for the next estimate of n(j+1) when recalculating P(j+1) and L(j+1) insures better stability. There are many variations (Turco and Whitten, 1974) that have particular advantages and shortcomings. Advanced implicit methods such as the Gear code (Gear, 1971) have been developed which insure stable solutions. These codes, however, require a lot of computer

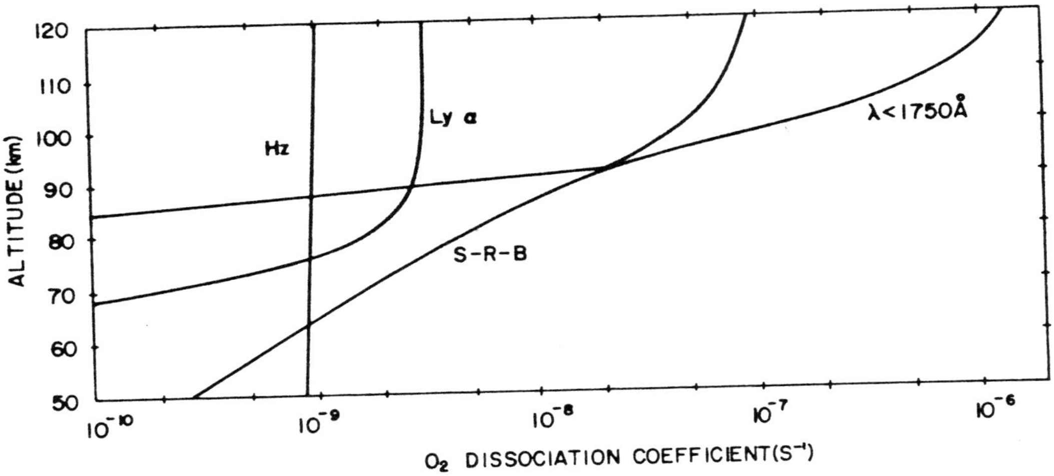

Figure 12. O_2 dissociation coefficients for an overhead sun. Contributions from four separate spectral regions are shown: (a) Hz, Herzberg continuum, S-R-, the Schumann-Runge bands, (c) Ly α the solar hydrogen Lyman-α line, (d) λ <1750 Å a region including mostly the Schumann-Runge continuum.

storage space and computer time. They are not very practical for large chemical models if one needs to make many calculations corresponding to changing parameters in the model, such as K_z or uncertain rate coefficients. In these cases faster, although approximate methods must be used. We will consider one such class next as we discuss the chemistry of the upper mesosphere.

The Chemistry of HO_x and O_x

In the mesosphere photodissociation of molecular oxygen is a source of atomic oxygen and ozone (collectively known as O_x). The dissociation coefficient of O_2 is shown in Figure 12 for an overhead sun which represents the maximum possible rate. Similarly the dissociation of water vapor leads to the formation of the hydrogen containing radicals H, OH, HO_2, and H_2O_2 (collectively known as HO_x), with H_2O_2 being of limited importance in the mesosphere. The O_x and HO_x groups are made up of species that are strongly coupled to other members of the group, and relatively weakly coupled to species outside their own group or family. The family technique is an approximation that takes advantage of this coupling in order to facillitate the numerical solution for the species densities. We will consider the O_x chemistry in more detail in order to illustrate the family technique.

The continuity equations for O_3 and O are

$$d[O_3]/dt = k_2[O][O_2][M] - k_4[O][O_3] - k_9[H][O_3] - k_8[OH][O_3]$$
$$-(J_{3D} + J_{3P})[O_3] + T_{O3} \qquad (2.5)$$

$$d[O]/dt = 2J_1[O_2] + J_{3P}[O_3] + k_6[O(^1D)][M] - k_2[O][O_2][M]$$
$$-k_4[O][O_3] - k_5[O][O][M] - k_7[O][OH]$$
$$-k_{11}[O][HO_2] + T_O. \qquad (2.6)$$

where the subscripts on rate coefficients correspond to the reactions listed in Table 3, and the $O(^1D)$ continuity equation is not used since $O(^1D)$ is always quickly quenched to the ^3P state with a time constant

$$\tau(O^1D) = (k_6[M])^{-1}$$

which below 90 km is less than 10^{-3} seconds. The $O(^1D)$ concentration quickly adjusts to the ozone concentration and its concentration can be calculated assuming photochemical equilibrium (PCE)

$$d[O(^1D)]/dt = 0 = J_{3D}[O_3] - k_6[O(^1D)][M]$$

$$[O(^1D)] = J_{3D}[O_3]/(k_6[M]). \qquad (2.7)$$

Equations (2.5) and (2.6) both have transport terms T_i that couple different levels of the atmosphere with time constants of the order of 10^5 seconds. However, (2.5) and (2.6) are strongly coupled by reactions k_2, k_6, J_{3P} and J_{3D}. This shown schematically in Figure 13. The ozone

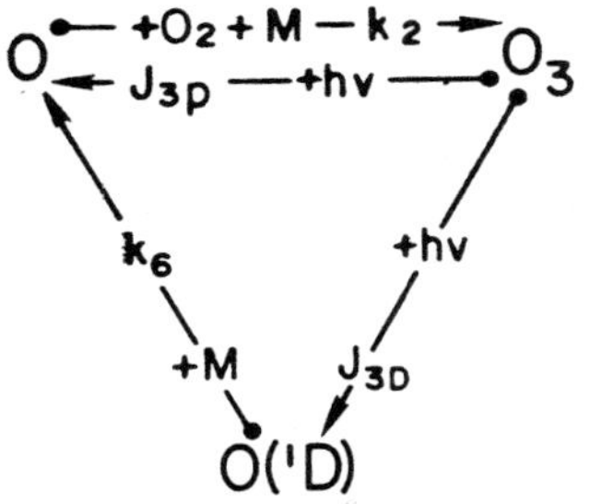

Figure 13. O_x coupling scheme in the mesosphere.

lifetime versus dissociation via J_{3P} and J_{3D} is about 100 seconds for most of the daylight atmosphere above 50 km. The atomic oxygen lifetime against recombining to form ozone is

$$\tau(O_{k2}) = (k_2[O_2][M])^{-1} \qquad (2.8)$$

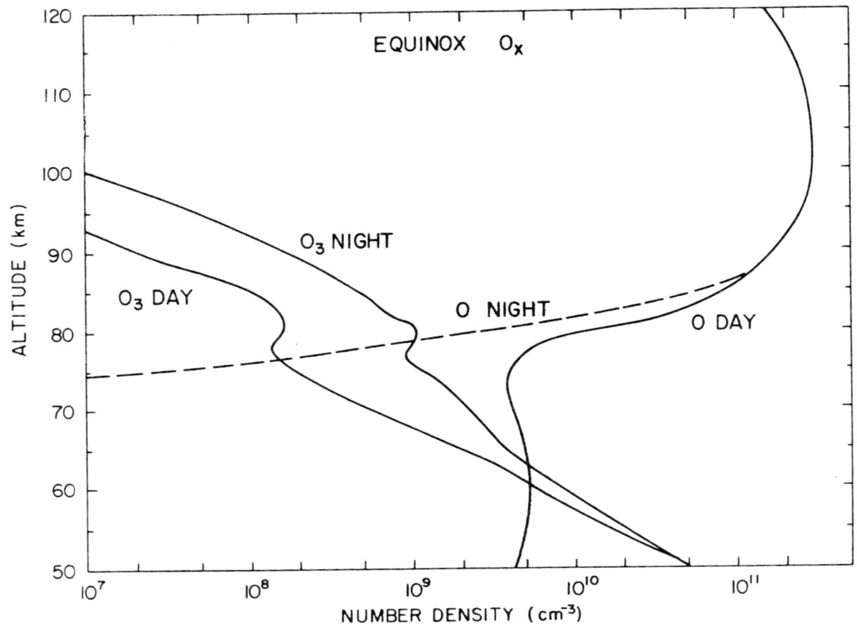

Figure 14. O_x profiles obtained from a one-dimensional diurnal model.

which varies from about $6x10^5$ seconds at 90 km to 1.7 seconds at 50 km. Thus, especially in the lower and middle mesosphere, the interconversion of O_3 and O via recombination and dissociation of ozone has a small time constant. We can obtain a continuity equation for the family O_x, by adding equations (2.5) and (2.6), in which the rapid interconversion terms cancel out, yielding (neglecting reaction 5)

$$\begin{aligned} d[O_x]/dt &= 2J_1[O_2] - 2k_4[O][O_3] - k_7[O][OH] - k_{11}[O][HO_2] \\ &\quad - k_9[H][O_3] - k_8[OH][O_3] \\ &= 2J_1[O_2] - 2k_4[O/O_x][O_3/O_x][O_x]^2 - \{k_7[OH][O/O_x] \\ &\quad + k_{11}[HO_2][O/O_x] + k_9[H][O_3/O_x] \\ &\quad + k_8[OH][O_3/O_x]\}[O_x]. \end{aligned} \tag{2.9}$$

We have neglected the slow transport terms here and have set $[O/O_x] \equiv [O]/[O_x]$. We now have an equation that has a longer chemical lifetime. The explicit difference equation corresponding to (2.9) may still tend to be unstable, but larger time steps can now be tolerated. Implicit solutions are complicated by the quadratic terms in O_x, which fortunately do not dominate. However, we must have expressions for the fractional terms $[O/O_x]$, and $[O_3/O_x]$ since we are no longer solving for [O] and $[O_3]$. These can be obtained by assuming O_3 interconversion with O is in a steady state

$$d[O_3]/dt = 0 = -(J_{3D}+J_{3P})[O_3] - k_9[H][O_3] + k_2[O][O_2][M], \quad (2.10)$$

then solving for

$$R_1 = [O_3/O] = k_2[O_2][M]/(J_{3D}+J_{3P} + k_9[H]), \quad (2.11)$$

$$[O/O_x] = [O]/([O] + [O_3]) = 1/(1 + R_1). \quad (2.12)$$

There is definitely error involved in the assumption (2.10), but during daylight R_1 is fairly accurate and the ratios do not change very much from one time step to another. At twilight J_{3D} and J_{3P} change rapidly as can R_1, but $[O_x]$ need not change rapidly, which is indeed the case.

Figure 14 shows the results of a one-dimensional diurnal calculation for O and O_3, which includes the reaction set of Table 3. The important points are the dominance of ozone at lower altitudes, and the abundance of atomic oxygen above 70 km during the day and above 80 km at night. The diurnal variation of O and O_3 is most noticeable in the 70-85 km region. At sunset O_3 is no longer photodissociated and atomic oxygen recombines to form O_3 with a lifetime τ (O,k_2) shown in equation (2.8). Ozone builds up after sunset and its lifetime is then determined by atomic hydrogen. The chemistry of odd oxygen (O_x) is strongly coupled with the odd hydrogen (HO_x) through the loss terms. Figure 15 shows typical ozone profiles obtained with only oxygen chemistry (Chapman scheme includes only reactions 1-5) and with inclusion of HO_x and NO_x which has little effect. The inclusion of the HO_x species is necessary to reduce the ozone amount to reasonable agreement with the measured profile. Thus the hydrogen radicals are important trace species which need to be simultaneously calculated.

Although the HO_x radicals are found in only small concentrations (see Figure 18), they are effective in destroying O_x via catalytic mechanisms such as

$$H + O_3 \longrightarrow OH + O_2 \qquad k_9 \qquad (2.13)$$

$$OH + O \longrightarrow O_2 + H \qquad k_7 \qquad (2.14)$$

$$O + O_3 \longrightarrow O_2 + O_2$$

at a rate that can be faster than the equivalent reaction with rate constant k_4. At 70km $k_9=1.65\times10^{-11}$ and $k_4=5.37\times10^{-16}$ and the catalytic cycle procedes much faster for concentrations of H and OH of about 1×10^7 cm^{-3}. Since the reaction scheme (2.13)-(2.14) is catalytic, HO_x is recycled and can recombine many O_x before it is lost by a slower process. Another catalytic cycle involving HO_x is;

$$O + OH \longrightarrow O_2 + H \qquad k_7$$

$$H + O_2 + M \longrightarrow HO_2 + M \qquad k_{10} \qquad (2.15)$$

$$HO_2 + O \longrightarrow OH + O_2 \qquad k_{11} \qquad (2.16)$$

$$O + O\ (+M) \longrightarrow O_2\ (+M)$$

which can recombine oxygen atoms at a faster rate than the single reaction with rate constant k_5. At 70 km k_5=8x10^{-33}, k_{10}=7.9x10^{-32} and for oxygen atom densities of about 1x10^{10}cm^{-3} the catalytic pathway proceeds, limited by reaction 11, at about 4x10^5 molecules cm^{-3} s^{-1} while reaction 5 removes only about 3x10^3 O atoms cm^{-3} per second.

The odd hydrogen species H, OH and HO_2 in the mesosphere are closely coupled to each other as seen in Figure 16. The source of HO_x in the mesosphere is dissociation of water vapor which is shown for an overhead sun in Figure 17. The family method is applied in a manner similar to the O_x case. If we add the individual continuity equations for the three species, assume

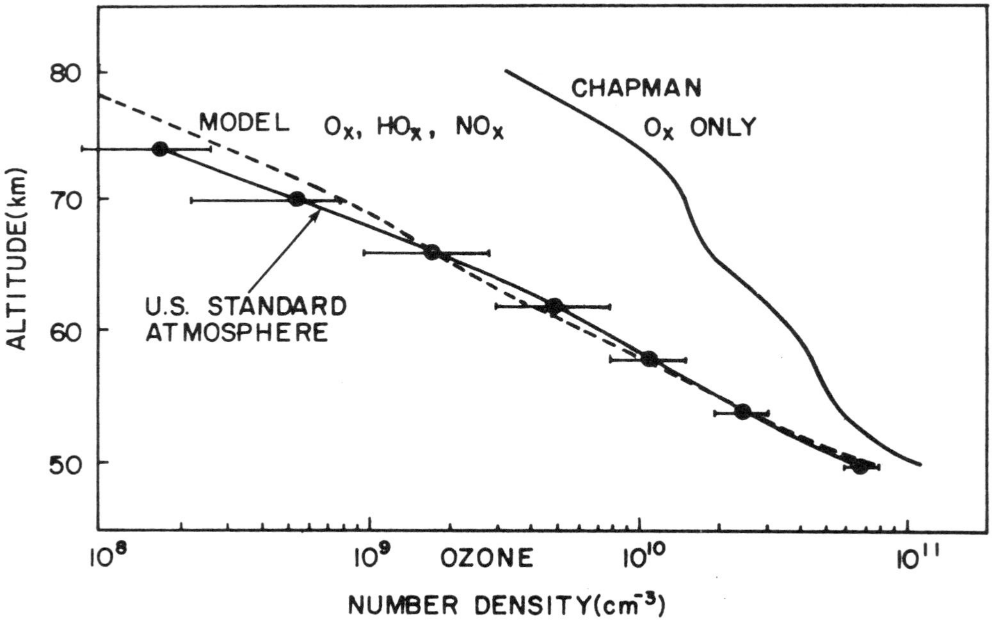

Figure 15. Comparison of ozone profiles obtained with one-dimensional model calculations and the mid-latitude ozone model from the U.S. Standard Atmosphere 1976. The Chapman scheme includes only O_x reactions.

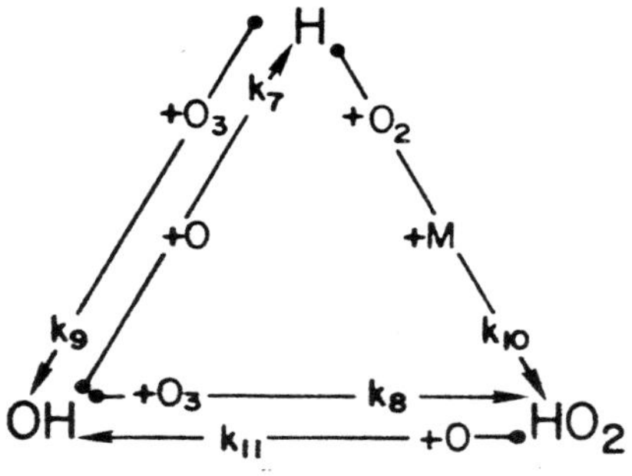

Figure 16. HO_x coupling scheme in the mesosphere.

$$d[H]/dt = 0 = d[HO_2]/dt, \quad (2.17)$$

and consider only the fast shuffling (interconversion) reactions, we can obtain expressions for the ratios

$$RH1 = [H]/[OH] = k_7[O]/(k_9[O_3] + k_{10}[O_2][M]), \quad (2.18)$$

$$RH2 = [HO_2]/[OH] = (RH1(k_{10}[O_2][M] + k_8[O_3])/Y, \quad (2.19)$$

$$Y = (k_{11}[O] + k_{26}[NO]).$$

Then the expressions for the ratios of HO_x species are:

$$[OH]/[HO_x] = 1/(1 + RH1 + RH2), \quad (2.20)$$

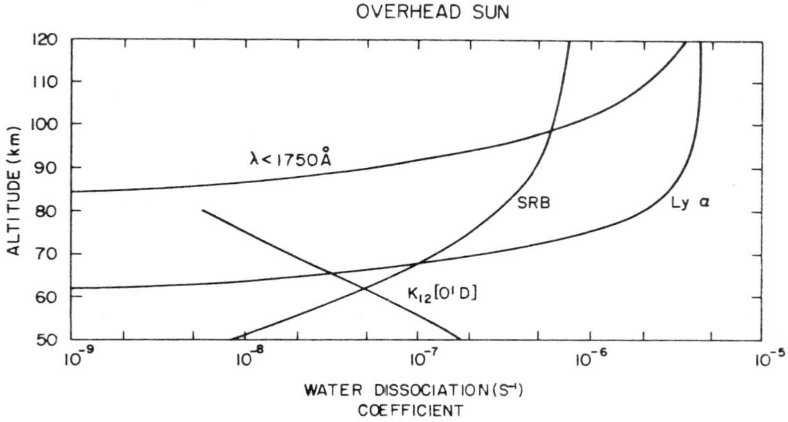

Figure 17. Water dissociation coefficients for an overhead sun. Contributions from three spectral regions and chemical dissociation by $O(^1D)$. Labels have the same meaning as in Figure 12.

$$[H]/[HO_x] = [OH]/[HO_x]\ RH1, \tag{2.21}$$

$$[HO_2]/[HO_x] = [OH]/[HO_x]\ RH2. \tag{2.22}$$

And the continuity equation for HO_x (neglecting transport) can be written as

$$\begin{aligned} d[HO_x]/dt = {} & 2\ J_{27}[H_2O] + 2\ k_{12}[O(^1D)][H_2O] \\ & - 2\left\{(k_{14a}+k_{14b})[H/HO_x][HO_2/HO_x] \right. \\ & \left. + k_{15}[OH/HO_x][HO_2/HO_x]\right\}[HO_x]^2 \end{aligned} \tag{2.23}$$

This equation is complicated by the fact that the loss term is a quadratic form. However, the loss terms are much smaller than the coupling terms between the HO_x species. The approximations involved in (2.17) that are used to obtain (2.18)-(2.23) in the family technique must be examined for their validity. Table 4 illustrates the lifetimes of the individual species for altitudes of 70, 80, and 90 km. The coupling of

Table 4. Lifetimes for the Coupled O_x - HO_x Chemistry in the Mesosphere

Species	Day	Night (below 80 km)
H	$(k_{10}[O_2][M]+k_9[O_3])^{-1}$	$(k_{10}[O_2][M])^{-1}$
HO_2	$(k_{11}[O])^{-1}$	$(k_{15}[OH]+k_{14a}[H])^{-1}$
OH	$(k_7[O])^{-1}$	$(k_{15}[HO_2])^{-1}$
O	$(k_2[O_2][M])^{-1}$	same as day
O_3	$(J_{3P}+J_{3D})^{-1}$	$(k_8[OH]+k_{24}[HO_2]+k_9[H])^{-1}$
HO_x	$\dfrac{0.5([H]+[OH]+[HO_2])}{(k_{14ab}[H][HO_2]+k_{15}[OH][HO_2]}$	same as day
O_x	$([O]+[O_3])/([O](k_7[OH]+k_{11}[HO_2])$	$0.5([O]+[O_3])/(k_4[O][O_3])$

Numerical estimates of lifetimes in seconds

Alt. z(km)	τ(H)	τ(HO_2)	τ(OH)	τ(O)	τ(O_3)	τ(HO_x)	τ(O_x)
			Day				
70	50	3	3	10^3	100	5×10^3	3×10^3
80	600	1	1	10^4	100	10^5	5×10^3
90	10^6	1	1	10^6	100	10^6	5×10^5
			Night				
70	50	10^3	10^3-10^4	10^3	very long	same	10^6
80	600	10^3-10^4	10^3-10^4	10^4	very long	as	10^6
90	10^3	1	1	10^6	10^3	day	3×10^5

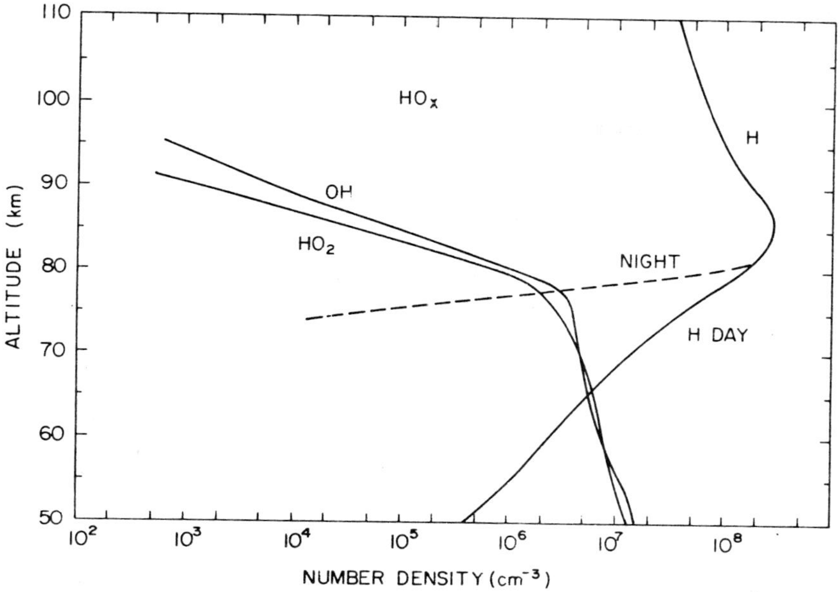

Figure 18. HO_x profiles from a one-dimensional diurnal model.

the HO_x species is dominated by the reactions of atomic oxygen. The lifetime of these reactions thus varies with altitude, but most significantly it is diurnally dependent through the O density. As O disappears rapidly at night below 75 km, the individual HO_x species become decoupled by the fact that the shuffling reactions are no longer much faster than the other loss reactions that are loss processes for HO_x as a whole. Thus when (2.17)-(2.22) are no longer valid, we cannot solve (2.23) for $[HO_x]$. Each species continuity equation must then be solved separately when atomic oxygen becomes less than about $1 \times 10^7 cm^{-3}$. However, this is not too cumbersome since the lifetimes of the individual species are then much longer than when [O] is large. Care must be taken to insure a smooth continuous transition from one scheme to another, since the O density falls below the cutoff value at different times for different altitudes. If the transition from family to individual solution is not smooth an error can be introduced which can be propagated by the transport term. However, twilight is usually a small fraction of the day, and since small time steps are necessary at twilight in order to follow the rapidly changing illumination, this possible error is controllable if a stable routine is used. However, slight discontinuities may appear at following the switching from one solution technique to another.

Profiles of HO_x are shown in Figure 18, obtained from the diurnal solution of (2.23) coupled with O_x and NO_x chemistry. The HO_x concentration is dependent upon two uncertain parameters: the H_2O concentration (the major mesospheric source) and the rate of reaction 15 which is an important loss of HO_x below 70 km. H disappears below about 80 km at night much like O does. In fact it is the decrease of O at night which eliminates formation of H by reaction 7. Hydrogen then combines with O_2 in reaction 10 to form HO_2. Reactions with O_3 continue to convert OH and HO_2 but they are much slower than the daytime reactions with atomic oxygen.

The water vapor profile in the stratosphere and mesosphere is not well known. Water is produced in the stratosphere from the oxidation of methane. In models of the mesosphere the mixing ratio at the stratopause (lower boundary) must be specified. The water loss rate, which is dominated by photodissociation, is slower than transport up to at least 75 km (see Figure 17). The model calculation shown in Figure 18 uses 5 ppm H_2O. Measurements have shown a large variability with some measurements as high as 12 ppm in the 60-65 km region. Photochemical models cannot yield more than about 5-6 ppm if there is 3-4 ppm H_2O in the lower stratosphere, and 1.6 ppm CH_4. Water can be transport-dominated below 70 km or at higher altitudes when solar radiation is very low, such as the polar night. Transport will tend to lead to smoothing of the mixingratio distribution, and not produce regions of enhanced mixing-ratio.

In the odd-hydrogen shuffling reactions there appears the reaction

$$NO + HO_2 \longrightarrow NO_2 + OH. \quad k_{26} \qquad (2.24)$$

This reaction is important in the stratosphere where NO, NO_2, NO_3, N_2O_5 and HNO_3 make up a group of odd nitrogen species. However, in the mesosphere NO dominates because there is much less O_3 to react with NO and NO_2 producing NO_2 and NO_3 respectively, which then proceed via three-body reactions to produce HNO_3 and N_2O_5. The major NO_x species in the mesosphere are NO and NO_2 which are coupled via the pathways illustrated

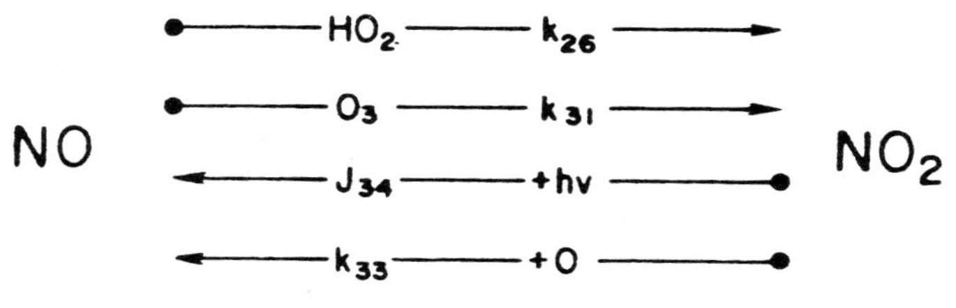

Figure 19. NO_x coupling scheme in the mesosphere.

in Figure 19. NO_2 can increase at night below 70 km since photodissociation ceases and atomic oxygen decreases so that the reaction

$$NO_2 + O \longrightarrow NO + O_2 \qquad k_{33} \qquad (2.25)$$

no longer destroys NO_2 rapidly. However,

$$NO + O_3 \rightarrow NO_2 + O_2 \qquad k_{31} \qquad (2.26)$$

is slow due to a fairly large activation energy in k_{31}. At 70 km $k_{31}=3.1\times10^{-15}$ and the NO lifetime is about 3.2×10^5 seconds versus this reaction if the ozone density is $1\times10^9 cm^{-3}$. For k_{26} equal to 1×10^{-11} and HO_2 1×10^7, NO is converted to NO_2 with a lifetime of about 10^4 seconds.

The NO_x profile is thus dominated by NO. However, there is no significant source of NO in the undisturbed mesosphere. The stratospheric source is from reaction of $O(^1D)$ with N_2O of biospheric origin (reaction 30a), and the N_2O has been depleted before diffusing to mesospheric altitudes. NO is produced when N_2 is dissociated, usually through ionic processes. We will consider this subject in the next section. However, it should be noted that if one puts in an NO_x lower boundary density at the stratopause and solves the continuity equation, the NO density decreases rapidly with increasing altitude above 60 km. The NO is being destroyed by photodissociation and reaction

$$NO + h\nu \rightarrow N(^4S) + O \qquad J_{34} \qquad (2.27)$$

$$NO + N(^4S) \rightarrow N_2 + O \qquad k_{35}, \qquad (2.28)$$

but only reformed by the slow reaction

$$N(^4S) + O_2 \rightarrow NO + O \qquad k_{36} \qquad (2.29)$$

because k_{36} is very slow at mesospheric temperatures due to the large activation energy. Thus photodissociation by reaction 34 also leads to $N(^4S)$ which can further "cannibalize" NO_x in reaction 35, a rate that is dependent upon the NO density. Figure 20 shows that a one-dimensional model NO density can be obtained in the 60-90 km region that roughly agrees with the scatter of measurements if a downward upper boundary flux at 120 km of $1\text{-}3\times10^8 cm^{-2}\ s^{-1}$ is imposed. This indicates that NO is transport dominated in the region below about 100 km where its production is very small. Large variations in transport can lead to large variablility in the density below about 100 km.

Production of Nitric Oxide

Production of nitric oxide proceeds by reactions 36 and 37 after nitrogen atoms are produced from dissociation or ionization of N_2,

$$\begin{aligned} N_2 + h\nu &\rightarrow N_2^+ + e \\ N_2 + h\nu &\rightarrow N + N^+ + 2e \\ e^* + N_2 &\rightarrow N_2^+ + 2e \\ e^* + N_2 &\rightarrow N + N^+ + 2e \\ e^* + N_2 &\rightarrow N + N + e, \end{aligned} \qquad (2.30)$$

where e* is an energetic photoelectron and N can be in either the ground state $\{(^4S)\}$ or excited states $\{N(^2D)$ or $N(^2P)\}$. The direct photodissociation of N_2 is only a weak source of N in the lower thermosphere

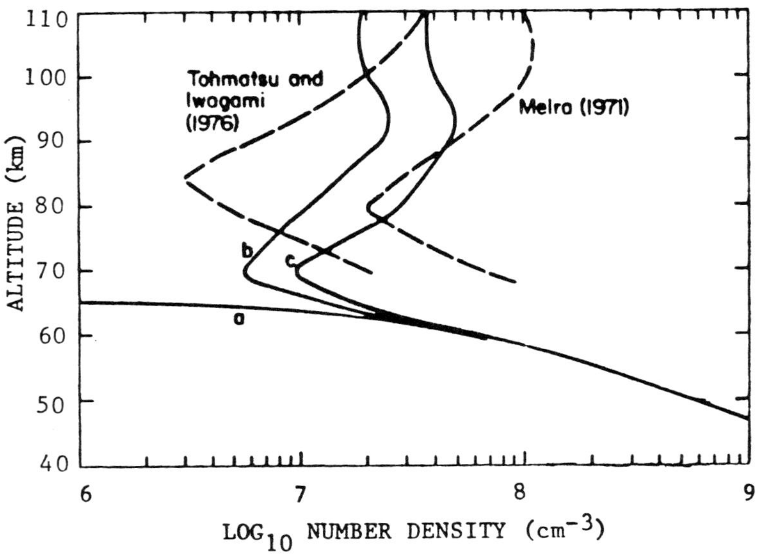

Figure 20. NO profiles obtained from one-dimensional model calculations for three different upper boundary conditions: (a) zero flux, (b) downward flux of 1×10^8 molecules cm^{-2} s^{-1}, and (c) downward flux of 3×10^8 molecules cm^{-2} s^{-1}. Two experimental observations are also shown. (Taken from S. Solomon, NCAR Cooperative Thesis NO. 62, 1981, with permission)

as was mentioned earlier. In the lower thermosphere photoionization of N_2 is mostly by X-rays and the resulting energetic photoelectrons which actually produce more ion pairs than the primary photon. The branching ratio for the atomic nitrogen states is a very important quantity due to the large difference between k_{36} and k_{37}. Ionization that produces N_2^+ leads to N dissociation by reaction 39 which produces NO^+ and N, the latter mostly in the $N(^2D)$ state. The NO-NO^+ reaction scheme is illustrated in Figure 21. NO^+ is a major precursor of NO since it is a terminal ion that is chemically destroyed by dissociative recombination (reaction 43). The $N(^2D)$ will react with O_2 to form NO, while $N(^4S)$ will react much more slowly, and thus be more likely to react with NO via reaction 35. Thus reaction 35 limits the amount of NO that can be formed even at the very high N_2 ionization and dissociation rates that occur in aurora. Since photodissociation of NO is a major source of $N(^4S)$ that destroys NO, NO can build up to higher levels in the polar night in the auroral zone. High NO densities are indeed measured in the auroral zone. The transport of NO down to the mesosphere (from the major auroral source region at 100 km and above) and equatorward is probably a major cause of the variations in the NO profile at mid-latitudes.

The NO density in the mesosphere is thus an uncertain quantity. Simple photochemical models are lacking not only because of the uncertain-

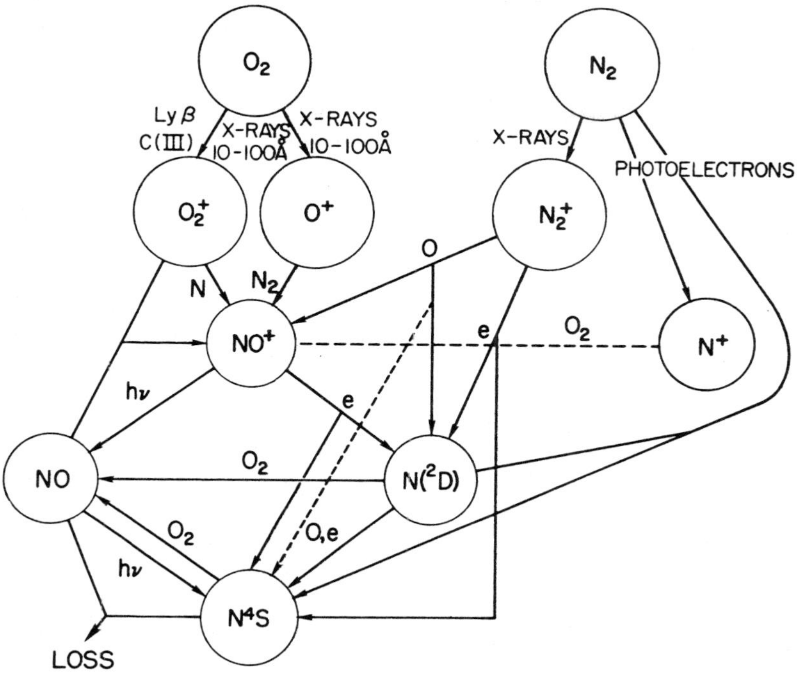

Figure 21. Chemical pathways for the formation of NO^+ and NO.

ty in the chemical source, but because of the difficulties in modelling three-dimensional transport. Measurements of NO below about 95 km (see Figure 20) are very difficult to make. NO densities determined by rocket measurements of the NO γ-band resonance scattering are subject to error because the NO density is decreasing as the altitude is decreased while the background atmospheric Rayleigh scattering is increasing, leading to small signal-to-noise. This error is magnified when one attempts to extract the density from the measured radiance. Modellers need a comprehensive determination of the branching ratios of N atom states to include in two-dimensional (Solomon et al, 1982) or three-dimensional models in order to simulate the nitric oxide distribution. Better experimental determination of the NO density is also needed before these simulations can be unambiguously verified.

Atomic Oxygen Profile

Atomic oxygen above 85 km is another constituent that is largely influenced by transport. At 90 km the chemical lifetime is the order of 10^6 seconds, nearly equal to or longer than the transport lifetime. The O profile is determined in the 80-100 km region by the large production rate in the lower thermosphere and downward transport. The chemical lifetime rapidly decreases as O is mixed downward, leading to a ledge in the profile around 80 km (Figure 14) where the three-body recombinatio reactions become sinks for O. In one-dimensional models changing K_z will change the O profile in the upper mesosphere and lower thermosphere. For smaller K_z there is slower downward mixing, the ledge will move upwards a little, and the O density will become larger in the 95-100 km region. Since K_z is quite uncertain, and only a crude parametrization of transport, the one-dimensional model is at best a semi-quantitative representation of the true atmosphere.

Measurements of atomic oxygen above 75-80 km do show great variability as indicated in Figure 22. However, it is difficult to determine whether the variation is real or is a systematic experimental error. Most recent determinations of O have been by in situ rocket measurements using a 1304 Å resonance lamp to scatter off the ambient atomic oxygen (Dickinson et al, 1980, Howlett et al, 1980). The absolute calibration of resonance lamps is a difficult procedure and may cause systematic errors between different experimenters and even between different flights by the same research group. However, the shape of the profile obtained should be correct, and it does show the ledge around 80 km, with a peak around 90-95 km and wavelike structure below 100 km (Figure 22). The O density in the 90-100 km region does seem to vary over too large a range, from 1×10^{11} cm^{-3} to $2\times10^{12} cm^{-3}$. The highest values cannot be obtained from models with "reasonable" transport rates. Values in the $1-3\times10^{11}$ range seem to be most consistent with airglow observations. High-latitude measurements, especially in winter tend to show larger peak atomic oxygen

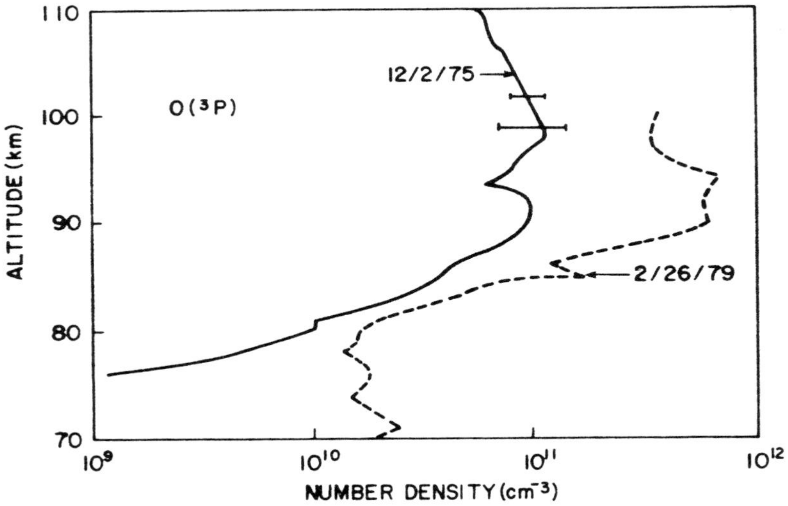

Figure 22. The atomic oxygen profiles measured by resonance fluorescence techniques by th Utah State University group. Profile dated 12/2/75 is a nighttime measurement at 33° N (Howlett et al, 1980). The profile dated 2/26/79 was obtained at 51° N during the early portion of a solar eclipse (Pendleton et al, 1982).

densities, often approaching $1x10^{12}$ cm^{-3} in the 90-95 km region. Most of the difference between the two profiles in Figure 22 is probably due to the difference in latitude and time of year.

4. ION CHEMISTRY

It was shown earlier that the ionization of N_2 is the major pathway in the production of NO and NO^+. In the remainder of this chapter we will give an overview of the ion chemistry in the 50-120 km region. First, we will give a brief discussion of the photochemistry of the lower E-region. The middle and upper ionosphere will be discussed in more detail elsewhere in this volume, so we will only outline the major photochemical processes that are characteristic of the region below 120 km. At altitudes below 90 km, three-body reactions become more important and trace species other than NO become important. The qualitative changes that characterize the D-region will be emphasized. The success of complex photochemical schemes in explaining the D-region compostition will be presented.

The Lower E-Region

The source of ionization in the lower E-region is solar EUV radiation at wavelengths below the O_2 ionization threshold at 1027 Å. The solar flux is dominated by chromospheric and coronal radiations, the Lyman series of hydrogen and its continuum as well as lines of other elements. O_2 is the major species with the lowest ionization potential, and it is the major absorber between 1027 Å and 910 Å which is the O ionization threshold. The two strong line sources in this region are hydrogen Lyman-β at 1025 Å with a flux of about $5x10^9$ photons cm^{-2} s^{-1}, and C(III) at 977 Å with a flux of about $4x10^9$ photons cm^{-2} s^{-1}. The flux in these lines probably varies by at least a factor of ±50% over the eleven year solar cycle. The total absorption and ionization cross sections of O_2 at Lyman-β are $1.55x10^{-18}$ cm^2 and $1.0x10^{-18} cm^2$, while at C(III) they are $4.0x10^{-18} cm^2$ and $2.5x10^{-18} cm^2$. These two line are the major reasons that O_2^+ is the dominant primary ion produced by solar EUV radiation.

Soft X-rays of 10-100 Å are absorbed by N_2 and O_2. The ionization cross sections vary from a few times 10^{-19} cm^2 around 10 Å to a few times 10^{-18} cm^2 at 100 Å, and thus the absorption occurs mostly below 120 km. These photons produce photoelectrons with energies greater than 100 eV. Energy degradation calculations show that about one ion pair is produced for each 35 eV of electron energy, and thus the photoelectrons produce more ionization than the primary X-ray itself. The soft X-rays produce O_2^+, N_2^+, N^+, and O^+. The solar X-ray spectrum is very weak so that this source is not very large for undisturbed solar conditions. Solar flares are accompanied by increases in the X-ray flux by orders of magnitude and produce sudden ionospheric disturbances in the sunlit hemisphere as ionization rates are suddenly increased.

The daytime ion production rate in the E-region increases rapidly from about 10 cm^{-3} s^{-1} at 90 km to close to $5x10^3$ cm^{-3} s^{-1} at 120 km for solar zenith angles near zero degrees. A simple-steady state model will yield the fact that NO^+ and O_2^+ are the dominant ions although O_2^+ and N_2^+ are the primary ion species. Charge transfer and ion-atom interchange reactions lead to the positive charge residing in the species with the lowest ionization potential, which is NO.

If we assume that the positive charge resides on molecular ions that are destroyed by dissociative recombination with a recombination coefficient of α=3.0x10^{-7} $cm^3 molec^{-1} s^{-1}$ then, setting the ionization rate equal to the loss rate

$$P = \alpha n_e^2, \tag{3.1}$$

the electron (and thus the molecular ion) density calculated is about $1x10^5$ cm^{-3} at 120 km and $6x10^3$ cm^{-3} at 90 km, which is in reasonable agreement with observations for daytime conditions. The lifetime of the molecular ions is then $(\alpha n_e)^{-1}$ which depends upon the electron density. Using the same α and n_e, the lifetime is 25 seconds at 120 km and 600 seconds at 90 km. These times are much shorter than transport times and thus for molecular ions the neglect of transport is justified. At night the major sources of ionization disappear and the lifetime increases as the ion density decreases. However, the E-region does not decay away. The electron density falls to about $5x10^2$ cm^{-3} at 90 km and $2x10^3$ cm^{-3} at 120 km due mostly to Lyman-β radiation transported into the dark hemisphere by resonance scattering in the earth's hydrogen geocorona. Thus the ion lifetime at 90 km may approach two hours, which is still shorter than most transport processes.

When calculating the ion densities and lifetimes we have assumed that all of the ions are molecular ions. N_2^+ ions are one of the major primary ions, but they react rapidly (see Figure 23)

$$N_2^+ + O_2 \longrightarrow N_2 + O_2^+ \qquad k_{44}=4.3x10^{-11} \tag{3.2}$$

$$N_2^+ + O \longrightarrow NO^+ + N \qquad k_{38}=1.4x10^{-10}. \tag{3.3}$$

If we use O from Figure 14 and O_2 equal to $1.25x10^{13}$ cm^{-3} at 90 km and $4x10^{10}$ cm^{-3} at 120 km, the N_2^+ lifetime at 90 km is less than 2 ms against forming O_2^+ and less than 40 ms versus forming NO^+, both much less than the recombination time. At 120 km the corresponding lifetimes against production of O_2^+ is less than 600 ms and against NO^+ is less than 70ms. Thus the N_2^{+2} density will not be significant, with O_2^+ and NO^+ being the favored ions. O_2^+ can further react

$$O_2^+ + NO \longrightarrow NO^+ + O_2 \qquad k_{45}=4.4x10^{-10} \tag{3.4}$$

This reaction only slowly converts O_2^+ to NO^+,the lifetime of O_2^+ against reaction 45 is of the order of hundreds of seconds for NO conctrations of about $1x10^7$ cm^{-3}. This is about the same lifetime as O_2^+

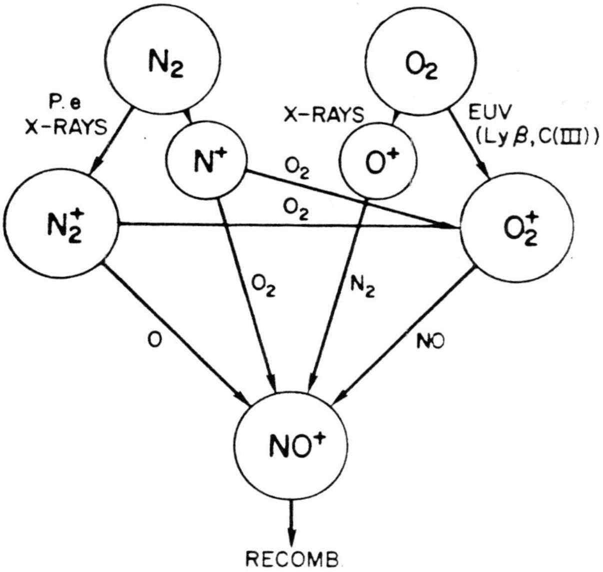

Figure 23. Schematic of E-region chemical pathways.

recombination in the 90 km region. The ${O_2}^+$ to NO^+ ratio will thus be sensitive to the NO concentration and will also depend upon the electron density.

The ionization sources for the lower E-region also produces atomic ions O^+ and N^+. Atomic ion can recombine only by very slow radiative processes with rate coefficients less than 1×10^{-11} cm^3 s^{-1}. Thus they have recombination lifetimes much longer than transport times and their profile will be determined by transport unless other loss processes are faster. In the E-region O^+ and N^+ are produced by X-rays, but they react quickly to form molecular ions

$$\begin{aligned}
&O^+ + N_2 \longrightarrow NO^+ + O && k_{39} = 1.0\times10^{-12} \\
&O^+ + O_2 \longrightarrow {O_2}^+ + O && k = 2.0\times10^{-11} \\
&N^+ + O_2 \longrightarrow NO^+ + O && k_{41a} = 3.0\times10^{-10} \\
&N^+ + O_2 \longrightarrow O_2{+} + N && k_{41b} = 3.0\times10^{-10}.
\end{aligned} \tag{3.5}$$

At 120 km where the lifetime is much longer than at 90 because the O_2 and N_2 densities are much less, the lifetime of O^+ is about 2.5 seconds and the lifetime of N^+ is about 0.04 seconds. Thus in the lower E-region the primary atomic ions are quickly converted to molecular ions before transport processes can effect their profile.

Atomic ions that do not undergo rapid ion-molecule reactions to form molecular ions will have long lifetimes and be subject to the influence of transport processes. The metallic ions that are derived from meteor ablation have very low ionization potentials and will not charge transfer to O_2 or NO. In fact the metallic ions are formed rapidly by photoionization and exothermic charge transfer

$$M + h\nu \longrightarrow M^+ + e \tag{3.6}$$

$$M + O_2^+ \ (NO^+) \longrightarrow M^+ + O_2 \ (NO) \tag{3.7}$$

where here M is a metal atom. These metallic ions are long-lived and can be transported into the E-region where local high concentrations cause increased E-region ionization associated with sporadic E. The metal atoms can also form oxides that have low ionization potentials,

$$M + O_3 \longrightarrow MO + O_2 \tag{3.8}$$

$$M + O_2 + N_2 \ (O_2) \longrightarrow MO_2 + N_2 \ (O_2). \tag{3.9}$$

These reactions occur at rates that are faster than or comparable to transport in the region below 90 km. The metal oxides often also have low ionization potentials and can react with photons or molecular ions,

$$MO + h\nu \longrightarrow MO^+ + e \tag{3.10}$$

$$MO_2 + h\nu \longrightarrow MO_2^+ + e \tag{3.11}$$

$$MO + O_2^+ \ (NO^+) \longrightarrow MO^+ + O_2 \ (NO) \tag{3.12}$$

$$MO_2 + O_2^+ \ (NO^+) \longrightarrow MO_2^+ + O_2 \ (NO) \tag{3.13}$$

The MO^+ and MO_2^+ recombine rapidly unlike the metallic atomic ions. The lifetime of the metallic ions is determined by the rate of conversion of M^+ to molecular ions in the reactions:

$$M^+ + O_3 \longrightarrow MO^+ + O_2 \tag{3.14}$$

$$M^+ + O_2 + N_2 \ (O_2) \longrightarrow MO_2^+ + N_2 \ (O_2). \tag{3.15}$$

However, formation of MO^+ is largely negated by its rapid destruction by

$$MO^+ + O \longrightarrow M^+ + O_2. \tag{3.16}$$

When the reaction in (3.14) is exothermic the rate constant can be estimated at 10^{-10} cm^3 s^{-1}. The rate coefficient of equation (3.15) is probably of the order of 10^{-30} cm^{6}molec^{-1}s^{-1}. With these estimates it is seen that the chemical lifetime of M^+ becomes increasingly large above 90 km and transport may control the M^+ density around 105 km. The chemistry of meteor metals and their ions is very complex and equations

(3.6)-(3.15) are a great simplification in part because the various metals react differently. In our qualitative look at the E-region chemistry, the meteor metals are interesting since the long-lived atomic ions that they produce can have a noticeable effect on the E-region electron density.

The D-Region

The ionizing solar flux of the E-region is rapidly attenuated below 95 km. Lyman-α radiation, however, penetrates below 80 km and is capable of ionizing NO, which has an ionization potential of 9.25 eV (1340 Å). Although the NO concentration is variable and not well characterized, any reasonable NO cncentration of the order of 5×10^6 cm^{-3} or more will supply the major source of ionization. Since O_2 has an absorption window at Lyman-α with a cross section of only about 1×10^{-20} cm^2 (see Figure 6), the maximum absorption occurs at about 75 km for an overhead sum where the optical depth is unity. The NO ionization rate can be quite accurately represented by the formula (Swider 1964)

$$q(NO^+) = 6 \times 10^{-7} [NO] \exp\{(-[O_2] H \, 10^{-20} \, Ch(x,\Theta)\} \qquad (3.17)$$

where H is the scale height in centimeters, and $Ch(x,\Theta)$ is the Chapman function defined in equation (1.46). The value 6×10^{-7} is derived from an NO cross-section of 2×10^{-18} cm^2 and a flux at Lyman-α of 3×10^{11} photons cm^{-2} s^{-1}. The Lyman-α flux may vary about ±50% about this value over the solar cycle, thus changing $q(NO^+)$ by a similar amount. The variation in NO concentration, however, is likely to be larger than this factor.

Solar X-rays of wavelength 1-10 Å penetrate into the mesosphere where they ionize N_2 and O_2, mostly by energetic photoelectrons. The intensity of these solar X-rays is extremely weak during quiet solar activity (about 200 photons cm^{-2} s^{-1} for 3-5 Å X-rays and about 10 photons cm^{-2} s^{-1} of 1-3 Å X-rays), but highly variable and may increase by three orders of magnitude between very quiet and disturbed solar conditions. Large solar flares can increase the flux by another factor of a hundred for brief periods. The relatively small variation seen in the D-region ionization at all but the most disturbed times indicates that the 1-10 Å X-ray flux is not a dominant ionization source there.

Another possible source of ionization is provided by the rather significant concentrations of metastable $O_2(^1\Delta_g)$ which can be ionized by solar ultraviolet radiation at wavelengths between 1027 Å and 1118 Å above the ground-state threshold. The calculation of the ionization rate involves the penetration of structured flux, most notably the Si(III) multiplet at 1108 Å through windows in the O_2 absorption cross section (Figure 5). An accurate expression (Paulsen et al, 1972) obtained from averaging over 75 wavelength intervals and nine solar lines is

$$q(O_2^+) = O_2(^1\Delta)\,\{.549\times10^{-9}\exp(-2.41\times10^{-20}[O_2]\,H\,Ch(x,\Theta)) + 2.61\times10^{-9}\exp(-8.51\times10^{-20}[O_2]\,H\,Ch(x,\Theta))\} \quad (3.18)$$

This source is important but not as large as the NO source unless $O_2(^1\Delta_g)$ is enhanced beyond what most models predict in the 80 km region.

Cluster Ions in the D-region

The mass spectrometer measurements of positive ions in the D-region indicated that ions of mass 19, 37, 55, and 73 were present below about 80 km as opposed to masses 32 (O_2^+) and 30 (NO^+) that dominated at higher altitudes. These ions were identified as proton hydrates ($H^+(H_2O)_n$) or hydronium ions ($H_3O^+(H_2O)_{n-1}$). Although these measurements are technically difficult and at first they were believed by many people to be artifacts, improved techniques have established that these cluster ions become the dominant positive ions below about 80 km. The height of the transition from molecular ions to cluster ions varies, but by 75 km they always seem to be dominant. The major problem of the photochemistry of the D-region was to identify the pathways that lead from the primary ions NO^+ and O_2^+ to the water cluster ions. This goal has been in large part accomplished, but there are still a few problems.

The proposed pathways are shown in Figure 24. There are two distinct pathways depending upon the primary ion, NO^+ or O_2^+. Lyman-α exclusively produces NO^+ while EUV ionization of $O_2(^1\Delta)$ produces O_2^+. Thus the quiet-time ionization favors the NO^+ source (Figure 25). X-rays or any source that ionizes N_2 and O_2 essentially in proportion to their concentrations will produce about 90% O_2^+ and 10% NO^+ taking into account the fast ion-molecule reactions. These are the same reactions that occur in the E-region [equations (3.1)-(3.5)], but at these low altitudes the molecular oxygen concentration is orders of magnitude larger than the atomic oxygen and thus reaction 44 proceeds much faster than reaction 38. Hence for disturbed conditions the situation can be reversed from the quiet-time situation with O_2^+ being the dominant primary ion.

The sharp transition from molecular NO^+ and O_2^+ ions to the cluster ions is a characteristic of the D-region. The actual height varies from as low as 77 km when ionization is high, to as high as 85 km during undisturbed nighttime. The transition near the mesopause region is close to the similar ledge in the atomic oxygen profile. This similarity is not a coincidence; this is the region where three-body reactions between a minor constituent and two major constituents become competitive with two-body reactions. The pathways in Figure 24 are complicated and involve more than fifty reactions. We will not discuss all of these (see Reid, 1976 for details), but we will emphasize the important features and the remaining uncertainties in a qualitative way.

The O_2^+ pathway, although less important for the undisturbed atmosphere is better established, and we will consider it first. The initial

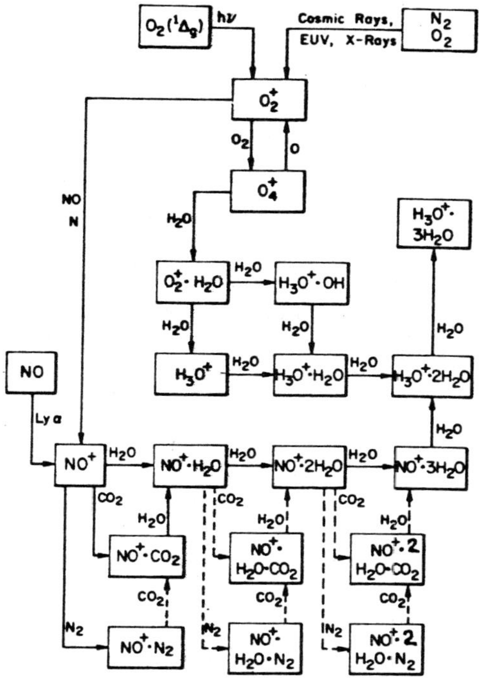

Figure 24. D-region positive ion chemical pathways (from Ferguson, 1979)

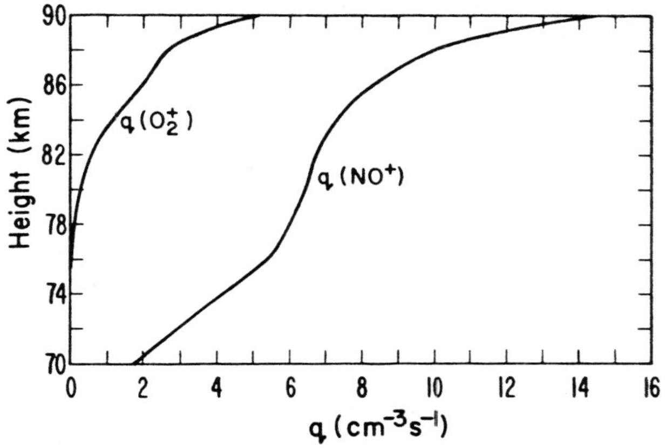

Figure 25. Quiet-time mid-latitude production rates of NO^+ and O_2^+ in the D-region (from Reid, 1979)

step for clustering is a three-body reaction with the primary ion to form the cluster ion O_4^+

$$O_2^+ + O_2 + M \longrightarrow O_4^+ + M. \qquad k_{46} \qquad (3.19)$$

This reaction proceeds rapidly since O_2 and M (either O_2 or N_2) are major constituents. Since water is found in only parts per million, the direct reaction proceeds at a slower rate

$$O_2^+ + H_2O + M \longrightarrow O_2^+(H_2O) + M. \quad k_{47} \qquad (3.20)$$

Although cluster ions recombine at a faster rate than molecular ions, the O_2^+ lifetime is much shorter than its recombination lifetime because it is rapidly destroyed by atomic oxygen regenerating the original O_2^+

$$O_4^+ + O \longrightarrow O_2^+ + O_3. \qquad k_{48} \qquad (3.21)$$

This is one of the major obstacles to forming cluster ions from O_2^+ above 80 km where the atomic oxygen concentration is about 1×10^{11} cm^{-3}, causing the O_4^+ lifetime to be less than a second. Reaction 46 becomes important below 80 km where it competes with recombination and conversion of O_2^+ to NO^+ as the major loss processes of O_2^+. For quiet or relatively quiet conditions reaction 45, charge exchange with NO, is the major O_2^+ loss above about 80 km. For disturbed conditions with high electron concentrations (the case where the O_2^+ pathway is dominant), the recombination lifetime will be shorter and thus the molecular ions will have a greater tendency to recombine before they form clusters. This will tend to lower the transition region somewhat. However, the rate of reaction 46 goes as the square of the pressure, and it is difficult to lower the transition altitude very much.

Reaction 48 competes with the reaction that continues down the pathway to $H^+(H_2O)_n$,

$$O_4^+ + H_2O \longrightarrow O_2^+(H_2O) + O_2 \qquad k_{49}. \qquad (3.22)$$

Reaction 46 followed by 49 is a good example of the efficient way of clustering a minor species, three-body clustering with a major species followed by a two-body switching reaction with the minor species. Exothermic cluster-ion switching reactions generally proceed near the limiting gas-kinetic rate with rate coefficients equal to about 1×10^{-9} $cm^3\ s^{-1}$.

The $O_2^+(H_2O)$ produced then can react with H_2O,

$$O_2^+(H_2O) + H_2O \longrightarrow H^+(H_2O)OH + O_2 \qquad k_{50}, \qquad (3.23)$$

followed by further reaction with water,

$$H^+(H_2O)OH + H_2O \longrightarrow H^+(H_2O)_2 + OH \qquad k_{51}, \qquad (3.24)$$

which yields the desired $H^+(H_2O)_2$ which can be further hydrated by three-body reaction with H_2O. It should be noted that the sequence (3.19)-(3.24) leads to production of HO_x. The recombination of $H^+(H_2O)_n$ will likely lead to H. Thus a cycle of ionization and recombination through the O_2^+ water-cluster pathway dissociates H_2O as a by-product.

Although the O_2^+ pathway is not dominant for quiet conditions, it is an efficient method for producing cluster ions. Once O_4^+ is produced in (3.19) faster than other losses of O_2^+, formation of water clusters is very rapid compared to recombination. The O_4^+ lifetime against forming $O_2^+(H_2O)$ is $(1.5\times10^{-9}[H_2O])^{-1}$ as opposed to the lifetime for recombination $(2\times10^{-6}\ n_e\ )^{-1}$. For $n_e \sim 1000\ cm^{-3}$ and 1 ppm H_2O at 75 km, the lifetime against forming water cluster ions is about 10^{-3} of the recombination lifetime.

The further clustering of $H^+(H_2O)_n$ for $n\geq3$ proceeds by three-body reactions with H_2O,

$$H^+(H_2O)_n + H_2O + M \rightleftarrows H^+(H_2O)_{n+1} + M, \qquad k_f,\ k_b$$

where the backward reaction (k_b) becomes increasingly important for large clusters. Measurements of the equilibrium constants $K_n=(k_f/k_b)$ as a function of temperature have shown these reactions to depend strongly upon the temperature. An example of this dependence is the case for n=5, where $k_b=3.9\times10^{-13}$ at 280 K, but at 150 K (possible at summer polar mesopause) $k_b=10^{-23}$ and thermal breakup is negligible. This is why the largest clusters occur near the high-latitude summer mesopause, which may enhance the formation of noctilucent clouds. Gravity-wave induced temperature oscillations can also lead to wavelike features in the degree of ionization due to the extreme temperature dependence of the cluster rates.

All of the reactions in the O_2^+ pathway have been well studied and the rate constants measured. The efficiency of this pathway is assured by the large rate of O_4^+ formation and the need of only two-body reactions to convert it to water clusters. In the case of the NO^+ precursor, the low ionization potential of NO^+ precludes conversion to water clusters until a later point in the hydration sequence. A direct method that has been established needs three consecutive direct three-body hydrations of NO^+,

$$NO^+ + H_2O + M \longrightarrow NO^+(H_2O) + M \qquad k_{53} \qquad (3.25)$$

$$NO^+(H_2O)_n + H_2O + M \longrightarrow NO^+(H_2O)_{n+1} + M, \qquad (3.26)$$

until $NO^+(H_2O)_3$ is produced. For $n\geq3$ a two-body reaction with H_2O is exothermic,

$$NO^+(H_2O)_3 + H_2O \longrightarrow H^+(H_2O)_3 + HNO_2 \qquad k_{54}. \qquad (3.27)$$

However, this pathway does not provide a route to clusters that is fast enough to compete with recombination. At 80 km with M about 4×10^{14} cm^{-3} and H_2O about 1 ppm the characteristic time for this path to water clusters is 5×10^4 seconds, while the recombination lifetime is a few times 10^3 seconds. Thus the water clusters will surely not dominate under these conditions. There would be mostly NO^+ and its first, or maybe second, hydrate present in the 80 km region . A mechanism which employs an initial three-body clustering with a species more abundant than water will allow the rate of clustering to be increased,

$$NO^+ + CO_2 + M \longrightarrow NO^+(CO_2) + M \qquad k_{55}. \qquad (3.28)$$

Although the rate coefficient is smaller than k_{53}, CO_2 is about a hundred times more abundant than H_2O. This is the rate-determining step, since switching reactions with H_2O are very rapid. This reaction still may not be of sufficient speed to overwhelm recombination and correctly yield the predominance of cluster ions.

Successive clustering of NO^+ ions and clustering with N_2 are proposed as a means of increasing the speed of the NO^+ pathway (Reid, 1977). The first step is

$$NO^+ + N_2 + M \longrightarrow NO^+(N_2) + M \qquad k_{56}. \qquad (3.29)$$

The rate coefficient is even smaller than k_{55}, but N_2 is about 3000 times more abundant than CO_2. The non-polar N_2 is very weakly bound to NO^+ with a binding energy of about 0.22 eV. The equilibrium between forward and backward (thermal breakup) reactions is very temperature dependent. The rate of $NO^+(N_2)$ thermal decomposition at 215 K is about the same as the rate of CO_2 switching at 80 km. At temperatures much above 200 K the N_2 clustering is not an important pathway, but for quiet times near the cold mesopause it can be important.

Only a few of the intermediate clusters of the NO^+ sequence have ever been measured, $NO^+(N_2)$ and $NO^+(CO_2)$ for example. The breakup of the weakly bound clusters in the sampling process may the major reason they have not been observed. However, it is not certain whether the N_2 clustering scheme is necessary. The mass-spectrometer measurements are difficult, and there have not been enough measurements made simultaneously with measurements of temperature, atomic oxygen, water, and NO.

The present scheme seems to be qualitatively correct, but there are still some unanswered problems. During disturbed times unclustered NO^+ is measured at much higher concentrations than is predicted with most models. During these disturbed times the O_2^+ pathway may be dominant and the recombination rate is faster due to the high levels of ionization. However, NO^+ can be more than an order of magnitude larger than the models predict. This means that during these disturbed times NO^+ clustering is being inhibited, and there is no specific process or processes in the photochemical scheme that simulates this effect.

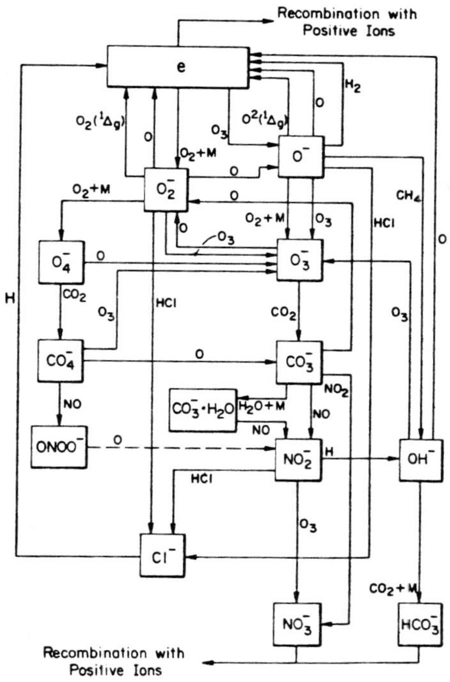

Figure 26. D-region negative ion chemical pathways (from Ferguson, 1979)

D-region Negative Ion Chemistry

A second characteristic of the D-region is the appearance of negative ions which replace electrons as the altitude decreases. The negative ion chemistry is more complex than the positive ion chemistry. The understanding of the negative ion chemistry is not on as firm ground. Part of this problem is due to experimental problems. Although techniques have improved in recent years, contamination and production of negative ions in mass-spectrometers are complications that are very hard to totally rule out. The experimentally obtained profiles seem to vary from one rocket measurement to another, and the variation is often difficult to explain using current photochemical concepts.

A simplified schematic of the negative ion pathways is shown in Figure 26. The initial formation of negative ions is a three-body process,

$$e + O_2 + M \longrightarrow O_2^- + M \qquad k_{57} \qquad (3.30)$$

in which electrons react with the major constituent O_2 to form the weakly bound (0.41 eV) O_2^-. N_2 does not form a stable negative ion. Dissociative attachment is a two-body mechanism for forming negative ions, but it important only when the dissociation energy is less than the electron affinity. This requirement is met for O_3, but the rate coefficient is small and the ozone is a trace constituent, so that the process

$$O_3 + e \longrightarrow O_2 + O^- \qquad k_{58} \qquad (3.31)$$

is not a major source of negative ions. Below 80 km reaction 57 becomes comparable to recombination in determining the chemical lifetime of electrons. The weakly bound O_2^- is easily detached by O and $O_2(^1\Delta_g)$

$$O_2^- + O \longrightarrow O_3 + e \qquad k_{59} \qquad (3.32)$$

$$O_2^- + O_2(^1\Delta_g) \longrightarrow O_2 + O_2 + e \quad k_{60} \qquad (3.33)$$

and in daytime is readily photodissociated by visible light

$$O_2^- + h\nu \longrightarrow O_2 + e \qquad J_{61}, \qquad (3.34)$$

with J_{61} equal to about 0.3 s^{-1}. Early measurements of the D-region at twilight during enhanced ionization events (Reid and Collins, 1959) indicated that the negative-ion/electron balance was responding to ultraviolet light that was absorbed by the ozone layer. This indicated that most of the negative charge resided on a species with a higher electron affinity than O_2^-.

The complicated scheme shown in Figure 26 is a series of paths that transfers charge until the negative ion with the largest electron affinity is formed, which is NO_3^- with an electron affinity of about 4 eV. The scheme shown is a simplification since photodetachment or photodissociation reactions are left out so that the figure remains legible. However, it should be remembered that each species can be photodetached and many can be dissociated into one or more product channels, although many of these paths have not been measured in the laboratory. Clustering of the ions is also neglected except for the clustering of CO_3^- with water which is well established. Clustering could complicate the reaction scheme because cluster formation stabilizes the negative charge and may prevent the clustered ion from continued reaction down the pathway of Figure 26. It should be noted, however, that there are measured rate coefficients for all of the steps in Figure 26.

The rate-determining step in the scheme is the reaction

$$CO_3^- + NO \longrightarrow NO_2^- + CO_2. \qquad k_{62} \qquad (3.35)$$

The lifetime of CO_3^- against this reaction is about 10^4 seconds for NO densities of about 10^7 cm^{-3}. CO_3^- is recycled back to the weakly bound ions O_2^- and O^- many times by the reactions

$$CO_3^- + O \longrightarrow O_2^- + CO_2 \qquad k_{63} \qquad (3.36)$$

$$CO_3^- + h\nu \longrightarrow CO_2 + O^- \qquad J_{64} \qquad (3.37)$$

before it reacts with NO. The daytime chemical lifetime of CO_3^- is only about a second even below 80 km. For the terminal ions NO_3^- and HCO_3^-, recombination determines the lifetime when hydration is neglected.

It is easy to estimate the ion density if a steady state is assumed. In the region where the electron attachment rate (reaction 57) is so large that most of the negative charge resides on negative ions and there are very few electrons (below 70 km), the ion production rate can be set equal to the neutralization rate between positive and negative ions. This ion-ion recombination is remarkably independent of the nature of the ions, with α varying from about 6×10^{-8} to 1×10^{-7} $cm^3 molec^{-1} s^{-1}$. Thus for a production rate of 2 cm^{-3}, the ion density is about $4\text{-}5 \times 10^3$ cm^{-3} and the terminal negative ions will have lifetimes of a few thousand seconds. Water clustering, especially if it proceeds by way of major species intermediates, can easily be competitive with recombination and thus the clustering of the terminal ions should be expected, especially at lower altitudes.

Mass-spectrometer measurements of the negative ions in the D-region do not give consistent results for the dominant ions. At times CO_3^- appears as the dominant species, especially above 60 km. At other times NO_3^- and its water hydrates are dominant. O_2^- often appears at much higher levels than photochemical models would predict. The fact that NO is quite variable in the mesosphere means that the rate-determining step of reaction 62 can vary widely, sometimes being faster than recombination and at other times being slower. The recombination rate is a function of the total ion density which can introduce more variability. Thus when the NO density is low, which is often the case for the low-latitude upper mesosphere, the CO_3^-/NO_3^- ratio should be large, and increase if the ionization rate increases. As NO_x increases at decreasing altitudes, NO_3^- should always become the dominant ion. For NO_3^- to dominate above 70 km the NO density must be large enough to make reaction 57 faster than recombination of the precursor negative ions. These qualitative ideas about the dominance of NO_3^- and CO_3^- seem to be correct, but experimental profiles show larger variations than the models predict for reasonable variations in the NO profile. Photodissociation and photodetachment of the terminal ions and possible cluster formation could change the ratios of the negative ion species, but most of the corresponding kinetic parameters are not known.

The D-region ion reaction schemes are dependent upon ion-molecule reactions between the ions and minor neutral constituents that lead to the formation of more stable ions. The positive ion scheme is sensitive to the NO, H_2O, and O concentrations as well as the temperature. The negative ion pathways depend upon the NO, O, $O_2(^1\Delta_g)$, and O_3 concentrations as well as the solar radiation through photodissociation and detachment processes. Thus accurate <u>in situ</u> ion measurements are a potentially sensitive probe of the minor species concentrations in the mesosphere. However, until the schemes are shown to be complete and quantitatively explain the variation observed in the ion composition, determination of minor neutral species from ion ratios should be done with caution.

A current unresolved issue in the negative ion chemistry is the mass-spectrometer measurements of high-mass negative ions in the region above 80 km. These negative ions of mass greater than 100 amu may be artifacts of the sampling process. If they are not, they represent a real challenge to the photochemical modeller. Three-body attachment by reaction 57 is too slow to start the scheme in Figure 26. The abundant atomic oxygen will quickly detach electrons, and the scheme will never get past CO_3^-, if that far. The formation of these heavy ions seems to require a two-body dissociative attachment to immediately form an ion stable enough to resist attack by O that would lead back to free electrons or less stable ions. Meteor ablation products could possibly play a role in formation of stable silicate ions, but further confirmation of the existence of the high mass ions is needed.

The elaborate photochemical schemes that have been developed to explain the ion composition of the D-region seem to be qualitatively correct with the possible exception of the large mass negative ions. The composition of the D-region illustrates the importance of chemical transformations. The positive and negative charge tends to reside in the most chemically stable form, even though many reactions with trace constituents must take place before the terminal ions are formed. The proposed schemes were developed in conjunction with improved in situ measurements, but the schemes often had to propose intermediates before they were measured in the atmosphere. All of the pathways have not yet been confirmed, and some of them may not turn out to be valid, but the study of the D-region has been most rewarding for the photochemical modellers, as well as the laboratory chemical kineticists and the experimentalists who made the in situ mass spectrometer measurements.

The Disturbed D-Region

The relatively low ionization rates that occur in the D-region can be increased by orders of magnitude following large solar flares. The greatly increased X-ray flux in the 1-10 Å wavelength region that accompanies the flares leads to suddenly increased ionization in the sunlit lower E-region and D-region. The flares are relatively short-lived events with lifetimes less than an hour. The X-rays ionize N_2 and O_2 producing 90% O_2^+ and 10% NO^+ in the D-region. The electron density increases markedly in the upper D-region. The increased ionization decreases the recombination lifetime of electrons and thus slows the conversion of electrons to negative ions. Thus during disturbed times the maximum altitude of conversion of electrons to negative ions decreases. The increased electron density can be remotely monitored since it causes increased radio-wave absorption.

The X-ray flares are of short duration and, although the increased ionization can increase the flow through the positive and negative ion pathways by orders of magnitude, the shortlived effects on radio communication are the most noticeable consequences. Solar proton events (SPE) that follow large solar flares also lead to large enhancements of the ionization rate. However, in this case the enhancement may last as long as days. During SPE's the ionization source is mostly MeV protons that can penetrate down to stratospheric levels. These charged particles are confined by the earth's magnetic field to the polar-cap regions. Although the polar-cap absorption region moves southward during these events, the precipitating particles are still confined to a small portion of the earth's surface. The very high energy protons produce high energy secondary electrons as well as Bremsstrahlung that ionize O_2 and N_2 yielding mostly O_2^+. The increased rate of ionization and thus ion chemistry reactions produces notable amounts of NO at altitudes well below the usual 100-115 km source region. The increased ionization leads to shorter recombination times and a slight lowering of the level at which clustering occurs. However, below 75 to 80 km most of the positive ions formed proceed through the O_2^+ pathways in Figure 24 to form water clusters before they recombine. When reaction 51 followed by recombination of $H^+(H_2O)_n$ with electrons occurs, it is equivalent to

$$H_2O \longrightarrow H + OH. \quad (3.38)$$

Where electrons dominate negative ions this will lead to the production of odd hydrogen at twice the ionization rate. Where negative ions are more numerous, neutralization of cluster ions with NO_3^- probably will yield HNO_3

$$NO_3^- + H^+(H_2O)_n \longrightarrow HNO_3 + nH_2O,$$

with HNO_3 being subject to dissociation into OH and NO_2 with a lifetime of hours in the mesosphere. For disturbed times CO_3^-, O_2^-, and O_3^- may also occur in noticeable amounts, and their neutralization with proton hydrates will likely lead to OH or HO_2. Recombination of clusters before the formation of $H^+(H_2O)(OH)$ is unlikely, but it tends to lower the yield of HO_x per ion pair to below two.

Very intense SPE's can increase the ionization rate to larger than 10^3-10^4 ions per cm^3 per second. Thus water can be dissociated at this rate by the ion chemistry. This rate can become comparable to or even greater than the neutral dissociation rates that are shown in Figure 17. A simple calculation assuming 5ppm H_2O at both 60 and 70 km and 2 ppm H_2O at 80 km, and using the maximum overhead sun rates yields: at 60 km, $(1\times10^{-7})(5\times10^{-6})(6.4\times10^{15})=3.2\times10^3$ cm^{-3} s^{-1}, at 70 km $(5\times10^{-7})(5\times10^{-6})$ $(1.7\times10^{15})=4.3\times10^3$ cm^{-3} s^{-1}, and at 80 km $(3\times10^{-6})(2\times10^{-6})(3.8\times10^{14})$ = 2.3×10^3 cm^{-3} s^{-1}. Thus the SPE ion chemistry can dissociate water at an equal or greater rate. If this continues for a time of the order of the HO_x lifetime, the HO_x density can be noticeably increased. HO_x will not grow proportionally to the increased rate since its loss by reactions 14 and 15 is proportional to the HO_x density. The increase in HO_x leads to increased catalytic destruction of O_x (and thus O_3) in the mesosphere. If the increased water dissociation rate continues for many hours in the upper mesosphere, the conversion of H_2O to HO_x and then to H_2 by reaction 14a may eventually deplete H_2O and thus lead to a decreased HO_x concentration and an eventual increase of O_3 above the initial values. This scenario is quite dependent upon the background H_2O concentration, the time history of the ionization rate, and the altitude (Solomon et al, 1981).

Large solar proton events are rare occurrences. With the highly elevated ionization rates the ion chemistry becomes fast enough to influence the neutral chemistry in the region below 90 km. At all other times, except intense aurora, the D-region modeller can seek a solution for the neutral constituents in the absence of D-region ions (using E-region ion chemistry or the appropriate boundary conditions for NO) and then use these values in an ion code that solves for the ion densities. During intense SPE's the neutral and ion chemistry become more strongly coupled, and the water-cluster-ion cycle effects the minor neutral constituents. Confirmation of ozone depletion by the D-region ion chemistry tends to verify the correctness of both the neutral and ion models.

4. SUMMARY

In this paper we have given an introduction to the important photochemical processes that occur in the 50-120 km region. The absorption of solar radiation and the resulting ionization or dissociation produces reactive species that partake in a series of chemical reactions. In the mesosphere solar ultraviolet dissociates molecular oxygen. The atomic oxygen that is produced can combine with molecular oxygen to form ozone. The total O_x profile is largely controlled by catalytic destruction by HO_x radicals that are formed by the photodissociation of water vapor. The family technique has been introduced as a relatively efficient method for solving the coupled continuity equations for the constituent profiles. We have given a brief discussion of the limitations and difficulties of applying the family technique for diurnal calculations in the upper mesosphere. We have also seen that NO and O in the 80-100 km region are greatly effected by transport processes, which means that the one-dimensional continuity equations for these species can yield poor approximations for their profiles.

Photochemistry controls the ion composition of the lower E-region except for sporadic-E events. Ion molecule reactions, often involving minor neutral constituents, control the D-region ion composition. Complex chemical pathways in the D-region lead from the primary positive ions and electrons to the terminal ions. As a result the positive charge tends to reside on the species with a high proton affinity (proton hydrates), while the negative charge resides on a species with a large electron affinity (NO_3^-). During highly disturbed solar conditions the ionization rate in the D-region can be increased by orders of magnitude. This leads to an increase in the rates of the water cluster ion chemistry, producing HO_x. Under these conditions enough HO_x can be produced so that O_3 concentration is lowered. The measurements of ozone destruction in the mesosphere during SPE's confirms our understanding of the neutral and ion chemical processes.

REFERENCES.

Banks, P. M., and Kockarts, G.: 1973,"Aeronomy", Academic Press, New York.
Baulch, D. L., Cox, R. A., Hampson, R. F. Jr., Kerr, J. A., Troe, J., and Watson, R. T.: 1979, Phys. and Chem. Ref. Data 9, pp295-471.
Carver, J. H., Gies, H. P., Hobbs, R. I., Lewis, B. R., and McCoy, D. G.: 1977,J. Geophys. Res. 82, p1955.
Dickinson, P. H. G., Bain, W. C., Thomas, L., Williams, E. R., Jenkins, D. B., and Twiddy, N. D.: 1980, Proc. Roy. Soc. A 369, pp379-408.
Ferguson, E. E., in Kinetics of Ion-Molecule Reactions, P. Ausloos, Editor, Plenum Press, N. Y., pp377-403.
Ferguson, E. E.,: 1979, in Middle Atmosphere Electrodynamics, NASA CP-2090, U. S. Government Printing Office, Washington, D.C., pp71-88.
Frederick, J. E., and Hudson, R. D.: 1979, J. Atmos. Sci. 36, pp737-745.
Frederick, J. E., and Hudson, R. D.: 1980a, J. Atmos. Sci. 37, pp1088-1098.
Frederick, J. E., and Hudson, R. D.: 1980b, J. Atmos. Sci. 37, pp1099-1106.
Gear, C. W.: 1971, Numerical Initial Value Problems in Ordinary Differential Equations, Prentice Hall, Englewood Cliffs, N.J.
Howard, C. J., and Evenson, K.: 1977, Geophys. Res. Lett. 4, pp437-440.
Howlett, L. C., Baker, K. D., Megill, L. R., Shaw, A. W., Pendleton, W. R., and Ulwick, J. C.:1980, J. Geophys. Res. 85, pp1291-1296.
Johnston,H.S.: 1966, Gas Phase Reaction Rate Theory, Ronald Press, N.Y.
Kockarts, G.: 1971, in Mesospheric Models and Related Experiments, D. Reidel, Dordrecht, Holland, pp160-176.
Kockarts,G .: 1976, Planet. Space Sci. 24, pp589-604
Meira, C. G., Jr.: 1971, J. Geophys. Res. 76, p202.
Mount, G. H., and Rottman, G. J.: 1981, J. Geophys. Res. 86, pp9193-9198.
NASA: 1979, The Stratosphere: Present and Future, RP 1049.
NASA: 1981, Chemical Kinetic and Photochemical Data for Use in Stratospheric Modelling. Evaluation Number 4, JPL 81-3.
Paulsen, D. E., Huffman, R. E., and Larrabee, J. C.: 1972, Radio Sci 7, p51.
Pendleton, W. R., Jr., Baker, K. D., Howlett, L. C., and Stair, A. T.: 1982, submitted to J. Atmos. Terr. Phys.
Purcell, J. D., and Tousey, R.: 1960, J. Geophys. Res. 65, p370.
Reid, G. C.: 1976, Adv. At. Mol. Phys. 12, pp375-413.
Reid, G. C.: 1977, Planet. Space Sci. 25, pp275-290.
Reid, G. C.,and Collins,C.: 1959, J. Atmos. Terr. Phys. 14, p63.
Reid, G. C.: 1979, "The Middle Atmosphere", in Middle Atmosphere Electrodynamics, NASA CP-2090, pp27-42.
Solomon, S.: 1981, NCAR Cooperative Thesis No. 62, University of California and National Center for Atmospheric Research, NCAR/CT-62.
Solomon, S., Rusch, D. W., Gerard, J.C., Reid, G. C., and Crutzen, P.J.: 1981, Planet. Space Sci. 29, p885.
Solomon, S., Crutzen, P. J., and Roble, R. G.: 1982, J. Geophys. Res. 87, pp7206-7220.
Swider, W.: 1964, Planet. Space Sci., 12, p761.
Tohmatsu, T., and Iwagami, N.: 1976, J. Geomag. Goelect. 28, p343.

Turco, R. P., and Whitten, R. C.: 1974, J. Geophys. Res. 79, pp3179-3185.
U.S. Standard Atmosphere, 1976: 1976, U.S. Government Printing Office, Washington,D.C.

PHYSICS OF THE MESOPAUSE REGION

Jeffrey M. Forbes
Department of Physics
Boston College
Chestnut Hill, Massachusetts 02167

The mesopause is the minimum in the vertical temperature structure of the earth's atmosphere occurring near 85-90 Km, ranging in value from roughly 140K to 220K from summer to winter polar latitudes (see Figure 1). The altitude region about the mesopause, roughly between 70 and 120 Km, may well be termed a 'transition' or 'boundary' region in many respects. Basically, fundamental mechanisms relating to heat deposition, conductive and radiative transfer, wave propagation, etc., applicable in their respective upper or lower altitude regimes, often require special consideration across this region. Most numerical models of the lower atmosphere or the thermosphere do not comprehensively address these difficulties, and consequently do not place confidence in results obtained between 80 and 120 Km. Furthermore, the mesopause region exceeds the altitude limits of most meterological rockets and is too low for in-situ satellite measurements, so that it is the most scantily observed region of the atmosphere as well.

The purpose of this lecture is to provide a cohesive presentation of the physics of the mesopause region, particularly emphasizing the fundamental principles and assumptions which lie at the foundation of our current knowledge and perception of mesopause region processes and contemporary models. It is intended to form a framework and basis from which to critically pursue the literature and perhaps initiate research investigations into this most interesting yet neglected region of the earth's atmosphere.

1. CHARACTERISTICS OF THE MESOPAUSE REGION

1.1 Thermal and dynamical structure

As illustrated by the U.S. Standard 1976 temperature profile for typical midlatitude conditions (Figure 1), the mesopause is the boundary between the mesosphere (ca. 50-85 Km) and the thermosphere ($\gtrsim$ 85 Km), adjacent regions of respectively negative and positive temperature lapse rates. The stratosphere is a region of positive

R. L. Carovillano and J. M. Forbes (eds.), Solar-Terrestrial Physics, 733–752.

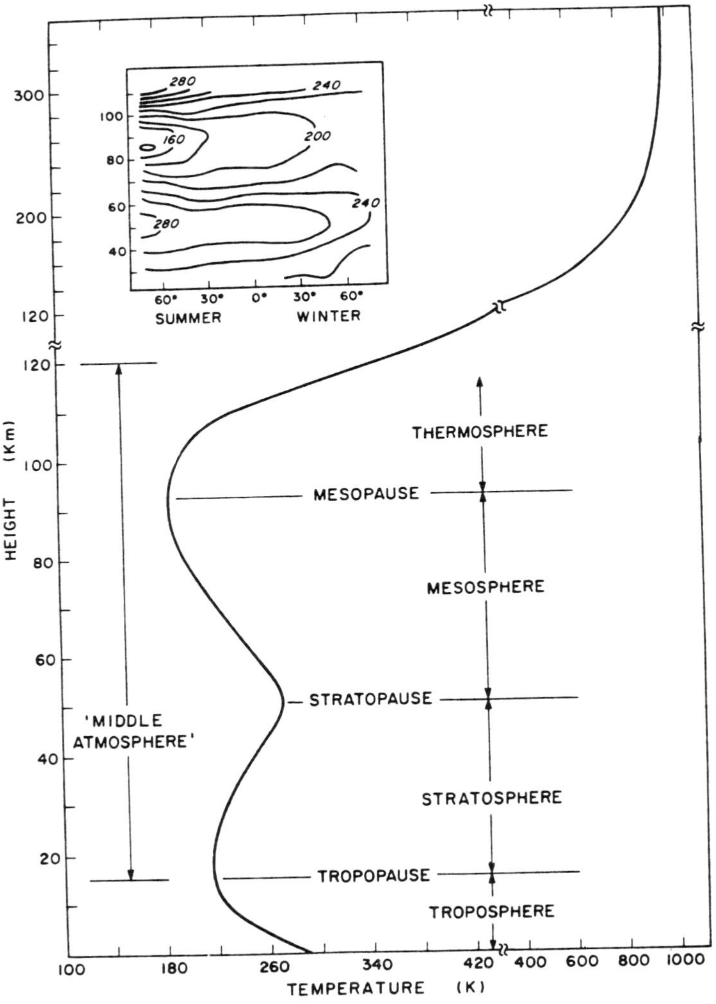

Figure 1. Average midlatitude vertical temperature structure (based on U.S. Standard Atmosphere, 1976) and (inset) zonally-averaged height-latitude temperature structure during solstice conditions (based on Cospar International Reference Atmosphere, 1972).

lapse rate bounded above by the stratopause (ca. 50 Km) and below by the tropopause (ca. 15 Km). In contemporary parlance the 15-120 Km altitude regime is termed the 'middle atmosphere'.

The zonally-averaged (i.e., around latitude circles) temperature structure as illustrated in the inset to Figure 1 reveals strong latitudinal temperature gradients in the stratopause and mesopause regions; temperatures increase from winter to summer in the stratosphere, but gradients are of the opposite sense above 70 Km. Often superimposed on this zonal mean structure are tidal (periods of 12 and 24 solar hours) and gravity wave (periods between 5 minutes and 3 hours) perturbations with vertical wavelengths typically in the range of 15-60 Km. Figure 2a illustrates a sequence of winter temperature profiles obtained from rocket measurements at Point Barrow, Alaska (71°N) during 1967 (Smith et al, 1968) which contain the signature of wave perturbations with 10-15 Km wavelength, periods of order 1-3 hours, and amplitudes increasing from roughly 5K at 40 Km to 25K at 90 Km. The background mesopause temperature is about 220 K. Summer temperature measurements at the same location (Figure 2b; Theon, 1968) depict comparatively quiescent behavior and cold mesopause temperatures (~ 130-150 K).

Figure 3 illustrates evidence of a diurnal (24-hour) propagating tide with vertical wavelength of about 25 Km from rocket measurements obtained over Natal (6°S; Smith et al, 1968). These profiles were obtained by differencing measurements taken 12 hours apart (sunrise minus sunset). Diurnal amplitudes increase from about 5K at 50 Km to 40K between 85 and 95 Km, and are consistent with the theoretical calculations of Lindzen (1967). (Note: This method of extracting tidal oscillations is fundamentally incorrect as contaminations from periods of less than 24 hours can render such results meaningless. However, in this particular case the diurnal propagating tide is known to dominate the equatorial daily temperature variations between 60 and 120 Km, and its phase and amplitude structure can reasonably be ascertained from a much more limited data sample.)

The temperature structures illustrated in Figures 1-3 are accompanied by winds of similar spatial and temporal scales. Mean zonal winds representative of solstice conditions (Groves, 1971) are depicted in Figure 4. The midlatitude stratopause region is characterized by strong easterlies (~ 50m sec^{-1}) in the summer and westerlies (~ 70m sec^{-1}) in the winter, whereas the midlatitude flow is basically westerly during both summer and winter above 80 Km. Note the midlatitude westerly jet streams peaking near 10 Km, and an easterly jet peaking near 105 Km over the equator.

1.2 Physical Processes

Figure 5 is a schematic of the physical processes which affect or occur in the mesopause region. The primary sources of heat input include:

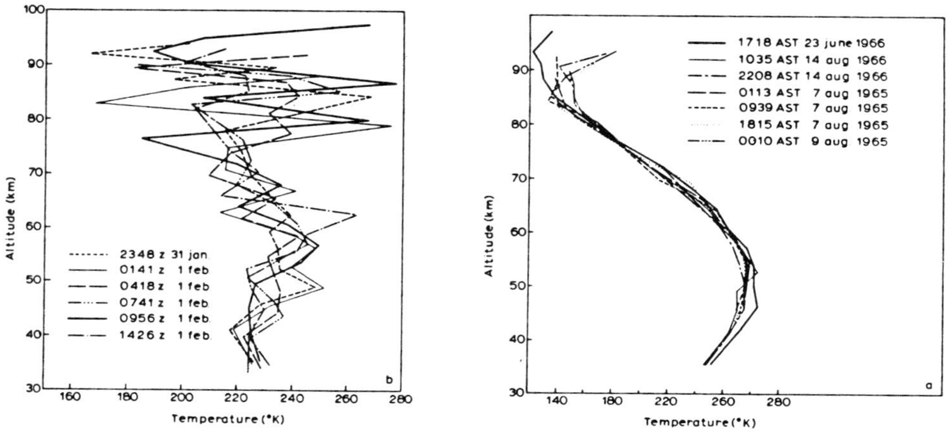

Figure 2. Temperature profiles from successive rocket measurements at Point Barrow, Alaska (71°N) during (a) 1967 winter (Smith et al, 1968) and (b) 1965 summer (Theon, 1968). Reprinted with permission of the American Meterological Society.

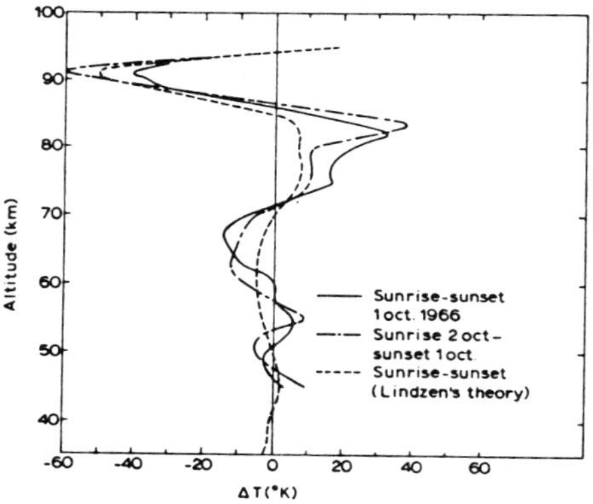

Figure 3. Sunrise-sunset differences in rocket measurements of temperature over Natal (6°S) illustrating presences of the diurnal propagating tide. Theoretical calculations of Lindzen (1967) are also shown. From Smith et al (1968) with permission of the American Meteorological Society.

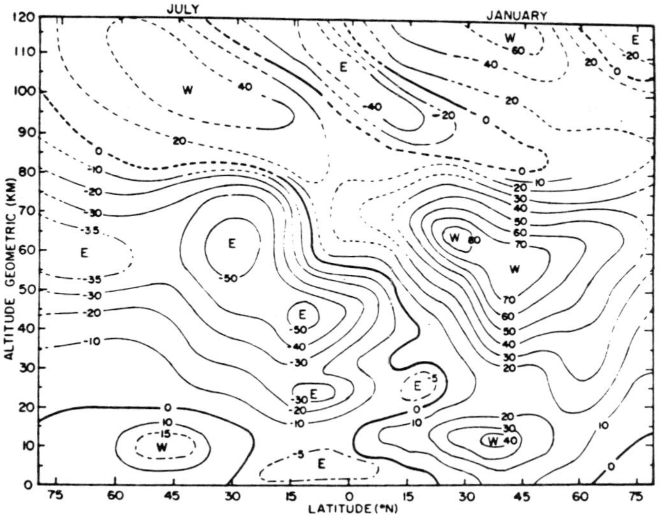

Figure 4. Zonally-averaged winds representative of solstice conditions. From Groves (1971).

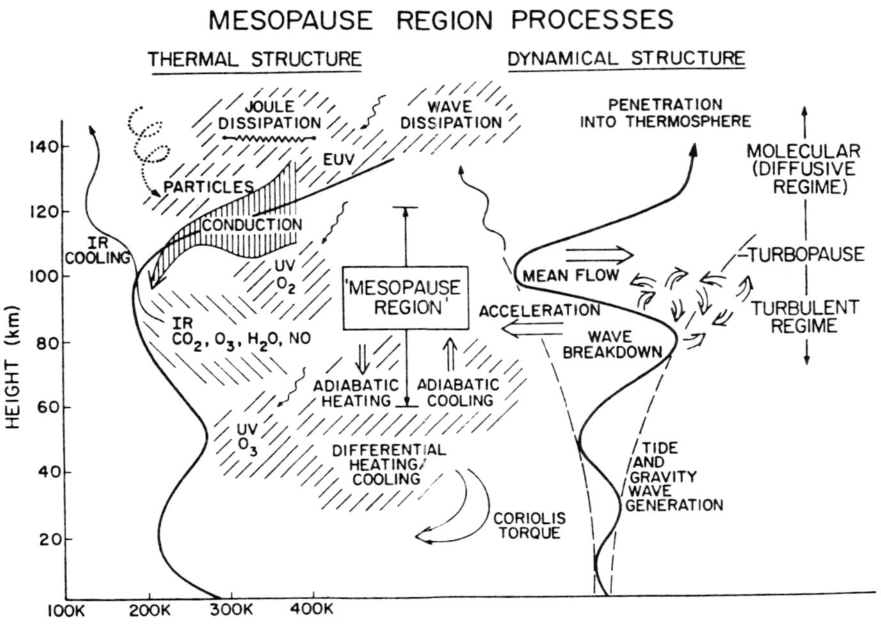

Figure 5. Schematic of physical processes which affect and/or occur in the mesopause region.

(a) EUV solar radiation absorption
(b) UV solar radiation absorption
(c) joule dissipation of ionospheric currents (high latitudes only)
(d) particle precipitation (high latitudes only)
(e) dissipation of wave and turbulent energy

The primary heat sink between 100 and 140 Km is downward heat conduction. (For further details of heat sources and thermal conduction above 100 Km, see lectures by A.D. Richmond in this volume.) Below 100 Km the densities and collision frequencies of polyatomic molecules CO_2, O_3, H_2O, and NO are sufficiently great that heat loss by infrared radiation plays a significant role in determining the thermal structure. The zonal mean structure is determined in part by a differential heating which also drives a meridional circulation which in turn feeds back to the temperatures via adiabatic cooling (heating) by rising (sinking) motions. A zonal (easterly/westerly) circulation is set up by Coriolis torques acting on the meridional flow. In addition, the deposition of mean momentum and heat by dissipating tides and gravity waves must also be taken into account in order to properly model the mesopause region mean zonal circulation. Gravity waves are thought to originate in such tropospheric sources as the adjustment of storm and circulation systems and flow over irregular topography. These waves propagate vertically, increase in amplitude with height at an exponential rate (e-folding distance of roughly 13 Km), often reach convectively unstable amplitudes in the mesopause region where they dissipate, generate turbulence, and deposit mean heat and momentum into the mean flow. Molecular diffusion, as its efficiency increases with decreasing density, usually becomes faster than turbulent diffusion rates above about 100 Km, the level sometimes referred to as the 'turbopause'. Similary, tidal oscillations of 24-hour ('diurnal') and 12-hour ('semidiurnal') periods are generated by H_2O and O_3 insolation absorption in the troposphere and stratosphere and propagate through the mesopause into the thermosphere. Tides represent a major source of temperature and wind variability in the mesopause region, and are also capable of generating turbulence and depositing mean momentum and heat above 85 Km.

2. SOME FUNDAMENTAL DYNAMICS

2.1 The Perturbation Approach (Linearization)

A.D. Richmond (this volume) documents the fundamental hydrodynamic and thermodynamic equations applicable to the upper atmosphere, and discusses many of the usual simplifications and assumptions. In the lectures on atmospheric waves he introduces a commonly utilized perturbation approach whereby the various meteorological variables (winds, temperature, pressure, density) are represented in the following form:

$$\psi = \overline{\psi} + \psi' \tag{1}$$

where ψ is any meterological variable, $\overline{\psi}$ represents the 'background' or 'unperturbed' component of ψ (usually the zonal mean (longitudinal average) of ψ), and ψ' represents the deviations of ψ from $\overline{\psi}$. Further, products of primed quantities are neglected. Generally this leads to separate (but coupled) sets of momentum, continuity, and thermal energy equations for the zonal mean and perturbation fields. As an example, consider the zonal (westerly) momentum equation in 'flux form' and pressure coordinates neglecting any type of friction (Holton, 1975):

$$\underline{\frac{\partial u}{\partial t}} + \underline{\frac{\partial u^2}{\partial x} + \frac{1}{\cos^2\theta}\frac{\partial}{\partial y}(uv\cos^2\theta) + \frac{1}{\rho_o}\frac{\partial}{\partial z}(\rho_o uw)} - \underline{2\Omega\sin\theta\, v} + \underline{\frac{\partial \Phi}{\partial x}} = 0 \qquad (2)$$

consisting of inertia, advective, coriolis, and geopotential gradient terms, respectively, where:

$$\frac{\partial \Phi}{\partial z} = \frac{RT}{H}$$

$$\frac{d\Phi_o}{dz} = \frac{T_o}{H}$$

$dx = a \cos\theta \, d\lambda$ = distance increment in westerly direction

$dy = a \, d\theta$ = distance increment in southerly direction

λ = longitude

θ = latitude

z = a measure of 'height' [$\equiv -H \ln (p/p_s)$]

H = scale height [$\equiv RT_s/g$]

R = gas constant for dry air

T_s = a constant reference temperature

g = gravitational acceleration

p = pressure

p_s = a constant reference pressure

u = westerly velocity

v = southerly velocity

w = a measure of 'vertical velocity' [$\equiv dz/dt$]

T_o = a basic state temperature [$\equiv T_o(z)$]

Φ_o = a basic state geopotential $[\equiv \Phi_o(z)]$

Ω = angular velocity of earth

a = radius of earth

Derivatives are taken at a level of constant pressure. One then obtains the zonal mean westerly momentum equation:

$$\frac{\partial \bar{u}}{\partial t} + \bar{v}\frac{\partial \bar{u}}{\partial y} + \bar{w}\frac{\partial \bar{u}}{\partial z} - \left(2\Omega \sin\theta + \frac{\bar{u}}{a\cot\theta}\right)\bar{v} = -\bar{F}_x \quad (3)$$

and the corresponding perturbation equation:

$$\frac{\partial u'}{\partial t} + \bar{u}\frac{\partial u'}{\partial x} + \bar{v}\frac{\partial u'}{\partial y} + \bar{w}\frac{\partial u'}{\partial z} - \left(2\Omega \sin\theta + \frac{\bar{u}}{a\cot\theta}\right)v'$$
$$- \frac{\bar{v}}{a\cot\theta}u' + \frac{\partial \bar{u}}{\partial y}v' + \frac{\partial \bar{u}}{\partial z}w' + \frac{\partial \Phi'}{\partial x} = 0 \quad (4)$$

where the so-called 'eddy flux' term or the 'convergence of the eddy momentum flux' is given by

$$\bar{F}_x \equiv \frac{1}{\cos^2\theta}\frac{\partial}{\partial y}\left(\overline{u'v'}\cos^2\theta\right) + \frac{1}{\rho_o}\frac{\partial}{\partial z}\left(\rho_o\,\overline{u'w'}\right) \quad (5)$$

The corresponding meridional momentum and thermal energy equations would similarly be obtained. It is common practice to investigate solutions to the zonal mean and perturbation fields separately (and most often by different investigators!). However, the zonal mean and perturbation equations are coupled either in their coefficients or the appearance of effective 'forcing' terms (e.g., $\bar{F}_x$). In order to obtain closed sets of equations, it is necessary to either neglect these terms, assume they are known a priori, or parameterize them in some way. It turns out that the radiative-dynamical balance of the mesopause region is inextricably involved with wave-mean flow interactions (see Section 4). It is towards this end that the current section, leading to an understanding of wave-mean flow interaction relationships in subsection 2.3, is devoted.

2.2 Thermal wind balance

For most of the middle atmosphere the large-scale horizontal quasi-steady zonal flow may be characterized by a balance between the coriolis and geopotential gradient terms in the meridional momentum equation:

$$f\,\bar{u} = -\frac{\partial \bar{\Phi}}{\partial y} \quad (6)$$

where $f = 2\Omega\sin\theta$ is the coriolis parameter. Using the definition of φ (6) may be rewritten in 'thermal wind' form:

$$\frac{\partial \bar{u}}{\partial z} = - \frac{g}{fT_s} \frac{\partial T}{\partial y} \tag{7}$$

which specifies how the mean zonal wind must change with height in the presence of a given meridional temperature gradient. Figures 1 (inset) and 4 illustrate the consistency of the observed zonal mean circulation with (7). Take the stratopause region for instance. During June solstice pole-to-pole temperature gradients exist which are positive near the stratopause region (40-70 Km) and negative near the mesopause region (70-100 Km). Therefore, we see that summer (winter) easterlies (westerlies) increase with height below 70 Km and decrease above 70 Km. The dynamical-radiative balances which drive the zonal mean circulation will be examined in more detail in Section 4.

2.3 Wave-mean flow interactions

The theory of atmospheric waves is treated in detail by A.D. Richmond (this volume). The following points are relevant to the present discussion:

(1) Gravity waves are atmospheric oscillations whose restoring forces arise from the buoyancy of displaced fluid parcels in a stably stratified atmosphere. Typical periods range between 10 minutes and 3 hours.

(2) Gravity waves propagating into the mesopause region are primarily excited in the troposphere by flow over topography or any other mechanism which causes vertical displacements of fluid elements.

(3) In the absence of dissipation the amplitudes of atmospheric gravity waves increase with height roughly as $1/\sqrt{\rho_o}$ where ρ_o is the background density. Further, Eliassen and Palm (1961) show that for linearized, plane, internal gravity waves in the absence of rotation, dissipation, and local excitation (cf. Lindzen, 1973):

$$\overline{p'w'} = -\rho_o\,(\bar{u} - c)\,\overline{u'w'} \tag{8}$$

and $$\frac{d}{dz}\,(\rho_o\,\overline{u'w'}) = 0 \text{ for } \bar{u} \neq c \tag{9}$$

where c = horizontal phase speed and $\overline{p'w'}$ (the overbar referring to the average over one cycle of the wave) may be associated with the vertical flux of energy and $\rho_o\,\overline{u'w'}$ with the vertical flux of horizontal momentum. Equation (8) implies that an upward moving wave ($\overline{p'w'} > 0$) carries easterly momentum if its phase speed is easterly relative to the mean flow, and (9) implies that none of the mean momentum carried by the wave is deposited in the mean flow. Eliassen and Palm (1961) do not consider the consequences of $\bar{u} = c$.

(4) Propagating tides can be viewed as gravity waves for which the rotation and sphericity of the earth must be taken into account (due to their long periods and horizontal length scales). Also, for tides $c \gg \overline{u}$.

There are two categories of problems connected with the interaction between the wave and zonal mean fields (see review by Lindzen, 1973). The first, which we will only touch upon briefly, involves the propagation of waves through regions of zonal flow comparable to the horizontal phase speed of the wave. Neglecting $\overline{v}$ and assuming solutions of the form $e^{i(kx + nz + \omega t)}$ in (3),

$$\frac{\partial}{\partial t} + \overline{u}\frac{\partial}{\partial x} \rightarrow \quad i\, k\, (\overline{u} - c)$$

where $c = -\omega/k$ = horizontal phase speed. The height at which $\overline{u} = c$ is called the critical level. Waves penetrate only very inefficiently through critical levels. In fact, the momentum flux $\rho_o\, \overline{u'w'}$ suffers a discontinuity at the critical level implying absorption of the wave and acceleration of the mean flow. The critical level may also be characterized by sharp unstable gradients leading to the production of thin turbulent layers.

Secondly, apart from critical level effects we wish to consider the possible role which waves might have on the zonal flow by virtue of other dissipation mechanisms. Recall from Section 2.2 that the zonal mean momentum and heat balance equations allow for the possible deposition of heat and momentum by virtue of the convergences of momentum and heat fluxes due to eddy (wave perturbation) motions. However the Eliassen and Palm relation (9) indicates that in the absence of dissipation or critical levels, mean momentum and heat will not be deposited into the mean flow. However, for some time it has been hypothesized that exponentially-growing gravity waves can reach amplitudes for which they are convectively unstable, leading to their breakdown and generation of sufficient turbulence to inhibit further growth (Hodges, 1969; Lindzen, 1981). This mechamism has often been suggested as the major source of mesospheric turbulence. Evidence of this type of behavior is in fact suggested by the temperature profiles illustrated in Figure 2a. Above the height of wave breaking (and cessation of exponential growth) the divergences of eddy heat and momentum fluxes (cf. Section 2.2) are non-zero and acceleration (or deceleration) of the mean flow can thus occur.

3. RADIATIVE COOLING NEAR THE MESOPAUSE

3.1 Simple Radiative Transfer

According to the familiar Beer's law, the fractional decrease of monochromatic radiation of intensity I_ν due to absorption by an atmospheric consituent of density n is

$$\frac{dI_\nu}{I_\nu} = -\sigma_\nu n\ ds = -\ d\tau \qquad (11)$$

where σ_ν is the absorption coefficient at frequency ν, ds is an incremental distance along the ray path, and λ is the optical depth. This formulation neglects the possibility that the absorbing species may re-emit radiation at the same wavelength or receive radiation from other levels in the atmosphere (we are neglecting scattering altogether in this discussion). For absorption of EUV and UV radiation in the upper atmosphere, the atmosphere is too cool to re-radiate at such short wavlengths. However, when considering infrared radiative transfer, this is not the case. Equation (11) must be rewritten as follows:

$$\frac{dI_\nu^+}{d\tau} = -I_\nu^+ + J_\nu \qquad (12)$$

$$\frac{dI_\nu^-}{d\tau} = I_\nu^- - J_\nu \qquad (13)$$

where J_ν is a 'source function', I_ν^+ and I_ν^+ are the upward and downward components of I_ν, and τ and its sign convention adopted here are given by (11). Under conditions of thermodynamic equilibrium J_ν is just the Planck (Black-Body) function B_ν:

$$B_\nu = \frac{2h}{c^2} \frac{\nu^3}{e^{h\nu/kT}-1} \qquad (14)$$

where h = Planck's constant, c= speed of light, and k=Boltzmann's constant. The integral of B_ν over all directions and frequencies gives the total flux of energy as σT^4 where σ is the Stefan-Boltzmann constant. This is the 'Stefan-Boltzmann Law'. The function B_ν plotted vs. ν for various temperatures yields the well-known family of black-body curves. For a given T, the black body emittance peaks at a wavelength given by 'Wiens Displacement Law':

$$\lambda_{max} = \frac{2897}{T} \text{ (microns)} \qquad (15)$$

and decreases as $|\lambda-\lambda_{max}|$ increases. For typical atmospheric temperatures below 100 Km ranging between 200 K and 300K, λ_{max} lies between 14.5μ and 9.7μ. Atmospheric species which possess strong vibration-rotation bands in this region, and which are therefore likely to strongly influence radiative transfer and heating in the earth's atmosphere (subject to adequate abundances), include CO_2 (15μ), H_2O (6.3μ), and O_3 (9.0μ, 9.6μ, 14μ).

The primary infrared radiative process of importance to the mesosphere and lower thermosphere involves cooling by the 15μ band of CO_2 (however, NO may also be important under some circumstances, especially at auroral latitudes; see Kockarts, 1980). Temperature changes are related to the divergence of the radiative energy flux:

$$\frac{\partial T}{\partial t} = -\frac{1}{\rho c_p} \frac{\partial}{\partial z} (F^+ - F^-) \tag{16}$$

where ρ=total density, c_p is the specific heat at constant pressure, and $F^{+,-}$, represent $\int\int I_\nu^{+,-} \cos\theta \, d\omega d\nu$ where integration is taken over downward facing and upward facing hemispheres, respectively, I_ν being the intensity at an angle θ to the vertical and $d\omega$ an element of solid angle (Houghton, 1979). Under conditions of thermodynamic equilibrium I may to a good approximation be replaced by F in (12) and (13) if $d\tau$ is replaced by 5/6 $d\tau$ and J_ν by the integral of B_ν or πB_ν (Houghton, 1979).

3.2 Non-LTE and Collisional Relaxation

The above discussion assumes that thermodynamic equilibrium prevails so that the source function J_ν in the equation of radiative transfer can be replaced by the black-body function, B_ν. The possibility and ramifications of deviations from thermodynamic equilibrium must be examined. Although the atmosphere as a whole is not in thermal equilibrium, a condition referred to as 'local thermodynamic equilibrium' (LTE) often prevails which says that the vibrational (rotational) energy levels are populated according to a Boltzmann distribution determined by the local kinetic temperature (Craig, 1965). Quantitatively, this means that the fraction of particles (n_i/n) in a level with energy E_i above ground level is proportional to $\exp(-E_i/kT)$. Radiative processes tend to inhibit the establishment of LTE. Thus, LTE can only be maintained if collisions are frequent enough so that the characteristic collisional relaxation time (τ_c) for transfer of vibrational (rotational) energy to kinetic energy is much less than the radiative lifetime (τ_r) of the energy state involved. Since collision frequencies decrease exponentially with height, for a given state there exists some altitude where $\tau_c = \tau_r$. This is called the level of vibrational (or rotational) relaxation (Craig, 1965). Since rotational collisional (radiative) relaxation times are generally much shorter (longer) than vibrational times, the level of vibrational relaxation is usually the relevant one for determining the validity of using the Planck function in the equation of radiative transfer (Curtis and Goody, 1956).

Non-LTE effects on the 15μ CO_2 cooling rate are discussed by Curtis and Goody (1956), Craig (1965), Houghton (1969, 1979), Allen et al (1979), among others. Assuming that cooling to space dominates, the 15μ CO_2 cooling rate is given approximately by

$$\text{cooling rate} \approx \frac{\text{LTE cooling rate}}{1 + \beta} \tag{17}$$

$$\beta = \frac{5}{12} \overline{\tau_\nu^*} \frac{\tau_c}{\tau_r}$$

$\overline{\tau_\nu^*}$ = the "transparency", or the probability of a photon emitted by CO_2 at level z getting to space

τ_r = radiative time constant (.74 sec)

τ_c = collisional relaxation time constant

The time constant for collisional relaxation varies inversely with the atmospheric density. For a τ_c at STP of 30μ sec, $\tau_r/\tau_c \approx 1$ at 73 Km; but, since $\overline{\tau\nu^*}$ is still relatively small at this altitude. LTE does not begin to break down ($\beta > 1$) until about 80 Km (Houghton, 1979). There has been considerable uncertainty as to what value to choose for τ_c. The most recent measurements indicate 15μ sec at 180 K and about 12μ sec at 300K, both at standard pressure. Figure 6 illustrates the effect of various values of τ_c on the non-LTE radiative heating/cooling due to the 15μ band of CO_2. Note that below 85 Km the cooling rate essentially follows the background temperature profile (approximately the same as that illustrated in Figure 1), yielding maximum cooling from the warmest temperature regions. In fact, a small amount of heating occurs near the mesopause which is primarily a result of slow cooling from this cold region, and radiative transfer from the warmer levels below.

3.3 The 'Cooling to Space' and 'Newtonian Cooling' Approximations

A widely used approximation is to assume that the exchange of radiation between different atmospheric layers is negligible compared to the loss of radiation to space. This is the 'cooling to space' approximation:

$$\frac{\partial T}{\partial t} = - \frac{1}{\rho c_p} \frac{\partial F^+}{\partial z} \tag{18}$$

However, this may not be a good approximation in cold regions of space like the polar summer mesopause, which cools slowly due to its low temperature yet receives non-negligible upward fluxes from the warmer layers below. The cooling to space term depends only on the temperature at a particular level, whereas the layer exchange term (omitted in (18)) depends on temperatures at all other levels (Dickinson, 1973). Often associated with the cooling to space approximation is the so-called 'Newtonian Cooling' parameterization of IR cooling which consists of the sum of a cooling for a globally-averaged reference temperature (T_o) and a linear approximation for departures (δT) from T_o:

$$Q_{IR}(T) = Q_{IR}(T_o) + \alpha\ \delta T \tag{19}$$

where $\alpha = \partial Q_{IR}/\partial T$ is the Newtonian Cooling coefficient. For the cooling to space approximation α could be analytically calculated directly from the Planckian function, which is a function only of the temperature at the level of interest. However, it is generally recognized (Dickinson, 1973) that this approximation is inadequate for mesospheric calculations, such that we must determine α (which is actually a function of temperatures at other levels, radiative exchange with other levels, as well as τ_c) empirically by performing accurate radiative culations for a range of temperature profiles. Actually α is perhaps

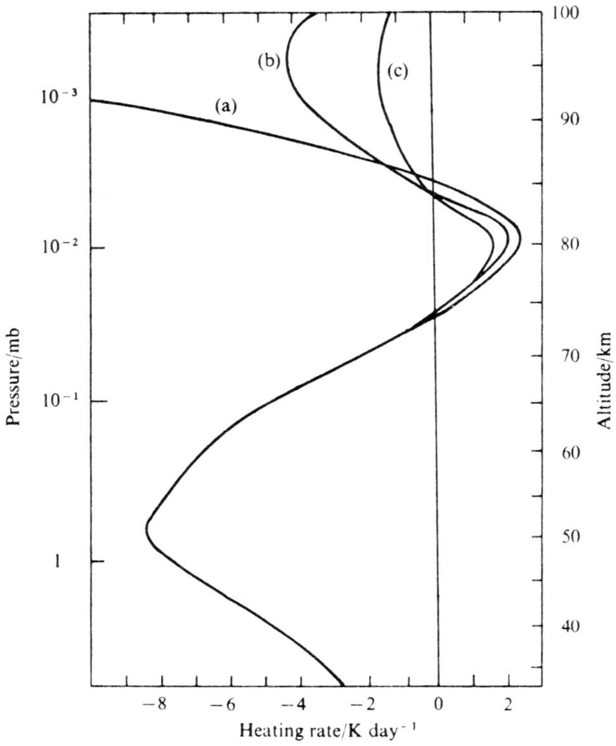

Figure 6. Heating rates for the 15μ band of CO_2 for a mean atmosphere with collisional relaxation times at STP of (a) 2μ sec, (b) 10μ sec, and (c) 30μ sec. (From Williams, 1971; see also Houghton, 1979).

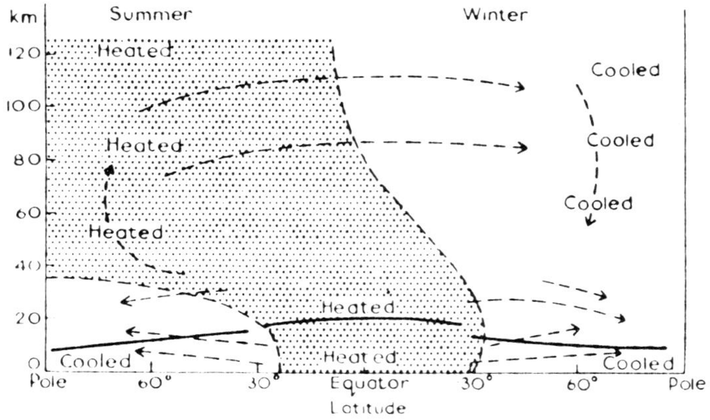

Figure 7. Heat transfers (dashed lines) due to the net radiation field (solar heating minus atmospheric cooling) during solstice conditions. From Murgatroyd (1971).

better termed the 'radiative relaxation rate' since α^{-1} is the time constant (radiative relaxation time) in which radiative processes tend to return a perturbed atmosphere to the mean state. The purpose of pursuing the Newtonian Cooling parameterization is that $Q_{IR}(T_o)$ and α, whose evaluation require extensive and time-consuming radiative transfer calculations, can be computed once and for all; the Newtonian cooling approximation allows the dependence of IR cooling on temperature to be easily included in dynamical calculations which require evaluation of Q_{IR} to be made for many numerical grid points in space and time (Dickinson, 1973). Dickinson's (1973) estimates of the Newtonian Cool ing coefficient did not extend above 70 Km due to uncertainties involved with non-LTE effects (see Section 3.2) and the adequacy of the cooling to space approximation in the cold mesopause region. While accurate specification of the radiative relaxation time has extremely important consequences with regard to dynamical modelling of the mesopaus region, values of α^{-1} used in such efforts have ranged between 10-20 days and 2-3 days. The dynamical consequences of this problem are dealt with in the following section.

4. ZONAL MEAN LATITUDE STRUCTURE

4.1 Radiative and Thermal Budgets

In the stratosphere and mesosphere the thermal structure is primarily accounted for by a balance between O_3 insolation absorption (peaking near 50 Km) and cooling at 15μ by CO_2. Cooling by H_2O (6.3μ, 80μ), O_3 (9.0μ, 9.6μ, 14μ) and other bands of CO_2 (2.7μ, 4.3μ) play a secondary role in the radiative balance over these regions. CO_2 cooling at 15μ is by far dominant above 80 Km. Schumman-Runge radiation absorption by O_2 adds a small but measurable amount of heating around the mesopause, but in fact becomes the dominant radiative heat source between 100 and 120 Km. As indicated schematically in Figure 7, the net radiative heating (heating minus cooling) primarily consists of maximum heating (cooling) at middle to high summer (winter) latitudes over most of the stratosphere and mesosphere. Net perturbation heating rates as calculated for instance by Murgatroyd and Goody (1958) are of order +4 deg day^{-1} and +4 deg day^{-1} at the summer polar stratopause and mesopause, respectively; and -12 deg day^{-1} and -4 deg day^{-1}, respectively, for winter conditions. The above 16 deg day^{-1} heating difference between the summer and winter polar mesopause levels translates to a radiative relaxation time α^{-1} of about 5 days for a corresponding temperature difference of 80K. A recent determination Allen et al (1982) of $\alpha^{-1} \approx$ 2-3 days using the latest measurements of τ_c in fact suggests pole-to-pole heating differences of 20-30 deg day^{-1} at the mesopause level.

It is evident that the atmosphere is not in radiative equilibrium, especially in the mesopause region. The internal transfer of this energy is accomplished by atmospheric motions. In fact the imbalance which results from differential heating and cooling has strong meridional and seasonal variations which drive a meridional circulation as depicted in Figure 7. In the stratopause region upward (downward) motions at the summer (winter) pole decreases (increases) the temperature from its radiative equilibrium value. In the mesopause region the vertical motions are sufficiently large to reverse the meridional temperature gradient imposed by radiative heating. As a point of reference, is useful to estimate the upward (downward) wind speed necessary for the adiabatic cooling (heating) to exactly balance a given imposed net radiative heating (cooling):

$$\frac{\partial T}{\partial t} \approx w \left(\frac{\partial T}{\partial z} + \frac{g}{C_p}\right) = w\Gamma \tag{20}$$

At the mesopause $\Gamma \approx 10\text{K Km}^{-1}$. Therefore, for a summer (winter) heating(cooling) rate of +(-) 10 deg day^{-1}, w is found to be +(-) 1 Km day^{-1} (~ 1.2 msec^{-1}).

4.2 Radiative-dynamical balance and the possible role of wave stress

Before investigating the radiative-dynamical balance of the mesopause region, it is instructive to first examine the stratopause (ca. 50 Km) region circulation. Referring to Figure 8, the net heating distribution during solstice drives a summer to winter circulation and maintains a positive winter to summer latitudinal temperature gradient. Examination of the zonal mean momentum equations (cf. Equation 3):

$$\frac{\partial \bar{u}}{\partial t} = f\bar{v} - \bar{F}_x + (\text{other terms}) \tag{21}$$

$$\frac{\partial \bar{v}}{\partial t} = - f\bar{u} - \frac{\partial \varphi}{\partial y} + (\text{other terms}) \tag{22}$$

and temporarily assuming $\bar{F}_x = 0$, we see that a northerly meridional motion gives rise to an easterly acceleration in the northern (summer) hemisphere (f>0), and a westerly acceleration in the southern (winter) hemisphere (f<0). In fact, the zonal winds are to a good approximation in thermal wind balance with the meridional temperature distribution (cf. Figures 1 and 4; Equation 7), indicating that the summer (winter) easterlies (westerlies) should increase with height for a winter to summer temperature increase. The same types of arguments and balances obtain for equinox conditions, except that the maximum (minimum) net heating occurs in equatorial (polar) regions, the temperature decreases away from the equator, and the zonal winds are therefore westerly in both hemispheres.

Observations (Figure 1 inset) indicate that above about 70 Km the latitudinal temperature gradient reverses in a sense opposite to that expected from radiative equilibrium. This distribution can only be dynamically maintained by strong adiabatic cooling (heating) due to rising (sinking) motions in the summer (winter) hemispheres. Based on the most up to date radiative heating calculations (Allen et al, 1979), Holton (1982) estimates that vertical motions of order 1.5 msec^{-1} are required, which implies (through continuity considerations) mean meri-

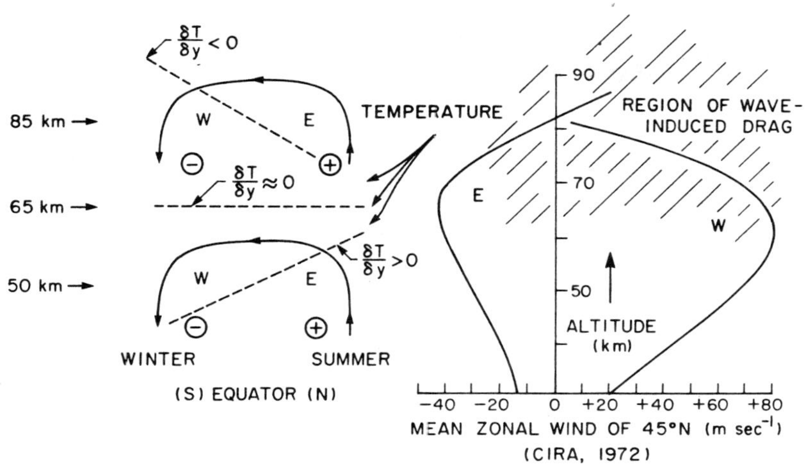

Figure 8. Schematic of radiative-dynamical balances in the mesosphere (left) and vertical profiles of mean zonal easterly (E) and westerly (W) winds at 45°N (right). Plus (+) and minus (-) signs indicate regions of net radiative heating and cooling, respectively. Also indicated on left are directions of net mean meridional flow (solid arrow) and meridional temperature gradients (dashed lines) at various heights.

dional velocities of order 5 msec^{-1} near the mesopause. This translates (cf. Equation 21) to a very large zonal acceleration induced by the Coriolis torque of order 50 m s^{-1} day^{-1}; yet, despite this tendency to accelerate the mean flow, zonal winds are observed to decrease with height above 70 Km, consistent with the thermal wind balance relationship. This suggests that the zonal wind is experiencing some type of mechanical damping (drag) which is not accounted for in the above formulation. As discussed at the end of Section 2.3, one possibility involves the divergence of eddy momentum flux associated with breaking gravity waves. This effect has commonly been parameterized in numerical simulation models as an ad-hoc 'drag' or 'Rayleigh friction' term, where $\overline{Fx}$ is approximated as being linearly proportional to the zonal wind:

$$\frac{\partial \bar{u}}{\partial t} - f\bar{v} = - \nu_R \bar{u} + \text{(other terms)}$$

and where ν_R is a Rayleigh friction coefficient chosen so that the calculated u field approximates observations. The degree to which such a parameterization is successful depends on the radiative relaxation time. For mesopause region radiative relaxation times of order 20 days, Leovy (1964) was able to obtain reasonable simulations of the observed mean zonal wind using a constant $\nu_R^{-1} \sim 15$ days, whereas Schoeberl and Strobel (1977) and Holton and Wehrbein (1980) obtained even better results with a strongly height-dependent Rayleigh friction with time scales (ν_R^{-1}) of order 50 days below 50 Km and $\nu_R^{-1} \sim 1$-2 days near 85 Km. However, on the basis of heating rates calculated by Murgatroyd and Goody (1958) and known temperature distributions a more appropriate choice of radiative relaxation time would have been closer to 5 days. In fact, given the more recently determined radiative relaxation times of order 2-3 days (Allen et al, 1979), even a more efficient and strongly height-dependent Rayleigh friction parameterization is incapable of preventing unrealistically large mean zonal winds near the mesopause (Holton, 1982). The basic problem is that a drag linearly proportional to u cannot possibly provide the force necessary to balance the momentum budget at the zero mean wind level (Holton, 1982).

A recent numerical investigation of this problem (Holton, 1982) uses a slight modification of the more realistic parameterization of Lindzen (1981) for the zonal drag and eddy diffusion effects generated by breaking internal gravity waves in the mesosphere. Lindzen (1981) uses a simple WKB formalism to quantitatively estimate the divergence of the eddy momentum flux and eddy diffusion coefficient between the level of wave breaking (Z_{break}) and the critical level (Z_{crit}) where the wave is absorbed into the mean flow. Further, based on an assessment of critical level filtering in the upper troposphere and stratosphere of gravity waves of tropospheric origin, he is able to show that seasonal variations in the lower level mean flow as reflected in Z_{crit} and consistent with summer/winter differences in turbulent echo intensities measured by the Poker Flat, Alaska MST radar (Balsley et al, 1980). Holton (1982) finds that use of such a formulation in a numerical model of the mean flow yields realistic summer and winter mean zonal wind profiles provided that a gravity wave spectrum is assumed which is characterized by phase speeds in the -20 m s^{-1} to $+20$ m s^{-1} range. Holton contrasts these results with the unrealistic mean wind profiles obtained when Rayleigh friction is used to parameterize the mean flow effects of breaking waves.

Our understanding of the radiative-dynamical balance of the mesopause region is by no means complete. The recent research described above undoubtedly represents the beginning of a surge of new knowlege in the area. New studies are expect to update/modify these results. For this reason, the present description has been kept at a conceptual rather than quantitative level. Rather, the primary purpose here has been to emphasize the essential interplay between radiative processes, the global circulation, and small-scale motions in determining the thermal and dynamical structure of the zonally-averaged mesopause region, and to illustrate the difficulties inherent in modelling the feedback between these processes.

ACKNOWLEDGEMENTS:

This work was supported under AFOSR Grant 81-0090 from the Air Force Office of Scientific Research to Boston College. The efforts of Mrs. Joan Feeney in typing this manuscript are greatly appreciated.

REFERENCES:

Allen, D.C., Haigh, J.D., Houghton, J.T., and Simpson, C.J.S.M.: 1979, Nature 282, pp. 660-661.

Balsley, B.B., Ecklund, W.L., Carter, D.A., and Johnston, P.E.: 1980, Radio Sci. 15, pp. 213-223.

Craig, R.A.: 1965, Meteorology of the Upper Atmosphere, Academic Press, New York, 509 pp.

CIRA: 1972, Cospar International Reference Atmosphere, Akademie-Verlag, Berlin, 450 pp.

Curtis, A.R., and Goody, R.M.: 1956, Proc. Roy. Soc. Lon. A236, pp. 193-206.

Dickinson, R.E.: 1973, J. Geophys. Res. 78, pp. 4451-4457.

Eliassen, A., and Palm, E.,: 1961, Geophys. Publ. 22, pp. 1-23.

Groves, G.V.: 1971, AFCRL Rept. No. 71-0410, Air Force Cambridge Research Laboratories, Bedford, Ma.

Hodges, R.R.: 1969, J. Geophys. Res. 74, pp. 4087-4090.

Holton, J.R.: 1975, The Dynamic Meteorology of the Stratosphere and Mesosphere, Amer. Meteor. Soc., 218 pp.

Holton, J.R.: 1982, J. Atmos. Sci. 39, pp. 791-799.

Holton, J.R., and Wehrbein, W.M.: 1980, Pure Appl. Geophys. 118, pp. 284-306.

Houghton, J.T.: 1969, Quart. J. Roy. Meteor. Soc. 95, pp. 1-20.

Houghton, J.T.: 1979, The Physics of Atmospheres, Cambridge University Press, 203 pp.

Kockarts, G.: 1980, Geophys. Res. Lett. 7, pp. 137-140.

Lindzen, R.S.: 1967, Quart. J. Roy. Meteor. Soc. 93, pp. 18-42.

Lindzen, R.S.: 1973, Bound. Lay. Met. 4, pp. 327-343.

Lindzen, R.S.: 1981, J. Geophys. Res. 86, pp. 9707-9714.

Murgatroyd, R.J., and Goody, R.M.: 1958, Quart. J. Roy. Meteor. Soc. 84, pp. 225-234.

Murgatroyd, R.J.: 1971, in Mesospheric Models and Related Experiments, Fiocco (ed.), pp. 104-121, D. Reidel Publishing Co., Dordrecht-Holland.

Schoeberl, M.R. and Strobel, D.F.: 1978, J. Atmos. Sci. 35, pp. 577-591.

Smith, W.S., Katchen, L.B., and Theon, J.S.: 1968, Meterol. Monogr. 9, Am. Meterol. Soc., pp. 170-175.

Theon, J.S.; 1968, Proc. 3rd Nat. Conf. Aerosp. Meteorol., New Orleans, Am. Meteorol. Soc., pp 449-456.

U.S. Standard Atmosphere: 1976, U.S. Government Printing Office, Wash., D.C.

Williams, A.P.: 1971, Unpublished Ph.D. Thesis, Univ. of Oxford, Oxford, England.

AURORAL EXCITATION AND ENERGY DISSIPATION

M.H. Rees
Geophysical Institute
University of Alaska
Fairbanks, Alaska 99701

1. INTRODUCTION

The spectroscopic aurora is the result of excitation of atomic and molecular constituents in the thermosphere. While the primary source of the aurora is the precipitation of energetic electrons and ions there are several processes that contribute to excitation. Electron impact excitation is a part (or a consequence) of the energy degradation suffered by the electron stream as it penetrates into the atmosphere. Many ionic and chemical reactions initiated by auroral ionization and dissociation lead to the production of excited states. The theoretical concepts and mathematical description relevant to these mechanisms are described in this chapter. A discussion of excitation processes in aurora cannot be divorced from the larger problem of various energy dissipation channels, and the resulting effects on the ionosphere and the neutral atmosphere. The coupling or interrelationship between various auroral processes is important. Theoretical concepts and models may be validated using various observable geophysical parameters. Several equations used in auroral modeling have already been given in preceding chapters on The Thermosphere by Richmond and on The Terrestrial Ionosphere by Schunk, because of similarities in the physical processes that are described. The derivation and general discussion of these equations will not be repeated here but the application to auroral problems is presented.

The penetration of auroral electrons is discussed in section 2. Details of inelastic scattering processes are given in section 3. An example of electron intensity profiles and energy deposition rate is presented in section 4. Excitation by electron impact is detailed in section 5. Auroral ion and neutral chemistry is explained in section 6 leading to chemical excitation processes in section 7. The auroral spectrum is briefly touched in section 8. In section 9 various mechanisms that dissipate auroral energy are discussed. The interaction of auroral protons with the atmosphere is the subject of section 10. The various components are brought together in section 11 in a summary flow chart.

R. L. Carovillano and J. M. Forbes (eds.), Solar-Terrestrial Physics, 753–780.

2. PENETRATION OF AURORAL ELECTRONS INTO THE ATMOSPHERE

Transport of auroral electrons in the magnetosphere is governed by the magnetic field topology, by currents and by waves. In the thermosphere, the region of auroral production, transport of the electrons is dominated by collisions. A transport equation can be derived from the Boltzmann equation,

$$\mu \frac{dI}{dz} = - n(z)\, \sigma\, I(z, E, \mu)$$

$$+ n(z) \iint \sigma(E', E; \mu' \mu)\, I(z, E' \mu')\, dE'\, d\mu' \qquad (1)$$

in which $I(z, E, \mu)$ is the electron differential intensity (cm^{-2} sec^{-1} eV^{-1} $ster^{-1}$), $n(z)$ is the thermospheric particle density (cm^{-3}), μ is the cosine of the pitch angle, and σ is the scattering cross section. The left side of the equation and the first term on the right side are simply Beer's Law for the attenuation of radiation passing through a medium with density n and absorption cross section σ. Indeed, the electron transport equation has the form of the basic equation of radiative transfer. The second term on the right side specifies the scattering of electrons from energy E' to E and angle μ' to μ and the production of secondary electrons. The following assumptions have been made in the formulation of the transport equation. 1. The magnetic field is parallel to the vertical direction, z, and is constant, 2. Azimuthal scattering is symmetric, 3. The atmosphere is plane parallel, 4. The gas is weakly ionized and the electrons are therefore not influenced by collective effects, i.e. by waves, 5. There are no external fields to influence the behavior of the energetic electrons. In addition, the temporal evolution of the electron intensity is governed by auroral processes other than tranport into the atmosphere and the time dependence of the Boltzmann equation need, therefore, not be included in the auroral electron transport equation.

The second term on the right side of equation (1) is decomposed into the elastic and inelastic components of scattering. The cross section for elastic scattering is rewritten in the form

$$\sigma_e(E', E; \mu', \mu) = \delta(E - E')\, P_e(E; \mu', \mu)\, \sigma_e(E) \qquad (2)$$

to show explicitly that it does not involve a change in energy but only angular deflection. P_e is the phase function for elastic scattering and σ_e the cross section. Elastic Coulomb scattering is described by the screened Rutherford cross section. While the electron basically moves in the Coulomb potential of the atom, at large distance the atoms appear effectively neutral. The screening parameter avoids the divergence of the integral obtained for a simple Coulomb potential varying as 1/r. The Rutherford cross section is valid under the same assumptions as the Born approximation for collisions, that the perturbation on the system due to the interaction is small. The differential cross section is

$$\sigma_{COUL}(E, \mu) = \frac{Z^2 e^4}{p^2 v^2 [1 - \mu + 2\varepsilon]^2} = P_e(\mu)\, \sigma_e(E) \tag{3}$$

where Z, e, p and v are the nuclear charge number, electronic charge, momentum and velocity of the electron, respectively, μ is the cosine of the deflection angle and ε is the screening parameter. For 10 keV electrons $\varepsilon \sim 10^{-3}$, increasing to 10^{-1} at 100 eV. It is therefore at low electron energy that screening tends to decrease the Coulomb cross section. Defining the phase function by

$$P_e(\mu) = \frac{\sigma_{COUL}(E, \mu)}{\sigma_e(E)} \tag{4}$$

where

$$\sigma_e(E) = \int_0^{\pi} \sigma_{COUL}(E, \mu) \frac{d\Omega}{4\pi} \tag{5}$$

one obtains

$$P_e(\mu) = \frac{4\varepsilon(1 + \varepsilon)}{(1 - \mu + 2\varepsilon)^2} \tag{6}$$

This scattering function is strongly peaked in the forward direction which causes some mathematical difficulties in the solution of the transport equation.

Inelastic scattering is divided into three processes. The production of excited states accompanied by a discrete energy loss equal to the excitation potential W of the state,

$$\sigma_{in}^{d} = \sum_j P_j^{d}(E', E; \mu', \mu)\, \delta(E' - E - W_j)\, \sigma_j(E) \quad , \tag{7}$$

ionization, a continuum process,

$$\sigma_{in}^{I} = P_I(E', E, \mu', \mu) \frac{d\sigma}{dW}(E', W = E' - E) \quad , \tag{8}$$

which is related to the production of secondary electrons

$$\sigma_{in}^{S} = P_s(E', E, \mu', \mu) \frac{d\sigma}{dW}(E', W = E + I) \quad , \tag{9}$$

where the energy loss W is the sum of the ionization potential and the energy of the secondary electron, E.

It is convenient to define the scattering depth, τ, analogous to the optical depth in radiation transfer,

$$\tau = \sigma_{tot} \int_z^\infty n(z)\, dz \,, \qquad (10)$$

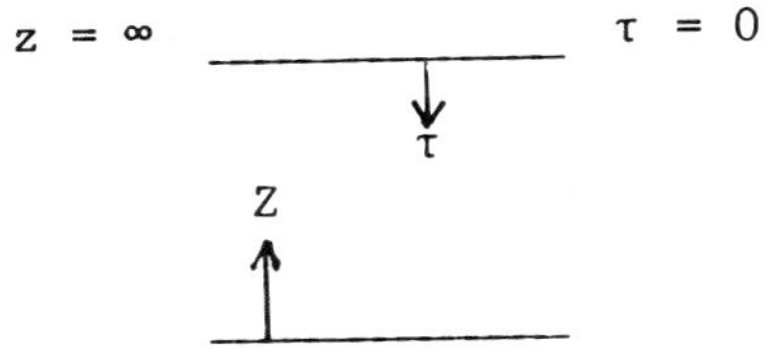

where

$$\sigma_{tot} = \sigma_e + \sigma_{in} \,, \qquad (11)$$

is the sum of the elastic and inelastic cross sections and the integral is the column density of scatterers above altitude z. The transport equation is rewritten in terms of the scattering depth, separating the elastic and inelastic scattering terms,

$$\mu \frac{dI}{d\tau} = I(\tau, E, \mu) - \frac{\sigma_e}{\sigma_{tot}} \int_{-1}^{1} P_e(\mu', \mu)\, I(\tau, E, \mu')\, d\mu'$$
$$- \frac{Q(\tau, E, \mu)}{\sigma_{tot}} \qquad (12)$$

Further simplifying assumptions can be made that are based on the behavior of various scattering processes. In excitation and ionization reactions the electron is scattered in the forward direction,

$$P_j^d \text{ and } P_i = \delta(\mu - \mu') \qquad (13)$$

while the secondary electrons that are produced have an isotropic angular distribution,

$$P_s = 1 \,. \qquad (14)$$

In addition, superelastic collisions may be neglected, i.e. the scattered electrons do not gain energy in any collision. This assumption decouples the transport and energy degradation aspects of the scattering process, and the source term can be written in the form,

$$Q(\tau, E, \mu) = \int_{E' > E} R(\tau, E', E)\, dE' \cdot \frac{1}{2} \int_{-1}^{1} P_{in}(\mu, \mu')\, I(\tau, \mu')\, d\mu' \qquad (15)$$

where

$$R(\tau, E, E') = \frac{\sigma_{in}(E, E')}{\sigma_{tot}} \qquad (16)$$

is the redistribution function for inelastic scattering. The lowest order solution of the transport equation (12) is the two stream solution in which the integrals over all angles are replaced by a two-point quadrature,

$$\int_{-1}^{1} P(\mu, \mu') \, I(\tau, \mu') \, d\mu' \Rightarrow (1 - \beta) \, I^+ + \beta \, I^- \tag{17}$$

where β is the backscatter ratio defined by

$$\beta = \frac{1}{2} \int_0^1 \int_0^1 P(\mu, -\mu') \, d\mu d\mu' \tag{18}$$

In terms of the upgoing, I^+, and downcoming, I^-, intensity the integro-differential equation (12) reduces to two coupled differential equations,

$$\pm\bar{\mu} \frac{dI^\pm}{d\tau} = I^\pm - \frac{\sigma_e}{\sigma_{tot}} [I^\pm - \beta(I^\pm + I^\mp)] - \frac{1}{\sigma_{tot}} Q^\pm \tag{19}$$

which are solved subject to the boundary conditions

$$I^\pm(\tau = \infty) = 0 \quad \text{and} \quad I^-(\tau = 0) = I^\infty \tag{20}$$

where I^∞ is the auroral electron intensity precipitating into the atmosphere. In the two-stream approximation detailed information on the angular distribution of the electron intensity is lost and can only be specified by an averaged cosine of the pitch angle, $\bar{\mu}$. Multistream solutions (4, 8, 16, etc.) are required to investigate the angular evolution of I. Equation (18) can be solved analytically for constant coefficients, as obtained for a single constituent gas. In a real multiconstituent atmosphere the mixing ratio changes with scattering depth, the total scattering efficiency $[= \Sigma \, \sigma_j \, n_j(z)]$ changes, and the coefficients in the equations are no longer constant. The equations must be solved numerically.

Detailed information on the angular distribution of scattered electrons is required for assessing the importance of acceleration mechanisms in the atmosphere, compared to collisional scattering. The two-stream solution does yield a backscatter albedo, i.e. the fraction of the incident intensity that is scattered out of the atmosphere again, and the values predicted by the theory are in reasonable accord with observations. The angular distribution of scattered electrons is not required for computing excitation and emission of auroral radiation. Only the total electron intensity,

$$I(E, z) = 2\pi(I^- + I^+) \tag{21}$$

is needed as a function of energy and altitude.

3. INELASTIC SCATTERING

The phase functions for various inelastic scattering processes, discussed in the preceding section, are simply described by forward scattering for ionization and excitation and isotropic production of secondary electrons. The energy dependent magnitude of various cross sections is derived from experimental measurements and theoretical computations.

Cross sections for ionization of N_2, O_2 and O, the major neutral constituents of the terrestrial atmosphere, are well known from laboratory measurements. The cross sections reach a maximum at an electron impact energy of 100 to 150 eV, slowly decreasing with increasing electron energy. Ionization is the major energy loss process above 100 eV. In addition to degrading the energy of the primary incident electron, ionization is the source of secondary electrons. The cross section for the production of secondary electrons can be expressed by an empirical formula based on laboratory measurements. Designating the energy of the incident primary electron E by E_p and of the secondary electron E by E_s the cross section is given by,

$$\sigma(E_p, E_s) = \frac{C(E_p)}{[1 + E_s/\overline{E}]^{2.1}} \quad , \quad \text{where} \tag{22}$$

$$C(E_p) = \frac{\sigma_I(E_p)}{\overline{E} \quad \tan^{-1}\left(\frac{E_p - I}{2\overline{E}}\right)} \quad ; \tag{23}$$

the ionization cross section is $\sigma_I(E_p)$, the ionization potential is I, and $\overline{E}$ is a spectral shape parameter. I and $\overline{E}$ have characteristic values for each gas; for N_2, $\overline{E}$ = 13.0 eV and I = 15.6 eV, while for O_2, $\overline{E}$ = 17.4 eV and I = 12.2 eV. Several excitation reactions contribute to energy degradation of electrons while producing excited states of atoms and molecules. They include a multitude of electronic states, vibrational and rotational states in molecules, and fine structure levels in the ground state of atomic oxygen. To illustrate the relative importance of various inelastic processes as a function of the electron energy the cross sections for N_2 are shown in Figure 1. Above 25 eV ionization and the excitation of many high-lying singlet states leading to dissociation of the molecule have by far the largest cross sections. The potential energy curves for N_2 and N_2^+ are shown in Figure 2. There are a multitude of electronic states to which the molecule can be excited but only states for which the cross section is appreciable and therefore contributes to energy degradation, or states that are upper levels of observable radiative transitions are included in the compilation of cross sections. Excitation of vibrational and rotational levels of the $X^1\Sigma_g^+$ ground state of the molecule has a large cross section and is a large energy sink

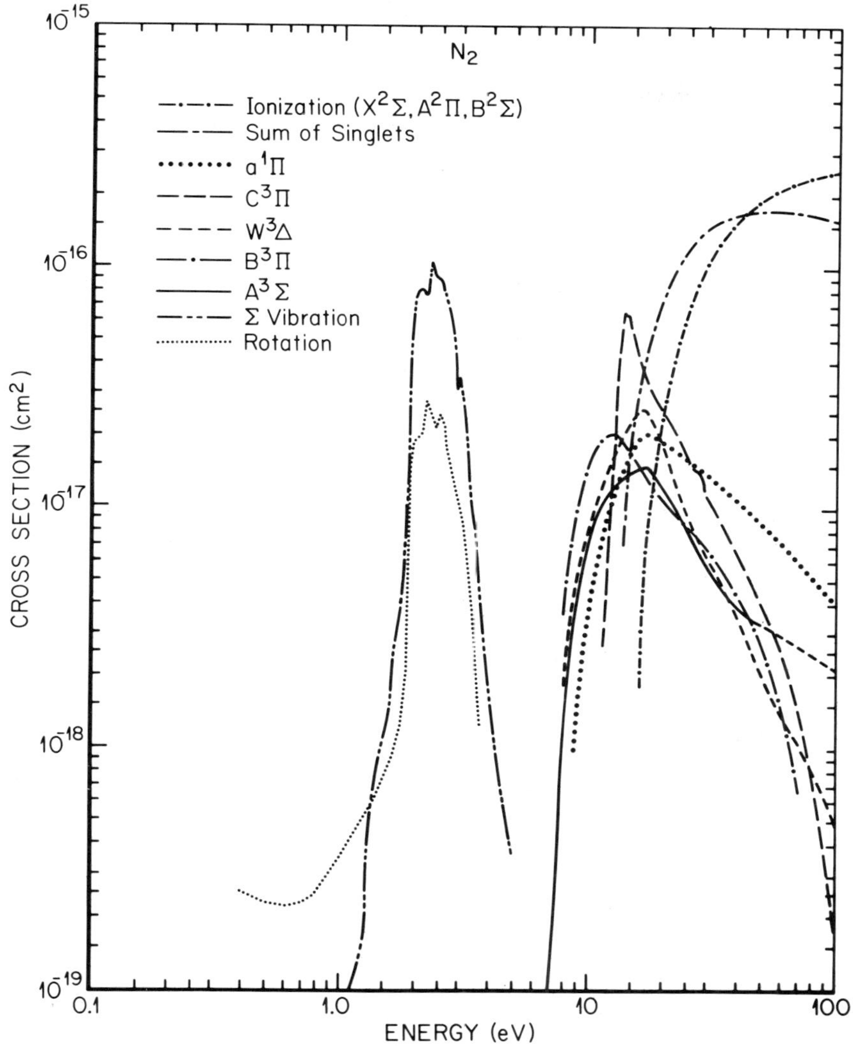

Figure 1. Inelastic cross sections for the production of several states in N_2 molecules by electron impact.

for electrons between about 1.5 and 4 eV. Molecules excited to high vibrational levels of the $C^3\pi_u$ state and to the $a^1\pi_g$ state predissociate. Likewise, the singlet states with excitation potential above 12 eV (e.g. $b'^1\Sigma_u^+$, $c^1\pi_u$, $b^1\pi_u$, etc.) lead to dissociation of the molecule, providing a source of odd nitrogen atoms in aurora.

The cross sections for molecular oxygen and atomic oxygen are not shown here. (See, however, bibliography at the conclusion of this chapter). They are obviously different, and become important in different energy intervals. For example, the vibrational cross section in O_2 has a maximum value of about 2×10^{-17} cm^2 at 0.6 eV,

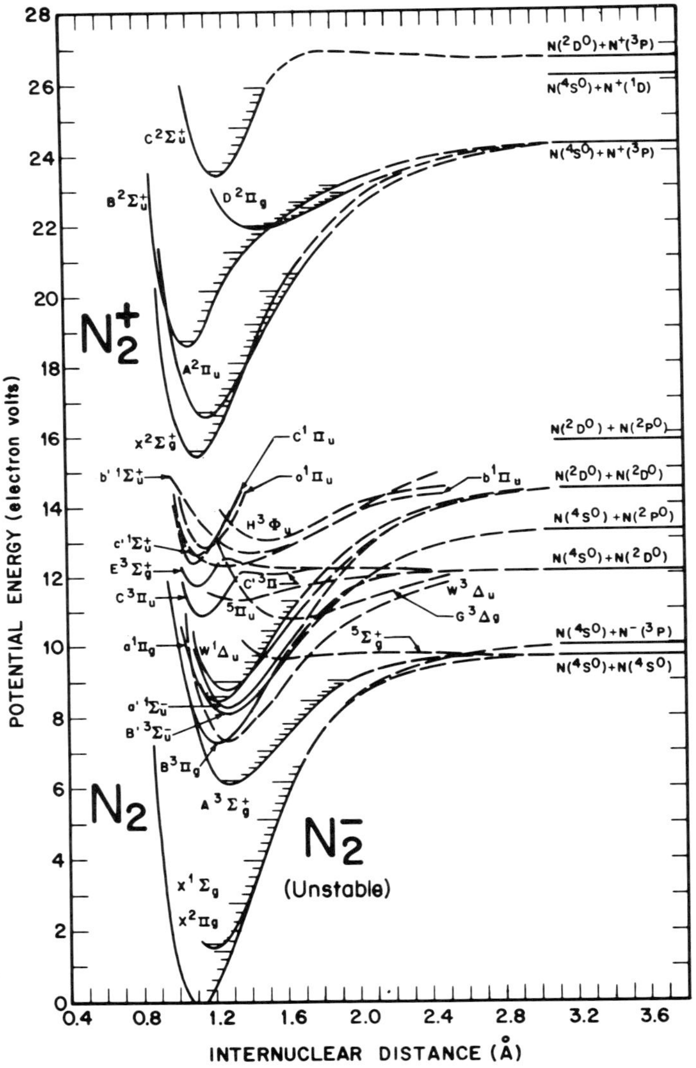

Figure 2. Potential energy curves for N_2 and N_2^+. (A. Lofthus and P.H. Krupenie, The Spectrum of Molecular Nitrogen, J. Phys. Chem. Ref. Data, 6, 113, 1977.

and the atomic oxygen fine structure excitation cross section is almost constant between 0.1 and 2.5 eV with a value of 4×10^{-17} cm^2.

Energy degradation of the auroral electrons is completed by sharing any remaining energy with the ambient thermal electron gas. This process is discussed in the chapter by Schunk in connection with energy degradation of photoelectrons. This loss channel becomes important at electron energies below about 5 eV. The net effect is to heat the electron gas. Further discussion is given in the section on energy dissipation.

4. ELECTRON INTENSITY AND ENERGY DEPOSITION

Using the elastic and inelastic cross sections described above and specifying a neutral atmosphere, the transport equations (19) can

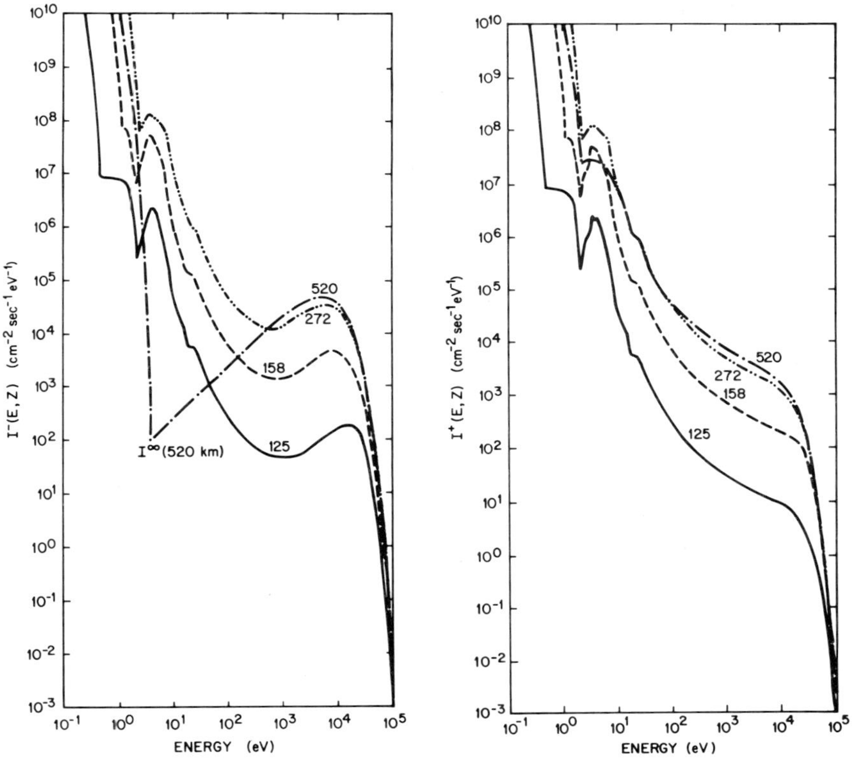

Figure 3: Electron intensity at several altitudes for a typical auroral spectrum characterized by a maxwellian energy distribution: (left) downward intensity, I^-, and (right) upward intensity, I^+.

be solved numerically (see bibliography). The results for a typical auroral electron stream are shown in Figures 3a and b. The intensity at the "top of the atmosphere" (520 km) is assumed to have an isotropic angular distribution and a maxwellian energy distribution, E exp(-E/5) with electron energy given in keV. The energy input rate is assumed to be 10 ergs cm^{-2} sec^{-1}. The electron intensity is shown at four altitudes as a function of electron energy. Comparison of the upward and downward intensity indicates that secondary electrons dominate over primaries below a few hundred eV energy since the secondaries have an isotropic angular distribution. The general shape of the electron spectrum does not strongly vary with altitude while the detailed structure in the spectra is a consequence of the structure in the total inelastic scattering cross section. The high energy tail of the thermal ambient electron gas determines the lowest energy to which the degradation computations must be carried out. As shown in Figure 3, the electron temperature varies with altitude.

The energy deposition rate at altitude z is given by

$$q(z) = \sum_j \sum_k n_j(z) \int_{W_{kj}}^{\infty} W_{kj}\, \sigma_j^{\,k}(E)\, I(E, z)\, dE \quad , \qquad (24)$$

where the summation extends over all excitation and ionization states and all neutral species. The altitude profile of energy deposition rate for the incident electron spectrum illustrated in Figure 3 is shown in Figure 4. Integration over altitude shows that energy is conserved, i.e. the energy absorbed by the atmosphere and the energy backscattered out of the atmosphere equal the energy incident on the

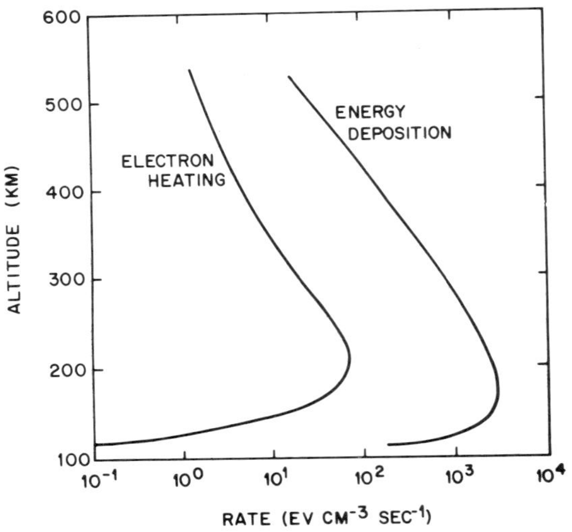

Figure 4. Altitude profiles of the energy deposition rate and electron heating rate for the typical aurora described in the text.

top of the atmosphere. Energy is conserved but particles are not since a copious amount of secondary electrons is produced.

5. EXCITATION BY ELECTRON IMPACT

There are several processes that lead to the production of excited states and it is convenient to distinguish between the direct process, electron impact, and indirect processes that involve chemical and ionic reactions. The electron impact reactions that lead to excitation are basically the same reactions accounting for electron energy loss discussed in the preceding sections as part of the degradation process. However, while previously only the energy lost by the electron was of consequence, in this section interest is focused on the excited state in which the atom, molecule, or ion is formed. The ionization cross section of N_2 shown in Figure 1 refers to the production of all ions, regardless of electronic state. Yet, reference to Figure 2 shows that the molecular ion could be in any of five electronic states or could dissociate to form an atomic ion. The majority of the ions are produced in excited states. For example,

$$e^* + N_2 \rightarrow e^* + N_2^+(B^2\Sigma_u^+) + e \quad (25)$$

The excited state radiates by an allowed transition to the ground state of the ion,

$$N_2^+(B^2\Sigma_u^+) \rightarrow N_2^+(X^2\Sigma_g^+) + h\nu \text{ (1 N.G.)} \quad (26)$$

producing the First Negative Bands of ionized molecular nitrogen with radiations at 3914Å, 4278Å, 4709Å, etc. Production of the various vibrational levels follows the Frank-Condon factors. Excitation of N_2^+ $(A^2\pi_u)$ leads to emission of the Meinel bands in the infrared portion of the auroral spectrum. Only examples of various processes are given here and a more extensive survey is available in references listed at the end of this chapter.

Dissociative ionization,

$$e^* + N_2 \rightarrow e^* + N^+(3d^3F^0) + N + e \quad (27)$$

may leave the N^+ ion in an excited state, as well as the product N atom. In the example shown the excited N^+ ion decays by radiation to a lower level through an allowed transition

$$N^+(3d^3F^0) \rightarrow N^+(3p^3D) + h\nu \text{ (5001Å)} \quad (28)$$

The product state is populated by cascading from a higher level and further emission results until the end product is in the ground state or a highly metastable state. Since the electron intensity generally increases with decreasing energy (Fig. 3) excitation of electronic states in N_2 is an efficient process if the cross section is not too small (Fig. 1). There are numerous states that are excited but we only give one example,

$$e^* + N_2(X^1\Sigma_g^+) \rightarrow e^* + N_2(C^3\pi_u) \qquad (29)$$

followed by the allowed radiative transition

$$N_2(C^3\pi_u) \rightarrow N_2(B^3\pi_g) + h\nu(2\ P.G.) \qquad (30)$$

with emission of the Second Positive Group bands. The $N_2(B^3\pi_g)$ level is thereby populated by cascading from a higher level, but it can also be excited by direct electron impact. Another excited state is produced by radiative decay,

$$N_2(B^3\pi_g) \rightarrow N_2(A^3\Sigma_u^+) + h\nu(1\ P.G.) \qquad (31)$$

with radiation of the First Positive Group bands. The $N_2(A^3\Sigma_u^+)$ state, which is also populated by direct electron impact, is a metastable state that decays radiatively to the ground state of the molecule,

$$N_2(A^3\Sigma_u^+) \rightarrow N_2(X^1\Sigma_g^+) + h\nu(V - K) \qquad (32)$$

with emission of the Vegard-Kaplan bands in the ultraviolet and near U.V. part of the spectrum, but is also subject to collisional deactivation, which will be discussed in a subsequent section.

Although dissociation of N_2 occurs through excitation of predissociation levels the process may be written as

$$e^* + N_2 \rightarrow e + N(^4S) + N(^2D) \qquad (33)$$

in which one or both product atoms may be left in an excited state such as the $N(^2D)$ state used as an example. This is a highly metastable state that predominantly takes part in chemical reactions or is collisionally deactivated before undergoing a radiative transition. At high altitude, however, where the density is low, emission of a doublet does occur,

$$N(^2D) \rightarrow N(^4S) + h\nu(5200Å) \qquad (34)$$

To summarize, there are several electron impact reactions that lead to excitation and emission. The excitation or production rate is computed from the general equation

$$P_j^k(z) = n_j(z) \int_W^\infty \sigma_j^k(E)\ I(E, z)\ dE \qquad (cm^{-3}\ sec^{-1}) \qquad (35)$$

where $n_j(z)$ is the density of species j at altitude z, $\sigma_j^k(E)$ is the energy dependent cross section for excitation of state k in species j, and I(E, z) is the electron intensity spectrum at altitude z.

Analogous electron impact processes occur in O_2 and in O (excepting dissociation in the latter), and the interested reader may consult the references at the end of this chapter.

6. AURORAL ION CHEMISTRY AND TRANSPORT

The indirect processes (those not involving direct electron impact) that lead to the production of excited states followed by emission of auroral spectral features require knowledge of the composition and density of many positive ions and odd nitrogen species. The concentration of each species is given by the continuity equation

$$\frac{\partial n_j}{\partial t} = - \nabla \cdot (n_j \vec{v}_j) + P_j - n_j L_j \qquad (36)$$

where the first term on the right side describes the transport of n_j, the second term is the local production and the third term represents local losses. This equation has already been given in the chapter by Schunk where the diffusive transport term is set forth in detail. Auroral ion chemistry includes a larger number of species than are considered in the solar case and the source term, P_j, may be subject to rapid temporal variations in response to active and variable auroral input. The transport term is included in the continuity equation for the major ions and for all species that have a long lifetime for reaction or radiation. These are $O^+(^4S)$, NO^+, O_2^+, $N(^4S)$, $N(^2D)$, and NO. Species that are assumed to be governed by chemical equilibrium are N_2^+, N^+, $O^+(^2D)$, $O^+(^2P)$, $O_2^+(a^4\pi)$, $N(^2P)$, and H^+. The ions that are initially produced by electron impact ionization (e.g. N_2^+) are not necessarily the most abundant species that are observed. For example, NO^+ is the dominant ion in the auroral E-region even though neutral NO is a minor constituent. Numerous chemical and ionic reactions operate, each with its own reaction rate coefficient which may be temperature (and therefore altitude) dependent. A reaction that represents a loss process for one species may be a production source for another. This circumstance couples the 13 continuity equations which must therefore be solved simultaneously.

The energy balance in a reaction is determined by the ionization and dissociation energies of the reactants and products. Current accepted values for these parameters that are used in auroral ion chemistry are given in Table 1.

Table 1
Ionization and Dissociation Energies

Species	Ionization (eV)	Dissociation (eV)
N_2	15.58	9.76
O_2	12.075	5.115
NO	9.25	6.50
O	13.61	
N	14.54	
H	13.59	
H_e	24.58	

Excited states in the ground configurations of O, O^+ and N play an important part in auroral ion and neutral chemistry, and the excitation potentials of these states, together with $O_2^+(a^4\pi)$ and $N_2(A^3\Sigma_u^+)$ are given in Table 2.

Table 2
Excitation Energies

Species	Excitation (eV)
$O(^1D)$	1.97
$O(^1S)$	4.12
$O^+(^2D)$	3.31
$O^+(^2P)$	5.0
$N(^2D)$	2.38
$N(^2P)$	3.57
$O_2^+(a^4\pi)$	4.05
$N_2(A^3\Sigma_u^+)$	6.2

To illustrate the various types of reactions that contribute to auroral chemistry, the sources and sinks (production and loss) for NO are listed, with their respective reaction rate coefficient (in units $cm^3\ sec^{-1}$).

Reactions leading to chemical production of NO

$N(^2D) + O_2 \rightarrow NO + O + 3.76$ eV $\quad 6 \times 10^{-12}$ (37)

$N(^4S) + O_2 \rightarrow NO + O + 1.385$ eV $\quad 4.4 \times 10^{-12} \left(\frac{-3220}{T_n}\right)$ (38)

$N(^2P) + O_2 \rightarrow NO + O + 4.95$ eV $\quad 5 \times 10^{-12}$ (39)

$N^+ + O_2 \rightarrow O^+ + NO + 2.31$ eV $\quad 2 \times 10^{-11}$ (40)

$O_2^+ + N_2 \rightarrow NO^+ + NO + 0.933$ $\quad 5 \times 10^{-16}$ (41)

Reactions leading to chemical loss of NO

$NO + N(^4S) \rightarrow N_2 + O + 3.25$ eV	$1.5 \times 10^{-12} \sqrt{T_n}$	(42)
$NO + N(^2D) \rightarrow N_2 + O + 5.63$ eV	7×10^{-11}	(43)
$NO + O_2^+ \rightarrow NO^+ + O_2 + 2.86$ eV	4.4×10^{-10}	(44)
$NO + N_2^+ \rightarrow NO^+ + N_2 + 6.33$ eV	3.3×10^{-10}	(45)
$NO + O^+ \rightarrow NO^+ + O + 4.36$ eV	8×10^{-13}	(46)
$NO + h\nu(\lambda < 1910Å) \rightarrow N + O$	8.3×10^{-6} sec^{-1}	(47)
$NO + h\nu(\lambda < 1340Å) \rightarrow NO^+ + e$	6×10^{-7} sec^{-1}	(48)

All the above reactions are exothermic and the excess energy is given. This excess energy may appear as kinetic energy of the products or as internal excitation of electronic or vibrational states. For example, $N(^2D) + O_2 \rightarrow NO + O(^1D) + 1.79$ eV produces an excited oxygen atom and there is still some excess energy remaining. Some reactions in the preceding list have reaction rate coefficients that are temperature dependent and the magnitude of the various coefficients vary over a wide range. It should be pointed out, however, that the rate at which a reaction proceeds is the product of the densities of the reactants and the rate coefficient and it is this product which determines the relative importance of various processes.

Similar sets of production and loss reactions have been compiled for the other twelve species of ions and minor neutrals. The interested reader should consult the references. For each species some reactions are more effective than others at a given altitude, but the same ones are not always dominant at all altitudes in view of changing composition, density and temperature. This accounts for the necessity of retaining so many reactions in deriving the auroral ion composition between 90 and 500 km.

A flow chart of ion production and major ion chemistry is given in Figure 5. The three major neutral species are ionized by energetic electron or photon impact. (Dissociative ionization is also indicated for O_2 to form O^+ ions, but the corresponding process for N_2 to yield N^+ is not shown in this simplified chart.) Reaction between an ion and a neutral to form a different ion and neutral is governed by the ionization and dissociation energies to insure a net exothermic process. Excited O^+ ions can also decay to the ground state of the ion by radiation or collisional deactivation. In general, atomic ions are converted to molecular ions, particularly NO^+ which has the lowest ionization potential. The molecular ions are destroyed by dissociative recombination to produce atomic oxygen and nitrogen. Excited states of O and all states of N participate in

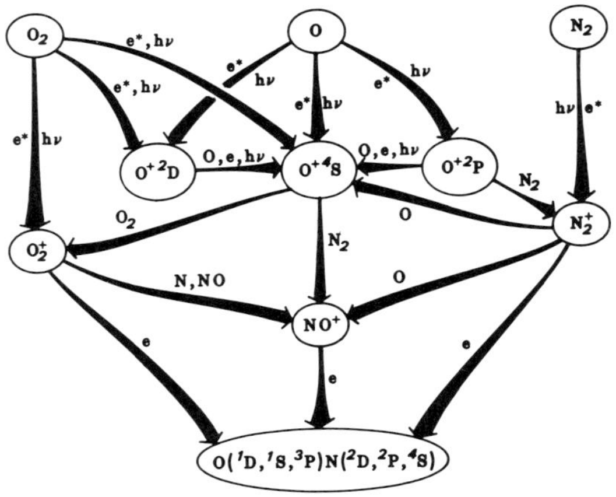

Figure 5: Schematic flow chart of auroral ion production and chemistry. Only the major reactions and species are shown.

the neutral chemistry. Ground state O becomes indistinguishable from the background atomic oxygen after thermalization of the hot atoms.

Atomic nitrogen is a minor neutral species in the terrestrial atmosphere and production by auroral processes has a major impact on the odd nitrogen chemistry and energy balance in the thermosphere. The effects of aurorally produced odd nitrogen are felt not only in the thermosphere but also in the mesosphere and upper stratosphere by downward transport of NO. The diagram in Figure 6 traces the produc-

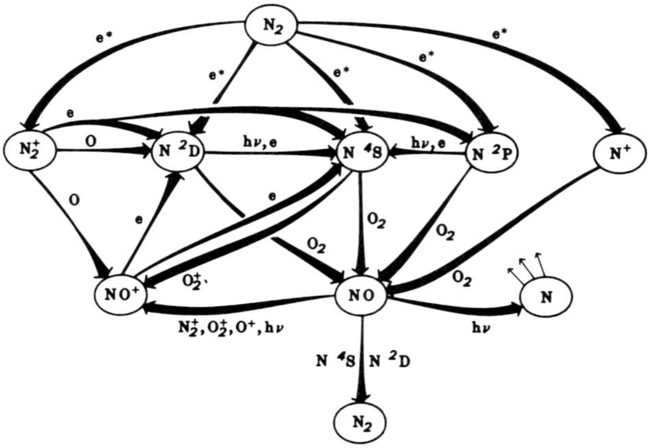

Figure 6. Schematic flow chart of the odd-nitrogen production and chemistry.

tion, conversion and destruction of odd nitrogen in the thermosphere. Electron impact dissociation of N_2 (eq. 33) is a direct source of N atoms; approximately one half are produced in the excited ^2D and ^2P states, the remainder in the ground ^4S state. About one fourth of the ionization of N_2 yields N^+ ions (eq. 27) in the ground or in excited states. This process is also a direct source of odd nitrogen. The N_2^+ ions react with O atoms or with electrons to form nitrogen atoms or NO^+ ions. The latter is both a source and a sink of N atoms. Both N^+ and N are converted to NO which reacts with N to form the major gases N_2 and O. This last process, $NO + N(^4S) \rightarrow N_2 + O$, (eq. 42) is slow and has a temperature dependent reaction rate coefficient allowing NO to be a longlived species. Photodissociation and photoionization destroy NO but not the resulting odd nitrogen which is then converted to NO, thereby prolonging the cycle.

7. CHEMICAL EXCITATION

Chemical reactions are an important source of excited species whence auroral radiation originates. An example of each type is given but numerous additional ones occur and have a major role in exciting the auroral spectrum.

(a) Dissociative recombination

$$O_2^+ + e \rightarrow O + O(^1D), \quad \text{followed by} \qquad (49)$$

$$O(^1D) \rightarrow O(^3P_2) + h\nu(6300\text{Å}) \qquad (50)$$

(b) Chemical reaction

$$N(^2D) + O_2 \rightarrow NO + O(^1D), \quad \text{followed by} \qquad (51)$$

$$O(^1D) \rightarrow O(^3P_2) + h\nu(6300\text{Å}) \qquad (52)$$

(c) Energy transfer

$$N_2(A^3\Sigma) + O(^3P) \rightarrow N_2(X^1\Sigma) + O(^1S), \quad \text{followed by} \qquad (53)$$

$$O(^1S) \rightarrow O(^1D) + h\nu(5577\text{Å}), \qquad (54)$$

which also populates the $O(^1D)$ state by cascading.

Only radiative decay has been shown for the excited state products, however, metastable states are also deactivated by collisions. The "auroral green line" is used as an example of excitation by several processes, the excitation rate, and the emission rate. The upper state for the green line is $O(^1S)$ and there are three important production mechanisms, (a) the energy transfer reaction (equ. 53), $N_2(A^3\Sigma) + O(^3P) \rightarrow N_2(X^1\Sigma) + O(^1S)$, with a production rate $\beta\ n(N_2 A\ \Sigma)\ n(O)$ ($cm^{-3}\ sec^{-1}$), where β is the reaction rate coefficient. (b) dissociative recombination, $O_2^+ + e \rightarrow O + O(^1S)$, with a production rate $\alpha\ n(O_2^+)\ n(e)$ ($cm^{-3}\ sec^{-1}$), where α is the appropriate reaction rate coefficient. Processes (a) and (b) require computation of the concentrations of $N_2(A^3\Sigma)$ molecules, of O_2^+ ions, and of the electron density as a function of altitude. Indeed, the concentration of all positive ions must be obtained to compute the dissociatve recombination rate because the electron density is equal to the sum of the densities of positive ions,

$$n(e) = \sum_j n_j(X^+) \tag{55}$$

Charge neutrality is assumed and the negative ion density is negligible in the auroral thermosphere. The third (c) source of $O(^1S)$ atoms is electron impact excitation, $e^* + O(^3P) \rightarrow e^* + O(^1S)$ with a production rate given by $n(O) \int_E \sigma(E)\ I(E)\ dE$, where $n(O)$ is the atomic oxygen density, $\sigma(E)$ the energy dependent excitation cross section, and $I(E)$ the differential intensity of energetic electrons. The preceding expression gives the volume excitation rate at each altitude, z, and both $n(O)$ and I are a function of height. Auroral excitation of $O(^1S)$ is independent of the mechanism whereby the excited state is produced in the airglow, energy transfer from electronically excited O_2, $O_2^* + O(^3P) \rightarrow O_2 + O(^1S)$, in which the O_2^* molecules are formed by three body recombination, $O + O + M \rightarrow O_2^* + M$. Airglow and auroral mechanisms add in the total production rate of $O(^1S)$ atoms.

Radiative decay of $O(^1S)$ produces the auroral green line, but radiation must compete with collisional deactivation at altitudes where the deactivating species has a large concentration. The emission rate of 5577Å is given by

$$q(5577) = \Sigma \text{ excitation rates} \times \frac{A_{5577}}{A_{^1S} + k\ n(O)} \tag{56}$$

The A's are the Einstein radiative transition probabilities, $A_{^1S}$ including both the $(^1D - {}^1S)$ and the $(^3P - {}^1S)$ transitions. The $O(^1S)$ state is deactivated by collisions with atomic oxygen, k being the appropriate rate coefficient. Collisional deactivation becomes important below an altitude of about 95 km.

8. THEORY OF THE AURORAL SPECTRUM

The spectrum of the aurora is rich in line and band emissions. Ground-based observations are limited to a spectral region in which the atmosphere is transparent, i.e. the visible and near infrared, and it was not until spectrometers were carried on board rockets and satellites that auroral spectra in the ultraviolet and infrared regions were measured. Observations have now been made at wavelengths as short as 400Å, limited only by instrument sensitivity to weak emission features. Identification of lines and bands has improved with higher spectral resolution of the measurements. Advances in technology continually provide improved instrumentation that produces higher quality auroral spectra. Selected references are given in the bibliography.

The excitation mechanisms discussed in the two preceding sections can account for the auroral spectrum qualitatively, but a quantitative theoretical prediction can only be made for a few selected features for which the emission cross sections have been measured or reliable excitation cross sections have been computed. It is appropriate, therefore, to provide more general theoretical guidelines which are applicable to a wide range of emission features, using several lines of atomic oxygen for discussion purposes.

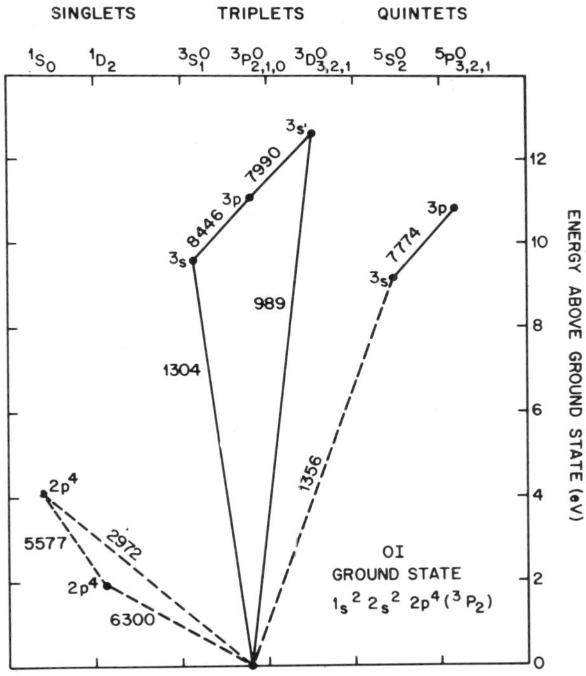

Figure 7. Partial energy level diagram of atomic oxygen showing the relationship between several auroral emissions in the visible and ultraviolet regions of the spectrum.

Figure 7 shows a partial energy level diagram of atomic oxygen. The ground state is $^3P_2$, but the $2p^4$ ground configuration also has the metastable $^1S_0$ and $^1D_2$ states that lie 4.12 eV and 1.97 eV above the $^3P_2$ state. The excitation energy of the next lowest state, 3s $^5S^0_2$, is 9.11 eV with many other states at still higher potentials. There are no auroral chemical reactions, discussed in sections 6 and 7, that have an exothermicity as high as 9 eV, but many have excess energy up to about 6 eV. Thus, if the excess energy in a reaction goes into electronic excitation it will be metastable states in the ground configuration that are populated. This will occur not only in OI but also in OII, NI and NII (the spectroscopic notation of roman numerals referring to the neutral atom, I, and singly ionized species, II). It is for this reason that emission rates from metastable states are higher than those resulting from allowed transitions. The allowed states are populated by electron impact excitation of atoms or by dissociative ionization of molecules. Transitions from excited states (3s, 3p, 3d) to the ground state are accompanied by radiation in the ultraviolet (see Fig. 7); these lines are prominent features of EUV and FUV auroral spectra. Higher levels are also populated, and allowed transitions between such levels occur in auroral spectra. Simultaneous measurements of several lines provides information on the efficiency of populating various states by electron impact. Atomic oxygen lines at 7990Å, 989Å, 8446Å and 1304Å are all prominent features in auroral spectra. Relative emission rates of these lines are used to compute the population rates of excited states, 3s $^3D^0$, 3p $^3P^0$, and 3s $^3S^0$. These states are populated by cascading from higher levels, in addition to direct electron impact. Knowledge of excitation cross sections, branching ratios and transition probabilities are required to compute the auroral emission spectrum, given the electron intensity, I(E, z) and the concentration of the gas, n(0). Conversely, observations of the auroral spectrum may yield data for deriving basic atomic parameters, or an unknown density.

Auroral spectra in the FUV were initially obtained with broad band detectors and the observed emission features were identified with band emissions of molecular nitrogen from high-lying excited states of the molecule. It was surprising, therefore, that at higher resolution the spectrum below 1500Å is dominated by atomic lines, and the N_2 band emissions are weak by comparison. In section 3 it was noted that many high-lying bound singlet and Rydberg states in N_2 are predissociation states that decay by dissociation of the molecule rather than radiatively. A fraction of the dissociation products, both N and N^+, are produced in highly excited states that are the upper levels for transitions with emission of NI and NII lines in the FUV spectrum. Another source of these lines is electron impact excitation or ionization of atomic nitrogen, however, the abundance of N atoms is too small for this process to be a major contributor, except perhaps in aurorae that originate at high altitude, where the relative abundance of atomic species is high compared to molecular gases. Interpretation of FUV auroral spectra is an important current research topic.

9. DISSIPATION OF AURORAL ENERGY

Precipitating auroral particles inject energy into the atmosphere at some rate that can vary from undetectably small values to a few hundred ergs cm^{-2} sec^{-1} over a limited area and for a short time interval. The processes described in the previous sections are all sinks of auroral energy. For example, the mean energy required for the formation of an electron-ion pair is about 34 eV. Part of this energy serves to maintain the enhanced ionization in the atmosphere due to aurora. The chemical and ionic reactions distribute much of the energy into other channels. For example, production of a long lived species such as NO is a sink of auroral energy in the form of stored chemical energy. Although eventually the NO recombines with N, during its long life it can diffuse downward out of the thermosphere into the mesosphere. Transport is therefore a sink of auroral energy. Auroral radiation at all wavelengths is a sink of energy. Emissions at wavelengths at which the atmosphere is optically thin radiate out of the region of excitation but radiation for which the atmosphere is optically thick is entrapped and much is dissipated locally by secondary processes. Optical emissions in the visible and near infrared spectrum, those that comprise the visual aurora, account for only a few percent of auroral energy dissipation.

By far the largest fraction of auroral energy input rate goes into heating of the neutral gas, the ions and the electrons. There are several processes that contribute to neutral gas heating:

(a) collisional deactivation of metastable states produced by electron impact. The two important excited states are $N_2(A^3\Sigma)$ and $O(^1D)$ and heating is the result of the following processes.

$$e^* + N_2 \rightarrow e^* + N_2(A^3\Sigma) \quad \text{followed by} \tag{57}$$

$$N_2(A^3\Sigma) + M \rightarrow N_2(X^1\Sigma) + M + 6.2\ \text{eV} \tag{58}$$

and

$$e^* + O \rightarrow e^* + O(^1D) \tag{59}$$

followed by

$$O(^1D) + N_2 \rightarrow O(^3P) + N_2 + 1.97\ \text{eV} \tag{60}$$

The excess energy goes into translational energy of the products.

(b) There are about thirty exothermic ion-chemical reactions that are sources of kinetic energy, i.e. of heating. Reactions such as

$$N_2^+ + e \rightarrow N + N + 5.82\ \text{eV} \tag{61}$$

are sources of neutral heating, the energy being shared equally between the products. Reactions such as

$$O^+ + N_2 \rightarrow NO^+ + N + 1.088 \text{ eV} \quad (62)$$

are a source of heating for ions as well as neutrals, and it is assumed that the excess kinetic energy is shared by the products in inverse proportion to their masses.

(c) The four neutral chemical reactions that involve atomic nitrogen are a major source of neutral heating. These are,

$$N(^4S) + O_2 \rightarrow NO + O + 1.4 \text{ eV} \quad (63)$$

$$N(^2D) + O_2 \rightarrow NO + O + 3.77 \text{ eV} \quad (64)$$

$$N(^4S) + NO \rightarrow N_2 + O + 2.68 \text{ eV} \quad (65)$$

$$N(^2D) + NO \rightarrow N_2 + O + 5.63 \text{ eV} \quad (66)$$

Although the oxygen atoms produced by reaction (64) may be in the excited ^1D state, collisional deactivation of this state is a source of neutral heating. The flow chart in Figure 6 shows that the four reactions involving odd nitrogen, (63) to (66), follow every N_2 ionization process.

Figure 6 illustrates another heating process.

(d) Dissociation of N_2 produces atoms with excess kinetic energy,

$$e^* + N_2 \rightarrow e^* + N + N + \text{k.E.} \quad (67)$$

It has already been noted that dissociation in N_2 occurs through predissociation levels. There are several electronic states involved, and many vibrational levels in each electronic state. Since each level of every state is at a different potential above the ground state of the molecule the hot nitrogen atoms acquire a distribution in energy rather than a single value. The hot atoms may enhance the rate at which reactions such as (65) proceed.

(e) The ion gas is frequently hotter than the neutral gas, i.e. $T_i > T_n$. Ions are heated by chemical reactions such as (62) and, more importantly, by orthogonal electric fields, as discussed below. Cooling of the ion gas by elastic collisions with the neutral gas is a source of heating for the neutrals.

(f) The electron gas, likewise, may be at a higher temperature than the neutral gas, i.e. $T_e > T_n$ and electron-neutral collisions transfer energy from the hotter to the cooler component, serving as a heat source for the neutral gas.

Although ions in the auroral atmosphere are heated by the ion chemistry and by cooling the electron gas ($T_i < T_e$), by far the largest source of ion heating is due to orthogonal electric fields in the ionosphere. An electric field will accelerate ions which then acquire a drift velocity with respect to the neutral gas, limited by collisions with the neutral molecules. Directed motion is thereby converted to random motion, or heating, by the collisional interaction. Both collision partners share in this energy and since the ion and neutral masses are about equal both ions and neutral are heated at a rate given by

$$Q_n \simeq Q_i = \frac{E_\perp^2}{B^2} \sum_i n_i \frac{\sum_n \mu_{in} \nu_{in}}{1 + (\sum_n \nu_{in})^2/\omega_i^2} , \quad \text{where} \tag{68}$$

μ_{in} is the reduced mass, ν_{in} the ion neutral collision frequency, ω_i the gyrofrequency, and n_i is the density of ion species i. The effect of electric fields on the ion temperature is discussed in the chapter by Schunk, who gives a simple equation (eq. 112) that relates the difference between the ion and neutral temperature to the effective electric field, in the local approximation. If transport effects are not negligible the energy conservation equation given by Schunk (eq. 3) must be solved. This includes conduction and advection terms in addition to local sources and sinks.

The principal heat source for the ambient electrons is the collisional interaction between auroral electrons and the thermal electron gas. Collisions between the suprathermal auroral electrons and the ambient thermal electrons are described by a term in the energy redistribution function, $R(\tau, E, E')$, (eq. 16). The complete expression for the electron heating rate, which includes Coulomb collisions and Cerenkov wave generation expressed as a function of the velocity, v, of the energetic electron is,

$$Q_e = \int \frac{dE}{ds} I(E)\, dE, \tag{69}$$

where

$$\frac{dE}{ds} = \frac{\omega_p^2 e^2}{v^2} \left[J_e(u)\, \ell_n \frac{2\Lambda}{3\gamma} + u\, G_e(u) \right] , \quad \text{where} \tag{70}$$

$$u = \frac{v}{v_e} , \quad v_e = \left[\frac{2\, k\, T_e}{m_e} \right]^{1/2} , \quad \omega_p^2 = \frac{4\pi\, n_e\, e^2}{m_e} , \quad \Lambda = \frac{3(k\, T_e)^{3/2}}{e^3 (4\pi\, n_e)^{1/2}} ,$$

and $J_e(u) = \frac{2}{\sqrt{\pi}} \int_0^u e^{-x^2} dx - \frac{2}{\sqrt{\pi}} 2 u e^{-u^2}$. Numerical values for $G_e(u)$ are listed in the appropriate reference given at the conclusion of this chapter. The full expression for the suprathermal electron energy loss rate is presented to show that it is a function of both the ambient electron density and temperature. Symbols not defined have their usual meaning. A simple empirical expression is given by Schunk (eq. 70). Expression (69) is valid at all values of electron energy, however, in the atmosphere electron heating becomes effective only at electron energies less than about 5 eV. The electron heating rate computed for the precipitating auroral intensity shown in Figure 3 is given in Figure 4 together with the total energy deposition rate discussed previously. Integrating over the entire atmospheric column it is found that less than 2% of the deposited energy goes into electron heating, but this small amount is sufficient to raise substantially the electron temperature above the neutral temperature. The electron energy equation (see Schunk, eq. 3) is solved to derive the electron temperature. But the heating rate is a function of T_e and n_e and therefore, the energy, ion continuity and electron transport equations are coupled and must be solved simultaneously. In addition, energy exchange between ions and electrons (through Coulomb collisions) requires a coupled solution of the electron and ion energy equations.

The electron gas is also heated by orthogonal electric fields and by parallel currents, but these are small sources by comparison with suprathermal electron cooling.

10. AURORAL PROTONS

Energetic protons are a part of auroral particle precipitation. Although the energy flux carried by the proton component is generally less than that carried by electrons there are occasions when precipitating protons account for nearly all the energy. Precipitating protons have a unique optical signature in the auroral spectrum in the form of Doppler broadened and shifted hydrogen lines, Balmer α and β in the visible. The shape and wavelength shift of the lines are a measure of the energy and line of sight velocity of the radiating hydrogen atoms. The process by which excited hydrogen atoms are formed is part of the reaction cycle undergone by the incident proton stream:

Electron capture

$$H^+ + N_2 \rightarrow H^{o*} + N_2^+ \qquad (71)$$

in which the proton captures an electron from the gas (using N_2 as an example) to form a hydrogen atom that may be in an excited state and an ion, N_2^+. This may be followed by stripping,

$$H^o + N_2 \rightarrow H^+ + e + N_2 \qquad (72)$$

in which a proton is reformed, or by ionization.

$$H^o + N_2 \rightarrow H^o + N_2^{+*} + e \qquad (73)$$

where the N_2^+ ion may be left in an excited state. The H atom may also have an excitation collision,

$$H^o + N_2 \rightarrow H^{o*} + N_2^* \qquad (74)$$

where one or both products may be in excited states. Excited hydrogen atoms are the source of the auroral Balmer lines. Both stripping and ionization reactions are sources of ejected electrons with energy up to several tens of eV. These energetic electrons become indistinguishable from those associated with the electron stream, and they take part in the same processes already described.

The fraction of H^+ and H^o in a charge state equilibrated beam varies with energy; protons are more abundant above about 40 keV while hydrogen atoms dominate below this energy. Hydrogen atoms are not confined by the magnetic field, contrary to protons and electrons, resulting in substantial lateral spreading of an initially collimated beam of protons. This theoretical prediction has been verified by observations of proton aurora.

11. SUMMARY

The response of the atmosphere and ionosphere to auroral electron and proton precipitation and to orthogonal electric fields that map from the magnetosphere down into the ionosphere has been set forth in the preceding sections. The physical concepts and the equations that are required for quantitative investigations of auroral processes have been described, with more attention devoted to recent ideas and mathematical techniques. In this section the interrelationship between the various processes is outlined, using the flow chart shown in Figure 8.

Energetic electrons and protons ionize the atmospheric gases. Electrons also cause dissociation of molecules, and those with energy greater than about 10 keV produce bremsstrahlung x-rays that have been measured in the stratosphere and higher. The quantities in the ovals in Figure 8 are observable parameters while processes (not measurable) are indicated by rectangles. The proton-hydrogen flux produces excitation, in addition to ionization. Ionization and dissociation products may be formed in excited states. The ions and neutrals that are produced undergo a multitude of chemical reactions, some of which lead to excitation, to ion heating and to neutral heating. The observable quantities are the ion density and composition and the electron density. Ionization produces secondary

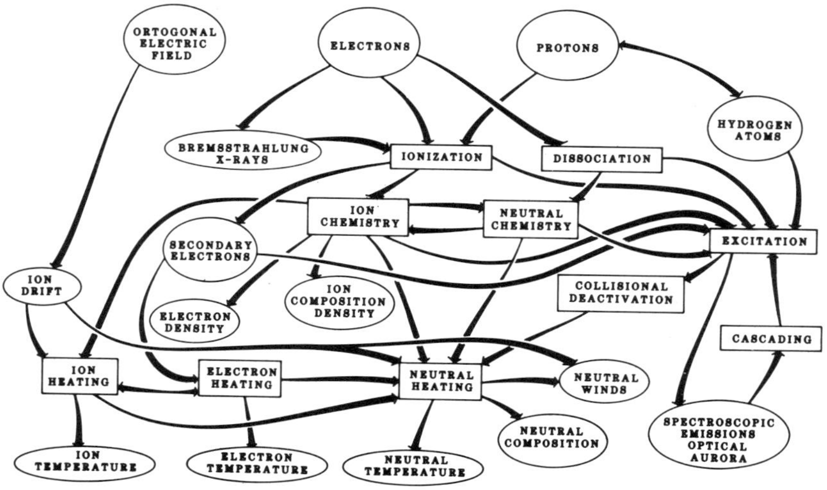

Figure 8. Flow chart of auroral processes and observables discussed in this chapter.

(and tertiary, etc.) electrons that can be measured. The secondary electrons cause excitation, some additional ionization and dissociation, and heating of the ambient electron gas. Excited states can be collisionally deactivated, resulting in neutral heating, or radiate spectroscopic features that comprise the optical aurora. Radiative cascading may cause additional excitation. Indeed, radiation in the EUV and FUV may be sufficiently energetic to cause additional ionization and dissociation (not shown in the diagram). Orthogonal electric fields cause observable ion drifts. Collisions with the neutral gas produce ion and neutral heating, and there is heat exchange between electrons, ions, and neutrals as well. The enhanced temperatures are all measureable. Neutral heating affects the neutral composition by enhanced vertical mixing. Neutral winds are also caused by heating as well as by ion-neutral momentum exchange.

Further coupling that is not shown in the diagram involves the dependence of the ion and neutral chemistry on the electron, ion and neutral temperature. Molecular dissociation produces atoms with excess kinetic energy that are an additional source of neutral heating. Changes in the neutral composition affects almost all the processes shown in Figure 8.

The interdependence of the various processes and physical quantities given in Figure 8 requires a coupled solution to the equations that govern each process. Those physical interactions that involve the transport of charged particles can be adequately described by a one-dimensional model but processes that involve the neutral gas and ion drift must be examined in three dimensions.

Current auroral models of electron and proton precipitation are one-dimensional, and the dynamical response of the thermosphere is decoupled from the remainder of the flow chart. Parameterized auroral inputs, in turn, are used in three-dimensional dynamical models, as discussed in the chapter by Richmond.

Bibliography

Section 2. The formulation and solution to the electron transport equation are given in,

Stamnes, K.: 1980, Analytic approach to auroral electron transport and energy degradation, Planet. Space Sci., 28, pp.427-441.

Stamnes, K.: 1981, On the two-stream approach to electron transport and thermalization, J. Geophys. Res., 86, pp. 2405-2410.

Another approach to this problem is discussed in,

Strickland, D.J., Book. D.C., Coffey, T.P., and Fedder, J.A.: 1976, Transport equation techniques for the deposition of auroral electrons, J. Geophys. Res., 81, pp.2755-2764.

Section 3. A comparison of various data sets is given in

Stamnes, K., and Rees, M.H.: 1983, Inelastic scattering effects on photoelectron spectra and on ionospheric electron temperature, J. Geophys. Res., 88.

Rees, M.H., and Stamnes, K.: 1983, Inelastic scattering effects on auroral electron transport and energy degradation, J. Geophys. Res., 88.

Section 4. The numerical solution of equation (19) is discussed in

Stamnes, K.: 1981, On the numerical solution to the two-stream equation, Scientific Report, Geophysical Institute, University of Alaska, UAG R-286.

Section 5. The references to section 3 also discuss electron impact excitation, and a more detailed exposition is presented in,

Vallance Jones, A.: 1974, "Aurora", D. Reidel Publishing Co., Dordrecht/Holland.

Section 6. Several basic concepts are discussed in the chapter by Schunk. A more extensive list of chemical reactions is given in,

Roble, R.G., and Rees, M.H.: 1977, Time-dependent studies of the aurora: Effects of particle precipitation on the dynamic morphology of ionospheric and atmospheric properties, Planet. Space Sci., 25, pp. 991-1010.

Section 7. References given for sections 5 and 6 also expand on the subject discussed in this section.

Section 8. New observations are reported at frequent intervals. The reference listed under section 5 presents spectra available up to about 1973. More recent results appear in,

Paresce, F., Chakrabarti, S., Bowyer, S., and Kimble, R.: 1983, The extreme ultraviolet spectrum of the aurora: 800 to 1400Å, J. Geophys. Res., 88.
Huffman, R.E., LeBlanc, F.J., Carreba, J.C., and Paulsen, D.E.: 1980, Satellite vacuum ultraviolet airglow and auroral observations, J. Geophys. Res., 85, pp. 2201-2215.
Gattinger, R.L., and Vallance Jones, A.: 1981, Quantitative spectroscopy of the aurora. V The spectrum of strong aurora between 10,000 and 16,000Å., Canad. J. Phys., 59, pp. 480-487 (and references therein).

Section 9. An early attempt at sorting out auroral energy dissipation is presented in,

Rees, M.H.: 1975, Magnetospheric substorm energy dissipation in the atmosphere, Planet. Space Sci., 23, pp. 1589-1596.

A recent quantitative treatment of auroral heating is given in,

Rees, M.H., Emery, B., Roble, R.G., and Stamnes, K.: 1983, Heating by auroral particle precipitation, J. Geophys. Res., 88.

A detailed discussion of electron heating is given by,

Schunk, R.W., Hays, P.B., and Itikawa, Y.: 1971, Energy losses of low-energy photoelectrons to thermal electrons, Planet. Space Sci., 19, pp. 125-126.

Section 10. The classic review paper on this topic is,

Eather, R.H.: 1967, Auroral proton precipitation and hydrogen emissions, Rev. Geophys., 5, pp. 207-285.

A recent article on the subject is,

Rees, M.H.: 1982, On the interaction of auroral protons with the Earth's atmosphere, Planet. Space Sci., 30, pp. 463-472.

Section 11. The standard textbooks on auroral excitation are,

Chamberlain, J.W.: 1961, "Physics of the aurora and airglow", Academic Press, New York, (out of print) and
Vallance Jones, A.: 1974, "Aurora", D. Reidel Publishing Co., Dordrecht/Holland.

SMALL-SCALE STRUCTURE IN THE EARTH'S IONOSPHERE: THEORY AND NUMERICAL SIMULATION

Steven T. Zalesak
Geophysical and Plasma Dynamics Branch
Plasma Physics Division
Naval Research Laboratory
Washington, D.C. 20375

Abstract. We describe in qualitative terms the cause and nonlinear evolution of the gradient drift and collisional Rayleigh-Taylor instabilities in the earth's ionosphere, by using the examples of ionospheric barium clouds and equatorial spread F "bubbles" respectively. We then derive the nonlinear differential equations governing these instabilities. Finally, we discuss the numerical solution of these differential equations.

1. Introduction

It is generally believed that the existence of ionospheric structures with scale sizes of tens of kilometers or smaller can be attributed primarily to the onset and evolution of instabilities of one sort or another. These instabilities can be thought of as being superimposed on, and indeed evolving from, the larger scale ionospheric configuration. Among the numerous such structures it is usually only those that are of reasonably large amplitude or those which cause problems (e.g. communications interference) that attract interest and study. Still, this number is greater than we can reasonably treat here. We shall therefore limit our discussion to two such structures whose physics and evolution we believe we understand reasonable well: 1) the steepening and subsequent recursive splitting of barium clouds released in the ionosphere, driven by the gradient drift instability; and 2) the formation and buoyant rise of low density "bubbles" of plasma in the nighttime equatorial ionosphere, known as equatorial spread F (ESF), driven by the collisional Rayleigh-Taylor instability. Each of these instabilities derive from the same set of plasma fluid equations and the same set of physical approximations, differing only in geometry and in the identity of the driving terms; hence we shall attempt here to unify their description as much as possible. We shall find that one of the characteristics of structures resulting from these instabilities is their tendency to be "field aligned", that is, for the plasma gradients and velocities parallel to the magnetic field to be much smaller than those perpendicular to the magnetic field. Our

R. L. Carovillano and J. M. Forbes (eds.), Solar-Terrestrial Physics, 781–804.

discussion will therefore focus on plasma motion perpendicular to the ambient magnetic field.

In Figure 1 we show a photograph of the Spruce event, a barium cloud released at 188 km altitude in February of 1971, 24 minutes after release. The cloud was originally released as a gaussian distribution of neutral barium which was subsequently photoionized by sunlight. In the very center of the photograph, our line of sight is parallel to the magnetic field lines at the cloud altitude, revealing the fine scale structure (termed "striations") that has evolved from this originally nearly gaussian distribution of plasma. In Figure 2 we show a sketch of what we believe to be the typical evolution of a barium cloud like Spruce, dervived from experimental observations and numerical simulations. The inital steepening of the top of the two-dimensinal cloud is caused by the buildup of polarization charge on its sides, causing the high density center of the cloud to $\underset{\sim}{E} \times \underset{\sim}{B}$ drift in the direction of the neutral wind to a greater extent than the low density

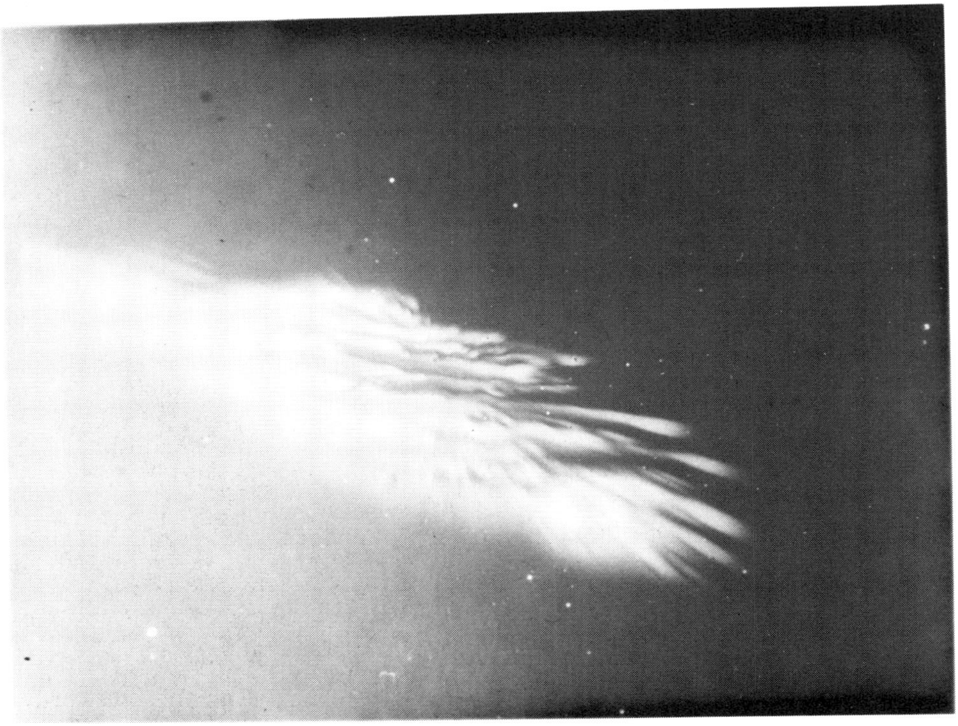

Figure 1. Photograph of the Spruce barium cloud 24 minutes after release. Bright areas are ionized barium. The line of sight near the center of the picture is parallel to the magnetic field at the cloud altitude.

edges. As the plasma gradient on the top of the cloud becomes steeper, the growth rate of the gradient drift instability (to be described later) active there becomes larger and eventually small perturbations on this gradient are amplified into visible ripples, which in turn evolve into finger-like structures. Each of the strucutes emerging from the steepened edge of the cloud then evolve into smaller clouds, and the process begins again, resulting in a cascade of recursively decreasing scale sizes until the instability is stopped by dissipation or other mechanisms which act more effectively on the smaller space scales.

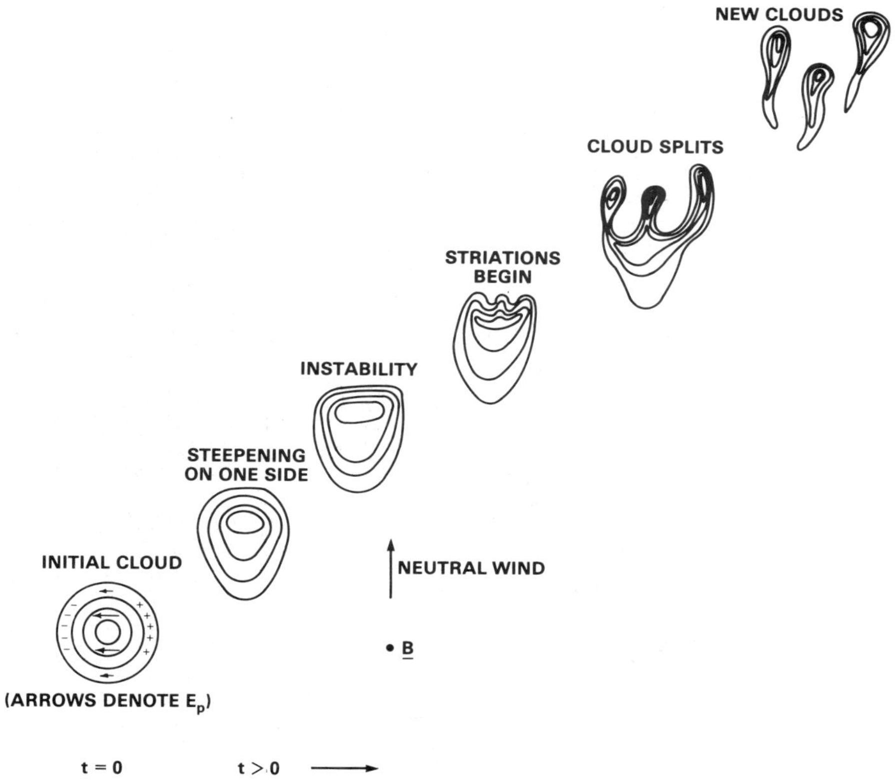

Figure 2. Sketch of the time evolution of a typical barium cloud in a plane perpendicular to the magnetic field, subject to an upward directed neutral wind. Lines demarking the cloud denote plasma density contours, with the highest plasma density in the center of cloud.

In Figure 3 we show maps of 1-m irregularities taken from Tsunoda (1981) at the earth's magnetic equator during equatorial spread F (ESF). These irregularities have been shown to be closely associated with "bubbles" or regions of large electron density depletion in the equatorial ionosphere, and can be thought of as at least a partial map of the locations of severe electron density depletion. In Figure 4 we show the results of a numerical simulation from Zalesak, et al. (1982), showing the time evolution of electron density contours at the earth's equator. The equatorial ionosophere was originally laminar with a maximum in electron density at 430 km altitude. A sinusoidal perturbation was applied in the east-west direction. The results show that the observed "bubbles" consist of low density plasma which has been transported from very low altitudes up through the F2 peak and beyond by the nonlinear evolution of the collisional Rayleigh-Taylor instability. The westward and eastward tilts of the bubble are due to an eastward neutral wind blowing at the equator coupled along magnetic field lines to background ionization (e.g., E regions) at higher and lower latitudes. Note that the various tilts of the bubbles in Figures 3 and 4 are consistent when allowance is made for the reversed abscissae in the two plots.

In Section 2 we shall present a qualitive, physical description of the instabilities active in ionospheric barium cloud and equatorial

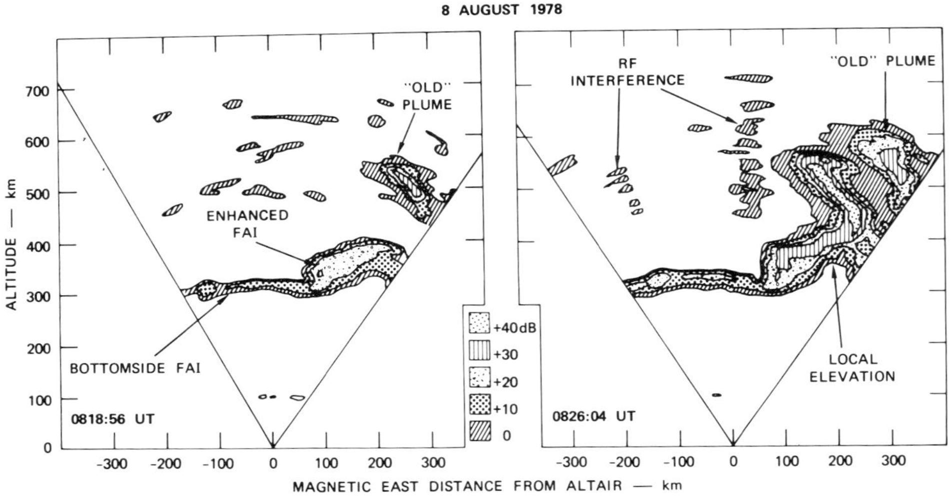

Figure 3. Two sequential maps of 3 meter radar backscatter at the earth's magnetic equator. Regions of intense backscatter have been shown to be associated with regions of severe electron density depletion. From R.T. Tsunoda, J. Geophys. Res., 86, 139, 1981, copyrighted by the American Geophysical Union.

spread F (ESF) cases. In Section 3 we derive the set of equation describing the motion of ionospheric plasma in general, and the evolution of barium clouds and ESF bubbles in particular. In Section 4 we discuss the simplifications made in constructing a mathematical representation of the physical system. In Section 5 we derive and summarize the equations describing the "simplest case" geometries and assumptions for each of the instabilities. Finally, in Sections 6 through 8, we treat the numerical integration of these differential equations.

2. The Gradient Drift/Collisional Rayleigh-Taylor Instability

In this section we shall attempt to give a qualitative physical picture of the gradient drift and collional Rayleigh-Taylor instabilities, both of which are caused by the differential motion of ions and electrons perpendicular to the magnetic field. We consider a two-dimensional x-y plane perpendicular to the ambient magnetic field $\underset{\sim}{B}$. A plasma species α in this plane embedded in a neutral gas will respond to an external force perpendicular to $\underset{\sim}{B}$, $\underset{\sim}{F}_{\alpha\perp}$, in two

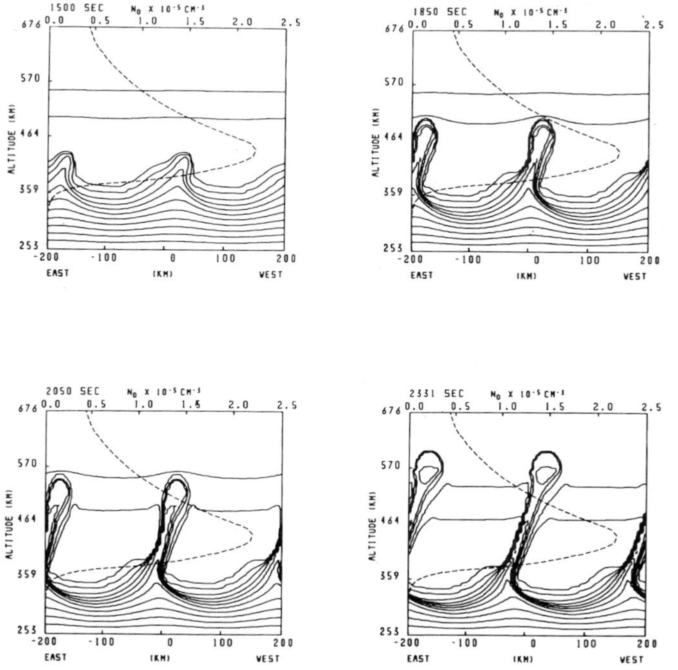

Figure 4. Four sequential plots of contours of electron density at the earth's equator depicting the formation and subsequent buoyant rise of an ESF "bubble", taken from a numerical simulation of Zalesak et al. (1982).

ways: 1) by drifting in a direction perpendicular to both $\mathbf{B}$ and $\mathbf{E}_{\alpha\perp}$ (Hall mobility) and 2) usually to a lesser extent, by drifting in a direction parallel to $\mathbf{E}_{\alpha\perp}$ (Pedersen mobility). We shall explicity derive these drifts in Section 3. We shall take our plasma to consist of a single ion species, denoted by subscript i, and of electrons, denoted by subscript e. The instabilities under discussion result from the fact that the ions and electrons drift with different velocities and directions in response to the same external force. In regions where plasma density gradients exist, this difference in velocities causes polarization charge to be created in our originally neutral plasma, which in turn produces a polarization electric field. The plasma drift associated with the eletric field will cause growth of a perturbation when the plasma gradient is properly aligned.

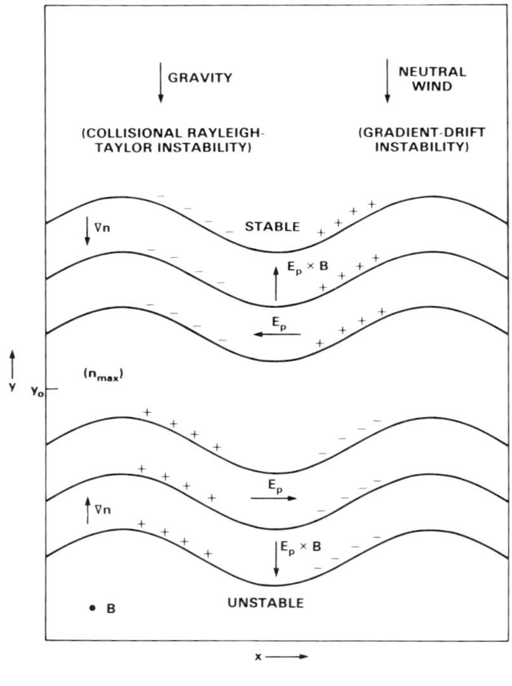

Figure 5. Contours of constant plasma density n for in an x-y plane perpendicular to the magnetic field. n is a function of y which maximizes at $y = y_o$, superimposed on which is a perturbation of the form sin kx, where k is a wavenumber. Either a downward-directed gravity or a downward-directed neutral wind causes ions to shift slightly leftward relative to the electrons, which results in an $\mathbf{E}_p \times \mathbf{B}$ velocity which either damps or enhances the perturbation, depending on the sign of $\partial n/\partial y$.

In Figure 5 we show contours of contant plasma density n in the two-dimensional x-y plane, where we assume the magnetic field $\underset{\sim}{B}$ to be aligned along the positive z axis. Depicted is a one-dimensional "slab" of plasma n(y) such that n maximizes at $y=y_o$, superimposed on which is a sinusoidal perturbation proportional to sin kx, where k is a wavenumber. Either a downward-directed gravational acceleration (in the collisional Rayleigh-Taylor instability) or a downward-directed neutral wind (in the gradient drift instability) will cause the ions to drift leftward relative to the electrons, leaving polarization charge where the relative drift has components parallel to the density gradient, as indicated in Figure 5. This polarization charge induces a polarization electric field $\underset{\sim}{E}_p$, which in turn induces an additional plasma drift in the $\underset{\sim}{E}_p \times \underset{\sim}{B}$ direction. This drift is such as to enhance the perturbation for $y < y_o$ (instability), as seen in Figure 5. In their most simplified geometries the linear growth rates γ for the gradient drift and collisional Rayleigh-Taylor instabilities are

$$\gamma = - \underset{\sim}{U}_n \cdot \frac{\nabla n}{n} \qquad \text{(gradient drift)} \tag{1}$$

$$\gamma = - \underset{\sim}{g} \cdot \frac{\nabla n}{n} \nu_{in}^{-1} \qquad \text{(Rayleigh-Taylor)} \tag{2}$$

where ν_{in} is the ion-neutral collision frequency. We note here that the gradient drift instability may be thought of as being driven by an ambient electric field simply by performing a Lorentz transformation into the rest frame of the neutral gas.

The above picture of the early (linear) stage of the instability evolution is quite informative, but unfortunately falls short of illuminating the complex nonlinear evolution of barium clouds and equatorial spread F "bubbles". In the next section we derive the equations necessary for a complete nonlinear description of these phenomena, which in general require numerical techniques for their solution.

3. The Motion of Ionospheric Plasma

We shall be concerned here with the motion of plasma consisting of ions and electrons in the presence of a neutral gas and magnetic field $\underset{\sim}{B}$, subject to an external force. We shall also be interested in the electric current $\underset{\sim}{J}$ arising from the differential motion of the various species comprising the plasma. In the course of deriving the equations we shall make some assumptions which are crucial to the model:

1) We shall assume the plasma can be adequately described by the fluid approximation. This assumes that the effective collision rate of each plasma species with itself is sufficiently high to maintain near Maxwellian distribution functions on time scales short compared to the times of interest, and is well satisfied for the plasmas we treat here.

2) We shall assume that the electric fields $\underset{\sim}{E}$ are electrostatic (i.e., $\nabla \times \underset{\sim}{E} = 0$) and hence can be described using a scalar potential ϕ such that $\underset{\sim}{E} = - \nabla \phi$. Note that this implies $\partial \underset{\sim}{B}/\partial t = 0$. The validity of this assumption can be related to the fact that the Alfven velocity is much larger than any other propagation speeds of interest for the plasmas we treat here. The assumption is also checked a posteriori by verifying that the calculated currents and displacement currents produce negligible time variations in $\underset{\sim}{B}$ which in turn produce negligible $\nabla \times \underset{\sim}{E}$.

3) We assume plasma quasi-neutrality; that is,

$$\sum n_i q_i \approx n_e e \tag{3}$$

where n is number density, q is ion species charge, e is the electron charge, the subscripts i and e refer to ions and electrons respectively, and the sum is taken over all ion species. This assumption is a statement that the Debye length is small compared to all length scales of interest, and again can be verified a posteriori by evaluating $\nabla \cdot \underset{\sim}{E}$. Note that this assumption implies that $\nabla \cdot \underset{\sim}{J} = 0$, where $\underset{\sim}{J}$ is the electric current. In addition to the above there are some other assumptions which, while they are not essential to the basic model, are nonetheless valid for many of the physical situations which we shall treat and impart a simplicity which we shall find convenient here:

4) We shall assume the electrostatic potential ϕ to be constant along magnetic field lines. As we shall see later, the electrical conductivity along magnetic field lines is much greater that that perpendicular to magnetic field lines, meaning that appreciable differences in potential along a field line will quickly be reduced by the resultant current. This assumption will break down for sufficiently small scale lengths perpendicular to the magnetic field, and for sufficiently large distances along the magnetic field.

5) We shall assume that the inertial terms in the plasma species momentum equations, i.e., the left hand side of Equation (5), are negligible with respect to the other terms in the equation. This assumption is justified whenever the time scales of interest are longer than the mean time between collisions for ions.

6) We shall neglect all collisions between species except those between ions and the neutral gas. This is justified simply by an evalution of the magnitudes of the terms involved.

7) We shall ignore production and loss terms which may appear as sources and sinks in the plasma continuity equations as a result of chemistry, photoionization, etc.

Assumptions (4) through (7) above, although made in this paper, are not necessary within the theoretical and computational framework we have developed, and adequate means exist to delete them, if necessary.

The continuity and momentum equations describing plasma species α are:

$$\frac{\partial n_\alpha}{\partial t} + \nabla \cdot (n\, \underset{\sim}{v}_\alpha) = 0 \tag{4}$$

$$\left(\frac{\partial}{\partial t} + \underset{\sim}{v}_\alpha \cdot \nabla\right) \underset{\sim}{v}_\alpha = \frac{q_\alpha}{m_\alpha}\left(\underset{\sim}{E} + \frac{\underset{\sim}{v}_\alpha \times \underset{\sim}{B}}{c}\right) - \nu_{in}\,(\underset{\sim}{v}_\alpha - \underset{\sim}{U}_n)$$

$$- \frac{\nabla P_\alpha}{n_\alpha m_\alpha} + \underset{\sim}{g} \tag{5}$$

where the subscript α denotes the plasma species (i for ions, e for electrons, for example), n is the species number density, $\underset{\sim}{v}$ is the species fluid velocity, P is pressure, $\underset{\sim}{E}$ is the electric field, $\underset{\sim}{g}$ is the gravitational acceleration, q is the species charge, $\nu_{\alpha n}$ is the species collision frequency with the neutral gas, $\underset{\sim}{U}_n$ is the neutral wind velocity, c is the speed of light, and m is the species particle mass. We can rewrite this equation as

$$\underset{\sim}{F}_\alpha / m_\alpha + \frac{q_\alpha}{m_\alpha c}(\underset{\sim}{v}_\alpha \times \underset{\sim}{B}) - \nu_{\alpha n}\, \underset{\sim}{v}_\alpha = 0 \tag{6}$$

where

$$\underset{\sim}{F}_\alpha \equiv q_\alpha \underset{\sim}{E} + m_\alpha g + \nu_{\alpha n}\, m_\alpha \underset{\sim}{U}_n - \nabla P_n / n_\alpha$$

$$- \left(\frac{\partial}{\partial t} + \underset{\sim}{v}_\alpha \cdot \nabla\right) \underset{\sim}{v}_\alpha\, m_\alpha \tag{7}$$

If we place ourselves in a Cartesian coordinate system in which $\underset{\sim}{B}$ is aligned along the z axis, and if we treat $\underset{\sim}{F}_\alpha$ as a given quantity then a componentwise evaluation of Equation (6) yields a set of three equations in three unknowns, the three components of $\underset{\sim}{v}_\alpha$. The formal solution is

$$\underset{\sim}{v}_{\alpha\perp} = k_{1\alpha}\, \underset{\sim}{F}_{\alpha\perp} + k_{2\alpha}\, \underset{\sim}{F}_{\alpha\perp} \times \hat{z} \tag{8}$$

$$\underset{\sim}{v}_{\alpha\parallel} = k_{o\alpha}\, \underset{\sim}{F}_{\parallel} \tag{9}$$

where

$$k_{1\alpha} = \frac{\nu_{\alpha n}}{\Omega_\alpha} \frac{c}{|q_\alpha B|}\left[1 - \frac{(\nu_{\alpha n}/\Omega_\alpha)^2}{1 + (\nu_{\alpha n}/\Omega_\alpha)^2}\right] \tag{10}$$

$$k_{2\alpha} = \frac{c}{q_\alpha B}\left[1 - \frac{(\nu_{\alpha n}/\Omega_\alpha)^2}{1 + (\nu_{\alpha n}/\Omega_\alpha)^2}\right] \tag{11}$$

$$k_{o\alpha} = (m_\alpha \nu_{\alpha n})^{-1} \tag{12}$$

$$\hat{z} \equiv \underset{\sim}{B}/|B| \tag{13}$$

$$\Omega_\alpha \equiv \left|\frac{q_\alpha B}{m_\alpha c}\right| \tag{14}$$

The vector subscripts $\perp$ and $\parallel$ refer to the components of the vector which are perpendicular and parallel respectively to $\hat{z}$. The quantities k_1, k_2, and k_o above are referred to as the Pedersen, Hall, and direct mobilities respectively. It should be pointed out that Equations (8) and (9) are only truly closed form expressions when the inertial terms (the last term on the right hand side of Equation (7)) are neglected, an assumption we have made previously. Typical ranges for collision frequencies are: $\nu_{in} \sim 30\ sec^{-1}$, $\nu_{en} \sim 800\ sec^{-1}$ at 150 km altitude; and $\nu_{in} \sim 10^{-1} sec^{-1}$, $\nu_{en} \sim 1\ sec^{-1}$ at 500 km altitude.

As we will see later, we will use the concept of "layers" to distinguish the various ion species, so for the moment we can consider only a single ion species, denoted by subscript i, and the associated electrons, denoted by subscript e. We will also consider only singly charged ions so that $q_i = e$ and $q_e = - e$. Noting that $\nu_{en}/\Omega_e \approx 0$ we obtain

$$k_{1i} = \frac{\nu_{in}}{\Omega_i} R_i \frac{c}{e|B|} \tag{15}$$

$$k_{1e} = 0 \tag{16}$$

$$k_{2i} = R_i \frac{c}{eB} \tag{17}$$

$$k_{2e} = - \frac{c}{eB} \tag{18}$$

where

$$R_i \equiv (1 + \nu_{in}^2/\Omega_i^2)^{-1} \tag{19}$$

We now define the perpendicular current

$$\underset{\sim}{J}_\perp \equiv \sum_\alpha n_\alpha q_\alpha v_{\alpha\perp} \tag{20}$$

Substituting Equations (15) through (18) and (8) into Equation (20), and using the quasi-neutrality approximation

$$n_i \approx n_e \equiv n \tag{21}$$

we obtain

$$\underset{\sim}{J}_\perp = \frac{\nu_{in}}{\Omega_i} R_i \frac{nc}{|B|} \underset{\sim}{F}_{i\perp}$$

$$+ \frac{nc}{B} (R_i \underset{\sim}{F}_{i\perp} + \underset{\sim}{F}_{e\perp}) \times \hat{z} \tag{22}$$

For the barium cloud and equatorial spread F (ESF) problems we shall treat here, we shall only consider neutral winds, electric fields, and gravity as external forces. Hence

$$\underset{\sim}{F}_{i\perp} = e\,\underset{\sim}{E}_{\perp} + m_i\,\underset{\sim}{g}_{\perp} + \nu_{in}\,m_i\,\underset{\sim}{U}_{n\perp} \tag{23}$$

$$\underset{\sim}{F}_{e\perp} = -\,e\,\underset{\sim}{E}_{\perp} + m_e\,\underset{\sim}{g}_{\perp} \tag{24}$$

Note that we have neglected the small term $\nu_{en}\,m_e$ in Equation (24). We obtain

$$\begin{aligned}\underset{\sim}{J}_{\perp} &= \frac{\nu_{in}}{\Omega_i} R_i \frac{nc}{|B|} (e\,\underset{\sim}{E}_{\perp} + m_i\,\underset{\sim}{g}_{\perp} + \nu_{in}\,m_i\,\underset{\sim}{U}_{n\perp}) \\ &+ R_i \frac{nc}{B} \left[e\,\underset{\sim}{E}_{\perp}\,(1 - R_i^{-1}) + m_i + \frac{m_e}{R_i})\,\underset{\sim}{g}_{\perp} + \nu_{in}\,m_i\,\underset{\sim}{U}_{n\perp}\right] \times \hat{z}\end{aligned} \tag{25}$$

Since $0.01 \leqslant R_i \leqslant 1.0$ we may neglect m_e/R_i with respect to m_i. Defining the Pedersen conductivity

$$\sigma_p \equiv R_i \frac{\nu_{in}}{\Omega_i} \frac{nce}{|B|} \tag{26}$$

and noting that $1 - R_i = -\,\nu_{in}^{\,2}/\Omega_i^{\,2}$ we obtain

$$\begin{aligned}\underset{\sim}{J}_{\perp} &= \sigma_p \left[\underset{\sim}{E}_{\perp} + \frac{m_i}{e}\,\underset{\sim}{g}_{\perp} + \nu_{in} \frac{m_i}{e}\,\underset{\sim}{U}_{n\perp}\right. \\ &\left. + \left(-\frac{\nu_{in}}{\Omega_i}\,\underset{\sim}{E}_{\perp} + \frac{\Omega_i m_i}{\nu_{in} e}\,\underset{\sim}{g}_{\perp} + \Omega_i \frac{m_i}{e}\,\underset{\sim}{U}_{n\perp}\right) \times \hat{z}\right]\end{aligned} \tag{27}$$

Our need for an expression for $\underset{\sim}{J}_{\perp}$ stems from our need for its divergence to evaluate $\nabla \cdot \underset{\sim}{J}$ (= 0 by quasi-neutrality), as we shall see in the next section.

4. Model Simplification and Mathematical Representation

We shall model our physical system using a simplified model as depicted in Figure 6. The magnetic field lines are assumed to be straight, to be aligned along the z axis of our cartesian coordinate system, and to terminate in insulators at $z = \pm\infty$. The plasma of interest is threaded by these magnetic field lines, and is divided into thin planes or "layers" of plasma perpendicular to the magnetic field. Since we have neglected collisions between different plasma species, we may use the device of layers to treat multiple ion species at a single point in space simply by allowing multiple layers to occupy the same plane in space, one for each ion species. In this way a "layer" consists only of a single ion species and its associated electrons.

Our quasi-neutrality assumption demands that

$$\nabla \cdot \underset{\sim}{J} = \frac{\partial}{\partial x} J_x + \frac{\partial}{\partial y} J_y + \frac{\partial}{\partial z} J_z = 0 \tag{28}$$

Integrating Equation (28) along z and noting from Figure 6 that J_z vanishes at $z = \pm \infty$ we obtain

$$\int_{-\infty}^{+\infty} \nabla_{\perp} \cdot J_{\perp} \, dz = 0 \tag{29}$$

where

$$\nabla_{\perp} \equiv \hat{x} \frac{\partial}{\partial x} + \hat{y} \frac{\partial}{\partial y} \tag{30}$$

From our model as depicted in Figure 6 we may approximate the integral in Equation (29) by a discrete sum

$$\sum_{k=1}^{N} \nabla_{\perp} \cdot \underset{\sim}{J}_{\perp k} \, \Delta z_k = 0 \tag{31}$$

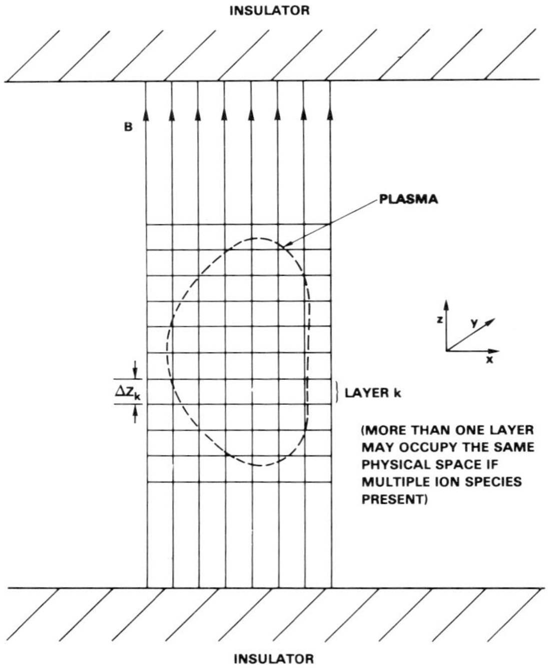

Figure 6. Model of plasma and magnetic field geometry used in this paper. Field lines terminate on insulators at $z = \pm \infty$. Plasma is divided into "layers" along z for mathematical and numerical treatment. Each layer consists of a single ion species and its associated electrons. Multiple collocated ion species are treated by having multiple collocated "layers".

where the subscript k refers to the layer number, N is the total number of layers in the system, and Δz_k is the thickness of layer k measured along the magnetic field line. By our assumption of equipotential magnetic field lines and electrostatic electric fields

$$\underset{\sim}{E}_{\perp k}(x,y) = -\nabla_{\perp}\phi(x,y) \text{ for all } k \tag{32}$$

Then Equation (31) becomes

$$\nabla_{\perp} \cdot \left[\sum_{k=1}^{N} (\Sigma_{pk})\nabla_{\perp}\phi\right] + \sum_{k=1}^{N} H_k = \sum_{k=1}^{N} \nabla_{\perp} \cdot \underset{\sim}{J}^{ext}_{\perp k} \tag{33}$$

where

$$\Sigma_{pk} \equiv \sigma_{pk}\, \Delta z_k \tag{34}$$

$$\begin{aligned} H_k &\equiv -\frac{\partial}{\partial x}\left(\frac{\nu_{in}}{\Omega_i}\Sigma_p \frac{\partial\phi}{\partial y}\right)_k + \frac{\partial}{\partial y}\left(\frac{\nu_{in}}{\Omega_i}\Sigma_p \frac{\partial\phi}{\partial x}\right)_k \\ &= -\frac{\partial\phi}{\partial y}\frac{\partial}{\partial x}\left(\frac{\nu_{in}}{\Omega_i}\Sigma_p\right)_k + \frac{\partial\phi}{\partial x}\frac{\partial}{\partial y}\left(\frac{\nu_{in}}{\Omega_i}\Sigma_p\right)_k \end{aligned} \tag{35}$$

$$\begin{aligned} \underset{\sim}{J}^{ext}_{\perp k} &\equiv \Sigma_{pk}\Big[\frac{m_i}{e}\underset{\sim}{g}_{\perp} + \nu_{in}\frac{m_i}{e}\underset{\sim}{U}_{n\perp} \\ &+ \left(\frac{\Omega_i m_i}{\nu_{in} e}\underset{\sim}{g}_{\perp} + \Omega_i \frac{m_i}{e}\underset{\sim}{U}_{\perp}\right) x\, \hat{z}\Big]_k \end{aligned} \tag{36}$$

and the subscript k denoting layer number on terms within parenthesis operates on all terms within those parentheses. Equation (33) is a second order elliptic partial differential equation for $\phi(x,y)$, subject to boundary conditions on ϕ. Our reason for writing Equation (33) in the form we did is related to the following picture of the physics. The external forces acting on a plasma, in this case gravity and a neutral wind collision term, will induce a current to flow. In general this current will not satisfy $\nabla \cdot \underset{\sim}{J} = 0$, meaning that in certain regions there will be a build-up of polarization charge, resulting in an electric field which causes secondary currents to flow. Over time scales much shorter than those of interest here, a quasi-steady state is reached such that subsequent plasma motion is well described by $\nabla \cdot \underset{\sim}{J}_{\perp} = 0$. In this physical picture the electric field represents the response of the plasma to a given externally driven current system. Thus the right hand side of the Equation (33) may be regarded as the "known" divergence of the external current, which we shall denote below by R, and the left hand side regarded as a differential operator L operating on ϕ:

$$L\phi = R \tag{37}$$

The operator L is a hermitian operator in the limit as the "Hall terms" H_k may be neglected, as is often the case at higher altitudes where ν_{in}/Ω_i is small.

5. The simplest Case Equations for Barium Clouds and for ESF

The simplest case equations for each of our physical systems are for one level only, i.e., N=1, and for altitudes such that terms of order $(\nu_{in}/\Omega_i)^2$ may be neglected with respect to terms of order (ν_{in}/Ω_i). We also treat only one external force for each case, and align that force along one of the coordinate axes. Since we have only one level, we drop the subscript k.

For barium clouds, we assume $\underset{\sim}{B}$ to be aligned along the z axis, that the only external force is a neutral wind $\underset{\sim}{U}_n \equiv U_n\hat{y}$.
Then

$$\underset{\sim}{J}_\perp^{ext} = \Sigma_p \left(\nu_{in} \frac{m_i}{e} U_n \hat{y} + \Omega_i \frac{m_i}{e} U_n \hat{x}\right) \tag{38}$$

Since Σ_p is already of order (ν_{in}/Ω_i), we may neglect the first term with respect to the second. Then

$$\nabla_\perp \cdot \underset{\sim}{J}_\perp^{ext} = \frac{\partial}{\partial x}\left(\Sigma_p \frac{BU_n}{c}\right) \tag{39}$$

where we have used Equation (14).
Noting that H in Equation (33) is of order $(\nu_{in}/\Omega_i)^2$ we obtain

$$\nabla_\perp \cdot (\Sigma_p \nabla_\perp\phi) = \frac{\partial}{\partial x}\left(\Sigma_p \frac{BU_n}{c}\right) \tag{40}$$

For the equatorial spread F case we assume a single plane of plasma located at the magnetic equator such that $\underset{\sim}{B}$ is along the z axis and y is "up". Our only external force is gravity $\underset{\sim}{g}_\perp = -\ g\hat{y}(g = +\ 980\ \text{cm/sec}^2)$.
Then

$$\underset{\sim}{J}_\perp^{ext} = \Sigma_p \left[-\frac{m_i}{e} g \hat{y} - \frac{\Omega_i}{\nu_{in}}\frac{m_i}{e} g \hat{x}\right] \tag{41}$$

The first term is of order ν_{in}/Ω_i times the second and may therefore be neglected. Then

$$\nabla_\perp \cdot \underset{\sim}{J}_\perp^{ext} = -\frac{\partial}{\partial x}\left(\Sigma_p \frac{Bg}{c\nu_{in}}\right) \tag{42}$$

Again neglecting H in Equation (33) we obtain

$$\nabla_\perp \cdot (\Sigma_p \nabla_\perp\phi) = -\frac{\partial}{\partial x}\left(\Sigma_p \frac{Bg}{c\nu_{in}}\right) \tag{43}$$

For both the one-layer barium cloud and ESF cases, one may solve either the electron or the ion continuity equations, since quasi-neutrality makes them equivalent (but not identical). For simplicity we choose the electron equation since we may neglect the Pedersen terms there $(\nu_{en}/\Omega_e = 0)$.

Summarizing the equations we must solve for each case we get

$$\frac{\partial n_e}{\partial t} = \nabla_\perp \cdot (n_e \underset{\sim}{v}_{e\perp}) = 0 \tag{44}$$

$$\nabla_\perp \cdot (\Sigma_p \nabla_\perp \phi) = \partial S/\partial x \tag{45}$$

$$\underset{\sim}{v}_{e\perp} = -\frac{c}{eB}\underset{\sim}{F}_{e\perp} \times \hat{z} \tag{46}$$

$$\underset{\sim}{F}_{e\perp} = \begin{cases} e\nabla_\perp \phi & \text{for barium clouds} \\ e\nabla_\perp \phi - m_e g\hat{y} & \text{for ESF} \end{cases} \tag{47}$$

$$S = \begin{cases} \Sigma_p U_n/c & \text{for barium clouds} \\ -\Sigma_p Bg/(c\nu_{in}) & \text{for ESF} \end{cases} \tag{48}$$

$$\Sigma_p = \Delta z \, (\nu_{in}/\Omega_i) nce/B \tag{49}$$

Solution of these equations requires the use of two-dimensional numerical simulation techniques.

6. Numerical Simulation: General

We saw in the previous section that in the simplest case for the barium cloud and equatorial spread F (ESF) problems, we can reduce our system to two partial differential equations posed on a two dimensional plane:

$$\frac{\partial n}{\partial t} + \nabla_\perp \cdot (n \, \underset{\sim}{v}_{e\perp}) = 0 \tag{50}$$

$$\nabla_\perp \cdot (\Sigma_p \nabla_\perp \phi) = \partial S/\partial x \tag{51}$$

where Σ_p and S are explicity given functions of n and $\underset{\sim}{v}_{e\perp}$ is an explicity given function of ϕ. Equation (50) is hyperbolic while Equation (51) is elliptic. Both require the imposition of physically relevant boundary conditions. Conceptually one solves this coupled system of equations as follows. At any given time t, we assume that we know n(x,y,t) and therefore $\Sigma_p(x,y,t)$ and S(x,y,t). We can then solve Equation (51) for its single scalar unknown $\phi(x,y,t)$, given properly specified boundary conditions on ϕ and/or its derivatives. Knowing ϕ we can compute $\underset{\sim}{v}_{e\perp}(x,y,t)$ explicity. We can then solve Equation (50) for $n(x,y,t + \Delta t)$ where Δt is a small time increment. The process is repeated recursively until the solution is advanced to the desired time.

Within the above context several numerical approaches exist for solving this system of coupled partial differential equations: spectral methods, finite element methods, Galerkin methods, and finite difference methods, among others. We have chosen finite difference methods here for reasons of simplicity, computational efficiency, and most importantly because acceptable techniques for solving Equation (50) in the presence of large gradients in n presently exist only within the finite difference domain. Fundamental to finite difference techniques is their use of a "grid", that is, a discrete set of points in space and time denoted by (x_i, y_i, t^m), $1 \leqslant i \leqslant NX$, $1 \leqslant j \leqslant NY$, $1 \leqslant n < \infty$ where i, j, m, NX and NY are integers, on which the solution is computed. For instance, the electrostatic potential $\phi(x,y,t)$ at $x = x_i$, $y = y_j$, and $t = t^m$ would be denoted ϕ^m_{ij}. In Figure 7, we show an example of a finite difference grid in space, and we also show how the grid would look in the case of multiple layers of plasma, although we shall treat only a single layer here. Note that the four "nearest neighbors" of the grid point (x_i, y_j) are the grid points (x_{i+1}, y_j), (x_{i-1}, y_j), (x_i, y_{j+1}), and (x_i, y_{j-1}). Many finite difference techniques employ what is known as a staggered grid, meaning that different dependent variables (n and ϕ for instance) are evaluated on different grids in space and possibly time. We do not employ staggered grids here; all dependent variables are evaluated on exactly the same grid.

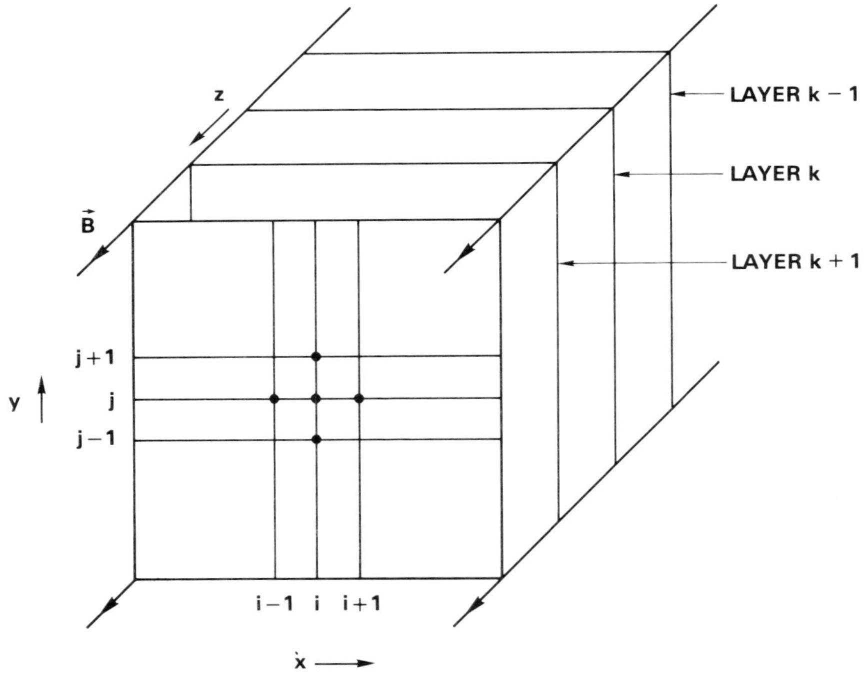

Figure 7. Spatial finite difference grid used in the numerical simulation codes, showing the correspondence between i and x and between j and y. Several layers of plasma are shown, even through the discussion in the text assumes only one layer. Grid points are shown as dots.

Looking at Equations (50) and (51) we see that there is more to distinguish them than just their hyperbolicity and ellipticity respectively. Both equations require the evaluation of spatial derivatives; but Equation (50) requires in addition the evaluation of temporal derivatives. Precisely because we do not yet know the solution at a time later than it has thus far been computed, the treatment of temporal derivatives is qualitatively different from that of spatial derivatives. More importantly, it has been found empirically that it is the numerical treatment of Equation (50) which will "make or break" the solution to the total system of equations. Specifically, Equation (51), once properly discretized (i.e., once the spatial derivatives are properly represented in finite difference form) simply yields a system of linear equations, albeit a very large system. Our experience has been that a number of algorithms will successfully yield a solution to this linear system, although it may be difficult to find one algorithm that will solve the system for all possible physical parameters. Accordingly we shall discuss the numerical treatment of Equation (51) only briefly here, in the next section, and reserve the bulk of our discussion for Equation (50).

7. The Numerical Solution of the Potential Equation

As was mentioned in the previous section, the numerical solution of Equation (51) takes place in two stages: 1) the discretization of the spatial derivatives and boundary conditions in finite difference form, resulting in a large linear system of NX • NY equations for the NX • NY unknowns ϕ_{ij}; and 2) the solution of this large linear system. Equation (51) is discretized as follows

$$[\partial\ (\Sigma\ \partial\phi/\partial x)/\partial x]_{i,j} = \frac{\Sigma_{i+1/2}\,\phi'_{i+1/2} - \Sigma_{i-1/2}\,\phi'_{i-1/2}}{(1/2)(x_{i+1} - x_{i-1})} \tag{52}$$

$$[\partial\ (\Sigma\ \partial\phi/\partial y)/\partial y]_{i,j} = \frac{\Sigma_{j+1/2}\,\phi'_{j+1/2} - \Sigma_{j-1/2}\,\phi'_{j-1/2}}{(1/2)\ (y_{j+1} - y_{j-1})} \tag{53}$$

$$[\partial s/\partial x]_{i,j} = \frac{s_{i+1,j} - s_{i-1,j}}{x_{i+1} + x_{i-1}} \tag{54}$$

where

$$\Sigma_{i+1/2} \equiv (1/2)\ (\Sigma_{i+1,j} + \Sigma_{i,j}) \tag{55}$$

$$\Sigma_{j+1/2} \equiv (1/2)\ (\Sigma_{i,j+1} + \Sigma_{i,j}) \tag{56}$$

$$\phi'_{i+1/2} \equiv (\phi_{i+1,j} - \phi_{i,j})/(x_{i+1} - x_i) \tag{57}$$

$$\phi'_{j+1/2} \equiv (\phi_{i,j+1} - \phi_{i,j})/(y_{j+1} - y_j) \tag{58}$$

The above expressions can be evaluated only for $2 \leq i \leq NX - 1$ and for $2 \leq j \leq NY - 1$, leaving $(NX-2) \cdot (NY-2)$ equations in $NX \cdot NY-4$ unknowns (note that the corner points of the grid do not appear in the above equations). The missing $2(NX+NY)-8$ equations are derived from the boundary conditions imposed on ϕ. For instance, the simplest boundary conditions that could be imposed would be Dirichlet, i.e., specification of known values of ϕ for the $2(NX+NY)-8$ grid points comprising the perimeter of our grid. Another possibility would be Neumann boundary conditions, which would specify known values of the normal derivative of ϕ at the boundary. For instance, the equation

$$(\phi_{N,j} - \phi_{N-1,j})/(X_N - X_{N-1}) = BX_{N-1/2,j} \tag{59}$$

can be thought of as imposing the condition that the normal derivative at the right boundary point $x = (1/2)\,(X_N + X_{N-1})$, $y = y_j$ be equal to $BX_{N-1/2,j}$, the value of which is presumably given.

The solution to this linear system of equations, while by no means a trivial exercise, can be accomplished by a number of algorithms. We have found one and only one algorithm which will yield a solution in all cases, the Stabilized Error Vector Propagation (SEVP) algorithm of Madala (1978). This is a direct solver and can be expensive on a large grid. Iterative solvers with which we have had success include the Chebyshev semi-iterative method of McDonald (1980), and the vectorized incomplete Cholesky conjugate gradient (ICCG) algorithm of Hain (1980), which is an extension of the work of Kershaw (1978).

8. The Numerical Solution of the Continuity Equation

The continuity equation is ubiquitous in all of physics. It is simply a statement of the fact that a conserved quantity (mass for instance) can only appear somewhere in space if it comes from somewhere else. As we have noted previously, Eq. (50) is distinguished by the appearance of both spatial and temporal derivatives. We have also noted previously that we intend to treat these spatial and temporal derivatives in distinctly different ways numerically. The formal distinction of spatial and temporal derivatives takes the form of a general numerical technique which has come to be known as the Method of Lines (MOL). In the Method of Lines one simply treats the entire spatial differential operator as some nonlinear operator H operating on the operand or operands of the temporal derivative operator, this case n:

$$\frac{\partial n}{\partial t} = H(n) \tag{60}$$

where

$$H(n) = \nabla_\perp \cdot (n\, \underset{\sim}{v}_{e\perp}) \tag{61}$$

Note that $\chi_{e\perp}$ is a function of n by Eq. (51) and the definitions of Σ_p, S, and $\chi_{e\perp}$. H is therefore a very complicated nonlinear operator acting on n which involves all of the spatial discretization and definitions implicit in solving Eq. (51), as well as (as we shall see) as the spatial finite difference discretization needed to represent the operator $\nabla_\perp$ for Eq. (50). Nonetheless this formalism considerably simplifies our task, for it allows us to cleanly separate out our treatment of the temporal derivatives. We note that now Eq. (60) is simply an ordinary differential equation (ODE) for which a wide variety of numerical integration techniques, known as "ODE solvers", exist. Actually, as we shall see later Eq. (60) and (61) actually represents a system of ODE's, one for each spatial grid point, which are coupled to each other through spatial finite differences and through the solution of the elliptic equation (51). We are fortunate here in that our system of ODE's never becomes stiff (i.e., there are no solutions with time scales much shorter than those of physical interest), and hence we have no need of the more sophisticated numerical techniques available for such situations. The solvers actually in use in the present versions of our simulation codes are as follows:

Leapfrog - Trapezoidal:

$$n' = n(t-\Delta t) \quad + 2H(n(t))) \Delta t \tag{62a}$$

$$n(t+\Delta t) = n(t) + (1/2)\ (H(n') + H(n(t)))\Delta t \tag{62b}$$

Modified Euler:

$$n' = n(t) \qquad + H(n(t))\Delta t \tag{63a}$$

$$n(t+\Delta t) = n(t) + (1/2)\ (H(n') + H(n(t)))\Delta t \tag{63b}$$

Note that each of these schemes consist of a predictor (62a, 63a) followed by a corrector (62b, 63b), and that the corrector stages are identical. Both schemes are of second order accuracy, meaning that if $n(t)$, $n(t-\Delta t)$, $H(n(t))$, and $H(n(t-\Delta t))$ are known exactly then the error $E(t+\Delta t)$ in the solution $n(t+\Delta t)$ decays as some constant times Δt^2 as $\Delta t \to 0$:

$$E(t+\Delta t) \to C\ \Delta t^2,\ \Delta t \to 0;\ C = \text{constant} \tag{64}$$

Restating this in the so-called O - notation:

$$E(t+\Delta t) = O(\Delta t^2) \tag{65}$$

The advantage that the modified Euler scheme enjoys is that only $n(t)$ need be known to advance the solution to time $t+\Delta t$, while the leapfrog-trapezoidal scheme requires in addition a knowledge of $n(t-\Delta t)$. However this advantage is outweighed by the fact that the modified Euler scheme is actually weakly unstable for the case $n(t) = e^{i\theta}$, θ a real number, $H(n) = ikn$, k a positive real number. That is, $|n(t+\Delta t)| > 1$

whereas analytically $|n(t+\Delta t)| = 1$ for all Δt. This form of H(n) is of great interest since if we set $n(x,t) = e^{i(kx-\omega t)}$ then the convective derivative for unit velocity is $\partial n/\partial x = ikn$. For the continuity equation this instability has the effect of amplifying the short spatial wavelength components of the density field slightly. The leapfrog-trapezoidal scheme does not have this defect, and is therefore the one we have chosen for use in our simulation codes. The modified Euler scheme is used in our codes only to start the calculation from the initial conditions, or to change the time step, which must be done occasionally, since even the leapfrog-trapezoidal scheme is stable only when $\Delta t \leqslant \Delta t^{min}$, where Δt^{min} depends on the effective value of k produced by the spatial operator H.

Our problem has now been reduced to that of evaluating the spatial operator H on the finite difference grid shown in Fig. 7. First we note that

$$H(n) = \nabla_{\perp} \cdot (n \, \underset{\sim}{v}_{e\perp}) = \partial f/\partial x + \partial g/\partial y \tag{66}$$

where

$$f(n) = n \, v_x(n) \tag{67}$$

$$g(n) = n \, v_y(n) \tag{68}$$

$$\underset{\sim}{v}_{e\perp} = v_x \hat{x} + v_y \hat{y} \tag{69}$$

and $\underset{\sim}{v}_{e\perp}$ is given by Eq. (46). As we stated earlier, n and ϕ are given on the mesh points (x_i, y_j) and are denoted by n_{ij} and ϕ_{ij} respectively. We shall also evaluate v_x and v_y on these same grid points, using centered finite difference formulae to be given presently. Therefore the quantities f and g above are also known on these grid points. We shall assume for the moment that our mesh is uniform, i.e., that $\Delta x_{i+1/2} \equiv x_{i+1} - x_i$ is independent of i and that $\Delta y_{j+1/2} \equiv y_{j+1} - y_j$ is independent of j, and denote these grid spacings by simply Δx and Δy respectively. Modifications necessary for a nonuniform mesh will be given later. Then we can approximate the quatity $\partial f/\partial x$ to various orders of accuracty:

$$\left(\frac{\partial f}{\partial x}\right)_{ij} = (f_{i+1,j} - f_{ij})/\Delta x + O(\Delta x) \tag{70}$$

$$\left(\frac{\partial f}{\partial x}\right)_{ij} = (f_{i+1,j} - f_{i-1,j})/(2\Delta x) + O(\Delta x^2) \tag{71}$$

$$\left(\frac{\partial f}{\partial x}\right)_{ij} = 2(f_{i+1,j} - f_{i-1,j})/(3\Delta x) - (f_{i+2,j} - f_{i-2,j})/(12\Delta x) + O(\Delta x^4) \tag{72}$$

Similar expressions exist for approximating $\partial g/\partial y$. For instance

$$\left(\frac{\partial g}{\partial y}\right)_{ij} = (g_{i,j+1} - g_{i,j-1})/(2\Delta y) + O(\Delta y^2) \qquad (73)$$

Recall that earlier we had assumed that v_x and v_y were known on grid points (x_i, y_j). Looking at Eq. (46) and (47) we see that this requires a knowledge of $\nabla_\perp \phi = \partial\phi/\partial x\ \hat{x} + \partial\phi/\partial y\ \hat{y}$ on grid points, which are obtained using the above formulae by substituing ϕ for both f and g.

If we simply choose an order of accuracy desired or required for our problem, we have apparently completely specified our solution algorithm; and indeed, for many kinds of problems this would be completely sufficient. However, if one attempts to solve even the simplest of continuity equations ($\partial n/\partial y = 0$, $v_y = 0$, v_x = constant) in the presence of very steep gradients of n in the x direction, the numerical solution is soon seen to be contaminated by the appearance of spurious nonphysical oscillations or "ripples" which can grow in time and eventually destroy all of the information content of the calculation. The reasons are many and varied, but in the final analysis are directly caused by the error associated with the finiteness of Δx, Δy, and Δt: the "discretization error". Often this error can be reduced to acceptable levels simply by using formally more accurate finite difference approximations for spatial and temporal derivatives, for instance by using Eq. (72) instead of Eq. (71), or using a fourth-order Runge Kutta solver to integrate in time instead of our leapfrog-trapezoidal scheme. However, if the spatial or temporal gradients are such that the Taylor series expansion implicit in all finite difference formulae is either nonconvergent or slowly convergent, then this technique will not improve matters appreciably, and may even increase the error. The brute force approach, of course, is to keep reducing Δx, Δy and Δt until we resolve all spatial and temporal structure sufficiently well to get a convergent solution. However there are many physical systems, among them the barium cloud and equatorial spread F system, which allow "shock-like" solutions, i.e., solutions which contain regions where the gradient scale length is orders of magnitude smaller that that of the other features in the problem. On the scale of the overall structure of the solution, these regions are well approximated by discontinuities. These discontinuities effect the rest of solution in time solely through their propagation speed ("shock speed") and the change in the physical characteristics and velocity of the plasma across the discontinuity ("jump conditions"). It is obviously impossible in a situation like this to reduce Δx, Δy and Δt to the point where the actual structure inside the shock is resolved on our finite difference mesh. Fortunately, it is also unneccessary. In their classic paper, Lax and Wendroff (1960) showed that when these shock-like solutions appeared within the context of a system of conservation laws (mass, momentum, and energy, for example), then any finite difference scheme which could represent the shock as a stable propagating entity, regardless of the computed internal shock structure, would recover the correct shock speed and jump conditions (and thus the correct influence of the shock on the rest of the solution) if it were in conservation form, a term we shall define momentarily. Thus it is sufficient to utilize a scheme which is both in conservation form and which has the property of representing a shock as a stable propagating entity. Within this class of schemes one is usually confronted with a choice between schemes which allow numerical oscillations near the shock front, which may be severe and which may in fact destroy the accuracy of the entire calculation if not carefully controlled, and schemes which artifically smear the shock front over large numbers of grid points. The oscillatory schemes in general are of second or higher order accuracy in time or space, while the dissipative, non-oscillatory schemes are all first order accurate in time and space. We shall therefore use the terms "high order" and "low order" to describe the above oscillatory and non-oscillatory schemes respectively. The choice between high and low order schemes is a particularly unpleasant one. The inherently high numerical dissipation of the low order schemes tends to excessively smooth the other physical structures in the problem as well as the shock front, and the low convergence rate ($O(\Delta x, \Delta y, \Delta t)$) may mean that almost as many grid points may be required for sufficient accuracy as would have been required to actually resolve the shock structure to begin with. On the other hand the numerical oscillations associated with the high order schemes often propagate into the entire domain of the solution, destroying all of the accuracy of the calculation. Again we are fortunate in that we do not have to make this choice, since we can have the best of both schemes by utilizing a technique known as flux-corrected transport (FCT), which was originally developed by Boris and Book (1973) and later generalized by Zalesak (1979).

Consider our continuity equation

$$\partial n/\partial t + \partial f(n)/\partial x + \partial g(n)/\partial y = 0 \tag{74}$$

We shall say that a finite difference approximation to Eq. (74) is in conservation (or "flux") form when it can be written in the form

$$n_{ij}(t+\Delta t) = n_{ij}(t) - \Delta V_{ij}^{-1} [F_{i+(1/2),j} - F_{i-(1/2),j} + G_{i,j+(1/2)} - G_{i,j-1/2)}] \tag{75}$$

where $\Delta V_{ij} = \Delta x_i \, \Delta y_j$ is an area element centered on grid point (x_i, y_j). The $F_{i+(1/2),j}$ and $G_{i,j+(1/2)}$ are called transportive fluxes and are functions of f and g respectively at one or more of the time levels. The functional dependence of F on f and of G on g defines the numerical scheme. For instance, if we choose the trapezoidal corrector step Eq. (62b) combined with fourth order accurate spatial derivatives Eq. (72) then

$$F_{i+(1/2),j} = [\frac{7}{12}(f''_{i+1,j} + f''_{ij}) - \frac{1}{12}(f''_{i+2,j} + f''_{i-1,j})]\Delta y_j \tag{76}$$

$$f''_{ij} = \frac{1}{2}\left(f(n'_{ij}) + f(n_{ij}(t))\right) \qquad (77)$$

The essence of the FCT method is as follows. For each time step one computes two distinct sets of F and G: one set by a low order scheme (the "low order fluxes") and the other set by a high order scheme (the "high order fluxes"). Then at each cell interface $(i + (1/2), j)$ and $(i, j+(1/2))$ one uses a weighted average of the high and low order flux as the final flux. This weighting is done in a manner which insures that the high order flux is utilized to the greatest extent possible without introducing numerical oscillations in the solution. The solution which would have resulted if the low order scheme had been used alone is used as the standard by which to determine whether an oscillation is numerical or physical. The result is a family of schemes capable of resolving discontinuities over 2 – 3 grid points with very little smearing of other physical details and no numerical oscillations. For more details, see Boris and Book (1973) or Zalesak (1979).

Before closing this section, let us briefly describe our treatment of nonuniform spatial grids. The basic technique is to utilize a smooth mapping from our "grid space" (i,j) to real space (x,y). The mappings we use are especially simple in that $x = x(i)$ and $y = y(j)$. Since our nonuniform spatial mesh enters only in our evaluation of $\partial f/\partial x$ and $g/\partial y$ and since the treatment for each is the same we shall simply show our evaluation of $\partial f/\partial x$ here. Utilizing the dummy index k and treating it as a continuous variable, we simply use the chain rule:

$$\left(\frac{\partial f}{\partial x}\right)_{ij} = \left(\frac{\partial f}{\partial k}\right)_{ij}\left(\frac{\partial k}{\partial x(k)}\right)_{i} = \left(\frac{\partial f}{\partial k}\right)_{ij}\left(\frac{\partial x}{\partial k}\right)_{i}^{-1} \qquad (78)$$

Now f as a function of the index k is by definition given on a uniform mesh, so we can use all of our previously given formulae for spatial derivatives and fluxes. The derivative $(\partial x/\partial k)_i$ can be taken analytically if we have specified an analytic map form "k-space" to "x-space", or if this map is not given explicity but is still smooth we can again use the previously given formulae for spatial derivatives since x as a function of k is also by definition given on a uniform mesh. In terms of our flux formulation, this simply means that Δx_i is defined to be $(\partial x/\partial k)_i$, and the rest of the scheme remains intact.

7. Concluding Remarks

We hope to have given the reader an understanding of the basic physics of the plasma instabilities underlying the ionospheric irregularities treated here, as well as of some of the fundamental concepts involved in the numerical integration of the differential equations describing this physics. We cannot treat the subject in its entirety here, but have rather tried to give the reader enough information to get started on his own if he so desires. Both aspects of the subject, the physics and the numerical analysis, are extremely dynamic fields. Of particular interest to this author is the fact that the subject of numerical solutions to continuity equations has recently become an area of widespread intensive study by many researchers. The reader is strongly advised to monitor the relevant numerical and mathematical literature.

Acknowledgement

This work was supported by the Office of Naval Research.

References

Boris, J.P., and Book, D.L.: 1973, J. Comput. Phys., 11, 38.
Hain, K.: 1980, Nav. Res. Lab. Memo. Rep. 4264, Naval Research Laboratory, Washington, D.C.
Kershaw, D.S.: 1979, J. Comput. Phys., 26, 263.
Lax, P., and Wendroff, B.: 1960, Comm. Pure Appl. Math., 13, 217.
Madala, R.V.: 1978, Mon. Weather Rev., 106, 1735.
McDonald, B.E.: 1980, J. Comput. Phys., 35, 147.
Tsunoda, R.T.: 1981, J. Geophys. Res., 86, 139.
Zalesak, S.T.,: 1979, J. Comput. Phys., 31, 335.
Zalesak, S.T., Ossakow, S.L., and Chaturvedi, P.K.: 1982, J. Geophys. Res., 87, 151.

COMPARATIVE IONOSPHERES: I. THE INNER PLANETS

Thomas E. Cravens
Space Physics Research Laboratory, Ann Arbor, MI 48109

The basic physical and chemical processes controlling the thermospheres and ionospheres of the inner planets, Venus and Mars, are described. The neutral composition and temperature structure of Venus and Mars are compared with the earth's. Chemical and diffusion processes in the ionosphere are then considered. Ionospheric energetics and heat sources are treated briefly. The mechanisms responsible for the maintenance of the nightside ionosphere of Venus are reviewed.

1. INTRODUCTION

The terrestrial ionosphere has been studied and explored since the advent of radio communication, although most of our understanding of the underlying physical processes has come with the use of space vehicles during the last two decades. Naturally, the ionospheres of the other planets in our solar system have been studied much less intensively, although the various planetary missions have generated a great deal of interest and have provided some of the data needed to initiate comprehensive studies of the various planets.

Mariners 5 and 10 and several Soviet Venera spacecraft have observed the ionosphere of Venus remotely using the radio occultation technique; and in 1978, the Pioneer Venus spacecraft (orbiter and probes) encountered Venus and began the most intensive study ever undertaken of a planetary atmosphere other than earth's. Mariners 4, 6, 7, and 9 observed the atmosphere of Mars in the visible and ultraviolet part of the spectrum as well as making radio occultation measurements of the ionosphere, but the only in situ measurements were made by instruments on the Viking landers in 1976. Jupiter and Saturn, as well as many of their satellites were studied by some combination of Pioneers 10 and 11 and Voyagers 1 and 2. The effect of all these missions has been to allow us to formulate theories of the various atmospheres and ionospheres and to make meaningful comparisons of all the planets.

Each planetary ionosphere is unique in many respects. The composi-

R. L. Carovillano and J. M. Forbes (eds.), Solar-Terrestrial Physics, 805–844.

TABLE I: Physical Characteristics of Planets

Planet	Mass (10^{23}kg)	Equatorial Radius (km)	Equatorial Gravitational Acceleration (ms^{-2})	Average Heliocentric Distance (10^6 km)	Length of Year (days)	Period of Rotation (days)	Magnetic Dipole Moment ($G - cm^3$)
Venus	48.7	6050.	8.60	108.2	224.7	-243	$\lesssim 4.3 \times 10^{21}$
Earth	59.8	6378.	9.88	149.6	365.3	1	8.06×10^{25}
Mars	6.42	3398.	3.72	227.9	687.0	1.03	$\lesssim 1.4 \times 10^{22}$
Jupiter	18990.	71900.	22.88	778.3	4333.0	0.41	1.5×10^{30}
Saturn	5686.	60000.	9.05	1423.0	10759.0	0.43	2.2×10^{29}
Io	.892	1816.	1.80	----	(1.77)*	----	~ 0
Titan	1.36×10^{23}	2560.	1.37	----	(15.9)*	----	~ 0
Comets	$\sim 1.5 \times 10^{-9}$	~ 3	$\sim 10^{-3}$	----	----	0.3	~ 0

* sidereal period

tion, temperature, gravity, magnetic field all differ from one planet to the next. However, the basic physical and chemical processes controlling the behavior of these ionospheres are very similar.

The first lecture will be a survey of the inner planets, Venus and Mars, and the second lecture will cover the outer planets, Jupiter and Saturn, as well as the Jovian satellite, Io, the Saturnian satellite, Titan, and comets. I will rely on the earlier lectures, dealing with the terrestrial atmosphere and ionosphere, to provide most of the necessary background material. A good general reference for the inner planets is the review article by Schunk and Nagy (1980). The book, "Venus", is also informative, especially the chapters by Russell and Vaisberg; by Nagy, Cravens and Gombosi; and by Brace et al.

A comparison of some relevant physical characteristics appears in Table I. The heliocentric distance, gravitational acceleration, length of day, among other characteristics, all differ from planet to planet. For instance, Venus is closer to the sun than Saturn, and the solar flux is 174 times more intense on Venus than on Saturn. Another important planetary characteristic is the intrinsic magnetic field, which to a large degree can be represented by the planet's magnetic dipole moment. Planets with sizable magnetic fields, like earth, Jupiter, and Saturn, will have large magnetospheres in which the ionosphere is shielded from direct interaction with the solar wind. On the other hand for planets with very small dipole moments like Venus and Mars, the solar wind is able to interact more or less directly with the ionosphere.

The order of presentation in this lecture will be: (1) the introduction, (2) the neutral atmosphere, (3) the dayside ionosphere (including the basic equations, ion composition, and energetics), (4) the nightside ionosphere of Venus, and (5) a summary.

2. NEUTRAL ATMOSPHERE

The nature of an ionosphere not only depends on the parameters listed in Table I but also on the type of neutral atmosphere a planet possesses. Composition and temperature structure are the two basic aspects of a neutral atmosphere. N_2 and O_2 are the major constituents of the terrestrial atmosphere. Carbon dioxide is the major constituent on Venus and Mars, although there is some N_2 and some noble gases. The surface pressure on Venus is 60 bars, on earth it is 1 bar, and on Mars it is 6 mb; however, the atmospheric density at 120 km is comparable on all three planets due to a fortuitous combination of temperature structure in the lower atmosphere and gravitational acceleration.

The neutral temperature in the thermosphere of a planet is a consequence of the deposition of energy in the uppermost part of the atmosphere. Only far and extreme ultraviolet (EUV) solar radiation can be significantly absorbed above 100 km on the inner planets, all longer wavelength radiation penetrates deeper. Particle precipitation and frictional heating due to electric fields (Joule heating) can also deposit energy in the upper atmosphere. A significant fraction (the heating efficiency) of this absorbed or deposited energy is available to heat the neutral gas. This heat can be redistributed by: (1) being transported to other parts of a planet (like the nightside) by thermospheric winds which themselves are due to pressure gradients caused by the heating, and (2) being transported vertically by heat conduction to lower altitudes where the gas is denser and cooling occurs by the emission of infra-red radiation by CO_2. The terrestrial mixing ratio of CO_2 is only 3×10^{-4} by volume, but CO_2 is the major constituent on Venus and Mars. It is not too surprising then that on earth the thermospheric temperature is larger than on Venus or Mars. Typical daytime exospheric temperatures are:

Venus - 280 K   Earth - 1500 K   Mars - 150 K

The temperature on Mars is lower than on Venus primarily because Mars is further from the sun. For all these planets, the neutral temperature is lower during the night than the day and is maintained partly by the thermal capacity of the thermosphere, partly by transport of heat by winds, and partly by localized nighttime heat sources such as the aurora. The transport of heat by strong day to night winds (~400 m/s) is, in fact, the primary mechanism for maintaining the nightside temperature on Venus where the time from sunset to sunrise is about 60 earth days.

Once the temperature structure is known the atmospheric pressure can be obtained in terms of the pressure P_o at an altitude z_o by applying hydrostatic equilibrium. For an isothermal atmosphere,

$$P = P_o \, e^{-(z - z_o)/H_{Av}} \tag{1}$$

where the scale height is

$$H_{Av} = \frac{kT}{M_{Av} g} \tag{2}$$

T is temperature, k is Boltzmann's constant, g is the gravitational acceleration and M_{Av} is the mean mass. We already have some idea of what g and T are for the inner planets, but we also need to know the composition. In the lower atmosphere we know that Venus and Mars are mostly CO_2 and the earth, mostly N_2 and O_2; but in the thermosphere, photodissociation generates other species.

O_2 and N_2 in the earth's thermosphere are both dissociated and ionized by solar EUV radiation (and by particle precipitation in the auroral regions) thus producing atomic oxygen and odd nitrogen (N and NO). Similarly, CO and O are created in the Martian and Venusian thermospheres by photodissociation of CO_2:

$$h\nu + CO_2 \rightarrow CO + O \tag{3}$$

Ionization of CO_2, followed by ion chemistry also generates CO and O. CO and O are transported by diffusion (molecular diffusion at higher altitudes and eddy diffusion at lower altitudes) from the region where they are produced to a region below the thermosphere where they are destroyed. Eddies tend to keep the various atmospheric species uniformly mixed at lower altitudes and this is described by an eddy diffusion coefficient, K.

The density of a minor neutral constituent in an isothermal atmosphere can be calculated by solving the following continuity and momentum equations:

$$\frac{\partial n_s}{\partial t} + \frac{\partial \phi_s}{\partial z} = P_s - L_s \tag{4}$$

$$\phi_s = -D_s \left(\frac{\partial n_s}{\partial z} + \frac{n_s}{H_s} \right) - K \left(\frac{\partial n_s}{\partial z} + \frac{n_s}{H_{Av}} \right) \tag{5}$$

n_s is the number density of the sth species, ϕ_s is the vertical flux, P_s the production rate, L_s the rate of destruction, H_s the scale height of species s, g is the gravitational acceleration and k is Boltzmann's constant. H_s is found using equation (2) but with the mass of species s rather than the mean mass. The molecular diffusion coefficient for the diffusion of species s through the major species is D_s and is inversely proportional to the collision frequency and therefore to the neutral density of the major species ($D_s = D_{so}/n$).

In a steady-state situation where transport is not important, photochemical equilibrium is said to hold, and equation (4) states that $P_s = L_s$. That is, net production must equal net loss of a species, including chemical production and loss. This situation, where chemistry is more important than transport, arises when the chemical lifetime of a species is short compared to diffusion times. Transport, however, must be considered for those species and/or at those altitudes for which the chemical lifetime is long compared to the time it takes for a molecule to diffuse a typical distance such as the scale height. In the latter case, if the species is relatively inert, the net diffusive flux given by equation (5) will be negligible and the density, n_s, can be determined by setting either the first or second terms equal to zero depending on the relative magnitudes of D and K. If K is larger, as it is near the bottom of the thermosphere, then eddy mixing will keep all species well-mixed so that they will have the same scale height, H_{Av}. D_s will become larger than K above some altitude (called the homopause) and then each species will fall off with its own scale height - this is called diffusive separation. Heavy species like CO_2 (mass 44) will fall off more rapidly than lighter species like He (mass 4). The homopause for all the inner planets is in the altitude range 100 km - 140 km.

The neutral atmospheres of Venus and Mars are shown in Figure 1 and 2, respectively. The densities of long-lived species like N_2 and He fall off with the average scale height (basically the CO_2 scale height)

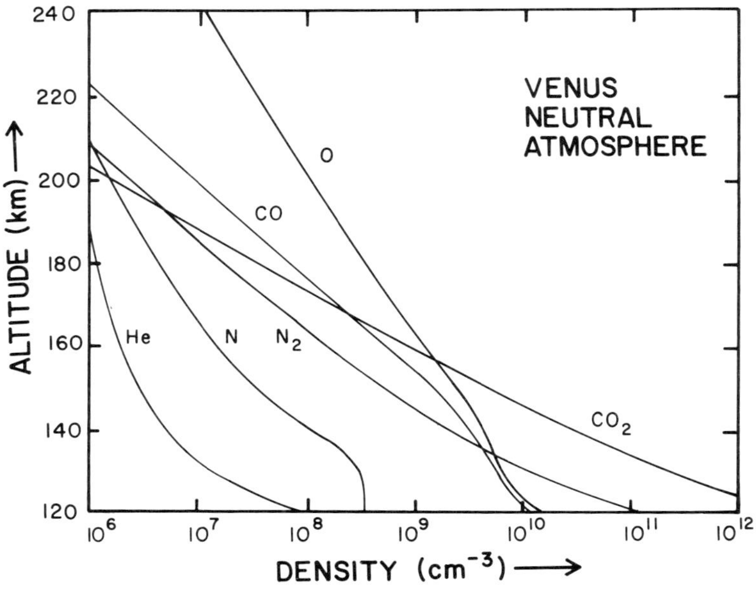

Figure 1. Altitude of neutral gas density profiles on Venus (adapted from Niemann et al., J. Geophys. Res., 85, 7817, 1980).

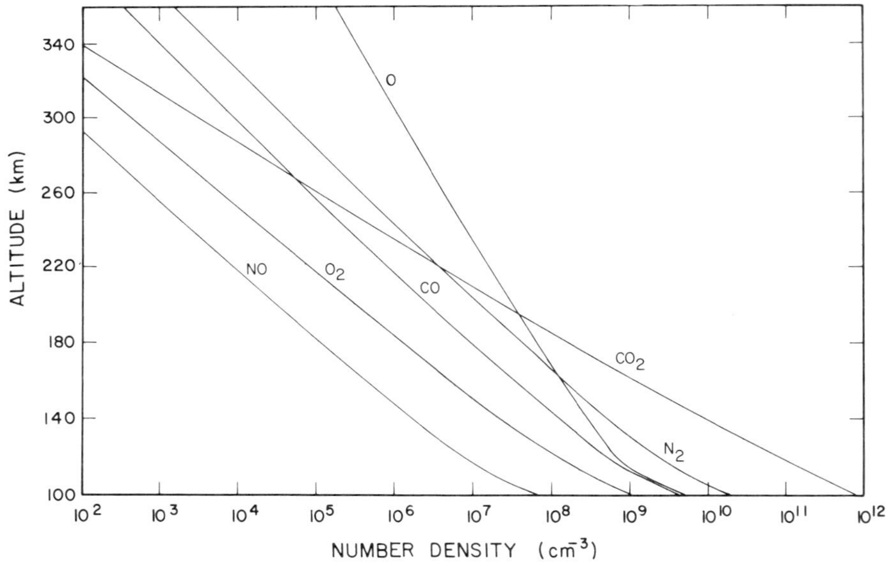

Figure 2. Altitude of neutral gas density profiles on Mars (from Chen et al., J. Geophys. Res., 83, 3871, 1978). Copyright 1978 by American Geophysical Union.

below the homopause and with their individual scale heights above the homopause. Chemical production and loss effects become important for species like CO, O, and N at lower altitudes, although in these figures these effects are not properly taken into account. These species should have a maximum density at altitudes between about 90 and 110 km, just as O does in the terrestrial atmosphere. Atomic oxygen becomes the major constituent on all three inner planets in the altitude range of 160-220 km. The scale height of O is about 14 km on Venus, 70 km on earth, and 22 km on Mars. Clearly, the thermosphere of the earth is hotter and therefore more extensive than those of Venus and Mars.

3. SOLAR WIND - IONOSPHERE INTERACTION

The earth has a strong intrinsic magnetic field generated by a dynamo deep within its interior. This strong field deflects the solar wind several earth radii above the atmosphere. There is virtually no intrinsic magnetic field on Venus and the solar wind is instead deflected by the highly conducting ionosphere.

A schematic diagram of the solar wind interaction with Venus is shown in Figure 3. The ionosphere resists the encroachment of the interplanetary magnetic field and field lines "pile up" in front of the planet More precisely, a current sheet forms at the solar wind-ionosphere interface (which has been given the name ionopause) and generates an induced magnetic field which enhances the field external to the ionopause and reduces the field to almost zero inside the ionosphere. This induced magnetic field acts as the actual obstacle to the solar wind. A bow shock exists in the flow around Venus, just as it does around the earth, because the solar wind is supersonic and super-Alfvénic.

Mars seems to require not only its ionosphere but also some intrinsic magnetic field, albeit weak, in order to be able to deflect the solar wind around it. However, the schematic (Figure 3) should still be a fair representation.

The ionopause boundary between the shocked solar wind (or ionosheath) and the ionospheric plasma is usually sharp - about 30 km in extent. It is located at an altitude where the ionospheric thermal pressure is sufficient to withstand the dynamic pressure of the solar wind. The induced magnetic field mediates this interaction.

Let us consider this interaction in more detail. The momentum equation for a plasma can be written:

$$\rho \frac{\partial \vec{V}}{\partial t} + \rho \vec{V} \cdot \nabla \vec{V} = \frac{1}{c} \vec{j} \times \vec{B} - \nabla p \qquad (6)$$

where ρ is the mass density, $\vec{V}$ is the bulk velocity, $\vec{j}$ the current density, $\vec{B}$ the magnetic field, and p is kinetic pressure. For a highly conducting medium,

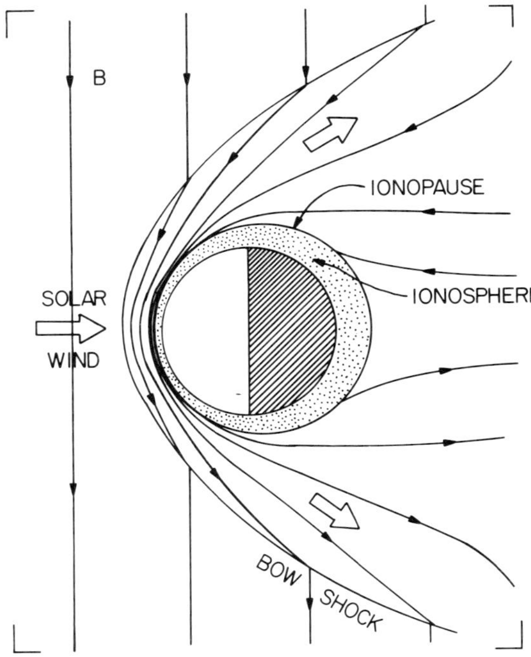

Figure 3. Schematic of solar wind-ionosphere interaction at Venus.

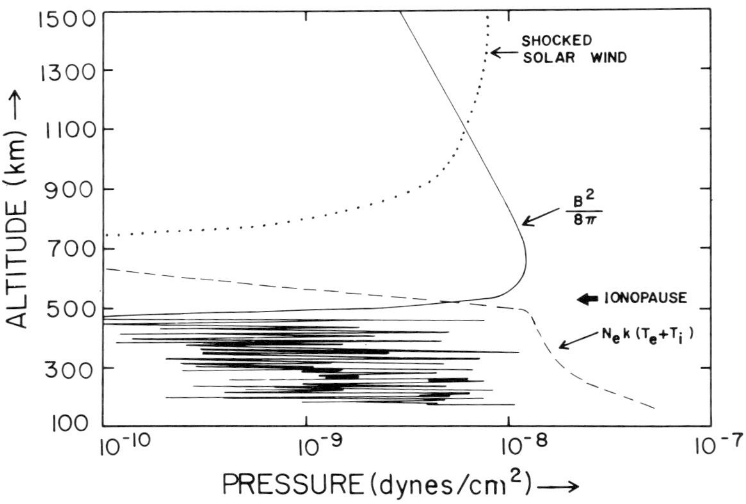

Figure 4. Pressure versus altitude.

$$\frac{4\pi}{c}\vec{j} = \nabla \times \vec{B} \tag{7}$$

$$\frac{1}{c}\vec{j} \times \vec{B} = \frac{1}{4\pi}(\nabla \times \vec{B}) \times \vec{B} = \frac{1}{4\pi}\vec{B} \cdot \nabla\vec{B} - \nabla \frac{B^2}{8\pi} \tag{8}$$

The first term on the right-hand side of (8) vanishes for a planar geometry like that near the ionopause surface. Equation (8) can be substituted in (6) and some further assumptions made to give the following approximate expression:

$$\frac{1}{2}\rho V^2 \cos^2\chi + \frac{B^2}{8\pi} + p = \text{constant in a direction normal to the ionopause surface} \tag{9}$$

χ is the solar zenith angle. The first term represents the dynamic pressure of the solar wind flow, the second term is the magnetic pressure, and the third term is kinetic pressure. p is largest within the ionosphere where $p = n_e k(T_e + T_i)$ and where n_e is the number density of electrons (or ions), T_e is the electron temperature, and T_i is the ion temperature.

Equation (9) without the dynamical pressure term (first term) is the standard static pressure balance relationship for a planar geometry which works very well near the ionopause. The dynamical pressure term is meant to represent the very complex effects of dynamics, but does so only crudely. What is usually done is to assume that the kinetic and magnetic pressures near the ionopause are balanced by the dynamical pressure of the undisturbed solar wind outside the bow shock, and this dynamical pressure is given by the Newtonian approximation (which is just the dynamical pressure term of equation (9) multiplied by a correction factor of order unity).

It is instructive to plot each of the above three pressure terms as a function of altitude for typical Venusian conditions (Figure 4). The ionospheric kinetic pressure decreases gradually with increasing altitude within the ionosphere (this will be discussed in the next section) but at the ionopause this pressure drops precipitously. The magnetic pressure builds up as p drops so as to keep the sum of all types of pressure constant. Thus, a pressure balance exists between the ionospheric pressure inside the ionopause and the magnetic pressure just outside the ionopause. The magnetic pressure then gradually decreases and the solar wind dynamic pressure increases. The magnetic "barrier" plays an important role in mediating the overall pressure balance between the shocked solar wind and the ionosphere.

Some actual in situ ion density data from the ion mass spectrometer on the Pioneer Venus Orbiter (PVO) is shown in Figure 5. The O^+ density is almost exactly equal to the electron density at higher altitudes in the ionosphere. T_e and T_i are almost constant at higher altitudes so that pressure is proportional to the O^+ density. The location of the ionopause is obvious and the height of the ionopause varies considerably

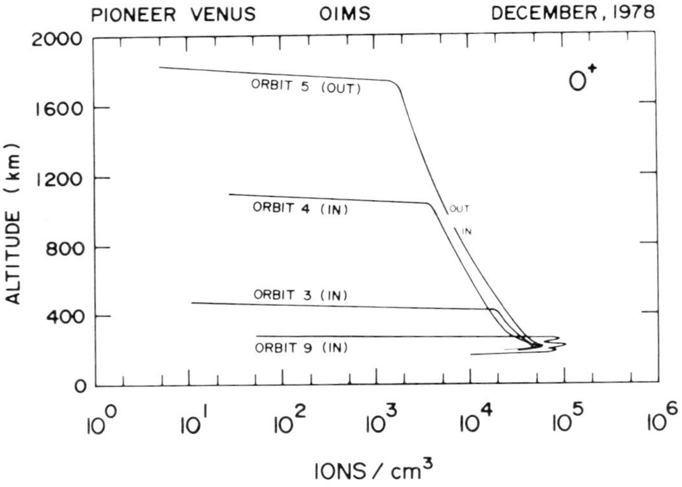

Figure 5. Altitude profiles of O^+ measured by the ion mass spectrometer on the Pioneer Venus Orbiter (from Taylor et al., Science, 203, 755, 1979). Copyright 1979 by the American Association for the Advancement of Science.

from orbit to orbit (the orbital period being ~24 h), reflecting the variability of the solar wind dynamic pressure. As the solar wind pressure increases, the magnetic barrier strength increases and the ionopause moves lower to an altitude where there is sufficient ionospheric pressure to match the external pressure. When the solar wind decreases, the ionopause moves higher.

The only measurement of a Martian ionopause was from the Viking 2 lander which showed an abrupt fall off of the electron density at 300 km. The ionospheric pressure at that altitude is insufficient by itself to balance the solar wind pressure, thus implying that some intrinsic magnetic field must exist in order to supply the extra pressure required.

The magnetic field within the ionosphere of Venus is highly variable and the average field is almost zero. However, there are localized (~10 km) structures within which the magnetic field is as large as 100 gammas.

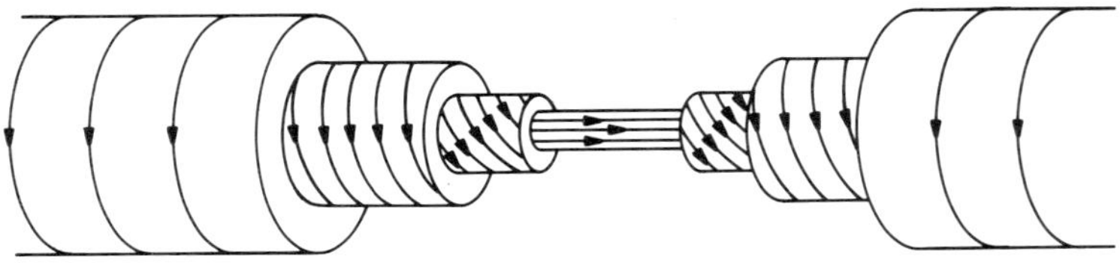

Figure 6. Magnetic flux rope structure as deduced from PVO magnetometer measurements (Elphic et al., J. Geophys. Res., 85, 7679, 1980).

The PVO magnetometer has determined that most of these structures are helical or rope-like in nature. Figure 6 is a schematic of a magnetic flux rope. The field configuration is such that:

$$\vec{j} \times \vec{B} \simeq 0 \tag{10}$$

where $\vec{j}$ is determined by equation (7). This means that the Lorentz force is almost zero, i.e., that the structure is force-free. Almost no external pressure is required to contain the magnetic field in these structures. A number of theories have been proposed to explain the origin of these flux-ropes, including the suggestion that the Kelvin-Helmholtz instability at the ionopause will entrain into the ionosphere and twist some of the magnetic field from just outside the ionopause. The flux-ropes plus even smaller scale magnetic irregularities within the ionosphere have an appreciable effect on the transport properties of the ionosphere.

4. THE DAYSIDE IONOSPHERE

4.1. Basic Equations

Conservation equations can be used to describe the flow velocities, electron and ion concentrations, and temperatures in an ionosphere. Schunk presented these equations as they apply to the terrestrial ionosphere in an earlier lecture. These basic equations are the same for all planetary ionospheres, but considerably simplified versions of these equations are usually used for planetary work. The simplified one-dimensional form of the continuity, momentum, and energy equations are:

$$\frac{\partial n_s}{\partial t} + \frac{\partial F_{sz}}{\partial z} = P_s' - L_s' \tag{11}$$

$$F_{sz} = n_s u_{sz} = -D_s n_s \left[\frac{1}{n_s} \frac{\partial n_s}{\partial z} + \frac{m_s g}{kT_i} + \frac{(T_e/T_i)}{n_e} \frac{\partial n_e}{\partial z} \right. \tag{12}$$

$$\left. + \frac{1}{T_i} \frac{\partial}{\partial z} (T_e + T_i) + \frac{\alpha_s}{T_i} \frac{\partial T_i}{\partial z} \right]$$

$$\frac{3}{2} n_m k \frac{\partial T_m}{\partial z} - \frac{\partial}{\partial z} (K_m \frac{\partial T_m}{\partial z}) = Q_m - L_m \tag{13}$$

where n_s is the number density of species s, m_s is the mass, F_{sz} is the vertical diffusive flux of the sth ion, D_s is the diffusion coefficient, α_s the thermal diffusion coefficient, P_s' is the ionization production rate (including chemistry), L_s' is the ionization loss rate (including chemistry), and u_{sz} is the drift velocity. In equation (13), m is an index denoting whether the electron or ion energy equation is being considered, K_m is the thermal conductivity for electrons or ions, Q_m is the heating rate, and L_m is the cooling rate.

One-dimensional equations such as these are usually solved along a field line in the case of the terrestrial ionosphere because of the preferred direction of transport parallel to the strong magnetic field. Venus and Mars do not have strong and steady magnetic fields, consequently these equations are solved as a function of altitude. The effects of the weak and variable field are incorporated into the transport coefficients.

The diffusion coefficient used in equation (12) was described in earlier lectures and is essentially,

$$D_s = \frac{kT_s}{m_s \nu_s} \tag{14}$$

where ν_s is an effective collision frequency for species s. The collision frequency becomes smaller and D_s larger as altitude z increases. The time for an ion to diffuse across one ion scale height ($\tau_D \sim H_s^2/D_s$) decreases with altitude, and therefore at higher altitudes diffusion is more important than local chemical processes. At lower altitudes, the chemical lifetimes of ions are typically smaller due to larger neutral densities, and where the chemical lifetime is less than the diffusion time chemistry dominates and ion concentrations can be obtained by assuming photochemical equilibrium ($P_s' = L_s'$).

4.2. Composition

Photoionization of the neutral gas by solar extreme ultraviolet radiation is the chief source of ions on the daysides of all the inner planets. The photoionization rate on Venus and Mars is calculated in the same way as in the terrestrial ionosphere except that instead of using photoabsorption and photoionization cross-sections for terrestrial neutral species like N_2, O_2, and O, cross-sections for CO_2, CO, and O are used instead. Naturally, the solar flux must be scaled for the relevant heliocentric distance. And the appropriate neutral atmosphere must be used into which the solar flux is deposited.

N_2^+, O_2^+, and O^+ are the most abundantly produced ions in the terrestrial ionosphere (although not the only ones) because N_2, O_2, and O are the most abundant neutral species in the thermosphere. Analogously, CO_2^+ and O^+ are the major products of photoionization on Venus and Mars; however, just as in the case of earth, chemistry influences the final outcome.

In the terrestrial E-region, O_2^+ and N_2^+ are produced, initially, but then the following ion-neutral chemical reactions take place:

$$N_2^+ + O_2 \rightarrow O_2^+ + N_2 \tag{15}$$

$$N_2^+ + O \rightarrow NO^+ + N \tag{16}$$

$$O_2^+ + NO \rightarrow NO^+ + O_2 \tag{17}$$

NO^+ becomes a major ion rather than N_2^+ which is a minor ion. O_2^+ and NO^+ recombine dissociatively.

In the terrestrial F_1 -region, O^+ is produced and is removed by reacting with N_2 to give NO^+. NO^+ again recombines dissociatively:

$$h\nu + O \rightarrow O^+ + e \tag{18}$$

$$O^+ + N_2 \rightarrow NO^+ + N \tag{19}$$

$$NO^+ + e \rightarrow N + O \tag{20}$$

In the F_2-region the chemistry is virtually the same (although at high enough altitudes H^+ is created by charge exchange of O^+ with H) as in the F_1-region but diffusion rather than chemistry becomes the dominant process. The terrestrial ionospheric peak is reached at this transition altitude (~300 km), and at even higher altitudes diffusive equilibrium is reached. Diffusive equilibrium is said to exist when the term in parentheses on the right-hand side of equation (12) is almost zero, as it is for conditions of small flux and large diffusion coefficient such as occur at higher altitudes. The density distribution for diffusive equilibrium can be shown to be exponential with a plasma scale height for the major ion of:

$$H_p = \frac{k(T_e + T_i)}{mg} \tag{21}$$

where m is the mass of the major ion. The temperature gradient is assumed to be zero in equation (21).

Now consider Venus and Mars again. The lower ionosphere ($z < 160$ km) on either Venus or Mars is analogous to the terrestrial E-region. Ion-neutral reactions convert the original photoionization products to other ions. The key reaction is the conversion of CO_2^+ to O_2^+ by atomic oxygen.

$$h\nu + CO_2 \rightarrow CO_2^+ + e \tag{22}$$

$$CO_2^+ + O \rightarrow O_2^+ + CO \tag{23}$$

$$O_2^+ + N \rightarrow NO^+ + O \tag{24}$$

NO^+ and O_2^+ recombine dissociatively. CO_2^+ is a minor ion and the major ion is O_2^+, although there is almost no neutral O_2 on Venus or Mars. And like the earth, the O_2^+ can be converted to NO^+. A significant difference between the terrestrial ionosphere and Venusian and Martian ionospheres is that for the former the peak electron density, n_e, is reached at ~300 km in the F_2-region and for the latter, the peak n_e is at ~140 km in the "E-region". The peak on either Venus or Mars is controlled chemically and is basically a Chapman layer.

O_2^+ is the major ion in the vicinity of the peak. The electron density for this photochemical situation can be found by setting the total ion production rate (i.e., P(z)) equal to the loss rate of O_2^+ via dissociate recombination (i.e., αn_e^2 where α is the O_2^+ dissociative recombination rate coefficient). The electron density is thus equal to $(P(z)/\alpha)^{1/2}$ and reaches a maximum value where P(z) reaches the maximum. The ion production rate, P(z), reaches a maximum at an altitude where the optical depth for the incoming solar EUV radiation is unity.

The region between about 160 km and 190 km on Venus and Mars is analogous to the terrestrial F_1-region in many respects. The major ion formed is atomic (O^+) rather than molecular because the major neutral at higher altitudes is O. O^+ is destroyed by reaction with the major neutral molecular species. On Venus and Mars, the O^+ produced by photoionization of O is destroyed by:

$$O^+ + CO_2 \rightarrow O_2^+ + CO \qquad k_{25} = 9.4 \times 10^{-10} \text{ cm}^3 \text{ sec}^{-1} \qquad (25)$$

The region above about 190-210 km is like the terrestrial F_2-region in that O^+ is the major (or at least an important) ion, and its density has a maximum at the transition altitude where diffusion becomes more important than chemistry. This peak/transition altitude is over 100 km higher on earth than on the other two inner planets because the neutral scale height is larger on earth so that an equivalent density (where the chemical lifetime of O^+ is equal to the diffusion time) is reached at a higher altitude. And on Venus and Mars the maximum O^+ density is not large enough to replace the O_2^+ peak at 140 km as the actual peak in the electron density profile.

Equations (11)-(13) have been solved for Venus and Mars with all the relevant production and loss terms, and a set of ion density profiles thus obtained for Venus and Mars (Figures 7 and 8). The calculated densities for Venus agree reasonably well with densities measured by the PVO ion mass spectrometer. PVO also saw the following ion species: C^+, N^+, H^+, He^+, N_2^+, CO^+, and O^{++}. Notice that the ionospheric peak is at ~140 km and is O_2^+, with CO_2^+ being a minor ion. The peak density on Mars is less than on Venus because the solar flux used for Mars was smaller (different heliocentric distances plus the Mars calculations were for solar minimum conditions and the Venus ones for solar maximum).

The topside profiles (Figure 7) exhibit a diffusive equilibrium type behavior with a scale height given by equation (21) for O^+. For Mars (Figure 8) O_2^+ actually remains the major ion in the topside region so a different mass is needed in equation (21). The O^+ density is an increasing function of altitude below the O^+ peak for almost the same reason it

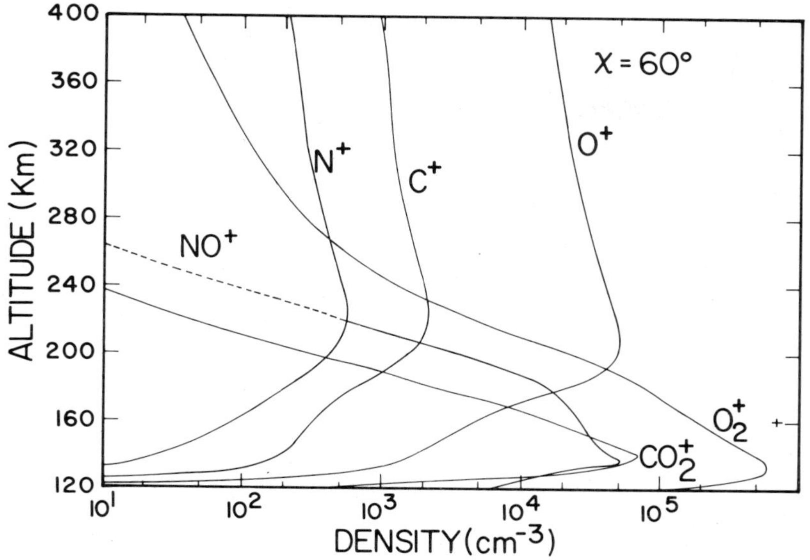

Figure 7. Calculated ion density profiles in the dayside of Venus (adapted from Nagy et al., J. Geophys. Res., 85, 7795, 1980).

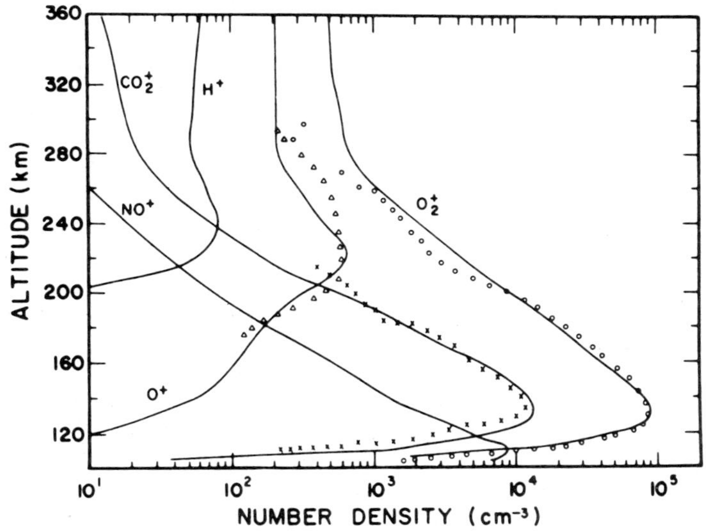

Figure 8. Ion density profiles for the daytime ionosphere of Mars. Solid lines and calculated densities. Circles are O_2^+ data, triangles O^+ data, and crosses CO_2^+ data, all from the Viking I retarding potential analyzer experiment (Hanson et al., J. Geophys. Res., 82, 4351, 1977). The figure is from Chen et al., J. Geophys. Res., 83, 3871, 1978. Copyright 1978 by the American Geophysical Union.

is on earth. The photochemical solution for the O^+ concentration can be found by equating the production rate of O^+ (photoionization of O with an ionization frequency at Venus, $I \simeq 10^{-6}\ sec^{-1}$) and loss rate of O^+ (destruction by CO_2 via reaction (25)):

$$[O^+] = \frac{I}{k_{25}} \frac{[O]}{[CO_2]} \tag{26}$$

The O^+ concentration increases because the neutral CO_2 concentration decreases more rapidly than the O concentration does. The O^+ density stops increasing at an altitude where diffusion "bends" the profile over into a diffusive equilibrium one.

4.3. Energetics

We have taken for granted, up to now, the electron and ion temperatures on Venus and Mars. Equation (13) for electrons and ions, or a more sophisticated version, can be used to find T_e and T_i. This particular equation includes the effects of heating, cooling, and heat transport by thermal conduction, but leaves out transport by bulk motions which is potentially important on the nightside of Venus. I will depend on earlier lectures to have treated this subject in some detail, and will present only a brief synopsis of the relevant issues. Figure 9 is a block diagram outlining the major sources and sinks of heat, and transfer processes, for the electron, ion, and neutral gases on Venus and Mars.

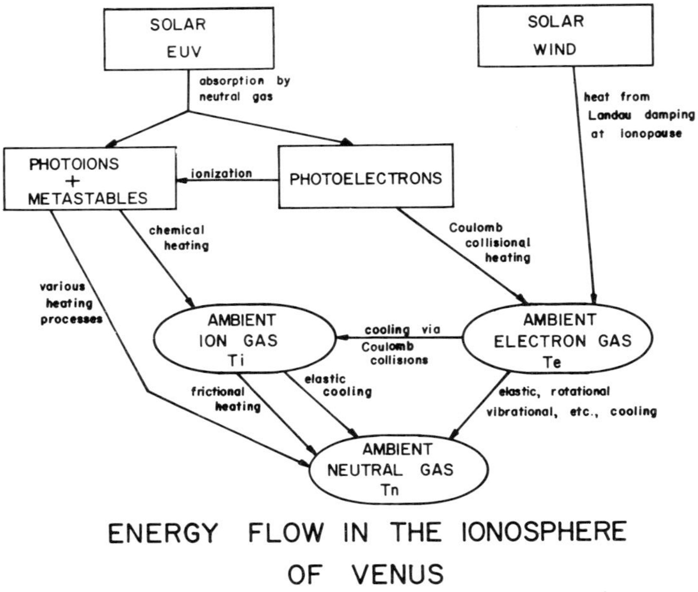

Figure 9. Block diagram of ionospheric energetics on Venus.

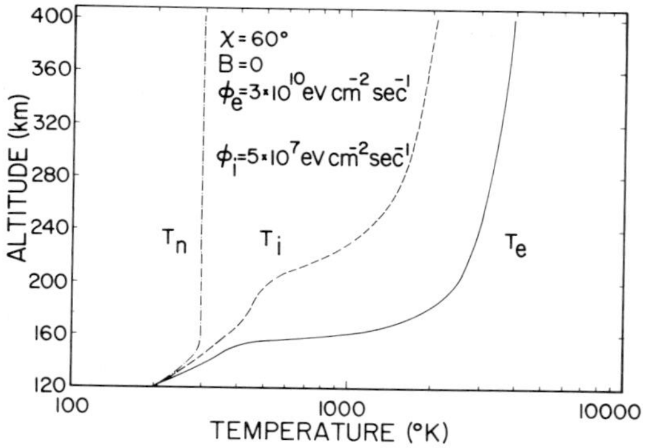

Figure 10. Calculated electron and ion temperatures for the dayside ionosphere of Venus (adapted from Cravens et al., Geophys. Res. Lett., 6, 341, 1979.) Downward heat fluxes into the electrons and ions at the ionopause are indicated.

A large part of the original energy in the "system" comes from the absorption of solar EUV radiation in the thermosphere which produces photoions, metastables, and photoelectrons. A photoelectron is initially produced with an energy equal to the photon energy minus the ionization threshold for a particular neutral constituent and a specific choice of ion final state. The photoelectrons can collide with neutrals and produce more ionization, metastables, or airglow. They can also interact with and heat the ambient thermal electrons via coulomb collisions, and they can travel from one place to another within the ionosphere. Eventually, a photoelectron loses most of its energy and joins the ambient thermal electron population.

The photoions and metastables can collisionally and chemically impart their stored energy to either the neutrals or ions. The relative drift of the neutral gas and ion gas can heat both gases by frictional heating. A major source of heat for the ions is coulomb collisions with the ambient electrons if T_e is greater than T_i - which it usually is. This is a major sink of heat for the electrons. The electrons can also cool by inelastic collisions with the neutrals (e.g., rotational and vibrational excitation of CO_2). The ions cool primarily by elastic collisions with the neutral gas. Ultimately, the neutral gas collects all this heat and gets rid of it by emitting infra-red radiation lower in the atmosphere. A significant difference between Venus and the earth is that most of the energy deposited into the electron gas at higher altitudes on Venus comes from the solar wind-ionosphere interaction and is injected into the ionosphere at the ionopause. Whistler-mode waves generated in the ionosheath get Landau damped when they reach the ionopause region, thus depositing their energy as heat into the ionospheric electrons at the top of the ionosphere.

Most of the heating takes place at altitudes higher than where the cooling occurs, consequently the heat is transported from higher to lower altitudes by thermal conduction. Some numerical solutions to equation (13) for electrons and ions are shown in Figure 10 for Venus. An appropriate heat flux attributed to the solar wind interaction is put in at the top of the ionosphere and heat is conducted down (opposite to the temperature gradient). The calculated temperatures agree rather well with those measured by PVO (not shown). T_e and T_i are much larger than the neutral temperature, T_n. These particular calculations were done for no magnetic field. The magnetic field on Venus is weak, irregular, but horizontal on the average. Such a magnetic field has the potential to inhibit the vertical transport of heat or particles. However, the field is weak enough that the ion gyrofrequency is usually less than the ion collision frequency, thus permitting the ions to move in a direction normal to the field. Although the field has little influence on the ions; even for a weak field the electron gyrofrequency is larger than the collision frequency, so that electrons must travel along field lines. The vertical electron heat conduction would therefore be severely inhibited for a strictly horizontal field, resulting in extremely high T_e because the upper ionosphere would be isolated from the lower ionosphere where cooling takes place. However, the actual instantaneous magnetic field on Venus can be in any direction and considerable vertical transport is thus permitted.

5. THE NIGHTSIDE IONOSPHERE

All three inner planets have substantial ionospheres on their nightsides. Peak electron densities of $\sim 10^4$ cm^{-3} at an altitude of $\sim$140 km have been observed on the nightsides of Venus and Mars. Mars, like earth, has a fairly rapid rotation rate, but on Venus night lasts about 60 earth days. An initial dayside ionosphere would decay to almost nothing in this span of time. Clearly, some ionization sources are required to maintain the nightside ionosphere of Venus. The two mechanisms which are now thought to be responsible for this maintenance are:

1. Impact ionization of the nightside atmosphere by low energy ($\sim$50 eV) electrons originating in the wake of Venus.

2. Horizontal transport of ions from the dayside, across the termination and onto the nightside.

Let us briefly consider the first mechanism. Fluxes of electrons ($\sim 10^7$ cm^{-2} sec^{-1}) have been observed in the wake of Venus outside the ionosphere and presumably were originally ionosheath electrons. These electrons precipitate into the neutral atmosphere interacting with (and ionizing) the neutral gas and losing their energy until they are stopped at an altitude of about 140 km. This has many features in common with electron precipitation in the auroral zones on earth. Equations (11) and (12) can be solved including this source of ionization (but obviously no photoionization) and the resulting peak electron densities are indeed $\sim 10^4$ cm^{-3}. However, Venus is known to have a substantial nightside ionosphere ($\sim 10^3$ cm^{-3}) at altitudes as high as 2000 km and the precipitation mechanism fails to generate sufficient ionization above 160 km.

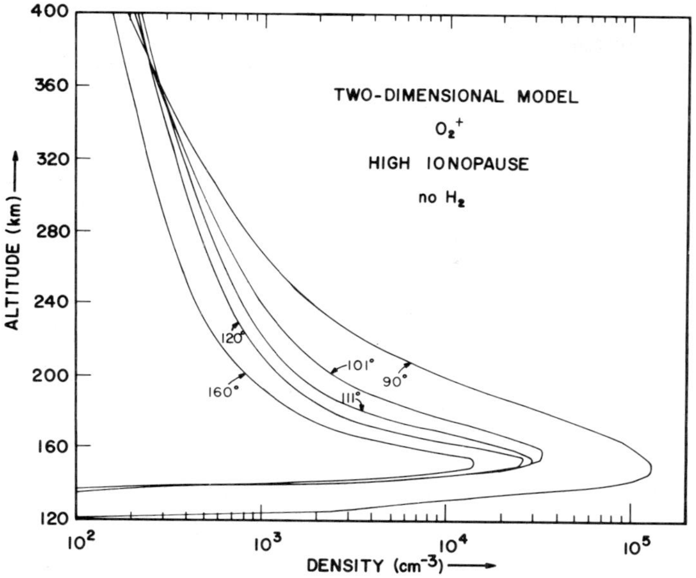

Figure 11. Calculated O_2^+ density profiles on nightside of Venus, labeled with solar zenith angle (from a preprint by Cravens et al., 1982). The initial dayside ionopause for this calculation was relatively high - 875 km.

The production of ions on the dayside is large enough to also maintain the nightside if these ions could be transported from the day to the nightside. PVO has measured horizontal ion drifts across the terminator with velocities ranging from less than 1 km/s at lower altitudes up to a few km/s at higher altitudes. Theoretically, one could calculate such velocities by solving a high-speed version of the momentum equation including horizontal pressure gradient terms, inertial terms, ion-neutral drag terms, and Lorentz force terms. Not surprisingly, this has not been done successfully up to now. However, the horizontal flow is undoubtedly caused by some combination of the following effects:

1. Day to night neutral winds can drag the ions along at lower altitudes where the ion-neutral collision frequency is large.

2. There is a large day to night gradient in ion density which means a large day to night pressure gradient force.

3. Near the ionopause, ionospheric plasma can be convected antisunward due to the interaction with the solar wind.

The nightside densities associated with horizontal transport can be calculated by solving the continuity and vertical momentum equations and by using empirical values for the horizontal drift velocity u_x. The appropriate continuity equation includes the horizontal contribution to the divergence of the flux and can be written in terms of altitude z and horizontal arc length from the subsolar point, x:

$$\frac{\partial n_s}{\partial t} + u_x \frac{\partial n_s}{\partial x} + n_s \frac{\partial n_x}{\partial x} + \frac{n_s u_x}{R} \operatorname{ctn}\left(\frac{x}{R}\right) + \frac{\partial F_{sz}}{\partial z} = P_s' - L_s' \qquad (27)$$

where R is the radial distance from the center of the planet and F_{sz} is given by equation (12). This equation can be solved rather easily for steady-state conditions. O^+ is transported horizontally fairly rapidly at higher altitudes and the nightside ionosphere at higher altitudes is thus maintained. But once over on the nightside the O^+ diffuses down until the CO_2 density is sufficient to rapidly destroy the O^+ by reaction (25) and thus form $O_2{}^+$. In this way, horizontal transport can also play an important role maintaining the $O_2{}^+$ density in the lower ionosphere. $O_2{}^+$ profiles calculated from equation (27) using measured horizontal velocities are shown in Figure 11. These calculations did not include electron impact ionization, yet even so the peak $O_2{}^+$ density deep on the nightside is greater than 10^4 cm^{-3}.

6. SUMMARY

Great advances have been made over the past decade or two both experimentally and theoretically in our understanding of the terrestrial ionosphere. Our understanding of the ionospheres of other planets is certainly less than of the earth, but has improved dramatically due to planetary missions and to the theoretical efforts these missions have encouraged.

ACKNOWLEDGEMENTS

I want to thank A. F. Nagy for useful discussions. I also wish to thank J. U. Kozyra for useful comments on the manuscript and E. K. Anagnost for preparation of the final typescript. This work was supported by NASA grants NAGW-15 and NGR23-005-015.

REFERENCES

Brace, L. H., Gombosi, T. I., Kliore, A. J., Knudsen, W. C., Nagy, A. F., and Taylor, Jr., H. A.: in press, 1982, "Venus", University of Arizona Press, Tucson, Chapter 23.
Nagy, A. F., Cravens, T. E., and Gombosi, T. I.: in press, 1982, "Venus", University of Arizona Press, Tucson, Chapter 24.
Russell, C. J., and Vaisberg, O.: in press, 1982, "Venus", University of Arizona Press, Tucson, Chapter 25.
Schunk, R., and Nagy, A. F.: 1980, "Rev. Geophys. Space Phys.", 18, 813.

COMPARATIVE IONOSPHERES: II. THE OUTER PLANETS

Thomas E. Cravens
Space Physics Research Laboratory, Ann Arbor, MI 48109

The upper atmospheres and ionospheres of the major planets, Jupiter and Saturn, are described. The temperature structure and composition of the thermospheres of the major planets are considered, including a description of the physical and chemical processes controlling the hydrocarbons and atomic hydrogen. The ionospheres of the major planets are then compared to the ionospheres of the inner planets. Io and Titan, satellites of Jupiter and Saturn, respectively, also have atmospheres and ionospheres and will be treated briefly in this lecture. Comets cannot be categorized as planets, but they do have atmospheres and ionospheres which are not gravitationally confined.

1. INTRODUCTION

All of the outer planets and their satellites are much further from the sun than the inner planets and consequently receive much less solar radiation. See Table I in the first lecture (Comparative Ionospheres - I. The Inner Planets). The gravitational acceleration on Jupiter is larger than on any of the other planets. On Saturn the gravitational acceleration is about the same as on earth, on Io and Titan it is much smaller than on earth, and on comets the gravitational acceleration is virtually zero. The length of the "day" for each of these objects is of the order of a few hours; none of them are slow rotators like Venus. Jupiter and Saturn have large intrinsic magnetic fields and therefore have sizable magnetospheres, whereas Io, Titan, and comets are thought to have weak intrinsic fields and will interact directly with the plasma external to their ionospheres. Consequently, comets will interact with the solar wind, Io with the magnetospheric plasma of Jupiter, and Titan with the magnetospheric plasma of Saturn sometimes or with the magnetosheath plasma or the solar wind at other times.

The order of topics will be: (1) the introduction, (2) Jupiter and Saturn (including composition and temperature structure of the thermosphere, hydrocarbons and atomic hydrogen, and the ionosphere), (3) Io, (4) Titan, and (5) comets.

2. JUPITER AND SATURN

A large body of literature now exists on the thermosphere and ionosphere of Jupiter and Saturn, due in large part to the Pioneer and Voyager missions to these planets. I will provide a few useful references here rather than clutter up the main body of the lecture. Many additional references can be found in these. The atmosphere and ionosphere of Jupiter were reviewed by Atreya and Donahue (1976; 1981). The important observations of Jupiter by the Voyager ultraviolet spectrometer experiments (UVS) were summarized by Broadfoot et al. (1982). Broadfoot et al. (1981) discussed UVS observations of Saturn and Titan. Yung and Strobel (1980) described the hydrocarbon photochemistry of the major planets, and Waite et al. (1982) provided an updated perspective on the ionosphere of Saturn.

The most striking feature of the Voyager UVS observations of Jupiter and Saturn was the intense airglow emissions seen at high latitudes, and which were identified as the Lyman and Werner bands of H_2 and the Lyman alpha line of atomic H. The natural interpretation of these features has been that they are produced by the precipitation of energetic particles - either energetic ions or electrons. These emissions were seen in a narrow latitude band near 65° on Jupiter and near 80° on Saturn and were about 40 and 10 times more intense, respectively, than the solar-produced dayglow features seen at lower latitudes on these two planets. The Saturnian aurora seems to be analogous to the terrestrial aurora and is associated with the plasmasheet. But the Jovian aurora appears to be linked by Jovian magnetic field lines with the Io plasma torus. This torus contains surprisingly dense plasma (electron densities $\sim 10^3$ cm^{-3}) in the vicinity of Io's orbit and originating from Io (Figure 1).

2.1. The Thermosphere - Composition and Temperature Structure

Molecular hydrogen is the major constituent of Jupiter and Saturn. The major planets are entirely gaseous except for small rocky cores, hence the altitude scales must be referenced to some arbitrary level - which is usually the 1 bar level (also the level of ammonia cloudtops on

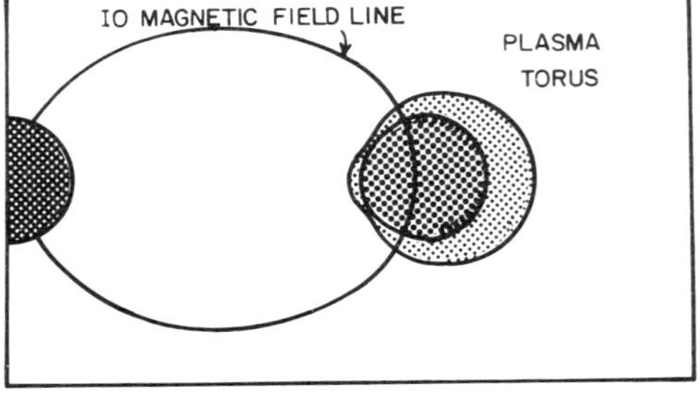

Figure 1. Schematic showing the Io plasma torus and a connecting Jovian magnetic field line.

Jupiter). Helium, methane, and ammonia are also present, as are more complex hydrocarbons and atomic hydrogen.

The temperature in the vicinity of the cloudtops is influenced not only by the absorption of solar radiation but also by the heat emanating from the interior due to the continuing gravitational self-collapse. Thermospheric temperatures are controlled by the same processes that operate on the inner planets. Solar ultraviolet (UV) radiation is absorbed by H_2 and some fraction of it is available as heat. This heat is then transported by winds and/or thermally conducted down to lower altitudes where the atmosphere is cooled by the emission of infra-red radiation. Although CO_2 is the infra-red emitter on the inner planets, C_2H_2 and C_2H_6 play this role on the major planets.

A solution of the standard neutral heat conduction equation gives ~130 K for Jupiter's exospheric temperature and ~105 K for Saturn's, if solar radiation is the only source of heat included. These temperatures are low compared to the exospheric temperatures on the inner planets because the solar flux at Jupiter is only 4% and at Saturn is only 1% of the flux at earth. Exospheric temperatures have been measured both indirectly by radio occultation measurements of the plasma scale height and more directly by Voyager UVS stellar occultation measurements of the neutral gas distribution. The exospheric temperature was found to be anywhere from 1100 K to 1600 K on Jupiter and ~900 K on Saturn.

Obviously, solar ultraviolet radiation is not a sufficient source of heat for the thermospheres of the outer planets. Some other possible sources of heat are:

1. Deposition of heat by gravity waves propagating upward into the thermosphere.

2. The precipitation of electrons or ions.

3. Joule (frictional) heating due to electric fields which are generated in the magnetosphere by departures of the magnetospheric plasma from corotation.

Both sources 2 and 3 are known to be very important in the high latitude terrestrial thermosphere, although on earth the convection electric field is associated with the solar wind. The relative importance of these heat sources remains unknown at the present time. However, source 2 is known to be important because of the intense auroral emissions observed by Voyagers 1 and 2.

The measured Jovian temperature profile is shown in Figure 2. Saturn's temperature profile is not shown here but is similar to Jupiter's although with a larger temperature gradient. These temperatures can be used to find that the atmospheric scale height (equation (2) of Lecture I) of H_2 is ~200 km in Jupiter's exosphere and ~400 km in Saturn's. These can be compared with the 14 km scale height on Venus, 70 km on earth, and

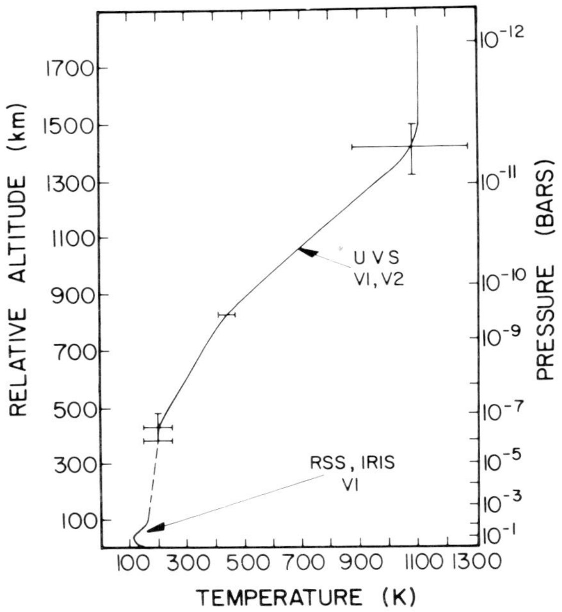

Figure 2. Thermospheric temperature profile for the equatorial region of Jupiter as determined from the Voyager stellar occultation experiment and other experiments (from Festou et al., J. Geophys. Res., 86, 5715, 1982).

22 km on Mars. The much more extensive nature of the atmospheres of the major planets is largely due to the lower molecular weight of H_2 compared with CO_2, N_2, or O.

2.2. The Thermosphere - Hydrocarbons and Atomic Hydrogen

The He/H_2 mixing ratio below the homopause is .11 by volume on Jupiter and the CH_4/H_2 mixing ratio is 8×10^{-4}. Diffusive separation takes place above the homopause (see Lecture I) which is at ~300 km on Jupiter and ~600 km on Saturn. He and CH_4 fall off very rapidly above the homopause because their molecular masses are larger than the average mass of ~2 amu. Thermospheric species other than H_2, He, or CH_4 are created either by photodissociation, or by dissociation associated with energetic particle impact. First consider the sources of atomic hydrogen:

$$(e_V \text{ or } h\nu) + H_2 \rightarrow H + H \quad (1)$$

$$h\nu + H_2 \rightarrow H_2^+ + e \quad (\sim 95\%) \quad (2a)$$

$$\rightarrow H^+ + H + e \quad (\sim 5\%) \quad (2b)$$

H_2^+ is rapidly removed by reaction with H_2, followed by dissociative recombination of H_3^+:

$$H_2^+ + H_2 \rightarrow H_3^+ + H \qquad k_3 = 2.1 \times 10^{-9} \text{ cm}^3 \text{ sec}^{-1} \tag{3}$$

$$H_3^+ + e \rightarrow H_2 + H \qquad \alpha_4 = 3 \times 10^{-7} \text{ cm}^3 \text{ sec}^{-1} \tag{4}$$

$$\rightarrow H + H + H$$

The exact yield of H atoms from ionization processes depends on the branching ratio for reaction (4), with the first branch currently being in favor.

The main sink of atomic hydrogen is three-body recombination with itself:

$$H + H + M \rightarrow H_2 + M \qquad k_5 = 8 \times 10^{-33} \left(\frac{T}{300}\right)^{.6} \tag{5}$$

It can also be destroyed catalytically by the hydrocarbon C_2H_2:

$$H + C_2H_2 + M \rightarrow C_2H_3 + M \tag{6a}$$

$$H + C_2H_3 \rightarrow C_2H_2 + H_2 \tag{6b}$$

$$2H \rightarrow H_2$$

The net effect of (6a) and (6b) is to destroy 2 H atoms and produce one H_2 molecule. There is no net destruction of either C_2H_2 or C_2H_3 so that this pair of reactions can be repeated for many cycles destroying 2 H's during each cycle. Destruction of H by CH_3 is of lesser importance. (6a) and (6b) are important below the homopause but the three-body recombination reaction (5) is more important overall.

The production of hydrocarbons other than CH_4 is initiated by the photolysis of CH_4. At this point, a fairly complicated set of chemical reactions takes place resulting in the production of many other hydrocarbon species. A schematic diagram of this chemistry is shown in Figure 3.

This concludes the discussion of the chemical sources and sinks of atomic hydrogen and hydrocarbons, but transport processes also play an important role in determining the density distributions of these species. Densities can be obtained by solving the continuity and momentum equations (equations (4) and (5) of Lecture I). The results of a numerical solution of these equations for the equatorial region of Jupiter are shown in Figure 4. The homopause height of ~300 km is apparent in the hydrocarbon profiles, and the eddy diffusion coefficient used in this calculation varied inversely as the square root of the total number density and was $\sim 10^6$ cm^2 sec^{-1} at the homopause. This eddy diffusion coefficient, K, is consistent with the Voyager measurements of the location of the He and hydrocarbon homopauses. C_2H_6 (ethane) and C_2H_2 (acetylene) are the most abundant hydrocarbons after CH_4 although CH_3 plays a pivotal role in the chemistry, being the precursor of C_2H_6. C_2H_2 is fairly abundant because it is only weakly photolyzed.

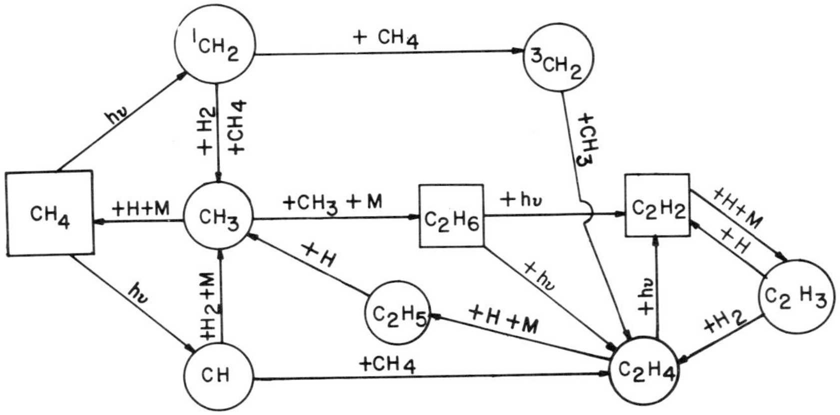

Figure 3. Hydrocarbon photochemistry. The squares indicate species that are relatively abundant (adapted from Strobel, Rev. Geophys. Space Phys., 13, 372, 1975).

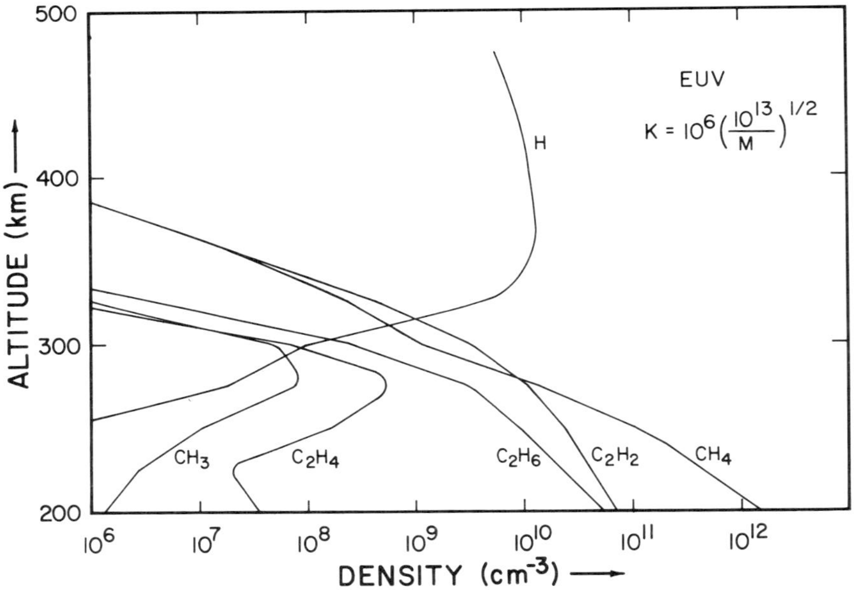

Figure 4. Theoretical hydrocarbon and atomic hydrogen concentrations for the Jovian equatorial region. The eddy diffusion coefficient used in the calculations is shown. M is the total number density. Adapted from Waite et al., manuscript submitted to J. Geophys. Res., 1982.

The hydrocarbon distribution affects the interpretation of H_2 Lyman and Werner band and H Lyman alpha measurements. These spectral features, with wavelengths between ~900Å and ~1600 Å, can be absorbed by CH_4 and by other hydrocarbons; consequently, only those emissions produced above the homopause are likely to escape and thus be observed by experiments like those on Voyager. One implication is that the energetic particles responsible for the Jovian aurora must deposit a large part of their energy above the CH_4 homopause altitude of 300 km - which means that if the particles are electrons, then they must be less energetic than ~30 keV.

The atomic hydrogen profile shown in Figure 4 has a maximum density of $\sim 1.5 \times 10^{10}$ cm^{-3} at an altitude of 400 km. Auroral sources of H were not included, therefore the maximum production rate of H should occur where the optical depth for incoming EUV radiation is unity (~650 km on Jupiter). The main production region of H is therefore more than 200 km higher than where the H density peaks which is also where most H recombines via reaction (5). In this and many other respects, this problem for Jupiter is similar to the O (or H) problem on the inner planets including earth. A simple and instructive analysis can be used to show how the H density behaves in the vicinity of the peak. This analysis presupposes that there is a diffusive flux of H down from a separate production region to a separate chemical loss region. The downward flux of atomic hydrogen, ϕ, is equal to the total production rate of H in a vertical column, and is about 2×10^9 cm^{-2} sec^{-1} in the Jovian equatorial region and 100 times larger than this in the auroral regions.

The diffusion equation ((5) of Lecture I) can be solved exactly for the above type scenario with a convenient choice of boundary conditions for the H density, though it cannot really give a totally realistic H profile since production and loss regions do overlap somewhat. Let n be the H density, let M be the H_2 density, H_1 the scale height of H (H_1 ~100 km at the H peak, although I will also use H_1 as a general characteristic length for this problem), and redefine z so that z = 0 occurs at some reference level where $M = M_o$. The molecular diffusion coefficient $D = D_o/M$ where for H diffusing in H_2, $D_o \simeq 5 \times 10^{19}$. First assume that the region of interest is above the homopause where D>>K, so that eddy diffusion can be neglected. Choose the reference level, z = 0, to correspond to the altitude where n is a maximum. In this case the exact solution for n in a region without sources and sinks is:

$$n = \frac{H_1 \phi M_o}{D_o} [2 - e^{-Z/H_1}] e^{-Z/H_1} \tag{7}$$

The "maximum" n, call it n_m, is at z = 0 and is:

$$n_m = \frac{H_1 \phi M_o}{D_o} \tag{8}$$

The "maximum" atomic hydrogen density depends on the flux and depends on the H_2 density at (or near) the maximum. The reference level (z = 0) is found by finding M_o. M_o can be found by requiring that the total downward flux of atomic hydrogen must all eventually recombine - hopefully, below the z = 0 level so that our original assumption of no sources or sinks holds. Assume that three-body recombination (equation (5)) is much more important than the hydrocarbon reactions (6). Setting column production of H (ϕ) equal to column loss, and then crudely approximating the integral gives:

$$\phi = 2 \int k_5 n^2 M \, dz \simeq k_5 n_m^2 M_o H_1 \qquad (9)$$

The solution of equation (9) for M_o is:

$$M_o = \left(\frac{D_o^2}{k_5 H_1^3 \phi} \right)^{1/3} \qquad (10)$$

Reasonable values for the parameters give $M_o \simeq 5\times10^{13}$ cm^{-3} which means that our reference altitude (which is the same as our approximate peak H altitude) is at a height of ~350 km. Substituting M_o into (8) we find a maximum H density of ~2×10^{10} cm^{-3}. Comparing these values with the numerical values from Figure 4, we see that the approximate analysis is not bad.

By substituting equation (10) into (8) one finds that n_m varies as $\phi^{2/3}$. One can now also estimate the maximum atomic hydrogen density in the auroral region using the fact that the auroral H flux is about 100 times larger than the H flux due to solar EUV. The peak H density is about 20 times larger or ~4×10^{11} cm^{-3}, which is within a factor of 2 of accurate numerical results. Of course, this assumes that none of the aurorally produced H is transported away from the auroral region horizontally by winds.

What happens when K is large and molecular diffusion can be ignored? One solution to the diffusion equation, again assuming no sources or sinks in the region of interest, and assuming constant K, is:

$$n = \frac{\phi H_{AV}}{K} \qquad (11)$$

where H_{AV} is the average scale height (which is just the H_2 scale height). The H density given by equation (11) is independent of altitude in this region. Although n is independent of M, (the same as being independent of altitude) the "peak" will occur at the bottom of the sourceless and sinkless region as determined by M_o. M_o can be found using equations (9) and (11):

$$M_o = \frac{2K^2}{k_5 \phi H_1^3} \qquad (12)$$

The distinction between H_1 and H_{AV} has been ignored. n (and n_m) is directly proportional to ϕ and inversely proportional to K, which makes intuitive sense if one recognizes that K/H_{AV} is an approximate diffusion or drift velocity. M_o increases as K increases and decreases as ϕ increases. If K becomes very large then M_o is large enough for the peak of H to be deep in the hydrocarbon layer where reactions (6a) and (6b) become more important than reaction (5).

There are observational determinations of the column density of atomic hydrogen above the hydrocarbon layer. Solar Lyman alpha is resonantly scattered by atmospheric atomic hydrogen and the Lyman alpha intensity re-emitted by the atmosphere is a function of the column density of atomic hydrogen, N_H. N_H can be crudely estimated by multiplying the maximum H density by the scale height. Figure 5 is a plot of N_H versus the eddy diffusion coefficient for solar production of H only. Molecular diffusion dominates for K less than about 10^6 cm^2 sec^{-1}, therefore N_H is given by equation (8) times the scale height and is independent of K. N_H varies inversely with K (equation (11)) for K greater than 10^6. The N_H deduced from the Voyager UVS Lyman alpha data from low latitudes is larger than either the calculated N_H shown in Figure 5 or the N_H from accurate numerical calculations. Apparently, at the time of Voyager, solar production of H was not sufficient and some additional source was required. Perhaps aurorally produced H is transported equatorward on Jupiter. On the other hand, the Pioneer Lyman alpha observations taken a few years before Voyager gave a column density of H less than 10^{15} cm^{-6} which implies (Figure 5) that the eddy diffusion coefficient was greater than 10^8 at that time.

The atomic hydrogen situation on Saturn is qualitatively the same as on Jupiter. However, the production rate of H, and therefore the flux, is less because Saturn is further from the sun, and also because the eddy diffusion coefficient at the time of the Voyager observations

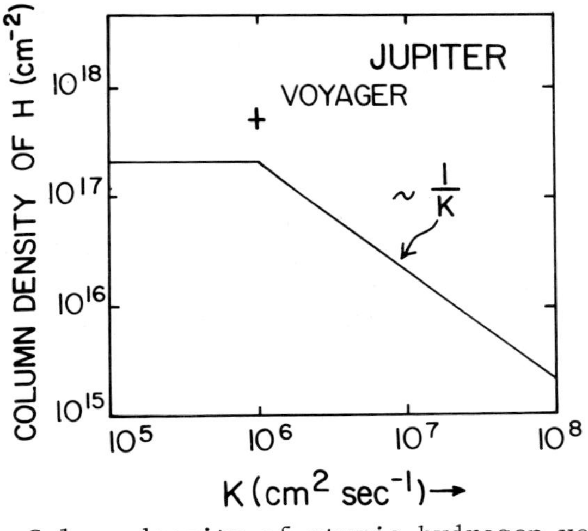

Figure 5. Column density of atomic hydrogen versus eddy diffusion coefficient for the simple model. The cross indicates the column density inferred from Voyager UVS data.

was $\sim 10^8$ cm^2 sec^{-1} at the CH_4 homopause. The maximum H density from equation (11) is about 3×10^8 cm^{-3} and a numerical calculation gives $\sim 10^9$ cm^{-3} for Saturn, as compared to $\sim 2 \times 10^{10}$ cm^{-3} for Jupiter.

2.3. The Ionosphere

The chemical and physical processes controlling the ionospheres of the major planets are basically the same as the ones discussed in earlier lectures (including Lecture I on the inner planets). For instance, the energetics can be described by a diagram like Figure 9 of Lecture I minus the solar wind-ionosphere electron heat source. The electron and ion temperatures (T_e and T_i) are almost equal to the neutral temperature, T_n, on the major planets because the considerable heliocentric distances of these planets imply small heating rates.

Solar EUV radiation and impact ionization by energetic particles are the sources of ionization (reactions (2a) and (2b)) on Jupiter and Saturn. Although the production rate of H_2^+ is much larger than the production rate of H^+, H^+ has a much higher concentration than H_2^+ because H_2^+ is rapidly removed by H_2 (reaction (3)) and the resulting H_3^+ is rapidly destroyed by dissociative recombination (reaction (4)). The main loss process for H^+ is radiative recombination which is very slow.

$$H^+ + e \rightarrow H + h\nu \qquad \alpha_{13} \sim 10^{-11}\ cm^3\ sec^{-1} \tag{13}$$

Three-body recombination of H^+ is also known to occur at lower altitudes. Direct ionization of H by solar photons or energetic particles is more important than the dissociative ionization of H_2 at higher altitudes and/or in the auroral regions where the mixing ratio of H is large.

The photochemical expression for the H^+ concentration can be found by equating the production rate of H^+, $P(z)$, to the loss rate $\alpha_{13}[H^+]\, n_e$. H^+ being the major ion, its concentration is the same as n_e, in which case the photochemical expression is:

$$n_e = \left(\frac{P(z)}{\alpha_{13}}\right)^{1/2} \tag{14}$$

The same expression also approximately describes the electron density in the vicinity of the peak on Venus and Mars except that instead of the radiative recombination coefficient, α_{13}, one has the dissociative recombination rate coefficient for O_2^+. In any case, n_e has a maximum where $P(z)$ is a maximum which is where the incoming solar EUV radiation reaches an optical depth of 1. This happens at 140 km on Venus and Mars but happens at ~ 650 km on Jupiter and at ~ 1200 km on Saturn because of their very extended H_2 atmospheres. The maximum production rate of H^+ is ~ 10 cm^{-3} sec^{-1} in the equatorial region giving a peak electron density of $\sim 10^6$ cm^{-3}.

Equation (14) presupposes that the chemical lifetime is shorter than the diffusion time so that transport of H^+ can be neglected. The chemical lifetime of H^+ ($\tau \sim 1/(\alpha_{13}\, n_e)$) on Jupiter and Saturn is fairly long

($\tau \sim 10^5$-10^6 sec); consequently, ambipolar diffusion becomes important at fairly low altitudes - in the vicinity of, or slightly above, the ionospheric peak. The diffusive equilibrium scale height is twice the chemical scale on the topside, which implies that diffusion causes the topside electron density to fall off more slowly than the photochemistry. The chemical scale height from equation (14) is twice the H_2 scale height because P(z) is proportional to $[H_2]$ above the peak where the atmosphere is optically thin to incoming solar EUV photons. The diffusive scale height (equation (21) of Lecture I) is 4 times the H_2 scale height when $T_e = T_i = T_n$.

Several electron density profiles for Jupiter and Saturn were measured by the Pioneer and Voyager radio occultation experiments. A low latitude Voyager profile for Jupiter is reproduced in Figure 6. There is a suggestion of a peak just below 2000 km. The Pioneer radio occultation data (not shown here) indicate the existence of a main peak near 1000 km as well as many narrow ionospheric layers below this which might be due to metallic ions like Na^+. Some results of an accurate numerical model of the Jovian ionosphere are shown in Figure 6. Expression (14), however,

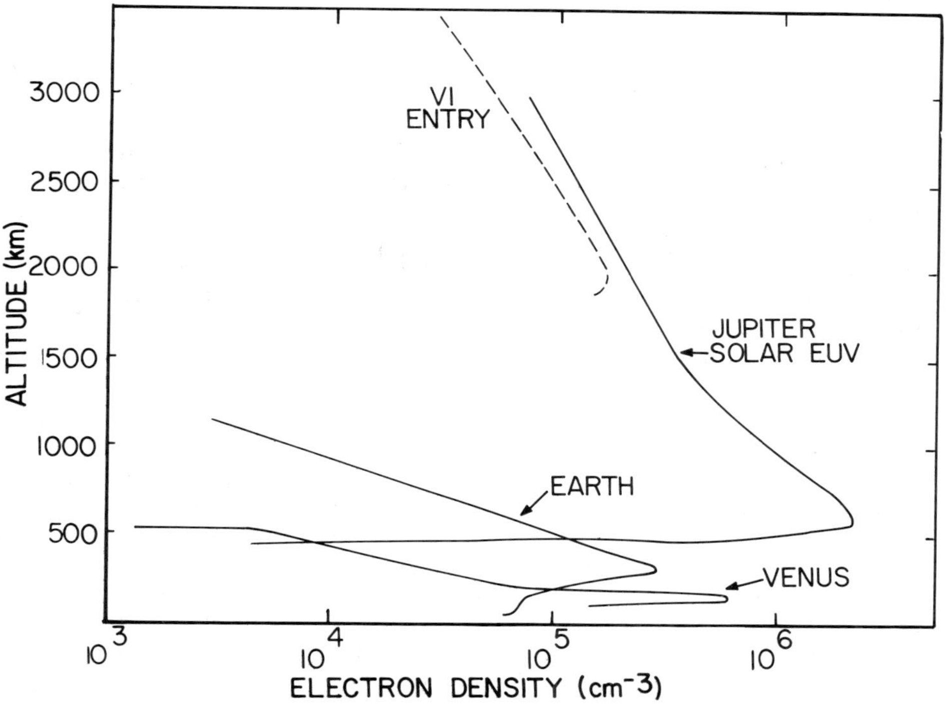

Figure 6. Theoretical electron density profile for low latitudes on Jupiter (adapted from Waite et al., manuscript submitted to J. Geophys. Res., 1982). Venus and terrestrial profiles are also shown, as is a Voyager profile (Eshleman et al., Science, 204, 976, 1979).

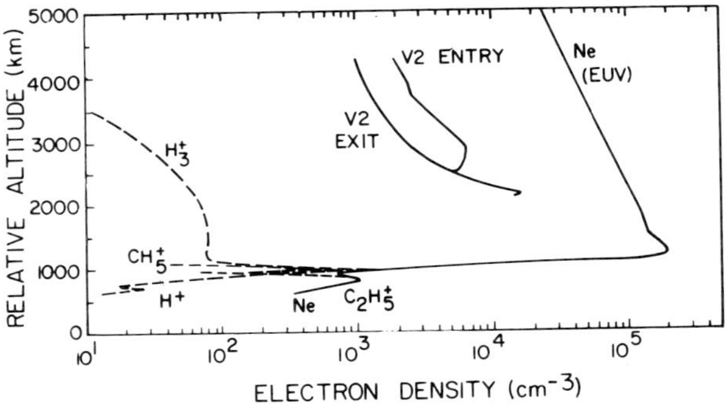

Figure 7. Theoretical electron density profile for Saturn (adapted from Waite, Atreya, and Cravens, manuscript submitted to J. Geophys. Res., 1982). Two Voyager profiles are also shown (Tyler et al., Science, 215, 553, 1982).

provides a reasonably good approximation for the electron density in the vicinity of the peak. Figure 6 also includes typical n_e profiles for Venus and the earth. The ionospheres of the inner planets are clearly more compact.

The theoretical n_e profile on Saturn has a peak density of about $2x10^5$ cm^{-3} at an altitude of ~1200 km (Figure 7). The electron densities measured by Pioneer or Voyager (for z > 2000 km) are about a factor of ten smaller than the calculated ones. Notice that the H_3^+ density is very small and that there are other ion species like CH_5^+ which have not been discussed.

The above ionospheric profiles were for low latitudes. What about the auroral regions? The auroral ionization rate on Jupiter is about 100 times the photoionization rate and theoretical auroral electron densities are therefore about a factor of ten larger than low or mid-latitude electron densities. However, the measured auroral n_e is even smaller than the low latitude n_e on Jupiter. It seems clear that the theoretical models are predicting electron densities which are considerably larger than they should be on Saturn and Jupiter, especially in the auroral regions. Another sink of H^+ is required to reduce the calculated densities. One popular possibility is the removal of H^+ by vibrationally excited H_2:

$$H^+ + H_2\ (v'' \geq 4) \rightarrow H_2^+ + H \qquad (15)$$

The H_2^+ is rapidly removed by reaction (3). Reaction (15) will not work with ground-state H_2 because then the reaction would be endothermic by about 2eV. Vibrationally excited H_2 can be created by recombination of H_3^+, by direct excitation of H_2 by particle impact, and by the cascading in energy levels associated with the excitation of the Lyman and Werner bands.

3. THE JOVIAN SATELLITE IO

Planets are not the only objects with ionospheres and atmospheres. Pioneer 10 measured an electron density of almost 10^5 cm^{-3} on Io and the Voyager infra-red experiment detected SO_2 gas on Io. The Voyager television camera detected extensive volcanic activity on Io, and the UVS experiment detected sulfur and oxygen ions in the Io plasma torus. In addition, solid SO_2 was seen on Io's surface by ground-based measurements. To quote Kumar and Hunten (1980), "Thus it seems likely that an SO_2 atmosphere in approximate equilibrium with the vapor pressure of solid SO_2 gives a surface pressure of 10^{-7} bars at a temperature of 130 K," and Kumar demonstrated that a thermosphere will exist with an exospheric temperature of about 1000 K.

An ionosphere can be formed by photoionization of SO_2:

$$h\nu + SO_2 \rightarrow SO_2^+ + e \tag{16}$$

SO_2^+ will recombine dissociatively and the electron density can be described by equation (14) but with the rate coefficient for SO_2^+. The peak n_e derived in this manner is $\sim 1.5 \times 10^4$ cm^{-3} at an altitude of ~80 km. The peak ionospheric density measured on the wake side of Io (Figure 8) is 4 times larger than this theoretical value. Kumar demonstrated that ionization by 20-100 eV electrons is required in addition to photoionization. Presumably, these electrons originate in the surrounding Io plasma torus.

The ionosphere on the upstream (or ram side) of Io appears to be compressed with a scale height of only 25 km (Figure 8). The corotating magnetospheric plasma of Jupiter is moving at a relative velocity of 57 km/sec with respect to Io, and there is a strong interaction between the two plasmas. The ionospheric pressure is certainly incapable of withstanding the external pressure (see equation (9) of Lecture I) so that there might not be any upstream ionosphere in the classic sense but only mass loaded convecting magnetospheric plasma. In either case, strong electrical currents are expected to flow in the ionosphere and maybe even the body of Io.

4. THE SATURNIAN SATELLITE TITAN

Voyager confirmed what had long been suspected - that Titan has a dense N_2 atmosphere. The atmospheric pressure at the surface of Titan is at least 1.4 bars and the temperature is 156 K. The temperature drops to a tropopause value of about 90 K at 50 km and then increases again to at least 160 K. The CH_4/N_2 mixing ratio measured by Voyager is about .05 and many other hydrocarbon species were also detected. There are also extensive cloud and haze layers. Volume 292 of the journal, "Nature", and Volumes 212 and 215 of the journal, "Science", contain many interesting articles on Titan and its interaction with Saturn's magnetosphere.

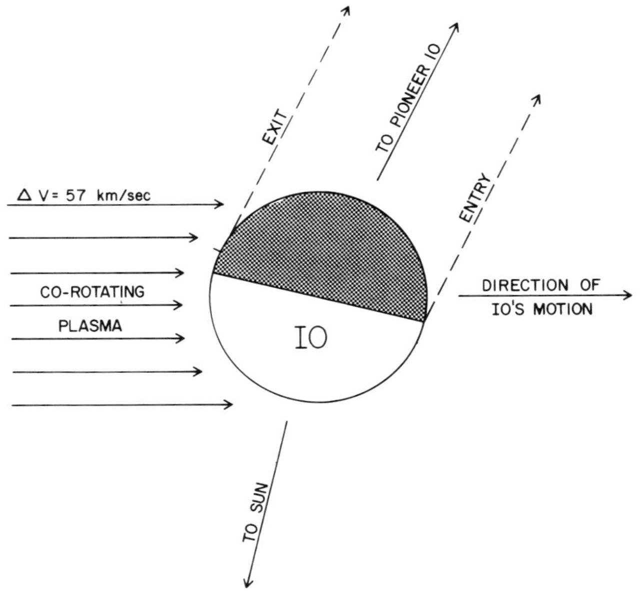

Figure 8a. Schematic of geometry of the Pioneer 10 radio occultation of Io.

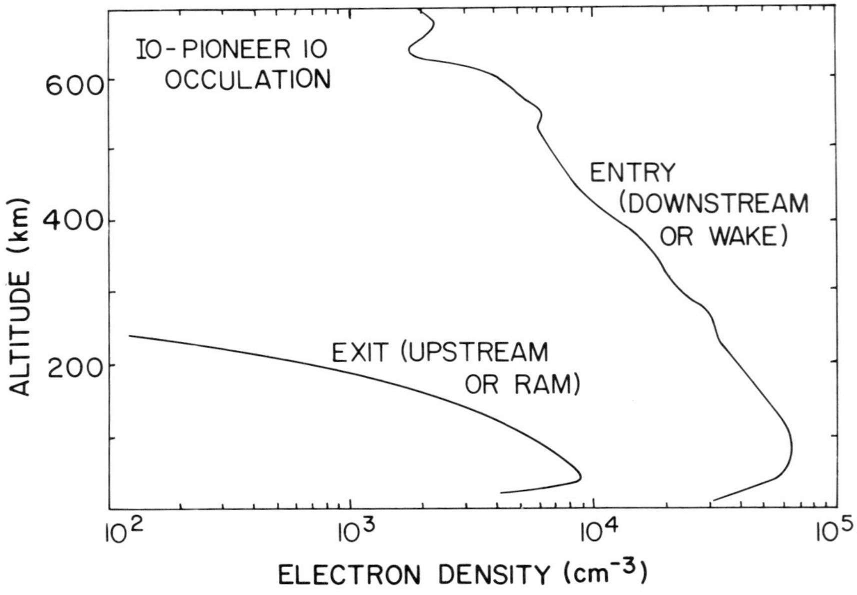

Figure 8b. Electron density profiles from the radio occultation on the upstream and downstream sides of Io. Adapted from Cloutier et al., Astrophys. Space Phys., 55, 93, 1978; and from Kliore et al., Icarus, 24, 407, 1975.

S. K. Atreya (private communication) has developed a chemical reaction scheme for the ionosphere of Titan of which I will present a brief synopsis. The dominant species, N_2, is either photoionized or possibly ionized by energetic electrons from Saturn's magnetosphere. The latter is plausible because the Voyager UVS experiment measured a spectrum from Titan that resembles laboratory spectra for electron impact excitation of N_2. N_2^+ and N^+ are produced by:

$$(h\nu \text{ or } e) + N_2 \rightarrow N_2^+ + e \qquad (17a)$$

$$\rightarrow N^+ + N + e \qquad (17b)$$

N_2^+ and N^+ are then removed by various reactions including,

$$N_2^+ + CH_4 \rightarrow CH_2^+ + H_2 + N_2 \qquad (18)$$

$$N_2^+ + N \rightarrow N^+ + N_2 \qquad (19)$$

$$N^+ + H_2 \rightarrow NH^+ + H \qquad (20)$$

$$N^+ + CH_4 \rightarrow H_2CN^+ + H_2 \qquad (21)$$

H_2CN^+ should be the major ion because CH_2^+ and NH^+ are rapidly destroyed by other ion-neutral reactions. H_2CN^+ recombines dissociately. Photoionization alone results in an electron density profile with a peak density between about 10^3 cm^{-3} and 10^4 cm^{-3} at an altitude of approximately 1500 km.

Titan's orbit at 20 Saturnian radii is such that Titan sometimes resides inside Saturn's magnetosphere and sometimes outside it. Thus, the ionosphere of Titan will sometimes interact with the solar wind and at other times with the magnetospheric plasma. The maximum ionospheric pressure calculated using the above estimates of the electron density from photoionization is such that it should often be able to withstand the solar wind and/or magnetospheric pressure in which case a Venus-like ionopause should exist. At other times, when the solar wind dynamic pressure is large the ionospheric pressure might be insufficient to hold off the external pressure. In the latter case the external plasma will penetrate quite deeply into the upper atmosphere of Titan and will be severely mass loaded by ions of Titan origin. Voyager 1 passed through the wake of Titan (within a few thousand km of it) when Titan was in Saturn's magnetosphere. Several plasma and magnetic field signatures of a strong magnetosphere-Titan interaction were seen, including a neutral sheet in the magnetic field indicative of the draping of Saturnian field lines about Titan.

5. COMETS

Large numbers of comets are thought to populate the outer solar system well beyond the orbits of Neptune and Pluto. Orbital perturbations will occasionally occur that put a comet into an orbit with a small perihelion, in which case the comet will brighten dramatically as it approaches close to the sun. The nucleus of a comet is a small (1-10 km) object thought to be made of a mixture of water, ice, dust, and small amounts of other volatiles -- they have been referred to as "dirty iceballs". The ice is sublimated by the increasingly intense solar radiation as the nucleus approaches the sun, and an expanding spherical cloud of gas and dust is carried off into space. It is this envelope of gas (called the coma) extending across distances of 10^5-10^6 km, which is actually observed from earth. The dust is usually spread out into a tail by the action of the radiation pressure of solar photons. The neutral gas can be photoionized thus forming an ionosphere and the ionosphere can interact with the solar wind. Cometary ions also form a second type of comet tail due to the solar wind interaction. Some useful references for this material are the book "Comet" (1982), a review article by Mendis and Houpis (1982), and a recent article by Gombosi et al. (1982).

The total gas production rate, Q (molecules sec^{-1}), from the surface of the nucleus can be determined from the rate of sublimation of water ice, Z(T), which is a function of the surface temperature, T.

$$Q = 4\pi R_n^2 Z(T) \tag{22}$$

where R_n is the radius of the nucleus. T can be found by balancing the solar radiation absorbed by the surface with the thermal infra-red radiation emitted by the surface as well as the latent heat lost in the sublimation process. This energy balance for a rapidly rotating nucleus can be written as:

$$\frac{1}{4}(F/d^2)(1-A) = \varepsilon\sigma T^4 + L(T)Z(T) \tag{23}$$

where F is the solar flux at 1 astronomical units (AU), d is the heliocentric distance in AU, A is the surface albedo, ε is the emissivity, σ is the Stefan's law constant, Z is the sublimation rate in molecules cm^{-2} sec^{-1}, and L is the latent heat of sublimation per molecule. Z(T) can be determined from the Clausius-Clapeyron equation.

The surface temperature, T, increases, and thus Z and Q increase, as the comet approaches the sun. The flow of gas from the nucleus out into the comet is described by the following continuity and momentum equations in the simple one-fluid approximation:

$$\frac{1}{r^2}\frac{d}{dr}(r^2 \rho V) = 0 \tag{24}$$

$$v\frac{dv}{dr} + \frac{1}{\rho}\frac{dp}{dr} = 0 \tag{25}$$

where r is the radial distance from the center of the comet, ρ is the mass density, v is the expansion velocity, and p is the kinetic pressure. More

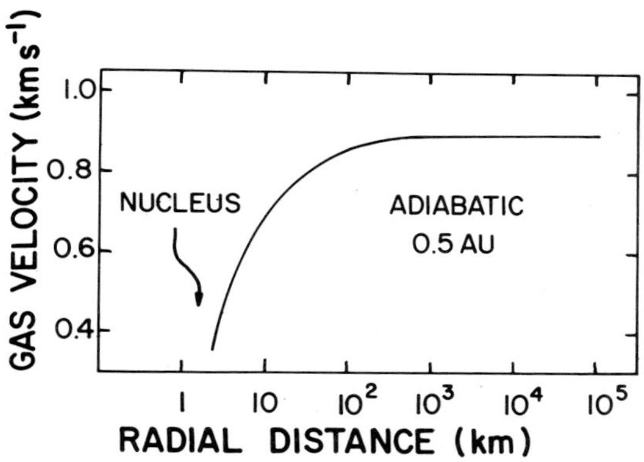

Figure 9. Theoretical neutral expansion velocity versus radial distance from the nucleus at a heliocentric distance of 0.5 AU (adapted from the chapter by Delsemme in "Comets", 1982).

accurate and complicated equations result if one includes dust or the effects of photodissociation. An energy relation is also required, but a polytropic relation with p proportional to ρ^γ with polytropic index γ is usually employed. A typical solution to these equations is shown in Figure 9. The expansion velocity is supersonic and levels off at radial distances not too far from the nucleus with a velocity of about 1 km/s.

Equations almost like equations (24) and (25), but with a gravitation term, describe the solar wind flow outward from the solar corona. Constant velocity is a good approximation for most purposes, in which case the continuity equation gives the following expression for the H_2O number density:

$$n = \left(\frac{Q}{4\pi V R_n^2}\right)\left(\frac{R_n}{r}\right)^2 \tag{26}$$

n is the number density of H_2O molecules. A typical gas production rate of a bright comet is $\sim 10^{30}$ sec^{-1} and a typical nucleus radius is ~ 3 km. The number density is then about 10^{13} cm^{-3} at the nucleus. The cometary atmosphere is formed by sublimation from a solid, like Io's atmosphere, but unlike Io's atmosphere, the cometary atmosphere is not gravitationally bound, instead expanding supersonically into the surrounding space.

The cometary gas will be photodissociated and photoionized as it expands. H and OH will be produced from H_2O and other dissociation products and radicals will be produced from other "parent" volatiles such as CO_2, CO, and HCN. Now let us consider the formation of the ionosphere for a pure H_2O comet. The water vapor is first photoionized by solar EUV photons:

$$h\nu + H_2O \rightarrow H_2O^+ + e \tag{27}$$

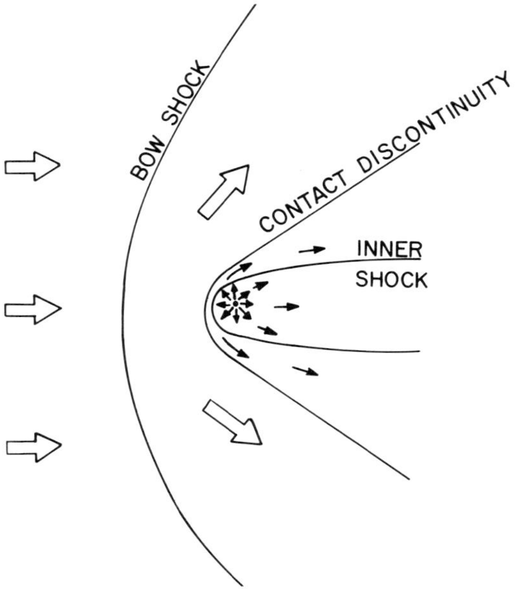

Figure 10. Schematic of the solar wind-cometary interaction.

The H_2O^+ then reacts with H_2O to produce H_3O^+:

$$H_2O^+ + H_2O \rightarrow H_3O^+ + H \tag{28}$$

$$H_3O^+ + e \rightarrow H_2O + H \qquad \alpha_{29} \simeq 10^{-6}\ \text{cm}^{-3}\ \text{sec}^{-1} \tag{29}$$

$$\rightarrow OH + H_2$$

Ionization processes other than photoionization such as electron impact might operate in a comet. The major ion is H_3O^+, which recombines dissociatively. This is similar to the major planets where H_2^+ is rapidly converted to H_3^+ by H_2. In comets with species other than H_2O there will be other ions like CO^+, which is often observed in comet tails.

The electron density, n_e, can be found by solving the following continuity equation for the optically thin regions of a comet:

$$\frac{1}{r^2}\frac{d}{dr}(r^2 v\, n_e) = I\, n - \alpha_{29}\, n_e^2 \tag{30}$$

where I is the photoionization frequency of H_2O ($I \sim 10^{-6}$ sec^{-1} at 1 AU), and n is the neutral density. The right-hand side of (30) is the net production rate - that is, the production rate minus the loss rate due to dissociative recombination. Two central assumptions here are that all H_2O^+ ions created by (29) are almost immediately converted to H_3O^+, and that the ions are swept along at the same expansion velocity as the neutrals. In this case, an approximate solution to (3), for constant velocity v and for r not too close to the nucleus, is:

$$n_e \simeq \frac{1}{r} \left\{ \sqrt{\left(\frac{v}{2\alpha}\right)^2 + \frac{IQ}{4\pi v\alpha}} - \frac{v}{2\alpha} \right\} \tag{31}$$

Notice that the electron density falls off as 1/r, whereas the neutral density falls off more rapidly as $1/r^2$.

Comets have no intrinsic magnetic field so that the solar wind interacts directly with their ionospheres. Figure 10 is a schematic diagram of the currently popular conception of this interaction, although there is virtually no experimental or observational evidence to verify it. The contact discontinuity separates the solar wind and the ionosphere, and is analogous to the ionopause on Venus. A bow shock lies outside the contact discontinuity. On Venus it is known that neutrals in the neutral exosphere beyond the ionopause are photoionized and can also charge exchange with solar wind protons, so that the solar wind (or ionosheath) is both attenuated and mass loaded. The same type of mass loading should occur in comets, but on a much larger scale because of the very extensive cometary atmosphere, and perhaps to such an extent that the cometary bow shock is either very weak or nonexistent. A significant difference between the cometary and Venus solar wind interactions is that thermal pressure balances the solar wind dynamic pressure on Venus, but in comets the ionosphere is expanding supersonically. This supersonic expansion means that there will almost certainly be an inner shock in the ionospheric flow which diverts this flow tailward. It also means that the dynamic pressure of the ionospheric flow, as well as the kinetic pressure, will contribute to the pressure balance with the solar wind dynamic pressure. The actual interaction will no doubt be mediated by an induced magnetic field like it is on Venus (see Lecture I).

6. SUMMARY

This concludes the second lecture on planetary ionospheres. It has been demonstrated that the basic physical and chemical processes controlling the atmospheres and ionospheres of the planets and their satellites are the same as on the earth, although the detailed application of these processes to individual planets or objects yield rather different results.

ACKNOWLEDGEMENTS

I want to thank S. K. Atreya for information on Titan. I also wish to thank J. U. Kozyra for comments on the manuscript and E. K. Anagnost for preparing the final typescript. This work was supported by NASA grants NAGW-15 and NAGW-312.

REFERENCES

Atreya, S. K., and Donahue, T. M.: 1976, "Jupiter", University of Arizona Press, Tucson.
Atreya, S. K., and Donahue, T. M.: 1981, "Vistas in Astronomy", Pergamon Press, Oxford.
Broadfoot, A. L., Sandel, B. R., Shemansky, D. E., Holberg, J. B., Smith, G. R., Strobel, D. F., McConnell, J. C., Kumar, S., Hunten, D. M., Atreya, S. K., Donahue, T. M., Moos, H. W., Bertaux, J. L., Blamont, J. E., Pomphrey, R. B., and Linick, S.: 1981a, "Science", 212, 206.
Broadfoot, A. L., Sandel, B. R., Shemansky, D. E., McConnell, J. C., Smith, G. R., Holberg, J. B., Atreya, S. K., Donahue, T. M., Strbel, D. F., and Bertaux, J. L.: 1981b, "J. Geophys. Res., 86, 8259.
Wilkening, L. L. (Ed.): 1982, "Comets", University of Arizona Press, Tucson.
Gombosi, T. I., Horanyi, M., Kecskemety, K., Cravens, T. E., and Nagy, A. F.: submitted, 1982, "Astrophys. J."
Kumar, S., and Hunten, D. M.: 1981, "The Satellites of Jupiter", University of Arizona Press, Tucson.
Mendis, D. A., and Houpis, H. L. F.: in press, 1982, "Rev. Geophys. Space Phys."
Waite, Jr., J. H., Atreya, S. K., and Cravens, T. E.: submitted, 1982, "J. Geophys. Res."
Yung, Y. L., and Strobel, D. F.: 1980, "Astrophys. J., 239, 395.

SUBJECT INDEX